Terrorism and Homeland Security

Timothy A. Capron dedicates this book to his lovely, gracious wife and best friend, Rhonda Allison Capron, to Rolando del Carmen, always an inspiration, and to the Foundation for Defense of Democracies, for making me a Fellow and allowing me to see the impact of terrorism firsthand.

Stephanie B. Mizrahi dedicates this book to her parents, Edward and Francis Mizrahi, for their unwavering support and love all her life and for making it possible for her to follow her dreams wherever they led.

SAGE was founded in 1965 by Sara Miller McCune to support the dissemination of usable knowledge by publishing innovative and high-quality research and teaching content. Today, we publish more than 850 journals, including those of more than 300 learned societies, more than 800 new books per year, and a growing range of library products including archives, data, case studies, reports, and video. SAGE remains majority-owned by our founder, and after Sara's lifetime will become owned by a charitable trust that secures our continued independence.

Los Angeles | London | New Delhi | Singapore | Washington DC

Terrorism and Homeland Security

A Text/Reader

Timothy A. Capron

California State University, Sacramento

Stephanie B. Mizrahi

California State University, Sacramento

Los Angeles | London | New Delhi
Singapore | Washington DC

Los Angeles | London | New Delhi
Singapore | Washington DC

FOR INFORMATION:

SAGE Publications, Inc.
2455 Teller Road
Thousand Oaks, California 91320
E-mail: order@sagepub.com

SAGE Publications Ltd.
1 Oliver's Yard
55 City Road
London EC1Y 1SP
United Kingdom

SAGE Publications India Pvt. Ltd.
B 1/I 1 Mohan Cooperative Industrial Area
Mathura Road, New Delhi 110 044
India

SAGE Publications Asia-Pacific Pte. Ltd.
3 Church Street
#10-04 Samsung Hub
Singapore 049483

Printed in the United States of America

ISBN 978-1-4129-9712-6

This book is printed on acid-free paper.

SUSTAINABLE FORESTRY INITIATIVE
Certified Chain of Custody
Promoting Sustainable Forestry
www.sfiprogram.org
SFI-01268

SFI label applies to text stock

Publisher: Jerry Westby
eLearning Editor: Robert Higgins
Editorial Assistant: Laura Kirkhuff
Production Editor: David C. Felts
Copy Editor: Kristin Bergstad
Typesetter: C&M Digitals (P) Ltd.
Proofreader: Lawrence W. Baker
Indexer: Joan Shapiro
Cover Designer: Janet Kiesel
Marketing Manager: Terra Schultz

15 16 17 18 19 10 9 8 7 6 5 4 3 2 1

BRIEF CONTENTS

Detailed Contents

READING

SECTION VIII. Southwest Asia 245

READINGS

SECTION IX. Europe, Turkey, and Russia 283

FOREWORD

The Criminal Justice System

You hold in your hands a book that we think is something new. It is billed a "text/reader." What that means is we have attempted to take the two most commonly used types of books, the textbook and the reader, and blend the two in a way that will appeal to both students and faculty.

Our experience as teachers and scholars has been that textbooks for the core classes in criminal justice (or any other social science discipline) leave many students and professors cold. The textbooks are huge, crammed with photographs, charts, highlighted material, and all sorts of pedagogical devices intended to increase student interest. Too often, however, these books end up creating a sort of sensory overload for students and suffer from a focus on "bells and whistles" such as fancy graphics at the expense of coverage of the most current research on the subject matter.

Readers, on the other hand, typically include recent and classic research articles on the subject matter. They generally suffer, however, from an absence of meaningful explanatory material. Articles are simply lined up and presented to the students, with little or no context or explanation. Students, particularly undergraduate students, are often confused and overwhelmed.

This text/reader represents our attempt to take the best of both the textbook and reader approaches to criminological theory. It can serve either as a supplement to a core textbook or as a stand-alone text. The book includes a combination of previously published articles and textual material introducing these articles and providing some structure and context for the selected readings. The book is broken up into a number of sections. The sections of the book track the typical content and structure of a textbook on the subject. Each section of the book has an introduction that serves to explain and provide context for the readings that follow. The readings are a selection of the best recent research that has appeared in academic journals, as well as some classic readings. The articles are edited as necessary to make them accessible to students. This variety of research and perspectives will provide the student with a grasp of the development of research, as well as an understanding of the current status of research in the subject area. This approach gives the student the opportunity to learn the basics (in the text portion of each section) and to read some of the most interesting research on the subject.

An introductory chapter explains the organization and content of the book and provides a context and framework for the text and articles that follow, as well as introducing relevant themes, issues, and concepts. This will assist the student in understanding the articles.

Each section will include a summary of the material covered and some discussion questions. These summaries and discussion questions should facilitate student thought and class discussion of the material.

It is our belief that this method of presenting the material will be more interesting for both students and faculty. We acknowledge that this approach may be viewed by some as more challenging than the traditional textbook. To that we say: Yes! It is! But we believe that if we raise the bar, our students will rise to the challenge. Research shows that students and faculty often find textbooks boring to read. We believe that many criminal justice instructors would welcome the opportunity to teach without having to rely on a "standard" textbook that covers only the most basic information and that lacks both depth of coverage and an attention to current research. This book provides an alternative for instructors who want to get more out of the basic criminal justice courses/curriculum than one can get from a basic textbook that is aimed at the lowest common denominator and filled with flashy but often useless features that merely serve to drive up the cost of the textbook. This book is intended for instructors who want to go beyond the ordinary, basic coverage provided in textbooks.

We also believe students will find this approach more interesting. They are given the opportunity to read current, cutting-edge research on the subject, while also being provided with background and context for this research. In addition to including the most topical and relevant research, we have included a short entry, "How to Read a Research Article." The purpose of this piece, which is placed at the beginning of the book, is to provide students with an overview of the components of a research article. It also serves to help walk them through the process of reading a research article, lessening their trepidation and increasing their ability to comprehend the material presented therein. Many students will be unfamiliar with reading and deciphering research articles; we hope this feature will help them to do so. In addition, we provide a student study site on the Internet. This site has additional research articles, study questions, practice quizzes, and other pedagogical material that will assist students in the process of learning the material. We chose to put these pedagogical tools on a companion study site rather than in the text to allow instructors to focus on the material, while still giving students the opportunity to learn more.

We hope that this unconventional approach will be more interesting, and thus make learning and teaching more fun. Criminal justice is a fascinating subject, and the topic deserves to be presented in an interesting manner. We hope you will agree.

Craig Hemmens, JD, PhD, Series Editor
Department of Criminal Justice and Criminology
Washington State University

PREFACE

Why this text, you ask? Yes, there are a number of excellent texts on violence and terrorism, as well as readers readily available to students and professors. But the texts and the readers are expensive and they do have difficulties. Texts may cover a great deal of material but in a very broad manner, often neglecting a more in-depth treatment of important topics. Readers may have an article with an in-depth treatment of a single issue or topic but there may be little explanation about why the reading was selected, or how it should be interpreted and understood in conjunction with the primary text and other readings. The authors hope to expose students to the key research and literature in this area, and encourage them to think critically and analytically about the state of studies in terrorism and homeland security. In addition, one of the key benefits of the text/reader approach is the ability to combine solid text materials and readings into the same textbook, thereby eliminating the need for students to buy multiple—often expensive—books for one course. Many of the journal articles were written for professional audiences and sometimes are lengthy and go into much detail. Therefore, where appropriate, the authors have edited, abridged, and explained the import of the articles selected. Certainly any instructor or any student may read entire articles by using the SAGE website access provided with this text.

The idea for this book stemmed from two catalysts. First was our very positive experience in using a text/reader in previous courses. As a result, the authors quickly saw the benefits of this text/reader series in engaging students with the course material. Second, we found ourselves searching, somewhat fruitlessly, for course materials that would work for our own course and the current texts we were using. This approach addresses this shortcoming as the authors married a thorough coverage of topics with readings that challenge students.

Too, many terrorism texts focus only on terrorism or only on homeland security or both, but none met our expectations. Our vision was to devise a text that lends itself to a course divided equally between the topics about terrorism (groups, incidents, types of attacks) and homeland security. In addition, the text takes a regional approach in the first half of the text/reader; a tour around the world of terrorism if you will. Each region is approached uniformly and has the following structure: (1) overview of the historical, religious, political, and social issues in the region: (2) the main terrorist groups operating (or having operated) in the region: and (3) a review of the main terrorist attacks attributed to those groups. This text/reader uses a regional approach for several reasons. First, many students are woefully short on the geographic knowledge and context needed to understand the actions or motivations of different terrorist groups whether based on religion, ideology, or nationalism. Second, because there is a great deal of overlap across types of terrorist groups, the regional approach provides students with a useful "hook" with which to categorize and distinguish groups. Within the discussion of each region, this book addresses whether the relevant groups can be typed as political, ideological, religious, or as having some other focus.

Finally, at the end of the textual material for each section there is a summary of key points, a listing of key terms that students should be able to define, a listing of review questions to serve as a study guide for students, and a list of useful websites to encourage further exploration of the topics covered in the section. Review questions available after each of the readings are designed to get students to think critically about the article they just read. Finally, each article/reading begins with a brief introductory paragraph that places the article in context with the rest of the book and suggests issues or questions for students to pay particular attention to. Each of the regional sections includes a map of the region. The homeland security sections include federal organizational charts prior to, and after, 2001, respectively.

The appendix describes the main tenets of Judaism, Christianity, and Islam to help students place the current conflicts in some context.

Structure of the Book

The text uses a typical outline for a violence and terrorism course and a course on homeland security, and it covers topics roughly equally divided between them. The sections are as follows:

Part I: Issues in the Study of Terrorism

Section 1: Introduction to the Study of Terrorism

This section introduces the book, provides an initial definition of terrorism, and details the types of terrorism covered in the text and what is not covered in the text. It discusses how the various disciplines study, research, and write about terrorism and homeland security. Examples of federal agencies and think tanks that study this are also presented. Lastly, this section introduces the concept of homeland security and the relatively new Department of Homeland Security. An example of a research article on terrorism is provided along with suggestions for how to read an article on terrorism research. Another reading for this section highlights how some take a critical terrorism approach that argues *for* including states in any definition and study of terrorism.

Section 2: Terrorism Definition and Typologies

This section provides students with an overview of the challenges of studying and understanding terrorism. Topics include the problems in defining terrorism, examples of other definitions of terrorism from various U.S. agencies and from other countries and perspectives, and definitions of terrorism from the international community. Terms frequently used in connection with terrorism studies are introduced. The authors note the difficulty with efforts to characterize terrorist groups and examine typologies of terrorist groups. This section concludes by addressing the distinction between state and non-state actors. The two readings selected for this section illustrate how difficult it is to actually define and study terrorism or terrorists.

Section 3: The History of Terrorism

This section provides a brief history of terrorism with a focus from the French Revolution forward. The authors use Rapoport's "Four Waves" as a guided organizational framework to present the history of terrorism by non-state actors. Rapoport's article is included in the readings.

Section 4: Methods, Patterns, and Trends of Terrorists

This section details the various methods of terror attacks and their (general) use by different types of terrorist groups, including bombings, assassinations, hostage taking, hijackings, suicide bombings, and the increasing use of remotely detonated Improvised Explosive Devices (IED) and Vehicle Borne Improvised Explosive Devices (VBIEDs).

Part II: Terrorism Around the World

These sections address terrorist groups by the region in which they operate or have operated (including issues regarding state-sponsored terrorism). Each regional section follows this format:

- Historical context of region
- Key groups operating in the region, including how and when the groups were established and their goals and motivations
- Major incidents conducted by the region's groups, with a focus on those involving U.S. citizens and property and the United States' national interest.
- Each regional section has two to four readings. The readings for the regional sections are generally articles that evaluate the success or failure of the groups in that region and/or evaluate government response to those groups. In some cases, articles that "explain" terrorism or the actions of terrorist groups or at least their stated justifications are presented. For the Middle East in particular, many articles assess conflicts over either land or religion (or both) and discuss how such conflicts generate terrorism in some places and more "conventional" unconventional warfare such as guerrilla actions or insurgencies in other places.

Section 5: Israel, Lebanon, and Palestine

Section 6: The Middle East and Africa

Section 7: Southeast Asia, South and North Korea, and China

Section 8: Southwest Asia

Section 9: Europe, Turkey, and Russia

Section 10: The Western Hemisphere, Including Canada, the United States, South America

Part III: Homeland Security

Section 11: Homeland Security: Before and After 9/11

This section is designed to give a broad but fairly detailed overview of the functions existing within the homeland security rubric prior to 2001. The section's organizational framework then follows with a comparison between pre- and post-2001 so that students can better understand how the current structure in the United States has evolved and assess the impact of the attacks on the United States on September 11, 2001.

Section 12: The Special Case of Weapons of Mass Destruction (WMD) Terrorism

This section covers the debate about the likelihood of an attack and the types of agents, including chemical, biological, radiological, and nuclear. It also addresses the special homeland security concerns in dealing with these types of attacks and the necessity for public-private partnership.

Section 13: Law and Terrorism:
Domestic and International Legal Regimes

This section covers basic legal provisions in place in the United States to deal with terrorism, especially the Patriot Act and the impact it has on civil liberties. International regimes, international organizations, and the numerous international treaties and agreements will be covered. Finally, the international cooperation against terrorism in practice will be presented.

Section 14: The Future of Terrorism

No one can know the future, but a consensus exists that allows prediction of future contentious issues that may lead to terrorism, and these are presented and discussed. These include rapid population growth, poverty, severe drought, limited opportunities, and unemployment.

Finally, the authors believe that they have provided a survey of the current state of violence and terrorism and a compilation of significant journal articles weighing in on issues. The authors recognize that violence and terrorism are dynamic and quickly changing significant phenomena that must be dealt with for some time to come.

 # Acknowledgments

We would first like to thank publisher Jerry Westby. He was tireless and supportive, and we are grateful for his efforts to enlist excellent reviewers making this text the best it could possibly be. Our copy editor, Kristin Bergstad, was amazing and thorough, and the manuscript benefited from her scrutiny. Senior Project Editor David Felts managed superbly to finish the process of proofing as well, and we thank him.

We are also very grateful to the reviewers who devoted so much time, effort and expertise during the writing/rewriting phase of production. Their efforts no doubt made this a much better text/reader:

Mark H. Beaudry, Anna Maria College

Michael M. Berlin, Coppin State University

Pamela M. Everett, Wayne State College

Shannon Hankhouse, Tarleton State University

Richard N. Holden, University of North Texas at Dallas

Linda Kiltz, Texas A&M University-Corpus Christi

John R. Michaud, Husson University

Robert G. Pastula, University of North Alabama

Brian L. Royster, Saint Peter's University

Shawn Schwaner, Miami Dade College

I

Issues in the
Study of Terrorism

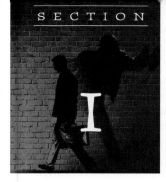

Introduction to the Study of Terrorism

Learning Objectives

At the end of this section, students will be able to:

- Define terrorism and the elements common to most definitions.
- Identify the challenges of studying terrorism.
- Discuss the debate of the role of the state in terrorism definitions.
- Discuss why it is important to define terrorism and measure its scope.
- Detail the response to the 9/11 attacks and the events that subsequently resulted in the Department of Homeland Security.
- Go to the CIA *World Factbook* site: https://www.cia.gov/library/publications/the-world-factbook/.
- Pick a country and take a look at the summary of the relevant facts and write a terrorism definition for that country.

 ## Introduction

This section introduces the concept of terrorism and discusses how it is studied and the various types of academic disciplines and organizations that may study terrorism. It details the limitations of the text to studying terror organizations, state-sponsors of terror, and some states that are clearly terrorist states, while recognizing that nation's and genocides as well as natural disasters have clearly wreaked more death, destruction, and chaos than any current modern terrorist organization. The concept of homeland security is introduced as well as the relatively new Department of Homeland Security. But the authors begin by providing an initial definition of terrorism. Terrorism should be a term

3

or concept that should be easily defined, right? Wrong. Defining the term is very difficult. Imagine that you are a corrupt, brutal dictator. You have managed to hang on for decades by rigging elections, generously taking care of those who facilitate you remaining in power, intimidating the press, and using the police and military to your advantage. But the economy has been destroyed through your incompetent efforts and the currency after a time of hyper-inflation is now worthless. Only U.S. dollars or similar currencies are accepted. Unemployment remains very high, probably around 80 percent. Through your bumbling efforts, the population tired and despite your unsurpassed tenacity for remaining in power, the last election was so narrow you were forced to share power with your opponent. Now, imagine your joy when WikiLeaks hands you a gift. Leaked cables from the U.S. Department of State may provide you with enough information to "prove" that your detractor and opponent actually encouraged other governments to continue sanctions on the country and on you personally. Imagine your joy as you immediately call your attorney general and ask him to peruse the statutes and determine if your country's definition of terrorism is sufficient to charge your opponent with terrorism. Unfortunately, the foregoing is not fiction. Consider this from the *Guardian* newspaper.

> Zimbabwe's Morgan Tsvangirai's call to public service has been a tortured one, punctuated by death and indignity. His numerous arrests and brushes with death began in 1997, when he emerged as the unlikely face of opposition to President Robert Mugabe. That year, Mugabe's henchmen nearly threw Tsvangirai from the window of his tenth floor office. He would be arrested on four separate occasions in the years to follow. During one such arrest, in 2007, he was severely beaten and tortured by Zimbabwean special forces at the behest of the ruling political party.
>
> After Zimbabwe's 2008 presidential contest—featuring incumbent Mugabe, Tsvangirai, and independent Simba Makoni—failed to award any candidate with the majority necessary to claim victory, the election defaulted to a runoff between the two highest vote-getters, Mugabe and Tsvangirai.
>
> In the days succeeding the first round of balloting, Tsvangirai was the alleged target of an assassination plot and subsequently taken into the custody of Mugabe's police, for which American and German diplomats demanded his immediate release. After initially committing to pursuing a second challenge to Mugabe, Tsvangirai withdrew in protest, lambasting the election as a "violent sham" in which his supporters were risking their lives to cast ballots in his favour. Indeed, it is estimated that over 100 Movement for Democratic Change supporters met an untimely demise in the period following the election.
>
> Following intense negotiations, the two parties agreed in February 2009 to a coalition government, in which Mugabe would remain head of state—a post he had held uninterrupted for 30 years—and Tsvangirai would assume the premiership. Not one month later, Tsvangirai and his wife were involved in a suspicious collision with a lorry. Though the prime minister survived, his wife for 31 years died.
>
> The *Guardian* last week published a classified US state department cable relating a 2009 meeting between Tsvangirai and American and European ambassadors, whose countries imposed travel sanctions and asset freezes on Mugabe and his top political lieutenants on the eve of Zimbabwe's 2002 presidential election. Though western sanctions don't prohibit foreign trade and investment or affect international aid—it's said that Zimbabwe's 2009 cholera epidemic topped 100,000 cases, registering some 4,300 deaths—the Mugabe administration effectively characterised the sanctions as an affront to the common Zimbabwean, further crippling the nation's already hobbled economy. (Zimbabwe's national unemployment figure hovers somewhere near 90%.)
>
> Publicly, Tsvangirai opposed the measures out of political necessity. In private conversations with western diplomats, however, the ascendant Tsvangirai praised its utility in forcing Mugabe's hand in the new unity government.
>
> Now, in the wake of the WikiLeaks' release, one of the men targeted by US and EU travel and asset freezes, Mugabe's appointed attorney general, has launched a probe to investigate Tsvangirai's involvement in sustained western sanctions. If found guilty, Tsvangirai will face the death penalty. (Richardson, 2011)

What Is Terrorism?

It seems as if the answer should be pretty straightforward but the authors suggest that if you put 20 professors in a room, those who study and even teach in this area would very likely have difficulty agreeing on a definition. The notion of **terrorism** and even discussions on defining terrorism bring out passions, emotions, biases, and anger. Consider what Gary LaFree, Director, National Consortium for the Study of Terrorism and Responses to Terrorism (START), University of Maryland, writes in 2009 about the difficulty in tabulating terror incidents:

> To begin with the term "terrorism" yields varying definitions, often loaded with political and emotional implications. As PLO Chairman Arafat famously noted in a 1974 speech before the United Nations, "One man's terrorist is another man's freedom fighter." Defining terrorism is no less complex for researchers. Researchers have identified dozens of different definitions of terrorism and it is not unusual for academic conferences to dedicate hours of discussion to exploring and defending competing definitions. Beyond the challenge of arriving at a defensible definition of terrorism are considerable challenges in collecting valid data on terrorism. In criminology, data on illegal violence come traditionally from three sources, corresponding to the major social roles connected to criminal events: "official" data collected by legal agents, especially the police; "victimization" data collected from the general population of victims and non-victims; and "self-report" data collected from offenders. However, each of these three data sources is problematic for examining terrorism: Official data on terrorism is incomplete because most individuals convicted of terrorism-related crimes are actually sentenced on non-terrorism charges such as homicide or weapons possession. Victimization data are difficult because often victims are randomly chosen, have little specialized knowledge about the attack, or do not survive the attack. And gaining access to active offenders through self-report surveys is both difficult and dangerous. (National Counterterrorism Center, 2009)

And why do organizations and others want to agree on a definition of terrorism? Researchers and scientists operate using a scientific method and the very first step is defining the problem. They want to define what constitutes terrorism because only then with documented cases or instances of terrorism do they get an idea of the scope of the problem. This can result in an allocation of resources devoted to doing something to diminish terrorism. How does one know what to do about terrorism or plan to deal with it if it cannot be defined? Too, many definitions of terrorism have a purpose, such as ensuring that the definition meets the need of clearly stating that it is a "crime" and that certain elements are met to allow prosecution. And some, including the subject and the author of one of your readings in the next section, take the position that defining terrorism and terrorists is a risky business, fraught with difficulties similar to labeling gangs criminal with the intent to marginalize them, persecute them, or even destroy them, as the introduction to this section clearly demonstrates.

While the next section explores defining terrorism in more detail, discusses issues and problems with definitions, and provides examples of various definitions, to set up a discussion of what exactly constitutes terrorism, let us examine a very satisfactory definition of terrorism.

Professor Cynthia C. Combs (2006) presents a definition of terrorism that has four major elements: "(1) It is an act of violence, (2) it has a political motive or goal, (3) it is perpetrated against innocent persons, and (4) it is staged to be played before an audience whose reaction of fear and terror is the desired result" (p. 19). What the good Professor Combs has done seems simple enough. Who could argue with that definition? But that definition is written from a perspective of an individual from the United States of America who enjoys immense freedom. What kind of definition of terrorism would someone like Morgan Tsvangirai want from a Zimbabwe government? There is the rub.

Is It Terrorism?

In 2009, Major Nidal Hasan opened fire at an Army base in Texas and killed 13 people. Review the definition of terrorism by Professor Combs. Is it terrorism? Not according to the federal government, which classified it as workplace violence, not as terrorism.

Is It Terrorism?

On September 12, 2014, Eric Frein, a survivalist, killed one Pennsylvania state trooper and wounded another. After a lengthy manhunt he was captured. He has been charged with criminal homicide of a law enforcement officer and terrorism.

To be fair, Professor Combs knows the difficulties in defining the term and wrestles with them, as do all. For perspective, let us take a look in detail at her definition. Terrorism must be an act of violence. That may seem pretty straightforward, but imagine you are living in a country, say Myanmar (Burma), where you have a brutal military dictatorship. They have sham elections; they brutalize with contempt any effort to protest, even peacefully. Would it be terrorism in that setting to consider an act of violence to overthrow that government?

The second point, having a political goal or motive seems pretty straightforward and it generally is as a matter of course. Terrorists engage in terrorism for a reason and are not shy about sharing those goals. Osama bin Laden periodically issued threats and called for killing Americans and their allies, stating it is an individual duty for every Muslim who can do it, in order to liberate the Al Aqsa mosque (Jerusalem) and the Holy Mosque (Mecca). And then, there is this phrase, "perpetrated against innocents," but what does that mean? Civilians are innocent surely, but what about civilians who work for the government and that government is Myanmar? And are military personnel innocents or acceptable targets? One of the authors of this text retired from the military and definitely has the opinion that they are not acceptable targets, but again, the military brutally enables the leadership in Myanmar and suppresses even peaceful demonstrations.

Next, terrorism is staged or played out before an audience, and clearly terrorism has a purpose and that is to instill fear and terror. Think back to the images of the attack on 9/11 and ask yourself if those acts of terrorism produced the desired result. Why do Americans remember those events so vividly? Is it because the media shamelessly and repeatedly ran countless days of scenes of the horror? This was followed by in-depth analyses by the media, and then more "specials" that guaranteed this act of terror was seared into your brain. One of the authors had a guest speaker in a class who worked in television, and he was asked how long the media could be responsible and respectful following a horrific act of terror. He answered that would be for about 30 minutes. And this is problematic, as the most successful terrorist act is not the one that kills the most people but gets the most publicity.

How to Study Terrorism?

Typing the above question into the Google search engine resulted in over 22 million hits. Not wanting to deal with that number, the authors went to the publisher of this text and their journals and entered the same question. Results were still large; over 8,000 journal articles were identified. So what does that tell us? There are many different attempts and efforts to study the topic, just as there are many definitions of terrorism. Browsing the journal articles, one finds that terrorism is studied as a political problem, a policing problem, or a criminal problem, just to name a few. And how does someone study terrorism? As an example, on the bookshelf of one of the authors is an ethnography about burglars. An ethnography is an observational, long-term study where researchers accompany, observe, and report what they see—in this case, burglars (with the blessing of law enforcement who said they welcomed any information they could get on burglars). But when you study terrorists, would this be a legitimate method?

Austin Turk (2004), a sociologist, also suggests that terrorism should be studied as (a) a social construct, meaning something that exists as a product of human social interaction instead of by virtue of objective, human-independent existence, or (b) as political violence, (c) communication, (d) organizing terrorism, (e) socializing terrorists, . . . social control of terrorism, and (g) theorizing terrorism. Researchers also spend a great deal of time and effort trying to understand how terrorism is financed and, how terrorists are organized, trained, and recruited, and they also spend a great deal of both time and effort studying terrorists' methods, such as suicide bombing.

To understand the breadth of studies and research devoted to terrorism, the authors visited a number of sites and present several that provide the reader with an idea of how varied and vast research is in this area. For example, the RAND Corporation devotes a great deal of time on this issue. Their National Security Research Division maintains a database, the **RAND Database of Worldwide Terrorism Incidents**, that tracks with some detail terrorism incidents around the world and then it provides an analysis of the incidents. Reports include evaluations of counterinsurgency efforts, piracy, and of course terrorism (RAND Corporation, 2011). RAND also has a Homeland Security and Defense Center; a RAND Infrastructure, Safety, and Environment center; and a Center for Terrorism Risk Management Policy. Obviously, the nation of Israel exists in a hostile area so it has a number of centers studying terrorism. Two examples are offered here. First, the International Institute for Counter-terrorism (defined as a range of efforts aimed at preventing terrorism) is a non-governmental organization that among other things tracks Islamic radicalization by country, monitors Jihadi Websites, maintains a database, and provides expertise in terrorism, counter-terrorism, homeland security, threat vulnerability and risk assessment, intelligence analysis, and national security and defense policy (International Institute for Counter-Terrorism, 2011). The other Israeli example is the Meir Amit Intelligence and Information Center. It writes bulletins on terrorism, hate, and anti-Semitic propaganda and incitement to violence, and it has an information center and a website on terrorism (Meir Amit Intelligence and Information Center, 2011).

Finally, the U.S. government has a new organization to put everything together under one organization to specifically counter the terror threat, the **National Counterterrorism Center** (NCTC). According to the site, the NCTC

> was established by Presidential Executive Order 13354 in August 2004, and codified by the Intelligence Reform and Terrorism Prevention Act of 2004 (IRTPA). NCTC integrates foreign and domestic analysis from across the Intelligence Community (IC) and produces a wide-range of detailed assessments designed to support senior policymakers and other members of the policy, intelligence, law enforcement, defense, homeland security, and foreign affairs communities. Prime examples of NCTC analytic products include items for the President's Daily Brief (PDB) and the daily National Terrorism Bulletin (NTB). NCTC is also the central player in the ODNI's [Office of the Director of National Intelligence] Homeland Threat Task Force, which orchestrates interagency collaboration and keeps senior policymakers informed about threats to the Homeland via a weekly update.
>
> NCTC leads the IC in providing expertise and analysis of key terrorism-related issues, with immediate and far-reaching impact. For example, NCTC's Radicalization and Extremist Messaging Group leads the IC's efforts on radicalization issues. NCTC's Chemical, Biological, Radiological, Nuclear Counterterrorism Group pools scarce analytical, subject matter, and scientific expertise from NCTC and CIA on these critical issues. (National Counterterrorism Center [NCTC], 2011)

The NCTC was created following recommendations from the 9/11 Commission Report. It is made up of some 500 personnel from various agencies of the federal government and is the primary organization for integrating and analyzing all intelligence pertaining to counterterrorism. It utilizes the World Wide Incidents Tracking System, which details all terrorist incidents worldwide and even now has a mapping feature (NCTC, 2011). Additionally it maintains a database on terrorist identities.

Ironically, criminologists, the logical folks to study this area, have, until recently, been absent from studying this. Many recently have begun studying this area and they provide a good framework to see what constitutes the study of terrorism. Lafree and Dugan (2004) provide the basis for a table outlining the study of crime and the study of terrorism:

Comparing the Study of Terrorism With the Study of Crime

Both Terrorism and Crime

- are interdisciplinary
- are social constructs
- are selectively prosecuted and show the discrepancy between law in books and law in action
- are disproportionately committed by young males
- undermine social trust, when they appear on a sustained level

Methodological

For the study of terrorism as for the study of crime, similar kinds of analysis are relevant:

- patterns, distributions, and trends
- geographic mapping
- time series analysis
- causal analysis
- life course analysis

Data Collection

For common crime, a wealth of different empirical data exist (official records, victimization and self-report surveys). For terrorist acts there are mainly "terrorism event" statistics (e.g., PGIS, ITERATE, RAND-MIPT and secondary data) (LaFree & Dugan, 2004, p. 56).

Who Else Studies Terrorism?

Criminologists are not the only academics studying terrorism. A quick browsing through a database of theses and dissertations yielded the following results of terrorism research done by various disciplines:

- Political Science
- Public Policy
- Education
- Philosophy
- Criminal Justice
- Security Studies
- Management
- American Studies
- International Relations
- Government
- Geography

This does not include those studying specific regions, areas, or countries, and those studying some languages. These are no doubt just a small sampling and one could easily add sociologists, cultural anthropologists, psychologists, and theologians as well. Many individuals from the various disciplines study the issues and problems created by terrorism at universities across the world, but many also work in the various intelligence agencies of the U.S. government and other nations' intelligence agencies. The United States alone has 17 agencies involved fully or in part in gathering intelligence with regard to terrorism, including the Central Intelligence Agency and the National Security Agency (Intelligence.gov, 2011).

Issues Terrorism Researchers Do Not Study

Terrorism researchers study the threats faced today, terrorist states, **state sponsors of terrorism**, terrorists groups, and individuals inspired and acting with support or inspired to act by terrorist groups. They do not study criminals or crimes, including horrible ones with scores of victims. They do not study, as a rule, the brutality that comes with war, leaving that to historians. This is not to minimize the impact that individuals and countries have had when it comes to slaughtering their citizens. In just the last century, the world witnessed mass murder and genocide on scales that are beyond imagining. Writing about concentration camps during the Second World War and the Nazis' extinction efforts, Theodore Abel (1951) coined the term *democide,*

> of which genocide is a sub-form pertaining specifically to the extermination of ethnic or racial groups. The broader term democide pertains to extermination procedures against a population selected on the basis of any kind of social attribute, racial, religious, educational, political, cultural, and so forth, including even distinctions on the basis of age. (p. 151)

R. J. Rummel, a professor emeritus, and political science scholar from the University of Hawaii, has an entire website dedicated to remembering the horror of 20th-century democide events. Some of the statistics are stunning and beyond comprehension. For example, he now estimates that between the years 1928 and 1987, in what is now the People's Republic of China, 76,702,000 individuals died (Rummel, 1994).

Interestingly, when you visit the CIA *World Factbook* Site regarding China, it waters down the numbers to "tens of millions" (CIA *World Factbook,* n.d.), but even that term is beyond comprehension. These events are surely worth studying but are not the focus of this text.

Finally, it is useful to jettison the notions of multiculturalism, tolerance, and diversity when studying terrorism and moral equivalency that "values" all peoples equally. They are not particularly useful when dealing with terrorists who have vowed to kill you. As an example, this occurred in 2011:

> The Imad Mughniyeh squad has claimed responsibility for the killing of five members of an Israeli family in the Jewish community of Itamar on March 12. The squad, part of the Fatah-sponsored Al Aqsa Martyrs Brigades said one of its fighters infiltrated Itamar and knifed the Jewish parents and three children, one of them a three-month-old. ("Iran-Backed Group," 2011)

Oh, and in the Gaza strip there were celebrations and the passing out of candy. That is terrorism.

For some perspectives, realize that many researchers and academics in the United States, Western Europe, and other countries maintain that it is foolish of terrorism scholars to neglect studying terror by the "state," and they include the United States, Great Britain, and France, among many as worthy of study. Ruth Blakeley has this to say:

State terrorism, along with other forms of repression, has been an ongoing feature of the foreign policies of democratic great powers from the North and the United States (US) in particular. The use of repression by the US was particularly intense during the Cold War, and we are seeing a resurgence of its use in the "war on terror." State terrorism, of which torture can sometimes be a tool, is defined as threats or acts of violence carried out by representatives of the state against civilians to instill fear for political purposes. (Blakeley, 2007, p. 258)

The thrust of her argument is that terrorist definitions in the West are used to focus analysis only on groups, not on states, and that is an interesting point, worthy of debate, and the authors are happy to do so. The authors do not believe that as part of a text on terrorism that time needs to be devoted to studying the West and democracies. Clearly this view is not shared by some, particularly those busy rewriting history. One example of this revisionism seems to occur on a regular basis on dates that mark the dropping of the atomic bombs on Japan in 1945. Should it have been done? This rethinking began almost as soon as the war was over. As one author notes:

Since then, there have been periodic eruptions of revisionism, uninformed speculation and political correctness on this subject, perhaps the most offensive of which was the Smithsonian Institution's plan for an exhibition of the Enola Gay for the 50th anniversary of the bombing of Hiroshima. In a particularly repugnant exercise of political correctness, the exhibit was to emphasize the "victimization" of the Japanese, mentioning the surprise attack on Pearl Harbor only as the motivation for the "vengeance" sought by the United States. (The exhibit as originally conceived was eventually cancelled.) (Miller, 2012)

One of the author's fathers was a Marine Raider in World War II and he would not think much of this as he spent 18 months in the hospital after being severely wounded on Okinawa. And note that the United States, which some of the above folks regard no doubt as "the Great Satan," has gone to war to assist Muslims in Kosovo and Bosnia, remove the Taliban in Afghanistan, get the Iraqi people free from a dictator, and as this is being written, the United States, along with others, is engaged in bombing Libya. One cannot think of any other country that would do that. Would China or Russia?

The authors do recognize that some regimes are terrorists; they are addressed in the text, and fortunately for us, Ambassador Mark Palmer has pointed them out and his call for a proposed Community of Democracies should be supported. This effort would form and pressure dictators, even friendly ones, to change or depart:

The United Nations, while it has global coverage, has a crippling flaw: dictators have full membership status, are treated as equals, and have representation at the top. China regularly deflects even the most trivial condemnations of other dictatorships, usually on the basis of "interference in internal affairs." Russia, while a tenuous democracy, maintains relationships with a host of brutal dictators. (Palmer, 2003, p. 46)

Palmer goes on to call for a Community of Democracies and an expanded North Atlantic Treaty Organization that will aim targeted sanctions against dictators. Imagine if all democratic countries cooperate in this effort and we no longer tolerate dictators.

⊠ What Is Homeland Security?

Homeland security is a term that simply refers to all efforts, offensive and defensive, to keep our country safe. It seems it was selected to be an all-encompassing term: "security was the umbrella term, incorporating local and national public-health preparedness for attack, the defense of the nation offered by the armed services, plus the intelligence and internal security activities of the C.I.A., F.B.I. and local police" (Safire, 2002, p. 12).

This text will present significant coverage of this topic, but it is introduced here since it is such a revolutionary change. Prior to the 2001 attack, many different agencies had some responsibility for keeping the United States secure. Following the attacks on the United States in 2001, President George W. Bush recommended a **Department of Homeland Security** in a 29-page document. His reasoning was clear—there were simply too many organizations (over 100) with some responsibility for homeland security. The president appointed Tom Ridge, the former governor of Pennsylvania, as his homeland security advisor to study this matter. Working with Congress, the Department of Homeland Security was formed by the **Homeland Security Act of 2002** (Proposal for Department of Homeland Security, 2002). It now has more than 240,000 employees and brought 22 agencies into it from many other federal departments as Chart 1.1 illustrates.

The following three directorates, created by the Homeland Security Act of 2002, were abolished by a July 2005 reorganization and their responsibilities transferred to other departmental components:

◆ Border and Transportation Security
◆ Emergency Preparedness and Response
◆ Information Analysis and Infrastructure Protection (Department of Homeland Security, 2008)

The second half of the text covers homeland security in more detail, but recognize that this has been a massive change, changing the federal government forever, in ways scholars will be attempting to understand for years. The department has the primary responsibility for counterterrorism, border security, preparedness, response and recovery, immigration, and cyber security.

Finally, the mission of the department, from its website, is below:

The Department's mission is to ensure a homeland that is safe, secure, and resilient against terrorism and other hazards.

We have five Departmental missions:

1. Prevent terrorism and enhance security

2. Secure and manage our borders

3. Enforce and administer our immigration laws

4. Safeguard and secure cyberspace

5. Ensure resilience to disasters

We also have taken significant steps to create a unified and integrated Department, focusing on accountability, efficiency, and transparency to enhance our performance and become a leaner, smarter agency better equipped to protect the nation.

Our efforts are supported by an ever-expanding set of partners, including federal, state, tribal, and local governments, law enforcement, the private sector, international allies, and individual citizens, communities, and organizations. Together, we are improving our awareness of risks and threats, working to build safer and more resilient communities and develop innovative approaches and solutions through advanced science and technology.

Every day, the more than 230,000 men and women of the Department contribute their skills and experiences to this important mission. Our duties are wide-ranging, but our goal is clear: a safer, more secure America. (Department of Homeland Security, n.d.)

Chart 1.1 Agencies in the Department of Homeland Security in 2003	
Original Agency (Department)	**Current Agency/Office**
U.S. Customs Service (Treasury)	U.S. Customs and Border Protection: inspection, border and ports of entry responsibilities
	U.S. Immigration and Customs Enforcement: customs law enforcement responsibilities
Immigration and Naturalization Service (Justice)	U.S. Customs and Border Protection: inspection functions and the U.S. Border Patrol
	U.S. Immigration and Customs Enforcement: immigration law enforcement; detention and removal, intelligence, and investigations
	U.S. Citizenship and Immigration Services: adjudications and benefits programs
Federal Protective Service	U.S. Immigration and Customs Enforcement
Transportation Security Administration (Transportation)	Transportation Security Administration
Federal Law Enforcement Training Center (Treasury)	Federal Law Enforcement Training Center
Animal and Plant Health Inspection Service (part) (Agriculture)	U.S. Customs and Border Protection: agricultural imports and entry inspections
Office for Domestic Preparedness (Justice)	Responsibilities distributed within FEMA
The Federal Emergency Management Agency (FEMA)	Federal Emergency Management Agency
Strategic National Stockpile and the National Disaster Medical System (HHS)	Returned to Health and Human Services, July, 2004
Nuclear Incident Response Team (Energy)	Responsibilities distributed within FEMA
Domestic Emergency Support Teams (Justice)	Responsibilities distributed within FEMA
National Domestic Preparedness Office (FBI)	Responsibilities distributed within FEMA
CBRN Countermeasures Programs (Energy)	Science & Technology Directorate
Environmental Measurements Laboratory (Energy)	Science & Technology Directorate
National Biological Weapons Defense Analysis Center (Defense)	Science & Technology Directorate
Plum Island Animal Disease Center (Agriculture)	Science & Technology Directorate
Federal Computer Incident Response Center (GSA)	US-CERT, Office of Cybersecurity and Communications in the National Programs and Preparedness Directorate
National Communications System (Defense)	Office of Cybersecurity and Communications in the National Programs and Preparedness Directorate
National Infrastructure Protection Center (FBI)	Dispersed throughout the department, including Office of Operations Coordination and Office of Infrastructure Protection
Energy Security and Assurance Program (Energy)	Integrated into the Office of Infrastructure Protection
U.S. Coast Guard	U.S. Coast Guard
U.S. Secret Service	U.S. Secret Service

Figure 1.1 U.S. Department of Homeland Security

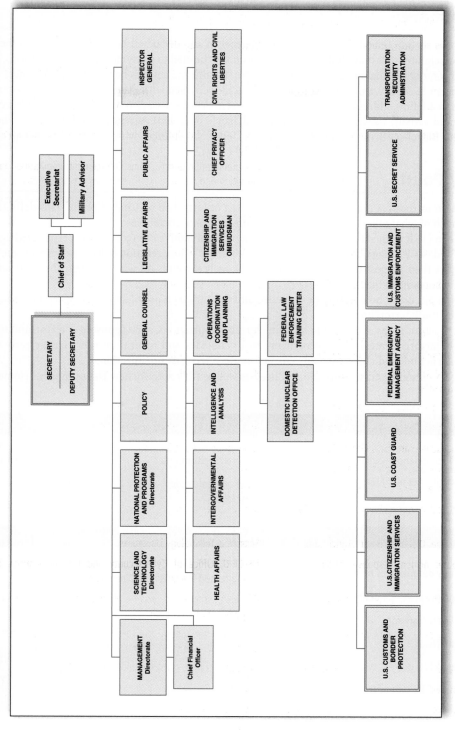

U.S. Department of Homeland Security

Summary

To reiterate, this section introduced the concept of terrorism and discussed how it is studied and the various types of academic disciplines that may study terrorism. This text is limited to studying terror organizations, state sponsors of terror, and some states that are clearly terrorist states. Finally, the authors have introduced readers to the Department of Homeland Security, a massive, complex bureaucracy charged with keeping the United States safe and secure. The readings that follow address separate issues. The first article discusses how to read a research article on the topic of terrorism and actually uses an abbreviated journal article by Robert Agnew as an illustration. The second is an article by Ruth Blakeley, "Bringing the State Back Into Terrorism Studies," which offers some perspective.

 Key Points

- This section introduces readers to a useful definition of terrorism by Cindy Combs. This definition is one that covers much of what other definitions dealing with terrorism encompass.
- To restate it: (1) It is an act of violence, (2) it has a political motive or goal, (3) it is perpetrated against innocent persons, and (4) it is staged to be played before an audience whose reaction of fear and terror is the desired result.
- The authors then suggest that defining terrorism largely depends on who one happens to be and the country where one happens to live.
- The section discussed how terrorism is studied and the various types of academic disciplines and organizations that might study terrorism.
- It details the limitations of the text to studying terror organizations, state sponsors of terror, and some states that are clearly terrorist states, while recognizing that genocides and natural disasters have clearly wreaked more death, destruction, and chaos than any current modern terrorist organization.
- The concept of homeland security is introduced as well as the relatively new Department of Homeland Security.

KEY TERMS

Department of Homeland Security

Homeland Security Act of 2002

National Counterterrorism Center

RAND Database of Worldwide
   Terrorism Incidents

State sponsor of terrorism

Terrorism

DISCUSSION QUESTIONS

1. What is your definition of terrorism?

2. What is the purpose of a definition?

3. Who studies terrorism?

4. What is the NCTC?

5. What does WITS do?

6. Should "states" be part of terrorism studies?

7. Does psychology have anything to offer terrorism studies?

WEB RESOURCES

American Civil Liberties Union: http://www.aclu.org

ASIS International: http://www.asisonline.org

Association of Former Intelligence Officers: http://www.afio.org

Center for Security Policy: http://www.centerforsecuritypolicy.org

DEBKAfile: http://www.debka.com (pro-Israeli)

Defense Advanced Research Projects Agency: http://www.darpa.mil

Defense Threat Reduction Agency: http://www.dtra.mil/

Democratic Underground: http://www.democraticunderground.com (very liberal)

Federation of American Scientists: http://www.fas.org

Free Republic: http://www.freerepublic.com (very conservative)

Freedom, Democide, War: http://www.hawaii.edu/powerkills/welcome.html

HIS Jane's 360: http://www.janes.com/security

Homeland Security Studies & Analysis Institute: http://www.homelandsecurity.org/

http://www.muhajabah.com/ (pro-Arab)

Johns Hopkins Center for Biodefense Strategies: http://www.hopkins-biodefense.org/

Middle East Media Research Institute: http://www.memri.org/index.html

Middle East Research and Information Project: http://www.merip.org

Middlebury Institute of International Studies at Monterey: http://www.miis.edu

Nuclear Waste and Weapons Solutions: http://www.nrdc.org/nuclear/default.asp

RAND Corporation: National Security: http://www.rand.org/natsec_area/

Sinn Féin: http://www.sinnfein.ie

Small Wars Journal: http://smallwarsjournal.com/ (Military, thoughtful)

Stormfront: White Nationalist Community: http://www.stormfront.org

U.S. Department of Defense: http://www.defenselink.mil/

U.S. Department of State: Bureau of Counterterrorism: http://www.state.gov/s/ct

Inspire: http://publicintelligence.net/inspire-al-qaeda-in-the-arabian-peninsula-magazine-issue-5-march-2011/ (this is for access to *Inspire,* the al Qaeda journal online); yes, it really exists with such great pieces as "How to Build a Bomb in the Kitchen of Your Mom." That would actually make a pretty good Rap song.

How to Read a Research Article on Terrorism

As you travel through your studies, you will soon learn that some of the best-known and emerging explanations, in any discipline, come from research articles in academic journals. This book is full of research articles, and you may be asking yourself, "How do I read a research article?" It is our hope to answer this question with a quick summary of the key elements of any research article, followed by the questions you should be answering as you read through the assigned sections. Do remember that terrorism studies is an area that stirs passion, debate, and agendas, and challenges scientific objectivity.

Most research articles published in a science or social science journal (admittedly some terrorism articles are sometimes way out there) will have the following elements: (1) introduction, (2) literature review, (3) methodology, (4) results, and (5) discussion/conclusion. In the introduction, you will find an overview of the purpose of the research. Within the introduction, you will also find the hypothesis or hypotheses. A hypothesis is most easily defined as an educated statement or guess. In most hypotheses, you will find that the format usually followed is "If X, Y will occur." For example, a simple hypothesis may be "If the price of gas increases, more people will ride bikes." This is a testable statement that the researcher wants to address in his or her study. Usually, authors will state the hypothesis directly, but not always. Therefore, you must be aware of what the author is actually testing in the research project. If you are unable to find the hypothesis, ask yourself what is being tested or manipulated and what the expected results are.

The next section of the research article is the literature review. At times, the literature review will be separated from the text in its own section, and at other times, it will be found within the introduction. In any case, the literature review is an examination of what other researchers have already produced in terms of the research question or hypothesis. For example, returning to our hypothesis on the relationship between gas prices and bike riding, we may find that five researchers have previously conducted studies on the increase of gas prices. In the literature review, the author will discuss their findings and then discuss what his or her study will add to the existing research. The literature review can also be used as a platform of support for the hypothesis. For example, one researcher may have already determined that an increase in gas prices causes more people to rollerblade to work. The author can use this study as evidence to support his or her hypothesis that increased gas prices will lead to more bike riding.

The methods used in the research design are found in the next section of the research article. In the methodology section, you will find the following: who/what was studied, how many subjects were studied, the research tool (e.g., interview, survey, observation), how long the subjects were studied, and how the data that were collected were processed. The methodology section is usually very concise, with every step of the research project recorded. This is important because a major goal of the researcher is reliability; describing exactly how the research was done enables other researchers to repeat it. Reliability is determined by whether the results are the same.

The results section is an analysis of the researcher's findings. If the researcher conducted a quantitative study, using numbers or statistics to explain the research, you will find statistical tables and analyses that explain whether or not the researcher's hypothesis is supported. If the researcher conducted a qualitative study, non-numerical research for the purpose of theory construction, the results will usually be displayed as a theoretical analysis or interpretation of the research question. The research article will conclude with a discussion and summary of the study. In the discussion, you will find that the hypothesis is usually restated, and there may be a small discussion of why this was the hypothesis. You will also find a brief overview of the methodology and results. Then the discussion section looks at the implications of the research and what future research is still needed.

Finally, the following section employs portions of an article from the journal *Theoretical Criminology*, which you have access to, by Robert Agnew, titled "A General Strain Theory of Terrorism," to provide an example of each aspect of a typical article. Agnew has proposed a theory suggesting that strain may produce crime. In his article he summarizes this:

A Brief Overview of General Strain Theory (GST)

GST states that certain strains or stressors increase the likelihood of crime (for overviews, see Agnew, 1992, 2001, 2006a, 2006b). Strains refer to events or conditions that are disliked by individuals. They involve negative or aversive treatment by others (receive something bad); the loss of valued possessions (lose something good), and/or the inability to achieve goals (fail to get what is wanted). Those strains most likely to increase crime are high in magnitude, are seen as unjust, are associated with low social control, and create some pressure or incentive for criminal coping.

Terrorism and Strain Theory

Robert Agnew

1. What is the thesis or main idea of this article?

The thesis or main idea is found in the introductory paragraph of this article. Agnew points out the main idea directly; you may read the introduction and summarize the main idea in your own words. For example, "The thesis or main idea is that general strain theory (GST) has been used to study and explain crime but should be extended to study sub-state terrorism and this article will do just that."

2. What is the hypothesis?

The hypothesis is found in the introduction of this article. It is first stated in the second paragraph: "it presents a general strain theory of terrorism, designed to explain why some people are more likely than others to form or join terrorist organizations and commit terrorist acts."

3. Is there any prior literature related to the hypothesis?

As you may have noticed, this article does not have a separate section called a literature review. However, you will see that Agnew devotes attention to prior literature under the heading "Current Strain-Based Explanations of Terrorism" and here, he offers literature regarding the history of GST research. This brief overview helps the reader understand the prior research.

4. What methods are used to support the hypothesis?

Agnew's methodology is known as a historical analysis or reviewing case studies. In other words, rather than conducting his own experiment, Agnew is using evidence from history to support his hypothesis regarding terrorism and GST. When conducting a historical analysis, most researchers use archival material from books, newspapers, journals, and so on. Agnew reviews case studies and the literature regarding terrorism and terrorist organizations and applies his previous work on GST to them.

5. Is this a qualitative study or quantitative study?

To determine whether a study is qualitative or quantitative, you must look at the results. Is Agnew using numbers to support his hypothesis (quantitative), or is he developing a nonnumerical theoretical argument (qualitative)? Because Agnew does not use statistics in this study, we can safely conclude that this is a qualitative study.

6. What are the results, and how does the author present the results?

Because this is a qualitative study, as we earlier determined, Agnew offers the results as a discussion of his findings from the historical analysis. The results may be found in the concluding section where Agnew suggests that he believes if researchers gain access to sufficient data to test the GST concerning terrorism, the data may yield some valuable information.

7. Do you believe that the author/s provided a persuasive argument? Why or why not?

This answer is ultimately up to the reader, but looking at this article, we believe that it is safe to assume that readers will agree that Agnew offered a persuasive argument.

8. Who is the intended audience of this article?

A final question that will be useful for the reader deals with the intended audience. As you read the article, ask yourself to whom the author wants to speak? After you read this article, you will see that Agnew is writing for researchers, professors, and policy makers. The target audience may most easily be identified if you ask yourself, "Who will benefit from reading this article?"

9. What does the article add to your knowledge of the subject?

One way to answer the question is as follows: This article helps the reader to understand that research on terrorism is not just about theoretical construction, but also about using theoretical construction and related research to conduct research into an area that is currently a major concern, terrorism.

10. What are the implications for public policy that can be derived from this article? Obviously, Agnew believes that policy makers should encourage research in this area and states this in his concluding paragraphs:

In addition to shedding light on the causes of terrorism, the general strain theory has important policy implications. The most obvious is to end or reduce collective strains of the above type.

Agnew then poses some examples, such as reducing strain by ensuring there are legitimate channels to air grievances.

Finally, we must again caution readers to remember that terrorism is a difficult subject to study and many individuals researching and writing in this area are passionate. The articles we will select come from all perspectives. Researchers try to maintain objectivity, but readers should realize that when it comes to terrorism, sometimes that is difficult.

READING 1

The reading presented next demonstrates how a theorist from criminology seeks to extend a useful theory on crime to terrorism. Agnew bases his explanation on the failure of individuals to achieve success through legitimate means and suggests that this may also explain why some turn to terrorism.

A General Strain Theory of Terrorism

Robert Agnew
Emory University, USA

It has been suggested that crime theories can shed much light on the causes of terrorism (Rosenfeld, 2002; LaFree and Dugan, 2004, 2008; Rausch and LaFree, 2007). Following that suggestion, this article applies general strain theory (GST) to the explanation of sub-state terrorism. The research on GST has focused almost exclusively on "common crimes," such as interpersonal assault, theft, and illicit drug use (although see Agnew et al., 2009). But as argued below, GST can contribute much to the explanation of terrorism, although the theory needs to be extended to account for this type of crime. The article is in three parts. First, it briefly reviews current strain-based explanations of terrorism. While promising, these explanations suffer from three major problems: they fail to describe the essential characteristics of those strains most likely to result in terrorism; they do not fully explain *why* such strains result in terrorism; and they do not explain why only a small percentage of those exposed to such strains turn to terrorism. Second, it provides a brief overview of GST, pointing to those key elements which can help address these problems. Third, it presents a general strain theory of terrorism, designed to explain why some people are more likely than others to form or join terrorist organizations and commit terrorist acts. In brief, this theory argues that terrorism is more likely when people experience "collective strains" that are: (a) high in magnitude, with civilians

affected; (b) unjust; and (c) inflicted by significantly more powerful *others*, including "complicit" civilians, with whom members of the strained collectivity have weak ties. These collective strains increase the likelihood of terrorism because they increase negative emotions, reduce social control, reduce the ability to cope through legal and military channels, foster the social learning of terrorism, and contribute to a collective orientation and response. These collective strains, however, do not lead to terrorism in all cases. A range of factors condition their effect, with these factors influencing the subjective interpretation of these strains; the emotional reaction to them; and the ability to engage in, costs of, and disposition for terrorism. Before applying GST to terrorism, however, it is first necessary to define terrorism. The many definitions of terrorism often disagree with one another, but several key elements are commonly mentioned (see National Research Council, 2002; LaFree and Dugan, 2004, 2008; Tilly, 2004; Goodwin, 2006; Hoffman, 2006; Post, 2007; Forst, 2009). Terrorism is defined as the commission of criminal acts, usually violent, that target civilians or violate conventions of war when targeting military personnel; and that are committed at least partly for social, political, or religious ends. Although not part of the formal definition, it is important to note that terrorist acts are typically committed by the members of subnational groups (LaFree and Dugan, 2004; Pape, 2005).

SOURCE: Agnew, R. (2010 May). A general strain theory of terrorism. *Theoretical Criminology, 14*(2), 131–153.

Current Strain-Based Explanations of Terrorism

Terrorism researchers commonly argue that strains or "grievances" are a major cause of terrorism (e.g. Gurr and Moore, 1997; Blazak, 2001; National Research Council, 2002; de Coming, 2004; Bjorgo, 2005; Pape, 2005; Victoroff, 2005; Callaway and Harrelson-Stephens, 2006; Goodwin, 2006; Hoffman, 2006; Robison et al., 2006; Piazza, 2007; Post, 2007; Smelser, 2007; Stevens, 2002; Freeman, 2008; LaFree and Dugan, 2008; Forst, 2009). Rosenfeld (2004: 23), in fact, states that "without a grievance, there would be no terrorism." Researchers, however, differ somewhat in the strains they link to terrorism. Terrorism is said to result from:

- absolute and relative material deprivation;
- the problems associated with globalization/modernization, such as threats to religious dominance and challenges to traditional family roles;
- resentment over the cultural, economic, and military domination of the West, particularly the United States;
- territorial, ethnic, and religious disputes resulting from postcolonial efforts at nation building and the breakup of the Soviet bloc;
- economic, political, and other discrimination based on race/ethnicity or religion;
- the problems encountered by certain immigrant groups, including unemployment,
- discrimination, and the clash between western and Islamic values;
- the denial of "basic human rights," including political rights, personal security rights, and the right to the satisfaction of basic human needs;
- harsh state repression, including widespread violence directed at certain groups;
- severe challenges to group identity or what Post (2007) calls "identicide";
- displacement or the loss of one's land/home;
- military occupation of certain types;
- threats to the status of working-class, white, male heterosexuals, including the loss of manufacturing jobs and the movements for civil, women's, and gay rights.

It should be noted that terrorists also explain their actions in terms of the strains they experience. This is apparent in the statements they make, the literature and videos they distribute, and on their websites (see Hoffman, 2006). The centrality of strain explanations for terrorists is frequently reflected in the names of their organizations, such as the Popular Front for the Liberation of Palestine and the Organization for the Oppressed on Earth (Hoffman, 2006: 21–2). Further, government figures frequently employ strain explanations when discussing terrorism. President George W. Bush, for example, stated that: "We fight against poverty because hope is an answer to terror" (quoted in Piazza, 2006: 160; also see Atran, 2003; Krueger and Maleckova, 2003; de Coning, 2004; Hoffman, 2006; Newman, 2006). Most of the academic research on strain and terrorism has involved case studies of terrorist groups. Such studies almost always conclude that strains played a central role in the formation of such groups (for overviews, see Callaway and Harrelson-Stephens, 2006; Hoffman, 2006; Post, 2007). It is possible, however, that similar strains do not lead to terrorism in other cases. Several quantitative studies have investigated this issue. Such studies should be interpreted with caution since they usually suffer from definitional, sampling, and other problems (Victoroff, 2005; Newman, 2007). Nevertheless, they provide the best test of the link between strain and terrorism. Surprisingly, such studies provide only mixed or weak support for strain explanations (see Gurr and Moore, 1997; Krueger and Maleckova, 2003; de Coning, 2004; Newman, 2006; Piazza, 2006; Robison et al., 2006; LaFree and Dugan, 2008). Most research has focused on the relationship between terrorism and material deprivation (absolute and, to a lesser extent, relative). Studies suggest that this relationship is weak at best (Atran, 2003; de Coning, 2004; Turk, 2004; Maleckova, 2005; Merari, 2005; Pape, 2005; Victoroff, 2005; Krueger and Maleckova, 2003; Newman, 2006; Piazza, 2006; Smelser, 2007; Araj, 2008; Forst, 2009). This is true at the individual level. For example, poor and poorly educated Palestinians are *not* more likely to support terrorism or engage in terrorist acts (Krueger and Maleckova, 2003; also see Maleckova, 2005). In some regions, terrorists are more often drawn from the ranks of the middle class and educated—including college students (Maleckova, 2005; Victoroff, 2005; Post, 2007). The weak link between deprivation and terrorism is also true at the macro-level. Most studies suggest that measures of material deprivation are unrelated or weakly related to the number of terrorist acts that take place in or originate in a country (Maleckova, 2005; Pape, 2005: 17–19; Newman, 2006; Piazza, 2006). It is a central contention of this article that this weak support stems from problems with the strain *explanations* that have been advanced, and not from the fact that strain plays a small role in terrorism. Drawing on GST, current strain explanations of terrorism suffer from three major problems. First, they fail to fully describe the core characteristics of strains likely to lead to terrorism. These theories typically focus on one or a few types of strain, such as material deprivation, threats to traditional values, and military occupation. But the characteristics of a given type of strain may differ greatly from situation to situation. For example, a type of strain such as material deprivation may differ in its magnitude (e.g. degree, duration, centrality, pervasiveness), perceived injustice, and source (e.g. is the source a more powerful *other*). Such differences, as argued below, have a major effect on whether the type of strain leads to terrorism. Second, most strain-based explanations of terrorism fail to fully explain *why* certain strains increase the likelihood of terrorism. Such explanations most commonly argue that the strains are intensely disliked and terrorism represents a desperate attempt to end them or seek revenge. The connections between strain and terrorism, however, are far more complex; and the failure to fully describe them significantly diminishes the completeness and policy relevance of existing strain explanations (more below). Finally, current strain explanations fail to explain why only a small portion of the individuals exposed to strains become involved in terrorism (see Victoroff, 2005: 19; Newman, 2006). For example, although 1.4 billion people lived in extreme poverty in 2005 (*New York Times*, 2008: A30), only a very small percentage turned to terrorism. The responses to strain are quite numerous, and include suffering in silence, legal challenge, common crime (e.g. theft, drug selling), political protest, and guerilla war. Current strain explanations of terrorism provide, at best, only limited information on those factors that influence or condition the response to strain. In sum, three problems account for the mixed or weak support for current strain-based explanations of terrorism. General strain theory (GST) holds the potential to correct for these problems. In particular, GST has much to say about the characteristics of strains most conducive to crime, the intervening mechanisms between strains and crime, and the factors that condition the effect of strains on crime. With some modification, these arguments can be used to construct a general strain theory of terrorism.

 # A General Strain Theory of Terrorism

Terrorism has certain special features that are in need of explanation. Terrorism is more extreme than most common crimes, since it often involves the commission of serious violence against civilians who have done nothing to directly provoke their victimization. Also, terrorists typically commit their acts with the support of sub-national groups, while most adult offenders act alone. Further, terrorism is committed wholly or in part for political, social, or religious reasons. Most common crimes, by contrast, are committed for reasons of self-interest. GST, then, must devote special attention to explaining the extreme and collective nature of terrorism.

Strains Most Likely to Contribute to Terrorism

Terrorism is most likely to result from the experience of "collective strains," or strains experienced by the members of an identifiable group or collectivity, most often a race/ethnic, religious, class, political, and/or territorial group. Only a small percentage of collective strains increase the likelihood of terrorism, however. These strains are: (a) high in magnitude, with civilian victims; (b) unjust; and (c) caused by significantly more powerful *others*, including complicit civilians, with whom members of the strained collectivity have weak ties. These arguments draw on GST (Agnew, 1992, 2001, 2006a, 2006b), but also take special account of the characteristics of and literature on terrorism (see especially Senechal de la Roche, 1996; Gurr and Moore, 1997; Black, 2004; Goodwin, 2006; Smelser, 2007).

Are High in Magnitude, With Civilian Victims (Nature of the Strain)

Collective strains are high in magnitude to the extent that they have the following characteristics: they involve acts which cause a *high degree of harm*, such as death, serious physical and sexual assault, dispossession, loss of livelihood, and major threats to core identities, values, and goals. They are *frequent, of long duration, and expected to continue into the future.* (However, strainful events—experienced in the context of persistent strains—may increase support for terrorism and precipitate terrorist acts (see Hamm, 2002; Oberschall, 2004: 28; Bjorgo, 2005; Newman, 2006; Post, 2007; Smelser, 2007: 34–5). And they are *widespread*, affecting a high absolute and/or relative number of people in the strained collectivity, including *many civilians* (defined as individuals not directly involved in hostile actions against the source of the collective strain). Case studies of terrorist organizations provide preliminary support for these arguments. Consider those strains associated with the emergence of several major terrorist groups: the Tamil Tigers, Basque Homeland and Liberty, Kurdistan Workers Party, Irish Republican Army, Shining Path, Hezbollah, Hamas, Revolutionary Armed Forces of Columbia, and al Qaeda. Such strains involved serious violence—including death and rape, major threats to livelihood, dispossession, large scale imprisonment or detention, and/or attempts to eradicate ethnic identity. Further, these strains occurred over long periods and affected large numbers in the collectivity, including many civilians (Callaway and Harrelson-Stephens, 2006; Hoffman, 2006; Post, 2007).

Are Seen as Unjust, Involving the Voluntary and Intentional Violation of Relevant Justice Norms by an External Agent (Reason for the Strain)

Collective strains may result from several sources other than the voluntary and intentional acts of an external agent. For example, they may result from the acts of members of the strained collectivity (e.g. some lower-class individuals victimize other lower-class individuals), from natural disasters (e.g. hurricanes, epidemics), or from "reasonable" accidents (e.g. airplane crashes, fires). In addition, collective strains may be seen as the result of "bad luck" (Merton, 1968) or supernatural forces, such as an angry God (see Smelser, 2007: 65). Terrorism is much less likely in these cases, even though the collective strain may be high in magnitude. Further, the voluntary and intentional infliction of collective strain by an external agent is unlikely to result in terrorism unless it also involves the violation of relevant justice norms. Several such norms appear to be applicable across a wide range of groups and cultures (Agnew, 2001, 2006a). In particular, the voluntary and intentional infliction of collective strain is more likely to be seen as unjust if:

(a) The strain is seen as undeserved. Strains are more likely to be seen as deserved if they result from the negatively valued behavior or characteristics of members of the strained collectivity that are deemed relevant in the particular situation. Further, the strain must not be seen as excessive given the behaviors or characteristics. To illustrate, members of a particular group may receive low pay for their work, but they may not view this as unjust if they believe they work in less demanding jobs and/or they have lower levels of education.

(b) The strain is not in the service of some greater good. Members of a collectivity, for example, may experience much loss of life during a war, but not view this as unjust if the war is seen as necessary.

(c) The process used to decide whether to inflict the strain is unjust. Among other things, victims are more likely to view the process as unjust if they have no voice in the decision to inflict the strain, they do not respect and trust those inflicting the strain, and no rationale is provided for the infliction of the strain.

(d) The strain violates strongly held social norms or values, especially those embodied in the criminal law.

(e) The strain that members of the collectivity experience is very different from their past treatment in similar circumstances and/or from the treatment of similar others (i.e. members of the collectivity are subject to discriminatory treatment).

Collective strains are likely to be viewed as unjust if conditions (a) and (b) are satisfied or if one of the other conditions is satisfied. Explanations of terrorism commonly make reference to the perceived injustice of the strains that are experienced. For example, Ahmed's

(2005: 5) account of Palestinian terrorism states that: "The fact is unmistakable and the message comes over loud and clear: a deep sense of injustice beyond the stage of profound frustration and despair stands at the heart of the issue."

Are Caused by More Powerful Others, Including "Complicit" Civilians, With Whom Members of the Strained Collectivity Have Weak Ties (the Relationship Between Those in the Strained Collectivity and the Source of Strain)

These "others" most commonly differ from members of the strained collectivity in terms of some salient social dimension, such as religion, race/ethnicity, class, territorial location, nationality, and/or political ideology. They are more powerful because of their greater resources, including numbers, military equipment and skills, and/or support from others. The strain they inflict may be partly attributed to civilians for several reasons (see the excellent discussion in Goodwin, 2006). Civilians may play a role in creating the Government or organization that inflicts the strain (e.g. through voting); they may support the Government/organization through acts such as paying taxes, public expressions of support, and service in government agencies; they may benefit from the infliction of the strain (e.g. occupying land formerly held by those in the strained collectivity); and they may fail to take action against those who inflict strain when such action is seen as possible (also see Pape, 2005: 137). Goodwin (2006) roughly measures civilian complicity in terms of whether the source of strain is a democratic state; the argument being that terrorists are more likely to believe that civilians in democratic states play major roles in electing and influencing their governments. Finally, members of the strained collectivity have weak emotional and material ties to the source of strain. These weak ties may stem from lack of contact, strong cultural differences (e.g. differences in language, values, beliefs, norms), and/or large differences in wealth/status/power, which tend to limit positive interaction and mutually beneficial exchange (Senechal de la Roche, 1996; Black, 2004; Goodwin, 2006). *In sum*, several characteristics related to the nature of the collective strain, reason for the strain, and the relationship between the recipients and source of strain influence the likelihood of terrorism. Most of these characteristics vary even when the focus is on a particular type of strain, such as material deprivation. Researchers sometimes take account of certain of these characteristics, but rarely

consider all of them. And this is a major reason for the weak quantitative support for strain theories of terrorism.

Why Do Strains of the Above Type Increase the Likelihood of Terrorism?

This section describes the intervening mechanisms between collective strains and terrorism. Examining such mechanisms not only provides a fuller explanation of terrorism, but suggests additional ways to prevent terrorism. Terrorism can be prevented not only by reducing or altering the strains that contribute to it, but also by targeting the intervening mechanisms below.

The Above Collective Strains Lead to Strong Negative Emotional States and Traits—Including Anger, Humiliation, and Hopelessness—Which Are Conducive to Terrorism

Strains of the above type contribute to a range of negative emotional *states*, and the persistent experience of these strains contributes to a heightened tendency to experience negative emotional states (referred to as an emotional *trait*). Negative emotions create much pressure for corrective action; individuals feel bad and want to do something about it. These emotions also reduce the ability to cope in a legal manner. Angry individuals, for example, are less able to accurately access their situation and effectively communicate with others. Further, these emotions lower inhibitions, reducing both the awareness of and concern for the consequences of one's behavior. Finally, certain of these emotions create a strong desire for revenge, with individuals feeling they must "right" the wrong that has been done to them (see Agnew, 2006a for an overview). There is much anecdotal data suggesting that negative emotions play a key role in the explanation of terrorism (Stern, 2003; Victoroff, 2005; Moghadam, 2006a, 2006b; Newman, 2006; Forst, 2009). A member of the Tamil Tigers, for example, stated that:

> In the late '90s when I was in school, the Sri Lanka military bombed my village. An elderly woman lost both legs, one person dies and two students were injured. I was angry with the [military] and joined the Tigers one year later (Post, 2007: 92).

Related to this, many terrorists state that revenge is a major motive for their acts. Araj (2008), in fact, argues that the desire for revenge is so strong that individuals will sometimes commit terrorist acts even when they believe that doing so will impede the achievement of their ultimate goals.

These Strains Reduce the Ability to Legally and Militarily Cope, Leaving Terrorism as One of the Few Viable Coping Options

The above strains reduce the ability of those in the strained collectivity to effectively employ such coping strategies as negotiation, lobbying, protest, appeals to external agents such as the United Nations, and insurgency. These strains frequently involve the massive loss of material and other resources, which facilitate these forms of coping. The weak ties between members of the strained collectivity and the source of strain further reduce the likelihood that many of these coping strategies will be effective, since the source has little emotional or material incentive to respond to the requests of those in the strained collectivity. In addition, these strains often involve exclusion from the political process and the brutal suppression of protest movements (Callaway and Harrelson-Stephens, 2006; Post, 2007). Finally, the significantly greater power of the source of strain reduces the effectiveness of these coping options (e.g. military campaigns by those in the strained collectivity are unlikely to be successful). Those in the strained collectivity may turn to common crimes in an effort to cope; for example, they may engage in theft to reduce their material deprivation. Common crimes, however, do little to end the collective strain and frequently do little to alleviate individual suffering. Common crimes, for example, are not an effective remedy for those collective strains involving violence or displacement. Further, common crime may not be a viable option in circumstances of massive deprivation. Terrorism, then, is often one of the few remaining coping options. While those in the strained collectivity may not have the resources to mount an effective military campaign, it is usually the case that they can easily target civilians. Civilians are generally more accessible and less able to resist attack than military targets. Further, there is some evidence that terrorism is an effective coping strategy in certain cases, ending or alleviating collective strain (see Pape, 2005; Victoroff, 2005; Hoffman, 2006; Moghadam, 2006b; Smelser, 2007). Also, terrorism serves other important functions for members of the strained collectivity (more below).

These Strains Reduce Social Control

In addition to reducing the ability to effectively cope through legal and military channels, the above strains also reduce most of the social controls that prevent terrorism (see Agnew, 2006a). In particular, these strains further weaken the emotional ties between members of the strained collectivity and the source of strain. They rob those in the strained collectivity of valued possessions, as well as hope for the future, leaving them with little to lose if they engage in terrorism. They weaken the belief that terrorism is wrong (more below). And they reduce the likelihood that members of the strained collectivity will sanction terrorists, since the experience of these strains tends to create tolerance, sympathy, or even support for terrorism. Again, there is much anecdotal evidence for these arguments in the terrorism literature, with terrorists frequently stating that they have weak/hostile ties to the source of their strain, that their strain has left them with little to lose, and that they no longer condemn terrorism (e.g. Post, 2007; Araj, 2008).

These Strains Provide Models for and Foster Beliefs Favorable to Terrorism

Collective strains of the above type frequently involve violent acts against civilians, thereby providing a model for terrorism. To illustrate, an aide to Arafat stated that Israeli civilians "are no more innocent than the Palestinian women and children killed by Israelis" (Hoffman, 2006: 26). These strains also foster beliefs that excuse, justify, or even require terrorism. Recall that these strains are high in magnitude, involve civilian victims, are seen as unjust, and are inflicted by more powerful *others*, including complicit civilians, with whom those in the strained collectivity have weak ties. Given these circumstances, it is not difficult for those in the strained collectivity to employ such techniques of neutralization as denial of the victim (the source of strain deserves punishment), appeal to higher loyalties (terrorism is necessary to protect those in the collectivity), and condemnation of the condemners (terrorism is no worse than the acts committed by the source of strain) (see Gottschalk and Gottschalk, 2004; Bloom, 2006; and Post, 2007 for examples).

These Strains Foster a Collective Orientation and Response

Members of the strained collectivity believe they are under serious assault by more powerful others with whom they have weak ties. This does much to foster a heightened sense of collective identity (see Hogg and Abrams, 2003; Stevens, 2002; Post, 2007). This identity amplifies the experience of vicarious strains, since we care more about those we closely identify with (Agnew, 2002). It creates a sense of "linked fate," or an "acute sense of awareness (or recognition) that what happens to the group will also affect the individual member" (Simien, 2005: 529). And it creates a sense of obligation to protect others in the collectivity, at least among those traditionally cast in the protector role. This collective orientation helps explain the terrorism of those who have not personally experienced severe strain. Such individuals strongly identify with others in the collectivity and, through this identification, they vicariously experience, feel personally threatened by, and feel responsible for alleviating the strain experienced by these others (see McCauley, 2002: 9). The literature on terrorism provides numerous illustrations of these points (e.g. McCauley, 2002; Gupta, 2005; Pape, 2005; Post, 2005; Victoroff, 2005: 21–2, 30; Loza, 2006; Forst, 2009).

> A Palestinian terrorist, for example, stated that she grieves for the loss of her homeland, for the loss of a whole people, the pain of my entire nation. Pain truly affects my soul; so does the persecution of my people. It is from pain that I derive the power to resist and to defend the persecuted. (Post, 2007: 26)

This collective orientation is also important because it contributes to the formation of "problem solving" groups that respond to the collective strain. When individuals confront shared problems that they cannot solve by themselves, they may develop a collective solution to their problems—one that sometimes takes the form of a criminal group (Cohen, 1955; Cloward and Ohlin, 1960). The Internet and media have come to play a critical role in facilitating the formation of such groups, since they publicize strains, allow strained individuals to (virtually) interact with one another, and facilitate the recruitment of individuals by terrorist groups (see especially Hoffman, 2006; Moghadam, 2006b; Forst, 2009).

Terrorist Groups, Once Formed, Promote Terrorism in a Variety of Ways

For reasons suggested above, those problem-solving groups that develop in response to collective strains of the above

type are often disposed to terrorism. These groups, in turn, play a critical role in the promotion of terrorism (Caracci, 2002; Hamm, 2002; McCauley, 2002; National Research Council, 2002; Victoroff, 2005: 30–1; Smelser, 2007). The members of such groups model terrorism; differentially reinforce terrorism—usually with social approval/status; promote the adoption of beliefs favorable to terrorism; and diffuse responsibility for terrorist acts. These effects are often heightened by isolating group members from others who might challenge the aims of the terrorist group. On a more practical level, terrorist groups provide informational, material, and other support necessary for the commission of many terrorist acts. It is important to note that while collective strains contribute to the development of terrorist groups that pursue collective goals, such groups also alleviate a range of individual strains. As suggested above, collective strains embody a host of individual strains; including feelings of anger, humiliation, and hopelessness; identity threats; and the loss of material possessions. Participation in terrorist groups allows for the alleviation of these strains (e.g. McCauley, 2002; Stevens, 2002; Stern, 2003; Victoroff, 2005; Moghadam, 2006a; Post, 2007; Forst, 2009). In particular, participation provides an outlet for one's rage, a sense of self-worth, and status. A Palestinian terrorist, for example, stated that: "An armed action proclaims that I am here, I exist, I am strong, I am in control, I am in the field, I am on the map" (Post, 2007: 61). Further, participation may alleviate material deprivation, since terrorist organizations frequently provide material aid to terrorists and their families (Hoffman, 2006). In addition, participation may address individual strains not directly linked to the collective strain. Abrahms (2008), for example, argues that many terrorists are socially alienated and that participation in terrorist groups allows them to develop close ties to others. Terrorist organizations, then, allow for the alleviation of a range of individual strains; some linked to the collective strain and some not. This fact helps explain the persistence of such organizations in the face of both repeated failure and full success (see Victoroff, 2005; Abrahms, 2008).

 ## Factors That Condition the Effect of Collective Strains on Terrorism

While collective strains of the above type are conducive to terrorism, they do not guarantee terrorism. The members of certain collectivities experiencing these strains have not turned to terrorism or have only turned to terrorism after many years (Gupta, 2005; Pape, 2005; Bloom, 2006; Moghadam, 2006a; Goodwin, 2007; Post, 2007). This is not surprising given the extreme nature of terrorism and its mixed effectiveness (see Victoroff, 2005; Hoffman, 2006; Moghadam, 2006b; Smelser, 2007; and Abrahms, 2008 for discussions on the effectiveness of terrorism). This section draws on GST and the terrorism literature to describe those factors that influence or condition the effect of strains on terrorism. These factors influence the *subjective* interpretation of strains; that is, the extent to which given strains are seen as high in magnitude and due to the unjust acts of others, including civilians. They also influence the emotional reaction to strains, the ability to engage in both non-terroristic and terroristic coping, the costs of terrorism, and/or the disposition for terrorism. It is important to note that while these factors are to *some* extent independent of the collective strains experienced, collective strains may alter them in ways conducive to terrorism. For example, the continued experience of collective strains may alter individual and group beliefs such that they come to excuse, justify, or require terrorism (see above).

Coping Resources, Skills, and Opportunities

The members of some collectivities may be better able to cope through nonterroristic means. This includes collectivities with extensive financial resources and legal and political skills. Also critical are the opportunities for coping provided by the larger political environment. In this area, some argue that terrorism is less likely in democratic states since there are more opportunities for legal coping (Crenshaw, 1995; National Research Council, 2002; Callaway and Harrelson-Stephens, 2006; Krueger and Maleckova, 2006; Piazza, 2007; Freeman, 2008). The research on the relationship between democracy and terrorism, however, is mixed (see the overview in Maleckova, 2005; also see Newman, 2006; Robison et al., 2006; Piazza, 2007; Abrahms, 2008; Freeman, 2008). This may reflect the fact that while democracies provide more opportunities for legal coping, they also provide more opportunities for terrorists—with democracies being less willing to harshly repress terrorists and more willing to negotiate with them (see Piazza, 2007). In addition to the ability and opportunity to engage in non-terroristic coping, it is important to consider the ability/opportunity to engage in terrorism. At the individual level, this ability includes certain physical skills and a willingness to engage in risky behavior; attributes which tend to favor the young males who most often engage in terrorism (LaFree and Dugan, 2004: 56; Forst, 2009: 22–3). At the group/collectivity level, this ability involves the knowledge, material resources (e.g. money, munitions), and organization to commit terrorist acts (see Gurr and Moore, 1997; Oberschall, 2004). It has been argued that there are more opportunities for terrorism in "failed states," since such states are less able to repress terrorist groups and often provide a base for such groups to operate. There is limited support for this view (Newman, 2007; Piazza, 2007; Forst, 2009: 409).

Social Support

Individuals, groups, and the collectivity itself may receive support for nonterroristic coping. Other individuals and groups, including foreign nations, may attempt to alleviate strain through the provision of such things as food, shelter, medical care, and military protection. They may attempt to end the strain through persuasion, sanction, and military intervention. And they may provide information, material assistance, and moral support in an effort to help those in the strained collectivity cope through non-terroristic means. Such support should reduce the likelihood of terrorism, particularly if it is believed to be effective and that terrorism will jeopardize it. For example, Goodwin (2007) argues that the African National Congress avoided terrorism partly because it feared alienating supportive groups. Individuals, groups, and collectivities may also receive support for terroristic coping, including information, moral support, material resources, and direct assistance (e.g. the provision of outside fighters). Such support may come from outside groups, including governments and foreign terrorist organizations. For example, the PLO and other terrorist organizations helped train members of the Tamil Tigers in the late 1970s and early 1980s (Pape, 2005: 73). And such support may come from internal sources. Individuals may receive support for terrorism from friends and family, terrorist groups, and members of the larger collectivity. And terrorist groups may receive support from members of the larger collectivity. Some researchers argue that it is unlikely that individuals and groups will engage in terrorism without such support (e.g. Merari, 2005; Pape, 2005; Smelser, 2007). There are rare cases of lone terrorists, but such terrorists are often loosely affiliated with others who support terrorism (see Hamm, 2002; McCauley, 2002; Smelser, 2007).

Social Control

While collective strains of the above type reduce most forms of social control, there may nevertheless be some independent variation in the control experienced by those in the strained collectivity. In particular, the source of strain may exercise high direct control over individuals and groups in the strained collectivity, thus reducing the likelihood of terrorism (see Gupta, 2005; Bloom, 2006; Callaway and Harrelson-Stephens, 2006; Robison et al., 2006; Smelser, 2007; Abrahms, 2008; Araj, 2008). This was the case in Cambodia under the Khmer Rouge, in Iraq under [Saddam Hussein], and in Germany under Hitler. Also, some strained individuals and groups may maintain their bonds with selected individuals associated with the source of the strain, again reducing the likelihood of terrorism. This is said to partly explain why the African National Congress avoided terrorism against white civilians; there were close ties between the ANC and whites involved in the antiapartheid movement (Goodman, 2007). Further, some strained individuals and groups may have valued possessions—including both material possessions and reputations—that would be jeopardized by terrorism. Finally, individuals and groups within the collectivity are less likely to engage in terrorism when it is condemned and sanctioned by others in the collectivity (see Pape, 2005).

Individual Traits

Terrorists are no more likely than comparable controls to suffer from psychopathology (McCauley, 2002; Atran, 2003; Pape, 2005; Victoroff, 2005; Post, 2007). Certain other traits, however, may increase the disposition for terrorism. Such traits include negative emotionality, low constraint, and cognitive inflexibility (Gottschalk and Gottschalk, 2004; Post, 2005; Victoroff, 2005: 27; Loza, 2006). Individuals with these traits are especially sensitive to strains, inclined to aggressive coping, attracted to risky activities, and prone to view the world in "black and white" terms. Also, those who are alienated and socially marginalized may be more inclined to terrorism, since they have less to lose through terrorism, terrorist groups may provide them with a sense of belonging, and terrorism may be seen as a solution to certain of their problems. Such individuals include young, unmarried males; widows; those not gainfully employed; and unassimilated immigrants (see National Research Council, 2002; Merari, 2005; Post, 2005; Smelser, 2007; Abrahms, 2008; Forst, 2009; see Victoroff, 2005 for a discussion of other traits that may contribute to terrorism).

Association With Close Others Who Support Terrorism

Associating with close others who support terrorism has a major effect on the disposition for terrorism. As indicated above, such others may model terrorism, reinforce terrorism, teach beliefs favorable to terrorism, and provide the training and support necessary for many terrorist activities. Anecdotal accounts and some research suggest that individuals whose family members and friends are involved in terrorism are much more likely to be involved themselves (Sageman, 2005; Victoroff, 2005; Post, 2007; Smelser, 2007; Abrahms, 2008; Forst, 2009).

Beliefs Favorable to Terrorism

The beliefs/ideology of those experiencing collective strains also influence the disposition for terrorism. Such beliefs may be learned from family members, friends, schools, neighbors, religious figures, and a variety of media sources (National Research Council, 2002; Victoroff, 2005: 18; Post, 2007; Forst, 2009). Beliefs favorable to terrorism have at least some of the following features: they emphasize the importance of collective identity (e.g. religious affiliation, ethnicity); increase the sensitivity to certain strains by, for example, placing much emphasis on "honor" and "masculinity"; claim that the collective strain being experienced is high in magnitude; provide an explanation for the strain, attributing it to the unjust acts of more powerful others, including complicit civilians; provide guidance on how to feel in response to the strain, with negative emotions such as rage and humiliation being emphasized; depict the source of strain as evil, subhuman, and otherwise deserving of a harsh response; encourage little or no contact with the source of strain; depict the source as both powerful but vulnerable to attack; point to the special strengths of those in the strained collectivity; excuse, justify, or require a terroristic response; provide a vision of a more positive, often utopian future that will result from such a response; promise rewards to those who engage in terrorism, including martyrdom and rewards in the afterlife; and create a history to support these views (e.g. emphasize the past victories of the collectivity over similar injustices) (Gupta, 2005; Loza, 2006; Moghadam, 2006a; Post, 2007; Smelser, 2007). While such beliefs are fostered by the experience of collective strains of the above type, they are also a function of other factors (see Smelser, 2007). As a result, some groups experiencing the above collective strains hold beliefs that discourage terrorism. For example, Goodwin (2007) argues that the emphasis of the African National Congress on nonracialism helped discourage terrorism against white South Africans. And the Dalai Lama's advocacy of the "middle way" has likely done much to prevent terrorism by the Tibetans against the Chinese (see Wong, 2008).

Anticipated Costs and Benefits of Terrorist Acts

Estimates of costs/benefits are partly a function of the success of prior terrorist acts, both those committed by the collectivity in question and by others (Gurr and Moore, 1997; Pape, 2005; Hoffman, 2006; Sedgwick, 2007). It is important to note, however, that it is difficult to objectively define "success" (see Pape, 2005; Goodwin, 2006; Hoffman, 2006; Moghadam, 2006a; Abrahms, 2008; Newman and Clarke, 2008). Terrorist acts seldom result in the end of the collective strain experienced. Such acts, however, may be deemed successful if they call greater attention to the collective strain, gain recruits or other support for the terrorist organization, boost the morale of those in the organization and sympathizers, inflict significant damage on the source of the collective strain, or result in the partial alleviation of the strain. Researchers must therefore consider the subjective views of those involved in terrorism when assessing success. Further, the anticipated costs and benefits of terrorism are influenced by a host of more immediate factors, including those having to do with the availability of attractive targets and the absence of capable guardians (Newman and Clarke, 2008).

 ## Conclusion

The general strain theory of terrorism presented in this article builds on current strain-based explanations of terrorism in three ways. First, it better describes the core characteristics of strains that contribute to terrorism. Terrorism is most likely in response to collective strains that

are high in magnitude, with civilian victims; unjust; and caused by more powerful others, including complicit civilians, with whom members of the strained collectivity have weak ties. Second, it more fully describes the reasons why such strains increase the likelihood of terrorism. In particular, such strains lead to negative emotional states and traits; reduce the ability to effectively cope through legal channels, common crime, and military means; reduce social control; provide models for and contribute to beliefs favorable to terrorism; and foster a collective orientation and response to the strain. Third, it provides the most complete description of those factors that condition the effect of the above strains on terrorism. Such factors include a range of coping resources, skills, and opportunities; various types of social support; level of social control; selected individual traits; association with others who support terrorism; beliefs related to terrorism, and the anticipated costs and benefits of terrorism. The general strain theory of terrorism extends GST in important ways; pointing to new strains, intervening mechanisms, and conditioning variables that are especially relevant to terrorism. It is important to note, however, that the general strain theory of terrorism is not a complete explanation of terrorism. Being a social psychological theory, it does not describe the larger social forces that contribute to the development of the above strains and help shape the reaction to them. Also, collective strains of the above type are likely only one of several causes of terrorism. Indeed, the final section of this article lists several factors that have been said to directly affect terrorism, such as social controls, beliefs/ideologies, association with others who support terrorism, and the anticipated costs and benefits of terrorism. A complete explanation of terrorism will require that we draw on a range of theories and describe the complex relations between them. The development of such an explanation is beyond the scope of this article, but this article does describe what will likely be a central variable in this explanation—collective strains of a certain type. In addition to shedding light on the causes of terrorism, the general strain theory has important policy implications. The most obvious is to end or reduce collective strains of the above type. For example, the source of strain may attempt to reduce civilian causalities. In addition, it may be possible to target those intervening mechanisms that link collective strains to terrorism. For example, governments may make it easier to address grievances via legal channels. Further, conditioning variables may be targeted. Outside groups, for example, may provide social support to members of the strained collectivity. These multiple points for intervention provide some hope for efforts to reduce terrorism.

At the same time, it is important to note that collective strains of the above type often set in motion a self-perpetuating process that is hard to interrupt. These strains gradually change members of the strained collectivity in ways that increase the likelihood of a terroristic response. Among other things, these strains foster individual traits conducive to terrorism, such as negative emotionality; further reduce social control, including ties to the source of strain; lead to the adoption of beliefs that favor terrorism; and contribute to the development of terroristic organizations. In addition, the terrorism carried out by such organizations frequently provokes a harsh response, which further increases support for terrorism in the strained collectivity (see Hamm, 2002; Smelser, 2007: 80–1; Araj, 2008; LaFree and Dugan, 2008). And, to further complicate matters, concessions by the source of strain may be seen as a success for terrorism—also prompting further terrorist acts. As LaFree and Dugan (2008) point out, however, it may be possible to escape this cycle of violence with a very carefully calibrated response to terrorism— one that does not reinforce terrorism or provoke a harsh reaction—along with efforts to address the types of root causes described above. Before proceeding further, however, it is critical to test the general strain theory. As indicated, most current tests of strain-based explanations are far too simplistic. They fail to measure the key dimensions of strain, including magnitude, injustice, and the nature of the source. Further, these tests do not examine intervening mechanisms, the subjective interpretation of strain, or conditioning variables. Unfortunately, most existing data sets do not permit anything close to a full test of the general strain theory. Agnew (2001, 2006a) provides suggestions on how to obtain both "objective" and subjective measures of many of the dimensions of strain that were listed, as well as measures of many of the intervening and conditioning variables. In the interim, researchers can draw on the theory to conduct better tests of strain explanations. One example of an approach that might be taken is provided by the cross-national research on criminal homicide (Agnew, 2006a). Material deprivation is often unrelated to violence in such research. Certain researchers, however, have attempted to roughly measure the perceived injustice of such deprivation. For example, they have estimated whether such deprivation is due to race/ethnic or religious discrimination (Messner, 1989; also see Gurr and

Moore, 1997). Deprivation resulting from discrimination is strongly related to violence. If the general strain theory is supported, it is critical to note that while collective strains may help *explain* terrorism, they do not *justify* terrorism. First, it is important to distinguish between the objective nature and subjective interpretation of such strains. In certain cases, there is good reason to believe that the members of terrorist groups exaggerate—often greatly—the strains they experience (e.g. members of certain white supremacist groups in the USA who claim they are being oppressed by the Zionist Occupation Government). Second, it is important to recognize that the members of the strained collectivity may sometimes contribute to the strains they experience through such things as attacks on the source of strain. Finally, the argument that collective strains contribute to terrorism is a causal one, not an ethical one. There are many responses to strain, some ethical and some not; being subject to strain does *not* justify any response to it.

✉ References

Abrahms, Max (2008) "What Terrorists Really Want," *International Security* 32(4): 78–105.

Agnew, Robert (1992) "Foundation for a General Strain Theory of Crime and Delinquency," *Criminology* 30(1): 47–87.

Agnew, Robert (2001) "Building on the Foundation of General Strain Theory: Specifying the Types of Strain Most Likely to Lead to Crime and Delinquency," *Journal of Research in Crime and Delinquency* 38(4): 319–61.

Agnew, Robert (2002) "Experienced, Vicarious, and Anticipated Strain: An Exploratory Study Focusing on Physical Victimization and Delinquency," *Justice Quarterly* 19(4): 603–32.

Agnew, Robert (2006a) *Pressured into Crime: An Overview of General Strain Theory.* New York: Oxford.

Agnew, Robert (2006b) "General Strain Theory: Current Status and Directions for Further Research," in Francis T. Cullen, John Paul Wright, and Michelle Coleman (eds) *Taking Stock: The Status of Criminological Theory, Advances in Criminological Theory*, Vol. 15, pp. 101–23. New Brunswick, NJ: Transaction.

Agnew, Robert, Nicole Leeper Piquero, and Francis T. Cullen (2009) "General Strain Theory and White-Collar Crime," in Sally S. Simpson and David Weisburd (eds) *The Criminology of White-Collar Crime*, 35–60. New York: Springer.

Ahmed, Hisham H. (2005) "Palestinian Resistance and 'Suicide Bombing,'" in Tore Bjorgo (ed.) *Root Causes of Terrorism*, pp. 87–101. London: Routledge.

Araj, Bader (2008) "Harsh State Repression as a Cause of Suicide Bombing: The Case of the Palestinian–Israeli Conflict," *Studies in Conflict & Terrorism* 31(4): 284–303.

Atran, Scott (2003) "Genesis of Suicide Terrorism," *Science* 299(5612): 1534–5, 1538.

Bjorgo, Tore (2005) "Introduction," in Tore Bjorgo (ed.) *Root Causes of Terrorism*, pp. 1–15. London: Routledge.

Black, Donald (2004) "Terrorism as Social Control," *Sociology of Crime, Law and Deviance* 5: 9–18.

Blazak, Randy (2001) "White Boys to Terrorist Men," *American Behavioral Scientist* 44(6): 982–1000.

Bloom, Mia (2006) "Dying to Kill: Motivations for Suicide Terrorism," in Ami Pedahzur (ed.) *Root Causes of Suicide Terrorism*, pp. 25–53. London: Routledge.

Callaway, Rhonda L. and Julie Harrelson-Stephens (2006) "Toward a Theory of Terrorism: Human Security as a Determinant of Terrorism," *Studies in Conflict & Terrorism* 29(7): 679–702.

Caracci, Giovanni (2002) "Cultural and Contextual Aspects of Terrorism," in Chris E. Stout (ed.) *The Psychology of Terrorism, Volume III: Theoretical Understandings and Perspectives*, pp. 57–81. Westport, CT: Praeger.

Cloward, Richard and Lloyd Ohlin (1960) *Delinquency and opportunity*. Glencoe, IL: Free Press.

Cohen, Albert K. (1955) *Delinquent Boys*. Glencoe, IL: Free Press.

Crenshaw, Martha (1995) "Thoughts on Relating Terrorism to Historical Contexts," in Martha Crenshaw (ed.) *Terrorism in Context*, pp. 3–26. University Park, PA: Pennsylvania State University Press.

De Coning, Cedric (2004) "Poverty and Terrorism: The Root Cause Debate?" *Conflict Trends* 3/2004: 20–9.

Eitle, David J. and R. Jay Turner (2002) "Exposure to Community Violence and Young Adult Crime: The Effects of Witnessing Violence, Traumatic Victimization, and Other Stressful Life Events," *Journal of Research in Crime and Delinquency* 39(2): 214–37.

Forst, Brian (2009) *Terrorism, Crime, and Public Policy*. Cambridge: Cambridge University Press.

Freeman, Michael (2008) "Democracy, Al Qaeda, and the Causes of Terrorism: A Strategic Analysis of U.S. Policy," *Studies in Conflict and Terrorism* 31(1): 40–59.

Froggio, Giancinto and Robert Agnew (2007) "The Relationship between Crime and 'Objective' versus 'Subjective' Strains," *Journal of Criminal Justice* 35(1): 81–7.

Goodwin, Jeff (2006). "A Theory of Categorical Terrorism," *Social Forces* 84(4): 2027–46.

Goodwin, Jeff (2007) "'The Struggle Made Me a Nonracialist': Why There Was So Little Terrorism in the Antiapartheid Struggle," *Mobilization: An International Quarterly Review* 12(2): 193–203.

Gottschalk, Michael and Susan Gottschalk (2004) "Authoritarianism and Pathological Hatred: A Social Psychological Profile of the Middle Eastern Terrorist," *American Sociologist* Summer: 38–59.

Gupta, Dipak K. (2005) "Exploring Roots of Terrorism," in Tore Bjorgo (ed.) *Root Causes of Terrorism*, pp. 16–32. London: Routledge.

Gurr, Ted Robert and Will H. Moore (1997) "Ethnopolitical Rebellion: A Cross-Sectional Analysis of the 1980s with Risk Assessment for the 1990s," *American Journal of Political Science* 41(4): 1079–103.

Hamm, Mark S. (2002) *In Bad Company: America's Terrorist Underground*. Boston, MA: Northeastern University Press.

Hoffman, Bruce (2006) *Inside Terrorism*. New York: Columbia University Press.

Hogg, Michael A. and Dominic Abrams (2003) "Intergroup Behavior and Social Identity," in Michael A. Hogg and Joel Cooper (eds) *The SAGE Handbook of Social Psychology*, pp. 407–31. Los Angeles, CA: Sage.

Krueger, Alan B. and Jitka Maleckova (2003). "Seeking the Roots of Terrorism," B10–13, *Chronicle of Higher Education: The Chronicle Review* 6 June.

LaFree, Gary and Laura Dugan (2004) "How Does Studying Terrorism Compare to Studying Crime?," *Sociology of Crime, Law and Deviance* 5: 53–74.

LaFree, Gary and Laura Dugan (2008) "Terrorism and Counterterrorism Research," unpublished document.

Loza, Wagdy (2006). "The Psychology of Extremism and Terrorism: A Middle-Eastern Perspective," *Aggression and Violent Behavior* 12(2): 141–55.

McCauley, Clark (2002) 'Psychological Issues in Understanding Terrorism and the Response to Terrorism', in Chris E. Stout (ed.) *The Psychology of Terrorism, Volume III: Theoretical Understandings and Perspectives*, pp. 3–30. Westport, CT: Praeger.

Maleckova, Kitke (2005) "Improverished Terrorists: Stereotype or Reality?," in Tore Bjorgo (ed.) *Root Causes of Terrorism*, pp. 33–43. London: Routledge.

Merari, Ariel (2005) "Social, Organizational and Psychological Factors in Suicide Terrorism," in Tore Bjorgo (ed.) *Root Causes of Terrorism*, pp. 70–85. London: Routledge.

Merton, Robert (1968). *Social Theory and Social Structure*. New York: Free Press.

Messner, Steven F. (1989) "Economic Discrimination and Societal Homicide Rates: Further Evidence of the Cost of Inequality," *American Sociology Review* 54(4): 597–611.

Moghadam, Assaf (2006a) "The Roots of Suicide Terrorism: A Multi-Causal Approach," in Ami Pedahzur (ed.) *Root Causes of Suicide Terrorism*, pp. 81–107. London: Routledge.

Moghadam, Assaf (2006b) "Suicide Terrorism, Occupation, and the Globalization of Martyrdom: A Critique of *Dying to Win*," *Studies in Conflict & Terrorism* 29(8): 707–29.

National Research Council (2002) *Terrorism: Perspectives from the Behavioral and Social Sciences*. Washington, DC: National Academies Press.

Newman, Edward (2006) "Exploring the Root Causes of Terrorism," *Studies in Conflict & Terrorism* 29(8): 749–72.

Newman, Edward (2007) "Weak States, State Failure, and Terrorism," *Terrorism and Political Violence* 19(4): 463–88.

Newman, Graeme R. and Ronald V. Clarke (2008) *Policing Terrorism: An Executive's Guide*. Washington, DC: Office of Community Oriented Policing Services, US Department of Justice.

New York Times (2008) "Failing the World's Poor," *New York Times*, 24 September, p. A30.

Oberschall, Anthony (2004) "Explaining Terrorism: The Contribution of Collective Action Theory," *Sociological Theory* 22(1): 26–37.

Pape, Robert A. (2005) *Dying to Win: The Strategic Logic of Suicide Terrorism.* New York: Random House.

Piazza, James A. (2006) "Rooted in Poverty? Terrorism, Poor Economic Development, and Social Cleavages," *Terrorism and Political Violence* 18(1): 159–77.

Piazza, James A. (2007) "Draining the Swamp: Democracy Promotion, State Failure, and Terrorism in 19 Middle Eastern Countries," *Studies in Conflict & Terrorism* 30(6): 521–39.

Post, Jerrold M. (2005) "The Socio-Cultural Underpinnings of Terrorist Psychology: When Hatred Is Bred in the Bone," in Tore Bjorgo (ed.) *Root Causes of Terrorism*, pp. 54–69. London: Routledge.

Post, Jerrold M. (2007) *The Mind of the Terrorist.* New York: Palgrave Macmillan.

Rausch, Sharla and Gary LaFree (2007) "The Growing Importance of Criminology in the Study of Terrorism," *The Criminologist* 32(6): 1, 3–5.

Robison, Kristopher K., Edward M. Crenshaw and J. Craig Jenkins (2006) "Ideologies of Violence: The Social Origins of Islamist and Leftist Transnational Terrorism," *Social Forces* 84(4): 2009–26.

Rosenfeld, Richard (2002) "Why Criminologists Should Study Terrorism," *The Criminologist* 27(6): 1, 3–4.

Rosenfeld, Richard (2004) "Terrorism and Criminology," *Sociology of Crime, Law and Deviance* 5: 19–32.

Sageman, Marc (2005) *Understanding Terror Networks.* Philadelphia, PA: University of Pennsylvania Press.

Sedgwick, Mark (2007) "Inspiration and the Origins of Global Waves of Terrorism," *Studies in Conflict & Terrorism* 30(2): 97–112.

Senechal de la Roche, Roberta (1996) "Collective Violence as Social Control," *Sociological Forum* 11(1): 97–128.

Simien, Evelyn (2005) "Race, Gender, and Linked Fate," *Journal of Black Studies* 35(5): 529–50.

Smelser, Neil J. (2007). *The Faces of Terrorism: Social and Psychological Dimensions.* Princeton, NJ: Princeton University Press.

Stern, Jessica (2003) *Terror in the Name of God: Why Religious Militants Kill.* New York: HarperCollins.

Stevens, Michael J. (2002) "The Unanticipated Consequences of Globalization: Contextualizing Terrorism," in Chris E. Stout (ed.) *The Psychology of Terrorism, Volume III: Theoretical Understandings and Perspectives*, pp. 31–56. Westport, CT: Praeger.

Tilly, Charles (2004) "Terror, Terrorism, Terrorists," *Sociological Theory* 22(10): 5–13.

Turk, Austin (2004). "Sociology of Terrorism," *Annual Review of Sociology* 30: 271–86.

Victoroff, Jeff (2005) "The Mind of the Terrorist: A Review and Critique of Psychological Approaches," *Journal of Conflict Resolution* 49(1): 3–42.

Wheaton, Blair (1990) "Life Transitions, Role Histories, and Mental Health," *American Sociological Review* 55(2): 209–24.

Wong, Edward (2008) "Tibetans Reaffirm a Conciliatory Approach to China," *New York Times*, 23 November, p. A13.

Robert Agnew is Professor of Sociology at Emory University. His research focuses on the causes of crime. Recent works include *Juvenile Delinquency: Causes and Control* (Oxford, 2009); *Pressured Into Crime: An Overview of General Strain Theory* (Oxford, 2006); and *Why Do Criminals Offend* (Oxford, 2005).

REVIEW QUESTIONS

1. Do you believe the article is persuasive and explains some terrorist behavior?

2. Would you fund further research in this area? Why or why not?

READING 2

The following article by Ruth Blakeley is presented with some abbreviation; the authors disagree with her conclusion but concede that if living in another country, Syria comes to mind, their opinion would likely be different. In it she argues that it is folly to allow western interpretations of terrorism to ignore the fact that a state, including the one readers likely dwell in, can be involved in terrorism. She seeks to explain why this is the case and argues for a critical or normative approach.

Bringing the State Back Into Terrorism Studies

Ruth Blakeley

State terrorism, along with other forms of repression, has been an ongoing feature of the foreign policies of democratic great powers from the North and the United States (US) in particular. The use of repression by the US was particularly intense during the Cold War, and we are seeing a resurgence of its use in the 'war on terror'. State terrorism, of which torture can sometimes be a tool, is defined as threats or acts of violence carried out by representatives of the state against civilians to instill fear for political purposes. According to dominant views in mainstream policy, media and academic circles, terrorism constitutes the targeting of Northern democratic states and their allies by non-state groups supplied and controlled by 'rogue' states or elements located in the South. This is only partially accurate. While such groups have carried out attacks against Northern democracies, including the devastating attacks of September 11th, 2001, it is also the case that Northern democracies have condoned and used terrorism, along with other forms of repression, against millions of citizens in the South over many decades.

There are three reasons for the notable absence of state terrorism—particularly that practised by Northern democracies—from scholarly debate within terrorism studies. The first has to do with the methods deployed by orthodox terrorism scholars. The second relates to their institutional affiliations. The third is connected to the marginalisation of explicitly normative approaches to foreign policy within international relations (IR) scholarship more broadly. I will outline some of the main flaws in the approaches of mainstream terrorism scholarship and show how these are exacerbated by the institutional affiliations of leading experts. I will then map out how this serious omission could be overcome.

The State of Contemporary Terrorism Studies

The way in which terrorism is theorised and defined in conventional terrorism studies is one of the main reasons why state terrorism by Northern democracies is largely absent from debate. This is correctly attributed to the way in which 'the term "terrorism" has been virtually appropriated by mainstream political discussion to signify atrocities targeting the West' (George, 1991: 1). It is in turn a consequence of the fact that most scholarship within terrorism studies is grounded in 'problem-solving theory'. As Robert Cox argues, problem-solving theory 'takes the world as it finds it, with the prevailing social and power relationships and the institutions into which they are organised, as the given framework for action' (Cox, 1981: 128). For orthodox terrorism scholars, the aim of their work is not to challenge these institutions and power relations, but to consider the problem of terrorism within the context of these existing institutions and power dynamics. Furthermore, the parameters of analysis for most terrorism scholars have been dictated by dominant neorealist approaches that tend to accept the benign character of the foreign policies of Northern democratic states, and the US in particular. When such states use force, it is assumed that this is in response to credible threats or as a means of protecting others. Yet as Alexander George accurately notes, 'on any reasonable definition of terrorism, taken literally, the United States and its friends are the major supporters, sponsors, and perpetrators of terrorist incidents in the world today' (George, 1991: 1).

A 'reasonable definition of terrorism' is offered by leading terrorism expert, Paul Wilkinson. He argues that terrorism has five main characteristics:

It is premeditated and aims to create a climate of extreme fear or terror; it is directed at a wider audience or target than the immediate victims of the violence; it inherently

> 'For orthodox terrorism scholars, the aim of their work is not to challenge these institutions and power relations, but to consider the problem of terrorism within the context of these existing institutions and power dynamics'.

SOURCE: Blakeley, R. (2007). Bringing the state back into terrorism studies. *European Political Science*, 6(3), 228–235. Copyright © 2007, Macmillan Publishers, Ltd. Reprinted with permission.

involves attacks on random and symbolic targets, including civilians; the acts of violence committed are seen by the society in which they occur as extra-normal, in the literal sense that they breach the social norms, thus causing a sense of outrage; and terrorism is used to try to influence political behaviour in some way (Wilkinson, 1992: 228–229).

Despite this, Wilkinson's only discussion of state terrorism is by Marxist–Leninist regimes and their client insurgencies (Wilkinson, 1992: 232). He makes no mention of the extensive terrorism used by right-wing states that during the Cold War, sought to repress left-wing movements across Latin America, often with US backing. Underpinning Wilkinson's work is an inbuilt assumption that Northern democracies are primarily victims and not perpetrators of terrorism. Importantly, it is not the content of Wilkinson's definition that precludes a focus on state terrorism by Northern democracies, but simply its inconsistent application in research.

The Centre for the Study of Terrorism and Political Violence (CSTPV) at St Andrew's University worked with the RAND Corporation to develop a database of international terrorism incidents between 1968 and 1997; it is widely recognised as the most authoritative source of data on international terrorism. The RAND Corporation is a non-profit-making research foundation with close links to the Pentagon. The largest private research centre in the world with an estimated annual budget of $160 million, it maintains close ties to the US government (Burnett and Whyte, 2005: 8). The RAND–St Andrew's data set defines international terrorism as 'incidents in which the perpetrators go abroad to strike their targets, select domestic targets associated with a foreign state, or create an international incident by attacking airline passengers or equipment'. From 1998, the data set was extended to include acts of domestic terrorism, which it defines as 'incidents perpetrated by local nationals against a purely domestic target' (RAND, 2007). Under both of these definitions, the assumption is that the perpetrators will not be the state itself, but sub-national individuals or groups acting against foreign or local interests. This is a crucial flaw. Explicitly excluded are acts of state terror committed by governments against their own citizens or acts of violence in warlike situations, even though such acts clearly fit Wilkinson's definition.

Terrorism is defined by the US State Department's Office of the Coordinator for Counterterrorism as 'premeditated, politically motivated violence perpetrated against non-combatant targets by subnational groups or clandestine agents, usually intended to influence an audience'. This means that rather than taking a literal approach to the study of terrorism, by which we determine what constitutes terrorism and then seek instances of the phenomenon to try and determine causes and remedies, the US government takes a propagandistic approach that focuses solely on actors seen as antithetical to US interests (Chomsky, 1991: 12). Importantly, the RAND–St Andrew's database also follows this pattern: the designated enemies are those non-state 'rogue' groups that seek to target foreign or domestic interests, and terrorist acts are those perpetrated by such groups against those targets. In this sense, the 'terrorist' label is used as a political tool to de-legitimise certain groups, rather than as an analytical category.

The selective ways in which terrorism is conceived and studied comes as no surprise considering the close connections between, first, RAND and the successive US administrations, and, second, between RAND and supposedly independent academic experts on terrorism, including Paul Wilkinson. Other leading academics associated with both RAND and the CSTPV are Bruce Hoffman who temporarily left the RAND Corporation in 1993 to found the CSTPV at St Andrew's and who remains an honorary senior researcher there, and Brian Jenkins, a senior analyst with RAND who is also a member of the CSTPV's advisory council (Burnett and Whyte, 2005: 8). Individuals associated with the CSTPV and RAND also retain key editorial positions in the two most prominent English language journals in the field of terrorism and political violence: Wilkinson as co-editor of *Terrorism and Political Violence;* Hoffman and Jenkins as members of its editorial Board and Hoffman as editor-in-chief of *Studies in Conflict and Terrorism,* a journal originally founded and editorially managed by RAND (Burnett and Whyte, 2005: 9). Burnett and Whyte correctly note that this means 'peer reviewed publications are dominated by academics connected with this nexus of influence', and while they are not in any way suggesting that the system of peer review is corrupt or less rigorous than it is in other publications, 'if we consider that two of the key journals are dominated by scholars from the RAND–St Andrew's nexus, then this does say something about their ability to impose their influence upon the field' (2005: 9). This may explain why there is so little scholarly literature published in the key journals that discusses the use of state terrorism by Northern democracies: it simply does not fit within the established frame of reference of dominant scholarship on terrorism.

Northern Democracies and Complicity in Repression

Northern democracies have a long history of complicity in repression, including state terrorism, often through providing military and financial support to highly repressive governments or to terrorist groups. For example, the US, Britain and Australia all backed Indonesia while it engaged in widespread repression against the people of East Timor (Chomsky, 2000: 51–61). Similarly, in Northern Ireland, British forces made extensive use of repression and torture and tacitly supported acts of Loyalist violence. For a long period, official British policy was to intern, without charge or trial, the suspected members of paramilitary groups. The British army also used torture as part of its interrogation of suspected Republican terrorists, as documented by Amnesty International (1972), which concluded that the British government had violated national and international law in relation to its treatment of fourteen Northern Irish men in 1972. These men were subjected to beatings with batons and kicking, often until they passed out; hooding; stripping; sensory assault, including being subject for a whole week to constant noise at various levels of intensity; food, water and sleep deprivation, and prolonged stress positions (Conroy, 2001: 5–11). It can be argued that these counter-terrorism measures themselves constituted a form of state terrorism.

The French also made extensive use of torture against large sectors of the Algerian population, both in Algeria itself by police forces and in France (Vidal-Naquet, 1963: 40–44). General Jacques Massu, Commander of the Tenth Parachute Division responsible for policing in Algiers from 1957, justified the use of torture on the grounds that the circumstances demanded its use and military necessity dictated it (Massu, 1997). The context of the counter-insurgency (CI) campaign saw French troops employing torture not simply as a means to secure

intelligence about imminent threats to French forces, although this was the justification used by Massu, but as an attempt to undermine the morale of the leaders and supporters of the Algerian insurgency. Used in this manner, torture is a tool of state terrorism.

US Repression in Context

The primary aims of US foreign policy are to maintain the dominant global position of the US and to ensure access to resources and markets in the South; these priorities are enshrined most openly in the Monroe doctrine, pronounced by US President James Monroe in 1923, when he declared the US the protector of the nations of the Americas from European states, whose efforts to extend their territory could undermine the security and the dominant position of the US in the Western Hemisphere (see Perkins, 1927; Shoup and Minter, 1977). During the Cold War, US foreign policy strategy was dominated by the use of repression (see Blum, 2003; Blakeley, 2006). As Chomsky and Herman (1979) demonstrated in their study of US relations with the South, the US was organising under its sponsorship a system of allied states, which ruled their populations primarily by terror.

US repression in the South has involved orchestrating or backing coups, as in Guatemala, Chile, Indonesia, Haiti and elsewhere, and in direct military intervention, such as in the Dominican Republic, Indochina, Panama and others (see Blum, 2003). One of the most significant ways in which the US has been complicit in repression in the South has been through the provision of training for military forces from the area, something that has had far-reaching consequences, not only in terms of human rights, but also in terms of the capacity of the US to achieve its foreign policy objectives. US military training of forces from the South since World War II has steadily increased and is now given to military personnel from over 150 countries each year. This has been most intense in Latin America: between 1950 and 1993, the US trained over 100,000 Latin American military and police personnel. A significant reason for the training is that the US prefers local elites to carry out its objectives in the South (Blakeley, 2006). This was particularly the case following the failure of the US in the Vietnam War, after which the American public had little sympathy for further US activities overseas (Klare, 1989: 97). As a consequence, during the Cold War the US provided covert military and intelligence assistance to elites from many Latin American states. Much of this involved support for CI operations and CI training, which advocated repression, including torture, of anyone suspected of being involved in or considered likely to become involved in activities that would threaten US interests (Blakeley, 2006). The forces trained would thereby act as US allies in pursuit of US objectives.

US involvement in state terrorism also included the use of torture as part of its CI strategy during the war with Vietnam. This occurred primarily through the Phoenix Program that was intended to improve intelligence and wipe out what was known among the CIA as the Vietcong Infrastructure (VCI). Valentine's (2000) definitive account shows that Phoenix had the effect not simply of destroying the VCI, but also of instilling terror among Vietnamese civilians. Large numbers of civilians, often not even members of the VCI but simply family members or neighbours of suspected members, were killed in their sleep by US and South Vietnamese military personnel:

> Phoenix was, among other things, an instrument of counter-terror—the psychological warfare tactic in which VCI members were brutally murdered along with their families or neighbours as a means of terrorising the neighbouring population into a state of submission. Such horrendous acts were, for propaganda purposes, often made to look as if they had been committed by the enemy (Valentine, 2000: 13).

As well as murder, torture was also widely practised, often at Province Interrogation Centres (PICs). Some of the documented atrocities included:

> Rape, gang rape, rape using eels, snakes, or hard objects, and rape followed by murder; electrical shock ('the Bell Telephone Hour') rendered by attaching wires to the genitals or other sensitive parts of the body, like the tongue; the 'water treatment'; the 'airplane', in which a prisoner's arms were tied behind the back and the rope looped over a hook on the ceiling, suspending the prisoner in midair, after which he or she was beaten; beatings

with rubber hoses and whips; the use of police dogs to maul prisoners (Valentine, 2000: 85).

According to CIA officer William Colby, who directed Phoenix between 1968 and May 1971, 20,587 alleged Vietcong cadres died as a result of Phoenix. The South Vietnam government places the number at 40,994. The true number will never be known, neither will the number of those killed under the programme's forerunners, operational from 1965 (Blum, 2003: 324).

The Way Forward: Critical Approaches to State Violence

In this article, I have tried to show that two main factors have contributed to the silence on state terrorism by Northern democracies from orthodox terrorism studies. The first relates to the theoretical framework of most terrorism scholarship and the way in which definitions of terrorism are applied in practice. Accepting an orthodoxy within IR that characterises the foreign policies of Northern democracies as largely benign, terrorism is understood to mean activities by non-state actors, often located in the South, against Northern democracies and their interests; state terrorism, when it is discussed, is assumed to constitute support for terrorists by 'rogue' states. The reality is that Northern democracies have been responsible for widespread terrorism against populations in the South. The second reason lies in the institutional affiliations of leading academic experts who are frequently tied to the institutions of state power. The exclusion of state terrorism from current usage of the term 'terrorism' means that terrorism studies scholars function to promote particular political agendas, such as those of the current US administration and its allies in the 'war on terror'. More specifically, by reinforcing certain political assumptions about what constitutes terrorism, they reinforce the

> 'critically oriented scholars need to reclaim the term 'terrorism' and use it as an analytical tool, rather than a political tool in the service of elite power'.

false notion that Northern democracies, especially the US, simply act to uphold liberal values and protect their populations from threats. In this sense, the approach taken by many terrorism studies scholars tends to serve particular national, sectional or class interests, which, as Cox notes, are comfortable within the given order (Cox, 1981: 129).

For these reasons, critically oriented scholars need to reclaim the term 'terrorism' and use it as an analytical tool, rather than a political tool in the service of elite power. There are several necessary steps in this reclaiming process. First, as Cox notes, critical approaches need to challenge institutions and approaches:

> Critical theory does not take institutions and social and power relations for granted but calls them into question by concerning itself with their origins and how and whether they might be in the process of changing. It is directed towards an appraisal of the very framework for action, or problematic, which problem-solving theory accepts as its parameters (Cox, 1981: 129).

This article has questioned the dominant interpretation of the foreign policies of the great powers—that it is benign in character—and has analysed the actual practices of those states and their outcomes. This has been with the specific, normative aim of offering suggestions for the emancipation of people in the South from the oppressive practices of Northern powers. Normative approaches of this kind are necessary for two main reasons. First, they enable us to overcome certain biases in the field, including the selective application of terms such as 'terrorism' that serves to fortify rather than confront illiberal practices. Second, they help to diversify and broaden debate beyond the narrow parameters set by the dominant, neo-realist and liberal approaches within IR.

It is also obvious that there is a pressing need to bring the state back into terrorism studies. Because terrorism is a tactic and not an ideology, states of any kind can be perpetrators of terrorism. Equally, the tactics that states use to combat terrorism can themselves resemble terrorism, as the cases of British, US and French counter-terror and CI efforts show. The field of terrorism studies therefore needs to reintegrate the state, as a potential instigator of terrorism, into the debate. This does not simply mean examining the role of the so-called 'rogue' states, but also that of states normally considered to be engaged in combating rather than perpetrating terrorism. Importantly, we must not focus solely on state terrorism by Northern powers; this would itself lead to

a further biasing of the debate. While the US was one of the greatest perpetrators of state terrorism in Latin America during the Cold War, it did not act alone, collaborating instead with authoritarian regimes which were themselves implementing state terror complexes before the provision of US support. In other words, the agency of other actors should not be ignored. We can better understand state terrorism when we examine the collaborations that are established between elites across state boundaries.

A number of issues present themselves as areas in need of further examination as part of the project to integrate state terrorism into critical terrorism studies. These include, among others: analysing the relationships between state terrorism and the use of torture; the nature of state responses to terrorism, including counter-terrorism and CI operations; the role of state military forces and, given their recent growth and increased use by states, private military companies, as potential agents of state terrorism; and finally, the degree to which the curtailments of civil liberties in the 'war on terror' may themselves run the risk of constituting state terrorism.

 # References

Amnesty International. (1972) *Report of an Inquiry into Allegations of Ill-Treatment in Northern Ireland,* London: Amnesty International.

Blakeley, R. (2006) 'Still training to torture? US training of military forces from Latin America', *Third World Quarterly* 27(8): 1439–1461.

Blum, W. (2003) Killing Hope: US Military and CIA Interventions Since World War II, London: Zed Books.

Burnett, J and Whyte, D. (2005) 'Embedded expertise and the new terrorism', *Journal for Crime, Conflict and the Media* 1(4): 1–18.

Chomsky, N. (1991) 'International Terrorism: Image and Reality', in A. George (ed.) *Western State Terrorism*, Cambridge: Polity Press, pp. 12–38.

Chomsky, N. (2000) *Rogue States,* London: Pluto Press Ltd.

Chomsky, N. and Herman, E. (1979) *The Washington Connection and Third World Fascism: The Political Economy of Human Rights,* Vol. I. Boston: South End Press.

Conroy, J. (2001) *Unspeakable Acts, Ordinary People: The Dynamics of Torture,* London: Vision Paperbacks.

Cox, R. (1981) 'Social forces, states and world orders: beyond international relations theory', *Millennium: Journal of International Studies* 10(2): 126–155.

George, A. (1991) *Western State Terrorism,* Cambridge: Polity Press.

Klare, M. (1989) 'Subterranean alliances: America's global proxy network', *Journal of International Affairs* 43(1): 97–118.

Massu, J. (1997) *La Vraie Bataille d'Alger,* Paris: Editions du Rocher.

Perkins, D. (1927) *The Monroe Doctrine 1823–1826,* Vol. I. Gloucester, MA: Harvard University Press.

RAND. (2007) 'RAND-MIPT Terrorism Knowledge Base, RAND Corporation (last updated April 2007), http://www.tkb.org/RandSummary.jsp?page ¼ about.

Shoup, L. and Minter, W. (1977) *Imperial Brain Trust: The Council on Foreign Relations and United States Foreign Policy,* New York: Monthly Review Press.

Valentine, D. (2000) *The Phoenix Program,* Lincoln: Authors Guild BackinPrint.com.

Vidal-Naquet, P. (1963) *Torture: Cancer of Democracy,* Middlesex: Penguin Books.

Wilkinson, P. (1992) 'International Terrorism: New Risks to World Order', in J. Baylis and N. Rengger (eds.) *Dilemmas of World Politics: International Issues in a Changing World,* London: Clarendon Press, pp. 228–257.

 ## About the Author

Ruth Blakeley is Lecturer in International Relations at the University of Kent, Canterbury. She has published articles on repression and state terrorism in *Third World Quarterly* and the *Review of International Studies* (forthcoming). Her research interests include US foreign policy, US–Latin American relations, state terrorism, political violence, torture and North–South relations.

REVIEW QUESTIONS

1. Do you agree or disagree with Blakeley that a definition of terrorism should include the possibility that there can be state terrorism? Why or why not? What about the country you live in?

2. If you lived in Syria (or any other country where a dictator was attempting to cling to power by killing opponents), what would your definition of terrorism be?

3. Does Blakeley adequately source her material, particularly some of the more inflammatory charges made? Was it convincing?

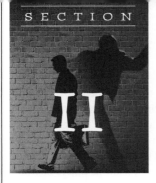

Terrorism Definitions and Typologies

At the end of this section, students will be able to:

- Discuss the differences in definitions of terrorism by various entities.
- Describe the terrorism definitions used by Great Britain and those of the United States.
- Explain why the United Nations does not have a definition of terrorism.
- Discuss the use of typologies when studying terrorism to include addressing the advantages and disadvantages.

Terrorists or Freedom Fighters?

In the late fall of 2014, several attacks were carried out in Jerusalem against Israelis. Kheir Hamdan attacked a police squad car with a knife; he was shot and killed by police. Al-Shaludi drove his car into a Jerusalem Light Rail station and killed 3-month-old Israeli-American Chaya Zissel Braun and Karen Yemima Muscara, 22, of Ecuador, and injured seven others. He was shot by the police when he attempted to flee the scene. Mu'taz Hijazi shot Rabbi Yehdah Glick, who was seriously wounded but survived. Hijazi was killed by the police during a gunfight. In response, Fatah officials (the ruling party the United States supports) praised their actions as heroic and called for "blood" to "purify" Jerusalem of Jews. Are these men **freedom fighters** or terrorists?

▲ Militant

©iStockphoto.com/zabelin

Introduction

This section introduces additional definitions of terrorism and students will see there is much variety among these definitions, and there will be a discussion of why it is so difficult to agree on a definition. Do not lose sight of the fact that defining something as terrorism or someone as a terrorist and then applying that label to a group or a person may trigger draconian actions on the part of a government intent on dealing with a serious matter it views threatening. For example, Great Britain has laws that make those of the United States seem timid. The **Anti-terrorism, Crime and Security Act of 2000** initially enabled the Home Secretary to indefinitely detain, without charge or trial, foreign nationals who are suspected of terrorism (Anti-terrorism, Crime and Security Act, 2001). That law has since been modified slightly and other revisions may follow, but indefinitely detained? Keep in mind that some definitions detailed are for specific purposes, such as a criminal prosecution, a persecution of a political opponent or individual protesters or groups, and some definitions presented will be from other countries and international organizations. Some individuals invoke the label of terrorism simply because it gets attention. For example, Robert Mugabe, already introduced in the first section, made the charge in a speech in 2011 that he regards the North Atlantic Treaty Organization (NATO) as a terrorist organization for its bombing campaign against Libya's leader, Muammar Qadhafi ("Mugabe Labels NATO a 'Terrorist Group' Over Libya," 2011). Mugabe alleged that NATO was bombing civilians and targeting the family of Qadhafi, ignoring the fact that NATO was operating under a United Nation's mandate and enforcing a no-fly zone. This demonstrates the looseness or applications of the term, and how capricious the uses of the term can be.

This will be followed by a presentation of attempts to "classify" terrorists using the concept of a typology. A typology is a method used to place actors or groups in categories for better understanding. For example, we had a horrific shooting in Norway, a normally placid country where police officers do not even carry weapons and must ask permission to obtain them for use. The shooter, Anders Behring Breivik, calmly spent an hour dispatching 76 individuals, some mere teenagers (Schwirtz, 2011). Immediately news reporters, pundits, and even experts struggled to explain this. They did so using such terms as a "lone wolf" or a "crazy person," which are examples of attempting to classify someone who engages in terrorism, hoping to explain the inexplicable and reprehensible behavior. And clearly, this was a case of terrorism, fitting the definition presented in the earlier section.

 What Is Terrorism?

So, what are other definitions of terrorism? The Federal Bureau of Investigation (FBI) in the United States defines terrorism as "the unlawful use of force or violence against persons or property to intimidate or coerce a government, the civilian population, or any segment thereof in furtherance of political or social objectives." The FBI further classifies terrorism as either domestic or international, depending on the origin, base, and objectives of the terrorist organization (Federal Bureau of Investigation, 2011). That definition is missing items from the previous definition provided by Professor Combs, such as against innocents (though it does use the term *civilian population*), and "staged before an audience to create fear," but it does emphasize the means or methods that would be the basis for an investigation and potentially a criminal prosecution. Great Britain has multiple definitions of terrorism as well. They have a very complex one similar to the one below and a much more concise one cited in the Reinsurance (Acts of Terrorism) Act 1993:

> acts of persons acting on behalf of, or in connection with, any organisation which carries out activities directed towards the overthrowing or influencing, by force or violence, of Her Majesty's government in the United Kingdom or any other government de jure or de facto.

The United Nations has been attempting for years to define terrorism and get on with doing something about it, but to no avail. Here is their proposed definition of a terrorist or of terrorism and one can see why it is slow going:

1. Any person commits an offence within the meaning of the present Convention if that person, by any means, unlawfully and intentionally, causes:

   (a) Death or serious bodily injury to any person; or

   (b) Serious damage to public or private property, including a place of public use, a State or government facility, a public transportation system, an infrastructure facility or to the environment; or

   (c) Damage to property, places, facilities or systems referred to in paragraph 1(b) of the present article resulting or likely to result in major economic loss, when the purpose of the conduct, by its nature or context, is to intimidate a population, or to compel a Government or an international organization to do or to abstain from doing any act.

2. Any person also commits an offence if that person makes a credible and serious threat to commit an offence as set forth in paragraph 1 of the present article.

3. Any person also commits an offence if that person attempts to commit an offence as set forth in paragraph 1 of the present article.

4. Any person also commits an offence if that person:

   (a) Participates as an accomplice in an offence as set forth in paragraph 1, 2 or 3 of the present article; or

   (b) Organizes or directs others to commit an offence as set forth in paragraph 1.2.3. of the present article; or

   (c) Contributes to the commission of one or more offences as set forth in paragraph 1, 2 or 3 of the present article by a group of persons acting with a common purpose. Such contribution shall be intentional and shall either:

      (i) Be made with the aim of furthering the criminal activity or criminal purpose of the group, where such activity or purpose involves the commission of an offence as set forth in paragraph 1 of the present article; or

      (ii) Be made in the knowledge of the intention of the group to commit an offence as set forth in paragraph 1 of the present article. (United Nations, 2010)

Can anyone diagram that definition? In a press release regarding the impasse that attempted to put a positive note on the meetings, note the following:

Another said the Committee had been wasting time over the past week because it was clear that negotiations on outstanding issues—chiefly a concrete definition of terrorism or terrorist acts—had stalled. (United Nations, 2011)

Cleary there must be something wrong if no one can agree on a definition after years.

A leading thinker on defining terror, and one of the authors of an early definition of terrorism, Brian Jenkins, weighed in on the definition problem in an early paper, first presented in 1978. Here are some highlights from that:

The term "terrorism" has no precise or widely-accepted definition.

It is pejorative.

It depends on one's point of view.

Use of the term implies a moral judgment.

Terrorism in the RAND chronology is defined by the nature of the act, not by the identity of the perpetrators or the nature of their cause.

All terrorist acts are crimes.

All involve violence or threat of violence.

Violence is mainly directed against civilian targets.

The motives are political.

"It" or "They"? carried out in a way to achieve maximum publicity.

And it produces effects beyond the immediate physical damage, including fear. (Jenkins, 1980, p. 1)

And in a much later report, Jenkins adds this:

The word terrorism dates from the eighteenth century, but as recently as 1971—when RAND published my essay "The Five Stages of Urban Guerrilla Warfare"—the term had not yet acquired its present currency. Nor did it then refer to a distinct mode of armed conflict. That meaning would be added subsequently, by terrorists themselves and by analysts of terrorism; and in my view the latter impart more coherence to the phenomenon than the former. (Jenkins, 2006, p. 117)

Jenkins addressed the main problems with defining terrorism and the authors agree with him. Bruce Hoffman, a contemporary of Jenkins, has spent a great deal of time and effort outlining the history of terror definitions and concludes that terrorism is:
Ineluctably political in aims and motives;

◆ Violent—or, equally important, threatens violence;
◆ Designed to have far-reaching psychological repercussions beyond the immediate victim or target;
◆ Conducted either by an organization . . . or by individuals . . . directly influenced, motivated, or inspired by the ideological aims or example of some existent terrorist movement and/or its leaders;
◆ Perpetrated by a subnational group or nonstate entity (Hoffman, 2006, p. 40).

It is obvious from reading both Jenkins and Hoffman that they exclude states from their definition of terrorism, but they certainly realize there are states out there that routinely terrorize their citizens. But their focus is on substate actors. Still, why is it so difficult to agree on a definition of terrorism? The answer can be found in the writing of Boaz Ganor. Dr. Ganor is the associate dean of the Lauder School of Government and founder and executive director of the International Institute for Counter-Terrorism (ICT). Often writing about terrorism, he has authored a paper called "Is One Man's Terrorist Another Man's Freedom Fighter?" and makes a compelling argument that terrorists are those who use violence or the threat of violence, have a political goal (the underlying ideology and motives are irrelevant), and the targets of terrorists are civilians. Thus, a freedom fighter from country x that used violence against civilians would be a terrorist if he or she had a political goal. Period. The confusion that reigns when it comes to defining terrorism is simply that some individuals and groups, often freedom fighters, the oppressed, or those seeking national liberty, regard civilians as targets (Ganor, 2010). One of the authors visited Israel a few years ago with a group of fellow professors. Taken to a prison for terrorists, the group interviewed a member of Hamas, a Palestinian terrorist organization. When asked what he wanted for his son in the future, he said he wanted his son to die as a martyr, killing Jews and liberating Palestine. Notice that he was very clear. He was a freedom fighter, fighting against oppressors and occupiers and wanted his son to be one, even become a martyr for the cause. But he was, in reality, a terrorist and his son would be a terrorist if his father's wishes come true. Dr. Ganor sums it up well:

> Surprisingly, many in the Western world have accepted the mistaken assumption that terrorism and national liberation are two extremes in the scale of legitimate use of violence. The struggle for "national liberation" would appear to be the positive and justified end of this sequence, whereas terrorism is the negative and odious one. It is impossible, according to this approach, for any organization to be both a terrorist group and a movement for national liberation at the same time. (Ganor, 2010)

Academics, policy makers, and yes, students need to understand that when the above elements are present there is terrorism. Those "freedom fighters" who blow up children in the name of their cause are terrorists. One may be sympathetic to the plight of the Palestinian people and not be sympathetic to their voting for Hamas as a government, but is very difficult to condone their support for terrorism. Another example of not regarding something as terrorism as long as it is carried out against a certain group is exemplified by this recent exhortation by a scholar in Egypt. Dr. Salah Sultan, a member of the Muslim Brotherhood and a lecturer at Cairo University, told a group of demonstrators that Egyptians meeting "Zionists" should kill them, and this comes on the heels of the Muslim Brotherhood threatening to kill the Israeli ambassador to Egypt if he does not leave and demanding the decades-old peace treaty with Israel be ended (Ronen, 2011).

To summarize, defining terrorism presents many difficulties. But when examining the facts objectively, Boaz Ganor gets it right. When individuals or groups have a political purpose, and use or threaten the use of violence against civilians, that is terrorism. This does bring up another issue. Is it terrorism when a group uses violence against an oppressive government, attacking members of that country's armed forces or government officials? One supposes that in some countries, it might be and in others, perhaps not. While writing this, President Bashar al-Assad of Syria is busy ordering the slaughter of his citizens, with reports that by the end of January 2015 there had been 200,000 deaths, 7.6 million persons displaced in the country, and another 3.2 million were registered refuges outside the country (Medecines Sans Frontiers, 2015). It is certainly fine with the authors if Syrians employ any means to fight back.

What Are Typologies of Terrorists?

A typology is a system of classification based on various characteristics or tactics of groups, in this case, those committing terrorism; it attempts to promote understanding and order. A typology may lead to development of a theory at some point in time, about a certain phenomenon. This has not yet happened when it comes to terrorism and there is much disagreement about what a terrorist typology should contain and there are many terrorist typologies; this is very similar to the multiple definitions of terrorism. Using a typology can be helpful for analysis, fashioning responses to terrorism, such as counterterrorism actions, but it can also be confusing and limiting. For example, an analyst may place a certain terrorist group in a category but lack enough information on the group to do so accurately. He or she may focus on its short-term goals and ignore the long-term ones. Finally, groups of terrorists may change radically with time.

One of the earlier **typologies of terrorists** was prepared by Frederick Hacker (1976), who suggested one could classify terrorists based on personalities, such as:

1. Crazies—strong survival attitude, but not based in reality; self-centered; goals clear only to perpetrator; irrational and unpredictable; strikes at random

2. Crusaders—sacrificial, death attitude; blends politics and religion; seldom willing to negotiate; task-oriented and indifferent to risk; seeks publicity and largest group possible

3. Criminals—strong self-preservation attitude; selfish; seeks gain and is task-oriented; avoids high risk; predictably targets small groups. (p. 12)

One wonders what Osama bin Laden would have thought examining that typology, since he and his group would have been classified as Crusaders, a name that is not to their liking, since Christian Crusaders pursued Muslims for many years. Too, understand that there are many bases for coming up with a classification or typology. Alex Schmid, Albert Jongman, and Michael Stohl (2005) produced the following ten bases: actor-based, victim-based, cause-based, environment-based, means-based, political-qrientation based, motivation-based, demand-based, purpose-based, and target-based. Groups could easily be in more than one category. For example, they could have an ideology that allows them to see themselves as victims or victim-based terrorists but have an environmental agenda or base. An example of a typical attempt at classifying terrorists into typologies was made in a federal government report by an individual following the al Qaeda attack on September 11, 2001.

The table focuses first on state actors and non-state actors. With regard to state actors, this typology specifies the type of action, who would perform this, likely victims, and the target audience. As mentioned earlier, Syria is in

Table 4.1	Terrorism Typology

State Actors			
Action Type	**Operational Group**	**Target Victims**	**Target Audience**
Internal Repression	Police, Military, Judicial, Vigilante	Internal individuals and groups considered subversive	Entire or segment of domestic population
State Sponsored	Foreign Affiliate Terrorist Group	Symbolic targets based on shared enemy	Population of shared enemy and allies
State Performed	Intelligence Service, Commando Unit	Symbolic targets based on foreign enemy	Population of foreign enemy and allies

	Non-state Actors		
Primary Motivator (Identity)	**Orientation**	**Target Victims**	**Target Audience**
Political Ideology	Anarchist	Symbolic targets based on relationship to "system"	Population of system
	Marxist	Symbolic targets based on relationship to capitalist-imperialist system	Population of capitalist-imperialist system
	Fascist	Symbolic targets based on opposition to Fascism (government, middle class, Marxists)	Population of nation-state
	Single Issue	Symbolic targets based on issue	Population of issue region (nation-state or region)
Ethno-nationalism	Pro-state	Symbolic targets based on relationship to anti-state revolutionaries	Population of nation-state
	Revolutionary	Symbolic targets based on relationship to support of the state	Population of nation-state
	Separatist	Symbolic targets based on relationship to state	Population of nation-state
	Autonomous	Symbolic targets based on relationship to state	Population of nation-state
	Ethnic Vigilante	Symbolic targets based on relationship to ethnic group	Population of issue region (nation-state or region)
Religious Extremism	Fundamentalist	Symbolic targets based on relationship to fundamentalist religious worldview	Population of worldview and competing systems (internal cohesion) and those outside worldview (external)
	Cults and Sects	Symbolic targets based on relationship to cult or sect	Population of sect region (nation-state or segment)

SOURCE: Cunningham (2003).

turmoil currently and the government urgently seeks to **internally repress** dissent; on a weekly basis following Friday prayers, the largely Sunni population is urged by Imams (religious leaders) to demonstrate against the state (state actors, again, will not be detailed in this text). This is followed by brutal crackdowns by the police and military resulting in many deaths, for the purpose of cowing the population at large and intimidating them. For non-state actors, the typology examines their motivation, such as political or religious, and their orientation, such as Marxist, and then who the targeted victims and audiences are. In this case, al Qaeda will serve as an example. Osama bin Laden clearly had a religious-based motivation for his actions; targets of al Qaeda were purposely symbolic and were intended for both an internal audience (fellow believers and sympathizers) and an external audience (the West and the world) that he hoped to intimidate.

The authors of this text reader use a simple typology when teaching a course on violence and terrorism. It breaks down terrorist groups by examining their background, ideology, religious motivation, and tactics:

Nationalist/Ethnic

Anti-colonial (vertical)

Separatist (horizontal)

Political/Ideological (leftist/communist; right wing)

Religious

Suicide

It should be noted that both religious and suicide terrorism are sometimes referred to as **New Terrorism.** In the above typology, groups may engage in terrorism due to nationalist motives or ethnic-based violence, or they may wish to separate from a state. Many may also have a political or ideological motivation. The religious-based terror groups are a genuine concern as the view violence is sanctioned by a higher power and their duty is to engage in violence to overthrow a state. This results in their seeing violence as "lawful" when used against non-believers or those not sympathetic to their cause.

Finally, typologies may be interesting and sometimes useful, but as Ross (2006) points out:

> There are a great number of problems with typologies. Some of the more important problems are: they can provide little connection to theory-building; they exclude exceptions; they can be conceptually confusing; they may be too general; and they're based on changing phenomena. If definitions are problematic, so then are typologies, which are context specific (bounded by time, culture, geography). (p. 10)

Summary

This section has presented additional material on the difficulty of defining terrorism. It is problematic, as the lack of consensus by the United Nations on a definition demonstrates. Readers should be mindful of any definitions presented and review the ideology of the individual, organization, and yes, even a nation, as well as the purpose. Additionally, the use of a typology is presented and how it can be helpful to determine a response to terrorism.

Key Points

- This section provides examples of other definitions of terrorism, including that of the Federal Bureau of Investigation and of the United Kingdom
- It introduces the concept of a typology and presents examples, while noting that there are advantages but limitations to their usefulness
- It notes the difficulty of any large body, such as the United Nations, ever arriving at a definition of terrorism

KEY TERMS

Anti-terrorism, Crime and Security Act of 2000	Freedom fighter	Typologies of terrorism
	New Terrorism	

DISCUSSION QUESTIONS

1. After reading two sections on how to define terrorism and the difficulties of defining the term, what would be your definition? Why?

2. Develop a typology that you believe would capture all current and active terrorist groups.

3. Do you believe one man's terrorist is another man's freedom fighter? Yes or no and explain your answer.

4. A government is currently shelling several cities with the civilian death toll over several thousand. Is that terrorism?

WEB RESOURCES

Chris Eskridge's terrorism page: http://www.unl.edu/eskridge/cj476index.html

Foundation for Defense of Democracy; includes opportunities for students and faculty to learn more about terrorism and what to do about it: http://www.defenddemocracy.org/

Global History of Terrorism Database: http://www.start.umd.edu/gtd/about/History.aspx

Israel site on suicide bombers: http://israelsmessiah.com/terrorism/suicide_bombers.htm

Middle East Media Research Institute; if you want to know what is really going on in the Middle East, this is the place: http://www.memri.org/

Rantburg (blog), a blog by a bunch of old retired folks that will terrify: http://rantburg.com/

Thomas Ricks's blog is essential for anyone following the various wars we are involved in at the moment. He is an amazing writer and author of *Fiasco: The American Military Adventure in Iraq,* and *The Gamble,* which any serious student should read: http://ricks.foreignpolicy.com/blog/2187?page=7

Tom O'Connor's pages on cults and terror: http://www.cultsandterror.org/sub-file/TOConnor%20Lecture.htm

READING 3

The reading presented here is from Brian Michael Jenkins, a scholar and researcher who has been studying terrorism before most readers were born. The reading is an almost stream-of-conscious essay, first presented at a conference in 1978, where he puzzles and discusses how difficult it is to define terrorism and how quickly it changes. Readers will no doubt recognize that they have had similar thoughts about defining terrorism after reading just the first two sections. It is presented in its entirety.

The Study of Terrorism: Definitional Problems*

Brian Michael Jenkins
The Rand Corporation, Santa Monica, California
November 1980

Terrorism has become part of our daily news diet. Hardly a day goes by without news of an assassination, political kidnapping, hijacking, or bombing somewhere in the world. As such incidents of terrorism have increased in the past decade, the phenomenon of terrorism has become one of increasing concern to governments and of increasing interest to scholars.

In the course of its continuing research on terrorism, The Rand Corporation has compiled a chronology of international terrorism incidents that have occurred since 1968. This chronology now contains over 1,000 incidents. In compiling the chronology, numerous problems of definition were encountered.

The term "terrorism" has no precise or widely-accepted definition. The problem of defining terrorism is compounded by the fact that terrorism has recently become a fad word used promiscuously and often applied to a variety of acts of violence which are not strictly terrorism by definition. It is generally pejorative. Some governments are prone to label as terrorism all violent acts committed by their political opponents, while anti-government extremists frequently claim to be the victims of government terror. What

is called terrorism thus seems to depend on one's point of view. Use of the term implies a moral judgment; and if one party can successfully attach the label *terrorist* to its opponent, then it has indirectly persuaded others to adopt its moral viewpoint. Terrorism is what the bad guys do.

The word *terrorism* is also an attention-getting word and therefore tends to be used, especially in the news media, to heighten the drama surrounding any act of violence. What we have, in sum, is the sloppy use of a word that is rather imprecisely defined to begin with. Terrorism may properly refer to a specific set of actions the primary intent of which is to produce fear and alarm that may serve a variety of purposes.But terrorism in general usage frequently is also applied to similar acts of violence—all ransom kidnappings, all hijackings, thrill-killings—which are not all intended by their perpetrators to be primarily terror-producing. Once a group carries out a terrorist act, it acquires the label *terrorist,* a label that tends to stick; and from that point on, everything this group does, whether intended to produce terror or not, is also henceforth called terrorism. If it robs a bank or steals arms from an arsenal, not necessarily acts of terrorism but common

*This paper was originally presented at the 1978 meeting of the Institute of Management Sciences and Operations Research Society of America, New York, May 3, 1978. It will be included as a chapter in a forthcoming book published by Pergamon Press.

SOURCE: Jenkins, B. M. (2006). The new age of terrorism. *RAND Corporation*. Retrieved August 11, 2011, from RAND Corporation: http://www.rand.org/pubs/reprints/RP1215. Reproduced with permission of The Rand Corporation, November 1980.

urban guerrilla tactics, these too are often described as terrorism. Eventually, *an* similar acts by other groups also come to be called terrorism. At some point in this expanding use of the term, terrorism can mean just what those who use the term (not the terrorist) want it to mean—almost any violent act by an opponent.

The difficulty of defining terrorism has led to the cliche that one man's terrorist is another man's freedom fighter. The phrase implies that there can be no objective definition of terrorism, that there are no universal standards of conduct in peace or war. That is not true.

Most civilized nations have identified through law modes of conduct that are criminal, among them homicide, kidnapping, threats to life, the willful destruction of property. Such laws may be violated in war, but even in war there are rules that outlaw the use of certain weapons and tactics.

The rules of war grant civilian noncombatants at least theoretical immunity from deliberate attack. They prohibit taking civilian hostages and actions against those held captive. The rules of war recognize neutral territory. Terrorists recognize no neutral territory, no noncombatants, no bystanders. They often seize, threaten, and murder hostages. One man's terrorist is everyone's terrorist.

Terrorism, in the Rand chronology, is defined by the nature of the act, not by the identity of the perpetrators or the nature of their cause. All terrorist acts are crimes—murder, kidnapping, arson. Many would also be violations of the rules of war, if a state of war existed. All involve violence or the threat of violence, often coupled with specific demands. The violence is directed mainly against civilian targets. The motives are political. The actions generally are carried out in a way that will achieve maximum publicity. The perpetrators are usually members of an organized group, and unlike other criminals, they often claim credit for the act. And finally the act is intended to produce effects beyond the immediate physical damage.

The fear created by terrorists may be intended to cause people to exaggerate the strength of the terrorists and the importance of their cause, to provoke extreme reactions, to discourage dissent, or to enforce compliance.

This definition of terrorism would not limit the application of the term solely to nongovernmental groups. Governments, their armies, their secret police may also be terrorists. Certainly the threat of torture is a form of terrorism designed to inspire dread of the regime and obedience to authorities. Some scholars make a semantic distinction

here, reserving the term "terrorism" for nongovernmental groups, while using the term "terror" to describe incidents of state terrorism. There are few incidents of state terrorism or terror in our chronology, not because it is considered to be less heinous, but because such terrorism tends to be internal rather than international. However, there are some international incidents of state terrorism: the assassination of a troublesome exile like Trotsky is an example. A more recent example may be the assassination in Washington, D.C. of a former Chilean cabinet minister, an action that was carried out by anti-Cuban extremists operating at the behest of the Chilean security services.

International terrorism comprises those incidents of terrorism that have clear international consequences: incidents in which terrorists go abroad to strike their targets, select victims or targets because of their connections to a foreign state (diplomats, executives of foreign corporations), attack airliners on international flights, or force airliners to fly to another country. It excludes the considerable amount of terrorist violence carried out by terrorists operating within their own country against their own nationals, and in many countries by governments against their own citizens. For example, Irish terrorists blowing up other Irishmen in Belfast would not be counted, nor would Italian terrorists kidnapping Italian officials in Italy:—"of course, such terrorism, although beyond the scope of our specific research task, is also of common interest and concern as it may lead to actions that will imperil foreign nationals, be carried abroad to other countries, be imitated by other groups, affect the stability of nations individually and collectively, strain relations between nations, or constitute intolerable violations of fundamental human rights, making it a matter of universal concern. Thus, while our research focuses on the specific problem of international terrorism, we find ourselves inevitably trespassing into an area of internal political violence as it bears upon the subject of international terrorism.

The Central Intelligence Agency, in its reports on the subject, makes a distinction between "transnational terrorism," which is terrorism "carried out by basically autonomous non-state actors, whether or not they enjoy some degree of support from sympathetic states," and "international terrorism" which is terrorism carried out by individuals or groups controlled by a sovereign state. This author, frankly, is somewhat skeptical about our ability to make such a distinction, as a growing number of terrorist operations seem to be virtually commissioned by governments. This trend will continue. The CIA also recognizes this problem and in a footnote goes on to say, "Given the element of governmental patronage that is common to both, the boundary line between transnational and international terrorism is often difficult to draw. To the degree that it can be determined, the key distinction lies in who is calling the shots with respect to a given action or campaign. Hence, groups can and do drift back and forth across the line. For example, even a one-time 'contract job' undertaken on behalf of a governmental actor by a group that normally acts according to its own lights qualifies as international terrorism."* In the Rand chronology, we stuck to the term "international terrorism", and attempted to make no distinction on the basis of government support.

This definition seemed pretty straightforward until we actually tried to use it in selecting incidents for our chronology of international terrorism. The chronology was to provide not only an historical record of international terrorism but was also to give some idea of the scope of the problem and allow the identification of trends.

We ran into several problems from the start. We decided that we would exclude incidents of terrorism that occurred in the middle of a war. There were potentially thousands of incidents of terrorism in Indochina, and in the Middle East, for example, during the civil war in Lebanon, some of international character. It would, however, be impossible to record all of these as they were submerged in a higher level of violence. Nor did we wish to engage in an unproductive debate as to whether the shelling of an Israeli kibbutz or the bombing of Hanoi constituted an act of international terrorism. The major incidents of obvious terrorism—the seizure of hostages in a border settlement, the murder of an official of the Palestine Liberation Organization in Beirut or West European capital—were picked up. When a Palestinian terrorist operation provoked an Israel military reprisal, we listed both.

Hijackings presented another problem. Would we include hijackings of airliners by people seeking political asylum? These certainly are not the same as hijackings by groups to publicize a political cause or coerce governments into making political concessions. Certainly the two are not in the same category except that the lives of innocent

bystanders are often jeopardized to satisfy basically political goals—asylum or revolution. The borderline separating political motives from highly personal motives and purely criminal motives is not always clear. We decided not to try to decipher motives. We would include all hijackings except those carried out for obvious criminal intent—individuals demanding cash and a parachute.

A further problem arose in deciding whether to include the activities of separatist groups. As we mentioned previously, our definition of international terrorism would exclude the Irish terrorists blowing up other Irishmen in Belfast. We would, of course, include IRA operations abroad such as the mailing of a letter bomb to a British official in Washington or the assassination of the British ambassador to the Republic of Ireland. We decided also to include IRA bombings in England; in a sense, this represented carrying the campaign abroad. To maintain consistency, we had to include bombings in New York and Chicago by Puerto Rican separatists. Would we then include actions by Corsican separatists if they took place on the French mainland? We have not done so to date but to remain consistent with our decision in the Irish case, I suppose we should. Must we then also include the terrorist activities of Basque and Breton separatists if they operate outside their own province? Even with a fairly precise definition, many decisions quickly become subjective. It becomes slippery around the edges.

Finally, we sometimes chose to list some incidents as one, for example, a single mailing of letter bombs, rather than list it as 40 to 50 separate acts of terrorism. This decision was made in order to avoid distorting the annual total of incidents. On the other hand, a bombing campaign over a period of time, carried on by a single group, was listed as a series of separate actions.

Despite these definition problems, which pertain to only a fraction of the total number of incidents, the chronology has been a useful tool in assessing the magnitude of the problem. The results are sometimes intriguing. We discovered, for example, that the level of international terrorism based upon the chronology does not exactly accord with the public's perception of the problem of terrorism nor with government reaction. To illustrate the point, the total number of incidents of international terrorism in 1972 was

---

*David L. Milbank, *International and Transnational Terrorism: Diagnosis and Prognosis,* Central Intelligence Agency, April 1976, iii, p. 9.

less than that of 1970, while the number of major incidents was about the same for the two years. Incidents with casualties and the number of deaths caused by terrorists were up in 1972. However, it was two particularly shocking incidents in 1972, the Lod Airport massacre in May and the Munich incident in September, that appalled the world and provoked many governments including the United States to undertake more serious measures to combat terrorism.

Similarly, the year 1975 was labeled by many in the news media as the "year of the terrorist". Certainly 1975 seemed to surpass previous years in the number of dramatic and shocking episodes that occurred. There were continued kidnappings in Latin America and in the Middle East, while in Europe two attempts to shoot down airliners at Orly Field in Paris, the kidnapping of a candidate for mayor in West Berlin, the seizure of embassies in Stockholm, Kuala Lumpur and Madrid, the Irish Republican Army's bombing campaign in London, the assassination of the Turkish ambassadors in Austria and France, the hijacking of a train in The Netherlands and the takeover of the Indonesian consulate in Amsterdam, and the seizure of the OPEC oil ministers in Vienna all combined to produce an enormous effect. Certainly, it seemed international terrorism had increased. However, measured by the number of incidents, by the number of major incidents, by the total number of incidents with casualties, and by the total number of casualties, it had in fact declined.

Some observers found encouragement in the seeming "downward trend" in 1976. In fact, however, more incidents of terrorism took place in 1976 and there were more casualties. There were more bombings, more assassinations, and even hijackings went up again after declining in earlier years.

Some continued to perceive a decline in the early months of 1977 but by the end of the year, judging by the number of news articles, television specials, and concern in government, virtually everyone agreed terrorism was on the rise. In fact, it was not. The figures for 1977 indeed show a slight decline.

How do we explain that terrorism often appears to be increasing when it is declining—appears down when it is up? Perhaps we count the wrong things. More likely, the things we can count do not reflect our perceptions of the phenomenon. Terrorism is not simply what terrorists do but the effect—the publicity, the alarm—they create by their actions.

Public perceptions of the level of terrorism in the world appear to be determined then not by the level of violence but rather by the quality of the incidents, the location, and the degree of media coverage. Hostage incidents seem to have greater impact than murder, barricade situations more than kidnappings. Hostage situations may last for days, possibly weeks. Human life hangs in the balance. The whole world watches and waits. By contrast, a death, even many deaths, are news for only a few days. They lack suspense and are soon forgotten. More people recall the hijacking of a TWA airliner by Croatian extremists in September 1976 than recall the bomb placed aboard a Cubana airliner three weeks later. No one died aboard the TWA airliner (although a policeman was killed attempting to defuse a bomb planted on the ground by the hijackers). Seventy-three persons died in the crash of the Cubana plane.

The location of the incident is also important. Incidents that occur in cities have more impact than those that occur in the countryside. Incidents in Western Europe and North America seem more important, at least to the American public, than incidents in Latin America, Africa, or Asia. It is a matter of communications. An unseen and un-heard terrorist incident produces no effect. The network of modern electronic communications laces Western Europe and North America more thoroughly than the rest of the world. We also tend to exhibit a higher tolerance for terrorist violence in the Third World. Terrorist violence in modern industrial societies with democratic governments jars this bias.

Finally, timing is important. Terrorist violence is easily submerged by higher levels of conflict. Individual acts of violence lose their meaning in a war. It is hard to say how many individual acts of terrorism there were during the war in Indochina or how many individual murders, how many kidnappings there were during the civil war in Lebanon. Even a war in another part of the globe can drown out an act of terrorism. There is only so much time and space for news. Terrorist acts themselves in succession produce the effect of a wave of terrorism but must now crowd each other too closely for world attention lest their impact be diluted.

The fact that each terrorist incident is in itself a complete episode—a bomb goes off, an individual is kidnapped and is either released, ransomed, or killed, plus the fact that there are now over 1,000 incidents in Rand's chronology of international terrorism, makes some type of quantitative analysis attractive. If complete chronologies of the terrorism in Argentina, Northern Ireland, Italy, and several other countries that have experienced high levels of terrorist violence were also available, potentially some quantitative analysis could be applied with even greater confidence.

Hostage situations have been examined quantitatively to determine their likely duration, probable outcomes, the risks to the hostages, and even the risks versus payoffs for the hostage-takers. This information has been used in examining the validity of certain policy assumptions and in actually dealing with such episodes.*

Enough airline hijackings occurred to permit the construction of a statistical profile of a typical hijacker, and this was used to reduce the crime.** Some work has also been done in constructing the demographic profile of a "typical terrorist"—a well-educated (although perhaps a university drop-out) male in his early twenties, coming from a middle or upper-class family, the son of a teacher, business executive, or professional, recruited in a university.*** Further analysis may enable us to understand more of his motivations and intentions.

Much more could be done if sufficient data bases were created. At the same time, some cautionary comments are in order. The term terrorism is slippery and also politically loaded. We have seen that it can be difficult to even grossly estimate the level and impact of terrorism by counting the number of incidents. The term "terrorist" is also a loose label applied to political extremists, common criminals, and authentic lunatics. Finally, we must recognize that we are dealing with a fast-moving subject. While there seem to be patterns to terrorist activity, we cannot assume that the historical record offers firm footing for predictions.

REVIEW QUESTIONS

1.   What does he mean when he says defining terror depends on one's point of view? Provide an example.

2.   Why does he characterize the use of the term *terrorism* as sloppy?

3.   How does he address the notion that one man's terrorist is another man's freedom fighter?

4.   Does Jenkins believe a government could engage in terror? How does he address this?

*Brian M. Jenkins, *Numbered Lives: Some Statistical Observations from 77 International Hostage Episodes,* The Rand Corporation, P-5905, July 1977.

**For a discussion of this research, see Evan Pickrel, "Federal Aviation Administration's Behavioral Research Program for Defense Against Hijacking," pp. 19–26, in U.S. Department of Commerce, National Bureau of Standards, *The Role of Behavioral Science in Physical Security: Proceedings of the First Annual Symposium, April 29–30, 1976,* NBS Special Publication 480–24, November 1977.

***Charles A. Russell and Bowman H. Miller, "Profile of a Terrorist," *Military Review,* August 1977, pp. 21–34.

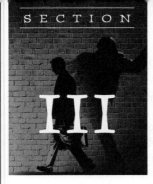

The History of Terrorism

At the end of this section, students will be able to:

- Trace and describe the role of religious terrorism from biblical times to the present day.
- Describe the origin of the word *terrorism* and trace its meaning from its first use to today.
- Identify and describe Rapoport's four waves of terrorism, including the defining characteristics of each wave.
- Trace and analyze the historical impact of the groups discussed in this section and use that historical information to analyze the outcome of modern-day terrorist conflicts.

Section Highlights

- Terrorism and Terminology in the Biblical Regions
- The French Revolution and the Emergence of "Terror"
- Rapoport's Four Waves of Terrorism

Introduction

With the death of Osama bin Laden in May of 2011, one may well ask: why study terrorism at all, let alone its early history? The answer is simply this: Terrorism did not begin with Osama bin Laden, nor will it end with him. Bin Laden had teachers and he was not their only pupil; the older groups have always served as models—directly or indirectly—for the newer groups. It is, therefore, important to understand the context in which terrorism developed over the centuries, before we can form any kind of coherent policy with which to deal with this particular student body.

Laquer (2003) has long argued that a historical approach to the study of terrorism is essential to understanding the phenomenon (p. 7). Indeed, as Laquer notes: Many of the ways we discuss terrorism are based on concepts and terms that have been around for millennia (p. 7). Laquer also sees a danger in current assumptions that radical Islamism is the only threat worth our current attention. He warns scholars not to engage in tunnel vision regarding current events, noting that "radical Islamism was not always the main threat and it may not always be in the future" (2003, p. 8).[1]

This section takes just such a historical approach as Laquer recommends. It will start with a review of some of the earliest terrorist groups that operated from the first to the 13th centuries (and in one case, even to the 19th century). The section then moves on to the development of the term *terrorism* during the French Revolution of the 18th century and then uses David Rapoport's four waves to explore the terrorism phenomenon from 1880 to the present. Readings discussing two of these waves in more detail follow this section.

 Terrorism and Terminology in the Biblical Regions

The following discussion centers on three groups whose names still reverberate in our language today. Indeed, these three groups are often used as case studies for the discussion of the early origins of terrorism. The **Zealots** were active in the latter half of the first century in the areas known today as Israel and Palestine. The **Assassins** were active in Persia and the Middle East from approximately 1090 to 1272. The Thugs (or **Thuggees**) were active in India from the 7th century to the 19th (for excellent summaries of the history of these groups see Rapoport, 1984; Hoffman, 2006).

In his seminal 1984 article on religious terrorism, Rapoport characterizes these groups as practicing "sacred terror" and highlights two reasons for their importance as case studies. First, while sacred terror did not form a significant part of the literature (at least in 1984), "it never disappeared altogether, and there are signs that it is reviving in new and unusual forms" (Rapoport, 1984, p. 659). Second, the study of these three groups allows scholars to challenge the conventional wisdom that technology and capabilities are the key factors on which to focus. Instead, Rapoport argues that "every society has weapons, transportation, and communication facilities and the clear meaning of our cases is that the decisive variables for understanding differences among the forms terror may take are a group's purpose, organization, methods, and, above all, the public's response to that group's activities" (Rapoport, 1984, p. 672).

The Zealots

The Zealots were a religious group existing early in the first century CE.[2] Their goals were to overthrow the Roman Empire in Judea and reform the practice of Judaism in their own image. They targeted both the Roman Empire and elite Jews that the Zealots believed were not pure enough in their practice of Judaism (Chailand & Blin, 2007; Laquer, 1999).

---

[1]One need only look at the events in Oklahoma City in March 1995—where the first assumptions were that the perpetrators were one of the Middle East terrorist groups—to find support for Laquer's warning.

[2]The abbreviations CE (Common Era) and BCE (Before Common Era) are used in this book to note chronological events.

Beliefs, Tactics, and Methods

The Zealots were one of the first messianic cults. According to Chailand and Blin (2007), "messianism postulates that one day in the not-too-distant future, the world will be completely transformed by an event marking the end of history" (p. 3). Those who believed in the coming Messiah and were appropriately pure in the practice of their faith (according to the rules of the cult) would be the ones most likely to survive this history-ending event. Therefore, "as a religious organization, they sought, often by force, to impose a degree of rigor in religious practice" (Chailand & Blin, 2007, p. 57). Indeed, the Zealots sought to purify Judaism in the same way that Assassins would seek to purify Islam some 1,000 years later, or the way al Qaeda seeks to in the 20th and 21st centuries (Chailand & Blin, 2007).

In addition to enforcing their religious beliefs, the Zealots also fomented numerous insurrections against Rome, far more than any other terrorist group in modern times. These insurrections served the parallel religious and political goals of the Zealots. The insurrections against Rome were meant to provide one sign among many of a pending messianic intervention (Rapoport, 1984). However, the insurrections were also meant to gain their country's independence from Roman rule. This close-knit relationship between politics and religion in terrorism is not relegated to Biblical times nor is it unusual. Many nationalistic and ethnic disputes in the 20th and 21st centuries reflect this relationship. Examples include the IRA (Irish Republican Army) against British control of Northern Ireland (Catholic vs. Protestant), the Palestinian conflict with Israel over the creation of a Palestinian state (Muslim vs. Jew), or the Sikh conflict with India over the Punjab region (Sikh vs. Hindu). As Chailand and Blin (2007) go on to note, "exclusively political terrorist organizations are rare in history, as are religious pressure groups with no political ambitions" (p. 57).

Like the Assassins to follow, the Zealots' favorite weapon was the dagger. Yet, their violence was not indiscriminate. Chailand and Blin characterize their strategy as complex (2007, p. 58). In order to create a sense of vulnerability in the general population, they attacked their victims—usually political or religious figures—in open, highly populated public places (such as marketplaces) by daylight and often during significant holidays (Laquer, 1999). They also used their targets to try to build support with various segments in the population. For example, they attacked buildings where loan documents were stored "with the aim of winning the support of a working class crushed by debt" (Chailand & Blin, 2007, p. 58).

Key Events in the History of the Zealots

The Zealots can be traced back as far as the 6th century BCE. (Chailand & Blin, 2007). Their efforts at revolt were soon put down by the Roman governor of Syria and nothing was heard from them for decades (Chailand & Blin, 2007). In fact, almost everything we know about the Zealots comes from the writings of a historian named Flavius Josephus, who chronicled their activities from about 60 CE to 70 CE. It was at this time that the Zealots engaged in their most significant efforts against the Roman Empire. However, while the Zealots are characterized as one of the more effective terrorist groups of their time, they could never match the strength, power, and resources of the Roman Empire and eventually fell to the greater force with catastrophic consequences. When Rome cracked down on the group in 70 CE, it not only went after the Zealots, but the entire Jewish state as well. Jerusalem was sacked, the Second Temple was destroyed, and the Jews were scattered to the far corners of the Roman Empire. At that point, the Jewish state ceased to exist and would not be seen again until 1947.

The final chapter of the Zealots ended on a now famous but tragic note. The few Zealots that survived the destruction of Jerusalem and the temple took refuge in a mountain fortress near the Dead Sea, known as Masada. Nine hundred and sixty men, women, and children held out for three years against a siege by Roman forces. Finally, knowing that the Romans would eventually breach Masada's walls—and rather than face slavery and death at the hands of the Romans—the men killed the women and children before taking their own lives. The myth of this tragic ending still resonates today as a call to never settle for anything less than sovereignty and independence (Chailand & Blin, 2007, p. 58). While the Zealots failed in their ultimate goal, their sustained efforts against a superior occupying force provided a model for many of the anti-colonial groups of the 20th century.

▲   The Ancient Fortress and Palace at Masada Overlooking the Dead Sea

The Assassins

The Assassins were founded by Hassan-i Sabbah in 1090 and lasted until 1275. The group was active in Persia, Syria, and Palestine. Their goal was to "purify Islam in both its religious and political institutions which they viewed as inseparable" (Rapoport, 1984, p. 664). Their philosophy dictated the nature of their attacks and led them to become one of the earliest religious/suicide groups, similar to Hamas and al Qaeda of today.

Beliefs, Tactics, and Methods

Over a period of almost two centuries, the Assassins, based in a stronghold in Northern Persia, targeted all whom they believed were enemies of Islam. This included both Arab leaders felt to be lax in the enforcement of and adherence to their ideal of Islam, but also prominent Crusaders and other foreigners whom they saw as a threat to Islam (Kronenwetter, 2004, p. 24). They often disguised themselves as devout emissaries to an official or an event, when, in fact, they were on a suicide mission, "in exchange for which [the bomber] was guaranteed the joys of paradise" (Laquer, 1999, p. 11).

The weapon of choice for the Assassins was most often a dagger, which mitigated the chance of going undetected or escaping, thus providing an early example of suicide terrorism. In fact, the term used to describe the Assassin suicide squads—the *fedayeen*—is still in use today by Palestinian guerrillas (Sinclair, 2003, p. 28). In addition, their attacks were in public places, which gave the group and their cause immediate publicity without any further efforts at propaganda (Rapoport, 1984). In a sense, this group recognized the strategy of "propaganda by deed" long before the 19th-century anarchists coined the term. As such, the Assassins could be viewed as the precursors to the suicide bombers of today.

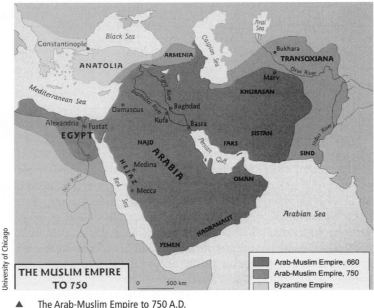

THE MUSLIM EMPIRE TO 750

0   500 km

Arab-Muslim Empire, 660
Arab-Muslim Empire, 750
Byzantine Empire

▲   The Arab-Muslim Empire to 750 A.D.

Key Events in the History of the Assassins

The roots of the Assassins being in Persia is not surprising, since their philosophy, according to Rapoport (1984) "emerged" from Shia elements of Islam. Yet the group also had a significant presence in Sunni Egypt and over time built a number of mountain strongholds. They even declared themselves an independent state, a feat few terrorist groups have ever been able to repeat until ISIS began holding territory in Iraq and Syria in 2014. As noted by Rapoport (1984), "the state provided a means for the creation of an efficient, enduring organization that could and did recover from numerous setbacks" (p. 666). The result of this foundation was to enable the Assassins to seriously threaten several governments over the years (Rapoport, 1984, p. 664). Still, Laquer (1999) notes that the overall impact of the Assassins was small. "They did not make many converts outside their mountain fortresses, nor did they produce any significant changes in Muslim thought or practice" (p. 11). Their "state" was eventually destroyed by Mongol and Arab armies, although their support in urban areas had diminished significantly almost 150 years earlier (Hodgson, 1955, cited in Rapoport, 1984, p. 666).

The Thuggees

The Thuggees, or Thugs, started long before the Assassins appeared in Persia, and were still active long after both the Zealots and the Assassins had been swept away in the desert sands. The Thuggees were a religious cult in India whose origins date back somewhere between the 7th and the 13th centuries. The cult was active well into the 19th century, when they were eventually defeated by British and colonial forces.

The Thuggees were a Hindu cult that worshipped and killed for Kali—often called the Hindu Goddess of Death. Kali ruled over both life and death by "representing the energy of the universe" (Rapoport, 1984, p. 662). As the story goes, Kali killed a monster that was devouring humans, but demons kept springing from the blood of the monster. So Kali created two men from her sweat and gave them cloth handkerchiefs with which to strangle the demons without shedding more blood from which more demons could spring. The men did so and were then bid to keep the handkerchiefs for their descendants (Rapoport, 1984). Based on this myth, Kali's Thuggee followers believed that Kali needed human sacrifice to maintain the balance between life and death.

The Thuggees attacked only travelers, and then only Indian travelers, never Europeans (Rapoport, 1984). They were often called the Deceivers because of their ability to hide within the households of their potential victims, sometimes for years. "The ability of the Thugs to deceive distinguished them radically from other related Hindu criminal associations, which also worshipped Kali" (Rapoport, 1984, p. 662). For the Thuggees, the kill was the name of the game. Although they would confiscate the property of their victims, robbery was never the point of the exercise.[3]

The Thuggees were a ritualistic and suspicious lot. There were rules for just about every activity, including the types of victims selected, the methods of the attack, how labor was to be divided, how the corpses were disposed of, how any booty was distributed, and how new members were trained.[4] Their activities were also dictated by religious omens. Good omens could doom travelers, while bad omens could save them regardless of the wealth they carried (Rapoport, 1984, p. 662).

Although there exists no precise number of Thuggee victims over the centuries, conservative estimates place the number at between 500,000 and 1 million (Rapoport, 1984, pp. 661–662). As Rapoport (1984) notes, some of this number may simply reflect the length of time the group lasted; even a marginally successful killing cult could amass a large number of victims over 1,200 years! Much more, however, is attributed to how well they kept their activities hidden and how extraordinarily difficult it was to see a pattern in their attacks.

That changed at the end of the 18th century. In 1799, the British, under Lord Mornington, captured or gained control of several of the kingdoms in southern India. It was then that "the scattered British soldiers and administrators began to discover that gangs of stranglers infested the roads of southern India in the winter season of travel" (Sinclair, 2003, p. 112). Still, there was no evidence that these murders were connected until Robert Sherwood—a British army surgeon—was able to find and build a network of informers among the

---

[3]Although, as Rapoport (1984, p. 663) notes, they were not above using the loot to bribe various princes for sanctuary.

[4]Membership in the order was by birth, although some captured male children were initiated (Sinclair, 2003, p. 15). As of age 10, the boys were taken under the wing of a near relative as mentor and tutor and allowed to accompany the murder bands (Sinclair, 2003).

Thuggees, prompting him to write a seminal report on the group (Sinclair, 2003, p. 112). Sherwood's report inspired another British officer—William Sleeman—to investigate Thuggee activity in his territory. He was eventually appointed to take charge of suppressing the cult across all of Central India (Sinclair, 2003, p. 117). This was easier said than done. It was extraordinarily difficult to distinguish between Thuggee activity and the more commonplace robbers and bandits that plagued all 19th-century travel. This was especially the case because the bodies of travelers were rarely found. In this, the lack of modern communication benefited the Thuggees greatly.

> It was essential for Thuggee to flourish that men should set out on six-month journeys through lands infested with cholera, cobras, flooded rivers, and ordinary bandits so that when travelers didn't return, no one worried for two years; then it was too late to find out; and the cobras got blamed again. (Masters, as cited in Sinclair, 2003, p. 123)

In addition, because of the close-knit, often familial, nature of Thuggee cells, it was extraordinarily difficult to place informants, adding to the difficulty of building any kind of an informant network. Eventually, however, the British triumphed.

Much of the downfall of the Thugees can and should be ascribed to the efforts of the British, as well as to the development of railways and modern communication. However, much can also be attributed to the actions of the Thuggees themselves. As noted by historian Andrew Sinclair (2003):

> The size of the Thuggee gangs had grown by Sleeman's time, particularly in the north of India. They moved about in groups of twenty or thirty and could combine quickly into greater gangs large enough to murder up to thirty travelers at a time. Greed had brought more Thugs onto the roads, and they had begun to kill indiscriminately for profit, selling off girl children [as] prostitutes and leaving too many live witnesses and unburied corpses behind them. (p. 121)

In the case of the Thugs, the reviews on their impact are mixed. On the one hand, their main goal was to please their Hindu goddess. On the other hand, for all of their violence and longevity, they achieved nothing of the goals attributed to terrorist groups. By focusing their attacks on individuals instead of institutions, they never posed a significant challenge to the authority or the nature of the society they were in (Rapoport, 1984, p. 660). In addition, any political effect they created was, according to Sinclair (2003), negative and divisive.

> In the chaos and anarchy that the British found in India, the British could easily apply their principle of "divide and rule" and use the troops of one petty state to conquer those of another. The Thugs helped to add to the general feeling of insecurity and terror in the subcontinent, so that the peasants often welcomed the British as their only guarantee of justice. (p. 122)

Summary

Although the terms *terror* and *terrorism* were not in use when the Zealots, Assassins, and Thugs were active, these three groups fall clearly into the terrorism category as we know it today. As Rapoport (1984) notes in his discussion of religion and terrorism: "as persons consciously committing atrocities, acts that go beyond the accepted norms and immunities that regulate violence, they were, according to one definition, clearly terrorists" (p. 660; see also Price, 1977). There is much in the old terrorism that we see today in the new. All three of the groups based their actions outside the mainstream in order to violently fulfill the obligations of their belief system (Rapoport, 1984). For both the Zealots and the Assassins, these obligations included purifying their respective religions of corrupt practices, much as al Qaeda and Hamas seek to create an Islamic state that reflects its own interpretations of the Islamic faith.

The record of success among these three groups is mixed. Although the Zealots never came close to overthrowing the Roman Empire, their short-term success in generating insurrections against, and tying up the resources of, a stronger occupying power provided a model for later anti-colonial terrorist groups. The Assassins targeted institutions of government and religious control and, as a result, came seriously close to threatening the existence of several states, something very few modern terrorist groups have achieved (Rapoport, 1984, p. 664). The Thuggees had the most limited effect of the three, even though they operated the longest, because of their targeting of individuals instead of institutions (Rapoport, 1984, p. 664). Finally, as Rapoport (1984) notes, both the Assassins and the Thuggees are early examples of modern international terrorism; both obtained and relied on foreign sanctuaries and support in a way similar to such modern groups as Hezbollah (backed by Iran and Syria) and al Qaeda (at one point backed by Afghanistan) (p. 664).

 ## The French Revolution and the Emergence of "Terror"

The terms *terror* and *terrorism* can be traced to **the Great Terror** of the French Revolution in the late 18th century, the brainchild of the infamous and fanatical Robespierre (Kronenwetter, 2004; Parry, 1976; see also Laquer, 1999). The Great Terror began in earnest four years into the French Revolution when the revolutionary army was defeated by the monarchies of Austria, Russia, and Prussia (Sinclair, 2003). The daily spectacle of the guillotine claiming its victims became a way of distracting the citizenry from looking too closely at the competency (or apparent lack thereof) of their new revolutionary government. In addition, Robespierre used the Revolution's most famous tool against all enemies, foreign and domestic, real or imagined, to ensure the failure of any opposition to the ruling Jacobins (Kronenwetter, 2004, p. 26). Robespierre justified the never-ending appetite of Madam Guillotine both "as a means of the survival of the French people threatened by their enemies" and "as a path to the people's welfare and virtue" (Parry, 1976, p. 47). Estimates of the terror place the number of dead at 17,000 with 400,000 imprisoned (Parry, 1976; Sinclair, 2003).

Conventional wisdom characterized the French Revolution as the masses rising up against the hateful nobility and sending them to their just deserts. Yet, as noted by Parry (1976), the noble victims of the guillotine represented only 1/4 of 1% of all French nobility (p. 59). Instead,

> in the most haphazard way, men and women of all classes were guillotined or otherwise slain not only on charges of wartime treason or other anti-state activities, be such charges true or not, but often because of the sheer personal greed or grudge of the executioners. (Parry, 1976, p. 48)

However, such revolutions cannot stand forever; eventually they run out of new victims and start feeding on their own. While a number of factors came together in the summer of 1794 to reign in the Reign of Terror, chief among them was a sense of self-preservation among the remaining members of the revolutionary government (Parry, 1976, pp. 59–60).

The impact of the French Revolution and the Great Terror is not strictly semantic, however. Sinclair (2003) and Parry (1976) both argue that the French Revolution represented a more rational and planned use of the tactics of terror than had been seen before. Sinclair argues that until the revolution, terror lacked a cohesive definition or meaning. It merely described atrocities used by any number of groups to get and maintain power (2003, p. 74). By contrast, the French Revolution used the Great Terror as an effective tool (among others) of revolutionary politics. According to Parry (1976), the Great Terror of 1793–1794 was "the first in history to attempt the elevation of primitive passion into a high-flown political philosophy, and to create an organization that tried to systemize murder and other lawlessness into a set of rules" (p. 39). In fact, the philosophical impact of the French Revolution and its guillotine weapon went far beyond its years of existence; its anti-monarchical orientation directly influenced such revolutionaries as Karl Marx, Friedrich Engels, and Vladimir Lenin (Parry, 1976). In turn, as we shall see in later sections, Marx and Lenin provided much of the theoretical inspiration for groups in Latin America, Asia, and Europe well into the late 20th century (Parry, 1976; see also Hoffman, 2006).

⊠ Rapoport's Four Waves of Terrorism

Following the French Revolution, modern terrorism can be divided into approximately four waves, first described by David Rapoport in 1984 and expanded upon in numerous writings since. The first wave lasted from 1880 to 1920 and was dominated by groups seeking to overthrow an existing system of governance and replace it with one using the anti-state and anti-monarchical ideology of 19th-century **anarchism**. The second wave lasted from 1920 to 1960 and was dominated by anti-colonial groups seeking independence from the old colonial empires. The third wave lasted from 1960 to 1980 and was dominated by left-wing terrorist groups who sought to overthrow capitalism in favor of a Marxist-Leninist ideology. The fourth and current wave started in 1980 and has been dominated by religious groups seeking to eliminate all separation of church and state in political, social, and economic institutions (Rapoport, 2002). A summary of these four waves can be seen in Table 3.1.

The Anarchist (First) Wave, 1880 to 1920

Inspired by the anti-monarchical nature of the terror of the French Revolution, several anarchist writers began to form a revolutionary justification for terrorism. These included such writers as Germans Karl Heinzen and Johann Most and Russians Mikhail Bakunin and Peter Kropotkin (Laquer, 1999; Rapoport, 2002). These writers were early advocates of "propaganda by deed"; the idea that simple propaganda without action would be ineffective, and that violent, but mostly targeted, actions were needed to "sow confusion among the rulers and mobilize the masses" (Laquer, 1999, p. 14; see also Rapoport, 2002).

Some of these writers, Most and Heinzen in particular, even hinted at the use of weapons of mass destruction (WMD), believing that science and technology could lead to more effective and more lethal weapons, granting terrorists a hitherto unknown advantage over the stronger state armies. Indeed, Heinzen argued strongly for the use of poison and other more deadly weapons as vital to the cause of freedom (Laquer, 1999, p. 13). In the end, however, neither of these early writers actually carried out such dastardly deeds and most of them mellowed considerably with

Table 3.1	David Rapoport's "Four Waves of Terrorism"		
Wave	**Name**	**Dates**	**Main Focus of Groups**
First	Anarchist Wave	1880–1920	To overthrow the existing system of governance and replace it with one using the anti-state and anti-monarchical ideology of the 19th century anarchists.
Second	Anti-colonial Wave	1920–1960	To gain independence from the old colonial empires, especially following the end of WWII.
Third	New Left Wave	1960–1980	To overthrow the Western capitalist system and replace it with one using a Marxist-Leninist ideology.
Fourth	Religious Wave	1980 to present	To eliminate all separation of church and state in political, social, and economic institutions.

The above descriptions of the Four Waves were drawn from Rapoport (2002) and Rapoport (2004).

age. For example, both Most and Heinzen settled in the United States and moved away from calls to violence, focusing instead on political and fund-raising activities. Heinzen, in fact, became a staunch advocate for both women's rights and abolition (Laquer, 1999, p. 13).

Nevertheless, their writings and teachings influenced later groups who did take action, sometimes with tragic consequences. One of the most famous of the anarchist groups in this first wave of terrorism was the Russian group Narodnaya Volya (the People's Will). The group sought to challenge the tsarist rule over Russia. They believed that the nature of the Russian masses, and the strong hold the tsarist regimes had over any political opposition, required that they resort to "daring and dramatic acts of violence designed to attract attention to the group and its cause" (Hoffman, 2006, p. 5; see also Rapoport, 2002). However, they did not go so far as to accept the indiscriminate bombing initially advocated by Most and other early anarchist writers. Instead, their targets were specifically addressed to individuals in the Tsarist regime or supporting it. As noted by Hoffman (2006), their targets were selected for their "symbolic value as the dynastic heads and subservient agents of a corrupt and tyrannical regime" (p. 5).

In the end, Narodnaya Volya chose one symbolic target too many. In 1881 they assassinated Tsar Alexander II himself. In one of the most well-known ironies in terrorism history, Alexander II was considered a reformer and probably the only one capable of creating and leading a transformed Russia.[5] In response, the Tsarist regime turned its full force on, and eliminated, Narodnaya Volya. As described by Laquer (1999), his assassination was entirely counterproductive:

▲　Picture of Anarchist

Hulton Archive/Getty Images

> The reformer Tsar Alexander II was replaced by the more repressive regime of Alexander III. The assassination helped to shut the door to a political solution of the constant Russian crisis and led to the revolution of 1917. The Tsarist regime bore principal responsibility for the events of 1917, but the activities of the terrorists, despite their political aims, had not helped to resolve the continuing political crisis. (p. 18)

The Anti-Colonial (Second) Wave, 1920 to 1960

Rapoport's second wave covers the years from 1920 to 1960. The legacies of both WWI and WWII played a significant role in the terrorism of this era (described in more detail in In Focus 3.1).

---

[5]Reforms proposed by Alexander II included freeing the serfs and providing money to buy land, giving more control to local governing institutions, relaxing controls over speech and education, and reforming the judicial system. His tragedy may have been to raise expectations too fast for actual reform to keep pace (Rapoport, 2002, p. 5).

IN-FOCUS 3.1

The Fall of the Ottoman Empire and Today's Middle East

The call of many of the Islamic fundamentalist groups of today is for an end to the nation state system imposed on the region by the European powers and the creation of an Islamic caliphate that would govern the entire region under a strict interpretation of Islamic law and would recognize no national or ethnic boundaries. A caliphate is "an institution defined by rightful succession to the earthly political authority that the Prophet [Muhammed] had exercised" (Kissinger, 2014, p. 421). The idea of an Islamic caliphate is not a new one but an old one, and one that existed—in a less idealized version—for centuries until its defeat in World War I (WWI). It was known as the Ottoman Empire.

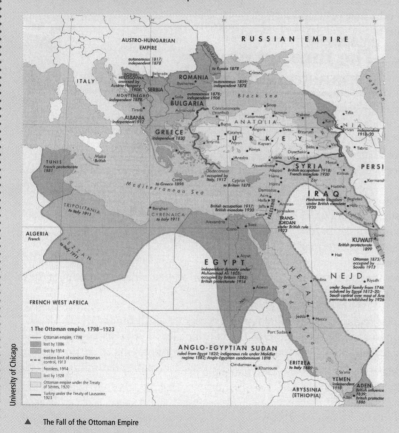

The Fall of the Ottoman Empire

In his book *A Peace to End All Peace: Creating the Modern Middle East, 1914–1922*, historian David Fromkin argues that the Middle East "as we know it from today's headlines, emerged from decisions made by the Allies during and after the First World War"—most importantly, decisions about how the territories of the fallen Ottoman Empire were divided (Fromkin, 2009, p. 7). What was, then, the Ottoman Empire?

In the 13th century the Turks of the Anatolian region began to expand their territory under their leader, Osman (hence the name Ottoman). At its height in the 15th and 16th centuries the Ottoman Empire unified a territory that stretched from the Mediterranean Sea to the borders of Central Asia and from Eastern Europe and the Balkans throughout much of what is now considered the Middle East, North Africa, and the Persian Gulf. Notes Kissinger (2014) in his book on the world order of nation states, "the Ottoman Empire was territorially larger than all of the Western European states combined and for many decades militarily stronger than any conceivable coalition of them" (p. 449). At various points in its history, the Empire was known for firm control over its vast territories as well as efficiency, prosperity, culture, and tolerance of non-Muslim communities within its borders.

In the 19th century, the balance of power between the Ottoman Empire and Western Europe began to shift. In part this was due to the growing military power and influence of Russia, Austria, Britain, and France as they sought to move into conquered Ottoman territories. But just as much, it was due to the internal decay in the Empire itself. Its lack of modernization in the face of powerful religious interests, increasing internal disturbances, and a gradual weakening of control over its territories made the Ottoman Empire vulnerable (Fromkin, 2009; Kissinger, 2014). The end came with the alliance of the Ottoman Empire with Germany against the Allied Powers in WWI. They lost.

To the victor go the spoils. In this case, the spoils involved almost the entirety of the Middle East. Between 1920 and 1923, the victorious European powers divided the Ottoman Empire based on their own national interests, with little regard for either natural boundaries or the ethnic, religious, or cultural identities and allegiances of the populations involved. The Middle East was reconceived "as a patchwork of states—a concept heretofore not part of its political vocabulary" (Kissinger, 2014, p. 464). France was given control over Syria and Lebanon; Britain was given control over both the Palestinian mandate (consisting of Palestine and Trans-jordan) and over Mesopotamia (now Iraq). Egypt became a nominally independent protectorate under Britain with Fuad I placed on the throne. Italy was given control of Libya, and the core of the original Ottoman Empire became modern Turkey. The Jews were promised a home in Palestine and the question of a homeland for the Kurds (considered in early negotiations) was ignored, setting the stage for years of conflict in Iraq, Turkey, and Syria. Fromkin describes the actions of the Allied Powers even more bluntly:

> It was an era in which the Middle Eastern countries and frontiers were fabricated in Europe. Iraq and what we now call Jordan, for example, were British inventions, lines drawn on an empty map by British politicians after the First World War; while the boundaries of Saudi Arabia, Kuwait, and Iraq were established by a British civil servant in 1922, and the frontiers between Muslims and Christians were drawn by France in Syria-Lebanon and by Russia on the borders of Armenia and Soviet-Azerbaijan. (Fromkin, 2009, p. 9)

"Until World War II (WWII), the European powers were sufficiently strong to maintain the regional order they had designed for the Middle East in the aftermath of WWI" (Kissinger, 2014, p. 472). The end of WWII saw a profound shift in the governance of the nations of the Middle East and in their relationship with the international community. As discussed in Part II, some of the nations under direct control, or a "mandate," of the colonial powers formed their own independent states. Others, especially in the 1950s and 1960s, displaced governments established or supported by the colonial governments in favor of more nationalist rulers. In addition, as both elder statesman Henry Kissinger and historian David Fromkin argue, the nation-state system was a concept imposed by non-Muslim foreigners on Muslim populations. As a result, it pasted over but never addressed millennia-old religious, ethnic, and tribal conflicts that continued to simmer beneath the surface—especially the split between the Sunni and Shia sects of Islam. Adding to the political and diplomatic challenges in the region was a growing Arab nationalism that sought to create secular national governments free from European control. These governments soon began to clash with forces that demanded a governing structure based on religious ideology. As noted by Fromkin, the European belief in secular civil sovereign-state government so taken for granted elsewhere is "an alien creed in a region most of whose inhabitants, for more than a thousand years, have avowed a faith in a Holy Law that governs all of life, including government and politics" (2009, p. 564).

In international forums, the Cold War between the United States and the Soviet Union also greatly impacted events in the Middle East. Both superpowers and their respective allies used the Middle East states

(Continued)

(Continued)

as pawns in the never-ending jockeying for power between the two camps. This often resulted in the West and the Soviet Union keeping in power tyrannical despots as a guarantee of stability and support for their policies—regardless of the consequences for the populations these despots ruled. The rulers in turn became masters at exploiting the Cold War rivalry for their own ends and profit.

In late 2010 this melting pot of artificial boundaries, despots with questionable legitimacy, simmering religious and ethnic tensions, and disputes over secular versus religious governing systems came to a head and then boiled over. Challenges to existing governments in Tunisia, Egypt, Syria, Yemen, and Bahrain (discussed in more detail in Part II) suggested for a brief moment the potential triumph of liberal democratic movements and governments. However, as of this writing, the Arab Spring has instead brought to the forefront all of the tensions and conflicts exacerbated by the treaties of the 1920s and the effort to impose from the outside a system of government that had never been a significant part of the region's history or experience. However, while much about today's conflicts in Lebanon, Syria, Iraq, Libya, Israel, and the Palestinian territories can be traced to the events of the 1920s, much also belongs to disputes and differences dating to the founding of Islam and even before. These disputes are not likely to be prevented or solved by any outside intervention, no matter how well intentioned.

This anti-colonial wave began just after the Versailles Peace Treaty of 1919, which ended the war between Germany and the Allied powers. In fact, this wave was sparked in part by some of the treaty's provisions. As noted by Rapoport (2002):

> The empires of the defeated states (which were mostly in Europe) were broken up by applying the principles of self-determination. Where independence was not immediately feasible, territories were understood to be "mandates" ultimately destined for independence. But the victors could not articulate the principle without raising questions about the legitimacy of their own empires. (p. 5)

While many of the colonies' aspirations for independence were placed on hold during WWII (anyone was better than Hitler), they returned full force soon after. Public comments in support of sovereignty in such documents as the Atlantic Charter, and the obviously weakened condition of the old colonial powers, sparked revolutionary movements around the globe (Hoffman, 2006; Laquer, 1999). Included in these movements were the Irgun against the British in Palestine, the National Organization of Cypriot Fighters (EOKA) against the British in Cyprus, and the National Liberation Front (FLN) against the French in Algeria. Each of these groups was shaped not only by its political agenda, but by the environment in which it operated.

> In urban societies such as Palestine and Cyprus, the action by necessity took place mainly in the cities. In Algeria, the struggle against the French proceeded both in the cities and in the countryside, and elements of terrorism and guerrilla warfare appeared side-by-side. (Laquer, 1999, p. 23)

Despite differences in environment, anti-colonial groups did share three common strategies:

◆ Do not try to overwhelm a larger military force. Instead, make it too economically and politically expensive for the occupying force to remain.
◆ Do not try to hold territory against the larger military force of the state
◆ Force an occupying power to overreact, by randomly targeting soft targets and generating a fear of attack in the general population (Hoffman, 2006)

Terrorist groups over the next several decades would use these strategies as a model for their own efforts.

All three of these groups can claim ultimate success based on the creation of the states of Israel, Cyprus, and Algeria, respectively. They also inspired nationalist uprisings later in the 1960s, including groups such as the Palestinian Liberation Organization (PLO) and the Irish Republican Army (IRA). Nevertheless, while many of the terrorist groups that will be discussed in later chapters of this book drew on these lessons, they were rarely able to repeat the successes of their earlier teachers. The fight by the Palestinian groups for a state of Palestine, and by the IRA for an independent Northern Ireland, continues today at varying levels of intensity. The Liberation Tigers of Tamil Eelam (Tamil Tigers) were eventually defeated by the government of Sri Lanka. The Basque Nation and Liberty Party (ETA) in Northern Spain have been rendered superfluous by political, economic, and social reforms in the region; their attacks are now sporadic and self-defeating.

The reasons for this lack of success are suggested by Hoffman (2006), who noted a key difference between the anti-colonial groups of the early decades after WWII and the nationalist groups that followed them:

> In sum, the anti-colonial terrorism campaigns are critical to understanding the evolution and development of contemporary terrorism. They were the first to recognize the publicity value inherent in terrorism and to choreograph their violence for an audience far beyond the immediate geographical loci of their immediate struggles (p. 62). . . . [However], the establishment of these independent countries [Israel, Cyprus, and Algeria] was confined to a distinct period of time and was the product, in some cases, of powerful forces other than terrorism. (p. 61)

The New Left (Third) Wave, 1960 to 1980

Rapoport (2002) called his Third Wave the New Left Wave. This wave drew on the ideology of Marx and Lenin (among others) to justify violent action needed to overthrow the Western capitalist system. Many of the Third Wave groups took hold in the Western capitalist democracies of Europe, including the Red Army Faction (RAF) in Germany,[6] Action Direct (AD) in France, the Communist Combatant Cells (CCC) in Belgium, the Red Brigades in Italy, and November 17 in Greece. Most of these groups died out with the death of the Soviet Union, a spectacular example of the failure of their ideology.

Longer lasting were the Third Wave groups that combined left-wing ideology with nationalist goals. These groups not only sought independence from a colonial power, or autonomy for a minority ethnic group, but sought to create the resulting state under a Marxist-Leninist system. Groups in this category include the Basque Nation and Liberty (ETA) in Spain, the Armenian Secret Army for the Liberation of Armenia (ASALA) in Armenia, and some of the Palestinian groups such as the Popular Front for the Liberation of Palestine (PFLP) and the Popular Front for the Liberation of Palestine-General Command (PFLP-GC).

Also emerging onto the scene in the Third Wave were activities deemed by many terrorism scholars to be "international terrorism." International terrorism refers to the tendency of the terrorist groups to choose targets with "international dimensions" and to conduct as many, if not more, of their attacks abroad, as they did on their home turf (Rapoport, 2002, p. 7). One of the earliest efforts at this form of terrorism took place in 1968, when the PLO hijacked several passenger planes from several countries and blew them up on the tarmac of an airfield in Jordan on national television. Attacks that gained prominence in this wave included hijackings, kidnappings, and hostage takings, and strikes on foreign embassies or other symbolic institutions (Rapoport, 2002). Some Third Wave groups even gained some international support for their cause. "As the United Nations (UN) grew by admitting new states, virtually all of which were former colonial territories, that body gave the anti-colonial sentiment more structure, focus, and opportunities. Significantly, UN debates regularly described anti-colonial terrorists as 'freedom fighters'" (Rapoport, 2002, p. 12).

While few of these groups, if any, achieved their ultimate goal of changing the system—and as of 2011 most of these groups have been defeated, disarmed, or reduced to sporadic attacks—the effectiveness of their international

---

[6]The RAF [Red Army Faction] was originally known as the Baader Meinhof Gang.

attacks in garnering publicity and attention would influence the Fourth Wave groups where such tactics would reach new heights. At the same time, there were some positive impacts resulting from efforts to deal with their activities. By the end of the 1980s, international cooperation had become much more prevalent and effective. Although not perfect by any means, these cooperative efforts laid the foundation for the significantly increased demands for international efforts following the events of September 11.

The Religious (Fourth) Wave, 1980 to the Present

Beginning in the early 1990s, religious terrorism began to demand more attention on the world stage. As the earlier discussion of the Zealots, Assassins, and Thugs demonstrated, there was nothing new about religious terrorism. What was new was the intersection of religious terrorism and modern technology (Chailand & Blin, 2007, p. 2). The marriage of religious fervor and technology enabled the groups in this wave to widen the killing zone considerably, both in terms of number of casualties and the distances over which they could operate.

While religious terrorism stretches across all religious traditions and often overlaps with nationalist aspirations, Hoffman (2006) did note some striking similarities among the different groups. First, the violence is the point of the exercise, the commission of which is seen as a sacramental or divine duty. Second, the actions of the group members are sanctioned by some type of religious authority. Third, the goal of the group is fundamentally to overthrow the current system and replace it with one that is based upon, and adheres strictly to, a rigid religious code. Finally, the groups in this category advocate and even applaud more and more deadly violence against any and all non-believers.

These characteristics have led to a concern hitherto unknown in the earlier waves. Prior to this "**new terrorism**," the conventional wisdom argued that terrorist groups would not resort to extremely high casualty attacks or attacks using weapons of mass destruction for fear of losing any international or domestic support they might have gained. This may no longer be true in a system where religious authority is equating the highest casualties possible with one's divine duty. Indeed, this fear was confirmed by the actions of a millennium cult in Japan known as Aum Shinrikyo. In March 1995, Aum Shinrikyo released sarin gas into the Tokyo subway system during morning rush hour, killing 12 and injuring or sickening over 1,000. As a result of the Tokyo attack, "a worldwide anxiety materialized over expectations that a new threshold in terrorist experience has materialized, various groups would be encouraged to use chemical and biological weapons soon, and each separate attack would produce casualties numbering tens of thousands" (Rapoport, 2002, p. 54). Still, it is important to remember that, despite the fact that numerous government policies and programs have been put in place to prevent or respond to WMD attacks, the most lethal attack from a religious or non-religious group was still a wholly conventional one—three fully fueled passenger jets that resulted in the death of over 3,000 people.

This section has used Rapoport's **Four Waves of Terrorism** to trace the evolution of terrorism from the middle of the 19th century to the present. It was a gradual evolution, each wave drawing on contributions from the others. The anti-monarchical tendencies of the Anarchists influenced a number of nationalist and anti-colonial groups, some of whose conflicts continue today. In the 1960s, this emphasis shifted toward one promoting the establishment of a left-wing governing system using the philosophies of Marx and Lenin, who were themselves influenced by the anarchists of the First Wave. By the 1990s the left-wing ideology gave ground to ideologies based on religion and calls for states where religious rather than secular law governs, in many ways reviving the inseparable nature of political and religious goals that were so much a part of the Zealots and Assassins of old. Perhaps, not such a "new" terrorism after all.

 ## Summary

Those who ignore history are doomed to repeat it. This oft-cited phrase is as true with regard to the study of terrorism as with anything else. This chapter has examined terrorist groups and the phenomenon of terrorism from early in the first century to the present, focusing on the impacts earlier eras and groups have on the terrorism of today and on the similarity of themes they presented over the centuries.

This section has examined these trends and impacts through the lens of Rapoport's Four Waves of Terrorism: the Anarchist Wave, the Anti-Colonial Wave, the New Left Wave, and the Religious Wave. Later sections in this book will examine specific regions where terrorism has been prevalent, starting with the New Left Wave.

In addition to the ideological and strategic examples provided by these older groups, they also provided a road map for tactics we see in use today, such as suicide bombings, assassinations and bombings in public places to attract the maximum attention, and the choice of targets most likely to intimidate a government or a population. These are now considered the norms of modern terrorism. Indeed, we have become so used to these tactics that terrorists of today often feel the need to use modern technology to ratchet up the body count just to gain the same amount of attention. In fact, the "new terrorism" of today may in actuality symbolize the return to something very old, making the old worth studying.

Key Points

- In the first century (CE) the Zealots provided an early lesson in how a weaker force can challenge a stronger occupying power—a lesson used in tactics and strategy by the anti-colonial terrorist groups of the 20th century.
- Both the Zealots and the Assassins (beginning in the 11th century) used terrorist activities to "purify" their respective religions of corrupt practices, likely providing a justification and a road map for the current Islamic fundamentalist groups such as Hamas or al Qaeda.
- The anti-monarchical flavor of the terrorism of the French Revolution (specifically the use of the guillotine by the Jacobins as a political tool) influenced the anti-state motivations of the anarchist groups of the 19th century.
- The anarchists significantly influenced the teachings of the left-wing ideologies of Marx and Lenin, who in turn provided the ideology of left-wing groups in Europe, Latin America, and even in the United States, particularly in the latter half of the 20th century.
- A number of terrorist groups in operation today have over time found vital support, funding, and shelter with foreign governments in much the same way as the Assassins and Thugs did.

KEY TERMS

Anarchism	the Great Terror	Zealots
Assassins	New Terrorism	
Four Waves of Terrorism	Thuggees	

DISCUSSION QUESTIONS

1. Why, if at all, is it useful to study the history of terrorism?

2. Briefly describe the motives and tactics of the Zealots, the Assassins, and the Thugs. What influence, if any, do you believe these ancient groups had on modern terrorism?

3. Trace the development of the term *terrorism* beginning with the French Revolution. What, if anything, remains of its original connotations today?

WEB RESOURCES

British Broadcasting Corporation (looks back at famous days and events in history). http://www.bbc.co.uk/history/recent/sept_11/changing_faces_01.shtml

RAND Corporation: http://www.rand.org

Maps as History: http://www.the-map-as-history.com

READING 4

The theme for both of the readings in this section is that "history matters." In his article "The Four Waves of Modern Terrorism," David Rapoport divides the history of terrorism beginning with the middle of the 19th century into four waves or eras: 1880 to 1920; 1920 to 1960; 1960 to 1980; and 1980 to the present. Rather than focus on individual groups or incidents, Rapoport uses the concept of a wave to study terrorism as a series of expansions and contractions over the years, describing the political, social, and cultural contexts that fashioned each wave.

Rapoport's work is important because it forces us to move from the more narrow focus of previous work in the discipline to a broader focus on the context in which terrorism takes place. In addition, by comparing the characteristics of waves gone by, we may be able to make more accurate assessments regarding the nature and longevity of the current terrorist threat. To quote a very old cliché, "One cannot get there if one does not know where one has been."

The Four Waves of Modern Terrorism

David C. Rapoport

September 11, 2001, is the most destructive day in the long, bloody history of terrorism. The casualties, economic damage, and outrage were unprecedented. It could turn out to be the most important day too, because it led President Bush to declare a "war (that) would not end until every terrorist group of global reach has been found, stopped, and defeated."[1]

However unprecedented September 11 was, President Bush's declaration was not altogether unique. Exactly 100 years ago, when an anarchist assassinated President William McKinley in September 1901, his successor Theodore Roosevelt called for a crusade to exterminate terrorism everywhere.[2]

No one knows if the current campaign will be more successful than its predecessors, but we can more fully appreciate the difficulties ahead by examining features of the history of rebel (non-state) terror. That history shows how deeply implanted terrorism is in our culture, provides parallels worth pondering, and offers a perspective for understanding the uniqueness of September 11 and its aftermath.[3] To this end, in this chapter I examine the course of modern terror from its initial appearance 125 years ago; I emphasize continuities and change, particularly with respect to international ingredients.[4]

 ## The Wave Phenomena

Modern terror began in Russia in the 1880s and within a decade appeared in Western Europe, the Balkans, and Asia. A generation later the wave was completed. Anarchists initiated the wave, and their primary strategy—assassination campaigns against prominent officials—was adopted by virtually all the other groups of the time, even those with nationalist aims in the Balkans and India.

Significant examples of secular rebel terror existed earlier, but they were specific to a particular time and country. The Ku Klux Klan (KKK), for example, made a striking contribution to the decision of the federal government to end Reconstruction, but the KKK had no contemporary parallels or emulators.[5]

The "Anarchist wave" was the first global or truly international terrorist experience in history;[6] three similar, consecutive, and overlapping expressions followed. The "anticolonial wave" began in the 1920s and lasted about forty years. Then came the "New Left wave," which diminished greatly as the twentieth century closed, leaving only a few groups still active today in Nepal, Spain, the United Kingdom, Peru, and Colombia. In 1979 a "religious wave" emerged; if

SOURCE: Copyright 2004 by Georgetown University Press. "The Four Waves of Modern Terrorism" by David C. Rapaport From *Attacking Terrorism: Elements of a Grand Strategy*, Audrey Kurth Cronin and James M. Ludes, Editors, pp. 46-73. Reprinted with permission, www.press.georgetown.edu

the pattern of its three predecessors is relevant it could disappear by 2025, at which time a new wave might emerge.[7] The uniqueness and persistence of the wave experience indicates that terror is deeply rooted in modern culture.

The wave concept—an unfamiliar notion—is worth more attention. Academics focus on organizations, and there are good reasons for this orientation. Organizations launch terror campaigns, and governments are always primarily concerned to disable those organizations.[8] Students of terrorism also focus unduly on contemporary events, which makes us less sensitive to waves because the life cycle of a wave lasts at least a generation.[9]

What is a wave? It is a cycle of activity in a given time period—a cycle characterized by expansion and contraction phases. A crucial feature is its international character—similar activities occur in several countries, driven by a common predominant energy that shapes the participating groups' characteristics and mutual relationships. As their names—"Anarchist," "anti-colonial," "New Left," and "Religious"—suggest, a different energy drives each.

Each wave's name reflects its dominant but not its only feature. Nationalist organizations in various numbers appear in all waves, for example, and each wave shaped its national elements differently. The Anarchists gave them tactics and often training. Third-wave nationalist groups displayed profoundly left-wing aspirations, and nationalism serves or reacts to religious purposes in the fourth wave. All groups in the second wave had nationalist aspirations, but the wave is termed anticolonial because the resisting states were powers that had become ambivalent about retaining their colonial status. That ambivalence explains why the wave produced the first terrorist successes. In other waves, that ambivalence is absent or very weak, and no nationalist struggle has succeeded.

A wave is composed of organizations, but waves and organizations have very different life rhythms. Normally, organizations disappear before the initial wave associated with them does. New Left organizations were particularly striking in this respect—typically lasting two years. Nonetheless, the wave retained sufficient energy to create a generation of successor or new groups. When a wave's energy cannot inspire new organizations, the wave disappears. Resistance, political concessions, and changes in the perceptions of generations are critical factors in explaining the disappearance.

Occasionally an organization survives its original wave. The Irish Republican Army (IRA), for example, is the oldest modern terrorist organization—emerging first in 1916, though not as a terror organization.[10] It then fought five campaigns in two successive waves (the fourth struggle, in the 1950s, used guerrilla tactics).[11] At least two offshoots—the Real IRA and Continuity IRA—are still active. The Palestine Liberation Organization (PLO}, founded in 1964, became active in 1967. When the Viet Cong faded into history, the international connections and activity of the PLO made it the preeminent body of the New Left wave, although the PLO pursued largely nationalist ends. More recently, elements of the PLO (e.g., Fatah) have become active in the fourth wave, even though the organization initially was wholly secular. When an organization transcends a wave, it reflects the new wave's influence—a change that may pose special problems for the group and its constituencies, as we shall see.

The first three waves lasted about a generation each—a suggestive time frame closest in duration to that of a human life cycle, in which dreams inspiring parents lose their attractiveness for children.[12] Although the resistance of those attacked is crucial in explaining why terror organizations rarely succeed, the time span of the wave also suggests that the wave has its own momentum. Over time there are fewer organizations because the enterprise's problematic nature becomes more visible. The pattern is familiar to students of evolutionary states such as France, the Soviet Union, and Iran. The inheritors of the revolution do not value it in the same way its creators did. In the anticolonial wave the process also seems relevant to the colonial powers. A new generation found it much easier to discard the colonial idea. The wave pattern calls one's attention to crucial political themes in the general culture—themes that distinguishes the ethos of one generation from another.

There are many reasons the first wave occurred when it did but two critical factors are conspicuous and facilitated successive waves. The first was the transformation in communication and transportation patterns. The telegraph, daily mass newspapers, and railroads flourished during the last quarter of the nineteenth century. Events in one country were known elsewhere in a day or so. Prominent Russian anarchists traveled extensively, helping to inspire sympathies and groups elsewhere; sometimes, as the journeys of Peter Prodhoun indicate, they had more influence abroad than at home. Mass transportation made large-scale emigrations possible and created diaspora communities, which then became significant in the politics of both their "new" and "old" countries. Subsequent innovations continued to shrink time and space.

A second factor contributing to the emergence of the first wave was doctrine or culture. Russian writers created a strategy for terror, which became an inheritance for successors to use, improve, and transmit. Sergei Nechaev was the leading figure in this effort; Nicholas Mozorov, Peter Kropotkin, Serge Stepniak, and others also made contributions. [13] Their efforts perpetuated the wave. The KKK had no emulators partly because it made no effort to explain its tactics. The Russian achievement becomes even more striking when we compare it to the practices of the ancient religious terrorists who always stayed within their own religious tradition—the source of their justifications and binding precedents. Each religious tradition produced its own kind of terrorist, and sometimes the tactics within a tradition were so uniform that they appear to be a form of religious ritual. [14]

A comparison of Nechaev's *Revolutionary Catechism* with Osama bin Laden's training manual, *Military Studies in the Jihad Against the Tyrants,* shows that they share one very significant feature: a paramount desire to become more *efficient* by learning from the experiences of friends and enemies alike. [15] The major difference in this respect is the role of women. Nechaev considers them "priceless assets," and indeed they were crucial leaders and participants in the first wave. Bin Laden dedicates his book to protecting the Muslim woman, but he ignores what experience can tell us about female terrorists. [16] Women do not participate in his forces and are virtually excluded in the fourth wave, except in Sri Lanka.

Each wave produces major technical works that reflect the special properties of that wave and contribute to a common modern effort to formulate a "science" of terror. Between Nechaev and bin Laden there were Georges Grivas, *Guerrilla War,* and Carlos Marighella, *Mini-Manual of the Urban Guerrilla,* in the second and third waves, respectively.

"Revolution" is the overriding aim in every wave, but revolution is understood in different ways. [17] Revolutionaries create a new source of political legitimacy, and more often than not that meant national self-determination. The anticolonial wave was dominated by this quest. The principle that a people should govern itself was bequeathed by the American and French revolutions. (The French Revolution also introduced the term *terror* to our vocabulary.) [18] Because the definition of "the people" has never been (and

perhaps never can be) clear and fixed, however, it is a source of recurring conflict even when the sanctity of the principle is accepted everywhere. Revolution also can mean a radical reconstruction of authority to eliminate all forms of equality—a cardinal theme in the first wave and a significant one in the third wave. Fourth-wave groups use a variety of sacred texts or revelations for legitimacy.

This chapter treats the great events precipitating each wave and the aims and tactics of participating groups. The focus, however, is the international scene. I examine the interactions of the five principal actors: terrorist organizations; diaspora populations; states; sympathetic foreign publics; and, beginning with the second wave, supranational organizations. [19]

 ## First Wave: Creation of a Doctrine

The creators of modern terrorism inherited a world in which traditional revolutionaries, who depended on pamphlets and leaflets to generate an uprising, suddenly seemed obsolete. The masses, Nechaev said, regarded them as "idle word-spillers." [20] A new form of communication (Peter Kropotkin named it "Propaganda by the Deed") was needed—one that would be heard and would command respect because the rebel took action that involved serious personal risks that signified deep commitment.

The anarchist analysis of modern society contained four major points. It noted that society had huge reservoirs of latent ambivalence and hostility and that the conventions society devised to muffle and diffuse antagonisms generated guilt and provided channels for settling grievances and securing personal amenities. By demonstrating that these conventions were simple historical creations, however, acts once declared immoral would be hailed by later generations as noble efforts to liberate humanity. In this view, terror was thought to be the quickest and most effective means to destroy conventions. By this reasoning, the perpetrators freed themselves from the paralyzing grip of guilt to become different kinds of people. They forced those who defended the government to respond in ways that undermined the rules the latter claimed to respect[21]. Dramatic action repeated again and again invariably would polarize the society, and the revolution inevitably would follow—or so the anarchists reasoned.

An incident that inspired the turbulent decades to follow illustrates the process. On January 24, 1878, Vera Zasulich wounded a Russian police commander who abused political prisoners. Throwing her weapon to the floor, she proclaimed that she was a "terrorist, *not* a killer."[22] The ensuing trial quickly became that of the police chief. When the court freed her, crowds greeted the verdict with thunderous applause.[23]

A successful campaign entailed learning how to fight and how to die, and the most admirable death occurred as a result of a court trial in which one accepted responsibility and used the occasion to indict the regime. Stepniak, a major figure in the history of Russian terrorism, described the Russian terrorist as "noble, terrible, irresistibly fascinating, uniting the two sublimities of human grandeur, the martyr and the hero."[24] Dynamite—a recent invention— was the weapon of choice because the assailant usually was killed too, so it was not a weapon a criminal would use.[25]

Terror was violence beyond the moral conventions used to regulate violence: the rules of war and punishment. The former distinguishes combatants from noncombatants, and the latter separates the guilty from the innocent. Invariably, most onlookers would label acts of terror atrocities or outrages. The rebels described themselves as terrorists, not guerrillas, tracing their lineage to the French Revolution. They sought political targets or those that could affect public attitudes.[26] Terrorism was a strategy, not an end. The tactics used depended upon the group's political objective and on the specific context faced. Judging a context constantly in flux was both an art and a science.

The creators of this strategy took confidence from contemporary events. In the Russian case, as well as in all subsequent ones, major unexpected political events dramatized new government vulnerabilities. Hope was excited, and hope is always an indispensable lubricant of rebel activity.[27] The turn of events that suggested Russian vulnerability was the dazzling effort of the young Czar Alexander II to transform the system virtually overnight. In one stroke of the pen (1861) he freed the serfs (one-third of the population) and promised them funds to buy their land. Three years later he established limited local self-government, "westernized" the judicial system, abolished capital punishment, and relaxed censorship powers and control over education. Hopes were aroused but could not be fulfilled quickly enough, as indicated by the fact that the funds available for the serfs to buy land were insufficient. In the wake of inevitable disappointments, systematic assassination strikes against prominent officials began—culminating in the death of Alexander himself.

Russian rebels encouraged and trained other groups, event hose with different political aims. Their efforts bore fruit quickly. Armenian and Polish nationalist groups committed to assassination emerged in Russia and used bank robbery to finance their activities. Then the Balkans exploded, as many groups found the boundaries of states recently torn out of the Ottoman Empire unsatisfactory.[28] In the West, where Russian anarchists fled and found refuge in Russian diaspora colonies and among other elements hostile to the czarist regime, a campaign of anarchist terror developed that influenced activities in India too.[29] The diaspora produced some surprising results for groups still struggling in Russia. The Terrorist Brigade in 1905 had its headquarters in Switzerland, launched strikes from Finland (an autonomous part of the Russian empire), got arms from an Armenian terrorist group Russians helped train, and were offered funds by the Japanese to be laundered through American millionaires.[30]

The high point of the first wave of international terrorist activity occurred in the 1890s, sometimes called the "Golden Age of Assassination"—when monarchs, prime ministers, and presidents were struck down, one after another, usually by assassins who moved easily across international borders.[31] The most immediately affected governments clamored for international police cooperation and for better border control, a situation President Theodore Roosevelt thought ideal for launching the first international effort to eliminate terrorism:

> Anarchy is a crime against the whole human race, and all mankind should band together against the Anarchist. His crimes should be made a crime against the law of nations . . . declared by treaties among all civilized powers.[32]

The consensus lasted only three years, however. The United States refused to send a delegation to a St. Petersburg conference to consider a German/Russian-sponsored protocol to meet these objectives. It feared that extensive involvement in European politics might be required, and it had no federal police force. Italy refused too, for a very different and revealing concern: If anarchists were returned to their original countries, Italy's domestic troubles might be worse than its international ones.

The first great effort to deal with international terrorism failed because the interests of states pulled them in different directions, and the divisions developed new expressions as the century developed. Bulgaria gave Macedonin nationalists sanctuaries and bases to aid operations in the Ottoman Empire. The suspicion that Serbia helped Archduke Franz Ferdinand's assassin precipitated World War I. An unintended consequences of the four terrible years that followed was a dampened enthusiasm for the strategy of assassination.

Second Wave: Mostly Successful, and a New Language

A wave by definition is an international event; oddly, however, the first one was sparked by a domestic political situation. A monumental international event, the Versailles Peace Treaty that concluded World War I, precipitated the second wave. The victors applied the principle of national self-determination to break up the empires of the defeated states (mostly in Europe). The non-European portions of those defeated empires, which were deemed not yet ready for independence, became League of Nations "mandates" administered directly by individual victorious powers until the territories were ready for independence.

Whether the victors fully understood the implications of their decisions or not, they undermined the legitimacy of their own empires. The IRA achieved limited success in the 1920s,[33] and terrorist groups developed in all empires except the Soviet Union (which did not recognize itself as a colonial power) after World War II. Terrorist activity was crucial in establishing the new states of Ireland, Israel, Cyprus, and Algeria, among others. As empires dissolved, the wave receded.

Most terrorist successes occurred twenty-five years after Versailles, and the time lag requires explanation. World War II reinforced and enlarged the implications of Versailles. Once more the victors compelled the defeated to abandon empires; this time the colonial territories were overseas (Manchuria, Korea, Ethiopia, Libya, and so forth) and were not made mandates. The victors began liquidating their own empires as well, and in doing so they generally were not responding to terrorist activity, as in India, Pakistan, Burma, Ceylon, Tunisia, Egypt, Morocco, the Philippines, Ghana, and Nigeria—which indicated how firmly committed the Western world had become to the principle of self-determination. The United States had become the major Western power, and it pressed hardest for eliminating empires. As the cold war developed, the process was accelerated because the Soviets were always poised to help would-be rebels.[34]

The terror campaigns of the second wave were fought in territories where special political problems made withdrawal a less attractive option. Jews and Arabs in Palestine, for example, had dramatically conflicting versions of what the termination of British rule was supposed to mean. The considerable European population in Algeria did not want Paris to abandon its authority, and in Northern Ireland the majority wanted to remain British. In Cyprus, the Turkish community did not want to be put under Greek rule—the aim of Ethniki Organosis Kyprion Agoniston (EOKA)—and Britain wanted to retain Cyprus as a base for Middle East operations.

The problem of conflicting aspirations was reflected in the way the struggles were or were not settled. The terrorists did get the imperial powers to withdraw, but that was not the only purpose of the struggle. Menachem Begin's *Irgun* fought to gain the entire Palestine mandate but settled for partition.[35] IRA elements have never accepted the fact that Britain will not leave Northern Ireland without the consent of the territory's population. EOKA fought to unify Cyprus with Greece (*enosis*) but accepted an independent state that EOKA tried to subvert for the sake of an ever-elusive *enosis*. Algeria seems to be the chief exception because the Europeans all fled. The initial manifesto of the Front de Liberation Nationale, Algeria (FLN) proclaimed, however, that it wanted to retain that population and establish a democratic state; neither objective was achieved.[36]

Second-wave organizations understood that they needed a new language to describe themselves because the term *terrorist* had accumulated so many negative connotations that those who identified themselves as terrorists incurred enormous political liabilities. The Israeli group *Lehi* was the last self identified terrorist group. Begin, leader of the *Irgtm* (*Lehi's* Zionist rival)—which concentrated on purpose rather than means—described his people as "freedom fighters" struggling against "government terror."[37] This self-description was so appealing that all subsequent terrorist groups followed suit; because the anti-colonial struggle seemed more legitimate than the purposes served in the first wave, the "new" language became attractive to potential political supporters as well. Governments also appreciated the political value of "appropriate" language and began to describe all violent rebels as terrorists. The media, hoping to avoid being seen as blatantly partisan, corrupted language further. Major American newspapers, for example, often described the same individuals alternatively as terrorists, guerrillas, and soldiers in the same account.[38]

Terrorist tactics also changed in the second wave. Because diaspora sources contributed more money, bank robberies were less common. The first wave demonstrated that assassinating prominent political figures could be very counterproductive, and few assassinations occurred in the second wave. The Balkans was an exception—an odd place especially when one considers where World War I started.[39] Elsewhere only *Lehi* (the British renamed it the Stern Gang) remained committed to a strategy of assassination. *Lehi* was much less effective than its two competitors, however, which may have been an important lesson for subsequent anti-colonial movements. Martyrdom, often linked to assassination, seemed less significant as well.

The new strategy was more complicated than the old because there were more kinds of targets chosen, and it was important to strike them in proper sequence. Second-wave strategy sought to eliminate the police—a government's eyes and ears—first through systematic assassinations of officers and/or their families. The military units replacing them, second-wave proponents reasoned would prove too clumsy to cope without counter-atrocities that would increase social support for the cause. If the process of atrocities and counter-atrocities were well planned, it could favor those perceived to be weak and without alternatives.[40]

Major energies went into guerrilla-like (hit-and-run) actions against troops—attacks that still went beyond the rules of war because weapons were concealed and the assailants had no identifying insignia.[41] Some groups, such as the Irgun, made efforts to give warnings in order to limit civilian casualties. In some cases, such as Algeria, terror was one aspect of a more comprehensive rebellion that included extensive guerrilla forces.

Compared to terrorists in the first wave, those in the second wave used the four international ingredients in different and much more productive ways. Leaders of different national groups still acknowledged the common bonds and heritage of an international revolutionary tradition, but the heroes invoked in the literature of specific groups were overwhelmingly national heroes.[42] The underlying assumption seemed to be that if one strengthened ties with foreign terrorists, other international assets would become less useful.

Diaspora groups regularly displayed abilities not seen earlier. Nineteenth century Irish rebels received money, weapons, and volunteers from the Irish-American community, but in the 1920s the exertions of the latter went further and induced the U.S. government to exert significant political influence on Britain to accept an Irish state.[43] Jewish diaspora communities, especially in the United States, exerted similar leverage as the horror of the Holocaust was finally revealed.

Foreign states with kindred populations also were active. Arab states gave the Algerian FLN crucial political support, and those adjacent to Algeria offered sanctuaries from which the group could stage attacks. Greece sponsored the Cypriot uprising against the British and against Cyprus when it became a state. Frightened Turkish Cypriots, in turn, looked to Turkey for aid. Turkish troops then invaded the island (1974) and are still there.

Outside influences obviously change when the purpose of the terrorist activity and the local context are perceived differently. The different Irish experiences illustrate the point well. The early effort in the 1920s was seen simply as an anti-colonial movement, and the Irish-American community had its greatest or most productive impact.[44] The diaspora was less interested in the IRA's brief campaigns to bring Northern Ireland into the Republic during World War II or later during the cold war. Conflicting concerns weakened overseas enthusiasms and influences.

As the second wave progressed, a new, fifth ingredient—supranational organization—came into play. When Alexander I of Serbia was assassinated in Marseilles (1934), the League of Nations tried to contain international terror by drafting two conventions, including one for an international court (1937). Neither came into effect. Two League members (Hungary and Italy) apparently encouraged the assassination and blocked the antiterror efforts.[45] After World War II the United Nations inherited the League's ultimate authority over the colonial mandates—territories that were now scenes of extensive terrorist activity. When Britain decided to withdraw from Palestine, the UN was crucial in legitimizing the partition; subsequently all anti-colonial terrorists sought to interest the UN in their struggles. The new states admitted to the UN were nearly always former colonial territories, and they gave the anticolonial sentiment in that body more structure, focus, and opportunities. More and more participants in UN debates regularly used Begin's language to describe anti-colonial terrorists as "freedom fighters."[46]

 ## Third Wave: Excessive Internationalism?

The major political event stimulating the third, or "New Left," wave was the organizing Vietnam War. The effectiveness of the Viet Cong's "primitive weapons" against the American goliath's modern technology rekindled radical hopes that the contemporary system was vulnerable. Groups developed in the Third World and in the Western heartland itself, where the war stimulated enormous ambivalence among the youth about the value of the existing system. Many Western groups—such as American Weather Underground, the West German Red Army Faction (RAF), the Italian Red Brigades, the Japanese Red Army, and the French Action Directe—saw themselves as vanguards for the Third World masses. The Soviet world encouraged the outbreaks and offered moral support, training, and weapons.

As in the first wave, radicalism and nationalism often were combined, as evidenced by the struggles of the Basques, Armenians, Corsicans, Kurds, and Irish.[47] Every first-wave nationalist movement had failed, but the linkage was renewed because ethnic concerns always have larger constituencies than radical aspirations have. Although self-determination ultimately obscured the radical programs and nationalist groups were much more durable than other groups in the third wave, none succeeded, and their survivors will fail too. The countries concerned—Spain, France, the United Kingdom, and Turkey—simply do not consider themselves to be colonial powers, and the ambivalence necessary for nationalist success is absent.

When the Vietnam War ended in 1975, the PLO replaced the Viet Cong as the heroic model. The PLO originated after the extraordinary collapse of three Arab armies in the six days of the 1967 Middle East war—its existence and persistence gave credibility to supporters who argued that only terror could remove Israel. Its centrality for other groups was strengthened because it got strong support from Arab states and the Soviet Union and made training facilities in Lebanon available to the other groups.

The first and third waves had some striking resemblances. Women in the second wave had been restricted to the role of messengers and scouts; now they became leaders and fighters once more.[48] "Theatrical targets," comparable to those of the first wave, replaced the second wave's military targets. International hijacking is one example. Terrorists understood that some foreign landing fields were accessible. Seven hundred hijackings occurred during the first three decades of the third wave.[49]

Planes were hijacked to secure hostages. There were other ways to generate hostage crises, however, and the hostage crisis became a third-wave characteristic. The most memorable episode was the 1979 kidnapping of former Italian Prime Minister Aldo Moro by the Red Brigades. When the government refused to negotiate, Moro was brutally murdered and his body dumped in the streets. The Sandinistas took Nicaragua's Congress hostage in 1978—an act so audacious that it sparked the popular insurrection that brought the Somoza regime down a year later. In Colombia the M-19 tried to duplicate the feat by seizing the Supreme Court on April 19, 1985, but the government refused to yield and in the struggle nearly 100 people were killed; the terrorists killed eleven justices.

Kidnappings occurred in seventy-three countries—especially in Italy, Spain, and Latin America. From 1968 to 1982 there were 409 international kidnapping incidents yielding 951 hostages.[50] Initially hostages gave their captors political leverage, but soon another concern became more dominant. Companies insured their executives, and kidnapping became lucrative. When money was the principal issue, kidnappers found that hostage negotiations were easier to consummate on their terms. Informed observers estimate the practice "earned" $350 million.[51]

The abandoned practice of assassinating prominent figures was revived. The IRA and its various splinter organizations, for example, assassinated the British ambassador to Ireland (1976) and Lord Mountbatten (1979) and attempted to kill prime ministers Thatcher (1984) and Major (1991).[52] The Palestinian Black September assassinated the Jordanian prime minister (1971) and attempted to assassinate Jordan's King Hussein (1974). Black September killed the American ambassador when it took the Saudi embassy in Khartoum (1973). Euskadi ta Askatasuna (Basque

Nation and Liberty; ETA) killed the Spanish prime minister in the same year.

First- and third-wave assassinations had a different logic, however. A first-wave victim was assassinated because he or she held a public office. New Left-wave assassinations more often were "punishments." Jordan's prime minister and king had forced the PLO out of their country in a savage battle. Similarly, the attempt against British Prime Minister Margaret Thatcher occurred because she was "responsible" for the death of the nine IRA hunger strikers who refused to be treated as ordinary criminals.[53] Aldo Moro was assassinated because the Italian government refused to enter hostage negotiations. The German Red Army Faction provided a second typical pattern: 15 percent of its strikes involved assassination. Although the RAF did not seek the most prominent public figures, it did kill the head of the Berlin Supreme Court and a well-known industrialist.[54]

For good reason, the abandoned term "international terrorism" was revived. Again the revolutionary ethos created significant bonds between separate national groups—bonds that intensified when first Cuban and then PLO training facilities were made available. The targets chosen reflected international dimensions as well. Some groups conducted more assaults abroad than on their home territories; the PLO, for example, was more active in Europe than on the West Bank, and sometimes more active in Europe than many European groups themselves were. Different national groups cooperated in attacks such as the Munich Olympics massacre (1972) and the kidnapping of OPEC ministers (1975), among others.

On their own soil, groups often chose targets with international significance. Strikes on foreign embassies began when the PLO attacked the Saudi embassy in Khartoum (1973). The Peruvian group *Tupac Amaru*—partly to gain political advantage over its rival *Sendero Luminoso* (The Shining Path)—held seventy-two hostages in the Japanese Embassy for more than four months (1996–97) until a rescue operation killed every terrorist in the complex.

One people became a favorite target of most groups. One-third of the international attacks in the third wave involved American targets—a pattern reflecting the United States' new importance. American targets were visible in Latin America, Europe, and the Middle East, where the United States supported most governments under terrorist siege.[55]

Despite its preeminent status as a target, cold war concerns sometimes led the United States to ignore its stated distaste for terror. In Nicaragua, Angola, and elsewhere the United States supported terrorist activity—an indication of how difficult it was to forgo a purpose deemed worthwhile even when deplorable tactics had to be used.

Third-wave organizations discovered that they paid a large price for not being able to negotiate between the conflicting demands imposed by various international elements.[56] The commitment to a revolutionary ethos alienated domestic and foreign elements, particularly during the cold war. The IRA forfeited significant Irish American diaspora support during the third wave. Its initial goal during the third wave was a united socialist Ireland, and its willingness to accept support from Libya and the PLO created problems. Most of all, however, the cold war had to end before the Irish diaspora and an American government showed sustained interest in the Irish issue again and assisted moves to resolve the conflict.

Involvement with foreign groups made some terrorist organizations neglect domestic constituencies. A leader of the 2nd of June, a German anarchist body, suggested that its obsession with the Palestinian cause induced it to attack a Jewish synagogue on the anniversary of *Kristall Nacht*—a date often considered the beginning of the Holocaust. Such "stupidity," he said, alienated potential German constituencies.[57] When the power of the cooperating terrorist entities was very unequal, the weaker found that its interest did not count. Thus, the German Revolutionary Cells, hijacking partners of the Popular Front for the Liberation of Palestine (PFLP), could not get help from their partners to release German prisoners. "(D)ependent on the will of Wadi Haddad and his group," whose agenda was very different from theirs after all, the Revolutionary Cells terminated the relationship and soon collapsed.[58]

The PLO, always a loose confederation, often found international ties expensive because they complicated serious existing divisions within the organization. In the 1970s Abu Iyad, PLO founding member and intelligence chief, wrote that the Palestinian cause was so important in Syrian and Iraqi domestic politics that those states felt it necessary to capture organizations within the PLO to serve their own ends. That made it even more difficult to settle for a limited goal, as the Irgun and EOKA had done earlier.

Entanglements with Arab states created problems for both parties. Raids from Egyptian-occupied Gaza helped precipitate a disastrous war with Israel (1956), and the *fidayeen* were prohibited from launching raids from that territory ever again. A Palestinian raid from Syria brought Syria into the Six-Day War, and ever after ward Syria kept a tight control on those operating from its territories. When a PLO faction hijacked British and American planes to Jordan (1970) in the first effort to target non-Israelis, the Jordanian army devastated the PLO, which then lost its home. Finally, an attempted assassination of an Israeli diplomat in Britain sparked the 1982 invasion of Lebanon and forced the PLO to leave a home that had given it so much significance among foreign terrorist groups. (Ironically, the assassination attempt was organized by Abu Nidal's renegade faction associated with Iraq—a group that had made two previous attempts to assassinate the PLO's leader Yasser Arafat.) Subsequently, Tunisia—the PLO's new host—prohibited the PLO from training foreign groups, and to a large extent the PLO's career as an effective terrorist organization seemed to be over. Paradoxically, the Oslo Accords demonstrated that the PLO could achieve more of its objectives when it was less dangerous.[59]

To maintain control over their own destiny, states again began to "sponsor" groups (a practice abandoned in the second wave), and once more the sponsors found the practice costly. In the 1980s Britain severed diplomatic relations with Libya and Syria for sponsoring terrorism on British soil, and France broke with Iran when it refused to let the French interrogate its embassy staff about assassinations of Iranian emigres. Iraq's surprising restraint during the 1991 Gulf War highlighted the weakness of state-sponsored terror. Iraq did threaten to use terror—a threat that induced Western authorities to predict that terrorists would flood Europe.[60] If terror had materialized, however, it would have made bringing Saddam Hussein to trial for crimes a war aim, and the desire to avoid that result is the most plausible explanation for the Iraqi dictator's uncharacteristic restraint.

The third wave began to ebb in the 1980s. Revolutionary terrorists were defeated in one country after another. Israel's invasion of Lebanon (1982) eliminated PLO facilities to train terrorist groups, and international counterterrorist cooperation became increasingly effective.

As in the first wave, states cooperated openly and formally in counter-terror efforts. The United States, with British aid, bombed Libya (1986) because of its role as a state sponsor, and the European Community imposed an arms embargo. The international cooperation of national police forces sought at St. Petersburg (1904) became more significant as Trevi—established in the mid-1970s—was joined in this mission by Europol in 1994. Differences between states remained, however; even close allies could not always cooperate. France refused to extradite PLO, Red

Brigade, and ETA suspects to West Germany, Italy, and Spain, respectively. Italy spurned American requests to extradite a Palestinian suspect in the seizure of the *Achille Lauro* cruise ship (1984), and Italy refused to extradite a Kurd (1988) because Italian law forbids capital punishment whereas Turkish law does not. The United States has refused to extradite some IRA suspects. Events of this sort will not stop until that improbable day when the laws and interests of separate states are identical.

The UN's role changed dramatically in the third wave. Now "new states"—former colonial territories—found that terrorism threatened their interests, and they particularly shunned nationalist movements. Major UN conventions from 1970 through 1999 made hijacking, hostage taking, attacks on senior government officials, "terrorist bombing" of a foreign state's facilities, and financing of international activities crimes. A change of language is some indication of the changed attitude. "Freedom fighter" was no longer a popular term in UN debates, and the term *terrorism* actually was used for the title of a document: "International Convention for the Supression Terrorist Bombing" (1997).[61] Evidence that Libya's agents were involved in the Pan Am Lockerbie crash produced a unanimous Security Council decision obliging Libya to extradite the suspects (1988), and a decade later when collective sanctions had their full effects Libya complied; this episode will continue to shape Libya's terrorist activities.

Yet very serious ambiguities within the UN remained, reflecting the ever-present fact that terror serves different ends—and some of those ends are prized. Ironically, the most important ambiguity concerned the third wave's major organization: the PLO. It received official UN status and was recognized by more than 100 states as a state that is entitled to receive a share of the Palestine Mandate.

 ## Fourth Wave: How Unique and How Long?

As its predecessor began to ebb, the "religious wave" gathered force. Religious elements have always been important in modern terror because religious and ethnic identities often overlap. The Armenian, Macedonian, Irish, Cypriot, French Canadian, Israeli, and Palestinian struggles illustrate the point.[62] In these cases, however, the aim was to create secular states.

Today religion has a vastly different significance, supplying justifications and organizing principles for a state.

The religious wave has produced an occasional secular group—a reaction to excessive religious zeal. Buddhists in Sri Lanka tried to transform the country, and a terrorist response among the largely Hindu Tamils aims at creating a separate secular state.

Islam is at the heart of the wave. Islamic groups have conducted the most significant, deadly, and profoundly international attacks. Equally significant, the political events providing the hope for the fourth wave originated in Islam, and the successes achieved apparently influenced religious terror groups elsewhere.[63]

Although there is no direct evidence for the latter connection, the chronology is suggestive. After Islam erupted, Sikhs sought a religious state in the Punjab. Jewish terrorists attempted to blow up Islam's most sacred shrine in Jerusalem and waged an assassination campaign against Palestinian mayors. One Jew murdered twenty-nine Muslim worshippers in Abraham's tomb (Hebron, 1994), and another assassinated Israeli Prime Minister Rabin (1995). Aum Shinrikyo—a group that combined Buddhist, Hindu, and Christian themes—released nerve gas on the Tokyo subway (1995), killing 12 people and injuring 3,000 and creating worldwide anxiety that various groups would soon use weapons of mass destruction.

Christian terrorism, based on racist interpretations of the Bible, emerged in the amorphous American "Christian Identity" movement. In true medieval millenarian fashion, armed rural communes composed of families withdrew from the state to wait for the Second Coming and the great racial war. Although some observers have associated Christian Identity with the Oklahoma City bombing (1995), the Christian level of violence has been minimal—so far.

Three events in the Islamic world provided the hope or dramatic political turning point that was vital to launch the fourth wave. In 1979 the Iranian Revolution occurred, a new Islamic century began, and the Soviets made an unprovoked invasion of Afghanistan.

Iranian street demonstrations disintegrated the Shah's secular state. The event also was clear evidence to believers that religion now had more political appeal than did the prevailing third-wave ethos because Iranian Marxists could only muster meager support against the Shah. "There are no frontiers in Islam," Ayatollah Khomeini proclaimed, and "his" revolution altered relationships among all Muslims as well as between Islam and the rest of the world. Most immediately, the Iranians inspired and assisted Shiite terror movements outside of Iran, particularly in Iraq, Saudi Arabia, Kuwait, and Lebanon. In Lebanon, Shiites—influenced by the self-martyrdom tactic of the medieval Assassins—introduced suicide bombing, with surprising results, ousting American and other foreign troops that had entered the country on a peace mission after the 1982 Israeli invasion.

The monumental Iranian revolution was unexpected, but some Muslims had always believed that the year would be very significant because it marked the beginning of a new Islamic century. One venerable Islamic tradition holds that a redeemer will come with the start of a new century—an expectation that regularly sparked uprisings at the turn of earlier Muslim centuries.[64] Muslims stormed the Grand Mosque in Mecca in the first minutes of the new century in 1979, and 10,000 casualties resulted. Whatever the specific local causes, it is striking that so many examples of Sunni terrorism appeared at the same time in Egypt, Syria, Tunisia, Morocco, Algeria, the Philippines, and Indonesia.

The Soviet Union invaded Afghanistan in 1979. Resistance strengthened by volunteers from all over the Sunni world and subsidized by U.S aid forced the Soviets out by 1989—a crucial step in the stunning and unimaginable disintegration of the Soviet Union itself. Religion had eliminated a secular superpower, an astonishing event with important consequences for terrorist activity in that the third wave received a decisive blow. Lands with large Muslim populations that formerly were part of the Soviet Union—such as Chechnya, Uzbekistan, Kyrgyzstan, Tajikistan, and Azerbaijan—became important new fields for Islamic rebels. Islamic forces ignited Bosnia. Kashmir again became a critical issue, and the death toll since 1990 has been more than 50,000.[66] Trained and confident Afghan veterans were major participants in the new and ongoing conflicts.

"Suicide bombing," reminiscent of anarchist bomb-throwing efforts, was the most deadly tactical innovation. Despite the conventional wisdom that only a vision of rewards in paradise could inspire such acts, the secular Tamil Tigers were so impressed by the achievement in Lebanon that they used the tactic in Sri Lanka to give their movement new life. From 1983 to 2000 they used suicide bombers more than all Islamic groups combined, and Tamil suicide bombers often were women—a very unusual event in the fourth wave.[67] Partly to enhance their political leverage at home, Palestinian religious groups began to use suicide bombers, compelling secular PLO elements to emulate them.

The fourth wave has displayed other distinctive international features. The number of terrorist groups declined dramatically. About 200 were active in the 1980s, but in the next decade the number fell to 40.[68] The trend appears to be related to the size of the primary audiences (nation versus religion). A major religious community such as Islam is much larger than any national group. Different cultural traditions also may be relevant. The huge number of secular terrorist groups came largely from Christian countries, and the Christian tradition has always generated many more religious divisions than the Islamic tradition has.[69] Islamic groups are more durable than their third-wave predecessors; the major groups in Lebanon, Egypt, and Algeria have persisted for two decades and are still functioning.[70] These groups are large organizations, and bin Laden's al-Qaeda was the largest, containing perhaps 5,000 members with cells operating in seventy-two countries.[71] Larger terrorist groups earlier usually had nationalist aims—with a few hundred active members and a few thousand available for recruitment. The PLO was a special case at least in Lebanon, where it had about 25,000 members and was trying to transform itself into a regular army. Likewise, most al-Qaeda recruits served with the Taliban in the Afghan civil war.

The American role too changed. Iran called the United States the "Great Satan." Al-Qaeda regarded America as its chief antagonist immediately after the Soviet Union was defeated-a fact not widely appreciated until September 11.[72] From the beginning, Islamic religious groups sought to *destroy* their American targets, usually military or civilian installations, an unknown pattern in the third wave. The aim was U.S. military withdrawal from the Middle East. U.S. troops were driven out of Lebanon and forced to abandon a humanitarian mission in Somalia. Attacks on military posts in Yemen and Saudi Arabia occurred. The destroyer USS *Cole* experienced the first terrorist strike against a military vessel ever (2000). All of the attacks on the U.S. military in the Arabian Peninsula and Africa drew military responses; moreover, Americans did not withdraw after those incidents. The strikes against American embassies in Kenya and Tanzania (1998) inflicted heavy casualties, and futile cruise missile attacks were made against al-Qaeda targets—the first time missiles were used against a group rather than a state. As Peter Bergen has noted, "The attacks, however, had a major unintended consequence: They turned bin Laden from a marginal figure in the Muslim world to a global celebrity."[73] Strikes on American soil began in 1993 with a partially successful effort on the World Trade Center. A mission to strike on the millennial celebration night seven years later was aborted.[74]

Al-Qaeda was responsible for attacks in the Arabian Peninsula, Africa, and the American homeland. Its initial object was to force U.S. evacuation of military bases in Saudi Arabia, the land containing Islam's two holiest sites. The Prophet Muhammed had said that only one religion should be in the land, and Saudi Arabia became a land where Christians and Jews could reside only for temporary periods.[75] Al-Qaeda's aim resonates in the Sunni world and is reflected in its unique recruiting pattern. Most volunteers come from Arab states—especially Egypt, Saudi Arabia, and Algeria—and the Afghan training camps received Sunnis from at least sixty Muslim and non-Muslim countries. Every previous terrorist organization, including Islamic groups, drew its recruits from a single national base. The contrast between PLO and al-Qaeda training facilities reflects this fact; the former trained units from other organizations and the latter received individuals only.

Beyond the evacuation of bases in Islam's Holy Land, al-Qaeda later developed another objective—a single Islamic state under the Sharia. Bin Laden gave vigorous support to Islamic groups that were active in various states of the Sunni world—states that many Muslims understand to be residues of collapsed colonial influence. Just as the United States refused to leave Saudi Arabia, it helped to frustrate this second effort by aiding the attacked states. The United States avoided direct intervention that could inflame the Islamic world, however. The support given to states attacked had some success, and perhaps September 11 should be understood as a desperate attempt to rejuvenate a failing cause by triggering indiscriminate reactions.[76]

The response to September 11 was as unprecedented as the attack itself. Under UN auspices, more than 100 states (including Iran) joined the attack on Afghanistan in various ways. Yet no one involved expected the intervention to be so quick and decisive. Afghanistan had always been difficult for invaders. Moreover, terrorist history demonstrates that even when antiterrorist forces were very familiar with territories containing terrorists (this time they were not), entrenched terrorists still had considerable staying power.

There are many reasons why al-Qaeda collapsed so quickly in Afghanistan. It violated a cardinal rule for terrorist organizations, which is to stay underground always. Al-Qaeda remained visible to operate its extensive training operations,[77] and as the Israelis demonstrated in ousting the PLO from Lebanon, visible groups are vulnerable. Moreover, al-Qaeda and the PLO were foreign elements in lands uncomfortable with their presence. Finally, al-Qaeda did not plan for an invasion possibility. The reason is not clear but there is evidence that its contempt for previous American reactions convinced it that the United States would avoid difficult targets and not go to Afghanistan.[78]

The PLO grouped in Tunisia, on condition that it would abandon its extensive training mission. Could al-Qaeda accept such limits and if it did, would any state risk playing Tunisia's role? Pakistan's revolving-door political suggests a much more likely reaction. Once al-Qaeda's principal supporter, Pakistan switched under U.S. pressure to give the coalition indispensable aid.

As of this writing, the world does not know what happened to al-Qaeda's leadership, but even if the portion left can be reassembled, how can the organization function without a protected sanctuary? Al Zawahiri, bin Laden's likely successor, warned his comrades before the Afghan training grounds were lost that "the victory . . . against the international alliance will not be accomplished without acquiring a . . . base in the heart of the Islamic world."[79] Peter Bergen's admirable study of al-Qaeda makes the same point.[80]

The disruption of al-Qaeda in Afghanistan has altered the organization's previous routine. Typically, al-Qaeda sleeper cells remained inactive until the moment to strike materialized, often designated by the organization's senior leadership. It was an unusual pattern in terrorist history. Normally cells are active and, therefore, need more autonomy so that police penetration in one cell does not go beyond that unit. Cells of this sort have more freedom to strike. They generally will do so more quickly and frequently, but the numbers and resources available to a cell constantly in motion limit them to softer or less protected targets. If direction from the top can no longer be a feature of al-Qaeda, the striking patterns will necessarily become more "normal."[81] Since the Afghan rout, strikes have been against "softer," largely unprotected civilian targets. As the destruction of tourist sites—such as the ancient synagogue in Tunisia and the nightclubs in Bali, Indonesia—suggests, however, the organization displays its trademark by maximizing casualties.

 ## Concluding Thoughts and Questions

Unlike crime or poverty, international terrorism is a recent phenomena. Its continuing presence for 125 years means,

however, that it is rooted in important features of our world. Technology and doctrine have played vital roles. The latter reflects a modern inclination to rationalize activity or make it efficient, which Max Weber declared a distinctive feature of modern life. A third briefly noted factor is the spread of democratic ideas, which shapes terrorist activity in different ways—as suggested by the fact that nationalism or separatism is the most frequently espoused cause.[82]

The failure of a democratic reform program inspired the first wave, and the main theme of the second was national self-determination. A dominant, however confused, third-wave theme was that existing systems were not truly democratic. The spirit of the fourth wave appears explicitly antidemocratic because the democratic idea is inconceivable without a significant measure of secularism.

For many reasons, terrorist organizations often have short lives; sometimes their future is determined by devastating tactical mistakes. A decision to become visible is rare in the history of terror, and the quick success of the coalition's Afghan military campaign demonstrates why. If al-Qaeda successfully reconstructs itself, it may discover that it must become an "ordinary" terrorist group living underground among a friendly local population. That also suggests but, alas, does not demonstrate that its strikes will become more "ordinary" too.

No matter what happens to al-Qaeda, this wave will continue, but for how long is uncertain. The life cycle of its predecessors may mislead us. Each was inspired by a secular cause, and a striking characteristic of religious communities is how durable some are. Thus, the fourth wave may last longer than its predecessors, but the course of the Iranian revolution suggests something else. If history repeats itself, the fourth wave will be over in two decades. That history also demonstrates, however, that the world of politics always produces large issues to stimulate terrorists who regularly invent new ways to deal with them. What makes the pattern so interesting and frightening is that the issues emerge unexpectedly—or, at least, no one has been able to anticipate their tragic course.

The coalition assembled after September 11 was extraordinary for several reasons. September 11 was not only an American catastrophe: The World Trade Center housed numerous large foreign groups, and there were many foreign casualties. The UN involvement climaxed a transformation; it is hard to see it as the same organization that regularly referred to terrorists as freedom fighters forty years ago.

The only other coalition against terrorism was initiated a century ago. It aimed to make waves impossible by disrupting vital communication and migration conditions. Much less was expected from its participants, but it still fell apart in three years (1904). Will the current coalition last longer? September 11 will not be forgotten easily,[83] and the effort is focused now on an organization—a much easier focus to sustain.

When the present campaign against al-Qaeda and the small groups in Asia loosely associated with it concludes, what happens next? No organization has been identified as the next target, and until that happens one suspects that the perennial inclination for different states to distinguish groups according to the ends sought rather than the means used may reappear. Kasmir, and Palestine are the two most important active scenes for terrorist activity. In Kashmir, Islamicins urgents are seriously dividing two important members of the coalition. India considers them terrorists but Pakistan does not. War between those states, both possessing nuclear weapons, will push the coalition's war against terror aside. Successful outside mediation may produce a similar result because that would require some acceptance of the insurgents' legitimacy. The Israeli-Palestinian conflict has a similar meaning; so many important states understand the issue of terror there differently.

Islam fuels terrorist activity in Kashmir, but the issue—as in Palestine, where religious elements are less significant—is a local one. To what extent are other organizations in the fourth wave local too? How deeply can the coalition afford to get involved in situations where it will be serving the interests of local governments? Our experience supporting governments dealing with "local" terrorists has not always served our interests well, especially in the Islamic world.

The efforts of Aum Shinrikyo to use weapons of mass destruction has made American officials feel that the most important lesson of this wave is that those weapons will be used by terrorists against us.[84] September 11 intensified this anxiety even though suicide bombers armed with box cutters produced that catastrophe, and the history of terrorism demonstrates that cheap, easy to produce, portable, and simple to use weapons have always been the most attractive.

The fourth wave's cheap and distinctive weapon is suicide bombing. The victory in Lebanon was impressive, and suicide bombers have been enormously destructive in Sri Lanka and Israel. Driving foreign troops out of a country is one thing, however; compelling a people to give up a portion of its own country (Sri Lanka) or leave its own land (Israel) is another. In the latter case, the bombers' supporters seem to be suffering a lot more than their enemies are.

How does September 11 affect our understanding of foreign threats? This is a serious question that needs more discussion than it has received. Nechaev emphasized that the fear and rage rebel terror produced undermined a society's traditional moral conventions and ways of thinking. He was thinking of the domestic context, and indeed the history of modern terrors shows that domestic responses frequently are indiscriminate and self-destructive.[85] Can the same pattern be observed on the international scene?

The 2003 invasion of Iraq suggests that Nechaev's observation is apt for the international scene as well. The justifications for the war were that Iraq might give terrorists weapons of mass destruction or use them itself against the West—considerations that are applicable to a variety of states, as the "axis of evil" language suggests. After September 11 the United States scrapped the deterrence doctrine, which we developed to help us cope with states possessing weapons of mass destruction and served us well for more than fifty years. Preemption seemed to fit the new age better. Deterrence worked because states knew that they were visible and could be destroyed "if they used the dreaded weapons. Underground terron·st groups do not have this vulnerability, which is why preemption has been an important part of police counterterrorist strategy since the first wave. Deterrence is linked to actions, whereas preemption is more suitable when intentions have to be assessed—a task always shrouded in grave ambiguities. Is there any reason to think the crucial distinction between states and terrorist groups has disappeared, however, and that we should put decisions of war and peace largely in the hands of very imperfect intelligence agencies?

The significance of the Iraqi war for the war against terrorism remains unclear. The coalition's cohesion has been weakened, and the flagging fortunes of Islamic groups could be revived. Both possibilities are more likely if preemption is employed against another state or if the victory in Iraq ultimately is understood as an occupation.

Notes

An earlier version of this essay was published in *Current History* (December 2001): 419–25. Another version was delivered at the annual John Barlow Lecture, University of Indiana, Indianapolis. I am indebted to Jim Ludes, Lindsay Clutterbuck, Laura Donohue, Clark McCauley, Barbara Rapoport, and Sara Grdan for useful comments, even those I did not take. The problems in the essay are my responsibility.

1. On September 20, 2001, the president told Congress that "any nation that continues to harbor or support terrorism will be regarded as a hostile regime. [T]he war would not end until every terrorist group of global reach has been found, stopped, and defeated."

2. See Richard B. Jensen, "The United States, International Policing, and the War against Anarchist Terrorism," *Terrorism and Political Violence* (hereafter *TPV*) 13, no. 1 (spring 2001): 5–46.

3. No good history of terrorism exists. Schmid and Jongman's monumental study of the terrorism literature does not even list a history of the subject. See *Political Terrorism: A New Guide to Actors, Authors, Concepts, Theories, DataBases, and Literature,* rev. ed. (New Brunswick, N.J.: Transaction Books, 1988).

4. I lack space to discuss the domestic sphere, which offers important parallels as well. The unusual character of terrorist activity made an enormous impact on national life in many countries beginning in the latter part of the nineteenth century. Every state affected in the first wave radically transformed its police organizations as tools to penetrate underground groups. The Russian *Okhrana,* the British Special Branch, and the FBI are conspicuous examples. The new organizational form remains a permanent, perhaps indispensable, feature of modern life. Terrorist tactics, *inter alia,* aim at producing rage and frustration, often driving governments to respond in anticipated, extraordinary, illegal, socially destructive, and shameful ways. Because a significant Jewish element, for example, was present in the several Russian terrorist movements the *Okhrana* organized pogroms to intimidate Russian Jews, compelling many to flee to the West and to the Holy Land. *Okhrana* fabricated *The Protocols of Zion,* a book that helped stimulate a virulent anti-Semitism that went well beyond Russia. The influence of that fabrication continued for decades and still influences Christian and Islamic terrorist movements today.

Democratic states "overreacted" too. President Theodore Roosevelt proposed sending all anarchists back to Europe. Congress did not act, but more than a decade later President Wilson's Attorney General Palmer implemented a similar proposal and rounded up all anarchists to ship them back "home," regardless of whether they had committed crimes. That event produced the 1920 Wall Street bombing, which in turn became the justification for an immigration quota law that for decades made it much more difficult for persons from southern and eastern European states (the original home of most anarchists) to

immigrate—a law Adolph Hitler praised highly. It is still too early to know what the domestic consequences of September 11 will be. The very first reactions suggested that we had learned from past mistakes. The federal government made special efforts to show that we were not at war with Islam, and it curbed the first expressions of vigilante passions. The significance of subsequent measures seems more problematic, however. Our first experience with terror led us to create important new policing arrangements. Now Congress has established a Department of Homeland Security with 170,000 employees—clearly the largest change in security policy in our history. No one knows what that seismic change means. One casualty could be the Posse Comitatus law, which prohibits the military forces from administering civil affairs—a law that ironically was passed because we were unhappy with military responses to KKK terrorist activity after the Civil War! A policy of secret detentions, a common reaction to serious terrorist activities in many countries, has been implemented. Extensive revisions of immigration regulations are being instituted. Prisoners taken in Afghanistan are not being prosecuted under the criminal law, reversing a long-standing policy in virtually all states including our own. Previous experiences suggest that it will take time for the changes to have their effect because so much depends on the scope, frequency, and duration of future terrorist activity.

5. David M. Chalmers, *Hooded Americanism: The History of the Ku Klux Klan,* 3d ed. (Durham, N.C.: Duke University Press, 1987), 19.

6. The activities of the Thugs and Assassins had international dimensions but were confined to specific regions; more important, there were no comparable groups operating at the same time in this region or elsewhere. See David C. Rapoport, "Fear and Trembling: Terror in Three Religious Traditions," *American Political Science Review* 78, no. 3 (1984): 658–77.

7. The lineage of rebel terror is very ancient, going back at least to the first century. Hinduism, Judaism, and Islam produced the Thugs, Zealots, and Assassins, respectively; these names still are used to designate terrorists. Religion determined every purpose and each tactic of this ancient form. See Rapoport, "Fear and Trembling."

8. By far most published academic articles on terrorism deal with counterterrorism and with organizations. Judging by my experience as an editor of *TPV,* the proportions increase further in this direction if we also consider articles that are rejected.

9. See note 1.

10. The rebels fought in uniform and against soldiers. George Bernard Shaw said, "My own view is that the men who were shot in cold blood . . . after their capture were prisoners of war." Prime Minister Asquith said that by Britain's own standards,

the rebels were honorable, that "they conducted themselves with great humanity . . . fought very bravely and did not resort to outrage." The *Manchester Guardian* declared that the executions were "atrocities." See my introduction to part III of David C. Rapoport and Yonah Alexander, eds., *The Morality of Terrorism: Religious Origins and Ethnic Implications*, 2d ed. (New York: Columbia University Press, 1989), 219–27.

11.   Guerrillas carry weapons openly and wear an identifying emblem—circumstances that oblige a state to treat them as soldiers.

12.   Anyone who has tried to explain the intensity of the 1960s experience to contemporary students knows how difficult it is to transmit a generation's experience.

13.   Nechaev's "Revolutionary Catechism" is reprinted in David C. Rapoport, Peter Kropotkin Revolutionary *Pamphlets* (New York: Benjamin Bloom, 1927); Nicholas Mozorov, *Terroristic Struggle* (London, 1880); Serge Stepniak, *Underground Russia: Revolutionary' Profiles and Sketches from Life* (New York, 1892).

14.   See Rapoport, "Fear and Trembling."

15.   It took time for this attitude to develop in Islam. If one compares bin Laden's work with Faraj's *Neglected Duty*—a work primarily written at the beginning of the fourth wave to justify the assassination of Egyptian President Sadat (1981)—the two authors seem to different worlds. Faraj cites no experience outside the Islamic tradition, and his most recent historical reference is to Napoleon's invasion of Europe. See David C. Rapoport, "Sacred Terror: A Case from Contemporary Islam," in *Origins of Terrorism*, ed. Walter Reich (Cambridge: Cambridge University Press, 1990), 103–30. I am grateful to Jerry Post for sharing his copy of the bin Laden treatise. An edited version appears on the Department of Justice website www.usdoj.gov/ag/trainingmanual.htm.

16.   Bin Laden's dedication reads as follows:

Pledge, O Sister

To the sister believer whose clothes the criminals have stripped off:

To the sister believer whose hair the oppressors have shaved.

To the sister believer whose body has been abused by the human dogs. . . .

Covenant O Sister . . . to make their women widows and their children orphans. . . .

17.   Ignore right-wing groups because more often than not they are associated with government reactions. I also ignore "single issue" groups such as the contemporary antiabortion and Green movements.

18.   The term *terror* originally referred to actions of the Revolutionary government that went beyond the rules regulating punishment in order to "educate" a people to govern itself.

19.   Vera Figner, the architect of Narodnaya Volya's foreign policy, identifies the first four ingredients. The fifth was created later. For a more extensive discussion of Figner, see David C. Rapoport, "The International World as Some Terrorists Have Seen It: A Look at a Century of Memoirs," in *Inside Terrorist Organizations*, 2d ed. (London: Frank Cass, 2001), 125ff.

20.   Nechaev, "Revolutionary Catechism."

21.   An equivalent for this argument in religious millennial thought is that the world must become impossibly bad before it could become unimaginably good.

22.   Adam B. Ulam, *In the Name of the People* (New York: Viking Press, 1977), 269 (emphasis added).

23.   Newspaper reports in Germany the next day interpreted the demonstrations to mean that a revolution was coming. See *New York Times*, 4 April 1878.

24.   Stepniak, *Underground Russia*, 39–40.

25.   The bomb was most significant in Russia. Women were crucial in Russian groups but sometimes were precluded from throwing the bomb, presumably because bombers rarely escaped. Other terrorists used the bomb extensively but chose other weapons as well.

26.   A guerrilla force has political objectives, as any army does, but it aims to weaken or destroy the enemy's military forces first. The terrorist, on the other hand, strikes directly at the political sentiments that sustain the enemy.

27.   Thomas Hobbes may have been the first to emphasize hope as a necessary ingredient of revolutionary efforts. The first chapter of Menachem Begin's account of his experience in the lrgun contains the most moving description of the necessity of hope in terrorist literature. Menachem Begin, *The Revolt: Story of the Irgun* (Jerusalem: Steinmatzky's Agency, 1997).

28.   There were many organizations: the Internal Macedonian Revolutionary Organization, Young Bosnia, and the Serbian Black Hand.

29.   See Peter Heehs, *Nationalism, Terrorism, and Communalism: Essays in Modern Indian History* (Delhi: Oxford University Press, 1998), chap. 2.

30.   The Japanese offer to finance Russian terrorists during the Russo-Japanese War (1905) encouraged Indian terrorists to believe that the Japanese would help them too. Heehs, *Nationalism, Terrorism, and Communalism*, 4. The Russians turned the Japanese offer down, fearing that knowledge of the transaction during a time of war would destroy their political credibility.

31.   Italians were particularly active as international assassins, crossing borders to kill French President Carnot (1894), Spanish Premier Casnovas (1896), and Austrian Empress Elizabeth (1898). In 1900 an agent of an anarchist group in Paterson, New Jersey, returned to Italy to assassinate King Umberto.

32. Jensen, "The United States, International Policing, and the War against Anarchist Terrorism," 19.

33. The IRA's success in 1921 occurred when the British recognized the Irish state. Nonhern Ireland remained British, however, and the civil war between Irish factions over the peace settlement ended in defeat for those who wanted to continue until Northern Ireland joined the Irish state.

34. For an interesting and useful account of the decolonialization process, see Roben Hager, Jr., and David A. Lake, "Balancing Empires: Competitive Decolonization in International Politics," *Security Studies* 9, no. 3 (spring 2000): 108–48. Hager and Lake emphasize that the literature on decolonization "has ignored how events and politics within the core (metropolitan area) shaped the process" (145).

35. Begin said that his decision was determined by the fact that if he pursued it, a civil war among Jews would occur, indicating that most Jews favored partition. Begin, *The Revolt*, chapters 9 and 10.

36. Alistair Horne, *A Savage War of Peace* (London: Macmillan, 1977), 94–96.

37. Begin, *The Revolt*.

38. For a more detailed discussion of the definition problem, see David C. Rapoport, "Politics of Atrocity," in *Terrorism: Interdisciplinary Perspectives*, ed. Yonah Alexander and Seymour Finger (New York: John Jay Press, 1987), 46.

39. Alexander I of Yugoslavia (1934) was the most prominent victim, and historians believe that Hungary and Italy were involved in providing help for Balkan terrorists. Begin points out in *The Revolt* that it was too costly to assassinate prominent figures.

40. The strategy is superbly described in the film "Battle of Algiers," based on the memoirs of Yaacev Saadi, who organized the battle. Attacks occur against the police, whose responses are limited by rules governing criminal procedure. In desperation, the police set a bomb off in the Casbah, inadvertently exploding an ammunition dump and killing Algerian women and children. A mob emerges screaming for revenge, and at this point the FLN has the moral warrant to attack civilians. There is another underlying element that often gives rebel terrorism in a democratic world special weight. The atrocities of the strong always seem worse than those of the weak because people believe that the latter have no alternatives.

41. See note 11.

42. See Rapoport, "The International World as Some Terrorists Have Seen It."

43. Irish Americans have always given Irish rebels extensive support. In fact, the Fenian movement was born in the American Civil War. Members attempted to invade Canada from the United States and then went to Ireland to spark rebellion there.

44. World War I, of course, increased the influence of the United States, and Wilson justified the war with the self-determination principle.

45. Martin David Dubin, "Great Britain and the Anti-Terrorist Convetions of 1937, *TPV 15*, no. 1 (spring 1993): 1.

46. See John Dugard, "International Terrorism and the Just War," in Rapoport and Alexander, *Morality of Terrorism*, 77–78.

47 Basque Nation and Liberty (ETA), the Armenian Secret Army for the Liberanon Armenia (ASALA), the Corsican National Liberation Front (FNLC), and the IRA.

48. The periods of the first and third waves were times when the rights of women were asserted more strenuously in the general society.

49. Sean Anderson and Stephen Sloan, *Historical Dictionary of Terrorism* (Metuchen, N.J.: Transaction Press, 1995), 136.

50. Although bank robbing was not as significant as in the first wave, some striking exampIes materialized. In January 1976 the PLO, together with its enemies the Christian Phalange, hired safe breakers to help loot the vaults of the major banks in Beirut. Estimates of the amount stolen range between $50 and a $100 million. "whatever the truth the robbery was large enough to earn a place in the *Guinness Book of Records* as the biggest bank robbery of all time"; James Adams, *The Financing of Terror* (New York: Simon and Schuster, 1986), 192.

51. Adams, *Financing of Terror,* 94.

52. The attack on Major actually was an attack on the cabinet, so it is not clear whether the prime minister was the principal target (Lindsay Clutterbuck, personal communication to author).

53. The status of political prisoner was revoked in March 1976. William White law, who granted it in the first place, ranked it as one of his "most regrettable decisions."

54. Anderson and Sloan, *Historical Dictionary of Terrorism*, 303.

55. Sometimes there was American support for terrorist activity (e.g., the Contras in Nicaragua).

56. When a disappointed óffice-seeker assassinated President Garfield, Figner's sympathy letter to the American people said that there was no place for terror in democratic states. The statement alienated elements of her radical constituency in other countries.

57. Michael Baumann, *Terror or Love* (New York: Grove Press, 1977), 61.

58. Interview with Hans J. Klein in Jean M. Bourguereau, *German Guerrilla: Terror, Rebel Reaction and Resistance* (Sanday, U.K.: Cienfuegos Press, 1981), 31.

59. Abu Nidal himself was on a PLO list of persons to be assassinated.

60. W. Andrew Terrill, "Saddam's Failed Counterstrike: Terrorism and the Gulf War," *Studies in Conflict and Terrorism* 16 (1993): 219–32.

61. In addition to four UN conventions there are eight other major multilateral terrorism conventions, starting with The Tokyo Convention of 1963, dealing with the aircraft safety. See http://usinfo.state.gov/topical/pol/terror/conven.htm and http://untreaty.un.org/English/Terrorism.asp.

62. Khachig Tololyan, "Cultural Narrative and the Motivation of the Terrorist," in Rapoport, *Inside Terrorist Organizations*, 217–33.

63. See David C. Rapoport, "Comparing Militant Fundamentalist Movements and Groups," in *Fundamentalisms and the State*, ed. Martin Marty and Scott Appleby (Chicago: University of Chicago Press, 1993), 429-61.

64. To those in the West the most familiar was the nineteenth-century uprising in the Sudan, which resulted in the murder of legendary British General "Chinese" Gordon.

65. This was not the first time secular forces would he!p Iaunch the careers of those who would become religious terrorists. Israel helped Hamas to get started, thinking it would compete to weaken the PLO. To check left-wing opposition, President Sadat released religious elements from prison that later assassinated him.

66. Peter Bergen, *Holy War Inc.: Inside the Secret World of Osama Bin Laden* (New York: Free Press, 2001), 208.

67. In the period specified, Tamil suicide bombers struck 171 times; the combined total for all thirteen Islamic groups using the tactic was 117. Ehud Sprinzak cites the figures compiled by Yoram Schweitzer in "Rational Fanatics," *Foreign Policy* (October 2001): 69. The most spectacular Tamil act was the assassination of Indian Prime Minister Rajiv Gandhi. (Religion did not motivate the notorious Kamikaze attacks during World War II either.) The example of the Tamils has other unusual characteristics. Efforts to make Sri Lanka a Buddhist state stimulated the revolt. Although Tamils largely come from India, there are several religious traditions represented in the population, and religion does not define the terrorists' purpose.

68. See Ami Pedahzur, William Eubank, and Leonard Weinberg, "The War on Terrorism and the Decline of Terrorist Group Formation," *TPV* 14, no. 3 (fall 2002): 141–47.

69. The relationship between different religious terror groups is unusual. Groups from different mainstream traditions (Christianity, Islam, etc.) do not cooperate. Even traditional cleavages within a religion—as in Shiite and Sunni Islam, for example—sometimes are intensified. Shiite terror groups generally take their lead from Iran regarding aid to Sunnis. Iran has helped the Palestinians and is hostile to al-Qaeda and the Saudi religious state.

70. I have no statistical evidence on this point.

71. Rohan Gunaratna, *Inside Al Qaeda: Global Network of Terror* (New York: Columbia University Press, 2002), 97.

72. The stated object of al-Qaeda is to recreate a single Muslim state, and one could argue that if the United States had withdrawn military units from the Muslim world, the attacks would have ceased. What if the issue really was the impact of American secular culture on the world?

73. Bergen, *Holy War Inc.*, 225.

74. Those attacks, as well as the expected attacks that did not materialize, are discussed in a special volume of *TPV* 14, no.1 (spring 2002) edited by Jeffrey Kaplan, titled *Millennial Violence*. The issue also was published as a book: *Millennial Violence: Past, Present, and Future* (London: Frank Cass, 2002).

75. Bernard Lewis, "License to Kill," *Foreign Affairs* (November/December 1998).

76. For a very interesting discussion of the circumstances that provoke American military responses to terrorist attacks, see Michelle Mavesti, "Explaining the United States' Decision to Strike Back at Terrorists," *TPV* 13, no. 2 (summer 2001): 85–106.

77. If the organization understood its vulnerability, it might have thought that an attack on the sovereignty of the state protecting it was unlikely. One reason the Taliban government refused a repeated UN demand to expel al-Qaeda was because without al-Qaeda support it could not survive local domestic opposition. Because most al-Qaeda recruits served in the Taliban forces in the ongoing civil war, the Taliban must have felt that it had no choice. Clearly, however, there must have been a failure to plan for an invasion possibility; the failure to resist is astonishing otherwise.

78. Gunaratna, *Inside Al Qaeda*.

79. Quoted by Nimrod Raphaeli, "Ayman Muhammad Rabi Al-Zawahri: The Making of an Arch-Terrorist," *TPV* 14, no. 4 (winter 2002): 1–22.

80. Bergen, *Holy War Inc.*, 234.

81. The Spaniards conquered the Aztecs and Incas easily, but the United States had more difficulty with the less powerful but highly decentralized native Americans. Steven Simon and Daniel Benjamin make a different argument, contending that bin Laden's group is uniquely decentralized and therefore less likely to be disturbed by destroying the center. See "America and the New Terrorism," *Survival* 42, no. 2 (2000): 156–57.

82. We lack a systematic comparison of the aims sought by organizations in the history of modern terror.

83. September 11 has had an impact on at least one terrorist group: The Tamils found diaspora financial support suddenly disappearing for suicide bombing—an opportunity the Norwegians seized to bring them to the bargaining table again.

84.   See David C. Rapoport, "Terrorism and Weapons of the Apocalypse," *National Security Studies Quarterly* 5, no. 3 (summer 1999): 49–69, reprinted in Henry Sokolski and James Ludes, *Twenty-First Century Weapons Proliferation* (London: Frank Cass, 2001), 14–33.

85.   See note 3.

REVIEW QUESTIONS

1.   Describe Rapoport's four waves of terrorism. Identify what you believe to be the key similarities and differences between the four waves.

2.   How does Rapoport characterize the wave phenomenon? Based on those characteristics, make a prediction as to how long the current (4th wave) will last and explain why.

3.   Discuss the influences of each wave on the subsequent waves. Identify those influences that are still the most potent today.

READING 5

In the chapter titled "The End of Empire and the Origins of Contemporary Terrorism" in his book *Inside Terrorism*, Bruce Hoffman—a well-known and long-standing terrorism scholar—focuses on what Rapoport would call the Anti-Colonial Wave. In part, Hoffman focuses on three case studies—Israel, Cyprus, and Algeria—from just before the start of WWII through the establishment of these respective states. He analyzes the nationalist terrorist groups in each of these states, discussing their development, strategies, tactics, leadership, and belief systems. Most importantly, Hoffman lays out the lessons in terms of tactics and strategies followed and still used by many modern terrorist groups. He also analyzes the reasons for the success of these groups in establishing their desired states and considers whether the success of these earlier groups can be replicated in current conflicts.

The End of Empire and the Origins of Contemporary Terrorism

Bruce Hoffman

Although terrorism motivated by ethno-nationalist/separatist aspirations had emerged from within the moribund Ottoman and Hapsburg empires during the three decades immediately preceding the First World War, it was only after 1945 that this phenomenon became a more pervasive global force. Two separate, highly symbolic events that had occurred early in the Second World War abetted its subsequent development. At the time, the repercussions for postwar anticolonial struggles of the fall of Singapore and the proclamation of the Atlantic Charter could not possibly have been anticipated. Yet both, in different ways, exerted a strong influence on indigenous nationalist movements, demonstrating as they did the vulnerability of once-mighty empires and the hypocrisy of war-time pledges of support for self-determination.

On February 15, 1942, the British Empire suffered the worst defeat in its history when Singapore fell to the invading Japanese forces. Whatever Singapore's strategic value, its real significance—according to the foremost military strategist of his day, Basil Liddell Hart—was as the

> outstanding symbol of Western power in the Far East. . . . Its easy capture in February 1942 was shattering to British, and European, prestige in Asia. No belated re-entry could efface the impression. The white man had lost his ascendancy with the disproof of his magic. The realisation of his vulnerability fostered and encouraged the postwar spread of Asiatic revolt against European domination or intrusion.

Indeed, within weeks Japan had also conquered the Dutch East Indies (Indonesia) and Burma. Hong Kong had already capitulated the previous Christmas, and more than a year earlier Japan had imposed its rule over French Indochina. Thus, when the American garrison holding out on Corregidor in the Philippines finally surrendered in May 1942, Japan's conquest of South-East Asia—and the destruction of the British, French, Dutch, and American empires there—was complete.

The long-term impact of these events was profound. Native peoples who had previously believed in the invincibility of their European colonial over-lords hereafter saw their former masters in a starkly different light. Not only had the vast British Empire been dealt a crushing blow, but American pledges of peace and security to its Pacific possessions had been similarly shattered. France's complete impotence in the face of Japanese bullying over Indochina had greatly undermined its imperial stature among the Vietnamese, and in Indonesia, Japanese promises of independence effectively negated any lingering feelings of loyalty to the Dutch. In the blink of an eye, the European powers' prewar arguments that their variegated Asian subjects were incapable of governing themselves were swept aside by Japan's policy of devolving self-government to local administrations and nominal independence to the countries they now occupied. Paradoxically, in many places it was the natives who now ruled the interned

Europeans—many of whom found themselves forced to perform the most menial and back-breaking tasks. It is not surprising, therefore, that even as the tide of war shifted in the Allies' favor over the following years, almost all these peoples resolved never again to come under European imperial rule.

It was not only the Asian subjects of these declining colonial powers who clamored for independence and self-determination. The litany of humiliating defeats had struck responsive chords in other places also, everywhere challenging the myth of European—indeed, Western—power and military superiority, if not omnipotence. In the Middle East as well as in Africa, India, the Mediterranean, and North Africa, indigenous peoples chafed at the prospect of returning to their prewar colonial status quo. They were encouraged, however unintentionally, by promises of independence and self-determination made by the Allies early in the Second World War. Even before the United States had entered the war in 1941, President Franklin D. Roosevelt had met with Britain's prime minister, Winston Churchill, on a warship off the coast of Newfoundland to formulate both countries' post-war aims. The result was an eight-point document known as the Atlantic Charter, whose main purpose, one historian has observed, was to "impress enemy opinion with the justice of the western cause." Its effects, however, went far beyond that lofty aim.

The charter's first point innocuously affirmed that both countries sought no "aggrandisement, territorial or other," from the war; it was the next two points that would be the source of future difficulties for the European powers. The second point declared unequivocally that neither Britain nor the United States desired to "see . . . territorial changes that do not accord with the freely expressed wishes of the peoples concerned," while the third point further pledged both countries to "respect the right of all peoples to choose the form of government under which they will live." These principles were embodied in the "Declaration of the United Nations," agreed to by Britain and the United States on January 1, 1942, and subsequently signed by all the governments at war with Germany, thus committing their countries to respect promises that in some instances they had no intention of keeping. Although on the first anniversary of the charter's signing Churchill attempted to qualify and restrict the terms of the original agreement—arguing that it was not intended that these principles should apply to either Asia or Africa, and especially not to India and

Palestine, but only to those peoples in hitherto sovereign countries conquered by Germany, Italy, and Japan—the damage had already been done. Indeed, all subsequent efforts by the European colonial powers to redefine or reinterpret the charter in ways favorable to the prolongation of their imperial rule fell for the most part on deaf ears.

The situation in postwar Algeria was perhaps typical of the bitterness, engendered by broken promises and misplaced hopes, that nurtured intractable conflict. In 1943, shortly after the Allied landings had liberated North Africa from Vichy rule, a delegation of Algerian Muslims sought an audience with the newly installed Free French commander, General Henri Giraud. Their request amounted to nothing more than recognition of the rights and freedoms that had been so loftily proclaimed in the Atlantic Charter, of which, of course, the French government-in-exile was a signatory. Giraud's reply was dismissively brusque. "I don't care about reforms," he thundered, "I want soldiers first." The ever-compliant Algerians obligingly provided these volunteers—much as they had some thirty years before, but they did so now in the expectation, shared by other European powers' colonial subjects elsewhere, that their loyalty would be rewarded appropriately at the war's end. As was the case in at least a dozen other colonial settings, however, their anticipation was to be disappointed. Indeed, by 1947 the future leader of the anti-British guerrilla campaign in Cyprus, General George Grivas (a Greek Cypriot who had fought beside the Allies), had already despaired in the face of repeated prevarications concerning Britain's own wartime promises of self-determination. "More and more," he recalled, "it seemed to me that only a revolution would liberate my homeland." The Algerians were rapidly coming to the same conclusion. They were further encouraged by the catastrophic defeat inflicted on the French at Dien Bien Phu in 1954 and by France's subsequent ignominious withdrawal from Indochina. Meanwhile, a revolt against British rule had broken out in Palestine, waged by two small Jewish terrorist organizations—the Irgun Zvai Le'umi (National Military Organization, or Irgun) and the Lohamei Herut Yisrael (Freedom Fighters for Israel, known to Jews by its Hebrew acronym, Lehi, and to the British as the Stern Gang). The Irgun's campaign was the more significant of the two, in that it established a revolutionary model that thereafter was emulated and embraced by both anticolonial- and postcolonial-era terrorist groups around the world.

Postwar Palestine

Palestine had, of course, long been the scene of numerous riots and other manifestations of intercommunal violence that between 1936 and 1939 had culminated in a full-scale rebellion by its Arab inhabitants. In 1937, a new element was added to the country's incendiary landscape when the Irgun commenced retaliatory terrorist attacks on the Arabs. The group expanded its operations to include British targets in 1939 following the government's promulgation of a White Paper in May that imposed severe restrictions on Jewish immigration to Palestine, thereby closing one of the few remaining avenues of escape available to European Jews fleeing Hitler. But the Irgun's inchoate revolt against British rule was short-lived. Less than three months after it began, Britain was at war with Germany. Confronted by the prospect of the greater menace of a victorious Nazi Germany, the Irgun declared a truce and announced the suspension of all anti-British operations for the war's duration. Like the rest of the Jewish community in Palestine, who had also pledged to support the British war effort, the Irgun hoped that this loyalty would later result in the recognition of Zionist claims to statehood.

In May 1942 a young private attached to General Wladyslaw Anders's Polish army-in-exile arrived in Palestine. Menachem Begin's journey had been a circuitous one. Born in 1913 in Brest Litovsk, Poland, the future prime minister of Israel (1977–83) had first become involved in Zionist politics as a teenager when he joined Betar, a right-wing nationalist Jewish youth group. By the time he had received his law degree from Warsaw University in 1935, Begin was head of the group's Organization Department for Poland. Three years later he was appointed its national commander. However, when Germany invaded Poland in September 1939, Begin was forced to flee to Lithuania. A year later, Russian secret police arrested him on the ironic charge of being "an agent of British imperialism." After spending nine months in a local jail, Begin was sentenced to eight years of "correctional labor." In June 1941, when Germany invaded Russia, he was on a Russian ship carrying political prisoners to a Stalinist labor camp in Siberia. A reprieve came in the form of an offer to join the Polish army or continue his journey. Begin chose the former and found himself in a unit ordered to Palestine. Shortly after his arrival, he established contact with the Irgun high command.

Since the suspension of its revolt, the Irgun had fallen into disarray. The deaths of its ideological mentor, Vladimir Jabotinsky, in August 1940 and its military commander, David Raziel, nine months later had deprived the group of leadership and direction at a time when its self-imposed dormancy required someone at the top with the vision and organizational skills necessary to hold it together. Throughout 1943, Begin met with the Irgun's surviving senior commanders to discuss the group's future. As the war against Germany moved decisively in the Allies' favor, they became convinced that the Irgun should resume its revolt. Four dominant considerations influenced this decision. First and foremost was news of the terrible fate that had befallen European Jewry under Nazi domination. Second, the expiration in March 1944 of the White Paper's rigidly enforced five-year immigration quota would likely choke off all future Jewish immigration to Palestine. Third, the Irgun's leaders agreed that the reasoning behind the self-imposed truce they had declared four years before—that harming Britain might help Germany—was no longer tenable since the course of the Second World War had now virtually assured an Allied victory. Finally, by renewing the revolt, the Irgun's revamped high command sought to position themselves and their organization at the vanguard of the active realization of the Jews' political and nationalist aspirations.

On December 1, 1943, Begin formally assumed command of the group and finalized plans for the resumption of anti-British operations. As a lowly enlisted man in an exile army with only the bare minimum of formal military training, Begin was an unlikely strategist. But he possessed an uncanny analytical ability to cut right to the heart of an issue and an intuitive sense about the interplay of violence, politics, and propaganda that ideally qualified him to lead a terrorist organization. Begin's strategy was simple. The handful of men and the few weapons that in 1943 constituted the Irgun could never hope to challenge the British Army on the battlefield and win. Instead, the group would function in the setting and operate in the manner that best afforded the terrorist with means of concealment and escape. Based in the city, its members would bury themselves within the surrounding community, indistinguishable from ordinary, law-abiding citizens. At the appropriate moment, they would emerge from the shadows to strike, then disappear back into the anonymity of Palestine's urban neighborhoods, remaining safely beyond the reach of the authorities. The Irgun's plan, therefore, was not to defeat Britain militarily but to use terrorist violence to undermine the government's prestige and control of Palestine by striking at symbols of British rule. "History and our observation," Begin later recalled, "persuaded us that if we could succeed in destroying the government's prestige in Eretz Israel [Hebrew: literally "the Land of Israel"], the removal of its rule would follow automatically. Thenceforward, we gave no peace to this weak spot. Throughout all the years of our uprising, we hit at the British Government's prestige, deliberately, tirelessly, unceasingly."

In contrast to other colonial rebellions that either had sought decisive military victories in actual battle or had relied on a prolonged strategy of attrition, the Irgun adopted a strategy that involved the relentless targeting of those institutions of government that unmistakably represented Britain's oppressive rule of Palestine. Thus the Irgun recommended operations in February 1944 with the simultaneous bombings of immigration department offices in Palestine's three major cities—Jerusalem, Tel Aviv, and Haifa. Subsequent attacks were mounted against the government land registry offices, from which the White Paper's provisions restricting Jewish land purchase were administered; the department of taxation and finance, responsible for collecting the revenue used to fund the government's repressive policies; and of course the security forces—the police and army—that were charged with enforcement of the White Paper.

The Irgun's most spectacular operation was without doubt its bombing in July 1946 of Jerusalem's King David Hotel. Although much has been written about this controversial incident, it is worth recalling that the King David was no ordinary hotel. On two floors of its southern wing (beneath which the explosives were placed), the hotel housed the nerve center of British rule in Palestine: the government secretariat and headquarters of British military forces in Palestine and Transjordan. The attack's target, therefore, was neither the hotel itself nor the people working or staying in it, but the government and military offices located there. Nor was its purpose random, indiscriminate carnage. Unlike many terrorist groups today, the Irgun's strategy was not deliberately to target or wantonly harm civilians. At the same time, though, the claim of Begin and other apologists that warnings were issued to evacuate the hotel before the blast cannot absolve either the group or its commander from responsibility for the ninety-one people killed and forty-five others injured—men and women, Arabs, Jews, and Britons alike. Indeed, whatever nonlethal intentions the Irgun may or may not have had, the fact remains that a tragedy of almost unparalleled magnitude was inflicted at the King David Hotel, so that to this day the bombing remains one of the world's single most lethal terrorist incidents of the twentieth century.

Despite—or perhaps because of—the tragic loss of life, so far as the Irgun was concerned the bombing achieved its objective: attracting worldwide attention to the group's struggle and the worsening situation in Palestine. Editorials in all the British newspapers focused on the nugatory results of recent military operations against the terrorists that had been previously trumpeted as great successes. Typical of these was the *Manchester Guardian's* observation that the bombing "will be a shock to those who imagined that the Government's firmness had put a stop to Jewish terrorism and had brought about an easier situation in Palestine. In fact, the opposite is the truth." These reactions accorded perfectly with Begin's plan to foster a climate of fear and alarm in Palestine so pervasive as to undermine confidence both there and in Britain in the government's ability to maintain order. Indeed, in these circumstances, the government could respond only by imposing on Palestine a harsh regimen of security measures encompassing a daily routine of curfews, roadblocks, snap checks, cordon-and-search operations and, for a time, even martial law. The failure of these measures to stop the Irgun's unrelenting terrorist campaign would, Begin hoped, have the effect of further underscoring the government's weakness. He also banked on the fact that the massive disruptions caused to daily life and commerce by the harsh and repressive countermeasures that the British were forced to take would further alienate the community from the government, thwart British efforts to obtain the community's cooperation against the terrorists, and create in the minds of the Jews an image of the army and the police as oppressors rather than protectors. Moreover, the more conspicuous the security forces seemed, the stronger the terrorists appeared.

At the foundation of this strategy was Begin's belief that the British, unlike the Germans who during the war had carried out wholesale reprisals against civilians, were incapable of such barbarity. "We knew," he explained, "that Eretz Israel, in consequence of the revolt, resembled a glass house. The world was looking into it with ever-increasing interest and could see most of what was happening

inside. . . . Arms were our weapons of attack; the transparency of the 'glass' was our shield of defence." By compelling a liberal democracy like Britain to take increasingly repressive measures against the public, the terrorists sought to push Britain to the limit of its endurance. In this respect, the Irgun did not have to defeat Britain militarily; it had only to avoid losing. Accordingly, British tactical "successes" did nothing to change the balance of forces or bring the security forces any closer to victory. Rather, measures such as massive cordon-and-search operations and the imposition of martial law proved to bring only ephemeral benefits, bought at the cost of estranging the population from the government. Nearly a quarter of a century later, the Brazilian revolutionary theorist Carlos Marighela would advocate the same strategy in his famous "Mini-Manual," the *Handbook of Urban Guerrilla War*.

In sum, this was not a war of numbers. Success was measured not in terms of casualties inflicted (between 1945 and 1947, the worst years of the conflict, just under one hundred fifty British soldiers were killed) or assets destroyed, but—precisely as Begin had wanted—by psychological impact. In place of a conventional military strategy of confrontation in battle, Begin and his lieutenants conceived operations that were designed less to kill than to tarnish the government's prestige, demoralize its security forces, and undermine Britain's resolve to remain in Palestine. Explaining his strategy, Begin argued, "The very existence of an underground must, in the end, undermine the prestige of a colonial regime that lives by the legend of its omnipotence. Every attack which it fails to prevent is a blow at its standing. Even if the attack does not succeed, it makes a dent in that prestige, and that dent widens into a crack which is extended with every succeeding attack." Thus, even though the British forces outnumbered the terrorists by twenty to one—so that there was, according to one account, "one armed soldier to each adult male Jew in Palestine"—even with this overwhelming numerical superiority, the British were still unable to destroy the Irgun and maintain order in Palestine.

Finally, an integral and innovative part of the Irgun's strategy was Begin's use of daring and dramatic acts of violence to attract international attention to Palestine and thereby publicize simultaneously the Zionists' grievances against Britain and their claims for statehood. In an era long before the advent of CNN and instantaneous satellite-transmitted news broadcasts, the Irgun deliberately attempted to appeal to a worldwide audience far beyond the immediate confines of the local struggle, beyond even the ruling regime's own homeland. In particular, the Irgun—like its nonviolent and less violent Zionist counterparts—sought to generate sympathy and marshal support among powerful allies such as the Jewish community in the United States and its elected representatives in Congress and the White House, as well as among the delegates to the fledgling United Nations, to bring pressure to bear on Britain to grant Jewish statehood. The success of this strategy, Begin claims, may be seen in the paucity of global coverage afforded to the civil war that had erupted in Greece after the Second World War, compared to that devoted to events in Palestine. Palestine, he wrote, had undeniably become a "centre of world interest. The revolt had made it so. It is a fact," Begin maintains,

> that no partisan struggle had been so publicized throughout the world as was ours. . . . The reports on our operations, under screaming headlines, covered the front pages of newspapers everywhere, particularly in the United States. . . . The interest of the newspapers is the measure of the interest of the public. And the public—not only Jews but non-Jews too—were manifestly interested in the blows we were striking in Eretz Israel.

In this respect, pro-Irgun Jewish American lobbyists were noticeably successful in obtaining the passage of resolutions by the U.S. Congress condemning "British oppression" and reaffirming American support for the establishment of a Jewish state in Palestine. These activities, which presaged the efforts undertaken more recently by Irish American activists on behalf of Sinn Fein and the IRA, had similarly corrosive effects on Anglo-American relations more than half a century ago.

By 1947 the Irgun had in fact achieved its objectives. Reporting on the situation to Washington, the American consul general in Jerusalem observed that

> with [British] officials attempting to administrate from behind masses of barbed wire, in heavily defended buildings, and with the same officials (minus wives and children evacuated some time ago) living in pathetic seclusion in "security

zones," one cannot escape the conclusion that the Government of Palestine is a hunted organization with little hope of ever being able to cope with conditions in this country as they exist today.

Indeed, each successive terrorist outrage illuminated the government's inability to curb, much less defeat the terrorists. Already sapped by the Second World War, Britain's limited economic resources were further strained by the cost of deploying so large a military force to Palestine to cope with the tide of violence submerging the country. Public opinion in Britain, already ill disposed to the continued loss of life and expenditure of effort in an unwinnable situation, was further inflamed by incidents such as the King David Hotel bombing and the Irgun's hanging in July 1947 of two sergeants in retaliation for the government's execution of three convicted Irgun terrorists. As the renowned British historian of the Middle East Elizabeth Monroe has noted with respect to the hangings: "The British public had taken Palestine in its stride . . . and had looked upon 'disturbances' and 'violence' there much as it viewed 'the troubles' in Ireland—as an unpleasant experience that was part of the white man's burden." All this changed, however, with the cold-blooded murder of the sergeants. Photographs of the grim death scene—depicting the two corpses just inches above the ground, the sergeants' hooded faces and bloodied shirts—were emblazoned across the front pages of British newspapers under headlines decrying their execution as an act of "medieval barbarity." As inured to the almost daily reports of the death and deprivation suffered by the army in Palestine as the British public was, the brutal execution of the two sergeants made a deep and unalterable impression on the national psyche. "All home comment on that deed," Monroe continued, was "different in tone from that on earlier terrorist acts, many of which caused greater loss of life—for instance, the blowing up of the officers' club or of the King David Hotel." For both the British public and the press, the murders seemed to demonstrate the futility of the situation in Palestine and the pointlessness of remaining there any longer than was absolutely necessary.

At the time, Britain was also, of course, coming under intense pressure from the United States and other quarters regarding the admission to Palestine of tens of thousands of Jewish displaced persons still languishing throughout liberated Europe and was itself trying to stem the flood of illegal Jewish immigrants attempting to enter Palestine. In addition, throughout the summer of 1947 the Special Committee on Palestine (UNSCOP) appointed by the UN General Assembly was completing its investigations regarding the country's future. It is a measure of the Irgun's success that Begin was twice granted audiences with the committee to explain the group's aims, motivations, and vision for a Jewish state in Palestine. The committee's unanimous recommendation calling for the immediate termination of British rule and granting of independence to Palestine finally forced the government's hand. In September the colonial secretary, Arthur Creech-Jones, announced that Britain would no longer be responsible for governing Palestine and that all civilian and military personnel would be evacuated as soon as was practicable.

A decade and a half after the event, Creech-Jones cited four pivotal considerations that influenced the government's decision. First, there were the irreconcilable differences between Palestine's Arab and Jewish communities; second, the drain on Britain's shrinking financial resources imposed by the country's heavy military commitment in Palestine; third, the force of international, American, and parliamentary opinion; and finally—and, he believed, most significant—the public outcry in Britain that followed the Irgun's hanging of the two sergeants. Describing the confluence of events that compelled the government to surrender the mandate, the former colonial secretary recalled specifically, "Terrorism was at its worst and the British public seemed unable to stand much more." Hence, with "accelerating speed," Creech-Jones explained, "the Cabinet was pushed to the conclusion that they could [no] longer support the Mandate." On May 15, 1948, Britain's rule over Palestine formally ended and the establishment of the State of Israel was proclaimed. In a communiqué issued that same day by the Irgun, Begin declared:

> After many years of underground warfare, years of persecution and suffering . . . [the] Hebrew revolt of 1944–48 has been crowned with success. . . . The rule of enslavement of Britain in our country has been beaten, uprooted, has crumbled and been dispersed. . . . The State of Israel has arisen. And it has arisen "Only Thus": through blood, fire, a strong hand and a mighty arm, with suffering and sacrifices.

 # The Anticolonial Struggles of the 1950s: Cyprus and Algeria

The Irgun's revolt provided a template for subsequent anticolonial uprisings elsewhere. Indeed, the most effective irredentist struggles of the immediate postwar era were those that emulated Begin's strategy and deliberately sought to appeal to—and thereby attract the attention and sympathy of—an international audience. "Our intention," explained General George Grivas, the founder and commander of EOKA (Ethniki Organosis Kyprion Agoniston, or National Organization of Cypriot Fighters), in his memoirs, "was to focus the eyes of the world on Cyprus and force the British to fulfil their promises." Similarly, the 1954 proclamation of revolt against French rule of Algeria by the FLN (Front de Liberation Nationale, or National Liberation Front) prominently cited the "internationalism of the Algerian problem" as among its principal goals.

In pursuit of this end, both groups also made a conscious effort to appeal directly to the United Nations for help. EOKA's own proclamation of revolt, for example, was specifically addressed to "Diplomats of the World." It called upon them to "Look to your duty. It is shameful that, in the twentieth century, people should have to shed blood for freedom, that divine gift for which we too fought at your side and for which you, at least, claim that you fought against Nazism." In Algeria, FLN offensives, general strikes, and other demonstrations were timed to occur when the UN General Assembly reconvened or was already scheduled to discuss the conflict. In January 1957 the fabled Battle of Algiers, immortalized by the 1966 Gillo Pontecorvo film of the same name, when the FLN unleashed its campaign of mass urban terrorism, was deliberately choreographed to coincide with the General Assembly's annual opening session. The FLN communiqué announcing the strike that accompanied the new terrorist offensive candidly admitted to this timing, announcing its desire to "bestow an incontestable authority upon our delegates at the United Nations in order to convince those rare diplomats still hesitant or possessing illusions about France's liberal policy." To a large extent, therefore, the success of both EOKA and the FLN in ending foreign rule of their respective countries was predicated upon their ability to attract external attention to their respective struggles, much as the Irgun had a decade before. In Cyprus, from the very start, Grivas appreciated the necessity of reaching out to an audience beyond the immediate geographical boundaries of his group's struggle. "The enlightenment of international public opinion," he recognized, "was bound to play an important part in bringing home to all concerned the Cypriot people's demand for self-determination. It is a fact that there were many foreigners and even United Nations representatives who were completely ignorant of why we were demanding our freedom." Accordingly, Grivas enshrined this principle in the "Preparatory General Plan" that he had formulated in 1953—two years before the campaign actually began—whose opening paragraph clearly states the fundamental objective "to arouse international public opinion . . . by deeds of heroism and self-sacrifice which will focus attention on Cyprus until our aims are achieved." Although there is no evidence that Grivas ever read Begin's book (an English-language translation of *The Revolt* had been published in London and New York in 1951) or studied the Irgun's campaign, the parallels between the two struggles are unmistakable. Grivas's strategy, like Begin's, was not to win an outright military victory against the numerically superior British forces but to rely on dramatic, well-orchestrated, and appropriately timed acts of violence to focus international attention on the situation in Cyprus and the Greek Cypriots' demand for *enosis*—unification with Greece.

"My small force was outnumbered by more than a hundred to one," Grivas later recalled, "but, as I have said, this made no difference to the type of subversive warfare I was planning." His plan was to deploy the majority of EOKA forces in the island's urban centers, where they were organized into individual terrorist cells numbering no more than eight to ten men each. Their mission was to tie up as many British troops as possible on static guard duties in the cities, thereby allowing EOKA to consolidate its control over the rest of the island. In words reminiscent of Begin's, Grivas explained in his treatise on guerrilla warfare that

> our strategy consisted in turning the whole island into a single field of battle in which there was no distinction between front and rear, so that the enemy should at no time and in no place feel himself secure. The enemy never knew where and when we might strike. . . . This strategy

achieved the dispersal, intimidation and wearing down of the enemy's forces and especially serious consequences resulting from our use of surprise.

By concentrating on urban operations, as the Irgun had in Palestine, EOKA also gained immediate access to the news media dispatched to the island to cover the escalating violence. Had EOKA, like traditional guerrilla forces, confined its operations to the island's rugged and isolated rural areas and mountain ranges, it would arguably have lost this access and forfeited the exposure so critical to Grivas's strategy. The urban terrorist campaign was thus pivotal in securing the propaganda platform required by the group to broadcast its cause to the world. In this respect, Grivas was able to derive great satisfaction from the fact that within five months of the revolt's proclamation, EOKA had succeeded in attaining what a decade of patient diplomacy and insistent lobbying had failed to achieve: UN consideration of the Greek Cypriots' nationalist claims. Hitherto, the General Assembly had upheld British arguments that the situation in Cyprus was an entirely *internal* matter and therefore outside the organization's purview. Now, for the first time, Cyprus had been placed on the UN's agenda. "This proved," Grivas later recalled, "that inside this international body the idea was beginning to penetrate that something must be done about the Cyprus question."

Indeed, by the end of 1955, Grivas had succeeded in plunging the island into complete disorder. An average of two British soldiers or policemen were being killed each week. The security forces had been thrown on the defensive, kept off balance by repeated EOKA hit-and-run attacks and unable to mount any effective offensive operations. Like Begin, Grivas also calculated that the unrelenting terrorist onslaught would sap the morale of British forces and compel them to overreact with counterproductive, self-defeating measures directed against the law-abiding Greek-Cypriot community.

"The 'security forces' set about their work in a manner which might have been deliberately designed to drive the population into our arms," Grivas recalled. "These attempts to frighten the people away from EOKA always had exactly the opposite effect to that intended: the population were merely bound more closely to the Organisation." Once again, the fundamental asymmetry between the terrorists' apparent ability to strike anywhere, at any time, and the security forces' inability to protect all conceivable targets, all the time, was glaringly demonstrated. As in Palestine, the more visible and pervasive the security forces in Cyprus became, the greater the public frustration caused by disruption to daily life, and the more powerful and omnipresent the terrorists appeared. At the height of the conflict, British security forces on Cyprus totaled nearly 40,000 men, arrayed against a hard core of fewer than 400 active terrorists, backed by some 750 "auxiliaries." Again, as in Palestine, this was not a war of "numbers." The massive deployment of British troops had little overall impact on the situation. As Grivas later reflected about his opponent commanding the British forces, Field Marshal Sir John Harding: "He underrated his enemy on the one hand, and overweighted his forces on the other. But one does not use a tank to catch field mice—a cat will do the job better."

Throughout the campaign, Grivas coordinated his underground campaign with the above ground diplomatic efforts of Archbishop Makarios III (Michael Christodoros Mouskos). As the appointed head of the Ethnarchy (Church Council) of Cyprus, Makarios was also the Greek-Cypriot community's de facto political leader. He and Grivas had met as early as 1950 to plan the general outline of the revolt, and prelate and soldier worked closely together throughout the struggle to achieve their shared goal of *enosis*. Thus the symbiotic relationship that today exists between Sinn Fein ("We Ourselves," an Irish nationalist political party) and the IRA has a historical parallel in that between the Ethnarchy and EOKA forty years ago, with Makarios playing the role allegedly performed today by Gerry Adams, Sinn Fein's president. Like Adams, Makarios was interned, and in 1956 he was exiled to the Seychelles, being allowed to return to Cyprus two years later as a condition for Greek-Cypriot participation in British-sponsored multiparty talks on the island's future. It was not until February 1959 that agreement was reached.

Under intense pressure from Greece, Makarios reluctantly accepted the proposal for the creation of an independent republic of Cyprus, with Britain being allowed to retain two strategic bases on the island. Fears that Turkey would otherwise forcibly impose partition (as in fact occurred fifteen years later) led Makarios to acquiesce in the arrangement despite Grivas's vehement opposition. The revolt officially ended a month later, when EOKA surrendered a large enough quantity of arms to satisfy the government that the peace agreement could be implemented. Although

enosis was never achieved, British rule was forcibly ended and Cyprus was granted its independence. The fruits of the terrorists' labors were also apparent in the new republic's general election: Makarios was elected the country's first president, polling 67 percent of the popular vote.

At the other end of the Mediterranean, the revolt against French rule over Algeria between 1954 and 1962 was the last of the immediate postwar anticolonial struggles. For that reason, perhaps, it had the most direct and discernible impact on many later ethno-nationalist terrorist campaigns. Yasir Arafat, for example, in his authorized biography, cites the pivotal influence that the FLN had on the PLO's struggle and the critical material assistance that Algeria later provided to the Palestinians. "I started my contacts with the Algerian revolutionaries in the early 1950s," he recalled. "I stayed in touch with them and they promised they would help us when they had achieved their independence. I never doubted for one moment that they would win, and that their victory would be very important for us." In his own candid memoir, Nelson Mandela similarly identifies the seminal influence that the FLN's struggle had on the decades-long effort of the African National Congress (ANC) to end minority white rule of South Africa. "The situation in Algeria," Mandela wrote, "was the closest model to our own in that the rebels faced a large white settler community that ruled the indigenous majority." Accordingly, the ANC studied the Algerian conflict closely, deriving the main lesson that a pure military victory was impossible to achieve in such circumstances. Instead, Mandela assiduously applied the advice given to him by an Algerian revolutionary that "international opinion . . . is sometimes worth more than a fleet of jet fighters." It was a lesson that the FLN itself had learned only belatedly. By the middle of 1956, the rebellion against French rule had been raging for nearly two years; the FLN, however, had precious few tangible achievements to show for its efforts, and recent advances by the security forces in the countryside had seriously undermined the group's rural insurgent strategy. Accordingly, the FLN embarked on a new strategy that would, for the first time, focus on the country's capital, Algiers, and thereby apply pressure to France by appealing directly to international opinion. The architect of this new strategy, first unveiled during the August 1956 summit convened at Soummam, Morocco, in hopes of reversing the FLN's declining fortunes, was Ramdane Abane, a leading figure in the movement until his execution the following year during an internecine power struggle. As the group's chief theoretician, he was also its most potent intellect. From an impoverished background and entirely self-educated, he was as completely unsentimental as he was ruthless, maintaining an unalterable faith in the efficacy of violence. This was clearly evident in his famous directive number nine, wherein he succinctly explained not just the purpose of the FLN's new strategy but the elementary logic behind urban terrorism. "Is it preferable for our cause to kill ten enemies in an oued [dry riverbed] of Telergma when no one will talk of it," he rhetorically asked, "or a single man in Algiers which will be noted the next day by the American press?"

The chain of events that led to the FLN's full-scale urban terrorist campaign, however, had actually begun two months earlier, in June, with the execution by guillotine of two convicted FLN fighters. As had occurred countless times elsewhere (Palestine and Cyprus included), such attempts by the ruling regime to deter further violence with a particularly harsh exemplary punishment backfired catastrophically. The recipients of this lesson, rather than serving as abject examples, as often as not become martyrs: emblematic rallying points for the revolutionary cause around which still greater sacrifice and still further bloodshed and destruction are demanded and justified. Thus it was in Algeria, where the FLN announced that for every FLN fighter executed, a hundred Frenchmen would meet a similar fate. Hitherto, the group could boast that its campaign in the capital had been deliberately nonlethal, its bombs directed against inanimate "symbols" of French rule—government offices and buildings, military cantonments and police stations—but not deliberately against people. This now changed with Abane's instructions to the FLN's urban cadres to unleash a reign of unprecedented bombings and terror. Within seventy-two hours, forty-nine French civilians had been gunned down. Then, in August, as a result of the new strategic direction approved that same month at Soummam, the bombings began.

The campaign was spearheaded not by the group's hardened male fighters but by its attractive young female operatives—whose comely bearing and European looks, as Saadi Yacef, the group's operations officer in Algiers (who later reprised his real-life role in the Pontecorvo film), correctly guessed, would arouse far less suspicion than their male counterparts. Their targets, moreover, were neither military nor even governmental, but the crowded seaside milk bar frequented by *pied noir* (French colonist) families after a day at the beach: a cafeteria particularly favored by European university students; and the downtown Air France passenger terminal. The coordinated operations killed three people and injured some fifty others—including several children, some of them among the dozen or so victims requiring surgical amputation of mangled limbs. Throughout it all, Abane was unmoved. Drawing the same analogy between the terrorist bomb in the dustbin and the "poor man's air force" cited in chapter 1, Abane is said to have dismissively observed, "I see hardly any difference between the girl who places a bomb in the Milk-Bar and the French aviator who bombards a *mechta* [village] or who drops napalm on a *zone interdite* [interdiction, or free-fire, zone]."

The urban campaign continued throughout the remainder of the year, climaxing on December 28, 1956, with the assassination of the mayor of Algiers. Widespread anti-Muslim rioting broke out, only to be followed by a new round of FLN assassinations. This was the last straw. In despair over the deteriorating situation, the governor-general called out the army. On January 7, 1957, General Jacques Massu, commander of the elite Tenth Parachute Division, assumed complete responsibility for maintaining order in the city. The FLN responded by declaring a general strike for January 28, to coincide, as noted above, with the UN's annual opening session. Its purpose—and that of the terrorist attacks that accompanied it: to focus international attention on Algeria. Once again, Yacef's bombers set about their work with startling efficiency. The FLN's target set now expanded to include popular bars and bistros, crowded city streets and sports stadiums packed with spectators. Within two weeks 15 people had been killed and 105 others wounded.

Massu went on the offensive. Having fought in Indochina, he and his senior commanders prided themselves on having acquired a thorough understanding of revolutionary warfare and how to counter it. Victory, they were convinced, would be entirely dependent on the acquisition of intelligence. "The man who places the bomb," declared Colonel Yves Godard, one of Massu's sector commanders, "is but an arm that tomorrow will be replaced by another arm"; the key was to find the individual commanding the arm. Accordingly, Godard and his men set out to uproot and destroy the FLN's urban infrastructure. Their method was to build up a meticulously detailed picture of the FLN's apparatus in Algiers that would home in relentlessly on the terrorist campaign's mastermind. Godard's approach, dramatically depicted onscreen by Pontecorvo, was described by the British historian Alistair Home as a "complex *organigramme* [that] began to take shape on a large blackboard, a kind of skeleton pyramid in which, as each fresh piece of information came from the interrogation centres, another name (and not always necessarily the right name) would be entered." That this system proved effective, there is no doubt. The problem was that it also depended on, and therefore encouraged, widespread abuses, including torture, for Massu and his men were not particularly concerned about *how* they obtained this information. Torture of both terrorists and *suspected* terrorists became routine. The French army in Algeria found it easy to justify such extraordinary measures, given the extraordinary conditions. The prevailing exculpatory philosophy among the Tenth Parachute Division can be summed up by Massu's terse response to complaints, that "the innocent [that is, the next victims of terrorist attacks] deserve more protection than the guilty."

The brutality of the army's campaign, however, completely alienated the native Algerian Muslim community. Hitherto mostly passive or apathetic, it was now driven into the arms of the FLN, swelling the organization's ranks and increasing its popular support. Domestic public opinion in France was similarly outraged, undercutting popular backing for continuing the struggle and creating deep fissures in French civil-military relations. Massu and his men stubbornly consoled themselves that they had achieved their mission and defeated the rebels' attempt to seize control of Algiers, but this military victory was bought at the cost of eventual political defeat. Five years later the French withdrew from Algeria and granted the country its independence.

For decades, Massu remained unrepentant, maintaining that the ends justified the means used to destroy the FLN's urban insurrection. In recent years, however, he has had second thoughts. In 2000, for instance, he told an interviewer that France should acknowledge that torture was routinely employed in Algeria and officially condemn it. The officer under Massu's command at the time, retired general Paul Aussaresses, who was directly responsible for implementing this policy, however, has never softened. "For my part, I do not repent," Aussaresses told the same interviewer. "Torture never pleased me but I was resolute

when I arrived in Algiers. At the time it was already widespread. If it were done again, it would piss me off, but I would do the same thing because I do not believe that one can do it differently." Respect for the rule of law and the niceties of legal procedure, much less international conventions governing the rights of combatants, he contends, were totally irrelevant given the crisis situation enveloping Algeria in 1957. "Only rarely were the prisoners we had questioned during the night still alive the next morning." Aussaresses explained in his memoir, *The Battle of the Casbah*:

> Whether they had talked or not they generally had been neutralized. It was impossible to send them back to the court system, there were too many of them and the machine of justice would have become clogged with cases and stopped working altogether. Furthermore, many of the prisoners would probably have managed to avoid any kind of punishment.

Thus the battle was won and the terrorists' indiscriminate bombing campaign ended. Extraordinary measures were legitimated by extraordinary circumstances—all of which Aussaresses maintains is besides the point. As torturers before and since have claimed, he too contends, "I don't think I ever tortured or executed people who were innocent. I was mainly dealing with terrorists who had been involved in attacks."

At the same time, there is no doubt that this "success" cut both ways. The FLN's tactical defeat in the city resulted in yet another complete reassessment of its strategy. Large-scale urban terrorism was now abandoned alongside the FLN's belief that France could be defeated militarily. The group's high command also concluded that the struggle could not be won inside Algeria alone; accordingly, the rebels relocated their operational bases to Tunisia, from which they pursued a rural hit-and-run strategy, making cross-border raids from their newly established sanctuaries. But the Battle of Algiers remains perhaps the most significant episode in bringing about the FLN's subsequent triumph, in that it succeeded in focusing world attention on the situation in Algeria, just as Abane had calculated. By provoking the government to overreact with torture, summary executions, and other repressive tactics, the FLN also revealed the bankruptcy of

French rule, thereby hastening the complete destruction of Algerie Francaise.

 Conclusion

The ethno-nationalist insurrections that followed the Second World War had a lasting influence on subsequent terrorist campaigns. Although governments throughout history and all over the world always claim that terrorism is ineffective as an instrument of political change, the examples of Israel, Cyprus, and Algeria, and of Begin, Makarios, and Ahmed Ben Bella (the FLN's leader, who became Algeria's first president), provide convincing evidence to the contrary. Admittedly the establishment of these independent countries was confined to a distinct period of time and was the product, in some cases, of powerful forces other than terrorism. At the same time, however, it is indisputable that, at the very least, the tactical "successes" and political victories won through violence by groups like the Irgun, EOKA, and the FLN clearly demonstrated that—notwithstanding the repeated denials of the governments they confronted—terrorism does "work." Even if this "success" did not always manifest itself in terms of the actual acquisition of power in government, the respectability accorded to terrorist organizations hitherto branded as "criminals" in

forums like the United Nations and their success in attracting attention to themselves and their causes, in publicizing grievances that might otherwise have gone overlooked, and perhaps even in compelling governments to address issues that, if not for the terrorists' violence, would have largely been ignored, cannot be disregarded.

In sum, the anticolonial terrorism campaigns are critical to understanding the evolution and development of modern, contemporary terrorism. They were the first to recognize the publicity value inherent in terrorism and to choreograph their violence for an audience far beyond the immediate geographical loci of their respective struggles. The Irgun directed its message to New York and Washington, D.C., as much as to London and Jerusalem. EOKA similarly appealed to opinion in New York and London as well as in Athens and Nicosia. And the FLN was especially concerned with influencing policy not only in Algiers but in New York and Paris as well. The ability of these groups to mobilize sympathy and support outside the narrow confines of their actual "theaters of operation" thus taught a powerful lesson to similarly aggrieved peoples elsewhere, who now saw in terrorism an effective means of transforming hitherto local conflicts into international issues. Thus the foundations were laid for the transformation of terrorism in the late 1960s from a primarily localized phenomenon into a security problem of global proportions.

REVIEW QUESTIONS

1. Briefly summarize the activities of the three terrorist groups described in this article. Identify the key differences and similarities among these groups.

2. Describe the significance of both the Japanese conquest of Singapore and the Atlantic Charter. Do you see any aspect of these two events mirrored in today's conflicts? Why or why not?

3. What strategic and tactical lessons can be drawn from the anti-colonial groups active in the 1950s and 1960s? How have these lessons been applied by terrorist groups operating in the latter half of the 20th century? With what level of success?

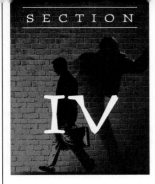

SECTION

IV

Methods, Patterns, and Trends of Terrorists

Learning Objectives

At the end of this section students will be able to:

- Describe the data available from the Global Terrorism Database.
- Discuss the implications of religion and terrorism.
- Describe the most common methods of terrorism.
- Detail the response to improvised explosive devices by the U.S. Department of Defense.

On the bright, cool, fall morning of October 4, 2003, the young woman left her parents' house promptly at 7:30 a.m., without even saying goodbye to them. Her parents assumed she was simply in a hurry to get to work. They were wrong. She met her handlers from the Palestinian Islamic Jihad and though hesitant and nervous, quickly filmed a brief last will and testament video. Later that day, in the early afternoon on the eve of the holiest Jewish festival, Yom Kippur, the same attractive female Palestinian and law school graduate rushed into a crowded, Jewish-Arab (Christian) owned restaurant, Maxim, in Haifa and set off a bomb. Hanadi Tayseer Jaradat avenged the death of her much-loved brother and a cousin at the hands of the Israeli military. In the process, she managed to kill not only herself but 20 others, including 4 children, and injured 50 more. Police do not know if she removed her headscarf and dressed in Western-style clothes or pretended to be pregnant and hid her explosive vest under maternity clothes. They do know that she hid over 15 kilograms or more than 33 pounds of explosives along with steel ball-bearings. She became a Shaheed, a martyr. In the video she made, we see Jaradat vowing revenge, and the next image the world would see of her would be of her severed head on the floor amid the devastation she created at Maxim. The Palestinian Islamic Jihad claimed credit and released the last will and testament video to Arab television networks.

Liberal Humanist

▲   Hanadi Tayseer Jaradat

Almost all definitions of terrorism begin with the phrase "is an act of violence" or something very similar to that. This section will present an overview of all the available violent actions that terrorists might use. At one time, cataloging them would have been difficult. Now, thanks to the Internet, the Global Terrorism Database, and the efforts of such groups as RAND, which produces and maintains the RAND Worldwide Database of Terrorism Incidents (introduced in Section 1), researchers and students have a great deal of information quickly at hand. It is very sophisticated, and users of many **databases** can filter this to reveal surprising results. Admittedly, the databases are not real time but do provide fairly current information and trends. For example, using the RAND database, anyone wishing to know all explosive or firearms attacks in a region would simply select the region, then the weapons filter and the time frame. Outputs are incident lists (date, perpetrator, and country), pie charts, or chronological graphs. This section will provide an overview of the types of attacks and the likely weapons or devices that have been used by terrorists. There will be a special section on **suicide bombing** and a presentation of the spectrum of weapons of mass destruction (WMD), or nuclear, radiological, biological, and chemical weapons. Too, these are considered low-probability high-consequence events that we hope never happen, but the possibility remains and several exercises, including the Dark Winter exercise of 2001, have demonstrated that we are not completely prepared (Dark Winter, n.d.).

A United Kingdom organization called the Community Security Trust, with a mission to protect members of the Jewish community, periodically publishes a report on terrorist incidents against Jewish communities and Israeli citizens abroad. A recent report notes that Jews have been targeted from 1968 through 2010 by a wide variety of groups from the left and right, by various groups of religious persuasions, and by terrorist groups. For our purposes here, an examination of the data indicates that of the 427 incidents, 80 were actually foiled or aborted, and 208 were bombs or **improvised explosive devices**. The next highest type of weapon used was some type of firearm (The Community Security Trust, United Kingdom, 2010).

The National Counterterrorism Center (NCTC) in the United States produces a report annually on terrorist incidents worldwide. Normally released in the spring for the previous year, it tallies information on attacks on specific countries and regions, identifying perpetrators and victims. For its purposes, NCTC uses the following categories of attacks:

> The WITS database contains a field that allows analysts to categorize an incident by **event type**. Event types are coded in the database as the following: armed attack, arson/firebombing, assassination, assault, barricade/hostage, bombing, CBRN, crime, firebombing, hijacking, hoax, kidnapping, near miss/non-attack, other, theft, unknown, and vandalism. While some incidents can easily be coded using this taxonomy, other kinds of attacks are more difficult to define. When it can be determined, incidents that involve multiple types of attacks are coded with multiple event types. Incidents involving mortars, rocket-propelled grenades, and missiles generally fall under armed attack, although improvised explosive devices (IED) fall under bombing, including vehicle-borne IEDs (VBIED). VBIEDs include any IED built into or made a part of a vehicle including cars, trucks, bicycles, and motorcycles. Suicide events are also captured, but the perpetrator must have died in the attack for the event type suicide to be included. (National Counterterrorism Center, 2011, p. 6)

When thinking about the attacks on one ethnic or religious community mentioned above, about 400 over a span of 40 years, the numbers do not seem large, though the authors do not want to minimize the trauma, pain, and suffering. But the numbers presented by the NCTC for just one recent year, 2010, are high. There were over 11,000 incidents or attacks resulting in over 50,000 victims and around 13,200 deaths. As the chart on p. 99 demonstrates, armed attacks were the most prevalent, with bombings the second most frequent, and kidnapping comes in third.

Although suicide bombings do not appear as a large number, 263, they accounted for around 13.5% of terror-related deaths (National Counterterrorism Center, 2011, p. 6).

 ## NCTC Chart: Primary Attack Types

Suicide bombers present a particular difficulty as they are the ultimate "smart bomb" and for a host of reasons are difficult to detect and stop. Governments, researchers, and academics have studied this extensively. Unfortunately, much of the research raises more questions than it answers and is the result of anecdotal evidence that cannot be refuted, small samples, and a host of methodological problems (Mintz & Brule, 2009). Here are the "facts" about suicide bombers.

Fact 1: Research about suicide bombers is limited and we do not know a great deal since suicide bombers who are successful are simply no longer with us and available to study. This means that we must infer from writings, last will and testament videos, and interviews with friends and relatives, and make some inferences or judgments regarding the reasons, motivations, or even ideologies.

Fact 2: Much of the research that has been done on intercepted or failed suicide bombers uses very small samples, and a failed suicide bomber is not the same in reality as a successful one. For example, Anat Berko and Edna Erez (2005) interviewed seven male and female Palestinian individuals who attempted suicide bombings and were serving sentences in an Israeli prison. They conclude that although they had a small sample there were indications that a decision to become a suicide bomber is essentially a social process, and using criminological theories is appropriate.

Fact 3: When more ambitious studies are attempted, and they are few, they have generated a great deal of controversy about their science, research, and methodology. One of the more ambitious research articles on the topic falls into this category. This 2009 report was a targeted article for *Political Psychology*. The article examines suicide bombers' taped death or last will and testament videos (made before they carried out their deadly act); coders listened to them and placed them in various categories using content analysis. The same process was used when viewing taped interviews with mothers. The findings, summarized by the authors below, are interesting and controversial, and they generated much amiable support, encouragement, and criticism:

> Recent analyses of the motivations for suicidal terrorism have identified a broad variety of motives (the "fatal cocktail") having to do with potential perpetrators' (1) personal traumas and frustrations, (2) ideological reasons, and (3) social pressures to which they may be subjected. In the present paper we introduced the notion of *significance quest* as an integrative concept tying these motivational categories together: Personal traumas and frustrations could encourage a "collectivistic switch" to a terrorism-justifying ideology because the latter may afford a means for restoring the lost significance occasioned by various unsettling events. Besides, terrorism-justifying ideologies may afford a relatively simple means of substantial *significance gain* and attainment of a hero or a martyr status in the eyes of one's community. (Kruglanski et al., 2009a)

Fact 4: Based largely on that study, it is apparent that those involved in suicide bombing have usually experienced some sort of trauma, often seek out an organization, may have ideological sympathy with a cause, and seek personal significance. However, critiques and reviews of the research were many. Two will be presented here. Margaret Crenshaw summarizes their efforts by suggesting that they determined:

> Individuals seek significance; ideologies provide a rationale, organizations channel and direct that search and society rewards the result. The process comes full circle as the individual believes (or is led to believe) that he or she will achieve transcendence through self-sacrificing participation in a collective endeavour of benefit to all, with the promise of living forever in the memories of successor generations (Crenshaw, 2009, p. 360).

Figure 4.1   Primary Attack Types

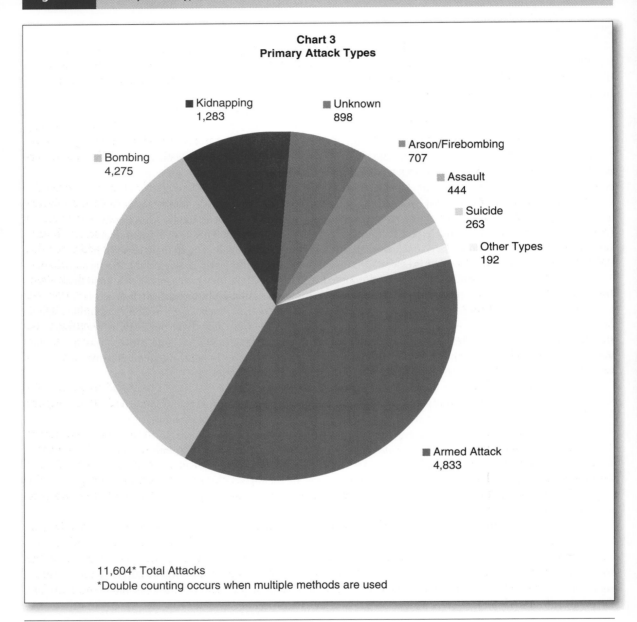

Chart 3
Primary Attack Types

■ Kidnapping
1,283

■ Unknown
898

■ Arson/Firebombing
707

■ Assault
444

■ Suicide
263

■ Other Types
192

■ Bombing
4,275

■ Armed Attack
4,833

11,604* Total Attacks
*Double counting occurs when multiple methods are used

National Counterterrorism Center 2010 Report on Terrorism, p. 13, April 30, 2011.

She then wonders if the quest for significance resulting in a suicide bombing is sufficient, questions how one explains the fact that suicide bombing campaigns vary in intensity, and debates if significance is uniform in all societies. Crenshaw suggests:

> that we need to ask when and why ideologies form that glorify suicide attacks? We need to know where these ideas come from and how they are presented and distributed as much as why they are attractive to the individuals who adopt them. We also need to ask about the process of adoption, which usually occurs in a group setting? (p. 360)

Answering the critics, Kruglanski et al. maintain that their approach explaining terrorism and suicide bombing remains plausible and is largely correct in that their main point is that there is a universal quest for significance; given intense conflicts, and an **ideology** such as radical Islam or even a dedication to Vishnu, a Hindu God, terrorism and suicide bombing is a rational response (Kruglanski et al., 2009a).

Fact 5: Today **religion** does play a role in suicide bombings, even among women suicide bombers, though the majority, almost 85%, were committed for secular organizations (O'Rourke, 2008). As noted by Capella and Sahliyeh (2007), "religion itself may be a necessary part of the explanation, but not a sufficient one" (p. 278). And as stated above, it is not just radical Islam, but other religions and other ideologies as well that are behind the suicide bombing and other terrorism that are predominant today. And how does one explain this? Kuntzel (2008) addressed this with regard to Islam and he places the blame squarely on Iran's late Ayatollah Khomeini, who articulated a worldview that life is not worth living and death performed as a martyr is preferable. Khomeini formed a force call the *Basiji* movement and enlisted the youth of Iran to serve as "mine sweepers" and suicide bombers during the Iraq-Iran war and soon, despite dissent from some experts, it was adopted by the other groups, including Hamas (Kuntzel, 2008). And clearly this continues today, as noted above by the National Counterterrorism Report, but what is chilling is the following quote: "Sunni extremists committed almost 60 percent of all worldwide terrorist attacks. These attacks caused approximately 70 percent of terrorism-related deaths, a significant increase from the almost 62 percent in 2009" (National Counterterrorism Center, 2011, p. 6).

Fact 6: Suicide bombing is clearly much more deadly when examining both non-religious and religious motivated suicide bombings and comparing it to other forms of violence, such as simply bombs and other weapons (Capella, 2007).

Religion sparks violence and terrorism, not just suicide bombing. Many weapons are in the hands of terrorists today and easily available. The world sees on a daily basis evidence of this. Suicide bombers are responsible for some of these events and while we may not be able to state with certainty the "cause," the research has provided sufficient information to suggest that they have suffered, want to do something significant, embrace an ideology (usually religious) that supports this, and they turn to an organization and a community that will support their efforts to succeed at self-destruction.

Fact 7: Suicide bombings are not a completely male-dominated activity. Females have been very effective suicide bombers in Sri Lanka, Israel, Afghanistan, India, Iraq, Lebanon, Pakistan, Russia, Somalia, Turkey, and Uzbekistan. In some regions and cultures where women dress in certain ways and may be segregated and/or respected, the use of a female suicide bomber is not only possible but encouraged by both secular and religious terror organizations. Female suicide bombers are older than male suicide bombers, a significant number have experienced the loss of someone close, and they are often unmarried and with unlikely prospects (O'Rourke, 2009).

 Summary

This section presented current tools used as methods to carry out terrorism and examined suicide bombings. Clearly this is a dangerous time, as the data demonstrate. In particular, improvised explosive devices (IED) and **vehicle born improvised explosive devices (VBIED)** were so deadly in the most recent Iraq war that they necessitated a

©iStockphoto.com/Rockfinder

▲ Photo of a Mine Resistant Ambush Protected (MRAP) vehicle produced at a cost of over $500,000 each in response to attacks in Iraq and Afghanistan by Improvised Explosive Devices.

concerted effort to do something about them. As one inspector general report states:

> DoD was aware of the threat posed by mines and improvised explosive devices (IEDs) in low-intensity conflicts and of the availability of mine-resistant vehicles years before insurgent actions began in Iraq in 2003. Yet DoD did not develop requirements for, fund, or acquire MRAP (mine resistant ambush-protected)-type vehicles for low-intensity conflicts that involved mines and IEDs. As a result, the Department entered into operations in Iraq without having taken available steps to acquire technology to mitigate the known mine and IED risk to soldiers and Marines. (Results in Brief: Marine Corps Implementation of the Urgent Universal Needs Process for Mine Resistant Ambush Protected Vehicles, 2008)

This report generated alarm and resulted in a concerted multiple-year effort by the entire Department of Defense, beginning with an ad hoc task force of 12 personnel and culminating with the formation of a Joint IED Defeat Organization in 2006 (Smith, 2011).

The response as reported by Smith was stunning. It encompassed not only the formation of high-level organizations to study this, but it necessitated research and development; upgrades in personal body armor; training, reconnaissance, and surveillance a special engineering modular force structure emphasizing route clearance; and ultimately, a number of sophisticated vehicles called **mine resistant ambush protected (MRAP)** (Smith, 2011). The price for these vehicles was staggering, over $22 billion dollars (Secretary of Defense Robert Gates, 2010). To put that dollar amount in perspective, it is almost half of the entire budget for the country of Cuba (*CIA World Factbook,* n.d.). Imagine if a nation not nearly as wealthy as the United States had to contend with sophisticated IEDS or VBIEDS. It would not go well. Bombs, assassinations, kidnappings, and general mayhem have been around for a long time, but with urbanization and advances in technology and explosives, dealing with them presents a challenge even for the most advanced and superior military. The MRAP and other solutions have proven effective but the search continues on how to deal with suicide bombers.

✕ Key Points

- Suicide bombing is an effective tactic that is difficult to detect and stop.
- Researchers are only now beginning to study suicide bombers and know little about them.
- Suicide bombers may not be highly motivated by religion.
- The U.S. military was unprepared to deal with improvised explosive devices and vehicle-born improvised explosive devices.

KEY TERMS

Databases

Ideology

Improvised explosive device(s) or IEDs

Mine Resistant Ambush Protected (MRAP)

Religion

Suicide bombing

Vehicle borne improvised explosive device(s) (VBIED)

DISCUSSION QUESTIONS

1. What are the databases used by researchers and government officials to examine terrorism?

2. Why do you believe individuals engage in suicide bombing? How can this be stopped?

3. What is the most common type of terrorist attack?

4. Do terrorists want large numbers of casualties or just publicity?

5. Is there a difference between male and female suicide bombers?

WEB RESOURCES

For a study by the military on female suicide bombers, go here: http://www.strategicstudiesinstitute.army.mil/pdffiles/pub408.pdf

To see what became of the response to IEDs and VBIEDs, go here: http://articles.latimes.com/2013/dec/27/world/la-fg-afghanistan-armor-20131227

The RAND Corporation has a wealth of information on suicide bombers, available here: http://www.rand.org/topics/suicide-attack.html

READING 6

Lindsey A. O'Rourke authored a very interesting journal article on female suicide terrorists, entitled "What's Special About Female Suicide Terrorism?" for the journal *Security Studies* in 2009. She also used it as a basis for an editorial for the *New York Times*. Both are so lengthy that they simply cannot be presented here but are well worth reading. Her abstract is presented below, along with a very thoughtful response from Michael Horowitz writing on the site of the Middle East Strategy at Harvard, National Security Studies Program: Weather head Center that includes a comment from Mia Bloom, a researcher on female suicide bombers.

What's Special About Female Suicide Terrorism?

Lindsey A. O'Rourke

This study analyzes the interaction between the motivations of individual attackers and terrorist group strategies. To do so, I combine a quantitative analysis of all known suicide terrorist attacks between 1981 and July 2008 with a strategic account of why terrorist organizations employ female suicide terrorism (FST) and case studies of individual female attackers. I advance five central claims. First, I reveal the superior effectiveness of FST from the perspective of the groups that employ women. Second, I explain that terrorist groups increasingly enlist women as suicide attackers because of their higher effectiveness. Third, I demonstrate that terrorist groups adapt their discourse, catering to the specific individual motives of potential female suicide attackers in order to recruit them. Fourth, I show that female attackers are driven by the same general motives and circumstances that drive men. Furthermore, and in contrast to the existing literature, women attackers uphold, rather than eschew, their societies' norms for gender behavior. Attempts to transform these societies into gender-neutral polities are therefore destined to increase FST. Finally, I conclude that, unless target states adapt their defensive strategies, we should expect an increase in FST.

Aug 11th, 2008 by MESH

From Michael Horowitz

Lindsey O'Rourke's recent op-ed in the *New York Times*, "Behind the Woman Behind the Bomb," is an interesting attempt to describe some of the issues surrounding the use of female suicide bombers in Iraq and elsewhere. As she points out, many of the groups that have utilized suicide terrorism have employed female suicide bombers. As such, her attempt to study the issue seriously is welcome and could significantly contribute to scholarship in this area.

Unfortunately, her piece contains a few misconceptions about suicide terrorism and the existing literature that deserve clarification. As someone also interested in questions surrounding suicide terrorism, I offer these comments in the spirit of helping build our knowledge in that area.

First, she states that "we are told" female suicide bombers are driven by "despair, mental illness, religiously mandated subordination to men, frustration with sexual inequality and a host of other factors related specifically to their gender." At least in the literature on suicide terrorism, this does not seem to be the case. Robert Pape's work on suicide terrorism, which she approvingly cites, does not come from this perspective. Neither does work by Mia Bloom, Bruce Hoffman, Assaf Moghadam, Ami Pedahzur,

Marc Sageman and others. So, while I agree with her argument that "feminine" motivations do not seem to be driving female suicide bombers and female suicide bombers have similar motivations to men, most other scholars of suicide terrorism agree as well.

Second, it is unclear whether her goal is to de-emphasize the "female" element of female suicide bombers or to argue they do deserve independent consideration. As many argue, she states that "there is simply no one demographic profile for female attackers," something true for male attackers as well. If there is no demographic profile and the motivations of female suicide bombers are similar to male suicide bombers, why do they deserve study as a separate category? Her answer is that female suicide bombers are used more frequently for a specific type of mission—assassinations—because they have an easier time getting close to hard targets due to cultural and societal norms about treating and handling women. This is a very interesting and an important finding, if true, for it points out a shortcoming in security screening procedures around the globe. However, that means we should not necessarily study female suicide bombers as an independent category, but as part of the larger category of suicide bombings designed to assassinate leaders.

Third, her focus on occupation as the cause of suicide terrorism is misplaced. Whether the feeling of occupation is accurate or not in the eyes of the West, perceptions of occupation likely play a powerful role in influencing the propensity for groups to engage in violent resistance. However, occupation is less likely to impact the choice of a particular tactic within the decision to engage in violent resistance. While Pape has shown that many of the groups that adopt suicide terrorism perceive themselves as occupied, many other groups that perceive themselves as occupied have not chosen to adopt suicide terrorism.

In fact, it makes more sense to think about suicide terrorism as a special case of a military innovation, one strongly influenced by diffusion dynamics. The extensive direct and indirect linkages between groups that have adopted suicide terrorism suggest that the probability of suicide terrorism is not an entirely independent choice, but one influenced by the knowledge and skills that groups gain from direct and vicarious learning. Moreover, we have to study both those groups and people that adopt suicide tactics and those that do not in order to gain the full picture. As Scott Ashworth et al. recently pointed out in the *American Political Science Review*, studying just the universe of suicide terror groups or

female suicide attackers selects on the dependent variable, making it hard to draw causal inferences from whatever correlations might exist. Things that are similar within the universe of suicide terror groups or the universe of female attackers might also be true of non-adopters as well, meaning those similarities do not actually predict behavior.

A more fruitful way to study the issue is to compare the groups that have adopted suicide terrorism and group members that have become suicide bombers with those that have not. Comparing adopters like Hamas, Al Qaeda, and the Tamil Tigers with non-adopters like the Provisional IRA and ETA, the Basque terrorist group, reveals the critical importance of organizational dynamics in driving adoption or non-adoption. Since, as O'Rourke points out, demographic profiling of potential suicide attackers does not seem promising, it makes more sense to evaluate group characteristics and focus on what makes adoption more or less likely.

Regardless of potential issues with her academic analysis, however, her policy prescription to improve screening of women at "key security checkpoints" is sensible. While I disagree that "occupation" is a primary cause of suicide attacks—as described above, it influences the probability that a group will adopt terrorism, not the choice of suicide tactics—hopefully ideas like the "Daughters of Iraq" can be more than a stopgap in the effort to decrease the number of suicide attacks against American and Iraqi forces, as well as ordinary Iraqis. I applaud O'Rourke's attention to this important topic, and hope to see more analysis of this kind in the future.

Comments are limited to MESH members and invitees.
Comment 1 on 11 Aug 2008 at 12:28 pm Mia Bloom

I have to agree with many of Michael Horowitz's assertions. I read the *New York Times* op-ed piece by Lindsey O'Rourke with great interest and felt that it emphasized foreign occupation at the expense of other competing and possibly complementary explanations. Asserting that there have been more secular attacks than religious attacks is only factually correct if we stop counting the events at the start of the Iraq war (as Robert Pape's *APSR* article did).

According to my own research (for a forthcoming book, *Bombshell: Women and Terror*), the best predictors of women's involvement in terrorist organizations continue to be association and especially relation to a male insurgent. Women's motivation to carry out a suicide attack increases exponentially if the male has been killed. This crosscuts radical Islamic extremist groups in addition to the secular organizations.

During my field work in Sri Lanka, most of the Liberation Tigers of Tamil Eelam (LTTE) women considered joining the organization as a family affair. Significantly, during my field research in Indonesia last year, I discovered that Jemaah Islamiyah (JI) uses its women to cement the linkages between different cells of the organization. This use of strategic marriage, akin to the European royal marriages of the 14th century that cemented ties between England and France or Spain and England, functions to keep the cell leaders within the group's orbit and control.

So in contrast to the ways in which we assume women and marriage moderate extremists (for example, when Yassir Arafat tried to marry off as many members of the Black September Organization after 1975 to de-mobilize them), women and marriage within the radical Islamic extremist organizations might have the reverse effect.

Lindsey O'Rourke is likely correct that the women may alternate their claims of motivation over time.

According to Yoram Schweitzer's interviews with failed female suicide bombers at Hasharon prison in Israel, their initial interviews reveal emotional reasons for their act. After spending time with the other prisoners in the jail (who are organized by political affiliation), they tend to parrot classic political propaganda. Schweitzer asserts that the women are in fact more motivated by the personal than the political. This might explain the motives for some of the women, but not someone like Ahlam Tammimi who was clearly motivated by political and not personal reasons.

The truth is likely a combination of motivations, which will include personal and political reasons—including occupation. If we consider suicide terrorism like any crime, we require both motive and opportunity to understand the event. Occupation may very well provide the opportunity (access to American or foreign troops) but the motive remains much more complex.

O'Rourke, like Pape, focuses on foreign occupation when, in fact, this is likely a necessary though insufficient condition. We might consider what it is specifically about occupation that causes intense levels of humiliation, outrage, and violent mobilization. But without the emotional content, and the religious justification, we would not see the literal explosion of suicide terrorism across the Islamic world.

Women, like men are motivated by a combination of reasons. The organizations now know that female operatives are more successful and less likely to be searched—and if women are invasively searched, this will only add to the population's anger and resentment. From the standpoint of the terrorist organizations, using women is a win-win strategy. Horowitz is correct in seeing suicide terrorism, especially by women, as a tactical adaptive innovation we will likely see much more of in the future.

Mia Bloom is assistant professor in the School of Public and International Affairs at the University of Georgia, Athens.

http://blogs.law.harvard.edu/mesh/2008/08/suicide_bombers_f/

REVIEW QUESTIONS

1. What does O'Rourke suggest motivates female suicide bombers? Do you agree or not and why?

2. What does O'Rourke believe we should do about the problem and what are her suggestions for doing this?

3. Summarize the criticisms of Horowitz with regard to the O'Rourke editorial.

4. What does he mean when he notes that "Scott Ashworth et al. recently pointed out in the *American Political Science Review*, studying just the universe of suicide terror groups or female suicide attackers selects on the dependent variable"?

❖

READING 7

The final reading for this section is one that directly explores the issue of Islam and suicide bombers. Matthias Küntzel maintains that true Islam does not condone this at all and provides a rationale for the rise of the Islamic suicide bomber.

Suicide Terrorism and Islam

Matthias Küntzel

 ## A Culture of Death

Suicidal terror is reaching every corner of the world and threatens the everyday life of European, American, Asian, and Middle Eastern citizens alike: Suicide killers hit cafés, hotels, restaurants, engagement parties, government institutions, the media, mosques, churches, means of transport, and funerals. Whereas war is supposed to be an extraordinary means that is limited in time and space, suicidal terror, in contrast, has no existence outside ordinary life. Its main aim is to make the exception of the "emergency rule" into the norm.

Suicide terror is frightening because its perpetrators behave like robots who seem to be devoid of that very instinct that normally unites all human beings: the survival instinct. Those who are ready to sacrifice their lives are able to perpetrate every conceivable crime. Moreover, it is frightening because every form of deterrence is inoperable, and the foundations of democracy—freedom and trust—are systematically undermined. Also, it is frightening because Al Qaeda's catch-phrase, "You love life; we love death," has already infected entire societies. The Hezbollah TV station recently broadcast a children's show in which a little girl thanks God for hearing her prayer and letting her father be killed in battle with the Israelis. The mother of the Palestinian who blew himself up in Eilat explained that she said goodbye to her son before he left and wished him success and that she was happy that God had heard her prayers.[1]

This culture of death, in which the child celebrates the loss of her father and the mother celebrates the loss of her son, is something beyond imagination: It is something George

SOURCE: *American Foreign Policy Interests*, 30: 227–232. Copyright © 2008 Routledge. Reprinted with permission.

Orwell was not able to write about. Nevertheless, 322 suicide bombings were carried out in Iraq from January to the end of August 2007, up from 179 in 2006. In Afghanistan, Taliban terrorists perpetrated 103 such attacks in the first 8 months of 2007, a 69 percent increase over the same period last year.

The type of world in which we are going to be living in the future depends on whether we defeat this surge of irrationality. In order to develop a successful strategy against suicide terrorism we need a correct understanding of its origins. Do hopelessness and desperation lie at the root of suicide murder, as is so often claimed? My answer is no.

Not Desperation but Joy

There are many people in the world who have every reason to feel desperate about their wretched lives. They do not, however, enter overcrowded buses or hijack planes with the sole purpose of blowing themselves up and killing as many innocent people as possible. That is definitely not a normal response to misery. Suicide attackers do not opt for paradise out of despair. If they did, their actions would be considered criminal and blasphemous. As Sheikh Qaradawi, one of the most prominent television personalities of the Muslim world and a member of the Muslim Brotherhood, reminds us: "These are heroic martyrdom operations, and the heroes who carry them out don't embark on this action out of hopelessness or despair."[2] Underscoring his judgment are the testamentary videos of the suicide bombers, which do not provide any evidence of desperation or hopelessness; on the contrary, they reveal an enormous amount of pride and even joy. Sheikh Qaradawi explains why: "He who carries out a martyrdom operation sells himself to Allah in order to buy Paradise in exchange."[3]

Not Islam Either

If the suicide bomber is not acting out of despair, then is Islam to blame? Once again my answer is no. Look at the example of Mali: This African Muslim country is one of the poorest nations in the world, but there have been no Malian suicide bombers. Look at Bosnia where the majority of the population is Muslim. Despite the experience of Srebrenica, Bosnian Muslims reject suicide terrorism for good reasons.[4]

Suicide bombing is contrary to Islam in three respects. First, since Abraham spared Isaac, it has been forbidden in all three religions for human beings to be sacrificed for any reason whatsoever. Yet those who advocate suicide terrorism are reducing human beings into tools of death. Second, no Jew, Christian, or Muslim is permitted to turn himself into a new god with the absolute control over life and death of civilians who just happen to be in a particular spot. Suicide killers, however, disregard the distinctions between a civilian and a soldier, between a minor and an adult, between a tank and an ambulance. Third, Islam strictly forbids suicide. Sura 2, verse 195 reads: "Cast not yourselves to destruction with your own hands." Sura 4, verses 29–30, still more explicit: "And do not kill yourselves. . . . Whoever does so in enmity and wrong, verily, we shall let him burn in Fire."

It is true that the Koran allows the killing of the unfaithful (4/89, 9/30, 47/4) and sometimes even demands it (4/74). Muslims can certainly find a religious justification for holy war in the Koran. It is true too that the Muslim doctrine of jihad advocates that a Muslim who finds himself in a hopeless situation in the struggle against the unbelievers should sacrifice his life as a *shahid* rather than surrender: "You must not think that those who were slain in the cause of Allah are dead," promises sura 3/169; "they are alive and well provided for by their Lord."

For a Muslim *deliberately* to be sent to *certain* death has been considered sacrilege within Islam. Even the founders of the Islamist movement—Hassan al-Banna, Abu Mawdudi, and Sayyid Qutb—never recommended that form of jihad. That is why in Soviet-occupied Afghanistan between 1979 and 1989 not a single suicide attack took place.[5] The systematic employment of Muslims as guided human bombs with the aim of killing as many people as possible was not seen in the first 1,360 years of Islam but was invented only 25 years ago.[6]

The Real Culprit

Suicide terror is a new development that is connected to the radicalization of a specific current within Islam—a current that we call Islamism. Here we come to the real nub of the matter. Suicide terror has little to do with Islam and still less to do with individual despair but a great deal to do with the ideology of Iran's Ayatollah Khomeini. Khomeini was the first to develop a full-blown death cult and a new interpretation of the aforementioned sura 3/169 of the Koran. According to his theological worldview, life is worthless and death is the beginning of genuine existence. "The natural world," Khomeini explained in October 1980, "is the lowest element, the scum of creation." What is decisive is the beyond: the "divine world—that is eternal."[7] According to Khomeini's mind-set, martyrs' deaths are nothing but the transition from this world to the world beyond where they will live on eternally and in splendor. Whether the warrior wins the battle or loses it and dies a martyr's death, in both cases, his victory is assured: either a mundane or a spiritual one.

Khomeini did not restrict himself to words: The first victims of his ideology of death were Iran's children, hundreds of thousands of whom were sent across minefields and so to a certain death between 1982 and 1988 in the war against Iraq—a crime that the world has yet to acknowledge. Those children formed part of the mass *Basij* movement that was called into being by Khomeini in 1979. They consisted of short-term volunteer militias and represented about 30 percent of the personnel on the battlefield. Most *Basij* members were between 12 and 18 years young. They went enthusiastically to their own destruction. Before every mission a small plastic key would be hung around each child's neck. It was supposed to open for all of them the gates to paradise. "The young men cleared the mines with their own bodies," a veteran of the Iran-Iraq War recalled: "It was sometimes like a race. Even without the commander's orders, everyone wanted to be the first."[8]

The human wave tactic was implemented in the following way: The barely armed children and teenagers had to move continuously forward in perfectly straight rows. It did not matter whether they fell as canon fodder to enemy fire or detonated the mines with their bodies. The important thing was that the *Basij* continued to move forward over the torn and mutilated remains of their fallen comrades, going to their death in wave after wave.[9] That tactic produced some undeniable initial successes for the Iranian side. "They come toward our positions in huge hordes with their fists swinging," an Iraqi officer complained in the summer of 1982. "You can shoot down the first wave and then the second. But at some point the corpses are piling up in front of you, and all you want to

do is scream and throw away your weapon. Those are human beings, after all."[10]

Nobody was more surprised by the effectiveness of his propaganda than Khomeini. "When Iranians go to war, they act as if they are going to a wedding," he exulted in September 1982. "Even in the earliest days of Islam, we didn't have that."[11] And indeed the history of Islam, although not lacking in atrocities, did not know acts like those of the *Basij*. Those children, honored as martyrs to this day by Ahmadinejad and the mullahs, were nevertheless the model for the first Islamically motivated suicide attacks against Israel.

Islamist Suicide Attacks Provoke Applause and Doubt

It is true that suicide attacks were launched against Israelis in the mid-1970s. But those were the work of Marxist-oriented groups like the Popular Front for the Liberation of Palestine (PFLP-GC). The first Islamist suicide murder took place in southern Lebanon on November 11, 1982. The perpetrator was 15-year-old Ahmad Qusayr, a follower of the then just emerging Shia militia, Hezbollah. He had been inspired by the model of the *Basij*. Khomeini personally consecrated the act of the 15 year old with a *fatwa*. Later he had a memorial built for Ahmad Qusayr in Tehran.[12]

Even in the jihadist camp, however, Khomeini's new instrument of jihad had to overcome considerable resistance to gain acceptance. The deviation from the Koran was too great and the break with tradition too sharp not to provoke a reaction. Even among Hezbollah's legal experts, suicide bombing was initially controversial. In 1993 the then spiritual leader of the group, Mohammad Husayn Fadlallah, expressed "reservations about resorting to suicidal tactics in political action" based on this reading of Islamic law. Is not the decision over life and death up to God alone? And was it not the case that inevitably an "innocent"—the bomber himself—would have to be killed?

More than 10 years passed before the Sunni Muslim Brothers in Palestine followed Ahmad Qusayr's example. Only in 1993 did Hamas's Al Qassam Brigades launch their first suicide missions. Two years passed before official approval was given. In 1995 Hamas founder, Sheikh Ahmad Yassin, declared "martyr operations" indispensable "because they confuse the Jews and fill them with fear and dread." But even now doubts about their religious legitimacy have not disappeared. In 2001 the mufti of Saudi Arabia issued a *fatwa* condemning suicide attacks as contrary to Islamic law. In April 2007 two senior Saudi religious scholars again came out against the use of religious edicts permitting suicide attacks in general and the use of explosive belts in particular.[13]

Khomeini's distortion of Islam has nevertheless become the calling card of today's Islamist movements throughout the world. In 2002 Iran's current Supreme Leader Ali Khamenei claimed: "A man, a youth, a boy, and a girl who are prepared to sacrifice their lives for the sake of the interests of a nation and their religion is the symbol of the greatest pride, courage, and bravery."[14] In one of his first television interviews, Iran's President Ahmadinejad enthused: "Is there an art that is more beautiful, more divine, more eternal than the art of the martyr's death?"[15] The "you love life; we love death" theme even appears in Ahmadinejad's letter to the president of the United States in May 2006, albeit in a somewhat watered down variant: "A bad ending belongs only to those who have chosen the life of this world. . . . A good land and eternal paradise belong to those servants who fear His majesty and do not follow their lascivious selves." In the same year, 2006, the Revolutionary Guard Corps announced that 40,000 Iranians were ready to carry out suicide missions against 29 identified Western targets.

bombings; 44 percent considered it justified in certain circumstances. In Indonesia, the largest Muslim country, 66 percent were totally opposed, and 33 percent approved, reflecting a decline in the level of support for suicide bombing over the past few years.[16]

There is a chance of ridding the world of the nightmare of suicide bombing within the next decades. For this to happen, however, three things have to be done. First, we must not ignore the religious aspect of suicide bombing. The roots of this kind of terrorism are in the preaching and sanctification that extol it.

There are many moral and political reasons to outlaw suicide terrorism. Accordingly, Muslim religious leaders are obliged to use theological arguments to discredit and condemn those who justify suicide bombing. "Muslims have to get to understand that a death cult has taken roots in the bosom of their religion, feeding off it like a cancerous tumor" wrote *New York Times* columnist Thomas Friedman. "If Muslim leaders don't remove this cancer—and only they can—it will spread, tainting innocent Muslims and poisoning their relations with each other and the world."[17] Second, we must open our eyes and vigorously support those forces within Islam that are rejecting and fighting suicide terrorism in all instances. Third, suicide terrorism presents the most imminent threat to the foundations of politics and law in the free world. It is high time for the West and Muslim countries as well to take the initiative and get the UN General Assembly to define suicide bombing as a crime against humanity and take resolute steps to punish its advocacy.

 ## The Struggle Between Islamists and Moderates

Suicide terrorism lies at the heart of a bitter struggle between two lines within Islam. The Islamists—Iran, Hezbollah, Al Qaeda, and Hamas—reject free speech and the modern world and support suicide bombing in pursuit of their aim of subjecting the whole globe to the will of Allah. The moderates reject suicide bombing and support modernity and reconciliation among religions.

In Pakistan, for example, 46 percent of the population in 2005 expressed their total rejection of suicide

 ## About the Author

Matthias Küntzel, a political scientist in Hamburg, Germany, is a research associate at the Vidal Sassoon International Center for the Study of Anti-Semitism at the Hebrew University of Jerusalem as well as a member of the Board of Directors of Scholars for Peace in the Middle East. He is the author of *Jihad and Jew-Hatred: Islamism, Nazism and the Roots of 9/11* (Telos Press, 2007). It was awarded the London Book Festival's grand prize in December 2007. His new book about the relationship between the Islamic Republic of Iran and (West) Germany will be published in 2009. Küntzel's essays about Islamism and anti-Semitism have been translated into 10 languages and published inter alia in *Policy Review, The New Republic, Telos,* and *The Wall Street Journal.*

 Notes

1. See columnist Hassan Haydar: "Iran Spreads a Culture of Death," in Al-Hayat (English edition), February 1, 2007, in Memri, Special Dispatch Series no. 1455, February 8, 2007.

2. Antidefamation League, "Sheik Yusuf al-Qaradawi: Theologican of Terror," August 1, 2005; see: www.adl.org.

3. MEMRI Special Dispatch Series, no. 542, July 24, 2003.

4. Scott Atran, "Who Wants to Be a Martyr," in *The New York Times,* May 5, 2003.

5. Waliullah Rahmani, "Combating the Ideology of Suicide Terrorism in Afghanistan," *Terrorism Monitor,* vol. IV, no. 21 (November 2, 2006). According to Rahmani, the first suicide attack in Afghanistan was in 1992 when an Egyptian fighter for Gulbuddin Hekmatyar in Kunar killed Maulvi Jamil Rahman, a Salafi leader who was against Hekmatyar.

6. The murders perpetrated by the Shia sect of the Assassins in the eleventh century were exclusively directed at individuals of the ruling elite.

7. Cited in Daniel Brumberg, "Khomeini's Legacy: Islamic Rule and Islamic Social Justice," in R. Scott Apple by, ed. *Spokesmen for the Despised. Fundamental Leaders of the Middle East* (Chicago and London, 1997), 56.

8. Cited in Christiane Hoffmann, "Vom elften Jahrhundert zum 11. September. Märtyrertum und Opferkultur sollen Iran als Staat festigen," *Frankfurter Allgemeine Zeitung,* May 4, 2002.

9. See the Basij report in Freidune Sehebjam, "Ich habe keine Tränen mehr," Iran: Die Geschichte des Kindersoldaten Reza Behrouzi, Reinbek: Rowohlt, 1988.

10. Cited in Erich Wiedemann, "Mit dem Paradies-Schlüssel in die Schlacht," *Der Spiegel,* no. 31 (1982): 93.

11. Cited in Dawud Gholamasad and Arian Sepideh, *Iran: Von der Kriegsbegeisterung zur Kriegsmüdigkeit* (Hannover: Internationalisms Verlag, 1988), 15.

12. Joseph Croitoru, *Der Märtyrer als Waffe. Die historischen Wurzeln des Selbstmord-attentats* (München: Hanser), 132.

13. Asaf Maliach, "Saudi Religious Scholars Come Out against Al-Qaeda's Use of Religious Edicts Permitting Suicide Attacks against Muslims," International Institute for Counter-terrorism, July 8, 2007.

14. Cited in: Ali Alfoneh, "Iran's Suicide Brigades Terrorism Resurgent," *Middle East Quarterly* (winter 2007).

15. *MEMRI, Special Dispatch* no. 945, July 29, 2005.

16. The Pew Global Attitudes Project, Support for Terror Wanes Among Muslim Publics. A 17-Nation Pew Global Attitudes Survey, July 14, 2005.

17. Thomas Friedman, "At a Theater Near You, . . ." *The New York Times,* July 3, 2007.

REVIEW QUESTIONS

1. Do you believe that this article is persuasive?

2. Can the United Nations make suicide terrorism and its adoption and use a crime against humanity when the UN cannot even agree on a definition of terrorism?

3. What do you think the current support for suicide bombings happens to be in various countries? For the answer, go here: pew research.org and search under global attitudes/foreign affairs.

PART

II

Terrorism Around the World

T he world is getting smaller. Anyone who has traveled from one country to another, played an online video game with someone located on a different continent, shared a college dorm room with an international student, stayed glued to the games of the XXXth Olympiad in London in 2012, or spent any time at all on the Internet or watching the news instinctively accepts this as true. Indeed, modern terrorism—the subject of this text/reader—thrives on this late 20th–early 21st-century phenomenon. Terrorism would not have nearly the same impact in an environment of international isolation. As the world has shrunk—metaphorically and culturally, if not physically—it has also come face to face with its own vast diversity of the human experience, sometimes with wonderfully positive results and sometimes with tragic and horrifying outcomes.

This human diversity leads to differences in things as simple as how we refer to the locations of this planet's different continents and regions. It is not as simple as our fifth grade mapping exercises assumed. For example, as discussed in the next section, the term *Middle East* is one developed from a Western perspective. If you were standing in Indonesia, referring to the "Middle East" is something of a misnomer, since from that position, the Middle East is actually to the west. Regional, continental, and national names and terms of references also often differ between the various ethnic, religious, and cultural groups within a boundary. In addition, continental, regional, and national terms of reference can reflect recent or even ancient changes in political structures. For instance, some of the former colonial territories changed their names upon independence to reflect the fact that they were no longer subservient to their old masters; some of the old names would be found offensive by the citizens of the new "state" even 50 to 100 years later.

Still, uniform terms of reference are necessary if for no other reason than, without them, a book of this nature would be unmanageable and probably unreadable. In addition, agreed-upon terms of reference provide a way of organizing the world. The world may be shrinking, but it still consists of six continents (not counting Antarctica) and somewhere between 189 and 196 independent countries (depending on the source being cited) (www.world atlas.com). They cannot all fit into one chapter.

Therefore, for simplicity and ease of reference, the authors have chosen to organize the regional part of this text into six different sections representing six different world regions. With one exception, these sections mirror the bureau divisions of the U.S. Department of State—perhaps a bit of a conceit given that both authors hail from California, U.S.A.—but a useful and easy-to-follow framework nonetheless. These divisions are also mirrored in a number of other U.S. government agencies, including the Central Intelligence Agency, for whom one of the authors once worked.

Because of the complexity of the issues involved in the conflict between Israel and the Palestinian territories, and the number of terrorist groups generated by or closely related to that conflict, Section V—Israel, Lebanon, and Palestine—deals specifically with Israel, Lebanon, Jordan, and Palestine. With those exceptions in mind, Section VI—The Middle East and Africa—encompasses the countries found in the Department's Bureau of Near Eastern Affairs (covering the Middle East and North Africa) and the Bureau of African Affairs. Section VII—Southeast Asia, South and North Korea, and China—mirrors those nations included in the Bureau of East Asian and Pacific Affairs. Section VIII—Southwest Asia—includes Afghanistan, Pakistan, India, and Sri Lanka, countries that can be found in the Bureau of South and Central Asian Affairs. Section IX—Europe, Turkey, and Russia—consists of the countries covered by the somewhat dated title of the Bureau of European and Eurasian Affairs, including a history of the Irish Republican Army and an overview of the conflicts in Chechnya. Finally, Section X covers the Americas—the continents of North and South America as well as the countries of the Caribbean found in the Bureau of Western Hemisphere Affairs. Issues surrounding domestic terrorism in the United States are also included in Section 10.

 ## The Problem of Terrorism Versus Insurgency: Which Terrorist Groups Do We Talk About?

In his introduction to *Patterns of Global Terrorism, 2003*, Ambassador Cofer Black, then coordinator for counterterrorism, noted that terrorist acts, which the U.S. State Department defines as "premeditated, politically motivated violence perpetrated against noncombatant targets by subnational groups or clandestine agents, usually intended to influence an audience," are part of the larger phenomenon of political violence and that drawing the important distinction between terrorism and other types of political violence can be difficult. This difficulty is exacerbated by the fact that the two strategies often overlap in the same group (U.S. Department of State, Patterns of Global Terrorism, 2003). Drawing this distinction has also proven to be a challenge for the authors, particularly in regions such as Central Africa where insurgencies and civil wars abound, sometimes crossing or ignoring "state" borders that have lasted less than half a century and made little if any sense when they were created.

Noted terrorism scholar Bruce Hoffman provides one of the best and most useful attempts at setting out the distinction between terrorism and, in particular, insurgency or guerrilla warfare—a distinction the authors have tried to draw on throughout this book. In identifying tactics particular to each strategy, Hoffman (2006) notes that insurgent groups engage in irregular military tactics found historically in guerrilla warfare such as: 1) numerically larger groups of armed individuals who operate as a military unit; 2) attacking enemy and military forces; 3) seizing and holding territory; and 4) exercising some form of control over a defined geographical area and its population" (p. 35). In contrast, terrorists "do not function in the open as armed units, generally do not attempt to seize and hold territory, deliberately avoid engaging enemy military forces in combat, are constrained both numerically and logistically from undertaking concerted mass political mobilization efforts, and exercise no direct control or governance over a populace at either the local or the national level" (Hoffman, 2006, p. 35). In addition, Hoffman has listed a number of characteristics that tend to be common to terrorist groups in particular. He argues that terrorism is:

- ineluctably political in aims and motives;
- violent—or, equally important, threatens violence;
- designed to have far-reaching psychological repercussions beyond the immediate victim or target;
- conducted *either* by an organization with an identifiable chain of command or conspiratorial cell structure (whose members wear no uniform or identifiable insignia) *or* by individuals or a small collection of individuals directly influenced, motivated, or inspired by the ideological aims or examples of some existent terrorist movement and/or its leaders; and
- perpetrated by a subnational group or non-state entity. (2006, p. 40)

However, in the same work, Hoffman also points out that these are not pure categories since groups that meet aspects of the criteria of both terrorists and insurgents abound, citing Hezbollah, the Liberation Tigers of Tamil Elam (LTTE), and the Revolutionary Armed Forces of Colombia (FARC) as examples.

There are hundreds of insurgencies and civil wars (of varying size and intensity) going on in the world at any given time. Many of them use tactics associated with terrorism such as assassinations, kidnappings and hostage-takings, and bombings of public places. There is no manageable way to discuss all of them in one publication. As a result, different authors will select different groups for their collection and the authors of this work are no different. We have tried our best to rely on uniform criteria for selection throughout this work. We have chosen to focus our efforts on non-state/subnational groups where a significant part of their attacks involve noncombatant targets, especially those groups emphasizing attacks against Western and U.S. civilians, or are likely to do so, or, have conducted, or are likely to conduct, attacks outside their country of origin (what many terrorism scholars identify in their typologies as international terrorism). We have also focused our efforts on groups operating within a still functional state as opposed to areas where a failed state or functioning government has devolved into conflicts between dueling insurgencies under relatively equal terms. Still, ultimately, the terrorist groups discussed in this work represent the groups the two authors simply felt should be highlighted. Perhaps there is a little channeling of Justice Potter Stewart at work here. Perhaps, that is inevitable.

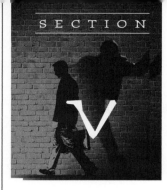

V

Israel, Lebanon, and Palestine

Learning Objectives

At the end of this section, students will be able to:

- Detail the major historical periods of Israel and Palestine, including periods when Israelis or Palestinians were marginalized, brutalized, or ruled fairly, before the 20th century.
- Describe the major events of the last century (and decade) of Israeli and Palestinian history, to include Israeli and Palestinian terror groups and attacks.
- Discuss the results of Israel becoming a nation.
- List and explain the wars, the opponents, and the outcomes that resulted.
- Explain the Arab and Palestinian responses to Israeli victories.
- Assess the impact of the First and Second Intifadas.
- Detail and evaluate the various attempts at peace made by Palestinians and Israel in the last six decades.
- Describe the major political organization that drew Palestinians toward a national identity.
- Describe the major terrorist organizations that exist in Palestine.
- Describe how the country of Lebanon has impacted the region and been impacted as well.
- Explain Syria's continued meddling in Lebanon.
- Describe the causes of the Lebanese Civil War and the solution that ended it.
- Describe the current threat to Israel from Iran.

Introduction

"No two historians ever agree on what happened, and the damn thing is they both think they're telling the truth."

—Harry S. Truman

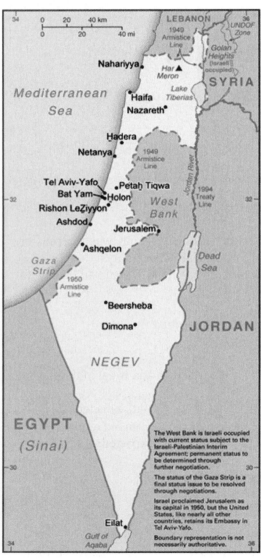

CIA World Fact Book

▲   Map of Israel

This section introduces readers to the lands and peoples of Israel and Palestine and makes a brief detour into Lebanese history. It clearly remains an historic place with rich and diverse stories that have many highs and lows. The authors will attempt to present objective information but realize that years, decades, and even centuries of bitter propaganda by various enemies of the Israelis and Palestinians have clouded the issues. This introduction will cover geography, the people, their economies, and governance and then a history of the conflicts between the two groups. The remainder of the section will present a more current review of Israeli and Palestinian conflict, with one view from an Israeli perspective and another from a Palestinian.

Israel and Palestine trace their history back thousands of years. Location is critical. Israel with a western border on the Mediterranean has a tough neighbourhood with a border on the north with Lebanon, on the east with Jordan and Syria, and Egypt lies to the south (*CIA World Factbook*, n.d.). It is not a large country, somewhat larger than the state of New Jersey, with some 20,330 square kilometers or just under 8,000 square miles, and it is stunning to actually go there and see how small it is, with a length of just 470 kilometers or 290 miles and a width of 135 kilometers or 85 miles as it widest point (Israeli Ministry of Foreign Affairs, n.d.). One of the authors spent some time there and stood on an Israeli military elevated lookout post on the West Bank, and it was very easy to see Jordan as well as Israeli cities. There is limited arable land, and water is a constant concern, with only about 15% of the land suitable for agriculture and some 2,000 square kilometers or just 868 square miles irrigated (*CIA World Factbook*, n.d.). Even providing details on the size of the country presents difficulties. The CIA acknowledges the West Bank and Gaza were occupied by Israel but treats the West Bank and Gaza as separate entities while simply stating the Golan Heights is occupied (*CIA World Factbook*, n.d.). It gets even more confusing, since Gaza was unilaterally abandoned by Israel, with all military and settlers leaving in 2005, but Israel still controls the land borders, water, and airspace. Why is this the case? The victory in the 1967 Six Day War provided significant territorial gains by Israel and gave it defensible borders, but these were not recognized as international borders until subsequent peace treaties with Egypt and later with Jordan; however, there has been no recognition of international borders or peace treaties with Iraq and Lebanon, and this is still a bone of contention for the Palestinians (Krämer, 2002).

The Palestinians have as their territory Gaza, north of Egypt, bordered by the Mediterranean Sea and Israel, and the West Bank or what is also known as the areas of Samaria and Judea. While Israel is not large, the areas of Gaza and the West Bank are even smaller, with Gaza a mere 360 square kilometers or 139 square miles (*CIA World Factbook*, n.d.). The West Bank area is larger, about 5,860 square kilometers or 2,260 square miles (*CIA World Factbook*, n.d.). For perspective, the county of Los Angeles in California is more than 4,000 square miles (U.S. Census Bureau, 2015).

Three fourths of Israel's approximately 7.5 million inhabitants are Jewish, with another 20% Arab, and 85% of these are Muslims while the remainder are Arab-Christians, Druze, and others (Israeli Ministry of Foreign Affairs, 2010). In Gaza, there is a population of 1,710,257 with the overwhelming majority Palestinian Arabs, 99% of them Sunni Muslims (*CIA World Factbook*, n.d.). The population of the West Bank is 2,622,544 with about 83% Palestinian Arabs (mostly Sunni Muslims) and about 17% Jewish settlers, and there are a small number of Christians as well (*CIA World Factbook*, n.d.).

▲ Map of Gaza

Briefly addressing the respective governments and economies demonstrates a stark contrast. Israel has a stable government with, interestingly, no constitution though they are working on that, a parliamentary system of government, and a thriving economy described by the CIA as a "technologically advanced market economy" (*CIA World Factbook*, n.d.); they have a biennial budget process; the budget for 2012 was $72.5 billion (France 24, 2010). The **Palestinian Authority**'s budget for 2014, the last year for which information is available (this includes Gaza), was $4.2 billion, much of that from donor nations (portlandtrust.org, 2014). To fully understand the disparity between Israel and the West Bank and Gaza, read the following explanation from the CIA on the fractious nature of relations between the **Fatah** Party and The Islamic Resistance Movement, **Hamas**, the parties that govern the West Bank and Gaza:

Israel unilaterally withdrew all of its settlers and soldiers and dismantled its military facilities in the Gaza Strip and withdrew settlers and redeployed soldiers from four small northern West Bank settlements. Nonetheless, Israel still controls maritime, airspace, and other access to the Gaza Strip. In January 2006, the Islamic Resistance Movement, HAMAS, won control of the Palestinian Legislative Council (PLC). HAMAS took control of the PA government in March 2006, but President ABBAS had little success negotiating with HAMAS to present a political platform acceptable to the international community so as to lift economic sanctions on Palestinians. Violent clashes between Fatah and HAMAS supporters in the Gaza Strip in 2006 and early 2007 resulted in numerous Palestinian deaths and injuries. In February 2007, ABBAS and HAMAS Political Bureau Chief

▲   Map of the West Bank

Khalid MISHAL signed the Mecca Agreement in Saudi Arabia that resulted in the formation of a Palestinian National Unity Government (NUG) headed by HAMAS member Ismail HANIYA. However, fighting continued in the Gaza Strip, and in June 2007, HAMAS militants succeeded in a violent takeover of all military and governmental institutions in the Gaza Strip. ABBAS that same month dismissed the NUG and through a series of presidential decrees formed a PA government in the West Bank led by independent Salam FAYYAD. Fatah and HAMAS in May 2011, under the auspices of Egyptian-sponsored reconciliation negotiations, agreed to reunify the Palestinian territories, but the factions have struggled to finalize details on governing and security structures. The status quo remains with HAMAS in control of the Gaza Strip and ABBAS and the Fatah-dominated PA governing the West Bank. FAYYAD and his PA government continue to implement a series of security and economic reforms to improve conditions in the West Bank. ABBAS, who on behalf of the Palestinians in September submitted a UN membership application, has said he will not resume negotiations with current Israeli Prime Minister Binyamin NETANYAHU until Israel halts all settlement activity in the West Bank and East Jerusalem. (*CIA World Factbook*, n.d.)

And, yes, Hamas, the elected government, remains a Designated Foreign Terrorist Organization by the United States (Laub, 2014).

⊠ The Israeli Narrative

Following the Six Day War the Israeli Defense Forces commissioned a song, "The Song of Peace." It ends thus:

Do not whisper a prayer

Better sing a song for peace

With a great shout.

The song was controversial among some military officers who took offense at the pacifist tone, but Yitzhak Rabin supported his education chief, Mordechai Bar-On, who took the view, "Israel will never reach peace unless it has a strong army, but the army will not be strong unless its combatants are convinced that the ultimate goal of all their endeavors is to reach peace" (Gilbert, 1998, p. 400). That tells you a great deal about the Israelis. They want peace and will do what they must to achieve it and survive as a nation.

Biblical accounts from the book of Genesis present Abraham as the Father of the Jews through his son Isaac, and Genesis 17:7 and 17:8 make it clear that they will reside in Canaan, where Abraham is during this exchange:

> 17:7 I will confirm my covenant as a perpetual covenant between me and you. It will extend to your descendants after you throughout their generations. I will be your God and the God of your descendants after you.

> 17:8 I will give the whole land of Canaan— the land where you are now residing—to you and your descendants after you as a permanent possession. I will be their God. (Net Bible, n.d.)

▲   Photo of an Israeli tank during Operation Cast Lead, the first of several incursions by the Israelis into Gaza after they gave the land back to the Palestinians. The Palestinians under Hamas took the opportunity to fire thousands of rockets and mortars into Israel, forcing Israel to go back in and respond.

In the Book of Exodus from the Bible, God instructs Moses, who led an exodus from Egypt to possess Canaan (what is now approximately Palestine), a land of well-developed city-states, ruled by various kings:

> 23:31 I will set your boundaries from the Red Sea to the sea of the Philistines, and from the desert to the River, for I will deliver the inhabitants of the land into your hand, and you will drive them out before you.

> 23:32 You must make no covenant with them or with their gods.

> 23:33 They must not live in your land, lest they make you sin against me, for if you serve their gods, it will surely be a snare to you. (Net Bible, n.d.)

Krämer (2002) reminds us that even though no one is completely certain of the actual boundaries the Israelis were supposed to possess, "the land actually settled by the Israelites formed only part of the land "promised" to them under the covenant (p. 9). The Israelis balk and spend another 40 years wandering in the desert, Moses dies, and Joshua eventually begins capturing Canaan, but their conquest is not complete and sets the stage for centuries of strife.

As one reads a history of Israel, over the centuries there have been times when the Jewish population in Palestine, which was the term the Romans used for the area, was not large and consisted of mere remnants (Krämer, 2002). Sachar (2010) details the sad history of the Jews from their slaughter at the hands of the Romans, to life under the Arab conquest, subjugation by the Turks, massacres by the Crusaders, better treatment under Salah-ed-Din (Saladin) after he defeated the Crusaders, and finally they end up in an often indifferent Ottoman Empire, many fleeing from the Spanish Inquisition and its Expulsion decree. However, by the mid-18th century, things deteriorated under Ottoman rule for the Jews in Palestine, as Sachar (2010) notes: "No more than 6,000 Jews altogether lingered in the four holy cities. Taxed and tyrannized as they were to within an inch of their lives, they regarded their ordeal essentially as a testimony of repentance" (p. 267).

Despite these many difficulties over the centuries, the Jews maintain that this is the land God gave them and they have a historical connection to it, but they were barely hanging on. Fortunately, things began to change, though slowly. By the 1870s, a movement began in Russia, where Jews were often subjected to persecution. This was Zionism, and the term *Zionism* was apparently coined in 1891 by Austrian, writer Nathan Birnbaum to describe the new ideology

(Sachar, 2010). This Zionist ideology stresses that the Jews are a people or nation like any other, and should gather together in a single homeland and that homeland should be Palestine. While the Zionist effort slowly begins to gain traction, aid comes from an unexpected front thanks to the revival of British Protestant evangelism. Influential individuals such as Anthony Ashley-Cooper, seventh earl of Shaftesbury, and Laurence Oliphant, a diplomat, who began campaigning enthusiastically for the British people to aid Jews in returning to the Holy Land (Sachar, 2010). In fact, it was Shaftesbury who coined the well-known phrase that would later cause problems for the Zionists, Anti-Zionists, and Palestinian Arabs, "a land without people for a people without land," not meaning that there were no people living in Palestine, because there were, but that they constituted a group with no history, culture, or territorial nationalism, which was a European notion; in short, there were no people identified as Palestinians (Garfinklle, 1991).

Sparking more support and interest in Zionism was an event that is now called the "Dreyfus Affair," a shocking display of anti-Semitism. In 1894, Captain Alfred Dreyfus, a promising French Army officer (Jewish and from Alsace, a German area) assigned to military intelligence, was accused of espionage and court-martialed, found guilty, and sentenced to public degradation, exile, and imprisonment for life on Devil's Island, French Guiana (Gueiu, 2000). Fortunately, both Captain Dreyfus's wife and his brother worked tirelessly to clear him, and they mobilized the services of Jewish writers Theodor Herzl and Emile Zola. Their combined efforts eventually resulted in finding the real culprit and reinstatement in the Army for Dreyfus, where he went on to serve honorably (Gueiu, 2000). Herzl was so incensed by the event that he produced a book, based on an earlier work, called *The Jewish State* that claimed the Jews had tried to live successfully in Europe only to be scorned. He called for the bringing of dispersed Jews from all over the world together, preferably in Palestine (Sachar, 2010). Herzl died in 1904 but he had smoothed the path and laid the foundations for those who succeeded him, including establishing a relationship with Arthur James Balfour, who as Lord Balfour, the British foreign minister, penned a letter to Lord Rothschild, president of the British Zionist Federation. The letter states in part, "His Majesty's Government view with favour the establishment in Palestine of a national home for the Jewish People" (Laquer, 2008, p. 16). Ultimately, due to the fact the British were on the winning side in the First World War and would control Palestine under a mandate, this would facilitate a return of the Jews to Israel.

Efforts to settle in the area had already begun in earnest at the end of the 18th century and by 1903, some 25,000 Jews managed to get to the area, some fleeing persecution in Russia and some simply wishing to study and live there (Sachar, 2010). This first immigration wave or First Aliyah (from the Hebrew word for ascent) was followed with a Second Aliyah between 1905 and 1914, with many more to follow with the Jewish population of Palestine now over 375,000 and cities becoming modernized (Sachar, 2010). Unfortunately for the Jews, the British were finding that governing was proving to be a difficult and thankless task, and the Palestine Arab population was frequently turning to violence. This came to a head after the riots of May 1921 and of August 1929 and culminated in Arab uprisings intermittently from 1936 to 1939, caused in part by increasing Jewish immigration and land purchases by well-capitalized Jews and suspicions that the Jews were increasingly arming themselves for protection, which they were (Krämer, 2002).

This state of affairs prompted the British to study the issues, and "The Palestine Royal Commission: Peel Commission Report" of July 1937 concluded that underlying causes of the unrest were the desire of the Arabs for national independence and a fear that the Jews were planning to make Palestine their Jewish National Home; the Commission recommended a two-state solution (Laquer, 2008). The British government rejected the recommendation as being too impractical

▲ Israeli security fence in the West Bank. These barriers, while deemed illegal, were determined to be necessary after an assault of suicide bombings.

and this set the stage for a miserable period in Palestine during World War II for the British, the Jews, and the Palestinians (Laquer, 2008). War found the British concerned that governing Palestine would become even more difficult and this proved to be the case, with the (illegal) security forces, the Hagana, now demanding a Jewish Brigade to aid the British. This was initially rejected (though one was finally allowed), so the Hagana encouraged a general mobilization and asked for men and women to enlist as volunteers in the British Army, with tens of thousands doing so (Krämer, 2002). Of course, the Hagana also cooperated with the British, volunteering to share intelligence with the British, all the while building and strengthening a secret army with its own illegal radio station, news service, and newspaper.

With victory in sight, the British, while happy with the support received by the Jews during the war, were now signaling they intended to focus on improving the

▲   Wailing at the Other Wall. A sad Arab Muslim woman is depicted passing near the Israeli separation barrier.

living standard of Arabs in Palestine. Jews were furious; the Arabs had done virtually nothing to contribute to the war effort, and after several leaders called the shift an outrage, members of Jewish underground organizations acted by stealing arms and ammunition from British military bases (Sachar, 2010). Following this, two serious terror groups formed, among them the "Fighters for the Freedom of Israel" or Lech'I (for its Hebrew initials), founded by Avraham Stern, with around 800 members (Sachar, 2010). Stern was shot dead by the British police in 1942 after a bombing by his group, but it continued and the group managed the murder of Lord Moyne, the British minister resident in the Middle East, in Cairo (Krämer, 2002). The Hagana was forced to act against the underground organizations, but they never went away and came back with a vengeance following the end of WWII. The immediate act that united Hagana, the Irgun (another group led by Menachem Begin, a future prime minister of Israel with around 2,000 members) to violence was Britain's refusal to accede to a request from the American president, Harry Truman, to accept immediate admission of 100,000 Jewish refugees (Krämer, 2002).

British civilian and military installations and personnel were under attack. On July 22, 1946, the Irgun bombed the south wing of the King David Hotel in Jerusalem, the headquarters of the British general staff, killing 91 people ("The Bombing of the King David Hotel," n.d.). Between September 1946 and May 1948, the Lech'I and the Irgun carried out hundreds of attacks while the British intensified efforts to stop them (Sachar, 2010). It was clear that Britain could not maintain a presence and on February 25, 1947, the British foreign minister told the world of the British decision to leave and hand the problem over to the United Nations (Krämer, 2002). At the end of 1946, the population of the Palestinian territories was estimated at 1.94 million, including 1.33 million Arabs, mostly Muslims, 603,000 Jews, and 16,000 "other" (Gideon, 2004). Governing the territory had proven to be expensive and dangerous, with a commitment of 100,000 soldiers trying to keep a fraying peace between Jews and Arabs (aish.com, n.d.). Deciding what to do with an angry Arab population and an increasingly armed and militant Jewish population was now up to a newly formed and untested organization, the United Nations. Despite being new, the United Nations General Assembly acted quickly and on November 29, 1947, voted (the shamed British abstained) to partition Palestine into two separate states, one Israeli and one Arab, with Jerusalem becoming an internationalized city (United Nations General Assembly Resolution 181, 1947).

With the United Nations' decision in 1947 and the subsequent declaration of independence, Israel was soon to be a nation, though immediately and often at war. The United Nations' decision initially resulted in an ugly period that eventually turned into a civil war, with the British ignoring brutality by both Palestinians and Israelis. A significant war (what the Israelis call their War of Independence and the Palestinians call "**Nabka**," meaning "catastrophe" or "disaster" in Arabic) followed Israel's Declaration of Independence on May 15, 1948 ("Modern History," n.d.). Facing the combined militaries of Egypt, Syria, Iraq, Saudi Arabia, Lebanon, and Jordan, Israel triumphed and added nearly 50% more land to its pre-war geography; Gaza fell under the jurisdiction of Egypt, and Jordan took possession of the West Bank of Jordan ("Israel War of Independence," n.d.). Thus, the military line, green line, or armistice line is simply where the fighting ended in 1949. It was the lack of defensible borders that really tempted the Arab nations to try another invasion in 1967:

> Israel's borders at the time were demarcated by the armistice lines established at the end of Israel's war of independence 18 years earlier. These lines left Israel a mere 9 miles wide at its most populous area. Israelis faced mountains to the east and the sea to their backs and, in West Jerusalem, were virtually surrounded by hostile forces. In 1948, Arab troops nearly cut the country in half at its narrow waist and laid siege to Jerusalem, depriving 100,000 Jews of food and water. . . .
>
> Forty-four years after Arab forces sought to exploit the vulnerable armistice lines, it remains clear that Israel cannot return to those lines. And 44 years after the United Nations, through Resolution 242, indicated that Israel would not have to forfeit all of the captured territories and must achieve "secure and recognized boundaries," the unsecure and unrecognized armistice lines must not be revived. Israel's insistence on defensible borders is a prerequisite for peace and a safeguard against a return to the Arab illusions and Israeli fears of June 1967. (Oren M., 2011)

To review, there were no pre-1967 borders, only a military line that left Israel in a precarious position militarily, which they corrected by winning the 1967 war. Following the 1948 war, no Arab states recognized Israel and two populations of refugees resulted from the conflict. Jews, as many as 800,000, were expelled from Arab states and quickly absorbed into the Israeli population while 750,000 (a disputed number) Palestinian refugees, some leaving voluntarily, some not, became pawns of the Arab states forever as well as victims (Herzog, 2004).

This section would be incomplete without a recap of some significant events that shaped and influenced Israel and Palestine. The early 1970s saw the tactic of hijacking airlines growing, as the Palestinian narrative will detail. But Israel did do something about it and one of the stellar examples would actually involve two future prime ministers. On May 8, 1972, Black September (later infamous for the Munich Olympic Massacre) hijacked a Belgium Sabena airliner and demanded the release of some 300 Palestinian prisoners held by Israel, or they would blow it up with all passengers. When the plane landed in Israel, commandos disguised as maintenance technicians took control of the plane, killed two of the hijackers, and captured two more in less than ten minutes. This was called Operation Isotope and was led by Ehud Barak, a future prime minister. One of his section leaders was slightly wounded; he is the current prime minister of Israel, Benjamin Netanyahu (RealClearWorld.com, n.d.).

Gaza Strip Update

In the summer of 2014, Israel Gaza Strip Update once again responded to kidnappings and the killing of individuals as well as rocket and mortar fire from Gaza. The resulting war greatly diminished the strength of Hamas, the power in Gaza. Following the summer of war, Hamas and the Palestinian Authority agreed to a new government in the Gaza Strip that will include both Hamas and Fatah members. The Palestinian Authority will begin to rebuild Gaza and pay Hamas officials who have not been receiving a salary. The war resulted in the deaths of both Israelis and Palestinians, but the results were lopsided with over 2,100 Palestinians and 72 Israelis killed.

SOURCE: http:ibttimes.com/fatah-weakened-hamas-agree-give-gaza-strip-palestinian-authority-1695297

Israel and several Arab nations, perhaps as many as nine Arab nations but primarily Syria and Egypt, would fight another war, called the Yom Kippur War since it began on the Israelis' holiest of days in 1973. While the Israelis did not win a decisive victory, they did not lose, and the reasons are similar to earlier wars: the Arabs did not fight with a unified command structure and despite massive support in military supplies from the Soviet Union, the aid to Israel, especially in intelligence provided by the United States, made the difference (HistoryLearningSite.co.Uk, n.d.).

Perhaps it seems strange to mention that the next significant historical event was a peace treaty signed, not between Israel and Palestine but between Israel and Egypt, the bitter enemies who had now fought three wars against each other. Under a great deal of pressure by then President Jimmy Carter, President Anwar Sadat of Egypt and Prime Minister Menachem Begin of Israel signed the treaty at a White House ceremony in Washington, D.C., on March 26, 1979; this shocked a number of conservative Israelis and Egyptians (Sachar, 2010). This would bring peace between the two countries for many years but would have significant ramifications. Following that event, Israel, convinced, with assistance from American intelligence, that Iraq was working on nuclear weapons, launched an attack on the Osiraq nuclear reactor in Tuwaituh, Iraq, destroying it with all Israeli planes returning safely and Israelis relieved and jubilant (Gilbert, 1998; Sacher, 2010). President Sadat continued dialog with the Israelis and they did make some progress on a number of fronts, but his last meeting occurred just three days before the Osiraq attack, and the Muslim Brotherhood and conservatives were angry. On October 6, 1981, at a military parade, a young Egyptian Army lieutenant and the three others he recruited attacked the reviewing stand by throwing grenades and firing automatic weapons, killing President Sadat. His vice president, Hosni Mubarak, was not injured and immediately took power, vowing to press on with peace initiatives with Israel (Gilbert, 1998; Sacher, 2010).

The next event that would shape Israeli relations with other nations for years to come is the invasion of Lebanon in 1982. Desperate to end attacks from southern Lebanon by the Palestinians, Israel made the decision to invade on June 6, 1982. Under the guidance of the minister of defense the Israeli military unleashed bombardments and heavily assaulted Palestinians positions, some even in the city of Beirut. Too, the Israeli military stood by as Phalangist militia (usually Druze or Christians and bitter enemies of the Palestinians) entered refugee camps and murdered over 2,000 individuals, including women and children. Israel remained in Lebanon until 2000, finally withdrawing after other countries, such as Syria, withdrew. The Palestinian threat was minimized but Israel still maintains a small presence in a disputed area called the Shebaa Farms, claimed by Lebanon though Israel maintains it won it from the Syrians, and it is part of the Golan Heights (Koekenbier, 2005).

Notable Terror Events

While detailing the events above, it is worth noting that Israel and the world were to observe a number of events carried out by enemies of Israel and the West; some would even be carried out by Israelis and damage the nation. What follows is a brief recap of some the more interesting or major actions. For example, on October 7, 1985, four terrorists, members of the Popular Liberation Front, an organization under the umbrella of the Palestinian Liberation Organization, hijacked a cruise ship, the M.S. *Achille Lauro,* en route to the Israeli port of Ashdod from Alexandra, Egypt (Berman, 2008; Sacher, 2010). Taking all of the crew and passengers as hostages, they killed an elderly, wheelchair-bound Jewish gentleman and threw him and his wheelchair overboard. Egyptian president Hosni Mubarak eventually convinced the terrorists to surrender to Egyptian authorities and they did so after releasing the remaining hostages. President Mubarak then lied to the Americans that the terrorists were no longer in Egypt, though they were. As they later flew later to Tunisia, their plane was forced down by U.S. Navy jets in Italy. Many were eventually tried and sentenced in Italian courts but the mastermind fled and never served prison time (Berman, 2008).

In 1985, the organization called Abu Nidal, founded by Sabri al-Banna, attempted to take over an airline counter in Rome, hijack an airplane, and blow it up over Tel Aviv, Israel, but were thwarted by Israeli security personnel. A similar attack was mounted at the Vienna airport and a total of 20 people died (Suro, 1987). This was then followed by the first of two brutal periods of Intifada or uprisings that caught the Israelis off guard both times and were generally devastating to both Israel and the Palestinians. The first began in 1987 and consisted of demonstrations, strikes, boycotts, suicide bombings, and almost a civil war that in the end, with the signing of the Oslo Accords in 1993, left more than 1,400 Palestinians and 185 Israelis dead (Gilbert, 1998; Jerusalem Media and Communications Centre, n.d.). While the **Oslo Accords** officially ended the First Intifada, they did not end the violence. Following the signing, a radical, violent Israeli physician, Baruch Goldstein, attacked a mosque in Hebron with automatic weapons on February 25, 1994, and killed 29 worshippers before being killed by other worshipers (Gilbert, 1998; Sachar, 2010). This resulted in the formation of Hamas and a new organization, Islamic Jihad, leading to a series of suicide bombings the next year that killed 86 Israelis and wounded 202 (Sachar, 2010).

The Oslo Accords provided the next events of note in Israel. Prime Minister Rabin, recognizing that Yasser Arafat and the Palestinian Liberation Authority represented the only chance for peace, signed not one accord but two treaties, one in 1993 and another in 1995, that empowered the Palestinian Liberation Authority to be given legislative, executive, and judicial powers, including police and internal security, as well as the responsibility for health, education, and welfare for the Palestinian people (Gilbert, 1998). At a rally in support of the peace process, after speaking and joining in with the singing of "The Song of Peace," Rabin, on the way to his car, was shot dead by an Israeli assassin, a religious student (Gilbert, 1998). Rabin was quickly replaced by Ariel Sharon as prime minister and the Israelis' struggle continued.

The Second Intifada began in September 2000 following a visit to Temple Mount and the Al Aqsa Mosque by Prime Minister Sharon; riots and violent demonstrations ensued and in the following years would claim thousands. Israelis killed 15 Israeli Arabs in riots in September and October 2000 and nearly 5,000 Palestinians in retaliatory raids thereafter. Palestinians killed over 1,000 Israelis. Violence continued for several years and abated, but did not end, following the death of Yasser Arafat in 2004 (MidEast Web.org, n.d.). To write in detail about this horrific period would take an entire book, but here is just a sample of the events that unfolded in just one month, March 2002. In various attacks, including suicide bombings, raids, and shelling by Palestinian terrorist groups, there were 117 Israelis killed. In response, the Israelis killed 247 Palestinians (MidEast Web.org, n.d.). Ultimately, as we cover more fully below in the Palestinian section, the Israelis would reoccupy Gaza and much of the West Bank before things improve.

Israel has been involved in more wars and military operations; has survived two periods of Intifada or Palestinian uprisings, waves of suicide bombers, and over 10,000 rockets; and has held decades of peace talks with no peace in sight. During a visit to Israel in 2007, one of the authors heard a briefing by a member of the Israeli National Security Council. Responding to a question of when, if ever, would there be peace, he suggested that there was no responsible, willing leadership on the Palestinian side so it would take a generation or more, 30 to 50 years. So where are we today? Recently numerous individuals, including President Barack Obama, insist that Israel should return to the pre-1967 borders, or give back part of Jerusalem, the Golan Heights, and the West Bank occupied as the result of the 1967 Six Day War, handily won by the Israelis, as a pre-condition for talks with the Palestinians, something that, as pointed out earlier, is doubtful they will ever do (Landler, 2011).

Finally, with regard to the future, not only must Israel worry about the Palestinians, Hamas, and **Hezbollah**, they have to worry about Iran, which will be covered in detail in another section. One of the authors was a Fellow for the Foundation for the Defense of Democracies and visited Israel in 2007. At the time, the Israelis told the Fellows their number one threat was Pakistan, as it had nuclear weapons and was unstable. A fellow professor attended the next year and returned saying the Israelis now believed Iran was the number one threat. Is Iran a threat to Israel? In a speech by Iran's Supreme Leader in 2012, the Ayatollah Ali Khamenei, the actual power in Iran, vowed to continue the pursuit of nuclear weapons and destroy Israel (Tobin, 2012). This is in line with the Ayatollah Khomeini's sentiments decades ago and reiterated in 2005 by Iranian president Mahmoud Ahmadinejad who called for Israel to be wiped off the map (Kirchick & Ahmari, 2012). Prime Minister Netanyahu has made it clear that he will never tolerate a nuclear Iran and in a speech in 2012 said if it must, Israel will act alone to stop Iran (Full Text of Netanyahu Speech to AIPAC, 2012). Israel has accomplished much, and in the years to come may have a bright future but as the rest of this section and the section on Iran demonstrate, Israel's existence and attempts to achieve peace have never been easy tasks and they will not be going forward.

The Palestinian Narrative

This narrative will provide a chronology of the Palestinian people. To do so, the authors follow the career of Yasser Arafat. Fatah is the political party that controls the Palestine Liberation Organization, and ultimately the Palestinian Authority, which may become a true Palestinian state. First, however, the authors provide some background on two major groups as one cannot understand what follows without some background on the actors; these would be Hamas and Hezbollah. The authors also stress that while these groups are the major organizations, this is a simplification of an almost overwhelming topic. If one examines the Global Terrorism Database from the National Consortium for the Study of Terrorism and Responses to Terrorism—better known as START—and searches for terror groups in the West Bank/Gaza, there are 35, and in Lebanon there are 54.

Hamas, which is an acronym of the name of the Islamic Resistance Movement, is an offshoot of the Muslim Brotherhood, a Sunni fundamentalist Islamic organization formed in Egypt. The chronology of Hamas is not long. It began as a social movement in the late 1970s and Israel supported its early growth as a counterbalance to Yasser Arafat's Palestine Liberation Organization. The founder and spiritual leader was Sheikh Ahmed Yassin, a blind Imam. Early activity concentrated on social and community issues, which consume most of their efforts even today. However, after the outbreak of the First Intifada or uprising, Hamas wrote a charter that made it very clear what their desires are:

- ◆ "Israel will exist and will continue to exist until Islam will obliterate it, just as it obliterated others before it" (The Martyr, Imam Hassan al-Banna, of blessed memory).
- ◆ The Islamic Resistance Movement is one of the wings of the Moslem Brotherhood in Palestine.
- ◆ The Islamic Resistance Movement is a distinguished Palestinian movement, whose allegiance is to Allah, and whose way of life is Islam.
- ◆ Allah is its target, the Prophet is its model, the Koran its constitution: Jihad is its path and death for the sake of Allah is the loftiest of its wishes.
- ◆ There is no solution for the Palestinian question except through Jihad. ("Hamas Covenant 1988," 2008)

Based on this they clearly wish to establish an Islamic theocracy in Israel, the West Bank, and Gaza, and they reject any compromise. Hamas determines to use any means possible to do so, including suicide bombing campaigns. They attack relentlessly and Israel reacts accordingly. As mentioned earlier, after the First Intifada ended in 1994 with some Israeli concessions, a second broke out in September 2000 that lasted four years, resulting in the deaths of some 3,500 Palestinians and 1,000 Israelis. In 2004, an Israeli missile strike killed Sheikh Ahmed Yassin, the Hamas founder, and four weeks later his successor, Abdel Aziz Rantisi (Challands, 2006). That was not the end of Hamas, as they became even more influential, acting as a shadow government, providing goods, health care, education, and recreation along with continuing harassment and military attacks on Israel. And they made a decision to enter the political arena, which would prove to be a game changing event (Challands, 2006).

 # Lebanon

This section includes information about Hezbollah, but an understanding of that organization requires some background on Lebanon, the country to the north of Israel. It is important to note that Lebanon has an eastern border with Syria; Syria is a country willing to intervene in Lebanon frequently. Lebanon is widely regarded as a bustling mercantile center. It has a major port and a highly educated population, though a relatively small population of about 4 million, nearly a tenth of whom are refugees (400,000 Palestinian refugees, 50,000 Iraq refugees, and 200,000 refugees as a result of the 2006 war with Israel); they live in camps, shanty towns, or ghettoes, enjoy few rights, and have been the source of much discord. Lebanon has some interesting politics. Given independence in 1943, the seeds

CIA World Fact Book

▲   Map of Lebanon

of dissent were sown when parliamentary seats were informally apportioned based on the 1932 census; this is the last one taken and the country at the time had a Christian majority (BBC News, 2012). This apportionment has been at the heart of most internal conflicts and the cause of much interference from neighbouring countries. Ninety-five percent of the population are Arab; almost 60% are Muslims, and most of the rest are Christian (*CIA World Factbook*, n.d.).

Lebanon endured a lengthy civil war characterized by brutality on all sides from 1975 until the early 1990s. The Lebanese government and military could only be characterized as weak and the war was fought by various factions and militias; sometimes it appeared to be Christians versus Muslim, and this was often the case but not always. In August 1990, parliament and the new president agreed on constitutional amendments with the Chamber of Deputies expanded to 128 seats and divided equally between Christians and Muslims (with Druze counted as Muslims). In March 1991, the parliament passed an amnesty law that pardoned all political crimes prior to its enactment. The amnesty was not extended to crimes perpetrated against foreign diplomats or certain crimes referred by the cabinet to the Higher Judicial Council. In May 1991, the militias (with the important exception of Hezbollah) were dissolved, and the Lebanese Armed Forces began to slowly rebuild as Lebanon's only major nonsectarian institution.

In all, it is estimated that more than 100,000 were killed, and another 100,000 left injured during Lebanon's 16-year civil war. A fifth of the pre-war resident population, or about 900,000 people, were displaced from their homes; perhaps a quarter of a million emigrated permanently.

Israel has invaded Lebanon twice, in 1978, withdrew the same year, and then invaded again in 1982, leaving only in 2000. Syria has a great deal of influence in Lebanon, despite the fact that most Syrian troops, sometimes as many as 16,000, were withdrawn in 2005 (*CIA World Factbook*, n.d.). In 2005 Lebanon held its first legislative elections since the end of the civil war, free of foreign interference. However, in July 2006, Hezbollah kidnapped two Israeli soldiers, leading to a 34-day conflict with Israel in which approximately 1,200 Lebanese civilians were killed. United Nations involvement ended the war in August 2006, and Lebanese Armed Forces (LAF) deployed throughout the country for the first time in decades, charged with securing Lebanon's borders against weapons smuggling and with maintaining a weapons-free zone in south Lebanon with the help of a larger UN Interim Force in Lebanon (UNIFIL). The LAF from May to September 2007 battled the Sunni extremist group Fatah al-Islam in the Nahr al-Barid Palestinian refugee camp, winning decisively and displacing 30,000 Palestinian residents. Lebanese politicians were unable to agree on a leader until the election of LAF commander Gen. Michel Sulayman in May 2008 and the formation of a new unity government in July 2008 (*CIA World Factbook*, n.d.). A national unity government was finally formed in November 2009 and approved by the National Assembly the following month. Inspired by the popular revolts that began in late 2010 against dictatorships across the Middle East and North Africa, marches and demonstrations in Lebanon were directed instead against sectarian politics. Although the protests gained some traction, they were limited in size and unsuccessful in changing the system. Opposition politicians collapsed the national unity government under Prime Minister Sa'ad Hariri in February 2011. After several months in caretaker status, the government named Najib Miqati its prime minister. Lebanon is stable at the moment and Syria, with problems of its own, is not interfering as much, though the Syrian situation has resulted in some demonstrations and violence. Only now are the militias all marginalized, with the exception, as noted later, of Hezbollah, and the LAF seems to have some degree of control ("Operation Cast Lead," n.d.). And Hezbollah? It is now a member of the government.

While Hamas is a Sunni Islamic organization, and a very sophisticated one at that, another Shia Islamic organization formed in Lebanon as a result of the Israeli incursions into Lebanon. These invasions had a severe negative impact on the Lebanese Shia population. This organization, now almost a quasi-governmental institution, Hezbollah, was originally founded by a religious scholar in 1975, Sayyid Musa al-Sadr, as the militia wing of al-Sadr's Harakat al-Muhrumin (Movement of the Deprived). Amal (Afwaj al-Muqawamat al-Lubnaniyya), an acronym for Lebanese Resistance Detachments, was formed in an attempt to arm as the Lebanese Civil War was approaching. After some time there was a split over ideology and purpose and Hussein Musawi, a leading figure in Amal, broke away from the organization and formed Islamic Amal, which would become Hezbollah, or the Party of God (Kennedy, 2009). He was the mentor of Hassan Nasrallah, now the general secretary. Hezbollah would soon deal with the occupation, and the group immediately engaged in a series of crude attacks, suicide bombings, and assassinations against Israel and the West; it soon began receiving funding and training from Iran and Syria (Fetini, 2009; Early, 2006). The world noticed them on October 23, 1983, following a suicide bombing of a U.S. Marine barracks in Beirut, which killed 241 American servicemen. This greatly increased their credibility in Lebanon and they were now a force to be reckoned with by the Israelis (Fetini, 2009). What does Hezbollah want? Very similar things to what Hamas wants. As Early (2006) notes:

> The foundational and immutable premise that united and defined Hezbollah at its outset, and continues to do so today, was the jihad that the organization's clerical leadership declared against Israel. Within Hezbollah's doctrine, Israel came to represent the ultimate oppressor and, therefore, had to be resisted by [the] Shiite community at all costs. (p. 119)

There is a slight difference from Hamas. Hezbollah has viewed Israel as an occupier and oppressor, not of Palestine, but of Lebanon. Despite the fact that today there is almost no Israeli presence in Lebanon (the Israelis hang on to ten square miles known as Shebaa Farms, claiming they took it from Syria) and Hezbollah is now arguably stronger, at least militarily and perhaps politically, than the Lebanese government, they maintain this fiction. They were responsible for the expulsion of Israel from Lebanese territory in 2000, further enhancing their prestige. They also endured an invasion in 2006 that while costly, resulted in the Israelis withdrawing after 34 days with little to show for their effort. They are also active politically and have substantial influence in the Lebanese government and especially many of the municipalities. Clearly Hezbollah is successful (Early, 2006):

> Paralleling the growth of Hezbollah's military prowess and strength against the state and other societal organizations have been its successes in penetrating into the realm of legitimate political authority and in displacing the state as primary social-welfare provider for a substantial portion of the Lebanese populace. (p. 125)

As a side note, United Nations peacekeepers have been present in Lebanon since 1978; they were completely ineffective during the 2006 Israeli incursion so their numbers have been strengthened to 11,780 peacekeepers from 39 different countries, plus a maritime contingent of nine ships, and they even have their own website and a magazine, *al-janoub* (UNIFIL, 2012).

Thus, when a Palestinian narrative is presented, the major figure is Yasser Arafat and the organizations he led, but the context provided by some background on these two major organizations leads to a better understanding of the situation. And yes, they are both on the list of Foreign Terrorist Organizations by the U.S. secretary of state (Department of State Bureau of Counterterrorism, 2012). The authors wish to present one final note. Israel is concerned with both Hamas and Hezbollah and should be. The West, and the United States in particular, does not have a great deal to worry about from Hamas. That is not the case with Hezbollah, which increasingly worries that Syria and Iran may reduce funding of Hezbollah as their purpose, thwarting Israel, has been achieved. Too, the Syrian government is threatened, so Hezbollah now engages in widespread criminal enterprises worldwide focusing especially on money laundering and cocaine smuggling (Hezbollah, 2012). And there is this cheery article in the *Jerusalem Post* that notes there may be thousands of Hezbollah donors in the United States and probably hundreds of operatives that could target Americans if the United States attacks Iran (Kreiger, 2012).

Palestinians maintain that the Jews are just a religious group (Kimmerling, 2006) and had only a very short history in Palestine as recounted by the Bible, and yes, they cite the Bible for this (Krämer, 2002). They lay claim to the fact that Israel did aggressively take the land of Canaan, from the Canaanites, or more specifically the tribe of the Jebusites, ancestors of the Palestinians (Net Bible, n.d.; Scham, n.d.). Too, they endorse the view that they have had a continuous presence in the area, since the nation of Israel was ordered by God to destroy the Jebusites but Israel disobeyed and did not, often intermarrying. David defeated Jebus in Jerusalem but there is no mention of their total destruction (Net Bible, n.d.). Why is this significant? As Wenkel (2007) notes, "Those asserting Jebusites heritage essentially argue that Jerusalem is rightfully theirs because Israel's own scriptures say that Jebusite possession predated the Jewish claim (p. 50). Wenkel (2007) also notes that this is a very recent development.

In fact, Zeine (1973) writes:

The world in which Arabs and Turks lived together was, before the end of the nineteenth century, politically a non-national world. The vast majority of the Muslim Arabs did not show any nationalist or separatist tendencies except when the Turkish leaders themselves, after 1908, asserted their own nationalism. (p. 127)

This linkage to the Canaanites and the Jebusites and a claim to a continued presence that pre-dates the Jews is not true and was even characterized by Buchanan (2000) as "perceptions which are ideologically motivated, history viewed emotionally, distortion becoming reality" (p. xiii). That has not stopped Palestinians from writing about their "history" from that perspective, and revisionist academics unquestionably join the charge. What this means is summarized by Kimmerling (2006), who asserts:

As such, if the Jews are not a nation, or even an ethnic group, they have no legitimate claim over Palestine. Their demand to "return" to their supposed fatherland is faulty and in any case not superior over the uprooted and disinherited Arab inhabitants of Palestine 48 years before. (p. 448)

And we find this nugget recently from the Hamas minister of the interior and of national security, Fathi Hammad:

"Every Palestinian, in Gaza and throughout Palestine, can prove his Arab roots—whether from Saudi Arabia, from Yemen, or anywhere. We have blood ties." More than that, Hammad stated that the true regional background of most "Palestinians" is not in "Palestine." "Brothers, half of the Palestinians are Egyptians and the other half are Saudis." (Jones, 2012)

Clearly he was not speaking to a Western audience; he was making an appeal for fuel for the Hamas-ruled Gaza, but it is telling.

While the Palestinian narrative does not begin coherently, they do have some history. The first stirrings of the Palestinian's actually considering something nationally and politically followed the unrest and riots that characterized much of the 1928 to 1939 time period and really ended when the United Nations imposed a two-state solution in 1948, not accepted by the Palestinians. In the Palestinian narrative, though, the United Nations had no right to give away Palestine's territory to the Jews and the resulting exodus of Palestinians was a planned expulsion, what they call "Nabkah" meaning a disaster or catastrophe, and it should be viewed as ethnic cleansing (Scham, n.d.). The Palestinians also condemn Arab countries that were very slow or failed to welcome Palestinian refugees or, in some cases, imposed huge restrictions and taxes on them, such as Jordan (Scham, n.d.; Zahran, 2012). Making things even worse was the United Nations. As May (2012) notes:

> Through two mechanisms: A refugee, by definition, lives on foreign soil but for Palestinians the definition has been changed so that a displaced Palestinian on Palestinian soil also receives refugee status. Second, the international organization responsible for resettling refugees, the United Nations High Commissioner for Refugees (UNHCR), was cut out from the start. A new organization was set up exclusively for Palestinians: the United Nations Relief and Works Agency (UNRWA). In 1950, UNRWA defined a refugee as someone who had "lost his home and his means of livelihood" during the war launched by Arab/Muslim countries in response to Israel's declaration of independent statehood. Fifteen years later, UNRWA decided—against objections from the United States—to include as refugees the children, grandchildren and great-grandchildren of those who left Israel. And in 1982, UNRWA further extended eligibility to all subsequent generations of descendants—forever.
>
> Under UNRWA's rules, even if the descendant of a Palestinian refugee has become a citizen of another state, he's still a refugee. For example, of the 2 million refugees registered in Jordan, all but 167,000 hold Jordanian citizenship. (In fact, approximately 80 percent of Jordan's population is Palestinian—not surprising since Jordan occupies more than three-fourths of the area historically referred to as Palestine.) By adopting such a policy, UNRWA is flagrantly violating the 1951 Convention Relating to the Status of Refugees which states clearly that a person shall cease to be considered a refugee if he has "acquired a new nationality, and enjoys the protection of the country of his new nationality." (May, 2012)

And yes, American taxpayers fund this, to the tune of $4.4 billion dollars since 1948 (Schanzer, 2012).

With a simmering, large refugee population and angry Arab nations the stage was set for the acts that followed. The first major event that would shape the Palestinian future occurred in East Jerusalem in 1964, when the Jordanian government allowed an "assembly" of Palestine Arabs to meet. This meeting established the Palestinian Liberation Organization with a goal of liquidating Israel and it also established the Palestinian Liberation Army (Gilbert, 1998). A small group of Palestinian students studying in Egypt in about the same time period established Fatah as a political party. That group would soon be dominated by Yasser Arafat, who soon would come to control the Palestinian Liberation Organization (PLO) when he was elected chairman in 1969 (Laquer, 2003; "Timeline: History of a Revolution," 2009). Following the chronology of the PLO provides an excellent path for following the history of Palestine. Feeling particularly bold, especially after witnessing the defeat of Arab nations in 1967, Palestinian Liberation Organization Fedayeens or freedom fighters and similar groups began attempting to take over Jordan, where King Hussein was still reeling from the pain of losing half his territory in the 1967 war.

Things came to a head in September 1970 when a group not under Arafat's control, the Popular Front for the Liberation of Palestine, became very active. The group, led by a Christian Arab physician with a Marxist-Leninist ideology, George Habash, decided to take major steps into the history books (Laqueur, 2003). The group hijacked four aircraft, landed them at an unused airfield in Jordan called Dawson's Field, and then destroyed them after removing all of the passengers. This forced King Hussein's hand and the Jordanian Army attacked the various Palestine factions viciously. This act was condemned widely by Arab nations; none responded except for Syrian Army units that entered Jordan to aid the Palestinians (the Syrian defense minister refused to provide air cover for the Syrian Army). This actually led King Hussein, worried about the Syrian Air Force joining the Syrian Army, to consider asking Israel for aid to beat back the Syrian Army units. Ultimately the king used his air force to expel the Syrians and defeat the Palestinians and force them from Jordan. This period became known as "Black September" and would serve as a stimulus for later terror (PLO: History of a Revolution, 2009).

© Can Stock Photo Inc./Golovniov

▲   Photo of a canceled stamp showing the leader of Syria, Hafez al-Assad, largely regarded by the world as a brutal monster, having killed an entire city that opposed him. He also regularly meddled in Lebanese affairs. No one suspected that his son, Bashar al-Assad, would exceed the slaughter of his father. He has presided over the deaths of some 150,000 Syrians thus far, and did so using chemical weapons among other means. He does seem to have little time to bother Lebanon though.

Smarting from this embarrassment, a new terror offshoot of the Palestinian Liberation Organization would soon emerge, known as Black September. The group acted quickly, assembling a hit squad and assassinating the Jordanian prime minister, Wasfi al-Tal, in Egypt and then the team gave themselves up to the authorities. It is clear now that Yasser Arafat knew of their plans and may have even directed the operation, including a later attack on Israeli athletes at the Munich Olympics in 1972 that resulted in the killing of 11 Israelis ("Timeline: History of a Revolution," 2009). The German police did kill five of the Palestinians and Mossad tracked down and killed all remaining perpetrators (Gilbert, 1998). The extraordinary lengths the Israelis took to do this may have included chocolate poisoning Wadi Haddad in Baghdad where he had fled ("Israel Used Chocs," 2006). Haddad had been expelled from the Popular Front for the Liberation of Palestine and may not have had anything to do with Munich but was clearly behind multiple hijackings, including one that ended up in Uganda where the Israelis carried out a spectacular rescue, killing all hijackers with only one casualty from the rescue force, Lieutenant Colonel Yonatan Netanyahu (brother of the current prime minister of Israel, Benjamin Netanyahu) (Gedalyahu, 2011; Sachar, 2010). Haddad was also instrumental in turning loose Illich Ramirez Sanchez, or Carlos the Jackal, first used in an attack on the Organization of the Petroleum Exporting Countries (OPEC) meeting of oil ministers, killing the security detail and taking 70 hostages. They were to be killed but he did not follow orders and after flying them to Algiers, released them, leading to a break with Haddad ("Timeline: History of a Revolution," 2009).

While defeated in Jordan, Arafat managed in 1974 to get the PLO recognized as the sole representative of the Palestinian people; the same year he also gave a fiery speech at the United Nations offering the world either an olive branch or a gun, and this received a standing ovation (PLO: History of a Revolution, 2009). The PLO and related organizations, such as the Popular Front for the Liberation of Palestine, next turned its sights on Lebanon as a base of operations. They arrived just in time for a civil war between Christians with a decided right-wing bent and Muslims with a socialist or left-wing leaning. Arafat for a time attempted to keep his forces out of the fray but inevitably they were sympathetic and soon joined the fighting on the side of the leftists. Ironically, fearing regional unrest, Syria with support from the United States and Israel, soon joined the conflict on the side of the Christians and prevailed, becoming a de facto peacekeeping force ("Timeline: History of a Revolution," 2009). Needing a haven after expulsion from Lebanon, Arafat and the PLO made their way to Tunisia. They remained briefly until fighting resumed in Lebanon between Syrian-backed Amal militia and anti-Arafat factions. The combat between the Palestinian camps in Beirut and southern Lebanon in an on-and-off onslaught would last three years and become known as the War of the Camps. This ended with a second expulsion of the PLO from Lebanon (PLO: History of a Revolution, 2009). While continuing efforts to work for a Palestinian state, Arafat saw the Palestinian people become more frustrated and the result was the First Intifada or uprising. Meanwhile the first Gulf War began with Iraq invading Kuwait. Arafat and the PLO supported Iraq, and the nearly 400,000 Palestinian refugees living in Kuwait were expelled ("Timeline: History of a Revolution," 2009). The Intifada continued and violence lasted for years until Arafat condemned terrorism, there is a mutual recognition of Israel and the PLO, and in 1993 the Oslo Accords, which provided some limited autonomy for the Palestinians, were signed (Sachar, 2010). Amazingly, King Hussein of Jordan even ceded the West Bank to the Palestinians, paving the way for them to actually have land, if they could claim it, and signed a peace treaty with Israel (Sachar, 2010).

Arafat returned to Gaza after a 27-year absence; he now headed the Fatah Party, as well as the Palestinian Liberation Organization, and was named the interim head of the fledging Palestinian Authority, but governing was not easy. He was later elected as president of the Palestinian Authority, but soon other groups, not nationalist and secular but nationalist and Islamic or simply Islamic, such as Hamas, began to challenge Arafat. Sachar (2010) notes this type of group was encouraged on the advice of Shin Bet, the Israeli Internal Security Service, as an option to check Arafat's organizations:

> With unofficial Israeli approval now, these right-wing religionists were authorized to build new mosques, Islamic schools and colleges, clinics and infirmaries, and thus presumably to function as a more "spiritual" alternative to Fatah and the PLO factions in Palestine. In fact, the gamble proved ill advised. Hamas soon revealed itself as irredeemably hostile to Israel. (p. 1027)

The end was near for Arafat and he was slowly losing his clout. Only one more significant, roaring acclamation and celebration from the crowds followed his return from negotiations with Prime Minister Ehud Barak of Israel and President Bill Clinton in 2000 at Camp David, rejecting a compromise, and as Carter (2006) notes, "There was no possibility that any Palestinian leader could accept such terms and survive" (p. 152).

Prime Minister Barak called early elections, which he then proceeded to lose to Ariel Sharon. Arafat refused to make any attempt to rein in the growing violence from the Second Intifada or Al Aqsa Intifada. It was named as such after Ariel Sharon, the leader of Israel's Likud party, visited al-Aqsa Mosque in Jerusalem, which is Islam's third holiest site. Arafat described the visit as a dangerous affront to Islam's holy places ("Timeline: History of a Revolution," 2009). It was soon clear that Arafat was unwilling and perhaps unable to stop the almost daily violence of bombings, shootings, and suicide bombers that spawned Israeli reprisals. Growing impatient, Sharon made a decision to end any discussion of peace with Arafat and to send military forces into the West Bank in 2001. This failed to stem the violence as Sachar (2010) notes; between October 2000 and December 2001 there were 685 Israeli deaths with over 4,500 wounded, and Palestinian casualties were 1,300 killed, mostly in firefights with the Israeli military, with a further 9,700 wounded. Sharon, after waiting many months and seeing violence that was approaching war, asked the cabinet for permission to mount "Operation Defensive Shield," a full-scale reoccupation of the West Bank, in March of 2002, and included the further step of closing all border entry points (Sachar, 2010). Included in this operation was the surrounding of the compound of the president of the Palestine Authority, Arafat, who was forbidden to leave, and to make certain, Israeli forces even destroyed the just developed runways of the Palestinian Authority's newly opened airport in the Gaza Strip. Israeli intelligence officers also discovered in the surrounding garages and warehouses extensive bomb-making facilities, plastic explosives, hundreds of automatic rifles, counterfeit Israeli currency, evidence of Iran supplying weapons, rockets, and explosives, and documents showing that the families of suicide bombers were compensated from Arafat's personal slush fund (Sachar, 2010).

Arafat and the Palestinian Authority were now effectively marginalized; they were not able to deliver any basic public services, protests against them were mounting by the Palestinian people, and terror groups such as Hamas and Islamic Jihad were supplanting them (Sachar, 2010). Arafat reluctantly, from his isolated compound, agreed to the election of a prime minister for the Palestinian Authority, something required as part of a roadmap for peace. Abu Mazen (Mahmud Abbas) was elected to the post with a security portfolio limited to primarily negotiating with Israel (mideastweb.org, n.d.). In June, Prime Minister Abu Mazen and Prime Minister Ariel Sharon met at a summit and pledged to work together for peace, with Prime Minister Abu Mazen calling for an end to violence; violenced escalated with rumors of an assassination attempt on him and even on members of Fatah Al Aqsa brigades. Arafat had lost control (mideastweb.org).

Ariel Sharon, tiring of dealing with the Gaza Strip, announced would that the Israelis would disengage from there and four West Bank settlements. This was an easier decision because Arafat had fallen ill on October 25, 2004, and died in a Paris hospital on November 11, 2004 (Sachar, 2010). Sharon was betting that building security fences, creating checkpoints, and leaving Gaza would deliver more peace and security, despite international criticism (Sachar, 2010). This was good news for former prime minister Abu Mazen, now the-president of the Palestinian Authority (mideastweb.org). In August of 2005, Sharon and the Israelis complete their disengagement from Gaza and some of the West Bank. This ushered in a quieter period of time and the security barriers and checkpoints worked, but uncertainty returned in January 2006 when Sharon suffered a massive brain hemorrhage and became comatose, suffering extensive brain damage, preparing the way for Ehud Barak to become acting prime minister (Sachar, 2010).

Further complicating the security situation, Hamas (the name is taken from the initials of the Arabic name, Islamic Resistance Movement; the word means *zeal* in Arabic and *violence* in Hebrew) not surprisingly, since it had been functioning as a shadow government providing goods and services, won the parliamentary elections in 2006 ("Who Are Hamas?" 2009). This surprised the United States and proved embarrassing for Secretary of State Condoleezza Rice and President George W. Bush as his administration's policy in the Middle East assumed that democratic elections would lead to pro-Western governments (Goldenberg, 2008). Hamas was viewed as a service provider, not corrupt and competent; the only problem is that they are a terrorist organization and deny that Israel has a right to exist ("Who Are Hamas?" 2009).

While worrying about the ramifications of the election of Hamas, and imposing an economic blockade on Gaza, Israel responded to the increasing number of rocket attacks and the kidnapping of two soldiers by Hezbollah (an Iranian-supported, well-organized terrorist group, long a proxy for Iran's terror attacks on Israel and the West) from Lebanon in the north by launching an invasion into southern Lebanon in July of 2006. This effort lasted for 34 days until a cease fire was signaled, but over 4,000 rockets were fired at northern Israeli cities; 1,187 Lebanese civilians were killed along with an unknown number of Hezbollah; Israel lost 161 and of that 44 were civilians (Sachar, 2010). It generally is conceded that while Israel may have "punished" Hezbollah, this effort was a failure and it achieved the destruction of perhaps only a fourth of the stockpile of enemy ballistic missiles.

As noted in an earlier section, Fatah and Hamas clashed in Gaza following the election win of Hamas and the inability of Fatah and Hamas to govern together. This resulted in the total defeat of Fatah as a force in Gaza and their complete expulsion from Gaza in 2007 (Goldenberg, 2008). Prime Minister Mahmoud Abbas sacked the Hamas members of government and formed a new Palestinian government in the West Bank where Fatah claimed some control (Kershner & El-Khodary, 2007). At last economic sanctions were lifted on the Palestinians, the future seemed brighter, and peace talks between Israel and Palestine began anew with a summit in Annapolis, Maryland, hosted by President Bush and attended by Prime Minister Ehud Olmert and President Abbas (Kershner & Graham, 2008; Myers, 2007).

The euphoria did not last long. Israel, tiring of endless mortars and rockets fired from Gaza and the kidnapping of an Israeli soldier (remember Gaza was not under the Palestinian Authority's control), launched on December 19, 2008, a week-long bombardment, after which Israeli troops and tanks moved in to once again destroy Hamas in what would become known as Operation Cast Lead (Kershner & EL-Khodary, 2009). While Israel regarded this as a successful operation, it was criticized due to the number of civilian deaths, including women and children; the number of dead at the close of the operation was 1,400 Palestinians, including more than 1,000 civilians, and more than 5,000 wounded. According to government figures, Palestinian deaths totaled 1,166, including 295 noncombatant deaths. There were 13 Israelis killed, including three civilians (Global Security.org, n.d.). The fighting ended on January 18, 2009, when Israeli forces withdrew.

▧ Coming Together, Maybe?

The Palestinian narrative does not end there, but it merges with the Israeli narrative after the election, again, of Prime Minister Benjamin Netanyahu, in March of 2009 ("Israel Today," 2009). Netanyahu, under pressure from the United States, for the first time ever endorsed the idea of a Palestinian state, with conditions such as a no right of return policy and noting that it must be a demilitarized state, something that the Palestinians immediately rejected, but it was an historic beginning (Federman, 2009). Meanwhile, Palestinian Authority president Abbas held a Fatah Party Congress to reinvigorate the Fatah Party, although Hamas refused to allow some Fatah members living in Gaza to attend (Gradstein, 2009).

Little occurred between Israel and the Palestinians for the next two years, but this report, based on leaked documents of Israeli and Palestinian negotiations reported by Al Jazeera television, provided hope that perhaps the Palestinians were finally getting serious about peace:

For one thing, the documents show that Palestinian leaders appeared to be far more willing to cut a peace deal than most Israelis, and even many Palestinians, believed. In contrast with Israelis' portrayal of Palestinian leaders as rejectionists, the Palestinians come across in the papers as the side better-prepared, with maps, charts and compromises, even broaching controversial trade-offs that went beyond what their own people were probably ready to accept. Though publicly Palestinians have insisted on a full right of return for refugees, Palestinian Authority President Mahmoud Abbas acknowledged in March 2009 that deep concessions would have to be made. "It is illogical to ask Israel to take 5 million [refugees] or, indeed, 1 million," Abbas is quoted as telling his team. Though other documents suggest that Abbas rejected a 2008 Israeli offer to accept only 10,000 refugees, one 2008 e-mail from a Palestinian attorney suggested that Abbas was prepared to accept an "extremely low proposal for the number of returnees to Israel." No figure was stated. (Sanders, 2011)

As well, the Palestinians offered in 2008 to allow Israel to annex most of the large Jewish housing developments built around Jerusalem on land seized during the 1967 Middle East War. As part of the offer, Israel would have had to give up comparable land around Jerusalem and agree to evacuate several large West Bank settlements.

Following that astonishing release, the bitter enemies, Fatah and Hamas, signed a reconciliation deal and agreed to work toward a unity government ("Fatah and Hamas sign landmark reconciliation deal," 2011). This was followed in 2012 with an agreement between the two to provide for elections and a unity government, allow the registration of voters in Gaza, and form an interim government (Rudoren & Fares, 2012). What does this portend? No one knows, but for the moment the situation in Israel and Palestine is generally stable. That does not, of course, mean it cannot change in a moment. It can.

Summary

This section presented overviews of the Israeli and Palestinian people, along with narratives from an Israeli and Palestinian viewpoint, as well as Lebanon and major terror organizations. Both Israel and Palestine claim the land of Palestine. Clearly it was a major mistake for the Palestinians not to accept a state and they have paid the price, suffering from hardships and living through wars and often corrupt, failed governance. As of this writing, efforts are under way for a united front between the Fatah and Hamas with the outcome uncertain. Israel continues to vigorously defend itself from terror attacks and is no doubt likely to do something about Iran. Will there be a Third Intifada, as this article notes: "Mideast experts warned Prime Minister Benjamin Netanyahu last week that construction policies in the settlements or the burning of a major mosque by extremists could help trigger a violent uprising in the West Bank" (Oren, 2012). Or will there be peace in the future?

Finally, it must be noted that the precarious situation in Egypt impacts Israel. Israel is busy building a security barrier on the border with Egypt as it no longer trusts the Egyptian military to do so because they are busy with other domestic concerns and there have been some attacks, some as recently as June 18, 2012, that left two infiltrators dead (Joshua, 2012). Infiltrations, rocket attacks, suicide bombings, and other unrest finally led Egyptian president Abdel-Fattah el-Sissi in 2014 to order attacks on terror bases in the Sinai and to construct a buffer zone between the Sinai and Gaza (Miller, 2014). Israel exists in a very tough neighborhood.

Key Points

- There is an Israeli narrative that provides an explanation of how they became a state, living surrounded by enemies.
- There is a Palestinian narrative that provides an explanation of how they became refugees with no country and they remain occupied by Israel, with no prospects for a state.
- Lebanon, Israel, and Palestine are volatile areas and the situation is not helped by the Syrian civil war.

KEY TERMS

Fatah	Hezbollah	Oslo Accords
Hamas	Nabka	Palestinian Authority

DISCUSSION QUESTIONS

1. Do you believe that Israel will agree to a peace deal with a united Hamas and Fatah? Why or why not?

2. What should Israel do about Iran?

3. Israel has been widely condemned for security checkpoints and barriers, but they have greatly reduced the number of attacks. Would you keep them or remove them?

4. What should the role of the United States be in this area?

5. Israeli soldiers have been kidnapped and held as hostages, exchanged years later for many Palestinian prisoners. What would you do in response if you were an Israeli prime minister?

6. Do you believe there will be peace between Israel and Palestine in your lifetime? Why or why not?

WEB RESOURCES

Al Jazeera, "Chronicling the PLO": Al Jazeera has a remarkably balanced history of the Palestine Liberation Organization and its period of time in Jordan: http://www.aljazeera.com/programmes/plohistoryofrevolution/2009/07/200974133438561995.html

CTV News: http://www.ctv.ca/CTVNews/Specials/20060127/hamas_chronology_060127/#ixzz1wqtMpRPW

Dreyfus Affair chronology: http://faculty.georgetown.edu/guieuj/DreyfusCase/Chronology%20of%20the%20Dreyfus%20Affair.htm

Israel Ministry of Foreign Affairs, report on measures taken by Israel to assist economic and social development of the Palestinians: http://www.mfa.gov.il/NR/rdonlyres/3EA95081-06AF-4FF7-977F-338D72FD8868/0/AHLCMarch2012.pdf

MidEast Web, Timeline of Second Intifada: http://www.mideastweb.org/second_intifada_timeline.htm

MidEast Web, Zionism: http://www.mideastweb.org/zionism.htm

Trans Arab Research Institute (TARI): incorporated in Massachusetts in 1998 as a nonprofit educational organization with the goal of providing focused research and public venues for analyzing, discussing, and presenting optional perspectives on solutions: http://tari.org/index.php?option=com_content&view=article&id=10&Itemid=10

READING 8

The following reading suggests one possible solution to the Palestinian state problem. Evaluate it and consider the ramifications for Israel, the Palestinians, and their political leaders.

Jordan Is Palestinian

Mudar Zahran

Thus far the Hashemite Kingdom of Jordan has weathered the storm that has swept across the Middle East since the beginning of the year. But the relative calm in Amman is an illusion. The unspoken truth is that the Palestinians, the country's largest ethnic group, have developed a profound hatred of the regime and view the Hashemites as occupiers of eastern Palestine—intruders rather than legitimate rulers. This, in turn, makes a regime change in Jordan more likely than ever. Such a change, however, would not only be confined to the toppling of yet another Arab despot but would also open the door to the only viable peace solution—and one that has effectively existed for quite some time: a Palestinian state in Jordan.

Abdullah's Apartheid Policies

Despite having held a comprehensive national census in 2004, the Jordanian government would not divulge the exact percentage of Palestinians in the kingdom. Nonetheless, the secret that everyone seems to know but which is never openly admitted is that Palestinians make up the vast majority of the population.

In his 2011 book, *Our Last Best Chance*, King Abdullah claimed that the Palestinians make up a mere 43 percent.

The U.S. State Department estimates that Palestinians make up "more than half" of Jordanians[1] while in a 2007 report, written in cooperation with several Jordanian government bodies, the London-based Oxford Business Group stated that at least two thirds of Jordan's population were of Palestinian origin.[2] Palestinians make up the majority of the population of Jordan's two largest cities, Amman and Zarqa, which were small, rural towns before the influx of Palestinians arrived in 1967 after Jordan's defeat in the Six-Day War.

In most countries with a record of human rights violations, vulnerable minorities are the typical victims. This has not been the case in Jordan where a Palestinian majority has been discriminated against by the ruling Hashemite dynasty, propped up by a minority Bedouin population, from the moment it occupied Judea and Samaria during the 1948 war (these territories were annexed to Jordan in April 1950 to become the kingdom's West Bank).

As a result, the Palestinians of Jordan find themselves discriminated against in government and legislative positions as the number of Palestinian government ministers and parliamentarians decreases; there is not a single Palestinian serving as governor of any of Jordan's twelve governorships.[3]

Jordanian Palestinians are encumbered with tariffs of up to 200 percent for an average family sedan, a fixed 16-percent sales tax, a high corporate tax, and an inescapable income tax. Most of their Bedouin fellow citizens,

Mudar Zahran is a Jordanian-Palestinian writer who resides in the United Kingdom as a political refugee. He served as an economic specialist and assistant to the policy coordinator at the U.S. Embassy in Amman before moving to the U.K. in 2010.

SOURCE: *Middle East Quarterly*, Winter 2012, pages 3–12.

meanwhile, do not have to worry about most of these duties as they are servicemen or public servants who get a free pass. Servicemen or public employees even have their own government-subsidized stores, which sell food items and household goods at lower prices than what others have to pay,[4] and the Military Consumer Corporation, which is a massive retailer restricted to Jordanian servicemen, has not increased prices despite inflation.[5]

Decades of such practices have left the Palestinians in Jordan with no political representation, no access to power, no competitive education, and restrictions in the only field in which they can excel: business.

According to Minority Rights Group International's *World Directory of Minorities and Indigenous Peoples of 2008*, "Jordan still considers them [Palestinian-Jordanians] refugees with a right of return to Palestine."[6] This by itself is confusing enough for the Palestinian majority and possibly gives basis for state-sponsored discrimination against them; indeed, since 2008, the Jordanian government has adopted a policy of stripping some Palestinians of their citizenship.[7] Thousands of families have borne the brunt of this action with tens of thousands more potentially affected. The Jordanian government has officially justified its position: Deputy Prime Minister and Minister of the Interior Nayef Qadi told the London-based *al-Hayat* newspaper that "Jordan should be thanked for standing up against Israeli ambitions of unloading the Palestinian land of its people," which he described as "the secret Israeli aim to impose a solution of Palestinian refugees at the expense of Jordan."[8] According to a February 2010 Human Rights Watch (HRW) report, some 2,700 Jordanian-Palestinians have had their citizenship revoked. As HRW obtained the figure from the Jordanian government, it is safe to assume that the actual figure is higher. To use the words of Sarah Leah Whitson, executive director of the Middle East and North Africa division of HRW, "Jordan is playing politics with the basic rights of thousands of its citizens."[9]

But Abdullah does not really want the Palestinians out of his kingdom. For it is the Palestinians who drive the country's economy: They pay heavy taxes; they receive close to zero state benefits; they are almost completely shut out of government jobs, and they have very little, if any,

> Palestinians in Jordan have no political representation, no access to power, and no competitive education.

political representation. He is merely using them as pawns in his game against Israel by threatening to make Jerusalem responsible for Jordanians of Palestinian descent in the name of the "right of return."

Despite systematic marginalization, Palestinians in Jordan seem well-settled and, indeed, do call Jordan home. Hundreds of thousands hold "yellow cards" and "green cards," residency permits allowing them to live and work in Israel while they maintain their Jordanian citizenship.[10] In addition, tens of thousands of Palestinians—some even claim hundreds of thousands—hold Israeli residency permits, which allow them to live in Judea and Samaria. Many also hold a "Jerusalem Residency Card," which entitles them to state benefits from Israel.[11] Yet they have remained in Jordan. Despite ill treatment by the Jordanian government, they still wish to live where most of their relatives and family members live and perhaps actually consider Jordan home.

◤ Playing the Islamist Card

The Hashemites' discriminatory policies against the Palestinians have been overlooked by the West, Washington in particular, for one main reason: the Palestine Liberation Organization (PLO) was the beating heart of Palestinian politics, and thus, if the Palestinians were empowered, they might topple the Hashemites and transform Jordan into a springboard for terror attacks against Israel. This fear was not all that farfetched. The Palestinian National Charter, by which the PLO lives, considers Palestine with its original mandate borders (i.e., including the territory east of the Jordan River, or Transjordan) as the indivisible homeland of the Palestinian Arab people.[12] In the candid admission of Abu Dawoud, Yasser Arafat's strongman in the 1970s, "Abu Ammar [Arafat] was doing everything then to establish his power and authority in Jordan despite his public statements" in support of King Hussein.[13] This tension led to the 1970 Black September civil war where the PLO was expelled from Jordan and thousands of Palestinians were slaughtered by Hussein's Bedouin army.

With the threat of Palestinian militants removed, the idea of having the Muslim Brotherhood entrenched in a Palestinian state with the longest border with Israel would naturally be of concern to Israel and its allies.

The only problem with this theory is that the Muslim Brotherhood in Jordan is dominated by Bedouins, not

Palestinians. The prominent, hawkish Muslim Brotherhood figure, Zaki Bani Rushiad, for example, is a native of Irbid in northern Jordan—not a Palestinian. Salem Falahat, another outspoken Brotherhood leader, and Abdul Latif Arabiat, a major tribal figure and godfather of the Brotherhood in Jordan, are also non-Palestinians. Upon President Obama's announcement of the death of Osama bin Laden, tribal Jordanians in the southern city of Ma'an mourned the terror leader's death and announced "a celebration of martyrdom."[14] Other cities with predominantly Bedouin populations, such as Salt and Kerak, did the same. The latter, a stronghold of the Majali tribe (which has historically held prominent positions in the Hashemite state) produced Abu Qutaibah al-Majali, bin Laden's personal aide between 1986 and 1991, who recruited fellow Bedouin-Jordanian, Abu Musab al-Zarqawi, head of al-Qaeda in Iraq who was killed in a 2006 U.S. raid.[15]

The Hashemite regime is keenly aware of U.S. and Israeli fears and has, therefore, striven to create a situation where the world would have to choose between the Hashemites and the Muslim Brotherhood as Jordan's rulers. To this end, it has supported the Muslim Brotherhood for decades, allowing it to operate freely, to run charitable organizations and youth movements, and to recruit members in Jordan.[16] In 2008, the Jordanian government introduced a new law, retroactively banning any existing political party unless it had five hundred members and branches in five governorates (counties). Since such conditions could only be fulfilled by the Muslim Brotherhood, most political parties were dissolved de jure because they did not meet the new standards, leaving the Islamic Action Front as the strongest party in the kingdom.

Both Jerusalem and Washington are aware of the Jordanian status quo yet have chosen to accept the Hashemite regime as it is, seduced by the conventional wisdom of "the devil you know is better than the devil you don't." The facts on the ground, however, suggest that the devil they think they know is in deep trouble with its own supposed constituency.

 ## The Bedouin Threat

Despite their lavish privileges, Jordanian Bedouins seem to insist relentlessly on a bigger piece of the cake, demanding more privileges from the king, and, in doing so, they have grown fearless about defying him. Since

▲   King Abdullah may find himself swallowed up by the growing force of the Muslim Brotherhood within his kingdom. In Jordan, the Brotherhood is dominated by Bedouins, not Palestinians, and although Bedouins remain the minority in the kingdom, they form the foundation on which Abdullah's regime is built.

2009, fully-armed tribal fights have become commonplace in Jordan.[17] Increasingly, the Hashemite regime has less control than it would like over its only ruling foundation—the Bedouin minority— which makes up the army, the police forces, all the security agencies, and the Jordanian General Intelligence Department. The regime is, therefore, less likely to survive any serious confrontations with them and has no other choice but to keep kowtowing to their demands.

What complicates the situation even further is that Bedouin tribes in Jordan do not maintain alliances only with the Hashemites; most shift their loyalties according to their current interests and the political season. Northern tribes, for example, have exhibited loyalty to the Syrian regime, and many of their members hold dual citizenships.[18] In September 1970, when Syrian forces invaded Jordan in the midst of the civil war there, the tribes of the northern city of Ramtha raised the Syrian flag and declared themselves "independent" from the Hashemite rulers.

Likewise, Bedouin tribes of the south have habitually traded loyalty for privileges and handouts with whoever paid better, beginning with the Turks, then replacing them with the better-paying Britons, and finally the Hashemites. This pattern has expanded in the last twenty years as tribesmen exchanged their loyalties for cash; in fact this is how they got involved in the British-supported Arab revolt

▲ The majority Palestinian population of Jordan bridles at the advantages and benefits bestowed on the minority Bedouins. Advancement in the civil service, as well as in the military, is almost entirely a Bedouin prerogative with the added insult that Palestinians pay the lion's share of the country's taxes.

▲ Ill feelings between the Palestinian and Bedouin citizens of Jordan took a turn for the worse in 2009 when riots broke out at a soccer match between the fans of the Palestinian favored al-Wihdat soccer club and those of the Bedouin favored al-Faisali team. Reports indicate that Jordanian police attacked Palestinian soccer fans without provocation soon thereafter.

of World War I, in which the Bedouins demanded to be paid in gold in advance in order to participate in the fighting against the Ottomans despite their alignment with the Ottoman Empire before joining the revolt.[19]

This in turn means that the Jordanian regime is now detested not only by the Palestinians but also by the Bedouins, who have called for a constitutional monarchy in which the king hands his powers to them.[20] Should the tribes fail to achieve their goals, they will most likely expand their demonstrations of unrest—complete with tribal killings, blockades, armed fights, robberies, and attacks on police officers—which the Jordanian state finds itself having to confront weekly. In 2010, an average of five citizens was killed each week just as a result of tribal unrest.[21]

The Hashemite regime cannot afford to confront the tribesmen since they constitute the regime's own servicemen and intelligence officers. In 2002, the Jordanian army besieged the southern Bedouin city of Ma'an in order to arrest a group of extremists, who were then pardoned a few years later.[22] Similarly, Hammam Balaoui, a Jordanian intelligence double agent was arrested in 2006 for supporting al-Qaeda, only to be released shortly thereafter, eventually blowing himself up in Afghanistan in 2009 along with seven senior CIA officers and King Abdullah's cousin.[23]

Palestinian Pawns

These open displays of animosity are of a piece with the Hashemite regime's use of its Palestinian citizens as pawns in its game of anti-Israel one-upmanship.

King Hussein—unlike his peace-loving image—made peace with Israel only because he could no longer afford to go to war against it. His son has been less shy about his hostility and is not reluctant to bloody Israel in a cost-effective manner. For example, on August 3, 2004, he went on al-Arabiya television and slandered the Palestinian Authority for "its willingness to give up more Palestinian land in exchange for peace with Israel."[24] He often unilaterally upped Palestinian demands on their behalf whenever the Palestinian Authority was about to make a concession, going as far as to threaten Israel with a war "unless all settlement activities cease."[25]

This hostility toward Israel was also evident when, in 2008, Abdullah started revoking the citizenship of Jordanian Palestinians. By turning the Palestinian majority in Jordan into "stateless refugees" and aggressively

> With the largest percentage of Palestinians in the world, Jordan is a logical location for a Palestinian state.

pushing the so-called "right of return," the king hopes to strengthen his anti-Israel credentials with the increasingly Islamist Bedouins and to embarrass Jerusalem on the world stage. It is not inconceivable to envision a scenario where thousands of disenfranchised Palestinians find themselves stranded at the Israeli border, unable to enter or remain in Jordan. The international media—no friend of the Jewish state—would immediately jump into action, demonizing Israel and turning the scene into a fiasco meant to burden Jerusalem's conscience—and that of the West. The Hashemite regime would thereby come out triumphant, turning its own problem—being rejected and hated by the Palestinians—into Israel's problem.

Palestinians in Jordan have also developed an intense hatred of the military as they are not allowed to join the army; they see Bedouin servicemen getting advantages in state education and health care, home taxes, and even tariff exemption on luxury vehicles.[29] In recent years, the Jordanian military has consumed up to 20.2 percent of the country's gross domestic product (GDP).[30]

Government spending does not end with the army. Jordan has one of the largest security and intelligence apparatuses in the Middle East, perhaps the largest compared to the size of its population. Since intelligence and security officers are labeled as "military servicemen" by the Jordanian Ministry of Finance, and their expense is considered military expenditure, Jordanian Palestinians see their tax dollars going to support job creation for posts from which they themselves are banned. At the same time, the country has not engaged in any warfare since 1970, leading some to conclude that this military spending is designed to protect the regime and not the country—a conclusion underscored by the Black September events.

A Pot Boiling Over

The Jordanian government's mistreatment of its Palestinian citizenry has taken a significant toll. Today, the Palestinians are a ticking bomb waiting to explode, especially as they watch their fellow Arabs rebelling against autocrats such as Egypt's Mubarak, Libya's Qaddafi, or Syria's Assad.

The complex relationship between the Palestinian majority and the Hashemite minority seems to have become tenser since Abdullah ascended the throne in 1999 after King Hussein's death. Abdullah's thin knowledge of the Arabic language, the region, and internal affairs, made him dependent on the Bedouin-dominated Jordanian Intelligence Department standing firmly between the king and his people, of which the Palestinians are the majority.[26] A U.S. embassy cable, dated July 2009, reported "bullying" practiced by the fans of al-Faisali Soccer Club (pre-dominantly Bedouin Jordanians) against the fans of al-Wihdat Soccer Club (predominantly Palestinians), with al-Faisali fans chanting anti-Palestinian slogans and going so far as to insult Queen Rania, who is of Palestinian descent.[27] Two days after the cable was released, Jordanian police mercilessly attacked Palestinian soccer fans without provocation, right under the eyes of the international media.[28]

A Path to Peace?

The desperate and destabilizing measures undertaken by the Hashemite regime to maintain its hold on power point to a need to revive the long-ignored solution to the Arab-Israeli conflict: the Jordanian option. With Jordan home to the largest percentage of Palestinians in the world, it is a more logical location for establishing Palestinian statehood than on another country's soil, i.e., Israel's.

There is, in fact, almost nothing un-Palestinian about Jordan except for the royal family. Despite decades of official imposition of a Bedouin image on the country, and even Bedouin accents on state television, the Palestinian identity is still the most dominant—to the point where the Jordanian capital, Amman, is the largest and most populated, Palestinian city anywhere. Palestinians view it as a symbol of their economic success and ability to excel. Moreover, empowering a Palestinian statehood for Jordan has a well-founded and legally accepted grounding: The minute the minimum level of democracy is applied to Jordan, the Palestinian majority would, by right, take over the political momentum.

For decades, however, regional players have entertained fears about empowering the Palestinians of Jordan.

While there may be apprehension that Jordan as a Palestinian state would be hostile to Israel and would support terror attacks across their long border, such concerns, while legitimate, are puzzling. Israel has allowed the Palestinians to establish their own ruling entities as well as their own police and paramilitary forces on soil captured in the 1967 war, cheek by jowl with major Israeli population centers. Would a Palestinian state on the other side of the Jordan River pose any greater security threat to Israel than one in Judea and Samaria?

> Abdullah's only "backbone" is Washington's political and financial support.

Moreover, the Jordan Valley serves as a much more effective, natural barrier between Jordan and Israel than any fences or walls. Israeli prime minister Benjamin Netanyahu confirmed the centrality of Israeli control over the western side of the Jordan Valley, which he said would never be relinquished.[31] It is likely that the area's tough terrain together with Israel's military prowess have prevented the Hashemite regime from even considering war with Israel for more than forty years.

It could be argued that should the Palestinians control Jordan, they would downsize the military institutions, which are dominated by their Bedouin rivals. A Palestinian-ruled Amman might also seek to cut back on the current scale of military expenditures in the hope that the U.S. military presence in the region would protect the country from unwelcome encroachments by Damascus or Tehran. It could also greatly benefit from financial and economic incentives attending good-neighbor relations with Israel. Even if a Jordanian army under Palestinian commanders were to be kept at its current level, it would still be well below Israel's military and technological edge. After all, it is Israel's military superiority, rather than regional goodwill, that drove some Arab states to make peace with it.

The Palestinians in Jordan already depend on Israel for water[32] and have enjoyed a thriving economic boom driven by the "Qualified Industrial Zones," which allow for Jordanian clothing factories to export apparel to the United States at preferred tariff rates if a minimum percentage of the raw material comes from Israel.[33] Hundreds of

▲   The storm of protests that have swept through the Middle East have left Jordan's King Abdullah on a shaky throne. Although not as violent as demonstrations in Egypt or Yemen, protesters have taken to the streets of Jordan calling for reforms.

Palestinian factory owners have prospered because of these zones. Expanding such cooperation between a future Palestinian state in Jordan and Israel would give the Palestinians even more reasons to maintain a good relationship with their neighbour.

Both the United States and Israel should consider reevaluating the Jordan option. Given the unpopularity of the Hashemite regime among its subjects, regime change in Amman should not be that difficult to achieve though active external intervention would likely yield better results than the wait-and-see-who-comes-to-power approach followed during the Egyptian revolution. After twelve years on the throne, and $7 billion dollars in U.S. aid, Abdullah is still running a leaky ship and creating obstacles to resolving the Palestinian issue.

Washington's leverage can come into play as well with the Jordanian armed forces which are, in theory, loyal to the king. With hundreds of troops undergoing training in the United States each year and almost $350 million handed out in military aid, the U.S. establishment could potentially influence their choices.

Recent events in the Middle East should serve as guidelines for what ought to be pursued and avoided.

U.S. diplomacy failed to nurse a moderate opposition to Egypt's Mubarak, which could have blocked Islamists and anti- Americans from coming to power. The current turmoil in Libya has shown that the later the international community acts, the more complicated the situation can get. An intervention in Jordan could be much softer than in Libya and with no need for major action. Abdullah is an outsider ruling a poor country with few resources; his only "backbone" is Washington's political and financial support. In exchange for a promise of immunity, the king could be convinced to let the Palestinian majority rule and become a figurehead, like Britain's Queen Elizabeth.

As further assurance of a future Palestinian Jordan's peaceful intentions, very strict anti-terrorism laws must be implemented, barring anyone who has incited violence from running for office, thus ruling out the Islamists even before they had a chance to start. Such an act should be rewarded with economic aid that actually filters down to the average Jordanian as opposed to the current situation, in which U.S. aid money seems to support mainly the Hashemites' lavish lifestyle.

Alongside downsizing the military, a defense agreement with Washington could be put in place to help protect the country against potentially hostile neighbours. Those who argue that Jordan needs a strong military to counter threats from abroad need only look again at its history: In 1970, when Syria invaded northern Jordan, King Hussein asked for U.S. and Israeli protection and was eventually saved by the Israeli air force, which managed to scare the Syrian troops back across the border.[34] Again in 2003, when Washington toppled Saddam Hussein, Amman asked for U.S. operated Patriot missile batteries and currently favors an extended U.S. presence in Iraq as a Jordanian security need.[35]

Should the international community see an advantage to maintaining the military power of the new Palestinian state in Jordan as it is today, the inviolability of the peace treaty with Israel must be reasserted, indeed upgraded, extending into more practical and tangible economic and political arenas. A mutual defense and counterterrorism agreement with Israel should be struck, based on one simple concept—"good fences make good neighbours"—with the river Jordan as the fence.

Conclusion

Considering the Palestinian-Jordanian option for peace would not pose any discrimination against Palestinians living in the West Bank, nor would it compromise their human rights: They would be welcome to move to Jordan or stay where they are if they so wished. Free will should be the determinant, not political pressure. Besides, there are indications that many would not mind living in Jordan.[36] Were the Palestinians to dominate Jordan, this tendency will be significantly strengthened. This possibility has also recently been confirmed by a released cable from the U.S. embassy in Amman in which Palestinian political and community representatives in Jordan made clear that they would not consider the "right of return" should they secure their civil rights in Jordan.[37]

Empowering Palestinian control of Jordan and giving Palestinians all over the world a place they can call home could not only defuse the population and demographic problem for Palestinians in Judea and Samaria but would also solve the much more complicated issue of the "right of return" for Palestinians in other Arab countries. Approximately a million Palestinian refugees and their descendents live in Syria and Lebanon, with another 300,000 in Jordan whom the Hashemite government still refuses to accept as citizens. How much better could their future look if there were a welcoming Palestinian Jordan?

> There are indications that many West Bank Palestinians would not mind living in Jordan.

The Jordanian option seems the best possible and most viable solution to date. Decades of peace talks and billions of dollars invested by the international community have only brought more pain and suffering for both Palestinians and Israelis—alongside prosperity and wealth for the Hashemites and their cronies.

It is time for the international community to adopt a more logical and less costly solution rather than to persist in long discredited misconceptions. It is historically perplexing that the world should be reluctant to ask the Hashemites to leave Jordan, a country to which they are alien, while at the same time demanding that Israeli families be removed by force from decades-old communities in their ancestral homeland. Equally frustrating is the world's silence while Palestinians seeking refuge from fighting in Iraq are locked in desert camps in eastern Jordan because the regime refuses to settle them "unless foreign aid is provided."[38]

The question that needs to be answered at this point is: Has the West ever attempted to establish any contacts with a pro-peace, Palestinian-Jordanian opposition? Palestinians today yearn for leaders. Washington is presented with a historical opportunity to support a potential Palestinian leadership that believes in a peace-based, two-state solution with the River Jordan as the separating border between the two countries. Such leadership does seem to exist. Last September, for example, local leaders in Jordanian refugee camps stopped Palestinian youth from participating in mass protests against the Israeli Embassy in Amman;[39] as a result, barely 200 protesters showed up instead of thousands as in similar, previous protests.[40] As for East Jerusalem, under Israel's 44-year rule, Muslims, Christians, and members of all other religions have been able to visit and practice their faith freely, just as billions of people from all over the world visit the Vatican or Muslim pilgrims flock to Mecca. Yet under the Hashemite occupation of the city, this was not done. Without claiming citizenship, Jerusalem would remain an open city to all who come to visit.

The Jordanian option is an overdue solution: A moderate, peaceful, economically thriving, Palestinian home in Jordan would allow both Israelis and Palestinians to see a true and lasting peace.

◪ Notes

1. "Jordan: Country Reports on Human Rights Practices, 2001," Bureau of Democracy, Human Rights, and Labor, U.S. Department of State, Mar. 4, 2002.

2. "The Report: Emerging Jordan 2007," Oxford Business Group, London, Apr. 2007.

3. "Jordan: Country Reports on Human Rights Practices, 2001," Mar. 4, 2002.

4. "Brief History," Civil Service Consumer Corporation, Government of Jordan, Amman, 2006.

5. Jordan News Agency (PETRA, Amman), Jan. 10, 2011.

6. "Jordan: Palestinians," *World Directory of Minorities and Indigenous Peoples*, Minority Rights Group International, 2008, accessed Sept. 20, 2011.

7. "Stateless Again," Human Rights Watch, New York, Feb. 1, 2010.

8. *The Arab Times* (Kuwait City), Jan. 13, 2011.

9. "Jordan: Stop Withdrawing Nationality from Palestinian-Origin Citizens," Human Rights Watch, Washington, D.C., Feb. 1, 2010.

10. "Jordan: Information on the right of abode of a Palestinian from the West Bank who holds a Jordanian passport which is valid for five years," Immigration and Refugee Board of Canada, Oct. 1, 1993, JOR15463.FE.

11. "Jordan's treatment of failed refugee claimants," Immigration and Refugee Board of Canada, Mar. 9, 2004, JOR42458.E.

12. The Palestinian National Charter, Resolutions of the Palestine National Council, July 1–17, 1968.

13. *Al-Jazeera* (Riyadh), Oct. 1, 2005.

14. *Amman News*, May 2, 2011.

15. Ibid., May 2, 2011.

16. Awni Jadu al-Ubaydi, *Jama'at al-Ikhwan al-Muslimin fi al-Urdunn wa-Filastin, 1945–1970* (Amman: Safahat Ta'arikhiyya, 1991), pp. 38–41.

17. Samer Libdeh, "The Hashemite Kingdom of Apartheid?" *The Jerusalem Post*, Apr. 26, 2010.

18. CNN, Nov. 28, 2007.

19. Michael Korda, *Hero: The Life and Legend of Lawrence of Arabia* (New York: Harper, 2010), p. 19.

20. *Hürriyet* (Istanbul), Mar. 4, 2011.

21. Libdeh, "The Hashemite Kingdom of Apartheid?"

22. PETRA, Aug. 6, 2011.

23. "Profile: Jordanian Triple Agent Who Killed CIA Agents," *The Telegraph* (London), Jan. 2010.

24. Al-Arabiya TV (Dubai), Aug. 3, 2004.

25. *The Jerusalem Post*, Sept. 24, 2010.

26. *Los Angeles Times*, Oct. 1, 2006.

27. *The Guardian* (London), Dec. 6, 2010.

28. *Qudosi Chronicles* (Long Beach, Calif.), Dec. 16, 2010.

29. "Assessment for Palestinians in Jordan," Minorities at Risk, Center for International Development and Conflict Management, University of Maryland, College Park, Md., Dec. 31, 2006.

30. "Jordan Military Expenditures—Percent of GDP," *CIA World Factbook*, May 16, 2008.

31. *Ha'aretz* (Tel Aviv), Mar. 2, 2010.

32. Lilach Grunfeld, "Jordan River Dispute," The Inventory of Conflict and Environment Case Studies, American University, Washington, D.C., Spring 1997.

33. Mary Jane Bolle, Alfred B. Prados, and Jeremy M. Sharp, "Qualifying Industrial Zones in Jordan and Egypt," Congressional Research Service, Washington, D.C., July 5, 2006.

34. Mitchell Bard, "Modern Jordan," Jewish Virtual Library, accessed Aug. 11, 2011.

35. *The Christian Science Monitor* (Boston), Jan. 30, 2003.

36. *The Forward* (New York), Apr. 13, 2007.

37. "The Right of Return: What It Means in Jordan," U.S. Embassy, Amman, to Bureau of Near Eastern Affairs, U.S. Department of State, Washington, D.C., Feb. 6, 2008.

38. "Non-Iraqi Refugees from Iraq in Jordan," Office of the United Nations High Commissioner for Refugees, Feb. 20, 2007.

39. Mudar Zahran, "A Plan B for Jordan?" Hudson Institute, Washington, D.C., Sept. 16, 2011.

40. *The Washington Post*, Sept. 15, 2011.

READING 9

This journal article suggests that Hamas is not the frightening terror group of the past and that it is capable of modifying and changing. If you were Israel, would you have a more hopeful view of Hamas after reading this? Why or why not?

Demystifying the Rise of Hamas

Wael J. Haboub
University of Illinois at Chicago, USA

Introduction

How can the literature on voting behavior and extreme political parties shed light on the electoral success of Hamas in the 2006 Palestinian parliamentary election? Drawing on expected utility theory, much of the literature on voting behavior assumes that the average voter selects parties and candidates in the center of the political spectrum. Prospect theory, on the other hand, holds that voters are risk-averse in their choice between gains and become risk seekers when choosing between losses. This implies that if voters perceive their conditions as worsening under the leadership of the incumbent party, they may be willing to take a chance and elect an extremist party, which they would not otherwise choose in relatively stable conditions. Most scholarly and popular writings portray Hamas as either extremist or anti-systemic and a threat to democracy (Alexander, 2002; Levitt, 2004, 2006). Since extremist and

SOURCE: Copyright © 2012 SAGE Publications www.sagepublications.com (Los Angeles, London, New Delhi, Singapore and Washington DC) Vol 28(1): 57–79. DOI: 10.1177/0169796X1102800103

anti-systemic parties, however, do not usually win the majority in parliamentary elections—except in cases in which the public is in domain of loss as predicted by prospect theory—how do we explain Hamas's electoral success? Which of the competing theories (expected utility or prospect theory) best explain the electoral victory of Hamas? In other words, did the Palestinian voters believe they were choosing an extremist or a centrist Hamas?

This research is important theoretically and has momentous policy implications. Theoretically, though counter intuitive, this study confirms the vast body of literature on voting behavior that assumes most voters would elect candidates in the center of the political spectrum. In terms of policy, understanding how voters cast their votes in various parts of the world, especially in the Muslim world, would help policy makers base their policies on sound theoretical foundations rather than untested ideological assumptions.

The rise of Hamas must be seen as part of a larger phenomenon of Islamic revivalism, which took a sharp turn in late 1970. In the post–Cold War era, especially after September 11, Islamic revivalism came to be seen as a threat not only to the West but also to the very concept of democracy. What became troubling to many in the West, not withstanding violent Islamist movements, is the participation of many Islamist groups in semi-democratic elections in many Muslim countries. When the Algerian Islamic Salvation Front, Front Islamique du Salut (FIS), for example, had won the first round of the double ballot votes in December 1991 and as a result the army cancelled the election, the West stood idly by. The US and Europe abandoned their proclaimed policy of the promotion of democracy because they simply did not want to see another Iran in the Middle East. More recently, despite the United States pressuring the then Fatah-led Palestinian Authority to open its system and allow for greater democratic participation, the democratically elected Hamas government came under sanctions. The US and Israel boycotted the Hamas government and pressured other governments to do the same until Hamas recognized Israel. Aside from Hamas's nonrecognition of Israel, many in the West fear that Islamist movements may participate in democratic systems in order to ultimately destroy them. Islamist movements, including Hamas, thus are viewed as anti-system parties.

I argue that in many Muslim countries, voters are increasingly choosing what they believe to be centrist Islamist parties, which may be viewed as extremist parties by the West. More importantly, the participation of Islamist movements in democratic political systems has a moderating effect on their domestic and foreign policies, as has been the case with Hamas, and their exclusion has destabilizing and radicalizing effects, as was the case with the Algerian FIS. Hamas suspended its attacks against Israel in the year preceding its participation in the parliamentary election of 2006.

I argue that the secret of Hamas's electoral success in 2006 lies not in its extremism but in its ideological centrist electoral message. The success of the electoral message of Hamas will be interpreted using voting behavior theory and the literature on extreme political parties. I will use the interpretive approach to investigate Hamas's ideological distance from other Palestinian political parties to determine whether expected utility or prospect theory of voting behavior best explains Hamas's electoral victory. I argue that the institutionalization of Fatah and Hamas, both centrist parties, made competition possible and enabled voters to evaluate their respective performance, competence, and electoral program.

Parties' Ideological Proximity and Voting Behavior

Hamas is an umbrella organization that encompasses three broad sub-divisions: civil society (charitable and educational institutions), political party (Al-Maktab al-Siyasi, or the Political Bureau), and a military wing (Ezz ed-Din el-Qassam Brigades). In order to understand the factors that led the Palestinian public to elect Hamas, despite the secularism of Palestinian society, this study focuses on the political wing of Hamas.

Expected utility theory posits that voters will choose candidates and parties after weighing the costs and benefits of their political decisions. Guided by expected utility theory, spatial theory (or the proximity model) believes that voters are risk-averse, and therefore, they select candidates that are closest to their ideal ideological point (Davis, Hinich, & Ordeshook, 1970; Downs, 1957; Enelow & Hinich, 1982, 1984; Grofman, 1985; Hinich & Pollard, 1981). Consequently, politicians will frame their electoral campaign to attract the median voter to increase their chances of winning.

The directional model expands on the proximity model in that it views voters' choices from the direction (that is, right vs left) of their and the candidate's position (Mathews, 1979; Rabinowitz & Macdonald, 1989). Rabinowitz and Macdonald (1989, p. 111) argue that instead of candidates moving to the center to match the median voter preferences (as argued by the proximity model), successful candidates are the ones that take "a position close to the boundary of the region of acceptability." In other words, voters select not centrist candidates but rather polarizing ones as long as they are still within the boundaries of the accepted political culture. Even though one may conclude from this finding that voters are more inclined to vote for extreme candidates, they, on the contrary, are still partial to the middle though closer to the border of accepted norms of political behavior.

Thus, regardless of whether we turn to the proximity or directional models for analysis, they each operate under the same basic assumption that voters will select candidates based on their relative position to the voters' preferences. We would expect the vast majority of the public, at least in the West, to be closer to the center of the ideological spectrum. Rabinowitz and Macdonald "incorporate into their model a penalty for extremism, on the grounds that voters will tend to find unacceptable a candidate or party whose stands are too 'far-out'" (Merrill, III, & Grofman, 1999, p. 38).

Under this assumption, extreme sides of the political debate are not able to mount an electoral victory. This observation should hold true under both two-party and multiparty systems. In a two-party system as in the US, both the Democratic Party and Republican Party appeal to the ideological center for electoral support. Similarly, in a multiparty system as in Austria, extremist parties may be able to gain a number of seats in the national parliament, but their success would be too small to gain either a majority or plurality of the votes to form a coalition. While multiparty systems allow extreme parties access to the political system, they do not afford them the opportunity to gain an electoral majority.

Extreme parties may be expected to rise only during periods of national loss, as during the rise of the National Socialist German Workers Party under the leadership of Adolf Hitler in Germany. On the basis of the empirical analysis of 103 elections in 16 western European countries from 1970 through 1990, Jackman and Volpert (1996) find that along with a multiparty system and electoral rules, higher unemployment is conducive to the rise of extremist parties. This argument is consistent with prospect theory, which was developed by Kahneman and Tversky (1979). Prospect theory offers an alternative to the expected utility theory that assumes people are utility maximizers. Kahneman and Tversky argue that when people are losing they tend to be risk-takers rather than risk-averse, as they are willing to take greater risks to get out of their loss.

The success of extremist parties is uncommon. The Swiss People's Party, Schweizerische Volkspartei (SVP), won an astonishing 29 percent of the popular vote in the 2007 election, placing it first in the lower chamber, Nationalrat, with 62 seats out of a total of 200. The SVP, however, is not categorically an extremist party. It is primarily a xenophobic nationalist conservative populist party that has benefited from the same conditions that animate extremist parties. Even though the SVP won during times of relatively stable economic conditions, its antiglobalization rhetoric attracted voters hurt by globalization, such as the elderly and voters in rural areas and small towns (Church, 2008, p. 618). Its extremism is seen in its xenophobic and Islamophobic stances as evidenced by its posters of white sheep kicking black sheep off the Swiss flag, and of a Muslim woman wearing *niqab* next to minarets shaped like missiles shooting out of the Swiss flag.

The success of the Austrian party Freiheitliche Partei Österreichs (FPÖ) in the 1999 election is one of the clearest examples of the rise of extremist parties in Europe. The electoral share of the FPÖ grew from 9.7 percent in 1986 to 26.9 percent in 1999, which made it "the greatest success experienced by an extreme right-wing party in Europe for the past fifty years" (Pedahzur & Brichta, 2002, pp. 31–32). Despite its apparent electoral success and its subsequent role in coalition building, the FPÖ received neither a majority nor first place to be able to form a government. By the 2002 election, FPÖ's share of votes dropped to 10 percent, less than half of its previous share, but it recovered slightly in 2008 to 17.5 percent of the total vote.

Similarly, Jewish religious political parties are a very small minority in Israel. Since no political party was ever able to gain a majority of the seats in the Knesset, Jewish religious political parties generally enter coalitions with larger secular parties that are unable to gain an electoral majority and thus form a government on their own (Kopelowitz, 2001, p. 170).

Chhibber (1996) attributes the electoral success of the Front Islamique du Salut (FIS) in Algeria in part to dire economic conditions that resulted in political realignments. The ruling Front de la Liberation Nationale (FLN) responded to fiscal crisis by partially opening its markets and downsizing its public sector. Large businesses benefited from privatization while small businesses bore the brunt of the cost. Since privatization led to the downsizing of the public sector, public sector employees shifted their support from the FLN to the FIS. Moreover, government spending was primarily put toward protecting the poor, while the middle class had to bear the cost of allocations, taxation, and housing. Consequently, an alliance of smaller business owners, low-level public employees, and the middle class lent their support to the FIS. If this coalition perceived the FIS as an extremist party, their electoral choices were influenced by their loss. Since they were losing, they had to take the risk of supporting the FIS rather than continue with the assured loss of the status quo.

Ascribing labels to political parties is not a simple conceptual endeavor, however. Rhetoric alone is not sufficient to help us conceptualize political parties along the spectrum of moderate versus extreme. Some extremist parties may seem moderate while moderate parties may appear extreme. Tepe (2000) argues that the electoral success of the radical right Nationalist Action Party (NAP) in Turkey in 1999 is due to the party's centrist message. Tepe argues that "it is not the party's extremism but rather, its message of ideological compromise that accounts for its electoral appeal" (Tepe, 2000, p. 59).

Along similar lines, Tepe (2005) maintains that it is a mistake to assign religious parties labels assuming that they have fixed policies, goals, and strategies. By far, political parties that sacralize secular institutions (such as Israel's Mafdal and Turkey's NAP) project a moderate image while endorsing authoritarian policies. On the other hand, religious parties that have been categorized as extremist (such as Shas in Israel and Prosperity and Justice and Development Parties in Turkey) are more likely to project a confrontational image while secularizing religious symbols. In other words, the projected image does not reflect the complex reality of religious parties.

Is Hamas more similar to FIS or does it resemble Shas in Israel and Prosperity and Justice and Development Parties in Turkey? Did Hamas gain the confidence of the Palestinian voter in 2006 because of its centrist message, or was Hamas seen as an extremist party that they chose because of national loss?

 ## Research Method

Is it Hamas's centripetal or centrifugal tendency that explains its success in the Palestinian parliamentary elections of January 2006? This question will be assessed using the interpretive approach. The primary goal of the interpretive approach is to interpret individuals' and communities' understanding of their realities. Regardless of whether there is an objective reality, individuals' subjective understanding of their environment is context-specific. The objects of reality are assigned meanings that become shared and color the way people understand their surroundings. These meanings, therefore, are socially constructed through discourses and interaction with various actors. Because of this, meanings assigned to objects tend to vary and therefore, different understandings of reality emerge (Creswell, 2003, pp. 8–9).

A comparison between expected utility theory and prospect theory must be undertaken to assess the reason behind Hamas's electoral success. To be able to assess the competing claims of both theories, I first must ascertain whether the conditions that led to the rise of extremist parties were present in Palestine. To test the merit of both the competing theories, I use certain measurable indicators derived from each theory's main assumption.

I will assess whether the empirical evidence is consistent with the predictions of either one of the competing theories. I will evaluate the content of Hamas's electoral message to see whether it fits within or outside of the "region of acceptability," to use the term of Rabinowitz and Macdonald (1989). In order to understand the reasons for the success of Hamas, it is essential to understand the specific historical, social, and political environment in which it was formed and developed. We must also understand the voters' attribution of meaning to their social and political environment, their perception of Hamas and other political parties, and how Hamas framed its electoral campaigns in a way that attracted the Palestinian public to its program.

For data collection, I rely on a combination of historical facts, field research in the Gaza Strip, and content analysis of statements made to the media by Hamas officials. The historical facts elucidate the cultural and political context of the Palestinian landscape, while field research illuminates the public views, feelings, and circumstances. Conducting field research in the Gaza Strip alone does not diminish the explanatory power of the findings. On the contrary, the Gaza Strip serves as a hard case since poverty is more severe there than in the West Bank, which strengthens the case of prospect theory. Moreover, statements and writings by Hamas officials provide the medium from which to interpret Hamas's electoral message. The success of the electoral message of Hamas will be interpreted using voting behavior theory and the literature on extreme political parties.

The Emergence of Hamas

Hamas emerged on the political scene during the Palestinian *Intifada* of 1987. The *Intifada* was primarily a form of civil protest in which civilians clashed with the Israeli military by throwing stones. By the early 1990s, Hamas had carried out its first military attack against Israeli targets. The Oslo Accord (1993) marked the signing by Palestinian Liberation Organization (PLO) and Israel of a "declaration of principles for interim Palestinian self-rule" in the Gaza Strip and the West Bank town of Jericho. Oslo II (1995) granted the establishment of a five-year period of limited autonomy in the occupied territories, beginning in Gaza Strip and Jericho and extending to the entire West Bank. The Jewish settlements would remain under Israeli control. It also provided for the election of a president of the Palestinian Authority and of a Palestinian parliament in the occupied territories, including East Jerusalem, in July 1994. The elections, however, did not take place until 1996.

The Palestinian Authority would establish a Palestinian police force to replace the Israeli troops that would withdraw from the population centers but retain control over all border crossings. Future negotiations would lead to more land transfer to the Palestinians as confidence elevated. Permanent status negotiations were intended to begin no later than May 4, 1996, and were meant to lead to the implementation of Security Council Resolutions 242 and 338, which called for the principle of "land for peace."

Later agreements led to more transfer of land under control of the Palestinian Authority. Progress, however, was very slow and economic conditions continued worsening. Opposition from both sides tried to obstruct the peace accord. On the Israeli side, for example, Baruch Goldstein killed 29 Palestinian worshippers in the Mosque of Abraham on February 24, 1994. On the Palestinian side, in October 1994, Hamas kidnapped an Israeli soldier who was later killed during an Israeli rescue mission. A subsequent suicide bombing in Tel Aviv killed 22 Israelis and injured 50. Yitzhak Rabin was assassinated on November 4, 1995, by Yigal Amir because Rabin pursued peace with the PLO. The subsequent election of Benjamin Netanyahu (1996–1999) slowed the progress towards peace.

The election of Ehud Barak brought with it a renewed hope for a permanent peace between Israel and the Palestinians, but these hopes were short lived with the collapse of the permanent status negotiations at Camp David Summit in 2000. Negotiations focused on reaching an agreement on Israel's security, the final borders of the Palestinian state, Palestinian refugees' right of return, and the status of Jerusalem. Yasser Arafat and Ehud Barak were unable to reach an agreement, especially on sovereignty over Jerusalem, and the peace process collapsed.

Shortly after, Ariel Sharon, surrounded by a massive police force, visited the holy sites in Jerusalem, antagonizing the Palestinians. Since Sharon was responsible for the whole enterprise of Jewish settlements in Gaza Strip and the West Bank, his visit to the holy sites was seen not as a neutral visit but as confirming Israel's claim to Jerusalem despite the fact that Jerusalem was occupied in the 1967 War. Palestinian demonstrators clashed with the Israeli forces, leading to many deaths, which led to a widespread uprising. The Al-Aqsa *Intifada* of 2000, as it came to be known, marked the official death of Oslo. Unlike the 1987 *Intifada*, Al-Aqsa *Intifada* was met with a heavy-handed Israeli military response since the Palestinians now had a police force. It was during this *Intifada* that Hamas was credited with resisting the Israeli military, which enhanced its popularity.

Conditions After Oslo

The conditions that led to the rise of extremist parties were clearly present in Palestine. Whether one chooses a scant look or a detailed investigation of people's circumstances after Oslo, it is very apparent that people's lives were getting worse. I travelled to Gaza in June 1998, July 2000, and July 2005, during which time I interacted with hundreds of people. Gathering data was not an issue since people are endlessly debating politics; everyone talked about politics all the time. It did not take much effort to get information. A simple question of "How are you?" would trigger a three-hour conversation about the life of the community and its individuals. In 1998, five years after Oslo, every person I talked to told me, without exception, that their lives were worse than they had ever been. The overwhelming majority of people I talked to were either unemployed or underemployed, and they told me they would not mind migrating to look for a better life. When I returned in 2000, two months before Al-Aqsa *Intifada*, conditions had reached a boiling point: the level of frustration was higher, the economy was worse, and political corruption had reached its climax. According to the United Nations Millennium Development Goals (MDG), employment-to-population ratio held at 30 percent between 1993 and 2005 (since the signing of the Oslo Accord until one year before the election of Hamas). In other words, roughly 70 percent of the population was either unemployed or, at best, underemployed (Table 1). Furthermore, undernourishment increased from affecting 8 percent of the population in 1996 to 15 percent in 2004.

Among the people with whom I interacted, support for Hamas was very low, though they considered Hamas as a part of mainstream Palestinian society. Hamas's base support was roughly 20 percent, and the then ruling Fatah's was between 40 and 50 percent. Arafat was in charge and made sure that Fatah continued to rule by controlling the treasury and blurring the lines between Fatah and the Palestinian Authority. Thus, even though people were unhappy, they blamed Israel—and not Fatah—for their miserable living conditions. On the other hand, 2005 marked a change. Al-Aqsa *Intifada* (2000) had been in existence for almost five years. According to citizens' testimonies, life was significantly worse than in 1998 and 2000. I saw posters in the city of Beit Lahia, a city north of Gaza, which read, "Hamas is the solution and Beit Hanon is the evidence." I was informed that, in the municipal election in Beit Lahia, this was a campaign slogan for Hamas, drawing on their previous success in Beit Hanon to attract voters. Hamas ultimately won the municipal election in Beit Lahia.

It became apparent that, at least in Gaza, Hamas was becoming powerful. (By and large, however, Palestinian society remains secular, though many people observe their daily prayers.) I also observed violent clashes between Hamas and Fatah-dominated preventive security forces. The word on the street was that "1996 would never come back." They refer to 1996 when the Fatah-dominated PA arrested and tortured members of Hamas as part of its commitment to Israel's security. Hamas had become powerful enough that it could no longer be pushed around. Along with the death of Arafat, the end of Oslo, and Hamas's success at the municipal level, Hamas was now ready for bigger-ticket parliamentary seats. They decided to run in the January 2006 election.

The success of Hamas, therefore, is directly associated with the decline of living conditions, as the above examples demonstrate, and the inability of Fatah to deliver better life opportunities; the conditions that normally lead to the rise of extremist parties were present in Palestine. Now that I have demonstrated the connection between worsening living conditions and the rise of

Table 1	Employment-to-population Ratio												
	1993	**1994**	**1995**	**1996**	**1997**	**1998**	**1999**	**2000**	**2001**	**2002**	**2003**	**2004**	**2005**
Total	28.8	28.7	29	29.3	30.7	33.2	34.3	33.4	29.1	26.4	29.2	29.5	31.2
Men	48.9	48.2	49.2	49.6	52.2	56.6	58.6	57	49	43.9	48.2	47.8	50.7
Women	8.2	8.8	8.4	8.7	9	9.5	9.6	9.3	8.9	8.6	9.8	10.7	11.2

SOURCE: United Nations Millennium Development Goals.[1]

extremist parties, I will assess whether Hamas fit the definition of an extremist party.

Hamas in the Palestinian Party System

It is also important to qualify Hamas's electoral victory. Even though Hamas's electoral victory was momentous to the extent that it was compared to a tsunami, the magnitude of the victory is due to the Palestinian electoral system, which combines features of both the plurality system and the party list proportional representations (PR). If we look at the national level of support for Hamas, even though it is larger than Hamas's usual political base, the numbers are comparable to that of Fatah. Hamas received 440,409 votes (44.45 percent) compared to Fatah's 410,554 votes (41.43 percent), a difference of only 29,855 votes (3.02 percent) (Table 2).

Thus, even though Hamas defeated Fatah in both PR and at the electoral district level, its massive victory was at the local level. At the district level, Hamas won 45 seats compared to Fatah's 17 seats (Table 3).

Candidates elected through party list proportional representation tend to be less responsive to voters than the ones elected in single-member districts (King, 1990). This is because party list candidates are typically more concerned with an abstract party platform, as they tend to be more nationally oriented. Single-member district candidates, on the other hand, are closer to the people and therefore tend to be more responsive to their constituents and more likely to deliver on their promises.

The Palestinian voters, therefore, gave their confidence to Hamas at the local level because they expected their elected candidates to deliver. Since Fatah squandered its chance to deliver for more than 10 years, people decided to give their vote to Hamas, which had proven itself in municipal elections though it did not participate in national

No.	Electoral lists	Number of votes	Percent (%)	Number of seats
Table 2	The Results for the Electoral Lists			
1.	Change and Reform (Hamas)	440,409	44.45	29
2.	Fatah Movement	410,554	41.43	28
3.	Martyr Abu Ali Mustafa (PFLP)	42,101	4.25	3
4.	The Alternative	28,973	2.92	2
5.	Independent Palestine	26,909	2.72	2
6.	The Third Way	23,862	2.41	2
7.	Freedom and Social Justice	7,127	0.72	0*
8.	Freedom and Independence	4,398	0.44	0*
9.	Martyr Abu al-Abbas	3,011	0.30	0*
10.	The National Coalition for Justice and Democracy (Wa'ad)	1,806	0.18	0*
11.	The Palestinian Justice	1,723	0.17	0*
	Total (95.05%)	990,873	100.00	66
	Total number of invalid papers (2.86%)	29,864		
	Total number of blank papers (2.08%)	21,687		
	Total number of electors	1,042,424		

SOURCE: Central Election Commission—Palestine.[2]

NOTE: *Less than the threshold percentage of 19,817 votes.

Table 3	The Final Distribution of PLC Seats in 2006		
No.	Political affiliation	Number of seats in party lists	Number of seats in electoral districts	Total seats
1.	Change and Reform (Hamas)	29	45	74
2.	Fatah Movement	28	17	45
3.	Martyr Abu Ali Mustafa	3	0	3
4.	The Third Way	2	0	2
5.	The Alternative	2	0	2
6.	Independent Palestine	2	0	2
7.	Independents	0	4	4
	Total	66	66	132

Source: Central Election Commission—Palestine.[3]

elections before. In the summer of 2005, I spoke with a secular Palestinian who believed that Hamas should participate in the election—even though he did not support it politically—in order to increase political competition and to end Fatah's political monopoly. I also spoke with a long-time Fatah activist who informed me that he voted for Hamas in the municipal election not because he supported Hamas—he was still a Fatah loyalist—but because he wanted to see a change.

The failure of Fatah-led PA to secure a lasting peace with Israel, along with the death of Arafat, made it politically possible for Hamas to join the parliamentary elections. Hamas previously refused to participate in Palestinian national politics because of its rejection of the Oslo Accord and its inability to justify working through Oslo institutions. Thus, the death of Oslo laid the groundwork for Hamas's entry into national politics. Furthermore, the death of Arafat transformed Fatah from a charismatic centrist party to an institutionalized party. Marwan al-Barguthi, a prominent Fatah leader, describes Arafat's personal style of leadership:

> Talk again of building democratic institutions, meaning decision making by an institution in a democratic way, and talk of collective leadership in the shadow of Yasser Arafat are hopes with no basis in reality: not in Fatah movement, not in the Palestinian people, not in the PLO, and not in

the Palestinian Authority. As long as Arafat exists, he is the alternative to institutions. Yasser Arafat is the institution, and with his existence there will be no institutions (quoted in Brown, 2007, p. 5).

Fatah's power, therefore, stemmed from its affiliation with Arafat as the political symbol of Palestine, which made it infeasible for any other party to compete. The death of Arafat and the lack of any charismatic personality in Fatah after him spurred the institutionalization of Fatah. Angelo Panebianco argues that a "charismatic party is institutionalized when loyalty shifts from the charismatic leader to the organization itself" (cited in Pedahzur & Brichta, 2002, pp. 32–33). In fact, Fatah lost its cohesion after Arafat's death; young activists started to demand that the "old guard" step down and allow them to play a bigger role in setting the party's agenda. The end of Fatah's reliance on Arafat's personality to hold the party together and the transformation of debate on setting the agenda and priorities of the party marked the beginning of the institutionalization of Fatah.

Charismatic centrist parties are more difficult to defeat because they are centered on the personality of the founder. The transformation of centrist Fatah from charismatic to an institutionalized party made it possible for Hamas, already institutionalized, to compete effectively with Fatah. Hamas has been an institutionalized party since its inception because it has always had bureaucratic, routinized process of governance that follows formal rules of conduct. Furthermore, Hamas has a grassroots approach of interacting with the public. Now institutionalized Fatah and institutionalized Hamas were on equal footing where they could be judged by the voters based on their respective performances. Hamas tested the ground by first running for municipal elections; once it proved to itself and its constituency that it was able to deliver, it participated in parliamentary elections and finally won a majority of the seats.

Hamas, therefore, does not fit any of the conditions of extreme parties. First, extreme parties tend to be protest parties that criticize the major parties for what they have failed to accomplish; however, extreme parties do not provide any alternative policy. Second, extreme parties obtain their support through the charismatic personalities of their leaders and tend to be more personalized with a less formal institutional structure (Van der Brug & Mughan, 2007, p. 30). Hamas was neither a protest nor a charismatic party, because its parent Al-Mujama' Al-Islami (or Islamic Congregation) and the present Hamas always had a network of social institutions. It had a set of formal structures that delivered educational and charitable services. Hamas did not see itself as merely protesting the policies of Fatah but also as providing a viable Islamic alternative. Until today, Hamas did not have a charismatic leader either. It has struggled through grassroots activities to have a place among the crowded Palestinian national movement. Because of these characteristics, Hamas does not fit the definition of extremist parties.

Further examination of Hamas's ideological distance from other Palestinian parties shows that Hamas is a centrist party—rather than an extremist or anti-system party—in the context of Palestinian politics. The term "anti-system" typically has been used to refer to parties that aim to destroy the system in which they are embedded (Capoccia, 2002, p. 11). Capoccia, however, recommends that we distinguish between "relational" anti-systemness and "ideological" anti-systemness. Relational anti-systemness is not defined in its own right, but rather it is defined relationally by comparing its ideological stances with other parties in the system. In this case, "an anti-system party will oppose some fundamental values of the regime, which, for its very salience, is shared by all other parties and constitutes a major basis for electoral competition." Because of its ideological distance, an anti-system party

will have a "polarizing" effect on the system (Capoccia, 2002, pp. 14–15).

Ideological anti-systemness, on the other hand, is reflected in an extremist ideology without regard to its relative ideological stance from other parties. This form of anti-systemness is considered a threat to the existing political regime and to democracy itself (Capoccia, 2002, p. 10). Capoccia further argues that a "threat to democracy can be still identified nowadays among the most radical of the Islamic fundamentalist parties and groups that are now active in a number of new democratic regimes" (Capoccia, 2002, p. 13).

Hamas, however, does not fit either the definition of "ideological" or "relational" anti-systemness. Hamas is not an ideological anti-system party, because its ideology is not a threat to the Palestinian democratic institutions. It is true that Hamas has an antagonist relationship with Israel, but anti-systemness is judged based on the threat to domestic political institutions and not on their foreign policies toward their adversaries. Mishal (2003, p. 585) argues that "Hamas, like other Islamic movements, tends to be reformist rather than revolutionary." He also warns against a simple classification of Islamic movements in general and Hamas in particular, which he believes hinders our understanding of these movements. Hamas, he argues, is not a monolithic entity with fixed goals and strategies. It is analytically important to distinguish between Hamas's "grassroots" activists and its military wing. Hamas has maintained a combination of "horizontally and vertically differentiated positions." Vertically, Hamas has a strict line of command to facilitate its various functions (security, military, political, and preaching), while horizontally, it has relied on a web of informal relations to maintain local support (Mishal, 2003, pp. 580–584).

Understanding and adapting to its environment, Hamas has "demonstrated rigidity within the formal Hamas doctrine while showing signs of political flexibility" (Mishal, 2003, p. 576). "Calculated policy based on pragmatic interpretation and negotiated profit/loss consideration rather than on bondage to a stated doctrine and rigid dogma," Mishal maintains, is "thus [what] characterized its mode of operation" (Mishal, 2003, p. 577). Hamas, therefore, is not an ideological anti-system party, because it does not intend to undermine Palestinian political institutions. Its antagonistic view toward Israel does not necessarily make it an anti-system party. Similarly, all Jewish political parties in Israel informally exclude Arab political parties when deciding on coalition partners. The case is much worse for the Palestinians in Gaza and the West Bank whose right to full sovereignty is yet to be recognized by any Israeli political party. Kopelowitz maintains that "there is little ambivalence over the status of Arabs [in Israel]—they are outsiders. With other Jews, however, there is 'no choice' but to negotiate" (Kopelowitz, 2001, p. 168).

Even though the overthrow of Fatah from Gaza by Hamas during the Palestinian civil war (June 7–15, 2007) may seem to reflect Hamas's antisystemic and extremist ideology, it is more likely a reflection on the weak Palestinian constitutional structure. Theoretically, the Palestinian Basic Law provides for the independence and impartiality of the Judiciary; however, neither the Basic Law nor the Judicial Authority Law of 1999 was able to organize properly the relations between the judicial and the executive authorities (Salem, 2006). When conflicts arose between the elected government of Hamas, led by Ismail Haniyeh, and President Abbas of Fatah over the control of security agencies, no sound judicial system was in place to arbitrate between the government and the presidency.

Consequently, the Haniyeh government established the Executive Force to control the lawless streets of Gaza. The existence of partisan rather than Palestinian security apparatuses led to the clash between the Executive Force and some Fatah groups, culminating in the civil war. Hamas insisted that its conflict was with certain factions within Fatah that aimed to undermine its elected government, not with Fatah as a whole or the Palestinian Authority institutions. Hamas continued after the civil war to work with mediators (for example, Egypt) to try to reconcile with Fatah. Even though Hamas overthrew Fatah, it does not seem that this act resulted directly from an anti-system ideology but rather as a result of actual political conflicts on the ground with no institutions capable of resolving conflict. The civil war thus was a result of institutional weakness.

Hamas is not a relational anti-system party either. A close examination of its ideological distance from other Palestinian parties shows that Hamas is close to the political center. I compare Hamas with the other major political parties on the main issues of the campaign: combating corruption, negotiation with Israel, the use of violence,

Jerusalem, refugees, and borders ("Main Issues," Aljazeera, January 24, 2006). All Palestinian parties vowed to fight corruption. Fatah and other political parties wanted to negotiate with Israel, while Hamas remained vague on the issue, including not ruling out a long-term truce with Israel. Fatah did not condone the use of violence even though its military wing, Al-Aqsa Martyrs Brigades, carried out attacks against Israel. Hamas stated that it would use violence, especially to counter Israeli attacks, but has honored a unilateral ceasefire that it announced in February 2005.

All parties claim Jerusalem as the future capital of a Palestinian state and that the refugees have the right to return home. As for borders, Fatah wants a Palestinian state in Gaza and West Bank, while Hamas sees this as a transitional phase in the recovery of historic Palestine. The ideological difference between Hamas and other Palestinian parties, therefore, is very slim (Table 4). The only major difference between both parties is recognition of Israel, which is a foreign policy issue; however, Hamas today is where the PLO was in 1988 when it struggled for international legitimacy. Furthermore, as was demonstrated earlier, Hamas has shown signs of accommodations even though they have maintained a rhetorical hard-line policy.

Hamas has long been successful in its charitable work; most people believe they are less likely to be corrupt since they are religious. They have successfully run some municipalities and were pragmatic in working with Israeli mayors on issues of common concerns such as water. The remaining issue for them was to convince people that they could deliver at the parliamentary level as well. Hamas's message was a mainstream message that it shared with other parties, including its primary rival Fatah. Fatah simply replied that they knew they made a mistake in not properly fighting corruption, and they asked for a second chance. Hamas claimed to have functioned free of corruption.

Hamas's centrism is also demonstrated in the following: its mainstream message (Hamas campaigned under the name Reform and Change List), the inclusion of a Christian candidate on their electoral ticket, distancing itself from Al-Qaeda and Al-Qaeda's beliefs, its grassroots campaign activities, and the inclusion of many female candidates and activists.

Table 4	Ideological Comparison of Palestinian Political Parties					
Main Issues	**Fatah**	**Hamas**	**Martyr Abu Ali Mustafa**	**The Third Way**	**The Alternative**	**Independent Palestine**
Peace with Israel	Yes	Favors a long-term hudna, or truce, instead of "peace"	Yes	Yes	Yes	Yes
Use of violence	No, but its armed factions carried out attacks against Israel	Yes, but honored with Israel	No	No	No	No
Israel's destruction?	No	Yes, but its campaign makes no mention of Israel's destruction	No	No	No	No
East Jerusalem as capital	Yes	Yes	Yes	Yes	Yes	Yes
Right of return for refugees	Yes	Yes	Yes	Yes	Yes	Yes
Borders	Accepts the 1993 Oslo peace accords	Rejects the 1993 Oslo peace accords	Accepts the 1993 Oslo peace accords	Accepts the 1993 Oslo peace accords	Accepts the 1993 Oslo peace accords	Accepts the 1993 Oslo peace accords
Fighting corruption	Yes	Yes	Yes	Yes	Yes	Yes

SOURCE: Aljazeera (2006).

A Christian candidate named Hosam al-Taweel ran on Hamas's electoral ticket, competing for one of six parliamentary seats designated for the Christian community in the Palestinian parliament. When asked by *Aljazeera.net* (January 25, 2006) about what he had in common with Hamas, he replied, "We are fighting for the right of return of displaced Palestinian refugees and fighting corruption in Palestinian governance."[4]

When Hamas was criticized by Ayman al-Zawahiri, Al-Qaida's second-in-command, on March 5, 2006, for its participation in the elections, Mahmoud Al-Zahar, the foreign minister in the elected Hamas government said, "We are a movement that neither brands other Muslims as infidels nor abandons them. We are a movement that lives with the people in a real world and try to attract them to Islam through wisdom and good advice." Al-Zahar continued, "By adopting this moderate approach, which is initiated by the prophet (Muhammad), Hamas movement has succeeded in attracting voters to its programme of Islam" ("Hamas Rejects al-Qaida Reprimand," Aljazeera, March 5, 2006).

Hamas's campaign program can be summarized from analyzing the speeches of various Hamas candidates on the following topics: political pluralism and partnership, Palestinian unity, consolidation of the Palestinian democratic system, social services, end of corruption in the PA institutions, and the empowerment of Palestinian women. As for negotiations with Israel, Hamas has maintained that the problem does not lie in Hamas's refusal to recognize Israel but rather in Israel's occupation of Palestinian lands.

In an interview with Aljazeera's program *With no Limits* on January 18, 2006, the Palestinian Prime Minister, Ismael Hanya of Hamas, said that the problem does not lie in Hamas's position but in the continuation of Israeli occupation, military incursion, arrests, and assassinations.

⊠ Conclusion

The secret of Hamas's electoral success lies not in its extremism but rather in its ideological centrist electoral message. Hamas neither fits the definition of an extremist party nor is its ideological distance from other Palestinian political parties wide enough to qualify as an extreme party. It is neither a protest nor a charismatic party because its parent Al-Mujama' Al-Islami (or Islamic Congregation) and the present Hamas always had a network of social institutions that proved capable of delivering services. Furthermore, Hamas does not see itself as merely protesting the policies of Fatah but also as providing a viable Islamic alternative. Until now, Hamas did not have a charismatic leader either. It has struggled through the efforts of grassroots activities to secure a place in the crowded Palestinian national movement.

In a public opinion survey conducted on April 3–7, 2003, by the Palestinian Center for Policy and Survey Research, before Hamas announced its intention of running for the 2006 parliamentary election, Fatah received 26 percent of public support, while Hamas received only 17 percent. A few days before the election on January 17–19, 2006, 42 percent of respondents said they would vote for the Fatah proportional representation list while 35 percent would vote for Hamas.

In the election of 2006, Hamas won 44.45 percent of the national vote, while Fatah won 41.43 percent. In the single-member electoral district portion of the seats, Hamas won 45 seats compared to Fatah's 17 seats. Thus, the primary difference in number of seats won was not in the proportional representation (Hamas's 29 compared to Fatah's 28), but in the single-member electoral districts seats: Hamas won 45 seats compared to Fatah's 17. The Party platform, which is more relevant in proportional representation, played little role in Hamas's electoral success. Instead, Hamas won decisively at the district level

because its individual candidates promised their constituency reforms and a more competent delivery of services.

Palestinian voters, therefore, elected what they saw as a centrist rather than an extremist Hamas. The institutionalization of both Fatah and Hamas made competition possible and allowed voters to evaluate their respective performances, competences, and electoral programs. Hamas won because voters wanted an alternative to Fatah, especially in light of Hamas's success at service delivery in both its social institutions and at municipal councils. This finding is consistent with expected utility theory of voting behavior that views voters' electoral choices in terms of their relative ideological distance from the parties they choose.

If Hamas pursued a more extremist domestic electoral message, its level of support would have remained at the same level as at the beginning of the campaign or before election time. Extremist parties tend to garner the support of only their narrow base and a small portion of disaffected voters, which impedes their ability to win a majority in parliamentary elections even during the direst economic conditions. For example, for the Nazi Party that ultimately gained control of Germany, the highest level of support it received was little over a third of total votes. King, Rosen, Tanner, and Wagner (2008) show that while the increase in the support for the Nazi Party was a result of extreme economic conditions, its support fluctuated "from less than 3 percent of eligible voters in the December 1924 election to 31.1 percent in July 1932, 26.5 percent in November 1932, and 38.7 percent in March 1933" (King et al., 2008, p. 957).

My analysis, therefore, provides support for Expected Utility Theory and only indirectly calls into question the Prospect Theory. None of the Palestinian political parties ran on an extremist electoral domestic agenda in the 2006 parliamentary election. The elections preemptively compelled these parties to appeal to the center of the political spectrum to be able to win sufficient number of votes. The Arab Spring that brought down the Tunisian president Zine El Abidine Ben Ali and the Egyptian president Hosni Mubarak in early 2011 and that still threatens many Arab autocrats has further moderated not only Hamas but many other Arab Islamist groups as well. Both the Tunisian al-Nahdha Party and Egyptian Muslim Brotherhood, for example, are calling for the establishment of a democratic and civil rather than an Islamic state. They claim to be similar to the Turkish Justice and Development Party, which is presumably modeled after Christian Democratic parties in Europe. In order to garner a sufficient number of potential votes in light of new hopes in Tunisia and Egypt, these Islamic groups had to appeal to the center of prospective electorate in their countries. Even the formerly violent Egyptian Islamic Group has established a political party called Building and Development Party.

Hamas is also increasing its tolerance of Fatah's political activities in Gaza after some period of tightening control following its overthrow of Fatah's political control in Gaza in 2007. The fear of the spread of Arab Spring into the Gaza Strip and the West Bank is forcing the Palestinian Authority and Hamas to start the reconciliation process, especially after the fall of Mubarak, which made an Egyptian neutral arbitration between both Palestinian factions a reality. Abbas is also seeking recognition from the United Nations for Palestinian statehood because the Palestinian public no longer tolerates endless negotiations with no end in sight. In short, in Palestine and other Arab countries, governments are becoming more sensitive to the demands of their people.

 Notes

1. Retrieved April 17, 2010, from http://mdgs.un.org/unsd/mdg/Data.aspx.

2. Retrieved April 17, 2010, from http://www.elections.ps/CECWebsite/events/ elections2006/results.aspx.

3. Ibid.

4. Hosam al-Taweel is listed as independent. He is one of the many independent candidates who is affiliated with and is endorsed by Hamas.

 References

Alexander, Yonah (2002). *Palestinian religious terrorism: Hamas and the Islamic Jihad*. Ardsley, NY: Transnational Publishers.

Aljazeera (2006, January 24). Main issues: Here are some of the main issues surrounding the Palestinian elections. *Aljazeera.com*. Retrieved July 5, 2007, from http://www.aljazeera.com/archive/2006/01/200849163938504228.html

Brown, Nathan J. (2007). *Requiem for Palestinian reform: Clear lessons from a troubled record*. Democracy and Rule of Law Books, No. 81. Washington, DC: Carnegie Endowment for International Peace. Retrieved July 5, 2007, from http://www.carnegieendowment.org/files/cp_81_palestine_final.pdf

Capoccia, Giovanni (2002). Anti-system parties: A conceptual reassessment. *Journal of Theoretical Politics*, *14*(1), 9–35.

Chhibber, Pradeep K. (1996). State policy, rent seeking, and the electoral success of a religious party in Algeria. *The Journal of Politics*, *58*(1), 126–148.

Church, Clive H. (2008). The Swiss elections of 21 October 2007: Consensus fights back. *West European Politics*, *31*(3), 608–623.

Creswell, John W. (2003). *Research design: Qualitative, quantitative, and mixed methods approaches* (2nd ed.). Thousand Oaks, CA: SAGE Publications.

Davis, Otto A., Hinich, Melvin J., & Ordeshook, Peter C. (1970). An expository development of a mathematical model of the electoral process. *American Political Science Review*, *64*(2), 426–448.

Downs, Anthony (1957). *An economic theory of democracy*. New York: Harper & Row.

Enelow, James M., & Hinich, Melvin J. (1982). Ideology, issues, and the spatial theory of elections. *American Political Science Review*, *76*(3), 493–501.

Enelow, James M., & Hinich, Melvin J. (1984). *The spatial theory of voting: An introduction*. New York: Cambridge University Press.

Grofman, Bernard (1985). The neglected role of the status quo in models of issue voting. *Journal of Politics*, *47*(1), 230–237.

Hinich, Melvin J., & Pollard, Walter (1981). A new approach to the spatial theory of elections. *American Political Science Review*, *25*(2), 323–341.

Jackman, Robert W., & Volpert, Karin (1996). Conditions favouring parties of the extreme right in western Europe. *British Journal of Political Science*, *26*(4), 501–521.

Kahneman, Daniel, & Tversky, Amos (1979). Prospect theory: An analysis of decision under risk. *Econometrica*, *47*(2), 263–292.

King, Gary (1990). Electoral responsiveness and partisan bias in multiparty democracy. *Legislative Studies Quarterly*, *15*(2), 159–181.

King, Gary, Rosen, Ori, Tanner, Martin, & Wagner, Alexander F. (2008). Ordinary economic voting behavior in the extraordinary election of Adolf Hitler. *The Journal of Economic History*, *68*(4), 951–996.

Kopelowitz, Ezra (2001). Religious politics and Israel's ethnic democracy. *Israel Studies*, *6*(3), 166–190.

Levitt, Mathew (2004). Hamas from the cradle to the grave. *Middle East Quarterly*, *11*(1), 1–12.

Levitt, Mathew (2006). *Hamas: Politics, charity, and terrorism in the service of Jihad*. New Haven: Yale University Press.

Mathews, Stevens A. (1979). A simple directional model of electoral competition. *Public Choice*, *34*(2), 141–156.

Merrill, Samuel, III, & Grofman, Bernard (1999). *A unified theory of voting: Directional and proximity spatial models*. New York: Cambridge University Press.

Mishal, Shaul (2003). The pragmatic dimension of the Palestinian Hamas: A network perspective. *Arms Forces and Society*, *29*(4), 569–589.

Palestinian Center for Policy and Survey Research (2003, April 3–7). Appointment of prime minister, political reform, roadmap, war in Iraq, Arafat's popularity, and political affiliation. Palestinian Center for Policy and Survey Research, Poll #7. Retrieved August 15, 2011, from http://www.pcpsr.org/survey/ polls/2003/p7a.html

Palestinian Center for Policy and Survey Research (2006, January 17–19). Special public opinion poll on the upcoming Palestinian elections. Palestinian Center for Policy and Survey Research. Retrieved August 15, 2011, from http://www.pcpsr.org/survey/polls/2006/preelectionsjan06.html

Pedahzur, Ami, & Brichta, Avraham (2002). The institutionalization of extreme right-wing charismatic parties: A paradox? *Party Politics*, 8(1), 31–49.

Rabinowitz, George, & Macdonald, Stuart Elaine (1989). A directional theory of issue voting. *American Political Science Review*, 83(1), 93–121.

Salem, Jamil (2006, July 6). *The Palestinian judicial system.* Paper presented at the annual meeting of the Law and Society Association, Baltimore, MD.

Tepe, Sultan (2000). A Kemalist–Islamist Movement? The Nationalist Action Party. *Turkish Studies*, 1(2), 59–72.

Tepe, Sultan (2005). Religious parties and democracy: A comparative assessment of Israel and Turkey. *Democratization*, 12(3), 283–307.

Van der Brug, Wouter, & Mughan, Anthony (2007). Charisma, leader effects and support for right-wing populist parties. *Party Politics*, 13(1), 29–51.

Wael J. Haboub is a PhD candidate in political science at the University of Illinois at Chicago. He is a lecturer at Northeastern Illinois University, where he teaches courses in comparative politics, international relations, and Middle East and Islamic politics. [email: whabou2@uic.edu]

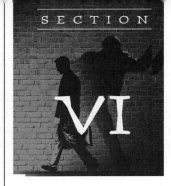

SECTION

VI

The Middle East and Africa

Learning Objectives

At the end of the section, students will be able to:

- Describe the major terrorist groups in the region and their associated attacks from the early 1970s to the present.
- Describe the role and importance of Saudi Arabia in promoting religious extremism in the region.
- Describe the impact of Iran and Libya on terrorist groups.
- Discuss the underlying tensions within and between the major Middle Eastern and African states that have led to modern-day terrorism in the region.
- Assess the threat posed by al Qaeda and its affiliates before and after the death of Osama bin Laden.
- Analyze the impact of the Arab Spring on terrorism in the region.

Section Highlights

- The Middle East
- North Africa
- Africa

✖ Introduction

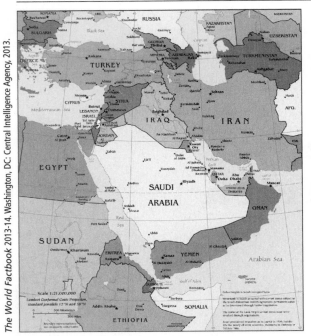

The World Factbook 2013-14. Washington, DC: Central Intelligence Agency, 2013.

▲   Map of Middle East

Calling this section the *Middle East and Africa* does not provide the reader with as much of a road map to the section as one might think. Scholars differ widely as to what, exactly, the term *Middle East* refers to. Most, however, agree that the term *Middle East* represents a Eurocentric and Western perspective. After all, the region is only in the East if you are standing in Europe or the United States. If you are standing in China or India, the Middle East is in an entirely different direction (Catherwood, 2006; Lewis, 1995; White, 2012). Even in the West, the region has not always been referred to as the Middle East. Other terms used over the past centuries include the Near East, the Levant, the Orient (a term now disfavored), and the Fertile Crescent (the area around the old Mesopotamia and now governed mostly by **Iraq**).

The term *Middle East* is also synonymous for many with the concept of the Muslim or Islamic world. This assumption ignores the fact that the vast majority of Muslims do not even live in the Middle East. Indeed, the largest Muslim population can be found in Indonesia—definitely not part of the Middle East no matter what definition is used. In addition, many of the residents of the region referred to as the Middle East are not Muslim at all. Finally, several countries often described in terms of terrorism in the Middle East—such as **Libya** and **Algeria**—are actually in North Africa.

As one can probably guess by now—like the term *terrorism*—the term *Middle East* is subject to different meanings for different people. Still, it is the term that seems to be most commonly used and will be for this text as well. For organizational purposes, this section will also draw on the divisions of the U.S. Department of State's regional bureaus. Using the State Department divisions, the first subsection of this topic will address terrorism stemming from the Middle East and North Africa (what the State Department refers to as the Bureau of Near Eastern Affairs). The discussion in this subsection will focus on terrorist groups emanating from the countries of **Egypt, Saudi Arabia**, and **Yemen** (the countries of Israel, Palestine, Lebanon, and Jordan were discussed in an earlier section for obvious reasons). The second subsection of this topic will cover the countries of North Africa—most notably Algeria and Libya. Finally, this section will conclude with a brief discussion of the region assigned to the Department of State's Bureau of African Affairs.

One should note that, while the terrorist groups described in this section are Islamic, not all terrorist groups are trying to create an Islamic state or are even based on Islamic ideology. As will be seen in later sections, other regions and countries have contributed to the extensive list of Foreign Terrorist Organizations—religious and non-religious alike—maintained by the U.S. Department of State. In addition, it is important to note that this section does not equate all Islamist movements in the Middle East and **Africa** with violent terrorist organizations or as active sponsors of such organizations. Indeed, as a recent essay for the Carnegie Endowment for International Peace notes, a number of Islamist movements have chosen to participate in the legal political processes of their countries and seek change within those legal political processes (Hamzawy, Ottoway, & Brown, 2007).

⬣ The Middle East

Saudi Arabia

With a land mass of over 2 million square miles inhabited by 26 million people, Saudi Arabia occupies the vast majority of the Arabian Peninsula and is the second largest Arab country behind Algeria. It is slightly more than one fifth the size of the United States, although most of its land is uninhabitable desert with only 1.45% of its land mass designated as arable (*CIA World Factbook,* n.d.). This is one of the key features that led this mysterious land to be dominated by nomadic or semi-nomadic Bedouin tribes well into the 20th century, and the influence of Bedouin culture, traditions, and perspectives can still be seen throughout modern Saudi Arabia. Ethnically, Saudi Arabia is 90% Arab and 10% Afro-Asian (*CIA World Factbook,* n.d.). Its official language is Arabic.

Saudi Arabia is also the birthplace of Islam and the home of Islam's two most holy sites: the mosques in Mecca and Medina. In fact, Islam requires its followers to make a pilgrimage to Mecca at least once in their lifetime. Not surprisingly, the official religion of the Kingdom of Saudi Arabia is Islam. Kingdom law forbids the public practice of any other religion except for very limited circumstances by non-citizens (Spindlove & Simonsen, 2013, p. 294).

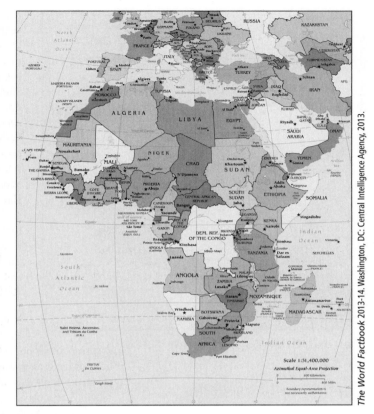

▲ Map of Africa

The World Factbook 2013-14. Washington, DC: Central Intelligence Agency, 2013.

Around 17% of the world's oil reserves lie under the soil of Saudi Arabia. It is the world's largest exporter of petroleum. Indeed, oil accounts for approximately 80% of the country's budget revenue, 45% of its gross domestic product (GDP), and 90% of its export earnings. At the same time, the Saudi oil industry is highly dependent on foreign workers, over 5 million of whom reside in the Kingdom (*CIA World Factbook,* n.d.). A considerable amount of the oil revenues have found their way into the personal fortunes of the Saudi royal family, which includes over 5,000 princes.

The modern-day Kingdom of Saudi Arabia was founded in September 1932 when the Arabian Peninsula was united under the control of **Ibn Saud**, who became the first Saudi king. However, events that ultimately led to the founding of the new kingdom can be traced back several centuries earlier. In the second half of the 18th century, the leader of the Al Saud family—Muhammad bin Saud—made an alliance with Muhammad ibn Abd al Wahhab, the leader of a religious movement that would become known as **Wahhabism**. Wahhabism is a strict, puritanical form of Islam that tolerates no dissent and no shared interpretations of Islam. It is based on the interpretations of Islam as it existed at its origins in 7th-century Medina. Wahhabism in particular targets the beliefs and practices of the Shia sect of Islam.

The support of the Wahhabis allowed the bin Saud family to gain control over a considerable amount of the Arabian Peninsula in the 18th and 19th centuries. The support of the descendants of al Wahhab—the Al ash-Sheikh family—and the military resources they provided were once again key when Ibn Saud captured the city of

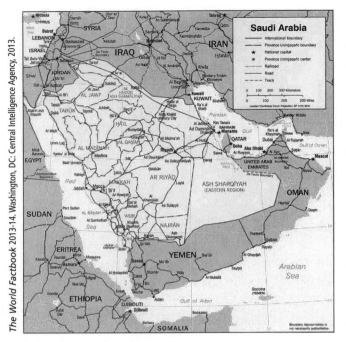

▲ Map of Saudi Arabia

▲ The United States Rewards for Justice poster for Osama bin Laden, founder of the al Qaeda terrorist organization.

Riyadh in 1902 and then spent the remaining 30 years uniting the rest of the peninsula, culminating in the creation of the Kingdom of Saudi Arabia in 1932. In return, Wahhabism became the defining religion of the new kingdom with the Al ash-Sheikh given considerable control over the Kingdom's clerical matters. Even though Ibn Saud eventually had to defeat the military wing of the Wahhabists to avoid losing British support for his position, the influence of Wahhabism over the religious, cultural, social, and political life of the Kingdom cannot be overestimated (Metz, 1992). Having the resources of such a wealthy power base as Saudi Arabia has also allowed Wahhabi clerics to expand the movement to other regions, and the influence of Wahhabism can be seen throughout the Middle East as well as in Southwest and Central Asia, and in the ideology of such groups as the Taliban. As noted by Lawrence Wright,

Not content to cleanse its own country of the least degree of religious freedom, the Saudi government set out to evangelize the Islamic world, using the billions of riyals at its disposal through the religious tax—*zakat*—to construct hundreds of mosques and colleges and thousands of religious schools around the globe, staffed with Wahhabi imams and teachers. Eventually Saudi Arabia, which constitutes only 1 percent of the Muslim population, would support 90 percent of the expenses of the entire faith, overriding other traditions of Islam. (2006, p. 149)

Saudi Arabia and al Qaeda

Perhaps the most famous, or infamous, terrorist group of the 20th and 21st century is that of **al Qaeda**—which in English translates as The Base—founded by the now notorious **Osama bin Laden**. Ironically, although bin Laden was shaped by the political, religious, and social life of a wealthy Saudi citizen, the main ideology of al Qaeda can be traced to the Islamic writer and activist **Sayyid Qutb** (discussed in more detail in the subsection on Egypt, below).

Al Qaeda was named for the idea that it forms both the physical and ideological base for global Islamic jihad that will one day bring all the world under the true and pure practice of Islam. There are, as of the writing of this book, hundreds of books and journal articles on al Qaeda. No text can keep up with an exhaustive review of the work available on this group—both bad and good. This subsection therefore represents a mere summary drawing in particular from works by Lawrence Wright, Brian Jenkins, Michael Scheuer, and assessments by the U.S. Department of State, the National Counterterrorism Center, and the Combating Terrorism Center at the U.S. Military Academy.

Born out of the fight against the Soviet Union in Afghanistan, al Qaeda first came to the attention of Western—and U.S. officials in particular—in 1988.[1] At first the U.S. and the Arab fighters were on the same anti-Soviet side. Ultimately, the long-term goals of al Qaeda would become several-fold: to overthrow all apostate Muslim regimes (defined as regimes that do not follow and require the purist form of Islamic practice), to remove Western—especially U.S.—influence from the Middle East and Asia, to defeat Israel, and to recreate a united Islamic Caliphate (an institution that had not been seen since the height of the **Ottoman Empire**) under strict Sunni interpretations of Islamic law and governance (National Counterterrorism Center, 2012; U.S. Department of State, Country Reports, 2011).

The founder of al Qaeda, Osama bin Laden, was born in January 1958, the 17th son of Mohammed bin Laden, a self-made billionaire with close personal ties to the Saudi royal family. By the time Osama bin Laden was born, the Saudi bin Laden Group had been responsible for most of the construction that followed Saudi Arabia's discovery of oil and the country's resulting explosion into modernity and modern infrastructure (Wright, 2006). Although a relatively minor player in the family hierarchy, bin Laden nevertheless inherited a significant fortune, which he put to a use his father probably never envisioned.

Early on in his adolescence, bin Laden demonstrated his well-known penchant for religious zeal and the ideas of violent jihad. He was strongly influenced by the teachings of Qutb and the Salafist (known in the West as Wahhabi) that dominated the religious life of Saudi Arabia. Early in the 1980s, he was drawn to the fight in Afghanistan. With others from Saudi Arabia and other Muslim countries, he established a Services Bureau in Peshawar, Pakistan—on Afghanistan's eastern border—to funnel fighters to the mujahedeen groups fighting in the Afghan mountains, drawing on his vast financial resources and contacts to provide support for the Arabs flowing into Pakistan. Eventually, he created his own training base and fighting group in a compound in the Afghan mountains known as the Lion's Den (Wright, 2006).[2] It was also at this time that he met **Ayman Zawahari**—an Egyptian physician and Islamist who had broken with the **Muslim Brotherhood** over ideological differences to establish his own group—**Egyptian Islamic Jihad** (EIJ). Originally, Zawahari's and bin Laden's long-term goals were somewhat different, with Zawahari more focused on an Egyptian revolution. By the end of the Soviet invasion, however, their goals had grown closer together and they remained in Pakistan and Afghanistan to form what became the current al Qaeda organization.

When bin Laden returned to Saudi Arabia from Afghanistan in 1989, his welcome and that of his new organization by the Saudi government was less than he had expected and certainly less than he saw as his due. His plans to use his returning Afghan fighters to overthrow the Marxist government of Yemen (the region of his father's birth) were quickly dismissed as a "bad idea" (Wright, 2006). The Saudi king also became increasingly annoyed at bin Laden's efforts to intervene in Saudi foreign policy and its relations with neighboring states. When bin Laden's offer to use his organization to repel any threat resulting from Iraq's 1990 invasion of neighboring Kuwait was flatly rebuffed and Saudi Arabia turned to the United States, relations between bin Laden and the Saudi royal family rapidly began to deteriorate.

Finally, in 1991, only two years after returning in "triumph" from jihad in Afghanistan, bin Laden was expelled from his home country. He settled with his remaining followers in Khartoum, **Sudan**, at the invitation of the new Islamic government there. Bin Laden poured considerable money and infrastructure into his host country—and the

---

[1]Former CIA bin Laden unit chief Michael Scheuer argues that the signs of the threat from al Qaeda were available long before 1988, but missed by the U.S. intelligence and policy communities (Scheuer, 2005).

[2]Wright argues that, in truth, the Afghan Arabs (as they were called) led by bin Laden and jihadists from Saudi Arabia and Egypt never had much of an impact on the actual fighting and were viewed by the Afghan tribal militias as something of a joke. Far more important to the Afghan mujahedeen were the funds that flowed from the United States through Pakistan's Inter-Services Intelligence service.

pockets of its leaders—and for a while it seemed as if both bin Laden and the Sudanese government had found the perfect partnership. There, bin Laden increased his rhetoric against the United States—rather than just the West in general—angered by the continued presence of American troops in Saudi Arabia even though Kuwait had been liberated months before.[3] Bin Laden's anger was increased by the U.S. effort to intervene in the famine in Somalia, which the U.S. viewed as a humanitarian effort and bin Laden saw as an effort to seize "control of the pressure points of the Arab world and [push] into al Qaeda's arena" (Wright, 2006, p. 170).[4] Even after he was expelled from Saudi Arabia, his activities such as those against the newly formed united Yemen proved embarrassing to the Saudis and contributed to the international pressure building on Sudan to distance itself from bin Laden. In 1996, Sudan caved, and bin Laden and al Qaeda were asked to leave. They returned to Pakistan and were eventually given safe haven by the Taliban government then controlling Afghanistan. Bin Laden would never again leave Southwest Asia.

From its base in Afghanistan, al Qaeda was able to launch its deadly crusade against its enemies (real and perceived). In February 1998, bin Laden publicly declared "war" against the United States—now his main adversary outside the Saudi royal family. He issued a statement under the guise of the "World Islamic Front for Jihad Against the Jews and Crusaders," calling for all Muslims to fulfill their duty of jihad by killing all U.S. civilians and their allies anywhere they could be found; this duty presumably included all other "nonbelievers" as well (U.S. Department of State, Country Reports, 2010; National Counterterrorism Center, 2012). In May, the statement was expanded to specifically call for holy war against all foreign forces in the Arabian Peninsula. In reality, though, al Qaeda was simply intensifying attacks against the West that had started during its time in the Sudan. These included:

- Three bombings in Aden, Yemen, in December 1992 targeting U.S. troops.
- Claims to have shot down U.S. helicopters and killed U.S. soldiers in Somalia in 1993.

Three major attacks followed the February 1998 pronouncement:

- Two simultaneous car bombings in August 1998 against the U.S. embassies in Nairobi, Kenya, and Dar es Salaam in Tanzania; over 500 people were killed and over 5,000 injured in the two bombings.
- The October 2000 suicide attack on the **USS Cole**, a guided missile destroyer docked in Aden's port; 17 U.S. sailors were killed and another 39 were injured (U.S. Department of State, 2012).
- The September 11, 2001, attacks against the Twin Towers of the World Trade Center in New York City and the Pentagon in Washington, D.C. (see Section 10); over 3,000 people were killed in the attacks.

Following the devastating attacks of 9/11, U.S. forces entered Afghanistan, overthrew the Taliban government, and forced al Qaeda out of the Afghan mountains back into Pakistan. Despite taking significant losses, al Qaeda was able to continue attacks against the United States and its allies, some suggest with the tacit knowledge if not support of Pakistan's intelligence service (see Section 9). The attacks included:

- A November 2002 suicide bombing of a hotel in Mombasa, Kenya, that left 15 dead.
- More than a dozen attacks in Saudi Arabia between 2003 and 2004 that left 90 dead, including 14 Americans.
- Claimed responsibility for attacks in 2004 on the Madrid train system and the attacks in 2005 against the London subway system. (U.S. Department of State, 2012)

In addition, al Qaeda or **al Qaeda affiliates** were likely behind several disrupted attacks, including plans to use liquid explosives to destroy several commercial airliners traveling from the UK to the United States, plans to attack

---

[3]American troops had remained at air bases in Saudi Arabia to monitor the cease fire agreement.

[4]Ironically, as Wright (2006) notes, "this thinking took place at a time when the United States had never heard of Al-Queda, the mission to Somalia was seen as an act of thankless charity, and Sudan was too inconsequential to worry about" (p. 170).

the New York subway system, an attempt to destroy a Northwest Airlines flight en route from Amsterdam to Detroit, and a plot to place explosive-laden packages on cargo flights coming into the United States.

Over the years, the United States has been successful in capturing or killing numerous members of the al Qaeda leadership and key operatives, but none more important than Osama bin Laden himself, who was killed in a daring May 2011 special forces raid on his compound in Abottabad, Pakistan. After careful identification, his body was buried at sea. The location of bin Laden's compound, only blocks away from Pakistan's version of West Point, has fueled considerable speculation that members of Pakistan's intelligence services and military were aware of bin Laden's location and were complicit in hiding him from their U.S. "allies." Ayman al-Zawahari, bin Laden's long-time number two, and as of this writing, its new leader, is still at large.

Al Qaeda no longer looks the way it did in 2001, and its core has been severely disrupted. But by no means is it less dangerous. In many ways, the danger from al Qaeda has become harder to deal with as it has become more of an ideological umbrella for affiliate groups that have spread across the Middle East, Africa, and Asia. These groups actively recruit within the United States and Europe (see Sections 9 and 10). As noted by the U.S. Department of State in its annual report on international terrorism in 2012:

> AQ [al Qaeda] serves as a focal point of "inspiration" for a worldwide network of affiliated groups—al Qa'ida in the Arabian Peninsula (AQAP) [Yemen], al-Qa'ida in Iraq (AQI), al-Qa'ida in the Islamic Maghreb (AQIM) [Algeria], al-Shabaab—and other Sunni Islamic extremist groups, including the Islamic Movement of Uzbekistan, the Islamic Jihad Union, Lashkar i Jhangvi, Harakat ul-Mujahadin, and Jemaah Islamiya. Tehrik-e Taliban Pakistan and the Haqqani Network also have ties to AQ. Additionally, supporters and associates worldwide who are "inspired" by the group's ideology may be operating without direction from AQ central leadership, and it is impossible to estimate their numbers. (U.S. Department of State, 2012, n.p.)

This assessment is supported by terrorism scholars such as Brian Jenkins of the RAND Corporation who notes that competing views of the threat still posed by al Qaeda:

> derive from the fact that Al-Queda is many things at once and must therefore be viewed in all of its various dimensions. It is a global terrorist enterprise, the center of the universe of like-minded fanatics, an ideology of violent jihad, and an autonomous online network. It is a virtual army. Increasingly, it is a conveyor of individual discontents. (Jenkins, 2012, p. 2)

Jenkins goes on to note that al Qaeda "survives best when it can attach itself to deeply rooted local movements, which it then proceeds to radicalize" (2012, p. 3).

Yemen

Yemen has long been both a haven and a place of operations for al Qaeda and/or al Qaeda affiliates and is included in this section for that reason.

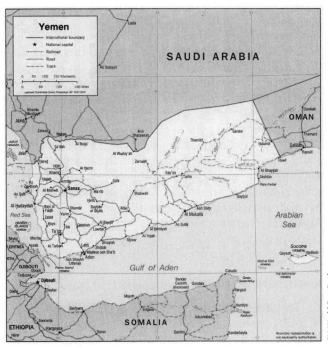

▲   Map of Yemen

CIA World Fact Book

Yemen occupies almost 528,000 square miles wedged between the southern border of Saudi Arabia, the western border of Oman, and the Gulf of Aden. In land mass, it is about twice the size of Wyoming. Its 25.4 million inhabitants are mostly Arab, with some Afro-Arabs, South Asians, and Europeans (*CIA World Factbook,* n.d.). The official language is Arabic and the official religion is Muslim (includes both Sunni and Shia sects). It is a republic with a president, a prime minister, and a bicameral legislature, although as of this writing, these institutions have collapsed before a threatening civil war. Its legal system is a mixture of Islamic law, Napoleonic code, English common law, and customary law.

Yemen started out as two separate countries: North Yemen and South Yemen. North Yemen gained its independence from the Ottoman Empire in 1918, while South Yemen remained a British protectorate until 1967. Three years later, South Yemen adopted a Marxist-style government and the remainder of the 1970s was characterized by a series of border wars between the two states. In 1990—partly spurred by the need to effectively exploit discoveries of oil in "the ill-defined border region between the two impoverished countries" (Wright, 2006, p.153)—North and South Yemen were able to form an agreement and became the united Republic of Yemen on May 22, 1990.

Along with many of its neighbors, the Yemeni government faced a challenge to its power as part of the Arab Spring in 2011. Under pressure from the UN and the Gulf Cooperation Council (GCC), President Saleh stepped down in November 2011 in favor of his vice president who then won the elections held in February 2012. Though fragile, the Yemeni government had been successful over the following four years in holding its newly unified territory together despite the existence of several southern secessionist groups, conservative tribal groups, and other anti-government militias. Until 2015, its most current threat stemmed from a particularly virulent al Qaeda affiliate known as **al Qaeda in the Arabian Peninsula** (AQAP).

Al Qaeda in the Arabian Peninsula (AQAP)

Al Qaeda elements had been in the Yemeni territories since at least the early 1990s at varying levels of intensity. But in 2009, the leader of al Qaeda in Yemen (AQY) announced that Yemeni and Saudi al Qaeda operatives were "working together under the banner of AQAP" (NCTC, 2012; U.S. Department of State, 2012, n.p.). In the years since that announcement, AQAP has been particularly active in its efforts against the United States and its allies, and what AQAP considers apostate Muslim regimes. Seventy-one percent of its attacks between 2004 and 2011 took place after the announced merger (START, 2011).

AQAP is responsible for several attempts to conduct terrorist attacks on U.S. soil, including (as noted above) an attempt to destroy a Northwest flight en route from Amsterdam to Detroit (conducted by Umar Farouk Abdulmutallah, known as the "underwear bomber" because he tried to blow up the plane with explosives concealed in his underwear), and a plot to place explosive-laden packages on cargo flights coming into the United States. Other countries targeted include Saudi Arabia, the United Arab Emirates, and the UK. An average of four fatalities per attack makes AQAP one of the more lethal groups at the time of this writing; over one third of their attacks are against Yemeni military targets while the remainder target private businesses, tourists, utilities, diplomatic and other government officials, and religious figures and institutions (START, 2011). Although the Yemeni military has had some significant successes against AQAP—especially in regaining previously AQAP-held territory—AQAP still retains a substantial capability to carry out lethal attacks. It continued to carry out attacks in 2012 and 2013 against Yemeni military and civilian targets, including a May 2012 suicide bombing targeting Yemeni soldiers rehearsing for a parade celebrating Yemen's Unification Day (U.S. Department of State, 2012).

And in 2015: AQAP, the Houthis, and Civil War?

The Sunni/Shia split underlying much of the fighting in Iraq and Syria has also intensified in Yemen. As of the spring of 2015, fighting among militant groups has brought Yemen to the brink of an all-out civil war; two of the main players in this tragedy are AQAP and the Houthis.

The Houthis belong to an offshoot of Shiite Islam known as the Zaydis and stem mostly from the northern regions of Yemen ("Behind the Houthi Insurgency," 2015). They have been fighting an on-again, off-again insurgency against the Saleh and the Hadi governments, as well as AQAP for over a decade. A critical turning point, however, came in September 2014, when they overran the airport of the capital of Yemen, San'a. The Yemeni government collapsed in February 2014 and President Hadi fled into hiding (Almasmari, Fitch, & Nissenbaum, 2015). The resulting power vacuum has provided yet another opportunity for Islamist groups such as AQAP to increase their strength in terms of personnel, resources, and even territory. As of this writing, AQAP and an ISIS affiliate are vying for a controlling role on the Sunni side as each tries to outdo the other in number and brutality of attacks (al-Mujahed & Morris, 2015). The ISIS affiliate claimed responsibility for a March 20 attack on two Shiite mosques that killed over 130 people, at least 13 of whom were children. This spurred an even more aggressive offensive by the Houthi rebels and by the end of March, they controlled Yemen's third largest city, Taiz, and were closing in on the port of Aden forcing Hadi to reportedly flee the country. As of March 2015, the United States had closed the embassy and pulled all remaining diplomatic and military personnel out of the country.

The Hadi government had been an important ally in U.S. counterterrorism efforts. The collapse of the Yemeni government and the inability of the United States to maintain a presence in the country will significantly risk the ability of the United States to counter the activities of AQAP, a group with the proven ability and motivation to conduct attacks in the United States and Europe ("Behind the Houthi Insurgency," 2015; Schmitt, 2015). In addition, the conflict in Yemen threatens to engulf its surrounding neighbors the way a black hole engulfs neighboring stars and planets. In fact, Bruce Reidel, a noted terrorism and intelligence expert, used the analogy of a black hole in a March 2015 interview with the *Wall Street Journal*, noting that Saudi Arabia—as well as the U.N. Security Council—supports President Hadi while Iran, although it denies the claim, is believed to be supporting the Shiite Houthis (Almasmari et al., 2015). On March 29, 2015, Egyptian president al-Sisi on behalf of the Arab League announced the creation of a joint Arab military force in response to the conflicts in Iraq, Syria, and Yemen. At the time of the announcement, a Saudi-led joint Arab force was already conducting operations against Houthi targets in Yemen ("Arab League Agrees," 2015).

Egypt

With 1,001,450 square miles of land mass, Egypt is about three times the size of New Mexico. Its population of over 85.3 million is 99.6% Egyptian, 90 % of whom are Muslim. It is bordered by Libya to the west and Sudan to the south, while its northern and much of its eastern boundaries are delineated by the Mediterranean and Red Seas, respectively. The official language is Arabic although English and French are widely understood by the educated classes (*CIA World Factbook*, n.d.). At the time of this writing, the government and legal systems of Egypt were in a state of flux after the overthrow of two presidents between 2011 and 2013.

Egypt is home to one of the world's greatest civilizations in history. The Egyptian dynasties ruled a unified kingdom for almost three millennia starting in 3200 BCE. Subsequently it was subjected over the centuries to conquest by the Persians, Greeks, Romans, Byzantines, Arabs (who introduced Islam to the region in the 7th century), Turks, and the British. Britain took control of Egypt's government in the late 19th century in order to protect British shipping interests through the Suez Canal. Egypt did not gain its full independence until 1952. In 1981, Hosni Mubarak became president after President Anwar Sadat was assassinated for his efforts to pursue peace with Israel, at the time viewed as Egypt's number one enemy. The harsh and autocratic regime of Mubarak that followed ended in 2012 as part of the Arab Spring.

Terrorist groups have operated out of Egypt with varying intensity since its independence, and some of the main ones are highlighted in this section. In addition, there have long been concerns about aid provided by Egyptian groups to Hamas and other groups operating in the Palestinian territories by smuggling arms and other provisions into Gaza from the Sinai. This has been exacerbated recently given the uncertain government presence in the area since the overthrow of Mubarak (Department of State, Country Report, 2010; Jenkins, 2012).

Highlights of Terrorist Groups in Egypt

This subsection will focus on the goals, motivations, capabilities, and activities of the three following Egyptian groups: the Muslim Brotherhood (MB), **Gama'a al-Islamiyya** (IG), and Egyptian Islamic Jihad (EIJ).

Muslim Brotherhood

The Brotherhood was founded in Egypt in 1928 by Hassan al Banna to return purity to the practice of Islam and to create a single Muslim nation, originally through education and reform (White, 2012). It has been at various times both a terrorist group and a "school of thought" that has influenced the development of Islamic groups all over the world with varying levels of radicalization (Vidino, 2011a). In common is a belief in Islam as the basis for a political system, but views differ widely among the groups the Brotherhood has inspired as to how that is to be achieved (Vidino, 2011a). Although the group has "officially" renounced violence and terrorism, its interpretations of Islam have served as inspiration for terrorist groups espousing nothing but violence, such as Hamas in the Palestinian territories and Jemaah Islamiya (JI) in Indonesia.

One particularly militant wing of the Brotherhood was founded by Sayyid Qutb, a radicalized Egyptian journalist and teacher (but not a recognized Muslim scholar) who spent a number of years living in the United States (Wright, 2006). He espoused violence to achieve the goals of the Brotherhood. He was jailed by the Egyptian government for attempting to overthrow the government in 1954; the Brotherhood was banned as an organization that same year. Qutb was released in 1964, published his most influential tract—*Milestones*—in 1965, was rearrested as a result, and was hanged in 1966. His influence far outreached his life, however, as his religious and militant views—including using violence to overthrow governments that did not follow pure Islamic practice, as well as their Western allies—provided much of the justification bin Laden used for al Qaeda's activities and goals (White, 2012).

The Arab Spring of 2012 changed much; at least much was hoped for. Once outlawed by the government, the Brotherhood was soon seen as a genuine player in the political process. Indeed, its representatives—through the Brotherhood's Freedom of Justice Party—won a majority of the parliamentary seats and the presidency in Egypt's first democratic elections following the fall of Mubarak (Newton-Small & Hauslohner, 2012). A Carnegie Endowment essay argued recently that the ability to create a legal political party separate from the Brotherhood (seen as a religious organization) is one of the key factors needed to establish the credibility of Islamist movements such as the Brotherhood as political and democratic actors on the world stage (Hamzawy et al., 2007).

According to Vidino (2011a) the Muslim Brotherhood had historical ties to al Qaeda but by 2011, those ties were fraying. Since the Brotherhood publicly rejected at least random violence against non-combatants,[5] its short-term goals at least were no longer compatible with those of al Qaeda (Vidino, 2011a). In addition, al Qaeda had little involvement with the uprisings of the Arab Spring in Egypt and elsewhere, although it has since taken advantage of the uncertainty created by the revolutions (Jenkins, 2012).

That is not to say that violence has not played a role in Muslim Brotherhood activities, nor that their activities were limited to Egypt. For instance, the Brotherhood has also been active in Syria since the 1930s, and in the early 1980s it engaged in a series of terrorist attacks against the Syrian government of Hafez al-Assad. The attacks backfired as the Syrian government response was harsh and thorough. In 1981, the Syrian army and security services crushed the Muslim Brotherhood in the towns of Aleppo and Hama, killing over 200 people and making membership in the Brotherhood a capital offense (Martin, 2011). One year later, the Syrian army responded to another Brotherhood revolt in Hama by killing approximately 25,000 civilians (Martin, 2011; see also Friedman, 1989).

Despite the hopes placed by Egyptians in their revolution and in the elected Muslim Brotherhood administration of President Mohammed Morsi, by 2013 things took a turn for the worse. Violent demonstrations and clashes between the original supporters of the revolution and the Morsi government broke out in the summer of 2013 over perceived incompetence of the Morsi government and its failure to live up to economic expectations. Over a million protesters—including many who had voted for Morsi and the Muslim Brotherhood just the previous year—now called for the military to oust the president. In July 2013, the military ousted Morsi and placed him under arrest, while the Muslim Brotherhood took to the streets in violent clashes with the military, vowing never to give up the fight (Vick, 2013). Morsi was replaced by interim president Adly Mansour, who once sat on Egypt's highest court. A

---

[5]The Brotherhood still supports the use of violence in conflict zones such as Iraq and Afghanistan where it sees Muslims as being under attack and violence as a necessary means of defense (Vidino, 2012).

new constitution was put in place in 2014, but many opponents argue that it leaves too much power in the hands of the military for Egypt to ever move toward true democracy. As of this writing, Egypt is once again ruled by a military dictatorship, and the Muslim Brotherhood is once again an outlawed organization.

Gama'a al Islamiyya (IG)

Indigenous to Egypt, IG was formed in the 1970s as a loosely organized, rather fluid organization whose goal it was to create an Islamic state, replacing the sovereign state of Egypt. The group targeted Egyptian security officials, Coptic Christians, and any Egyptians opposed to the fundamentalist view of Islam (U.S. Department of State, 1987). Although the group claimed responsibility for an attempt on the life of President Hosni Mubarak, its most famous attack took place against a major tourist attraction when IG bombed an Egyptian archeological site in 1997, killing 58 tourists and four Egyptians.

Today the IG is a shadow of its former self. Following the 1997 attack, security crackdowns—aided starting in 2001 by post-September 11 antiterrorism efforts—have left IG with only a relatively few number of fighters. In addition, its former spiritual leader, Sheik Umar Abd al-Rahman, is serving a life sentence in a U.S. prison for conspiracy in the first World Trade Center bombing in 1993. Ramzi Yousef, one of al-Rahman's disciples, is serving a 240-year sentence for his role in directing the bombing, as well as a life sentence for the attempted bombing of U.S. airliners in East Asia. However, according to the U.S. State Department, many former IG fighters have taken their experience to al Qaeda and al Qaeda affiliates (Country Report on Terrorism, 2010). In addition, supporters of Sheikh Abd al-Rahman still remain a possible threat to U.S. interests and have called for reprisal attacks in case of his death in prison (Patterns of Global Terrorism, 1997).

Egyptian Islamic Jihad (EIJ)

The EIJ can no longer be found on the list of foreign terrorist organizations. In fact, it no longer exists in its original conception. It is the form of its demise, however, that makes this group worth addressing. In June 2001, EIJ merged with al Qaeda through marriage and EIJ's leader—Zawahari—became bin Laden's number two and his successor upon bin Laden's death in 2011. As of the writing of this book, Zawahari is one of the most wanted men on the planet.

EIJ was originally formed out of Zawahari's concern that IG's tendency to target Christians and tourists would create a backlash against the fundamentalist jihadists. EIJ, by contrast, targeted only Egyptian government officials, at least initially. As White (2012) describes the distinction, Zawahari argued for targeting the "near enemy" (the Egyptian government) first, and then when the entire Islamic community was united under an Islamic governing structure, target "far enemies" such as Israel, the United States, and Western Europe (see Wright, 2006, for a description of Zawahari's transition from his original more limited goals to widespread and indiscriminate global jihad). However, his grand dreams for revolution in Egypt never materialized and he left Egypt for the Sudan in 1996, linking back up with bin Laden with whom he had worked closely in Afghanistan in the late 1980s and early 1990s to form the early structure of what would become the current al Qaeda.

North Africa

Sitting just to the west of Egypt and situated on the Mediterranean Sea and the North Atlantic Ocean is the region of North Africa. Two countries in particular stand out in our consideration of terrorism in this region: Libya and Algeria.

Libya

At just over 1.76 million square miles, Libya is slightly larger than Alaska. Its population of six million is 97% Sunni Muslim, and Arabic is the official language. Entering the 20th century, Libya—as with most of the region—was part of the Ottoman Empire. When the Empire was defeated in WWI and its regions divided among the victorious powers, Libya came under control of Italy. When Italy and its leader, Benito Mussolini, were defeated in World War II, Libya was placed under UN administration and gained its own independence in 1951. Any joy was short-lived, however, as Libya fell under the control of Colonel Muammar al-Qadhafi following a military coup in 1969.

Muammar Qadhafi was a vicious dictator who ruled by his own bizarre political ideology that combined parts of socialism and Islam. He used Libya's oil revenues to promote this ideology abroad and to support terrorist groups (including a number of the Palestinian groups discussed in Section 5) as well as conducting several high-profile attacks himself. In early 2011, the Arab Spring came to Libya. As protests against his brutal regime erupted across the country, Qadhafi engaged in one of the harshest crackdowns against rebel forces in the region. His attempts to use his air force to shell poorly armed protesters and militias prompted international military intervention under the auspices of the UN. Qadhafi's regime was toppled in mid-2011. Qadhafi himself was killed by a rebel mob that found him in a ditch after trying to escape from his hometown of Sirte. While a new parliament and prime minister were elected in 2012, as of this writing, Libya remains under a transitional government with no new constitution and a legal system that is "in flux and driven by state and non-state entities" (*CIA World Factbook,* 2013, n.p.). Much of this country is now divided among tribal militias, aided by foreign fighters from Al Qaeda, ISIS, and other Islamist affiliates.

Although this book focuses on the origins and activities of non-state actor terrorist groups, Qadhafi—and thus the Libyan government—was more directly involved with terrorist groups and clear acts of terrorism than almost any other state in the region, particularly outside of its own territory. Details of many of these groups and incidents are discussed in the other regional sections of this book (based on origin of the group or location of the attack) but are worth summarizing here.

Qadhafi's involvement with terrorism can be divided into two approaches. First, Qadhafi was a strong supporter of many of the Palestinian groups that splintered off from the PLO starting in the late 1960s and early 1970s (see Section 5). These included the Palestinian Liberation Front (PLF), led by Abu Abbas and responsible for the hijacking of the Italian cruise ship *Achille Lauro* (see Section 9) in October 1985, and the Abu Nidal Organization (ANO). Qadhafi provided money, safe havens, and even operational support to these groups. Indeed, in its later years ANO was considered by many to be nothing more than a mercenary group in Qadhafi's employ.

In addition, Libyan intelligence officials and agents were directly responsible for several high-profile attacks against U.S. and Western targets, and this is where he may differ the most from other state-supporters of terrorism. The attacks included:

- The bombing of the La Belle discotheque in 1986 in West Berlin (see Section 9).
- The bombing of Pan Am Flight 103 over Lockerbie, Scotland, in 1988 (see Section 9).
- The bombing of UTA Flight 772 over the Sahara Desert while en route to Paris, killing 159 passengers and 14 crew members.

The threat of terrorism against U.S. and Western targets did not end with the much-welcomed death of Qadhafi. On September 11, 2012, a violent assault was launched against the U.S. Consulate in Benghazi in which two U.S. security agents and two U.S. diplomats—including the ambassador to Libya, J. Christopher Stevens—were killed. Considerable debate and vitriol have been expended on the cause of the attack and the existence of a cover-up by the Obama administration to avoid responsibility for security lapses. In particular, two major questions were floated:

- First, was this a pre-planned attack by al Qaeda affiliates in Libya?
- Second, why were security levels at the consulate and access to reinforcements in the event of a crisis so limited?

In the end, a report by the Senate Intelligence Committee released in January 2014 found that the attack was opportunistic, involving Islamic militants, and that there was little evidence to categorize it as a pre-planned al Qaeda attack. The attack on the Consulate may have been—at least in part—spawned from protests that had broken out in Cairo earlier over a virulently anti-Muslim video shown in the United States (Goldman & Gearan, 2014; Drum, 2012). In addition, although the committee found no evidence of a cover-up, it did characterize the attack as preventable "based on extensive intelligence reporting on the terrorist activity in Libya—to include prior threats and attacks against Western targets—and given the known security shortfalls at the U.S. Mission" (Goldman & Gearan, 2014, n.p.).

Algeria

Algeria's 2.4 million square miles—an area about three times the size of Texas—sits between Morocco and Tunisia and overlooks the Mediterranean Sea. It is the largest country in Africa. Ninety-nine percent of its approximately 38 million people are Sunni Muslim. The official language is Arabic but French is widely spoken. Like many parliamentary systems, its constitution provides for a president, a prime minister, and a bicameral legislature. Its legal system is a mixture of French civil law and Islamic law (*CIA World Factbook*, n.d.).

As far back as the 16th century Algeria was a part of the Ottoman Empire. By 1710, however, Algeria had almost full autonomy with the sultan maintaining only minimal influence (Metz, 1994). Then Algerian history took a turn for the worse. In 1827, France blockaded Algeria in part as a response to a perceived insult to the French consul, but also because the French monarch at the time—Charles X—needed a distraction from his domestic unpopularity (Metz, 1994). When the blockade "failed" after three years, the French invaded and occupied Algeria, establishing a colonial administration four years later. Algeria was ruled as a French colony for close to 130 years.

Starting in the 1950s, Algeria began an often violent struggle for independence led by the **National Liberation Front** (FLN), which would become Algeria's dominant political party. At one point in the struggle, in 1956 and 1957, the FLN gained control of parts of the city of Algiers. The violence got so bad that the French army under the command of General Massau moved in to regain control of the city. The FLN in Algiers was defeated and destroyed in 1957, but the methods used by the French forces to do so were so brutal and harsh that a backlash against the French was the result. The French actions galvanized support for the FLN across the Muslim population of Algeria and five years later the French were gone and Algeria was an independent state.

In the early 1990s, Algeria was racked by a violent conflict between the more secular government and military forces and an Islamic fundamentalist group known as the Islamic Salvation Front (FIS). The government gained the upper hand by the late 1990s, and the military arm of the FIS was disbanded. Abdelaziz Bouteflika won the presidency in 1999 and has retained the office since (although many criticize most of his electoral victories as fraudulent). In 2011, in response to the uprisings of the Arab Spring, the Algerian government lifted the state of emergency that had existed since the end of the conflict with the FIS and instituted some minor political reforms. Still, Algeria continues to deal with the threat of Islamic extremism, especially with the formation of **al Qaeda in the Lands of the Islamic Maghreb** (AQIM) in 2006.

Highlights of Terrorist Groups in Algeria

This subsection focuses on two Islamist groups that developed from the Algerian conflict of the 1990s: GIA and AQIM.

Armed Islamic Group (GIA)

The **Armed Islamic Group** (GIA) formed in 1992 when the Algerian government voided the victory of the FIS (see above) in elections; at the time FIS was the country's largest Islamic party. The goal of GIA was to overthrow the Algerian government and replace it with an Islamic state. Targets across the country from 1992 to 1996 included municipalities, a press center, a school, cafes, and family members of the government security services. Also, in 1993, the group announced a campaign against all foreigners—mostly Europeans—living in Algiers. At one point, the group kidnapped seven French monks and then killed them when France refused to negotiate for their release (U.S. Department of State, 1996). In 1994, the GIA hijacked an Air France flight to Algiers and is suspected of bombings in France in 1995 and 1996. The group started to lose its support from Islamists abroad, however, when its constant targeting of civilians became seen as indiscriminate (U.S. Department of State, 1998).

Al Qaeda in the Lands of the Islamic Maghreb (AQIM)

Formed in 1998 as an offshoot of the GIA, AQIM was originally known as the Salafist Group for Preaching Combat. The Salafist group merged with al Qaeda in 2006 and officially changed its name to AQIM in 2007. At its inception, AQIM had close to 30,000 members, but as of 2012, government counterterrorism efforts reduced that to fewer than 1,000 (National Counterterrorism Center, 2012).

Originally, the GIA off-shoot concentrated mostly on Algerian targets in its effort to create an Islamic caliphate, but when it merged with al Qaeda, it turned to Western targets as well. Using IEDs similar to those seen in the conflict in Iraq, AQIM has gone after convoys of foreign energy workers as well as the UN building in Algiers. The group is also fairly sanguine about crossing borders it does not recognize; for example, it claimed responsibility for a small-arms attack on the Israeli embassy in Mauritania. In 2012, Algerian authorities claim to have disrupted a plot by AQIM to attack U.S. or European ships in the Mediterranean, and some of the militants involved in the attack (discussed above) on the U.S. Consulate in Benghazi might have had ties to AQIM (U.S. Department of State, 2012).

While focused on suicide bombings, AQIM also uses extortion and kidnapping both as sources of fund-raising and as part of its politically motivated activities. It continues GIA's practice of kidnapping Westerners for ransom and in 2009 claimed responsibility for the death of Christopher Leggitt, a U.S. citizen doing missionary work in Mauritania. According to the U.S. Department of State, the group is attempting to take advantage of volatile situations in other countries, such as Mali, to expand its membership, resources, and operations. Of additional and particular concern is AQIM's connections to the Algerian expatriate community in Europe—especially France. Efforts to broaden its operational network into Europe are addressed in Section 9 (see also Vidino, 2011b).

Rogue States and Terrorism: The Cases of Iraq and Syria

As noted in the introduction to Part II, this book focuses on terrorist groups classified as non-state actors. That being said there are three states whose involvement or connection to the region's terrorism issues are worth some mention.

Iraq

Iraq and Syria differ somewhat from the other countries discussed in this section for two reasons: First, at the time of this writing, they are both scenes of ongoing conflicts that make it difficult to distinguish insurgency groups from one another or insurgency groups from terrorist groups, despite the use of terrorist tactics on both sides of the conflicts. Second, although the presence of jihadist groups, such as al Qaeda, raises concerns about an increase in anti-state—and especially anti-Western terrorism—most of these groups are, for the moment, less likely (with one exception; see In-Focus 6.1) to conduct activities outside of the conflict region than those in Yemen or Algeria.

Iraq obtained its independence through a UN mandate in 1932 and declared itself a republic in 1958. In truth, however, the country has been ruled mostly by a series of dictators—the worst of which was **Saddam Hussein** who took control of the country in 1979. Hussein was a Sunni Muslim and leader of the Baath political party, resulting in a Sunni minority ruling a majority Shia country. After supporting Hussein in his eight-year war against **Iran**, relations between Iraq and the United States have been characterized by military conflict. In 1990, U.S.-led forces ousted Iraqi forces from Kuwait after Hussein's invasion of that country. U.S. forces then remained in the region to enforce a no-fly zone. Then in 2003, citing threats from Hussein's possession of WMD and his non-compliance with

UN inspection regimes, the United States invaded and occupied Iraq.[6] What followed was a years-long insurgency where both Sunni and Shia militias targeted both U.S. and Iraqi military and civilian personnel, including Sunni challenges to the newly elected Shia government. Thousands of U.S. military personnel and tens of thousands of Iraqi civilians were killed in the conflict.

The United States ceased military operations in 2011, but U.S. personnel remain in Iraq to assist in training and other security issues. Numerous militias remain armed across Iraq, however, including an al Qaeda affiliate known as al Qaeda in Iraq (now known as the Islamic State of Iraq and Greater Syria—see In-Focus 6.1). After losing the city of Fallujah to insurgents and regaining the city after a surge of U.S. forces in 2004, as of the time of this writing, the city of Fallujah is again at risk of being controlled by the group ISIS under the leadership of Abu Bakr al-Baghdadi. The group drove police and the Iraqi army units from the city in January 2014 (Baker, 2014). According to a 2014 feature in *Time* magazine, the presence of Baghdadi's group in other cities was so strong, they were collecting unofficial taxes (Baker, 2014, p. 32). Baghdadi has also challenged the authority of al Qaeda's current leader al-Zawahari and tipped the balance in the fight in Syria's civil war, using the conflict to draw new recruits into his group. ISIS also claimed responsibility for a January 2014 suicide bomb attack in Lebanon as Syria's conflict began to engulf its southern neighbor. In-Focus 6.1 provides more detail about this new and deadly player.

IN-FOCUS 6.1

The Emergence of the Islamic State of Iraq and Syria (ISIS): A New Chapter in a 1,300-Year-Old Fight

In 2014, in what appeared to catch most Western intelligence and diplomatic services by surprise, a particularly vicious terrorist group emerged onto the world stage from the morass of militant groups operating in Iraq and Syria. What makes this terrorist group different from most of its predecessors is that while it focuses as much on civilian as military casualties—and in especially heinous fashion—it also conquers and holds territory, governing its captured populations under a rigid and brutal interpretation of Islamic law. As of this writing, the self-styled Islamic State of Iraq and Syria (ISIS) controls territory that stretches across northern Syria and western Iraq. As such, it crosses the boundaries of insurgent versus terrorist group. Because of its brutal targeting of innocent civilians, its kidnapping and beheading of U.S. and Western journalists (as of this writing at least two American journalists and two British aid workers were beheaded and the videos of the beheadings posted on YouTube), its specific recruitment of U.S. and European "volunteers," and its calls for attacks by its members against the United States and West, the authors have chosen to include ISIS as a highlighted group for this text.

ISIS began life as al Qaeda in Iraq. Driven out of Iraq by the U.S. surge of 2007 and the increased cooperation of Iraqi Sunni tribes against fundamentalist groups targeting Muslim civilians, ISIS moved across the border to Syria where the chaos of the civil war allowed it almost unlimited freedom to operate without restriction and to take advantage of the porous and hard-to-police border between northern Iraq and southern Turkey. Working to ISIS's advantage was the inability of the Syrian Sunni opposition groups to organize and cooperate under the common goal of deposing Assad, or to provide sufficient indication of moderation to attract enough U.S. and Western European assistance to do any good. Instead, as former U.S. secretary of state Henry Kissinger describes the state of affairs in Syria by 2014:

(Continued)

---

[6]No weapons of mass destruction were ever found.

(Continued)

> The principal Syrian and regional players saw the war as not about democracy but about prevailing. They were interested in democracy only if it installed their own group; none favored a system that did not guarantee its own party's control of the political system. A war conducted solely to enforce human rights norms and without concern for the geostrategic and georeligious outcomes was inconceivable to the overwhelming majority of players. (Kissinger, 2014, p. 530)

Like most Sunni militant groups, ISIS calls for the supremacy of Sharia law and a rigid, seventh-century interpretation of Islam. Going further, however, ISIS has declared itself the representative of all Muslims worldwide and has declared the territory it currently holds to be the founding of the new Islamic Caliphate that will one day govern the Muslim world. Perhaps even more shocking, its brutal tactics against civilian Muslims ISIS considers to be apostate or blasphemous was too much even for al Qaeda, which has repudiated any connection with ISIS.

In June 2014, ISIS moved back into Iraq with stunning speed. The group claims to have executed 1,700 captured Iraqi soldiers as of late June (Crowley, 2014a). Captured territory includes the cities of Mosul, Tikrit, Tal Afar, and Kirkuk. ISIS has vowed to erase the "blasphemous" Shiites from existence and has targeted key worship sites for attacks designed to both destroy the sites and kill as many worshipers as possible. Any who refuse to convert to the ISIS version of Sunni Islam—whether Muslim or non-Muslim—are executed. As shocking to the West as ISIS's brutality was how quickly the Iraqi Army—which the United States had spent so many years and money rebuilding—disintegrated, abandoning its posts and its arms to the ISIS fighters with barely a challenge. Sunni populations in Iraq at first welcomed ISIS as a counter to a Shia-led government that was viewed as using its power to take revenge for decades of Sunni rule—until they began to realize the monster they had allowed into their midst.

In September 2014, with ISIS threatening the Kurdish territory in northern Iraq (and its oil reserves) and two Americans beheaded on YouTube, the United States and the international community moved in. The United States—although reluctant to include U.S. troops in actual combat—began air strikes in both Iraq and Syria in support of Iraqi and Kurdish troops and sent U.S. military advisors back into Iraq to help what was left of the Iraqi Army stabilize the situation. The United States and the international community also pressured the Iraqi Shia prime minister Malaki into resigning in favor of a leader—Haider al-Abadi—who at least recognized the need to try to heal some of the Sunni-Shia split Malaki had exacerbated since the United States relinquished control of Iraq's government.

As of this writing, the air strikes have been effective in reducing the speed with which ISIS can gain new territory and in protecting key strategic points in Iraq and Syria—including Baghdad. At the same time, however, ISIS targets have grown more limited as they have moved from targetable military convoys to harder to detect civilian vehicles (Vick, 2014a). In addition, while the international coalition spearheaded by the United States—consisting of over 40 countries, including Albania and South Korea—is holding, the hold is tenuous. For example, Saudi Arabia views ISIS as a significant threat and has agreed to pressure its Sunni clerics to denounce the group and to host a training camp for moderate Syrian rebels. But Saudi Arabia has demands of its own, including more U.S. efforts to depose Assad and to limit Iran's involvement—whose Iraqi Shia militias under Iraq's control are fighting alongside the United States, but which has also provided the bulk of the support that has kept Assad in power (Crowley, 2014b).

The fight against ISIS is not taking place only in Iraq and Syria. One of the key concerns of counterterrorism officials worldwide is the astounding ability of ISIS to manipulate modern media and social networks to recruit members from the United States and Western Europe. Many of these recruits participate in the fighting—including suicide attacks in the Middle East. What happens, however, ask counterterrorism officials, when these recruits—often with U.S. or European passports—return home with instructions to conduct attacks on their home soil? In November of 2014, the UK convicted and sentenced to prison its first citizen who returned home after fighting with ISIS. A few months earlier, Australian authorities arrested a number of suspects with links to ISIS for conspiracy to commit attacks in Australian cities.

Syria

The current incarnation of Syria is a relatively new one in the region. Under French control since the end of World War I, Syria first gained its independence in 1946. After a series of military coups characterized its first decade, it merged with Egypt to form the United Arab Republic in 1958. After only four years, the short-lived UAR dissolved in 1961, and Syria was once again its own state.

In 1970, Hafiz al-Assad seized power in a coup and took control as Syria's president. Assad brought political stability—if not political freedom—and upon his death in 2000, was succeeded by his son, Bashar. The Assad family comes from the minority Alawi sect, which results in the unusual specter of a Shia minority ruling a Sunni majority country; just the opposite of most of the region. The Shia connection to Syria's leadership has led directly to close ties with the Shia Islamic government in Iran and to the Shi'ite members of Hezbollah in Lebanon. In addition, Syria has long actively involved itself, including militarily, in the volatile politics of Lebanon, especially since the civil war and Israeli invasions of the late 1970s and early 1980s (Colello, 1987).

In early 2011, antigovernment protests calling for basic political reforms broke out in Syria's southern region. The unrest spread quickly to the rest of the country and the Assad government responded, at least initially, with both moderate concessions and force. By the end of 2011, force won out as the defining government reaction. At the time of this writing, a major civil war had been raging in Syria for close to four years. By the end of 2013, media reports had placed the death toll at over 125,000 (Solomon, 2013) and over 6 million pushed from their homes (Vick, 2014). In Jordan, 85,000 call the Za'atari refugee camp home and the refugees there are considered to be some of the lucky ones (Vick, 2014). Despite sanctions and calls from the Arab League, the EU, Turkey, the UN, and the United States for his ouster, Assad has refused to step down. His intransigence has been bolstered by support from Iran and Hezbollah, which has joined the fighting in Syria on the side of the Assad regime. In August 2013, Assad went so far as to use chemical weapons against civilians in and around Damascus. Reports on the death toll range anywhere from 300 to 1,300 (BBC, 2013). International pressure resulted in a negotiated settlement requiring Assad to turn over his chemical weapons cache to avoid international military intervention, but whether or not Assad is going to fully comply with the agreement is still an open question.

At the other end of the war sit a number of Sunni insurgent and militia groups, many under the umbrella name of the National Coalition of Syrian Revolution and Opposition Forces (*CIA World Factbook*, n.d.). Within the coalition, groups range from moderate, secular groups with the goal of establishing a democratic Syria to al Qaeda affiliates with the goal of creating a Sunni-Islamic state; the latter often the better funded, trained, organized, and experienced in fighting. Many of the Sunni and Shia fighters in Syria have been recruited from across the region and worldwide. Both sides of the conflict—government and rebel—have been accused of ongoing atrocities, including terrorist attacks against Sunni and Shia civilians (Baker, 2013). "The Obama Administration's reluctance to help arm the rebels is informed at least in part by the fear that the weapons may go to the wrong people" (Baker, 2013, p. 41). The extent of the involvement of al Qaeda, Hezbollah, and other terrorist groups in the Syrian conflict is hard to estimate, but clearly Syria is now seen as yet another battleground in the global jihad and the creation of an Islamic caliphate across the Middle East. Complicating the fight against Assad is the international coalition against the Sunni jihadist group ISIS, which is actually an opponent of Assad but is considered by most of the international community as an even worse option.

What now for the future of Syria (and Iraq, for that matter)? Lacking a crystal ball, a quote from former secretary of state Henry Kissinger might be apt, if somewhat pessimistic (or realistic):

A working regional or international security system might have averted, or at least contained, the catastrophe. But the perceptions of national interest proved to be too different, and the costs of stabilization too daunting. Massive outside intervention at an early stage might have squelched the contending forces but would have required a long-term, substantial military presence to be sustained. In the wake of Iraq and Afghanistan, this was not feasible for the United States, at least not alone. An Iraqi political consensus might have halted the conflict at the Syrian border, but the sectarian impulses of the Baghdad government and its regional affiliates were in the way. Alternatively, the international community could have imposed an arms embargo on Syria and the jihadist militias. That was made impossible by the incompatible aims of the permanent members of the [UN] Security Council. If order cannot be achieved by consensus or imposed by force, it will be wrought, at disastrous and dehumanizing cost, from the experience of chaos. (Kissinger, 2014, p. 539)

⊠ Rogue States and Terrorism: The Case of Iran

Unlike the countries of the Middle East discussed above, Iran is not Arab, but is in fact Persian—the remnant of the once grand Persian Empire; its heyday running from approximately 500 to 330 BCE. As late as 1935, Iran was still known as **Persia**. From 1941 to 1979, Iran was ruled by Mohammad Reza Pahlavi, the Shah of Iran. Pahlavi was an absolute monarch known for his vicious and brutal secret police. In 1979, the Shah was overthrown and sent into exile where he died in 1980. Power in Iran was claimed by its conservative Shia clergy. The clergy established an Islamic Republic with full political control vested in the religious leadership of the Supreme Leader Ayatollah Khomeini, whose regime and crackdown on political dissent turned out to be just as harsh as the Shah's.

As part of the 1979 revolution, a mob of Iranian students seized the U.S. Embassy in the capital of Tehran and held embassy personnel hostage for 444 days, from November 1979 until January of 1981. The United States broke off diplomatic relations with Iran in April of 1980. President Jimmy Carter's inability to gain the release of the hostages is widely credited as a factor in his defeat by Ronald Reagan in the 1980 presidential election.

Iran is a majority Shia country and its creation of an Islamic Republic governed by Shia clergy and Shia law continues to embolden Shia minorities across the region, as does its support for Shia extremist groups. The group Iran is most famous for, and the group that most directly connects Iran to international terrorism, is that of Hezbollah—the Shia terrorist group established in 1982 in Lebanon following the Israeli invasion (discussed in more detail in Section 5). Iran's monetary, training, operational, strategic, and political support was and is key not only to Hezbollah's success in carrying out a number of high-profile terrorist attacks against Lebanese, Israeli, and Western targets, but also its evolution into a significant military force and a major player in Lebanese politics. Since the mid-late 1980s, Iran has continued to support Islamic fundamentalist terrorists groups in the region—both Shia and Sunni—such as Hamas and Palestinian Islamic Jihad (also discussed in Section 5). Its support of these groups, as well as its ambitions and efforts to build a nuclear weapons program, has made it an international pariah in many circles and the target of substantial economic and other sanctions from the UN and many Western nations.

As of the time of this writing, Iran's support of terrorist groups has not waned, but it has moved to try to restore some of its international standing. In June 2013, the relatively moderate cleric Hassan Ruhani won election as Iran's president. His less confrontational approach to international relations and his willingness to come to the negotiating table over Iran's nuclear program in return for easing of economic sanctions has raised hopes of a "new" Iran that poses less of a threat to the stability of the region (Wright, 2014). However, much of the power still resides in the hands of Supreme Leader Ayatollah Khamenei and Iran's policies, such as its support for Hezbollah, for the Assad regime in Syria (see above), and presumably for the Houthis in Yemen (see above), are unlikely to change (Hosseinian, 2013; Wright, 2014).

Africa

The Region

When writing about terrorism in Africa—especially Central and West Africa—it is especially difficult to distinguish between terrorism and insurgency/guerrilla warfare. This is because there are, at the time of this writing, multiple civil wars or insurgencies going on in the region. While many of those involved will use terrorist tactics, it is next to impossible to attribute a specific attack to a specific group unless there is a claim of responsibility and often there is not. In addition, there is a limited law enforcement capability in the region. Indeed, some states lack any law enforcement or counterterrorism or even government presence as the states themselves have failed. In fact, a number of these contests are between tribes and clans with little interest in or respect for traditional "sovereign" boundaries. Exacerbating the problem is the emergence of Islamic fundamentalist groups in the region. While many of these conflicts did not start out as part of the global jihad of the Islamic fundamentalists (the disputes were often more about land and water than religion), the rampant and endemic poverty, violence, corruption, and lack of hope that things will change has provided a fertile and dangerous ground for recruitment by al Qaeda or numerous other Islamist terrorist groups. As noted by Lyman and Morrison (2004), "the terrorist threat in West and Central Africa comes less from religion and politics than from lack of sovereign control and general debility" (p. 83).

Nevertheless, in a book about terrorism around the world, some effort to address terrorism in this region is warranted. This subsection, therefore, will focus on two known terrorist groups to which specific terrorist attacks have been attributed, and which may be seeking to broaden their efforts beyond their original geographic boundaries.

Al-Shabbab in Somalia

Al-Shabbab is an al Qaeda affiliate that was designated a Foreign Terrorist Organization by the U.S. Department of State in 2008. It grew out of the militant wing of the Council of Islamic Courts, which was eventually defeated by the Somali government in its most recent civil war. The group uses bombings and assassinations to carry out disparate goals related to the overthrow of the Transitional Federal Government that was put in place at the end of Somalia's civil war in 2004.[7] Given its origins, most members are more concerned with Somali nationalist issues than worldwide jihad. Still, since its affiliation with al Qaeda in 2012, Al-Shabbab has also increased its targeting of Westerners—including journalists, peace activists, and international aid workers—as well as African Union peacekeepers (NCTC, 2012; U.S. Department of State, 2012). Concerns have also been raised about connections between Al-Shabbab and both AQAP and AQIM.

Attacks outside of Somalia include a 2002 attack on an Israeli-owned hotel that killed 15 people and a missile fired at Israeli jet at take off—both in Mombasa (Lyman & Morrison, 2004). In July 2010, an Al-Shabbab attack in Kampala, Uganda, during the World Cup killed over 70 people including an American citizen (U.S. Department of State, 2012). In 2011, the *New York Times* reported on another Al-Shabbab attack in Kenya where members of the terrorist group threw grenades into a church and then opened fire on the escaping attendees with weapons

---

[7]Somalia is still in the process of forming its long-term governmental institutions (*CIA World Factbook*, 2014).

taken from two security officers (Gettleman, 2012). Fifteen people were killed and 50 were wounded. Its most well-known attack so far is the highly publicized attack on the upscale Westgate Shopping Mall in Nairobi. The attack provided Al-Shabbab with four days of intensive media coverage (Anzalone, 2013). One analysis of the attack concluded that "despite facing increased political and military setbacks, [Al-Shabbab] remains adept at executing audacious attacks designed to attract the maximum amount of media attention" (Anzalone, 2013, p. 2). Seventy-two people, including five terrorists and six Kenyan soldiers, were killed in the Westgate attack (Anzalone, 2013).

In September 2014, a U.S. airstrike killed the leader of Al-Shabbab—Ahmed Ali Godane. The group's new leader—Ahmed Omar—has called for revenge attacks against Kenyan, U.S., and Western targets, stating that "avenging the death of our scholars and leader is a binding obligation on our shoulders that we will never relinquish no matter how long it takes" (Guled, 2014, p. A09).

Boko Haram in Nigeria

Boko Haram was initially viewed as a Nigerian group focused on Nigerian targets, emanating from its Northern region. Its goal is to establish an Islamic state in Nigeria based on strict Islamic law. Its recruits are local and its leader is a messianic individual whose followers resemble a cult more than a typical Islamist militant group (Nossiter & Kirkpatrick, 2014). However, in 2010, it issued a statement expressing solidarity with al Qaeda and its intent to target Westerners. It followed this statement with a car bomb attack in August 2011 against the UN Headquarters in Abuja, killing 23 and injuring 80 (NCTC, 2012). Its first act of violence, however, dates back to 2009 when Boko Haram fighters attacked a mosque and a police station, killing 55 people. Since 2011, its attacks have become increasingly indiscriminate (Nossiter & Kirkpatrick, 2014).

In April 2014, in an act that shocked even many Islamist militants for its audacity, Boko Haram kidnapped 200 girls from a school in northern Nigeria; it was an act even al Qaeda was unwilling to condone (Nossiter & Kirkpatrick, 2014). Since 2009, the group may be responsible for the kidnapping of more than 500 women and girls ("Explaining Boko Haram," 2014). Boko Haram's consistent killing of innocent Muslim civilians is the same basis for al Qaeda's break with ISIS. In fact, as argued in a feature article in the *New York Times,* groups like Boko Haram and its loose affiliates across Africa "may be far more brutally violent than even the acolytes of Bin Laden can accept" ("Explaining Boko Haram," 2014, p. 5).

The kidnapping of the girls made worldwide news and social media headlines for several months before being driven off the front page by ISIS. Considerable grassroots and international pressure was placed on the Nigerian government to find the girls and deal with Boko Haram. The United States and other Western countries as well as the African Union provided resources to help in the search. To no avail. In late September 2014, Boko Haram claimed that all of the girls had willingly converted to its brand of Islam and had been married off. As of this writing, Boko Haram presents a substantial threat to the stability of the Nigerian government. Elections that were scheduled for the fall of 2014 were delayed several months with security concerns cited as the reason.

Summary

This section has looked at the historical background and main terrorist groups of several countries spanning the Middle East and Africa, including Saudi Arabia, Yemen, Egypt, Libya, Algeria, Somalia, and Nigeria. A brief history of these countries was reviewed where appropriate, particularly the chronology following the fall of the Ottoman and colonial empires and the impact those events had on current governing structures and institutions. The key terrorist groups with origins in the region were traced, focusing on motivations, capabilities, and incidents. Special attention was paid to al Qaeda and its growing number of affiliates throughout the region, especially in Algeria, Yemen, Somalia, and Nigeria. The historical and current impact of state actors such as Libya, Iran, Iraq, and Syria on non-state actor terrorist groups was assessed. Highlights of this section also include:

- ◆ The impact of Saudi Arabia on Islamic fundamentalist ideology regionally and worldwide.
- ◆ The history of bin Laden and al Qaeda, tracing the movements of al Qaeda from Saudi Arabia to Sudan to Afghanistan.
- ◆ The changes in Egypt, Syria, and Yemen since the Arab Spring revolts.
- ◆ The emergence of ISIS and its impact on events in Iraq and Syria.

⬙ Key Points

- ◆ Saudi Arabia is the birthplace of Islam and is the most important proponent of the Wahhabi interpretation of Islam. It is also the birthplace of Osama bin Laden, the founder and leader of al Qaeda.
- ◆ Al Qaeda was formed following the Soviet Union's invasion of Afghanistan with the goal of overthrowing all apostate Muslim regimes (defined as regimes that do not follow and require the purist form of Islamic practice), removing Western—especially U.S.—influence from the Middle East and Asia, defeating Israel, and recreating a united Islamic Caliphate.
- ◆ Al Qaeda now has affiliates in numerous countries throughout the region including Yemen, Algeria, Somalia, Nigeria, Iraq, Syria, and Libya.
- ◆ Since the death of bin Laden in 2011, the danger from al Qaeda has in some ways become harder to deal with as it has become more of an ideology and umbrella group for affiliate groups that have spread across the Middle East, Africa, and Asia.
- ◆ Terrorist groups have operated out of Egypt with varying intensity since its independence; these include the Muslim Brotherhood, the Egyptian Islamic Jihad, and Gama'a al-Islamiyya. At the time of this writing, the government and legal systems of Egypt were in a state of flux after the overthrow of two presidents between 2011 and 2013.
- ◆ A civil war has been raging in Syria since 2011, with Iran and Hezbollah supporting the Assad family and groups ranging from moderates to al Qaeda affiliates supporting the opposition.
- ◆ Iran is a majority Shia country and its creation of an Islamic republic governed by Shia clergy and Shia law continues to embolden Shia minorities across the region; as does its support for both Shia and Sunni extremist groups.
- ◆ At the time of this writing, multiple civil wars or insurgencies are going on in Africa. While many of those involved will use terrorist tactics, it is next to impossible to attribute a specific attack to a specific group unless there is a claim of responsibility, and often there is not.

KEY TERMS		
Africa	Al Qaeda Affiliates	Al Qaeda in the Lands of the Islamic Maghreb
Algeria	Al Qaeda in the Arabian	
Al Qaeda	Peninsula	Al-Shabbab

Armed Islamic Group	Ibn Saud	Ottoman Empire
Ayman Zawahari	Iran	Persia
bin Laden, Osama	Iraq	Qadhafi, Muammar
Boko Haram	ISIS	Qutb, Sayyid
Egypt	Libya	Saudi Arabia
Egyptian Islamic Jihad	Middle East	Sudan
	Muslim Brotherhood	USS *Cole*
Gama'a Islamiyya	National Liberation Front	Wahhabism
Hussein, Saddam		Yemen

DISCUSSION QUESTIONS

1. Briefly describe the history of the formation of Saudi Arabia. How has the history of this country influenced the development of terrorism in the region, if at all?

2. How did the fall of the Ottoman Empire lead to the creation of the modern states of the Middle East and North Africa?

3. What are the implications of the defeat of the Ottoman Empire in WWII for the evolution of modern-day terrorism?

4. Trace and describe the evolution of al Qaeda. Identify three of the most defining events or influences in its history and explain your selection.

5. Briefly summarize the events transpiring in the Middle East and North Africa as a result of the Arab Spring. As a result of these events, do you think support and recruitment for terrorist groups in this region will increase or decrease over the next several years. Support your answer. (**Note: for this question, you will need to draw on and cite outside sources such as newspaper and magazine articles**).

WEB RESOURCES

Brookings Institution: http://www.brookings.edu

Central Intelligence Agency: http://www.cia.gov (publisher of *World Factbook*—available online)

Council on Foreign Relations: http://www.cfr.org

Human Rights Watch: http://www.hrw.org/middle-east/n-africa

Middle East and North Africa Financial Action Task Force: http://www.menafatf.org/

Middle East Media Research Institute: http://www.memri.org/middle-east-media-research-institute.html

RAND Corporation: http://www.rand.org

U.S. Department of State: http://www.state.gov (publisher of *Country Reports on Terrorism*—available online)

UNHCR: http://www.unhcr.org/pages/4a02db416.html

World Bank: http://www.worldbank.org/en/region/mena (analysis on Middle East and North Africa)

READING 10

In this article, Silke explores the underlying psychology that drives individuals to join jihadi-motivated terrorist groups, such as Al Qaeda, and engage in their violent activities. In particular, Silke explores the stereotype that terrorists suffer from mental disorders or illness. Instead, Silke finds that the decision to join such groups is often a rational one stemming from a gradual process of radicalization.

Holy Warriors

Exploring the Psychological Processes of Jihadi Radicalization

Andrew Silke

University of East London, UK

 ## Introduction

On 7 July 2005 the first Islamist suicide bombings in Europe were carried out in the UK. Four suicide bombers (three of whom were British born) detonated bombs in London during the morning rush hour; 52 people were killed by the bombers and more than 700 people were maimed and injured. Exactly two weeks later, on 21 July, more extremists attempted to carry out a second wave of suicide attacks on London's transport system but this time the devices failed to detonate and no one was killed.

Following in the wake of even more devastating attacks in Madrid in 2004, these events highlighted the severity of the threat now facing Europe from jihadi terrorists motivated by a radical interpretation of Islam. The fact that a growing number of the terrorists were home grown has represented a disturbing development and left the authorities and others struggling to understand the radicalization process that can produce such extremists within the relatively stable and prosperous states of Western Europe.

The first hurdle to be cleared in any effort for understanding is to clarify what exactly is meant by 'terrorism'. The questions of what constitutes terrorism and who is a terrorist are deeply problematic. There is still no precise and agreed definition of terrorism, and some writers have concluded that 'it is unlikely that any definition will ever be generally agreed upon' (Shafritz et al. 1991: 260). In this article, however, I follow the concise definition provided by Crenshaw (1992: 71), who described terrorism as 'a particular style of political violence, involving attacks on a small number of victims in order to influence a wider audience'. The claims as to what behaviours fit this definition still vary considerably, but the focus of this article is very much on 'insurgent' terrorism, which is essentially a strategy of the weak, adopted 'by groups with little numerical, physical or direct political power in order to effect political or social change' (Friedland 1992: 82).

In practical terms, 'insurgent' terrorists are members of small covert groups engaged in an organized campaign of violence. This violence is often extreme and frequently indiscriminate. The terrorists themselves tend to live isolated and stressful lives and enjoy varying levels of wider support. In the past, groups that fit within this framework have included the Irish Republican Army (IRA), the Basque separatist group ETA, the Red Army Faction and the Italian Red Brigades. In this article I primarily focus on al-Qaeda and the jihadi extremists affiliated with it.

The word '*jihad*' is often assumed to mean 'holy war' but the meaning is more complex than that. The phrase derives from the Arabic for 'struggle', and within Islam there are two forms of jihad: the Greater Jihad and the Lesser Jihad. The Greater Jihad refers to an individual's personal struggle to live a good and charitable life and adhere to God's commands as understood within Islam. This is a strictly personal and non-violent phenomenon. The Lesser Jihad refers to violent struggle on behalf of Islam. The *jihadis* then are literally 'those who struggle'. This term is typically used to describe individuals who have volunteered to fight in the Lesser Jihad, and the expression is used by members of groups such as al-Qaeda to describe themselves. (*Mujahideen*, meaning 'holy warriors', is another expression commonly used to refer to Muslims engaged in the Lesser Jihad).

 ## Quality of the Research Evidence

This article aims to provide an assessment of the psychological basis of jihadi radicalization. A great deal has been written on this subject since the attacks of 11 September 2001 (9/11), and much has been claimed, but critical issues remain concerning the quality of the evidence being used to justify many claims (Ranstorp 2006; Chen et al. in press).

Most research looking at jihadi radicalization has been limited, often relying on anecdotes and a small number of case studies (e.g. Mazarr 2004; Nesser 2006). Even the best available research on this subject is almost all based on secondary analysis of data, more specifically of archival records (e.g. Sageman 2004; Bakker 2006). These, however, are traditional features of research on terrorism where over 80 percent of all research is based either solely or primarily on data gathered from books, journals, the media (or media-derived databases) or other published documents (Silke 2001). An old failing of the field has been the very heavy reliance on literature review methods. Schmid and Jongman (1988) were very critical of the paucity of fresh data that researchers

were producing. In the 1990s, 68 percent of the published research essentially took the form of a literature review and did not add new information. Since 9/11, 65 percent of articles are still essentially just reviews (Silke 2006). Research providing new information is uncommon and where there is new information this is mostly gleaned indirectly from reports in the media (e.g. Pape 2005).

Currently, only about 20 percent of research articles provide substantially new knowledge that was previously unavailable to the field. The field thus is top heavy with what are referred to as pre-experimental research designs. Unfortunately, these are 'the weakest designs since the sources of internal and external validity are not controlled for. The risk of error in drawing causal inferences from preexperimental designs is extremely high and they are primarily useful as a basis . . . for exploratory research' (Frankfort-Nachmias and Nachmias 1996: 147). For example, in only 1 percent of research reports have systematic interviews been used to provide data and to date no such interview study has been carried out with jihadis.

The limitations of available research can be illustrated by reviewing two of the key studies so far published on jihadi extremism. Sageman (2004) studied 172 individuals who were or are members of extremist Islamist organizations. The data came from publicly available sources, mainly media reports and transcripts of court proceedings. These sources are not always reliable, and the study also suffers from the lack of a comparison with individuals who are not members of extremist groups. Nevertheless, the research still provides a useful analysis of the backgrounds of extreme Islamists, and many of the findings are very relevant to this paper. For example, the study provided a detailed description of the profiles of jihadis with respect to education, family background, childhood, socioeconomic background, religiosity, professions, and so forth. Sageman also explored the process of recruitment into the jihad, and the lives of the 172 revealed that a surprisingly large proportion joined in small groups (and not as isolated individuals). Even bearing in mind the acknowledged weaknesses, this study represented a major step forward in social science research on al-Qaeda and jihadi extremism, and was a significant improvement on the much more journalistic accounts that had largely dominated (e.g. Gunaratna 2002; Burke 2004; Vidino 2006).

A second significant study is the one by Bakker (2006), which in many respects is essentially a replication of the earlier study by Sageman (2004). It has a narrower focus in that it considers only jihadis active in Europe, but the data set is larger than Sageman's, with information on 242 individuals who had been involved in jihadi terrorist attacks and attempts. As did Sageman, Bakker relied on open source data for this study, again using court and media reports.

Currently, there is no published research where the researchers have had direct access to the jihadis. The result is that all of the analysis is done at a distance. This means either relying on media or court reports, as Sageman and Bakker have done, or else drawing inferences from other populations that are believed to be relevant in some way to the jihadis. For example, Ansari et al. (2006) distributed attitudinal questionnaires to 80 practising Muslims in the UK. The questionnaires contained Likert-type measures of attitudes towards religious martyrdom, terrorism, violence, suicide, jihad and 9/11. In-depth interviews were also carried out with 13 respondents. This research found that social identity had a major impact on attitudes: respondents who felt their primary identity was Muslim held more positive views towards jihad and martyrdom, whereas respondents with a dominant British identity did not. Such studies are helpful in highlighting the key role identity plays, but the work is limited because of the small sample size and the fact that the respondents were not actually jihadis. Nevertheless, even the pool of studies like this is currently very limited.

A different research strategy is to draw upon information about other terrorist groups that are thought to be similar or to generalize about new forms of terrorism. There is an increasing trend for commentators to argue that we are currently experiencing an age of New Terrorism (e.g. Laqueur 2000; Kegley 2002). This thesis argues that terrorism today is significantly different from terrorism in the past, primarily because it is now more lethal, more violent and more heavily motivated by fundamentalist religion. Al-Qaeda and its affiliates are held up as a good example of this new type of terrorist group. Implicit in the thesis is an assumption that any understanding gained from an analysis of the Old Terrorism is not very relevant to the understanding of the new form.

However, the New Terrorism thesis is overemphasized. At heart, the argument often shows a poor awareness of the history of terrorism and overlooks the many examples of lethal, violent and fundamentalist terrorist groups of the past. For example, the Muslim Moros campaign of suicide terrorism against US forces in the Philippines at the start of the 20th century closely mirrors many of the problems faced by the Americans in Iraq at the start of the 21st century (Woolman 2002).

Further undermining the idea that the New Terrorists are a breed apart are the reports from researchers who have been able to interview terrorists from many different groups and who have found more similarities than differences between the different factions (e.g. Horgan 2005). Ultimately, although context matters, this does not mean that research focused on the IRA, for example, is not relevant for understanding the experience of other terrorist groups, and vice versa. However, the New Terrorism thesis biases our understanding of terrorism. It encourages the belief that terrorism today is essentially a new phenomenon and that new ways of thinking are needed to understand it. Post (1990: 29) noted that every 'terrorist group is unique and must be studied in the context of its own . . . culture and history'. Although this is true, it is also the case that terrorist groups often have a great deal in common and terrorist campaigns often follow similar patterns. It would be wasteful to ignore entirely the insights and lessons from studies of other groups and conflicts (Silke 2003).

 ## The Psychology of Terrorism

When confronted with atrocities such as the Madrid and London bombings, it can be very difficult to consider the perpetrators as rational and mentally healthy individuals. On the contrary, it is much easier to view them as highly deviant personalities who are very likely to suffer from mental illness and psychopathological disorders. Indeed, it is not just the general public who feel this way. Many researchers on terrorism have shared this view, and some have argued that terrorist groups are made up of a mix of individuals suffering from psychopathic and paranoid personality disorders (e.g. Post 1986; Pearlstein 1991; DeMause 2002). Such views have had a powerful impact on government policy decisions.

As social science grapples to improve our current understanding of terrorism, much of the research it is drawing on was originally aimed at explaining how and why people were able to commit atrocities in past ages. For example, after the end of the Second World War a number of surviving Nazis, including Hermann Goering and Rudolf Hess, were tried for war crimes at Nuremberg. Prior to the trials, 16 of these Nazi leaders were assessed by an Allied psychologist. The psychologist concluded that their scores were those of violent, power-hungry personalities, obsessed with death and lacking in any real human feeling. This was pretty much what the world expected from the men responsible for the horror of the Holocaust. Years later, however, the same Nazi scores were inserted among a selection of scores from a group of average Americans. This mix was given to a panel of experts who failed entirely to identify anything unusual about the Nazi leaders, and instead concluded that *all* of the scores reflected stable and healthy personalities (Harrower 1976).

What had happened with the Nazis is an example of attribution theory. Attribution theory has shown that we tend to view our own behaviour as stemming from situational or environmental forces, but that we see the behaviour of other people as stemming from internal forces, such as their personality (Quattrone 1982). If people are involved in extreme and violent acts, we tend to assume that their personality must be similarly extreme and deviant. We then tend to make any available evidence fit in with our assumptions. This is exactly what happened with the captured Nazis, and the same effect can be seen in the way most people consider the psychology of terrorists.

Why does someone become a terrorist? The first answer offered to this question was to suggest that (like the Nazi leaders) terrorists must be psychologically different from everyone else. Their abnormality accounts for their involvement in terrorism. This view was very popular in the 1970s and it has survived in some form or other since then. As evidenced only too forcefully on 11 September 2001, terrorist violence can often be extreme, and extreme violence has a special ability to provoke extreme views about the perpetrator. Despite the indiscriminate and extreme violence of many terrorist attacks, the vast majority of research on terrorists has concluded that the perpetrators are not psychologically abnormal (Silke 2003; Horgan 2005). On the contrary, many studies have found that terrorists are psychologically much healthier and far more stable than other violent criminals (e.g. Lyons and Harbinson 1986). An act of extreme violence does not in itself show that the perpetrator is psychologically distinct from the rest of humanity. Although a few psychologists believe terrorists are mentally abnormal, their conclusions are based on very weak evidence (Silke 1998 provides a review of this literature). Psychologists who have met terrorists face to face have nearly always concluded that these people were in no way abnormal, and on the contrary that they had stable and rational personalities.

This is not to say that people suffering from psychological disorders are never found in terrorist groups. They are, but these are the exception and not the rule. Quite simply, such people do not make good terrorists. They lack the discipline, rationality, self-control and mental stamina needed if terrorists are to survive for any length of time (see Taylor 1988). When they are found, they tend to be fringe members of the group and not central characters.

Overall, terrorists are a very heterogeneous group and the range of people who become involved in terrorism is vast. They can vary hugely in terms of education, family background, age, gender, intelligence, economic class, and so on. Consequently, the manner in which they became a terrorist can also vary, and factors that played a pivotal role in one person's decision to engage in terrorist violence may play a peripheral role in the decision-making of others, or indeed may have played no part at all.

Despite this heterogeneity, four decades of research have shown that a number of factors appear to be relatively common in the background of terrorists (Silke 2003). Becoming a terrorist is for most people a gradual process (Horgan 2005), and progressing through the different stages is usually not something that happens quickly or easily. However, considering the factors in the following sections will help in gaining understanding and insight into this process. Not all of these factors will necessarily be present in the experience of every terrorist, but most will be there at least to some degree. Neither are their boundaries exclusive. They interact and mesh together in a complex manner that can often be very difficult to disentangle or differentiate in the case of any one person. Ultimately, it is the combined impact of a number of factors that pushes and pulls someone into becoming a terrorist, and these factors will vary depending on the culture, the social context, the terrorist group and the individual involved.

Age and Gender

As yet, there is no scientific evidence of any genetic role in determining why certain people become involved in terrorism, and specific biological approaches to explaining terrorism have tended to be very flawed (see Silke 1998).

The most important biological factors associated with joining a terrorist group are age and gender. Although a causative role for these factors is not entirely unambiguous, there is certainly a correlation between these two factors and most recruits to terrorist organizations. Ultimately, most people who join a terrorist group are young—by young here I am referring to teenagers and people in their early twenties. Further, most recruits to terrorist groups are male.

It is already well established in other spheres that young males are associated with a multitude of dangerous and high-risk activities (Farrington 2003). Statistics on violent crime consistently show that perpetrators are most likely to be males between 15 and 25 years of age (e.g. Budd et al. 2005). This is a very robust finding that is remarkably stable across cultures and regions (e.g. Schönteich 1999). More crime in general is committed by teenagers and young adults than by any other age category. Adolescence brings with it a dramatic increase in the number of people who are willing to offend, and cross-cultural studies tend to show that the peak age for male offending has generally been between 15 and 18 years of age, falling off quickly for most individuals as they grow older (Farrington 2003).

Research studies have found that between 54 percent and 96 percent of young men have been involved in some form of delinquent behaviour, with good international agreement on this finding (e.g. Fox and Zawitz 2006). Consider for example the research by Junger-Tas (1994), who compared delinquency rates in five countries and found that 64–90 percent of all young men surveyed admitted to having committed a criminal offence (with around 45 percent having committed at least one offence in the previous 12 months). These rates are far higher than those for other age groups. As Moffitt (1993: 675) noted, 'actual rates of illegal behaviour soar so high during adolescence that participation in delinquency appears to be a normal part of teen life'.

There is also widespread agreement that young men are more heavily involved in crime than are young women, in terms of both the quantity and the severity of the offences. Studies on violent crime for example show that the ratio of male to female offenders varies by at least 2:1 to 4:1 in western cultures, with this ratio generally climbing higher the more serious the offences become (e.g. Rutter et al. 1998).

Explaining why young men are so inclined to get involved in deviant behaviour is not straightforward. There has been a great deal of research on juvenile delinquency, but most of this has focused on examining life-course persistent offenders – the small minority of adolescents who continue to offend at significant levels into their adult life (Hollin et al. 2002). These individuals normally comprise no more than 4–7 percent of all juvenile offenders and a range of factors have been found to predict these persistent offenders (such as family criminality, poor school performance, family poverty, poor housing and high impulsivity). Yet, in focusing so much attention on life-course persistent offenders, surprisingly little effort has been devoted to the majority who cease deviant activity in their twenties.

Nevertheless, there is widespread recognition that most young men get involved in some form of criminal activity and deviancy during their teenage years. This involvement tapers off dramatically as individuals get older, having all but vanished for most by the time they reach 28. With terrorism, the same factors that attract young men to deviant activity in other spheres can also play at least a partial role in the attraction terrorism holds for a few. Higher impulsivity, higher confidence, greater attraction to risk-taking and a need for status can all work to give life as a terrorist a certain appeal for some young males. As discussed later, a desire for revenge and retribution is an extremely common motive for joining terrorist groups and again research indicates that young men hold the most positive attitudes toward vengeance and are the ones most likely to exhibit and approve of vengeful behaviour (Stuckless and Goranson 1992; Cota-McKinley et al. 2001).

Although young men make up the majority of terrorist recruits, some recruits are female and a few are much older. Out of 242 jihadi terrorists identified by Bakker (2006) in his review of jihadi terrorism in Europe since 2001, 5 were women. The review found a relatively wide age range, stretching from 16 to 59 at the time of their arrests, but most of the jihadis were in their teens to mid-twenties (Bakker 2006).

Further, it is also very evident that, even in troubled regions such as Northern Ireland, the vast majority of young males living in the affected communities do not in fact become terrorists. The unavoidable conclusion then is that other factors besides age and gender must be playing crucial roles in the process and decision to become a terrorist.

Education, Career and Marriage

In considering the factors associated with violent radicalization, especially towards jihadism, it is worth focusing on some surprising findings. Traditionally, factors such as good educational achievement, good socioeconomic background and marriage have been associated with a reduced likelihood of criminal offending. An analysis of the backgrounds of jihadis, however, suggests that such trends do not apply as strongly to them.

Sageman's (2004) survey of members of extremist Islamist groups found that they generally tend to be well educated as a group. Over 60 percent had some higher or further-level education. Such findings undermine the view that Islamic extremism can be explained as a result of ignorance or lack of education. Similarly, Sageman (2004) found that about three-quarters of Islamist extremists came from upper- or middle-class backgrounds. A relatively small proportion (27 percent) came from working-class or poor backgrounds. Poorer individuals accounted for a larger proportion of Bakker's (2006) sample, but there was still a very large middle-class element. This undermines explanations for involvement in extremism resulting from personal poverty or deprivation. Further, at the time of joining, the majority of Islamists had professional occupations (e.g. physicians, teachers) or semiskilled employment (e.g. police, civil service, students). Less than a quarter were unemployed or working in unskilled jobs.

Finally, Sageman (2004) found that 73 percent of members were married, and that most of these men had children. Bakker (2006) also found relatively high levels of marriage among jihadis. Family commitments have clearly not prevented individuals from embracing jihad.

It is often assumed that getting married and settling down is a significant factor in explaining desistance from crime in general. There is some research support for this view, but research findings show that the relationship is not entirely straightforward (Farrington and West 1995). Some criminological research has highlighted that marriage at a young age either can have no impact in terms of diminishing offending behaviour, or else can be associated with an increase in criminality. The quality of the relationship, for example, as opposed to the fact that it exists has

been found to be a crucial factor. West (1982) found that marriage had no effect among very young adults and that a preventive effect occurred only among older adults. Ouimet and LeBlanc (1993) found that cohabitation with a partner is positively associated with crime among very young adults (18–23 years old). In this study, marriage was found to contribute to desistance only after the age of 24. Ouimet and LeBlanc explained the negative impact of an early marriage or cohabitation in terms of seeing it as a possible sign of impulsiveness and that marriage at such young ages could lead to significant economic difficulties and family discord.

Building on such findings, Sampson et al. (2006) argued that marriage becomes a positive influence for desistance only once the relationship meets certain quality criteria. Strong social relationships need time and effort to develop. As a result, a marriage or other significant relationship cannot be expected to exert its fullest impact until a sufficient amount of time has passed. Laub and Sampson reviewed data from a series of longitudinal studies of criminal behaviour. The results showed that desistance from crime is facilitated by the development of *quality* marital bonds, and that this influence is gradual and cumulative over time. On average, a five-year-old marriage has more of an impact on desistance than a one-year-old marriage. Thus, the timing and quality of marriage matter: early marriages characterized by social cohesiveness lead to a growing preventive effect. Laub and Sampson (2001) concluded that the effect of a good marriage takes time to appear, and that the influence grows slowly until it inhibits crime.

In considering jihadi terrorism, a further important issue is whether there are different marriage patterns within Muslim communities and the possible impact of practices such as arranged marriages. Research on jihadis has found very high levels of marriage among members of such groups. For example, Sageman (2004) found that 70 percent of members of groups such as al-Qaeda were married. This included many individuals who carried out suicide attacks.

The UK's 2001 census found that young British Muslims (16–24 year olds) had the highest rates of marriage of all groups in the UK (National Statistics Online 2007a): for example, 22 percent of Muslims in these age ranges were married. (The Muslims were the third-lowest group to be co-habiting but not married; Sikhs, Hindus and Muslims were the groups least likely to co-habit.) Christians and those with no religion were the least likely to be married – just 3 percent of 16–24 year olds in each group. Divorce rates among Muslims also appear to be low: 17 percent versus 34 percent for Christians and 43 percent of those with no religion. Thus the high levels of marriage seen among jihadis may simply be a reflection of wider cultural practices and, as a result, the relationships between marriage and desistance from crime that have been observed for other groups may not apply to Muslims. Further, Sageman (2004) highlighted that many jihadi marriages are to wives who share strong ideological beliefs supporting jihadism (or that the wives' families share such beliefs). Thus the marriage provides an essentially endorsing environment for jihadist views as opposed to a restraining influence.

Also of relevance is the importance of peers in facilitating radicalization. Warr (1998) has argued that many of the benefits of marriage in reducing illegal behaviour relate to its impact on peer interaction. He found that marriage leads to a sharp decline in time spent with friends and hence reduces exposure to delinquent peers. On this argument, marriage is an important influence in favour of desistance when it reduces, weakens and severs ties with delinquent associates. In the case of married jihadis, this effect does not occur and peer interactions remain substantial in the lives of the extremists as they are radicalized.

The findings on education and economic background add to a growing body of research that has found that there is no clear link between poverty or deprivation and membership of extremist organizations (Maleckova 2005). The explanation proposed by Maleckova is that terrorists are motivated by belief in a political cause and not by economic factors. Thus, factors associated with desistance among other types of criminal offender (who often *are* motivated by economic factors) will have less of an impact on political offenders. As summarized by Maleckova (2005: 41):

> Just as political participation is much more typical of people who are wealthy enough to concern themselves with more than mere economic subsistence while the impoverished are less likely to vote, the poor are also less likely to become engaged in terrorist organizations.

Social Identity

Identity has been shown to play a vital role in explaining involvement in terrorism. Recruits always belong to the section of society that supports or shares the aims, grievances and ambitions of the terrorist group. In the case of jihadi extremism, individuals need to have a strong sense of Muslim identity and, equally, to identify strongly with the wider Muslim community – the *umma*. Recruits consistently report that, prior to joining, they perceived they had a very strong connection to other Muslims across the globe. This wider connection brought with it a sense of responsibility for these other Muslims, even when the individual had never met them or travelled to their lands. Research has shown that individuals who rate their Islamic identity as being more important than their national or ethnic identity express more positive views on topics such as jihad and martyrdom (Ansari et al. 2006). Further, the 2001 Home Office Citizenship Survey in the UK found that Muslims are more likely than any other religious groupings to rate their faith as their primary identity, and that this effect is particularly strong among young people aged 16–24 (Attwood et al. 2003).

The key aspects of social identity in the context of jihadism are (1) the role of religion and (2) group loyalties.

The Role of Religion

Religion is assumed to lie at the heart of Islamist terrorism and, to a degree, this is justified. As a movement, the global Salafi jihad has a strategic religious agenda with the ultimate goal of recreating past Muslim glory in a great Islamist state stretching from Morocco to the Philippines, eliminating present national boundaries. It preaches *salafiyyah* (from *salaf*, the Arabic word for 'ancient one' and referring to the companions of the Prophet Mohammed) – the restoration of authentic Islam – and advocates a strategy of violent jihad, resulting in an explosion of terror to wipe out what it regards as local political heresy. The global version of the movement advocates the defeat of the western powers that prevent the establishment of a true Islamist state (Sageman 2004).

Whereas the global jihad has a strategic religious aim, the religious backgrounds of the people who join the jihad is not as clear-cut. Indeed, Sageman (2004) found that only 18 percent of Islamist extremists have had an Islamic religious primary or secondary education. In contrast, 82 percent went to secular schools. These data undermine the view that Islamic extremism can be best viewed as resulting from brainwashing by teachers in madrassas as part of normal primary or secondary education.

Sageman (2004) also found that 45 percent of Islamist extremists were described by family and peers as being religiously devout as youth. The majority, however, were not described as being religious as youth. Further, 8 percent were raised as Christians and converted to Islam later in life. There was a definite shift in the degree of devotion, however, prior to joining: 99 percent of members were described as being very religious prior to joining – this represents a major change from the situation when they were younger. Sageman explained this change as originating from a desire for companionship, which was realized within the context of mosques. In the West, many individuals were recruited at a time when they were living away from their family home and old friends. Often the individuals were living in a foreign land and were even more isolated from the society around them. In the expatriate community from which most recruits come, especially in the West, the most available source of companionship with people of similar background is the mosque. The mosque provides a setting for non-radical individuals to socialize, but once groups form, in some cases a process leading to increased religious devotion started. These findings suggest that the mosque often provides the setting in which small groups become radicalized but that this radicalization does not typically result from the teachings of the official hierarchy within the mosque.

It is important to stress that not all Salafist Muslims support the global jihad. Indeed, such individuals remain a minority among Salafists. Thus, in order to understand the mind-set of Islamist terrorism, one needs to move beyond the limits of religious doctrine and explore other driving factors.

Group Loyalties

To a large extent, becoming involved in the jihad is a group phenomenon. Individuals tend not to join the jihad as isolated individuals. Rather, it is within small groups that individuals gradually become radicalized. This is a trend identified by both Sageman (2004) and Bakker (2006). Usually, members undergo a long period of intense social interaction with a small group of friends,

developing a strong mutual intimacy, which relieves their previous isolation.

In his analysis of 242 jihadis, Bakker (2006) found that these individuals tended to become involved in terrorism through networks of friends or relatives and that generally there were no formal ties with global Salafi networks. In short, the individuals were not becoming radicalized because of the efforts of an al-Qaeda recruiter, but rather the process was occurring almost independently of established jihadis.

Within the group context, individuals gradually adopt the beliefs and faith of the group's more extreme members (in a psychological process known as 'risky shift'). As both Sageman and Bakker found, individuals' new Salafi faith resulted in their becoming more isolated from older friends and family, and led to an ever-increasing dependence on, and loyalty towards, the group. With an increasing focus on this small group, their religious faith became more important and more intense. The polarization experienced within the group, combined with an increased sense of group identity and commitment, helped to radicalize individuals and facilitate their entry into the jihad in a way that was approved by their new social peers.

Marginalization

Social marginalization appears to be a common factor in the backgrounds of most jihadi recruits. Research has shown that most members of groups such as al-Qaeda joined the jihad while they were living in a foreign country or when they were otherwise isolated from older friends and family. Often these individuals were expatriates – students, workers, refugees – living away from home and family. Sageman (2004) found that 70 percent joined in a country where they had not grown up; 8 percent were second generation and might not have been fully embedded in their host country. In total, 78 percent of the recruits had been cut off from their social and cultural origins, far from family and friends.

Discrimination

If such marginalized groups are discriminated against or internal sections believe that there is discrimination, then there will always be sections within such communities who will be receptive to radical ideologies that advocate changing or reforming the established, mainstream social system. The aim of these changes will be to improve the lot of the disadvantaged group. People on the margins have less to lose if the current social order is maintained and conceivably a great deal to gain if it is radically changed. For example, a number of factors have been identified as important elements in driving the political conflict in Northern Ireland from 1969 and for increasing support within Catholic communities for extremists (O'Leary 2007). In the context of the Northern Irish conflict, these factors included:

◆ economic deprivation
◆ educational underperformance
◆ insufficient political representation

As a result, most Catholics in Northern Ireland had relatively little to gain by supporting the status quo of Protestant-controlled government, which ran the province until 1972, and potentially a great deal to gain if they could substantially reform or abolish that system of government. Among a disadvantaged Catholic population, ideologies advocating regime change would always have some support and, in the face of continuing discrimination, ideologies advocating even violent action to change the status quo attracted significant support from some quarters.

In the context of jihadi extremism, it is currently clear that Muslim populations in the West are experiencing considerable disadvantage. Indeed, statistics suggest that the Muslim population within the UK is far more disadvantaged than Catholics were in Northern Ireland in 1969. Compared with the UK population as a whole, Muslims have three times the unemployment rate; a higher proportion are unqualified; and a higher concentration are living in deprived areas (National Statistics Online 2007b). Muslims are significantly underperforming in secondary education and are also underrepresented in higher education: 36 percent of the population as a whole have no qualifications, compared with 43 percent of Muslims.

In proportional terms, the number of UK politicians from Muslim backgrounds is also out of balance. Approximately 3 percent of the UK population is Muslim but they only have 0.3 percent of the country's Members of Parliament (MPs) and 0.9 percent of district councillors. For levels to be proportionate, this would require 19 Muslim MPs (as opposed to the current 4) and 675 district councillors (as opposed to the current 217).

This is a problem repeated in other European countries. There are 15 million Muslims in the European Union, reflecting approximately 3 percent of the total population. However, Muslims hold only 0.8 percent of the seats in the European Parliament: out of 785 representatives only 6 are Muslim.

Given such levels of disadvantage, it is not surprising that many Muslim communities have come to view themselves as being unfairly marginalized. Other statistics, such as the fact that Muslims are grossly overrepresented in among prison populations, do not help (Joly 2005). The poor performance of Muslims on these measures can be interpreted in different ways – just as various interpretations of Catholic deprivation were offered in the 1960s – but the imbalance certainly creates a sense of dissatisfaction within the communities affected. Once a community and the individuals within that community have been largely excluded from mainstream society, they will lose much of their vested interest in maintaining that society.

If such deprivation is combined with exposure to an extremist ideology that advocates reforming or even replacing the current order, then an important step towards becoming an active extremist is taken. However, although many people are politically committed to a variety of causes, few are willing to commit acts of violence to further these ideals. To move into terrorism still requires something more.

Catalyst Events/ Perceived Injustice

One of the most important elements in understanding the psychology of why people become extremists is an appreciation of the psychology of vengeance. It has long been recognized that for most terrorists a key motivation for joining a terrorist organization ultimately revolves around a desire for revenge (Schmid and Jongman 1988).

Within the context of jihadi terrorism, the perception of a strong shared identity and link with the wider Muslim world – the *umma* – has serious consequences when the individual perceives that some Muslim communities are being treated brutally or unfairly. Perceived injustices are important drivers of individual decisions to become involved in militant activism. Catalyst events (i.e. violent acts that are perceived to be unjust) provide a strong sense of outrage and a powerful psychological desire for revenge and retribution (Silke 2003). Importantly, one does not need to experience these unjust events first hand in order to feel sufficiently motivated to become a terrorist. Indeed, the events do not even have to involve friends or family members. Many terrorists report that they first joined the organization after witnessing events on television or other media (e.g. O'Callaghan 1998). Although they did not come from the area where the events occurred – or indeed even know the people who lived there – at some level they identified with the victims. This identification, combined with the perceived injustice of the event, can provide a strong motivation to become involved in the jihad.

Such exposure is frequently facilitated through viewing extremist propaganda. This propaganda may be extremely biased, but usually there is a basis of truth and reality to the events portrayed. Islamist recruits tend to be extensively exposed to such propaganda material, with graphic images of abuse and violence being drawn from a wide range of conflicts involving Muslim populations, including Kashmir, Chechnya, Bosnia and the Palestinian Territories (Weimann 2006). US foreign policy is also heavily emphasized, with recent propaganda focusing on the US invasion and occupation of Afghanistan and Iraq. In a number of public statements (including the last statements of suicide bombers) Islamists specifically draw attention to these conflicts as a justification for their own violence.

Exposure to death-related imagery, such as that contained in the jihadi propaganda, results in what psychologists refer to as a 'mortality salience' effect. A variety of research studies have shown that mortality salience generally increases identification with and pride in one's country, religion, gender, race, etc. (for a review see Pyszczynski et al. 2002). Crucially, mortality salience can lead to an increase in support for extremism when it is linked to group identity. For example, one study found that under mortality salience conditions white Americans expressed more sympathy and support for other Whites who expressed racist views. Similarly, Pyszczynski et al. (2006) found that Muslim college students in the Middle East who were reminded of death showed increased support for other students who voiced support for suicide attacks. Significantly, the mortality salient group also indicated that they would be more willing to take part in a suicide attack themselves.

One important element of the desire for vengeance is the surprising willingness of individuals to sacrifice and suffer in order to carry out an act of revenge. As Cota-McKinley et al. (2001: 343) comment, 'vengeance can have many irrational and destructive consequences for the person seeking vengeance as well as for the target. The person seeking vengeance will often compromise his or her own integrity, social standing, and personal safety for the sake of revenge.' This observation is supported by a number of research studies (e.g. Fehr and Gachter 2002; Fowler 2005). In one Swiss study, researchers gave students a cooperative task of the 'prisoner's dilemma' kind: all students in the study will benefit provided each behaves honourably, but those who cheat will benefit more provided they are not caught (Tudge 2002). The students were rewarded with real money if they did well and fined if they did not. They were also able to punish fellow players who had cheated by imposing fines, but could do this only by forfeiting money themselves. This meant that those who punished others frequently would end up with considerably less than those who punished others only a little. Participants tended to punish cheats severely, even though they lost out by doing so. People seem to hate cheats so much that they are prepared to incur significant losses themselves in order to inflict punishment on the transgressors.

Cota-McKinley et al. (2001) show that revenge can fulfil a range of goals, including righting perceived injustice, restoring the self-worth of the vengeful individual and deterring future injustice. The vengeful individual 'sends the message that harmful acts will not go unanswered' (Kim and Smith 1993: 40). Not only is the goal to stop this particular form of maltreatment in the future, it is to deter this transgressor from wanting to commit similar crimes; additionally, vengeance may stop other potential offenders from committing similar crimes or from even considering similar crimes.

Significantly, some individuals are more vengeance prone than others. Men hold more positive attitudes towards vengeance than women do, and young people are much more prepared to act in a vengeful manner than are older individuals (Cota-McKinley et al. 2001). It is not surprising then to find that most recruits to terrorist groups are both young and male. In addition, some evidence exists to suggest that religious belief also affects one's attitude to vengeance, with more secular individuals showing less approval of vengeful attitudes (Greer et al. 2005).

Status and Personal Rewards

As well as providing an outlet for desires for vengeance, terrorist groups offer other inducements and rewards to outsiders. There are considerable dangers in becoming a terrorist. It can be an isolated, stressful and extremely dangerous existence. Recruits to the IRA, for example, were warned that they could expect only one of two things for certain from joining the organization: a lengthy prison sentence or a violent death (e.g. Collins 1997; McGartland 1998). Hardship and suffering are seen as inseparable aspects of life as a terrorist, yet there are still often benefits and advantages to being in a terrorist group. For example, in many communities and societies, terrorist groups and their members are regarded as courageous, honourable and important. As one Palestinian terrorist described it, '[r]ecruits were treated with great respect. A youngster who belonged to Hamas or Fatah was regarded more highly than one who didn't belong to a group, and got better treatment than unaffiliated kids' (Post and Denny 2002). It helps to remember that some communities see members of these organizations or movements not as 'terrorists' but rather as 'freedom fighters', 'rebels' or 'the resistance'. One cannot avoid the fact that applying the label 'terrorist' is often a value judgement (and a negative one) and is often a label imposed from outside of the communities and culture that the terrorists belong to. Those within that culture can reject the term or else reject such a clumsy effort to describe the actors in black and white terms. Ultimately, in many communities joining a terrorist group increases the standing of a teenager or youth considerably. As another Palestinian terrorist put it: 'After recruitment, my social status was greatly enhanced. I got a lot of respect from my acquaintances, and from the young people in my village' (Post and Denny 2002).

As well as increased status and self-esteem, life as a terrorist offers potential recruits excitement and danger. Becoming a jihadi is a dangerous, high-risk decision. Research has long shown that young males in particular are much more attracted to high-risk behaviour than are other segments of the population. The propaganda material developed by jihadi groups often attempts to portray the jihadi lifestyle as an exciting, dangerous and meaningful one. In recruitment videos seen by me, jihadis are shown training with a wide variety of weapons, including assault rifles, machine guns, rockets and hand grenades. There is also often footage of actual attacks against enemy targets.

Overall, such material paints a portrait of jihad as an exciting, dangerous and rewarding activity. This may be an unrealistic portrayal, but it will certainly have an appeal to some viewers. It is clear from other sources, however, that danger and excitement are a feature of life as a terrorist. Indeed, former terrorists, when asked what they miss about their old life as active members, often talk about the closeness they felt with group members, the sense of shared risk and common purpose. In their eyes, life as a terrorist had an intensity and purpose that life outside of the organization noticeably lacked. As a former IRA member described it:

> A part of me missed being in the IRA. I had spent six years leading an action-packed existence, living each day with the excitement of feeling I was playing a part in taking on the Orange State. At the very least, such activity gave a strange edge to my life: I lived each day in a heightened state of alertness. Everything I did, however trivial, could seem meaningful. Life outside the IRA could often feel terribly mundane. . . . I lived life with a weird intensity. I felt myself part of a large family whose members had powerful emotional links to each other. The idea of turning my back on the IRA had become as repugnant to me as turning my back on my own children. As soon as I left this intense environment I found myself missing my comrades: the dangers and risks we shared brought us close. (Collins 1997: 158, 363)

A former Italian terrorist, when asked was there anything they missed from their experience in a terrorist group, replied: 'The fact of being totally at risk' (De Cataldo Neuburger and Valentini 1996: 137). It is important then not to underestimate the appeal that this sense of risk and excitement will exert on potential recruits, especially on those who are otherwise living an ordinary but dull existence.

 ## Opportunity and Recruitment

People cannot become active terrorists unless they can find a terrorist group that is willing to let them join. The individual – now located in the appropriate social grouping and motivated by a desire for retribution – needs to identify an accessible avenue into a terrorist group. He or she is hampered in this task because terrorist groups are nearly always illegal and membership is a punishable crime. This presents difficulties for potential terrorists: they must try to identify current members in order to facilitate entry into the organization, yet they risk exposing themselves to the security services if they approach the wrong person at the wrong time.

For ethnic terrorist groups such as Hamas or the IRA, the problem is largely overcome by the use of legal political front organizations. Hamas, for example, is a very large group and many elements of it are engaged in nonviolent legitimate activity. The organization openly runs and is associated with many schools, hospitals, charities, businesses and mosques (Gunning 2007). For individuals interested in joining, it is not at all difficult to make enquiries about doing so with people involved in the organization and for them to be directed to the appropriate people. Similar situations exist elsewhere, particularly in the case of nationalist and ethnic conflicts.

Some terrorist groups will accept almost any individual who asks to join, but the vast majority of groups will closely vet all applicants. Other groups will first arrange for applicants to carry out some minor illegal acts to test their commitment. If the candidate carries these out successfully, then he or she is formally accepted into the terrorist group.

For the individual who cannot identify a route into an established terrorist group, or is refused membership after applying, the remaining options are to form their own group or to wage a one-person terrorist campaign. For example, the American far right's concept of 'leaderless resistance' explicitly endorses a philosophy of individuals (or small groups) mounting operationally independent campaigns of violence for ideologically similar reasons. The concept was first pushed in books such as *The Turner Diaries*, but has now been taken up and advocated very strongly on the Internet as well (Joyce-Hasham 2000).

Prior to 9/11, mosques played a key role in providing potential jihadis with a route into groups such as al-Qaeda. Radical imams acted as a magnet for potential recruits – for example, Omar Mahmoud Othman (alias abu Qatada) and Mustafa Kamel (alias abu Hamza al-Mazri), who were active in London. Indeed, a very high proportion of the UK's Islamist extremists can be linked by their attendance at one of two London mosques: Brixton mosque and Finsbury Park mosque. At both national and international levels these mosques established a reputation for hard-line extremist preaching that advocated the use of violence for furthering the global Salafi jihad.

It is not always the case that a radical imam actively recruits potential jihadis; rather, the social interactions between attendees listening to the sermons at these mosques builds and reinforces ideological commitment. The mosques offer opportunities for people to meet new friends and foster the development of an ideological commitment to the jihad, and (critically) provide links to the jihad through already connected members. Finsbury Park mosque was closed for an extended period following a major police raid on 20 January 2003. The police statement at the time of the raid stated that:

> [The raid] was aimed specifically at individuals who have been supporting or engaging in suspected terrorist activity from within the building. Police believe that these premises have played a role in the recruitment of suspected terrorists and in supporting their activity both here and abroad.

 Conclusions

The preceding discussion has highlighted that Islamist terrorists do not fit many of the stereotypes that shape public expectations. The individuals involved do not suffer from mental illness or disorders, but instead are generally ordinary and unremarkable in psychological terms. Their involvement in terrorism is usually the result of a gradual process – typically occurring over a period of years. The sense of personal identity and social networks of potential recruits are both extremely important factors. Most terrorists become radicalized as members of a small group of like-minded individuals. These groups do not start out as radical but become so gradually over time. Sometimes the initial contacts are facilitated at mosques, but the relative isolation of the individuals from the surrounding society beforehand appears to play an important role in the early bonding of the group.

As the members of a group become more interested in their faith, the concern about abuses experienced by Muslims elsewhere becomes a more prominent issue. Provocative and explicit propaganda can fuel these concerns, though it is worth noting that both Muslims and non-Muslims in the wider society can share many of these concerns.

Ultimately, ordinary psychological processes and small group dynamics play a major role in understanding Islamist radicalization. It can be very difficult, however, to see beyond the common myths and assumptions that are often offered as explanations for terrorist violence. Partly this is because the process of involvement is, in many respects, at odds with our understanding of the development and course of other types of offending. Farrington (2003: 224–5) summarized our existing understanding of other offenders:

> The main risk factors for the early onset of offending before age 20 are well known . . . individual factors (low intelligence, low school achievement, hyperactivity impulsiveness and risk-taking, antisocial child behavior, including aggression and bullying), family factors (poor parental supervision, harsh discipline and child physical abuse, inconsistent discipline, a cold parental attitude and child neglect, low involvement of parents with children, parental conflict, broken families, criminal parents, delinquent siblings), socioeconomic factors (low family income, large family size), peer factors (delinquent peers, peer rejection, and low popularity), school factors (a high delinquency rate school) and neighbourhood factors (a high crime neighbourhood).

Yet most of these factors are absent in the lives of jihadis, and indeed many terrorists appear to come from backgrounds that would normally protect against the onset of offending. This makes it clear that terrorists cannot simply be regarded as typical criminals; on the contrary, they are a distinct group, and in many ways the origins of their unconventional behaviour are exceptional. This distinctiveness also helps to explain why many of the factors that are normally associated with desistance, such as marriage, education and a career, often do not have the same influence on this group. Criminology will have to work hard to develop theories and models that can comfortably account for the distinctive patterns seen in the lives of terrorists.

 # References

Ansari, H., Cinnirella, M., Rogers, M. B., Loewenthal, K. M., and Lewis, C. A. (2006). Perceptions of martyrdom and terrorism amongst British Muslims. In M. B. Rogers, C. A. Lewis, K. M. Loewenthal, M. Cinnirella, R. Amlôt and H. Ansari (eds) *Proceedings of the British Psychological Society Seminar Series Aspects of Terrorism and Martyrdom*, COMMUNITY: *International Journal of Mental Health & Addiction*.

Attwood, C., Singh, G., Prime, D. and Creasey, R. (2003). *2001 Home Office Citizenship Survey: People, Families and Communities*. Home Office Research Study 270. London: Home Office.

Bakker, E. (2006). *Jihadi terrorists in Europe, their characteristics and the circumstances in which they joined the jihad: An exploratory study*. The Hague: Clingendael Institute.

Budd, T., Sharp, C. and Mayhew, P. (2005). Offending in England and Wales: First results from the 2003 Crime and Justice Survey. Home Office Research Study 275. London: Home Office.

Burke, J. (2004). *Al-Qaeda: The true story of radical Islam*. London: Penguin. Chen, H., Reid, E., Sinai, J., Silke, A. and Ganor, B. (in press). *Terrorism informatics: Knowledge management and data mining for homeland security*. New York: Springer-Verlag.

Collins, E. with McGovern, M. (1997). *Killing rage*. London: Granta Books.

Cota-McKinley, A., Woody, W. and Bell, P. (2001). Vengeance: Effects of gender, age and religious background. *Aggressive Behavior 27*, 343–50.

Crenshaw, M. (1992). How terrorists think: What psychology can contribute to understanding terrorism. In L. Howard (ed.) *Terrorism: Roots, impact, responses*, 71–80. London: Praeger.

De Cataldo Neuburger, L. and Valentini, T. (1996). *Women and terrorism*. London: Macmillan.

DeMause, L. (2002). The childhood origins of terrorism. *Journal of Psychohistory 29*, 340–8. URL (accessed 10 September 2007): http://www.psychohistory. com/htm/eln03_terrorism.html.

Farrington, D. P. (2003). Developmental and life-course criminology: Key theoretical and empirical issues – the 2002 Sutherland award address. *Criminology 41*, 221–55.

Farrington, D. and West, D. (1995). Effects of marriage, separation and children on offending by adult males. In J. Hagan (ed.) *Current perspectives on aging and the life cycle: Vol. 4. Delinquency and disrepute in the life course*, 249–81. Greenwich, CT: JAI Press.

Fehr, E. and Gachter, S. (2002). Altruistic punishment in humans. *Nature 415*, 137–40.

Fowler, J. (2005). Altruistic punishment and the origin of cooperation. *Proceedings of the National Academy of Sciences 102*, 7047–9.

Fox, J. and Zawitz, M. (2006). *Homicide trends in the United States*. Washington DC: Bureau of Justice Statistics.

Frankfort-Nachmias, C. and Nachmias, D. (1996). *Research methods in the social sciences*, 5th edn. London: Arnold.

Friedland, N. (1992). Becoming a terrorist: Social and individual antecedents. In L. Howard (ed.) *Terrorism: Roots, impact, responses*, 81–93. London: Praeger.

Greer, T., Berman, M., Varan, V., Bobrycki, L. and Watson, S. (2005). We are a religious people; we are a vengeful people. *Journal for the Scientific Study of Religion 44*, 45–57.

Gunaratna, R. (2002). *Inside al-Qaeda*. London: Hurst.

Gunning, J. (2007). Hamas: Harakat al-Muqawama al-Islamiyya. In M. Hieberg, B. O'Leary and J. Tirman (eds) *Terror, insurgency and the state*, 123–56. Philadelphia: University of Pennsylvania Press.

Harrower, M. (1976). Were Hitler's henchmen mad? *Psychology Today 6*, 76–80.

Hollin, C., Browne, D. and Palmer, E. (2002). *Delinquency and young offenders*. Oxford: BPS Blackwell.

Horgan, J. (2005). *The psychology of terrorism*. London: Routledge.

Joly, D. (2005). Muslims in prisons: A European challenge. Full Research Report. Swindon: Economic and Social Research Council.

Joyce-Hasham, M. (2000). Web offence. *The World Today 56*, 11–13.

Junger-Tas, J. (1994). *Delinquent behaviour among young people in the western world*. Amsterdam: Kugler.

Kegley, C. (2002). *The new global terrorism: Characteristics, causes, controls*. Upper Saddle River, NJ: Prentice Hall.

Kim, S. and Smith, R. (1993). Revenge and conflict escalation. *Negotiation Journal 9*, 37–43.

Laqueur, W. (2000). *The new terrorism: Fanaticism and the arms of mass destruction*. Oxford: Oxford University Press.

Laub, J. and Sampson, R. (2001). Understanding desistance from crime. In Michael Tonry (ed.) *Crime and justice*, vol. 28, 1–69. Chicago: University of Chicago Press.

Lyons, H. A. and Harbinson, H. J. (1986). A comparison of political and non-political murderers in Northern Ireland, 1974–84. *Medicine, Science and the Law 26*, 193–8.

McGartland, M. (1998). *Fifty dead men walking*, 2nd edn. London: John Blake.

Maleckova, J. (2005). Impoverished terrorists: Stereotype or reality? In T. Bjorgo (ed.) *Root causes of terrorism*, 33–43. London: Routledge.

Mazarr, M. J. (2004). The psychological sources of Islamic terrorism. *Policy Review 125*.

Moffitt, T. (1993). Adolescence-limited and life-course-persistent antisocial behaviour: A developmental taxonomy. *Psychological Review 100*, 674–701.

National Statistics Online (2007a). Marriage patterns. URL (accessed 4 July 2007): http://www.statistics.gov.uk/cci/nugget.asp?id=960.

National Statistics Online (2007b). Labour market. URL (accessed 4 July 2007): http://www.statistics.gov.uk/cci/nugget.asp?id=979.

Nesser, P. (2006). How does radicalization occur in Europe? Paper presented at the Second Inter-Agency Radicalization Conference, 10 July 2006, US Department of Homeland Security, Washington DC.

O'Callaghan, S. (1998). *The informer*. London: Granta.

O'Leary, B. (2007). IRA: Irish Republican Army (Oglaigh na hEireann). In M. Hieberg, B. O'Leary and J. Tirman (eds) *Terror, insurgency and the state*, 189–228. Philadelphia: University of Pennsylvania Press.

Ouimet, M. and LeBlanc, M. (1993). Life events in the course of the adult criminal career. *Criminal Behavior and Mental Health 6*, 75–97.

Pape, R. (2005). *Dying to win: The strategic logic of suicide terrorism*. New York: Random House.

Pearlstein, R. M. (1991). *The mind of the political terrorist*. Wilmington, DE: Scholarly Resources.

Post, J. M. (1986). Hostilité, conformité, fraternité: The group dynamics of terrorist behavior. *International Journal of Group Psychotherapy 36*, 211–24.

Post, J. (1990). Terrorist psycho-logic: Terrorist behavior as a product of psychological forces. In W. Reich (ed.) *Origins of terrorism: Psychologies, ideologies, theologies, states of mind*, 25–40. Cambridge, MA: Woodrow Wilson Center.

Post, J. and Denny, L. (2002). The terrorists in their own words. Paper presented at the International Society of Political Psychology Conference, 16–19 July, Berlin, Germany.

Pyszczynski, T., Solomon, S. and Greenberg, J. (2002). *In the wake of 9/11: The psychology of terror*. Washington DC: American Psychological Association.

Pyszczynski, T., Abdollahi, A., Solomon, S., Greenberg, J., Cohen, F. and Weise, D. (2006). Mortality salience, martyrdom, and military might: The great satan versus the axis of evil. *Personality and Social Psychology Bulletin 32*, 525–37.

Quattrone, G. A. (1982). Overattribution and unit formation: When behaviour engulfs the person. *Journal of Personality and Social Psychology 36*, 247–56.

Ranstorp, M. (2006). *Mapping terrorism research*. Oxford: Routledge.

Rutter, M., Giller, H. and Hagell, A. (1998). *Antisocial behavior by young people*. Cambridge: Cambridge University Press.

Sageman, M. (2004). *Understanding terrorist networks*. Philadelphia: University of Pennsylvania Press.

Sampson, R., Laub, J. and Wimer, C. (2006). Does marriage reduce crime? A counterfactual approach to within-individual causal effects. *Criminology 44*, 465–508.

Schmid, A. and Jongman, A. (1988). *Political terrorism*, 2nd edn. Oxford: North-Holland.

Schönteich, M. (1999). The dangers of youth? Linking offenders, victims and age. *Nedbank ISS Crime Index 3*.

Shafritz, J. M., Gibbons, E. F., Jr and Scott, G. E. J. (1991). *Almanac of modern terrorism*. Oxford: Facts on File.

Silke, A. (1998). Cheshire-cat logic: The recurring theme of terrorist abnormality in psychological research. *Psychology, Crime, and Law 4*, 51–69.

Silke, A. (2001). The devil you know: Continuing problems with research on terrorism. *Terrorism and Political Violence 13*, 1–14.

Silke, A. (2003). Becoming a terrorist. In A. Silke (ed.) *Terrorists, victims and society: Psychological perspectives on terrorism and its consequences*, 29–53. Chichester: Wiley.

Silke, A. (2006). The impact of 9/11 on research on terrorism. In M. Ranstorp (ed.) *Mapping terrorism research*, 175–93. Oxford: Routledge.

Stuckless, N. and Goranson, R. (1992). The vengeance scale: Development of a measure of attitudes toward revenge. *Journal of Social Behaviour and Personality 7*, 25–42.

Taylor, M. (1988). *The terrorist*. London: Brassey's.

Tudge, C. (2002). Natural born killers. *New Scientist 174*, 36–9.

Vidino, L. (2006). *Al Qaeda in Europe: The new battleground of international jihad*. London: Prometheus Books.

Warr, M. (1998). Life-course transitions and desistance from crime. *Criminology 36*, 183–216.

Weimann, G. (2006). *Terror on the Internet.* Washington DC: United States Institute of Peace Press.

West, D. (1982). *Delinquency: Its roots, careers and prospects.* London: Heinemann.

Woolman, D. (2002). Fighting Islam's fierce Moro warriors. *Military History 9*, 34–40.

Andrew Silke Andrew Silke currently holds a Chair in Criminology at the University of East London, where he is the Field Leader for Criminology and the Director for Terrorism Studies. Professor Silke has a background in forensic psychology and has worked both in academia and for government. He serves by invitation on both the European Commission's Expert Group on Violent Radicalisation and the United Nations' Roster of Terrorism Experts. His work has taken him to Northern Ireland, the Middle East and Latin America. a.silke@uel.ac.uk

REVIEW QUESTIONS

1. What is the main research question Silke is seeking to answer? How does this research fill a gap or gaps in the literature?

2. What is the main critique(s) Silke (2008) makes about the current research on jihadi radicalization? How does his article not suffer from the same defects? How would you solve this research dilemma?

3. What factors does Silke (2008) offer for why people become radicalized and join terrorist groups? Does he give more weight to one factor or set of factors over another? Do you agree with his conclusions? Why or why not?

READING 11

Plummer explores the connection between failed states and terrorism; asking the question "which failed states tend to promote what amounts of terrorism". Using the Failed States Index, Plummer examines the connection between how and why a state failed and the extent of terrorism generated by that state's failure. Case studies of Somalia and the Ivory Coast, combined with a least squares regression analysis of the variables in the Index, are used to address this question. Plummer finds that the relationship between state failure and terrorism is a complex rather than a linear one and depends heavily on the process by which the state failed.

Failed States and Connections to Terrorist Activity

Chelli Plummer[1]

 ## Introduction

In the current geopolitical structure, asymmetrical warfare (terrorism) has increased in lethality. Moreover, current research suggest that this trend will continue (Ellis, 2003; Hoffman, 1998) and have widespread consequences for a great number of people. The new age of religiously motivated terrorism (Bergesen & Lizardo, 2004; Ellis, 2003; Hoffman, 1998; Winkler, 2008) faces a highly complex new phenomenon, namely terrorists' goals have become more vague and ambiguous (Bergesen & Lizardo, 2004), while weapons have become more sophisticated (Ellis, 2003; Hoffman, 1998) and targets exceedingly indiscriminate (Bergesen & Lizardo, 2004).

SOURCE: *International Criminal Justice Review*, 22(4) 416–449. Copyright © 2012 Georgia State University

Previous terrorism research focuses on relationships between terrorism and political (Piazza, 2008), economic (Freytag, Kruger, Meierrieks, & Schneider, 2009; Schneider, Bruck, & Meierrieks, 2010), and social factors (Burgoon, 2006); yet does not consider the interplay of the three. Moreover, there is limited empirical research on the connections, relying heavily on case studies or theory building. The Failed States Index (Fund for Peace, 2009) has been used in its aggregate form to test for connections between state failure and terrorist activity; however, little research examines the variables that make up the Index. The present research employs quantitative analysis to examine the relationship between terrorist activity and failed states. First, ordinary least squares regression is utilized to ascertain which of the variables of social, economic, and political fragility are most closely associated with terrorist activity. Second, two case studies of Somalia and the Ivory Coast are utilized to explore the relationship between fragility and terrorism. These two countries were chosen because of similar scores on the Failed States Index while having very different outcomes of terrorist activity. The comparative case study analysis is used to contextualize how aspects of state failure differ in affects on terrorist activity.

The majority of prior research focuses primarily on the economic factors that foster terrorism. Therefore, examining the multifaceted components of state failure may shed light on the interdependency of economic factors on social and political forces. Research has been separated along political, economic, and social lines with little synthesis of the three. A broader view of terrorism's roots may lead to a better understanding of the qualities of states at risk for fostering terrorist activity. Through deeper exploration of state fragility, theory may emerge to help counterterrorism policies transnationally. This study is driven by the theoretical position that state failure is intimately connected to terrorist activity; however, it strives to go beyond the superficiality of state failure and delve into the processes of such failure.

 ## Literature Review

The first step in conducting a study on terrorism is defining the term terrorism. The present research employs the Global Terrorism Database (GTD) definition, "the threatened or actual use of illegal force and violence by a non-state actor to attain a political, economic, religious, or social goal through fear, coercion, or intimidation" (GTD, 2007; also see Schmid & Jongman, 1988, 2005). The GTD includes acts that meet two of the three following criteria:

1. The violent act was aimed at attaining a political, economic, religious, or social goal; the violent act included evidence of an intention to coerce, intimidate, or convey some other message to a larger audience other than the immediate victims; and the violent act was outside the precepts of International Humanitarian Law.

 ## The Influence of Social Welfare Factors on Terrorism

Burgoon's (2006) study of social welfare policies indicates that countries that put less effort into social welfare have more connections to transnational terrorism as well as terrorist incidents on their own land. Such policies as social security, unemployment, and health and education spending are said to discourage terrorism. Social welfare policies serve to reduce poverty, inequality, and insecurity (Burgoon, 2006; Schneider et al., 2010). Social welfare reduces economic insecurity, inequality, poverty, and religious extremism, which in turn reduce the likelihood of terrorism (Burgoon, 2006). Alesina and Perotti (1996) agree with Piazza (2006) that as populations increase, the likelihood of terrorism increases. Alesina and Perotti (1996) demonstrated that as democracy and social welfare increase, terrorism decreases.

Not all empirical research supports the social welfare argument (Freytag, Kruger, Meierrieks, & Schneider, 2011; Krueger, 2008; Krueger & Maleckova, 2003; Sageman, 2008). Terrorists predominately come from well-educated, higher socioeconomic statuses, making motivations more political than economic (Bergesen & Lizardo, 2004; Ellis, 2003; Hoffman, 1998; Winkler, 2008). Freytag, Kruger, Meierrieks, and Schneider (2011) demonstrate empirically that while terrorists themselves may be economically and educationally advantaged, the society that they come from is not. This environment, aspects of state failure, also fosters a large pool of economically disadvantaged and disenfranchised youth from which to pull potential recruits. Further, economic disparity between countries causes grievances against the current economic order and the disadvantage failing states experience (Blomberg, Hess, & Weerapana, 2004; Freytag et al., 2011; Harrison, 2006).

The presence of particular ethnic or religious communities with legacies of persecution or repression may also create black holes, where terrorists can hide. According to Korteweg (2008), terrorist groups can plug into the particular grievances of these local communities, gaining the advantage of popular support. Through community support, terrorists can hide, gain new recruits, and possibly have access to new resources. Important here is the fact that in tribal communities, a sense of duty and honor are paramount and, as such, obligations to help those in their tribe may be part of the moral code that terrorists exploit (Freytag et al., 2009; Kittner, 2007; Simons & Tucker, 2007). Ethnic divides and grievances along group lines also foster communities where terrorist activity thrives (Piazza, 2008).

Taking into consideration the research that has come before, and in particular, Piazza (2008), more exploration into aspects of state failure and their connections to terrorism is demanded. Through a comparison of case studies, this article examines exactly what the differences are in how a state fails and how these differences contribute to fostering terrorism.

The Influence of Economic Factors on Terrorism

Economic underdevelopment can be a catalyst for terrorist creation. Korteweg (2008) states that economic underdevelopment (e.g., areas of high poverty) may be advantageous for terrorist recruitment. Young men might have no other opportunities for employment than terrorist groups. Poverty is cited repeatedly as a main catalyst in terrorist group formation (Burgoon, 2006; Freytag et al., 2009; Li & Schaub, 2004; Piazza, 2008; Schneider et al., 2010). Even so, it may be the unequal distribution of wealth creating the conditions that spawn terrorist organizations (Burgoon, 2006; Li & Schaub, 2004; Piazza, 2008; Sageman, 2008). Much like Piazza (2011), Stewart (2000) found that increasing inequality fuels social discontent, political instability, and violence. Muller and Seligson (1987) found that misdistribution of land and income inequality are positively correlated to mass violence. Land reforms and rapid economic growth can further exacerbate this phenomenon.

Economic disparity between countries causes grievances against the current economic order and the disadvantage certain failing states experience (Blomberg et al., 2004; Freytag et al., 2011; Harrison, 2006). Ehrlich and Liu (2002) argue that Islamic rage is partially due to failure to achieve economic success. Upon examination of Pakistan's failure to take off economically, Looney (2004) found that high fiscal deficits, unsustainable public debt, sharp deterioration in the distribution of income, and a disturbing rise in the level of unemployment and poverty. He argues that these are the factors that shape the increasing terrorist activity in Pakistan (Looney, 2004).

Minority economic discrimination increases the propensity of terrorist activity (Piazza, 2011). Politically marginalized people, with no remediation for discrimination, are more likely to resort to terrorism. When there are mechanisms in place to address these issues, marginalized people are much less likely to resort to terrorism. Piazza (2011) found that countries with high economic development and a high Gini index are subject to more terrorist activity; however, he cautions that the overall country economic status is not as important as the minority group's economic status. While less developed countries seem to spawn terrorism, many factors contribute to the creation of a terrorist. Krueger and Maleckova (2003) found that although poor countries spawn terrorism, when controlled for civil liberties, the relationship disappears. While poverty may not breed terrorists, Freytag, Kruger, Meierrieks, and Schneider (2009) found that underdevelopment in the Middle East as well as other countries seems to be the bed from which terrorists do arise. The present research builds upon prior work by adding breadth by examining the economic indicators of state failure disaggregated from the Failed States Index. In addition, the present research adds depth by examining the economic failures of the two case studies.

The Influence of Political Factors on Terrorism

Terrorism is a form of asymmetrical warfare; military powers differ significantly between groups. As such, the targets, while not necessarily states themselves, are politically motivated and key to the analysis. Weber (1948) viewed the state as the legitimate source of force and, as such, states have a monopoly on the use of force. This is furthered by the states' ability to tax in support of a standing army (Weber, 1948). Legitimacy is paramount to state formation in that people must believe in the legitimacy of the state to exercise force in order to be compliant. Legitimacy allows the government the right to govern. Another aspect of the state is a particular geographical location or the right to property. This is the basis of unequal distribution of resources and, as such, the basis for conflict (Giddens, 1999). As the recognized state has certain rights to property and the ability to assign property in the manner it sees fit, certain unequal distributions of resources may occur that can spur conflict.

Weber (1948) states three resources where conflict can arise: economic, power, and status (cultural). Collins (1975) provides an important perspective to examine the historical aspects of terrorism in light of Weber's (1948) three resources. According to Collins, economic resources are those of material conditions. Power resources are those of social position within networks. Status or cultural resources are those that exert control over rituals that produce solidarity (Collins, 1975). Resources become important when groups mobilize in reaction to unequal distribution. Groups may mobilize in two ways, emotionally or materially (Collins, 1975). Emotional mobilization is important for terrorist groups because members must have a strong sense of group identity. The sense of group identity permits terrorist members to perceive their beliefs as morally right and to make sacrifices for the cause. Material mobilization is also important for terrorist groups, in that it encompasses communication and transportation as well as material and monetary supplies to sustain the conflict (Collins, 1975).

In opposition to the argument that failed states foster terrorism, Simmons and Tucker (2007), Sageman (2008), Bilgin and Morton (2004) as well as Patrick (2006) argue that the connection is tenuous at best. Failed states are too chaotic to promote terrorism, and there needs to be a degree of functioning that will allow the bare infrastructure to be in place so that terrorists can operate successfully (Patrick, 2006). Logistically, Simmons and Tucker (2007) argue that failed states are a nightmare for reliable operations. Simmons and Tucker (2007) propose that while people in failed states do gain skills that would be well utilized by terrorist organizations, the need for their skills locally is more critical. Simmons and Tucker (2007) posit

that few failed states are utilized as training camps (with the exception of Afghanistan). This could be because terrorists are now receiving on the job training in the wars in Afghanistan and Iraq (Simmons & Tucker, 2007).

Korteweg (2008) cites seven aspects of comparative advantage that lead to the creation of "black holes" or spaces in which terrorism may grow. The most important of these aspects are remote areas of countries where the governments of those countries have little ability to control what goes on in the area. It is physically impossible to govern such areas; as a result, the areas may be forgotten regions of no man's land. These are challenges to the state's ruling authority that erodes the confidence in the ability of the state to assert its control (Zartman, 1995). These are nontransparent areas where the central government has questionable legitimacy and groups are free to operate unnoticed that Piazza (2008) terms stateless regions.

In failing states, there is challenge to governmental authority as well as lack of confidence in the state to control the territory (Piazza, 2008; Zartman, 1995). Hehir (2007) posits that failed and/or failing states suffer administrative incapacity, where the government is unable to provide basic services that are expected from such an entity. Lambach (2004) notes that there is not a clear threshold of failure and that there are distinctions between weak states that may still function in some aspects and collapsed states that have no ability to effectively govern. Failed states may retain the appearance of sovereignty (Takeyh & Gvosdev, 2002). This type of functioning may best suit terrorist organizations in that outward manifestation of sovereignty limits outside intervention (Piazza, 2008). The present research builds upon prior work by examining political indices that compose the Failed States Index. In addition, the present research adds depth by examining the political failures of the two case studies.

 Hypotheses

This study is meant to examine the connections between failed states and terrorist activity. While the aggregate failed states score is statistically significant and linearly associated with terrorist activity, a deeper examination needs to be conducted on the individual components. Past research has focused on the linear relationship between state failure and terrorist activity; however, the relationship may not be linear. There may be a midrange of state failure that is most optimal for terrorist activities. The argument is informed by prior research (Li & Schaub, 2004; Mair, 2008) although in conflict with the most recent empirical findings (Piazza, 2008). There exists a range that fosters the most terrorist activity. While it has been put forth that the relationship between the total state failure score and terrorist activity is linear (Piazza, 2008), completely failed states are not hospitable to terrorism. Nor are states that experience the least amount of state failure immune from such activity.

Based on a review of the literature and theoretical understanding, the following hypotheses are explored:

Hypothesis 1: While the Failed States Index aggregate score has been significantly related to terrorist activity, a better model of prediction would be one that contains all 12 variables that comprise the Failed States Index.

Hypothesis 2: The predictive power of the 12 variable model from Hypothesis 1 will be significantly improved by the addition of the Global Peace Index (GPI) as an independent variable.

Hypothesis 3: There is a midrange of the total state failure scores that will be most strongly related to terrorist activity and a squared term of the overall Failed States score will be a better predictor of terrorist activity.

 Data and Methods

Quantitative

The data for these analyses came from the National Consortium for the Study of Terrorism and Responses to Terrorism (START). START is based at the University of Maryland, College Park, and is an open source. In conjunction with START, the GTD, also an open source, encompasses terrorist activities throughout the world from 1970 through 2008, with over 87,000 cases. The GPI (see Appendix A for list of indicators and operationalization of the overall GPT score) measures relative position of nations' peacefulness, created in 2007 by the Institute for Economics and Peace, with data from the Economist Intelligence Unit (EIU): the lower the GPI score, the more peaceful the country. It consists of 149 countries and examines a set of 23 internal as well as external violence factors. Also utilized

for this study is the Failed States Index of 184 countries measured on 12 variables and is an open source database.

Sample

The focus year of this study is 2008. The data for the study come from the START Center and consists of 187 countries, comprising 4,861 attacks and 104 terrorist organizations. The data were collapsed so that there is one entry for each country and the variable number of attacks was generated for each country. Once countries that had no information were eliminated, the total number of cases was 179 (see Appendix B for list of countries).

Dependent Variable

The dependent variable is the calculated rate of terrorist attacks within a country per 100,000 of the population. The calculation is a common measure of terrorism level in a given country (Piazza, 2008).

Independent Variables

The independent variables for Hypothesis 1 are the 12 indicators of the Failed States Index. The Failed States Index includes four social indicators, two economic indicators, and six political indicators (for the operationalization of the following measures, see Appendix C).

Social indicators. The Failed States Index uses measures of demographic pressure. These include the following: pressures from high population density relative to food supply, settlement patterns, border disputes, land ownership, and controls of religious or historical sites. The second social measure is massive movement of refugees and internally displaced peoples (IDPs). A legacy of vengeance seeking group grievances is a measure of atrocities committed against groups in forms of persecution, repression, or political exclusion. Chronic and sustained human flight, such as the "brain drain," is used to measure the emigration of the middle class as well as growth of exiled communities (Fund for Peace, 2009).

Economic indicators. Uneven economic development along group lines is determined by groupbased inequality in jobs, education and economic status, poverty levels, infant mortality rates, and education levels. Sharp and/or severe economic decline is a measurement of the society as a whole relying on per capita income, gross national product (GNP), debt, and business failures. Indicative of this variable is a collapse or severe devaluation of the national currency and increase in hidden economies (Fund for Peace, 2009).

Political indicators. Criminalization and/or delegitimization of the state is an indicator for government corruption with a lack of transparency, accountability, and political representation. Progressive deterioration of public services measures the disappearance of basic state services and protection of citizens. Widespread violation of human rights is a measure of the emergence of authoritarian, dictatorial, or military rule, along with large numbers of political prisoners. Security apparatus as a "state within a state" is the surfacing of private militias, which terrorizes opponents. The rise of factionalized elites demonstrates a fragmentation of ruling elites along group lines. Finally, intervention of other states or external factors indicates risk to the state's sovereignty (Fund for Peace, 2009).

For Hypothesis 1, an ordinary lease square (OLS) regression model with the 12 indicators of the Failed States Index was run with the dependent variable of rate of terrorist activity. This allows for comparison of the effects of the 12 variables on the dependent variable and comparison to the model with only the total Failed States score. Both standardized and unstandardized coefficients were generated.

For Hypothesis 2, the independent variables were the total Failed States Index scores and the GPI scores. It is necessary to run nested models, with the full model of the Failed States score, along with the GPI, and a restricted model, excluding the GPI, so that a comparison can be made between the two, using appropriate statistical significance tests. By comparison of the models, the importance of the GPI becomes apparent. For Hypothesis 3, the total Failed States Index score was to be squared to test my hypothesis that this is not necessarily a linear relationship and compared to the model with the linear term.

Control Variables

Control variables for demographics are the *population size* [*emphasis added*] and the *median age* [*emphasis added*] of the country. Because crime in general is most heavily represented in the age group 15–35 (Ehrlich & Liu, 2002),

controlling for the age structure of the country is important. Population size is a potential confounder if not included in the analysis, because as population size increases there is an increased potential for more criminal activity. The third control variable is a *religious diversity measure* [*emphasis added*], constructed based on the Herfindahl–Hirschman Index ([HHI] Hirschman, 1964).

The case study analysis provides depth by examining the interplay of the political, economic, and social factors to enhance the quantitative analysis. There is a need to look at countries whose failure scores are similar yet differ in levels of terrorist activity because processes involving the interplay of political, economic, and social factors are having differing outcomes. Feagin, Orum, and Sjoberg (1991) find that case studies permit the "grounding of observations and concepts about social action and social structures . . . [while]providing information from a number of sources and over a period of time."

Case Studies

Choice of countries was based on the total failed states score, focusing on finding two failed states, one with high levels of terrorist activity and one without. The scores on the 12 indices were very similar, as were their global peace score, median age, population density, and governance score. In focusing on the 10 most failed states, 2 states, Somalia and the Ivory Coast, most closely aligned on the independent variables, while still differing on the level of terrorist activity (see Table 1 for comparison of 2 countries on key variables).

Dependent Variable

The dependent variable is the calculated rate of terrorist attacks within a country, which is my unit of analysis, as was the prior section.

Independent Variables

Diversity. Because of long histories of colonial oppression and ethnic conflict, it is imperative to investigate the differences in population diversity. The two measures used were one of ELF and religious diversity. The religious diversity measure was constructed based on the HHI.

Table 1	Case Study Variables	
Country	**Somalia**	**Ivory Coast**
Terrorist attacks	172	0
ELF	0.082	0.82
Religious diversity	0.03	0.701
Voice and accountability	−1.87	−1.26
Political stability	−3.3	−1.88
Government effectiveness	−2.45	−1.24
Regulatory quality	−2.66	−0.90
Rule of law	−2.68	−1.46
Control of corruption	−1.92	−1.08
Colonial power	Great Britain	France

NOTE: ELF = Ethnolinguistic Fractionalization.

Table 2	Descriptive Statistics for Independent, Control, and Dependent Variables			
Variable	**Mean**	**SD**	**Minimum**	**Maximum**
Peace score (GPI)	2.050	0.486	1.176	3.514
Demographic pressures	6.390	1.987	1.000	9.800
Refugees/IDPs	4.971	2.372	0.900	9.800
Group grievance	5.896	2.043	1.000	10.000
Human flight	5.606	2.077	1.000	10.000
Uneven development	6.753	1.823	1.900	9.600
Economic decline	5.503	2.077	1.200	10.000
Delegitimization of the state	6.382	2.424	0.900	10.000
Public services	5.689	2.315	1.200	10.000
Human rights	5.943	2.260	1.400	9.900
Security apparatus	5.555	2.561	0.700	10.000
Factionalized elites	5.921	2.561	0.700	10.000
External intervention	5.697	2.202	0.900	10.000
Total failed states score	70.305	23.560	16.800	114.200
Rate of terrorist activity	0.0982	0.324	0.000	3.111

SOURCE: Failed States Index, $N = 179$.

NOTE: GPI = Global Peace Index; IDP = Internally Displaced People.

Table 3	Regression Predicting Terrorist Activity Rates by Failed States Indices Scores (N = 179)			
Variable	**Model 1**	**Model 2**	**Model 3**	**Model 4**
Adjusted R^2	.0588	.1385	.2234	.1851
Model F	3.16*	2.48*	3.48*	7.27*
Constant	−.516 (.245)*	.009 (.314)	−.323 (.310)	.149 (.269)
Demographic pressure		−.034 (.046)	−.031 (.044)	
Refugees/IDPs		.026 (.019)	.009 (.018)	
Group grievance		.059 (.031)	.040 (.030)	
Human flight		.025 (.024)	.033 (.023)	
Uneven development		−.078 (.032)*	−.071 (.031)*	
Economic decline		−.036 (.028)	−.008 (.028)	
Delegitimization of the state		−.033 (.040)	−.049 (.038)	
Public services		.025 (.042)	−.014 (.014)	
Human rights		.014 (.038)	.026 (.037)	
Security apparatus		.028 (.030)	.012 (.029)	
Factionalized elites		−.032 (.032)	−.019 (.030)	
External intervention		−.057 (.006)*	.041 (.027)	
FS total score	0.004 (.002)*			0.021 (.005)*
FS total squared				0.0002 (.00004)*
GPI score			.327 (.086)*	
Religious diversity	.025 (.119)*	0.21 (.12)	0.203 (.113)	0.266 (.111)*
Population	−1.17E−10 (1.85E−10)	7.37E−11 (2.13E−10)	3.98E−11 (2.02E−10)	−5.58E−11 (1.73E−10)
Median age	0.007 (.004)	−0.0003 (.006)	−0.002 (.006)	0.006 (.005)

NOTE: FS = Failed States; GPI = Global Peace Index; IDP = Internally Displaced People.

*p ≤ .05.

Governance. The worldwide governance indicators (WGIs) were analyzed for both countries. The WGI examines the following measures: voice and accountability, political stability and government effectiveness, regulatory quality, rule of law, and control of corruption (see Appendix D for operationalization of the variables).

Colonization. Countries who colonized Somalia and the Ivory Coast, Great Britain, Italy, and France, respectively, are included as independent variables.

Results

Table 2 presents the means, standard deviations, and minimum and maximum values for all variables in the model. For the dependent variable, rate of terrorist attacks per 100,000 of the population (activity), the mean is 0.098, with a minimum of 0, a maximum of 3.111, and a standard deviation of 0.324. The failed states' total score mean is 70.305. A total of 109 cases in the study have scores above this level, meaning a higher degree of state failure. The lowest score in the study is Norway, at 16.8, while the highest is Somalia at 114.201. The standard deviation for this variable is 23.561. The GPI score has a mean of 2.053, and 60 of the cases in the sample have a score greater than this. Iceland had the minimum score of 1.176 and Iraq had the maximum score of 3.154. The standard deviation is 0.486.

 Multivariate Analysis

The results of the multivariate modeling efforts are presented in Table 3. The 12 variables of the Failed States Index (Model 2) are compared to the total failed states score (Model 1), using the control variables population, median age, and religious diversity. Both of these models were statistically significant to the .05 *a* level, but note that the 12 indices of the Failed States Index make for a better predictive model than the total score model.

Adding the GPI improves the predictive power of the model for the 12 failed states (Model 3) while controlling for religious diversity, population, and median age. When examining the models for Hypothesis 2, nested models with the full Model 3 and the restricted Model 2 are compared. By adding the GPI, the model improves its predictive ability. There is approximately a 62% increase in explained variance between Models 2 and 3.

Finally, to test the hypothesis that the original relationship between the total Failed States Index score and rates of terrorist activity is not linear, a second-degree power polynomial was generated (Model 4). The second power polynomial is nonlinear in nature and provides for an examination of a nonlinear relationship (i.e., curvilinear). Model 4 has stronger predictive power than Model 1 as there is approximately a twofold increase in the explained variance between the models.

In this study, by disaggregating the Failed States Index, an improved model of prediction was constructed. On the surface, this supports Hypothesis 1 that a better model of prediction contains all 12 variables that comprise the Failed States Index. However, the items are strongly correlated, and their individual slopes are not significant, except for the slopes of uneven economic development and external intervention. Both have negative slopes, which intuitively does not seem right. While the model does seem to predict better than the original, this discrepancy leads me to believe that one must go deeper into this relationship. While it is important to look at the separate components that make up the Failed States Index, a deeper exploration into the variables involved in their individual construction may provide more explanation. This is the first step in a deeper understanding of how the factors of state failure influence terrorist activity. This may be indicative of a nonlinear relationship for the two statistically significant variables and will be explored in future research.

 Case Studies

Age distribution in both countries is very similar, although the Ivory Coast is slightly older, having a median age of 19.6 years to Somalia's 17.6. Birth and death rates are also slightly higher in Somalia as well. Literacy rates mirror these other differences, with Somalis faring slightly worse than Ivoirians. Somalia has over 1 million IDP, while the Ivory Coast has slightly less than 1 million (IDMC, 2010). The percentage of the population of Somalia living in urban areas is 37, while the Ivory Coast is 49. However, Somalia is seeing a growth in urbanization of 1½ times that of the Ivory Coast. Both countries have very young populations, which general research on violence demonstrates are more likely to commit violent acts.

Arable land is scarce in Somalia, which sees only 1.6% of its land usable. Struggle over land has fueled conflict within the country. Pastoralists have regularly crossed Kenyan and Ethiopian borders, while land scarcity fuels interclan rivalries (Dehez, 2009). The privileging of some clans, while disadvantaging others, further exacerbates this interclan conflict. Traditionally, clan elders handled such conflicts over land; however, with the government of Siyaad Barre introducing land registration laws that placed all land as state property until legal registration was completed, the privileging of some clans over others reinforced land conflict (Dehez, 2009). The land reform was meant to regulate agricultural economy as well as legalize inheritance claims through the registration process. Most small farmers who held inherited farms could not afford the bureaucratic process involved in registering claims. Pastoralists, largely not linked to government officials in a privileged manner, suffered difficulty in finding places to move herds.

The Ivory Coast's political crisis also had roots in land tenure. While the Ivory Coast has a much greater percentage of arable land (8.8%), legal access to land has privileged some groups over others. In particular, migrants' citizenship and land rights were pivotal to the civil conflict. The economic crisis of the 1980s fueled the discrimination of migrants and their rights to land, leading to the revocation of voting rights. The 1998 Rural Land Law (Bassett, 2009) was geared toward recognizing and formalizing customary land rights by procedurally setting out conditions to title land. This only served to highlight tensions between native Ivoirians and migrants because only citizens (native Ivoirians) could own land, while others could only gain long-term leases. Two percent of the rural land was legally registered. Land in protected forests served to further marginalize IDPs in that forests were deemed publicly held so those migrants who derived their livelihood in these areas could no longer use them. Further, the land reform threatened the mobility of pastoralists (Bassett, 2009).

Somalia and the Ivory Coast have very similar indicators on the Failed States Index; however, the most striking differences are in the areas of politics. This is perhaps predictable because Somalia has functioned without a government for a longer period than the Ivory Coast. Also operating in Somalia is the clan structure that pits groups against each other in very distinct separations. The warlord atmosphere is Somalia's state within a state, and the ad hoc administration of law by those with the most power at the time leads to instability. This is not as much of a problem in the Ivory Coast, mostly because of outside intervention from France and the United Nations. Because of the Ivory Coast's highly diverse population, there is not a clan structure in place to operate like there is in Somalia.

Progressive deterioration of public services gives the starkest difference between the two countries, with Somalia scoring 10 out of 10 and the Ivory Coast scoring 7.8. The ability of the Ivory Coast to continue to provide public services is heavily reliant on French intervention (Advameg, 2010) while Somalia has no such intervention. The other strong disparity worth noting is the rise of factionalized elites. Here the success of marginalized persons, and forced recognition of migrants by outside forces, helps the Ivory Coast's score. Through long-term external intervention and forced inclusion of minorities (Advameg, 2010), the Ivory Coast has less factionalized elites than the clan driven, warlord country of Somalia.

The potential for economic development is also important for a state's stability. As the difference in governance scores demonstrated, the Ivory Coast's perceived ability to promote private economic investment allows its people to accept its governance more readily. Somalia again suffers from its lack of central government in that even if a company wanted to do business there is no clear group with which to negotiate. This is related to foreign capital penetration and dependency theory. Dependency theory (Dixon & Boswell, 1996) states that foreign capital penetration into less developed states provides short-term economic benefits, however, in the long-term it is detrimental. This is demonstrated in the Ivory Coast in that while France's involvement has provided stability in times of chaos, the Ivory Coast is highly dependent on France and a French pullout would leave the country unable to provide basic utilities. The stability gained by French dependency lends itself to the perceived ability of the government to foster private economic development, leading Ivoirians to have more faith in their government. Conversely, Somalia's lack of any government apparatus prevents it from having the ability to promote economic development. Related to dependency is the concept of debt. The Ivory Coast's debt load is much higher than Somalia's. This increases its connections to the world polity. By engaging with the World Bank and other global structures, the Ivory Coast is better situated in the world than Somalia. Furthermore, while the Gini Index for the Ivory Coast is higher than Somalia's, indicating greater economic inequality, there is a greater amount of poverty in Somalia. It is not difficult for a country to have more equal distribution of wealth when there is little wealth. Somalia has no ability to provide the needed infrastructure and services to its people, while the Ivory Coast has benefited from outside intervention in these areas. Sen (2000) advocates development as freedom: thus, development may promote peace. The five essential freedoms for Sen (2000) are political, economic, social opportunity, transparency guarantee, and protective security. Development is an expansion of individual freedoms and as such may lead to a more stable state. If development is success, it is obvious where Somalia and the Ivory Coast are failing.

The diversity measures indicate that Somalia is a quite homogenous country, scoring .082 on the ELF and .03 on the religious diversity measure, indicating little diversity. In terms of religion, 97% of the population is Sunni Muslim. The Ivory Coast, by contrast, has an ELF score of .82 and a religious diversity score of .701, indicating a greater amount of diversity. Approximately a third is Muslim, a third is Christian, with the remaining one third comprising indigenous religions or no religion. While both countries have similar histories, their diversity scores are very different.

While ethnic diversity is very low in Somalia, conflicts are along clan lines. This is in part a reflection of the privileging of some clans by the ruling apparatus. Further, Somalis tend to be apprehensive of outsiders, having experienced such a violent and chaotic colonial history (Advameg, 2010). Contrasted with the Ivory Coast, which has over 60 different ethnic groups, Somalia would seem more capable of forging a national identity. On the contrary, the legacy of colonialism has only furthered divisions along clan lineage (Hayes & Robinson, 2010). And clan structure led to a massive power struggle between scores of rebel groups and subsequent civil wars (Homer-Dixon, 1999). WGIs were analyzed for both countries. The overall score for Somalia was −1.9 (scale −2.5 to 2.5), while the Ivory Coast's was −1.5. In looking at the components of governance examined, voice and accountability captures perceptions of the extent to which citizens are able to participate in selecting their government, exercising freedom of expression and association, and free media. The Ivory Coast, while still in the lowest 20%, is much better than Somalia, which ranks in the lowest 2% worldwide. Democracy may decrease terrorism in that it affords those with grievances other avenues of redress (Schneider et al., 2010). Components of political stability and government effectiveness have both countries in the lowest 10th percentile. These are measures of government stability and quality of public services as well as the government's credibility. With Somalia operating for over 20 years with no central

recognized government, and the Ivory Coast being in a state of transition for the last decade with its interim government continuously putting off elections, it is not difficult to understand these low scores.

Somalia scored −2.46 and the Ivory Coast scores just under the 20th percentile with a score of −.97 on the measure of regulatory quality. This measures the perception that the government has some ability to promote economic, private development. Because there is some sort of government in the Ivory Coast that is recognized by the international world and the French have continued their economic involvement. The Ivory Coast has weathered its civil wars and sharp economic declines much better than Somalia. France supplies many aspects of infrastructure, lessening pressure for these basics on the state. Somalia, with no internationally recognizable government and minute economic trade, has little hope of developing (Herblist, 2009). Also preventing quality of life in Somalia is the fact that without a recognizable government, international aid organizations have no coordination. This makes such agencies hesitant to provide aid to Somalia. For the measures of rule of law and control of corruption, both countries fell into the bottom 10th percentile (World Bank, 2010). Again, these scores are predictable because of the lack of governance each country has experienced.

Both the British and French colonized much of Africa; however, their methods of colonization were very different. The French emphasized cultural assimilation, replacing traditional African leadership with French bureaucracy (Hayes & Robinson, 2010). This left newly created postcolonial states with no governmental structure to rely on. The British, in contrast, relied on local elites to administer British rule. This created class divisions and thus a bureaucratic class was in place when independence was obtained. These crucial structural differences left nation-states with very different abilities for self-governing. Coupled with this is the haphazard way in which European powers created these nation-states, many times piecing together antagonistic groups in to one state (Cocodia, 2008). This is most clearly demonstrated in the case of Somalia, where independence left 3 million Somalis living in Kenya and Ethiopia once the lines were drawn. While neither Somalia nor the Ivory Coast was fully under British control, their colonial experiences were quite distinct. The French ruled the Ivory Coast and left them with a weak government apparatus once independence was attained in 1960; however, the French continued to maintain economic and military involvement through utilities and telecommunications as well as a strong military presence (ICEM, 2010). The British in the North, the French in the coastal region that is now Djibouti, and Italy in the South dominated Somalia. During World War II, Somalia was a hotbed for conflict between the British and Italians, extending the European-based war into the Horn of Africa. This speaks to McMichael's (2008) belief that how states pursue development has far-reaching effects on social, economic, and political development. Somalia was in play during the cold war and because of the power plays by the United States and the Soviet Union much of its policy was externally influenced, while the Ivory Coast avoided this situation. Further, at the end of the cold war, Somalia saw an abrupt change in external intervention, as it was no longer key in the fight between capitalism and communism. Complicating Somalia's state formation even more is that when independence was achieved and Northern and Southern Somalia were united, two distinctly separate states were brought together, having no common bureaucratic functions (Advameg, 2010). Further distinction between the two countries' experiences is that while the Ivory Coast still experiences intense French involvement; Somalia has not enjoyed the stabilizing effects of continuous outside intervention.

Colonial influence when these countries gained their independence reflects how the struggle between three countries for control of Somalia left it much less equipped for self-governing than the Ivory Coast. Differences in the histories of the two countries indicate that the continued presence of the French in the Ivory Coast may serve as a stabilizing factor in the rebuilding of the country. While the Ivory Coast experienced French colonialism, which overall left countries less adept at self-rule postcolonially, and may be a source for its overall high state failure score, continued involvement has made critical contributions to its disaggregated scores being lower in some areas than Somalia. Somalia was not only brutally colonized by three different empires, but it continued to experience violent struggles between two of these countries until the end of World War II. The form that colonialism took, along with the structure of the government it leaves behind, are important in creating an environment for state success.

Discussion

In this study, by disaggregating the Failed States Index, an improved model of prediction could be constructed. On the surface, this supports Hypothesis 1 that a better model of prediction contains all 12 variables that comprise the Failed States Index. However, the items are strongly correlated, and their individual slopes are not significant, with the exception of the slopes for uneven economic development and external intervention, both having negative slopes, which intuitively does not seem right. While the model does seem to predict better than the original, this discrepancy leads me to believe that one must go deeper into this relationship. While it is important to look at the separate components that make up the Failed States Index, a deeper exploration into the variables involved in their individual construction may provide more explanation. This is the first step in to a deeper understanding of how the factors of state failure influence terrorist activity. This may also be indicative of a nonlinear relationship for the two statistically significant variables and will be explored in future research.

The findings supported Hypothesis 2 that adding the score of the GPI the model would be improved. Not only is the adjusted R^2 larger, but the incremental F test demonstrates that the improvement is statistically significant. In this model, however, only uneven economic development and the GPI were statistically significant. Because this addition seemed to enhance the overall model, the GPI seems to be important in improving the model.

Hypothesis 3, that a regression equation with a squared term would have better predictive value than the original Model 1, was supported by the findings. Both independent variables were statistically significant to the .05 a, as was the control variable for religious diversity. This is important because debate within the literature has demonstrated that research is not settled as to the shape of the relationship. The findings, however, support a nonlinear relationship. While all three hypotheses were supported, the low adjusted R^2 necessitates further exploration into why some failed states promote terrorism while others do not. Because of this, case studies of two failed states, one with low levels of terrorism and one with high levels, have been conducted.

There is no current theory that definitively explains the connections between failed states and terrorism.

Asymmetrical or fourth-generation warfare is a very real threat globally today. Prior research has focused on political, economic, and social factors independently of each other and the effects on terrorist activity. The focus of this article, in examining state failure in detail for Somalia and the Ivory Coast, has worked to move away from looking at these factors (political, economic, and social) independently and tried to view the overall processes involved in state failure in regard to terrorism.

While Somalia and the Ivory Coast have very similar scores on most components of the Failed States Index, noticeable differences were found on political indicators of the ability of the state to provide public services and the rise of factionalized elites. In exploring diversity, both measures show a significant difference between Somalia and the Ivory Coast, with Somalia showing virtually no diversity, while the Ivory Coast is moderately diverse. In terms of governance, differences in voice and the perceived ability of the government to foster economic private development show vast differences between the countries. The history of colonialism has been explored and clear distinctions made between Somalia and the Ivory Coast's experiences. The study has been theoretically driven by the position that it is the process of state failure that is intimately connected to terrorism. In examining the 12 indices of state failure, along with diversity, governance, and colonial history, a clearer picture has emerged as to why Somalia and the Ivory Coast have experienced differences in terrorist activity.

 ## Conclusion

On first look, economic factors of state failure seem least informative of what differences are attributed to fostering terrorism because their economic histories are very similar. Both are poor countries that experienced severe economic decline prior to civil wars (Bassett, 2009; Dehez, 2009), with a heavy reliance on the informal sector (Aboygue, 1989; Guichaoua, 2007). As the quantitative analysis indicates, uneven economic development is predictive of terrorist activity. Since there is a significant difference between the two countries on this variable, the model would predict a difference in terrorist activity.

The differences between the Ivory Coast and Somalia go further than what can be measured by the Failed States Index. That is why, in extending the study of terrorism, focusing on the processes of diversity, governance and colonialism have helped to shed new light on the subject. Terrorism is a complex phenomenon and so are its root causes. Because of this, a simple OLS regression equation is not going to be as predictive as desired. It requires delving into the history that shapes a country spawning terrorism as well as critical analysis of the governance. This article has begun to look at these processes, but future research should focus on the colonial experiences and how they serve to shape postcolonial nation-states.

Further research should strive to look beyond failed states and perhaps a four-by-four study adding two nonfailed states, one with high levels of terrorism and one with none, would add a greater understanding of the processes involved. Also not addressed in this article are the different components of the GPI. By taking the index apart into its 23 components, further comparison of the actions involved could be done and a clearer picture of the operations involved would emerge. It is further noted that this is a simple snapshot in time from the year 2008. The states chosen are fragile and unstable, and as such, this must be taken into consideration in the analysis. Perhaps a time series or longitudinal approach would aid in bringing more stability to the data.

[*Appendices omitted.*]

 ## References

Aboagye, A. A. (1989). Generating Employment and Incomes in Somalia: Report of an Inter-disciplinary Employment and Project-identification Mission to Somalia Financed by the United Nations Development Programme and Executed by ILO/JASPA. United Nations Development Program, *International Labor Organization: Jobs and Skills Program for Africa*. Addis Ababa: Ethiopia. Accessed September 12, 2010, at Google Books.

Advameg, Inc. (2010). Every culture: Somalia. Retrieved September 12, 2010, from http://www.everyculture.com/Sa-Th/Somalia.html

Alesina, A., & Perotti, R. (1996). Income distribution, political instability, and investment. *European Economic Review*, 40, 1203–1228.

Bassett, T. J. (2009). Land, labor, livestock and (Neo) liberalism: Understanding the geographies of pastoralism and ranching. *Geoforum*, 40, 756–766.

Bergesen, A. J., & Lizardo, O. (2004). International terrorism and the world-system. *Sociological Theory*, 22, 38–52.

Bilgin, P., & Morton, A. D. (2004). From rogue to failed states? The fallacy of short-termism. *Politics*, 24, 169–180.

Blomberg, S. B., Hess, G. D., & Weerapana, A. (2004). Economic conditions and terrorism. *European Journal of Political Economy, 20,* 463–478.

Bossert, W., D'Ambrosio, C., & La Ferrara, E. (2005). A generalized index of fractionalization. Memeo, Bocconi University and University of Montreal.

Boylan, B. M. (2010). Economic development, religion, and the conditions for domestic terrorism. *Josef Korbel Journal of Advanced International Studies, 2,* 28–44.

Burgoon, B. (2006). On welfare and terror: Social welfare policies and political-economic roots of terrorism. *Journal of Conflict Resolution, 50,* 176–203.

Cocodia, J. (2008). Exhuming trends in ethnic conflict and cooperation in Africa: Some selected states. *African Journal on Conflict Resolution, 8,* 9–26.

Collins, R. (1975). *Conflict Sociology.* New York, NY: Academic Press.

Crenshaw, M. (1981). The causes of terrorism. *Comparative Politics, 13,* 379–399.

Dixon, W. J., & Boswell, T. (1996). Dependency, disarticulation, and denominator effects: Another look at foreign capital penetration. *American Journal of Sociology, 102,* 543–562.

Dehez, D. (2009). *The scarcity of land in Somalia: Natural resources and their role in the Somali Conflict.* Bonn, Germany: Bonn International Center for Conversation.

Ehrlich, P. R., & Liu, J. (2002). Some roots of terrorism. *Population and Environment, 24,* 183–191.

Ellis, B. (2003). Countering complexity: An analytical framework to guide counter-terrorism policy making. In R. D. Howard, & R. L. Sawyer (Eds.), *Terrorism and counterterrorism: Understanding the new security environment* (pp. 109–122). Guilford, CT: McGraw-Hill/Dushkin.

Feagin, J., Orum, A., & Sjoberg, G. Eds. (1991). *A case for case study.* Chapel Hill, NC: University of North Carolina Press.

Freytag, A., Kruger, J. J., Meierrieks, D., & Schneider, F. (2009). *The origins of terrorism cross-country estimates on socio-economic determinants of terrorism* (No. 2009, 009). Jena Economic Research Papers.

Freytag, A., Krüger, J. J., Meierrieks, D., & Schneider, F. (2011). The origins of terrorism: Cross-country estimates of socio-economic determinants of terrorism. *European Journal of Political Economy, 27,* S5–S16.

Giddens, A. (1999). *Runaway world: How globalization is reshaping our lives.* London, UK: Profile.

Guichaoua, Y. (2007). Access to informal sector employment in cote d'ivoire. *Development Research Reporting Service.* Retrieved September 12, 2010, from http://www.eldis.org/assets/Docs/46933.html

Harrison, M. (2006). An economist looks at suicide terrorism. *World Economics, 7,* 1–15.

Hayes, R. N., & Robinson, J. A. (2010). *An economic sociology of African entrepreneurial activity.* Conference on Entrepreneurship in Africa, Syracuse, NY.

Hehir, A. (2007). The myth of the failed state and the war on terror: A challenge to the conventional wisdom. *Journal of Intervention and State Building, 1,* 307–332.

Helman, G. B., & Ratner, S. B. (1992). Saving failed states. *Foreign Policy, 89,* 3–20.

Herblist, J. (2009). *Africa and the challenge of globalization* (Working Paper). Singapore: Singapore University Library.

Hirschman, A. O. (1964). The paternity of an index. *The American Economic Review (American Economic Association), 54,* 761.

Hoffman, B. (1998-99). Old madness new methods: Revival of religious terrorism begs for broader US policy. *Rand Review, 22,* 12–23.

Homer-Dixon, T. (1999). *Environment, scarcity and violence.* Princeton, NJ: Princeton University Press. IDMC (International Displacement Monitoring Centre). 2010. Retrieved November 21, 2010, from http://www. internal-displacement.org/

ICEM (International Federation of Chemical, Energy, Mine and General Workers' Unions). (2010). Creating union problems in cote d'ivoire. Retrieved November 30, 2010, from http://www.icem.org/en/78-ICEMInBrief/1544-French-based-Bouygues-Creating-Union-Problems-in-Cote-d%E2%80%99Ivoire

Kaufmann, D., Kraay, A., & Mastruzzi, M. (2003). *Governance matters III.* Washington, DC: World Bank.

Kittner, C. C. B. (2007). The role of safe havens in Islamist terrorism. *Terrorism and Political Violence, 19,* 307–329.

Korteweg, R. (2008). Black holes: On terrorist sanctuaries and governmental weakness. *Civil Wars, 10,* 60–71.

Krueger, A. B. (2008). What makes a homegrown terrorist? Human capital and participation in domestic Islamic terrorist groups in the USA. *Economics Letters, 101,* 293–296.

Krueger, A. B., & Maleckova, J. (2003). Education, poverty and terrorism: Is there a causal connection? *Journal of Economic Perspectives, 17,* 119–144.

Lambach, D. (2004). *The perils of weakness: Failed states and perceptions of threat in Europe and Australia.* Presented at the New Security Agendas: European and Australian Perspectives conference at the Menzies Center, Kings College, London.

Li, Q., & Schaub, D. (2004). Economic globalization and transnational terrorism: A pooled time-series analysis. *Journal of Conflict Resolution, 48,* 230–258.

Looney, R. (2004). Failed economic take-offs and terrorism in Pakistan: Conceptualizing a proper role for US assistance. *Asian Survey, 44,* 771–793.

Mair, S. (2008). A new approach: The need to focus on failing states. *Harvard International Review, 29,* 52.

McMichael, P. (2008). *Development and social change: A global perspective* (4th ed.). Los Angeles, CA: Pine Forge Press.

Muller, E. N., & Seligson, M. A. (1987). Inequality and insurgency. *American Political Science Review, 81,* 425–452.

Patrick, S. (2006). Weak states and global threats: Fact or fiction? *The Washington Quarterly, 29,* 27–53.

Piazza, J. A. (2006). Rooted in poverty?: Terrorism, poor economic development and social cleavages. *Terrorism and Political Violence, 18*, 219–237.

Piazza, J. A. (2008). Incubators of terror: Do fail and failing states promote transnational terrorism? *International Studies Quarterly, 52*, 469–488.

Piazza, J. A. (2011). Poverty, minority economic discrimination, and domestic terrorism. *Journal of Peace Research, 48*, 339–353.

Sageman, M. (2008). *Leaderless jihad terrorist networks in the twenty-first century.* Philadelphia: University of Pennsylvania Press.

Schneider, F., Bruck, T., & Meierrieks, D. (2010). *The economics of terrorism and counter-terrorism: A survey* (CESIFO Working Paper No. 3011). Retrieved September 12, 2010, from http://papers.ssrn.com/sol3/ papers.cfm?abstract_id¼1590148

Schmid, A. P., & Jongman, A. J. (1988). *Political terrorism.* New Brunswick, NJ: Transaction Books.

Schmid, A. P., & Jongman, A. J. (2005). *Political terrorism: a new guide to actors, authors, concepts, data bases, theories, & literature.* New Brunswick, NJ: Transaction Pub.

Sen, A. (2000). *Development as freedom.* New York, NY: Anchor Books.

Simons, A., & Tucker, D. (2007). The misleading problem of failed states: A socio-geography of terrorism in the post 9/11 era. *Third World Quarterly, 28*, 387–401.

START Global Terrorism Database. (2007). *START.* Retrieved September 12, 2010, from http://www.start. umd.edu/start/

Stewart, F. (2000). Crisis prevention: Tackling horizontal inequalities. *Oxford Development Studies, 28*, 245–262.

Takeyh, R., & Gvosdev, N. (2002). Do terrorist networks need a home? *The Washington Quarterly, 25*, 97–108.

The Fund for Peace. (2009). "Failed States Index." Retrieved September 12, 2010, from http://www.fundfor peace.org/web/

Weber, M. (1948). *From Max Weber: Essays in sociology.* London, UK: Routledge and Kegan Paul Ltd.

Winkler, T. H. (2008). The shifting face of violence. *World Policy Institute, Fall*, 29–36.

World Bank. (2010). Worldwide governance indicators. Retrieved October 26, 2010, from http://www.icem. org/en/78-ICEM-InBrief/1544-French-based-Bouygues-Creating-Union-Problems-in-Cote-d%E2%80%99 Ivoire

Zartman, W. I. (1995). *Collapsed states the disintegration and restoration of legitimate authority.* Boulder, CO: Lynne Reiner Publishers Inc.

Chelli Plummer graduated with her BA from California State University at Chico in 2006. She received her MS in sociology from North Carolina State University. She is currently a fourth-year graduate student in the Sociology and Anthropology Department at North Carolina State University. Her areas of interest are global social change and family and life course. In particular, Plummer conducts research on terrorism and women in the military.

REVIEW QUESTIONS

1. What does Plummer identify as the main problem with previous research on failed states and terrorism? How does she propose to solve this problem?

2. What two states does Plummer use for her case study? Why did she choose these two states? What other states might you have selected for the purpose of this research?

3. What are the main factors related to failed states and terrorism that Plummer focuses on? Why did she choose those factors?

4. What are the main conclusions drawn by Plummer. Do you agree with these conclusions? Why or why not? What other approaches might you use to explore similar questions?

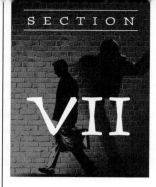

Southeast Asia, South and North Korea, and China

At the end of this section, students will be able to:

- Describe the countries that make up Southeast Asia.
- Discuss why it is appropriate to address North Korea, South Korea, Japan, and China in this section.
- Describe the impact of Jemaah Islamiya on the region.
- Detail how North Korea keeps it citizens in line.
- Describe how the Chinese deal with dissidents, such as the region of Tibet, the Uighurs, the Falun Gong, and others such as Christians and the rural poor moving to urban areas.
- Describe the threats to Indonesia and the Philippines with regard to terrorism.
- Discuss how the "one child policy" may well cripple China.
- Define Maritime Domain Awareness.
- Describe the issues Japan is currently dealing with in the 21st century.

Introduction

Southeast Asia consists of countries that stretch from eastern India to China, are dependent on seas and oceans for much of their transport and livelihood, and can be divided into those that are mainland countries or island countries. Those on the mainland, an extension of the Asian continent, are Burma, Thailand, Laos, Cambodia, and Vietnam; those that are essentially islands are Malaysia, Singapore, the Philippines, Brunei, Indonesia, and East Timor (Andaya, n.d.). It is worth noting that some of these countries are not only islands but many islands; Indonesia has more than 17,000 islands and the Philippines more than 7,100. In this section only Burma, Cambodia, Indonesia, and the Philippines will be covered as they have important lessons or they have to deal with terrorism, mostly from an al Qaeda affiliate, Jemaah Islamiya; terrorism is not a major concern for the other countries.

The population of Southeast Asia as of 2011 was 601,898,000 (statistics for China, South Korea, North Korea, and Japan will be included in the portions covering them) (Population Reference Bureau, n.d.). The area is diverse in ethnicity and religion, almost to the point of being overwhelming. For example, in Indonesia alone there are over 500 different ethnic groups speaking over 700 ethnic languages (Embassy of Indonesia, Oslo, 2012). Much of the region can be characterized as Islamic (over 40% are Muslim) and even in countries where Islam is not dominant, it still exerts influence; this should not be surprising since Muslim traders have been in Southeast Asia ports since the seventh century and the adoption of Islam was a peaceful process (Houben, 2003). Yet, as this section will spell out later, some terror groups in the region do use fundamentalist Islam as a basis for recruitment (Houben, 2003).

This section will also include China due to the influence it exerts in the region and the fact that it has nuclear weapons; and North Korea due to its destabilizing impact and the fact that it may have nuclear weapons. Finally, Japan and South Korea are presented as stabilizing forces against both China and North Korea; Japan has a history with terror groups as well, which will be addressed: the **Japanese Red Army** and **Aum Shinrikyo**, the terror group that managed to manufacture a nerve gas, **sarin**, and release it in a Tokyo subway.

Southeast Asia remains a mystery to many in the world but it is where several of the significant events of this century will occur. Europe was the focus of the world for much of the last century but the 21st century's focus will be on Southeast Asia as well as China, North Korea, and Japan as the Chinese in particular joust with the United States and India for influence, power, and resources (Kaplan, 2010). For example, consider this quote from Robert Kaplan (2010):

> In the foreseeable future the Chinese will help finance a canal across the Isthmus of Kra in Thailand that will provide another link between the Indian and Pacific Oceans—an engineering project on the scale of the Panama Canal and slated to cost $20 billion. (p. 288)

What this signifies is that China, realizing it will need much more crude oil, which is very expensive to continue to ship through the crowded and narrow Strait of Malacca, will then save a fortune and position itself militarily, economically, and politically to be a major player in the region and ultimately the world. China wishes to control the Pacific Ocean, the South China Sea, the Indian Ocean, and the Arabian Sea; if it does that, it will become the major power in the region if not the world.

This section will address countries individually, beginning with North Korea, South Korea, China, and Japan. This will be followed by discussions on Burma, Cambodia, the Philippines, and Indonesia. North Korea, Burma, and Cambodia are included because they are interesting in their history and brutality, not because they have a pressing terrorist problem; they are simply, as the Fund for Peace points out annually using very sophisticated measures, "Failed States" (Fund for Peace, 2012).

North Korea

Writing about terrorism and Southeast Asia is relatively straightforward when you address North Korea. It does not have any unless your definition of terrorism includes the possibility that a nation willingly terrorizes its own people. If you are unfortunate enough to be a North Korean, not a member of the few select individuals who control the country, then yes, you are a victim of terrorism. North Korea is an insignificant country, and discussion of it will be limited but readers should understand it can be a problem for stability in the region. More coverage of North Korea will be presented in Part III in a threat assessment of nuclear terrorism. If there were a prize for utter inhumanity and brutality it would necessarily go to North Korea where starvation and brutality are a way of life. Here is a recent *Wall Street Journal* book review of life in North Korea:

▲   An illustration of a North Korea nuclear launch pad. They may or may not have this capability; the fact that the United States does not know much about North Korea is a concern.

A Life Sentence, Then a New Life

"Escape From Camp 14" tells the story of one man's incarceration and personal awakening in North Korea's highest-security prison.

By Melanie Kirkpatrick

There is no dispute about the existence of the North Korean gulag. Anyone with a computer and access to the Internet can go to Google Earth and zoom in on a string of vast prison camps located in the unforgiving, mountainous center of the country. The U.S. State Department and international human-rights organizations put the number of inmates at about 200,000. As many as one million North Koreans are believed to have perished there. Only three people are known to have escaped.

One is Shin Dong-hyuk, a young man who defied the odds and managed to flee, first from the gulag and then from North Korea itself. He made it to China and eventually reached safety in South Korea in 2006. His remarkable story is told by Blaine Harden, a former *Washington Post* reporter, in "Escape From Camp 14." It is a searing account of one man's incarceration and personal awakening in North Korea's highest-security prison.

The book is also an indictment of the barbaric regime that rules North Korea, the world's most repressive totalitarian state. Mr. Shin is roughly the same age as Kim Jong Eun, who took over as North Korea's dictator after his father's death in December. Mr. Harden notes that the ruler and Mr. Shin "personify the antipodes of privilege and privation in North Korea, a nominally classless society where, in fact, breeding and bloodlines decide everything."

Mr. Shin was born in Camp 14, the offspring of two inmates who had been rewarded for good behavior. The prison authorities assigned his mother to his father and allowed them to sleep together five nights a year. The boy barely knew his father, living with his mother and older brother until he reached his early teens, when he was moved to a dormitory. Mr. Shin told the author that he had no experience of maternal love. He viewed his mother not as a source of affection but as a competitor for the limited amount of food that was available to them.

Prisons in North Korea are known for starvation-level rations, backbreaking work and brutal treatment. But unlike most prisoners, who, if they survive, at least have the possibility of release, everyone at Camp 14 is serving a life sentence. The camp ranks as a "total control zone," where prisoners are deemed "irredeemable."

Mr. Shin's unforgivable crime was being born of "bad seed." His father was sent to Camp 14 because two of his brothers had fled south during the Korean War. Under an edict laid down by Kim Il Sung, founder of the Democratic People's Republic of Korea, the crimes of such traitors must be paid for by their relatives, "through three generations." For most of Mr. Shin's life, Mr. Harden notes, he accepted that his tainted lineage meant that he deserved the suffering inflicted on him at Camp 14—"he believed the guards' preaching about original sin."

For children at Camp 14, schooling consisted mainly of memorizing the camp's 10 rules. Rule No. 3: "Anyone who steals or conceals foodstuffs will be shot immediately." When Mr. Shin was in the first grade, his teacher discovered five kernels of corn in a classmate's pocket. The girl was required to kneel down in front of the class. Mr. Shin and his classmates watched as their teacher beat her to death. She was 6 years old.

At 14, he was forced to witness two other executions: those of his mother and brother. They had been arrested for violating Rule No. 1: "Do not try to escape." His mother was hanged, and his brother was shot. The only emotion Mr. Shin felt was anger. He blamed them for his interrogation and torture after their arrests. Guards bound his hands and feet, hoisted him into the air by means of a hook pierced into his abdomen and dangled him over an open fire.

His escape from Camp 14—in 2005, when he was 22—came about because of a chance encounter with a new inmate, a man who had held a high-ranking government position. The two men shared a desperation to get out of the camp, whatever the risk, and plotted to reach China, a country Mr. Shin had never heard of but one where his friend had relatives. His friend, though, was killed during their escape over the electric fence that surrounded the camp; Mr. Shin crawled over his corpse to freedom. Thanks to luck as well as to skills he had honed at Camp 14—stealing, lying and fighting—he managed to travel to the border and cross into China by himself. In that country, he was helped by ethnic Koreans, local Christians and, eventually, a South Korean journalist who escorted him to the South Korean consulate in Shanghai.

Parts of "Escape From Camp 14" can be painful to read. Mr. Harden spares no detail of Mr. Shin's torment, physical or psychological. He writes in a direct, matter-of-fact style that puts the horrors he is relating in dark relief. He is equally explicit in describing Mr. Shin's difficulties in learning to succeed in a free society—his nightmares, his inability to hold down a job, his troubles making friends or placing trust in anyone. There is "no easy way for Shin to adapt to life outside the fence," he writes. Mr. Harden quotes him as saying, "I am evolving from being an animal." Today he is living in Seoul and trying to raise the world's awareness of his countrymen's plight.

The effects of North Korea's gulag extend beyond the lives of those unfortunate men, women and children who live and die there. The threat of being sent to the camps—and the monumental human suffering that implies—terrorizes every North Korean. It is one of the brutal control mechanisms by which the Kim family regime stays in power.

Meanwhile, Amnesty International reported last year that satellite photographs showed new construction inside the prison camps. As the world watches, North Korea apparently is planning to increase the number of inmates incarcerated there.

Ms. Kirkpatrick, a former deputy editor of the *Journal's* editorial page, is a senior fellow at the Hudson Institute and author of the forthcoming "Escape From North Korea: The Untold Story of Asia's Underground Railroad."

Following World War II, the powers that were victors, but primarily the Soviet Union and the United States, could not agree on what to do about Korea. Their clumsy Solomon-like solution was to divide the country along the 38th parallel. Encouraged by China and the Soviet Union, the north, now called the Democratic People's Republic of North Korea, attacked the Republic of Korea in a prolonged conflict from 1950 until 1953. Though a truce was declared, with the line of the truce drawn at the 38tth parallel, the conflict has never ended. President Kim Il Sung implemented a draconian policy of self-reliance and anti-Western philosophy that doomed the majority of the people of the country (*CIA World Factbook*, n.d.). Kim Jong II was the designated successor and took over after the death of his father in 1994. Kim Jong-un was the next to succeed following the death of his father in 2011 (*CIA World Factbook*, n.d.). All have followed the basic model of repression, brutality, military threats and provocation, and begging internationally for food, along with long-range missile development (*CIA World Factbook*, n.d.). They actively work on weapons of mass destruction (or WMD) and supposedly conducted tests of nuclear weapons or devices in 2006 and 2009 (*CIA World Factbook*, n.d.).

▲　Map of North Korea

One question the authors routinely field when presenting on North Korea is how its population of about 24 million people can put up with such an onerous situation. The response is twofold. First, from birth, children are indoctrinated with the glory of the leader and his omnipotence, and with hatred for the West, specifically Japan, their enemy and occupier from 1910 until the end of World War II, and the United States, their enemy since 1950. In early school experiences their play is to use toy guns and tanks to attack "Yankee Imperialists" ("How North Korean Children Are Taught", 2012). Second, Professor Bruce Bueno de Mesquita (2009) makes some startling observations that could apply to someone like the leaders of North Korea:

Virtually all long-lasting (read authoritarian) leaders, however, really depend only on a very small number of generals, senior civil servants and their own families for support. Because they rely on so few people to keep them in power, they can afford to bribe them handsomely. With such big paydays, those cronies aren't going to risk losing their privileges. They will do whatever it takes to keep the boss in power. They will oppress their fellow citizens; they'll silence a free press and punish protesters. They will torture, maim and murder to protect the incumbent as long as the incumbent delivers enough goodies to them. (p. xiii)

▲   Map of South Korea

North Korea is not a successful country due to decades of economic mismanagement; it may not be the worst country on the planet but it is working on it, and it is on *Foreign Policy Magazine's* 2012 Failed State Index at number 22 (Foreign Policy, 2012). It has not invested in its infrastructure, and people routinely suffer from malnutrition and starvation (*CIA World Factbook,* n.d.). North Korea has a gross domestic product (GDP) of only $40 billion, ranking 99th in the world; this is very significant because a fourth of that is spent on the military while the country starves (South Korea spends much more, nearly $37 billion, on its military but that is only 2.5% of its GDP) (Menon, 2012). North Korea has played a weak hand well by spending the last 20 years begging for "talks" and fuel and food in exchange for not sending missiles, testing nuclear weapons, or engaging in some sort of provocation, and further "talks" if it promises to behave itself (Menon, 2012). On paper North Korea has a large military but its equipment is dated and the technologically superior South Korean military would be an effective check on it, even if the United States did not have 24,000 military personnel willing to give

it pause. And Rajan Menon has some advice that the various governments should take when it comes to North Korea: We should just ignore it.

🏴 South Korea

▲   South Korea has a successful economy, currently number 12 in the world. This is a real contrast with North Korea which has little urbanization with the exception of the major city.

South Korea amazes and presents a sharp contrast to North Korea. Other than being constantly threatened by North Korea and with the capital city Seoul within artillery range from the North, it has a robust, democratic economy and has grown to become the 13th largest economy in the world. It has twice the population of North Korea and life expectancy is ten years longer than for someone living in North Korea (*CIA World Factbook,* n.d.; Menon, 2012). This rosy picture today has not always been the case. Democracy faltered early in the history of South Korea, with a military coup occurring in 1961, followed by the assassination of the primary member who fomented the coup, General Park Chung-hee, in 1979 (South Korea Profile, 2012). After strikes and demonstrations throughout the early 1980s, the government moved toward becoming a true democracy, electing its first civilian government in 1993 (*CIA World Factbook,* n.d.); it is indeed a first-tier country. A free trade agreement between the United States and South Korea was implemented in 2012; the future looks bright for South Korea.

China

China, with over 1.3 billion people, is an enigma for many. For much of the first half of the 20th century China was plagued by invasions and civil wars. Following WWII, under Mao Zedong, it became an autocratic communist state, with stability but at a price (*CIA World Factbook,* n.d.). China now embraces a market economy but it is still a country ruled by the Communist Party. As mentioned in an earlier section, the numbers of individuals killed by Chairman Mao are in the tens of millions and even more recently at Tiananmen Square in 1989, hundreds and perhaps a thousand were killed during protests (Saiget, 2012). When it comes to terrorism and China, the fact remains that there is not much, as the Chinese harshly crack down on any form of dissent, even perceived dissent. Three examples of Chinese dealings with dissent will be covered: China and Tibet, China and the Falun Gong, and China and the Uighurs living in the northwest Xinjiang area of northwest China, along with a discussion of *chengguan* urban management officers and the ongoing persecution.

▲   Map of China

Keep in mind too that China, though changing, is still a communist country that engages in propaganda and censorship. The spouse of one of the authors just returned from a trip to China. While FaceTime, Skype, magicJack Free, and of course cell phones all worked, there was no access to the Internet. Blogging on China required sending e-mails with attached photos for the spouse in the United States to post. The same spouse also met the U.S. Embassy person responsible for tweeting about the awful Beijing air quality; during the visit he described in a tweet that one day's pollution was so off the chart it was "crazy bad," resulting in a diplomatic protest from the Chinese (Capron, 2012).

The Chinese population does have some access to the Internet, though some sites are completely banned—such as Facebook, YouTube, and Twitter—and monitors and filters for other sites are in place (Chao, 2012). The Chinese military and government both have extensive cyber capabilities and have carried out attacks on U.S. government organizations and U.S. corporations. Both China and the United States are now developing cyber experts and commands that will focus on this area (Harrison, 2012). It should also be noted that the U.S. Department of Defense, the Department of Homeland Security, the Federal Bureau of Investigation, and the National Security Agency, to name

© Can Stock Photo Inc. / kentoh

▲ Pollution is also extremely bad in China, as this photo illustrates.

just a few, are all players in this game. It will be challenging to coordinate this and to include U.S. firms that are at risk (Corrin, 2012; Harrison, 2012). Cyber terrorism will be addressed in a later section.

While China may be a rising power, it has pursued one policy that may well prove disastrous. Alarmed about economic growth and its ability to feed a growing population, China mandated a "one child policy" in 1979 (there were exemptions for some minorities, rural Chinese, and those willing to pay a hefty fine) (Fried & Chinnareddy, 2012). The result was a bias in favor of males (females were routinely aborted). This policy has currently produced 32 million more boys under the age of 20 than girls, making marriage an unlikely prospect for many males (Fried & Chinnareddy, 2012). China will also have an increasingly aging population that it may find difficult to support given this policy, as one adult child may now be supporting a spouse and four parents. This is exacerbated by a declining birth rate; it was reported in 2011 that China now has an average fertility rate (the average number of children a woman can expect to have) of just 1.4, which is below the 2.1 rate need for a stable, not a growing, population ("China's Population," 2011).

China Has a Pollution Problem

China's pollution is so serious it is now impacting the growing of food. One of the author's spouses toured China recently and met some officials from the United States Embassy. One of them had an official protest filed against them for Tweeting that air pollution that day was so awful it was "crazy bad."

Tibet, once an autonomous region, came into the Chinese communist orbit by the direct command of Mao Zedong in 1951 with Chairman Mao claiming Tibet belonged to China and forcing Tibet's leader, the Dalai Lama, to agree to a 17-point plan guaranteeing religious freedom and some autonomy, which the Chinese have not honored (Tibet Profile Timeline, 2012). In 1959, growing tired of Chinese influence, a revolt broke out in Lhasa, Tibet's major city, with thousands killed there and in other areas of Tibet. The Dalai Lama and around 80,000 followers fled to Northern India (Tibet Profile Timeline, 2012). They remain there and the Dalai Lama has retired, giving up hope of ever returning. In 2012 the situation remained grim for Tibetans as policing had been stepped up and protesters hunted down (Mandel, 2012). Police even fired on a peaceful protest rally in 2014, wounding two ("Chinese police 'fire on Tibetan protesters,'" 2014). To truly understand the grip that the military and indeed the police have on the Chinese and especially the Tibetan people, it is worth noting that one of the more senior Chinese Communist Party members, Zhou Yongkang, presides over the Politics and Law Commission, an agency responsible for police, courts, and prosecution, that has an annual budget of $111.6 billion, more than the Chinese military ("China Leader Urges Resistance," n.d.).

China will no doubt stay in Tibet, because it sees doing so as strategically and economically beneficial. Tibet is positioned between China on the one side and India, Nepal, and Bangladesh on the other; it also has much needed water and mineral resources (Rastogi, 2008). The Chinese government is routinely criticized for its treatment of Tibetans.

Examining the reactions of the Chinese government to the Falun Gong presents a real challenge as the Falun Gong seems, to those in the West, as simply a group of people engaged in some form of yoga. Lum (2011) describes the Falun Gong as a group that

> combines an exercise regimen with meditation, moral values, and spiritual beliefs. The practice and beliefs are derived from qigong, a set of movements said to stimulate the flow of qi—vital energies or "life forces"—throughout the body, and Buddhist and Daoist concepts. The spiritual exercise reportedly gained tens of millions of adherents across China in the late 1990s. (p. 5)

▲ The Chinese are also building an increasingly larger Navy, and they have a number of submarines similar to the one in this photo, to control the seas regionally and then worldwide.

In 1999, thousands of practitioners quietly protested in Beijing and the group was outlawed as a dangerous cult. Overseas Falun Gong organizations reported that the government intensified its persecution of Falun Gong during the period of the 2008 Olympics and the 2009 Shanghai World Expo. Many practitioners who did not renounce their beliefs reportedly were held in labor camps for re-education and subjected to torture and other abuses (Jacobs, 2009). The Falun Gong leadership, or Falun Dafa as it also is known, and the founder, Li Hongzhi, moved to New York City where they are going strong ("Brief Introduction to Falun Dafa," n.d.).

Perhaps as serious as the situation with Tibet is the persecution of the Uighur, or Uygher, people, most of whom are Muslim. Information on the situation of the Uighurs is admittedly sketchy but they do have a presence in the United States and actively campaign to highlight the plight of their fellow Uighurs in China. The Chinese maintain that practitioners of Islam advocate violence and have made extreme efforts to diminish their influence in the region, including moving in Han Chinese, the dominant majority. Their location makes accurate information difficult but recently, on the third anniversary of riots, strikes, and violence in the area, northwestern China's Uighur minority claims it is close to war.

> Xinjiang Communist Party Chief Zhang Chunxian met with security forces in the regional capital Urumqi on Wednesday, urging them to use "iron fists" in crushing separatists, extremists and terrorists. Xinjiang is home to millions of Turkic-speaking and largely Muslim Uighurs who accuse Beijing of persecution and turning them into a minority in their homeland by flooding it with Han Chinese settlers. (Lipin, 2012)

Rebiya Kadeer (2012), the head of the World Uyghur Congress, writes that the Chinese, citing labor shortages, have moved Han Chinese into the area and forcibly relocated Uighurs to other parts of the country and have actively tried to eliminate those who practice Islam, which most Uighurs do. Some Uighur males remain in detention because the Chinese characterize them as terrorists (Lipin, 2012). Just as it is adamant about dealing harshly with Tibet and the Falun Gong, China is unlikely to change its policy toward the Uighurs, though the exemption from the one-child policy they enjoy may improve things in the future when China needs them.

There is a need to mention that two other populations receive harsh treatment from Chinese authorities on a regular basis. The first are rural peasants moving to cities seeking to better their lives, though some are forced to do so as officials take their land. Urban workers do earn three times as much as rural Chinese do, and there are now 690 million Chinese living in cities, more than those who live in rural areas ("China's Urban Population", 2011). The sheer numbers and density of the cities in China created a call for new policing and the development of the *chengguan* urban management officers, a para-police organization intended to control and enforce non-criminal regulations such as unlicensed street vendors, provide traffic control and monitor environmental practices (Kine, 2012). There are now chengguan urban management offices in over 300 cities, and Beijing alone has over 6,000 such officers. They have no uniform training, huge budgets, and little oversight (Kine, 2012).

Making things even worse is an outdated social system called the Hukou household registration system that makes registering for benefits and work permits possible only in the city or village a person is from (Simpson, 2012). What this means is that the rural poor come to the cities bereft of social benefits, including health benefits, education for children, or welfare (Simpson, 2012). Corruption remains a problem as well as a lack of planning; corruption in China occurs at levels that are unbelievable and is a way of life since the ability to incentivize remains the only way to get things done (Wang, 2012). It is estimated that more than 100 million more will soon come to the cities from rural areas and have to contend with this (Simpson, 2012).

The second group that must worry about persecution from Chinese officials are Christians, particularly members of "underground" churches or house churches not approved by the Chinese government; they meet in secret and have an underground network. No one knows the exact number of Christians in China as that is not a question asked during their census; it is true that to belong to the Communist Party and have a better life, one cannot belong to any religion (Green, 2011; "Religion and the Communist Party," 2011). It is possible that the number of Christians in China may number over 100 million, more than the 78 million people who belong to the Communist Party (Green, 2011). Christians wishing to worship in private with others play a cat-and-mouse game with the Chinese government. Perhaps the best insight into how China deals with Christians comes from an article in the *Wall Street Journal* on Bob Fu, a blind pastor who escaped from house arrest in China and immigrated to the United States:

> The news out of China these days is gripping, and there's no one more qualified to read the tea leaves than Bob Fu—who from a town in West Texas coordinates the most influential network of human-rights activists, underground Christians and freedom fighters in China. Since 2004, Midland (pop. 111,147) has been home to the spunky 44-year-old pastor and his nonprofit, China Aid. It's here that Mr. Fu and his staff of five use the Internet, telephone, letters—any means possible, though he's reticent to give specifics—to communicate with thousands of volunteers who promote religious freedom and the rule of law in China.
>
> Why Midland? "It's much safer here," Mr. Fu chuckles. After immigrating to the U.S. in 1997, he and his family settled in Philadelphia, where they were soon tailed by Chinese agents. Other agents "confronted" him in Washington, D.C., and invited him for tea. "The invitation for tea in China to dissidents means soft interrogation," Mr. Fu explains. When a Midland-based minister invited Mr. Fu and his wife to visit, they liked the place so much that they stayed.
>
> China Aid provides support for underground house churches, legal counsel for victims of forced abortions, financial help to prisoners in labor camps and their families, and more. It also sponsors the only nationally circulated underground church magazine in China. China Aid's three-pronged strategy, Mr. Fu says, consists of "exposing the abuses, encouraging the abused, and spiritually and legally equipping" the Chinese "to defend their faith and freedom."
>
> Excerpt from "Bob Fu: The Pastor of China's Underground Railroad," Mary Kissel, *The Wall Street Journal*, June 1, 2012. Copyright © 2012 Dow Jones & Company, Inc. All Rights Reserved. Reprinted with permission.

Too, the respected Christian group Voice of the Martyrs Canada reports that Chinese Christians have experienced more widespread and more severe persecution than at any time in the history of the Church, and Voice of the Martyrs report on Chinese plans to identify and investigate house churches. These house churches will then be "encouraged" to transition to the Three-Self Patriotic Movement (the Chinese approved and controlled Protestant Church, and there is one for every religion) and then in ten years, churches refusing to comply with be shut down (The Voice of the Martyrs Canada, 2015).

In summary, China is changing but it will take some time and change will be painfully slow. In the meantime, things will be very difficult for groups such as the people living in Tibet, the Falun Gong, the Uighurs, the migrating rural poor, and Christians.

Indonesia

Indonesia had a remarkably tough birth as a nation, but by 1999 it finally held democratic elections and has now grown to be the fourth most populous country in the world with 248,216,193 people; its chief geographic characteristics are islands, more than 17,500 of them, coastal plains, and mountainous interiors, and it is large, almost three times the size of Texas (*CIA World Factbook,* n.d.). Indonesia is plagued with serious income inequality, with wealth concentrated among a few and many not doing particularly well (von Luebk, 2011). It is not an Islamic theocracy but it does have the largest Muslim population of any country in the world so leaders are sensitive to the concerns of the majority (U.S. State Department, 2012). Indonesia does have to deal with terrorists, and it also has to contend with natural disasters of such magnitude that they put terrorism into the proper perspective:

> Natural disasters have devastated many parts of Indonesia over the past few years. On December 26, 2004, a 9.1 to 9.3 magnitude earthquake took place in the Indian Ocean, and the resulting tsunami killed over 130,000 people in Aceh and left more than 500,000 homeless. On March 26, 2005, an 8.7 magnitude earthquake struck between Aceh and northern Sumatra, killing 905 people and displacing tens of thousands. After much media attention on the seismic activity on Mt. Merapi in April and May 2006, a 6.2 magnitude earthquake occurred 30 miles to the southwest. It killed more than 5,000 people and left an estimated 200,000 people homeless in the Yogyakarta region. An earthquake of 7.4 struck Tasikmalaya, West Java, on September 2, 2009, killing approximately 100 people. (U.S. State Department, 2012)

The really dangerous and destructive tsunamis did have an upside; the Indonesian government decided the people of Aceh and the Free Aceh Movement (GAM) should have autonomy, something the GAM had been waging a terror campaign over for years. It is interesting to note that because the Indonesian government and the GAM did not trust each other, they asked the Common Security and Defense Policy (CSDP), of the European Union (EU), to facilitate negotiations and demobilization, which was done successfully (Kasmeri, 2011).

According to a 2009 press release from the National Consortium for the Study of Terrorism and Responses to Terrorism, there were 421 terror attacks in Indonesia from 1970 to 2007, with the GAM responsible for 113 attacks. Since signing the peace treaty in 2005, the GAM has not engaged in any terrorism (National Consortium for the Study of Terrorism and Responses to Terrorism, 2009). The major terror group that plagues Indonesia (and neighboring countries) is Jemaah Islamiya (JI). It has carried out 11 terrorist attacks in cities throughout Indonesia and has killed 265 people, most of them in the 2002 Bali night club bombing that resulted in 202 deaths and 150 injuries (Mapping Militant Organizations, Stanford University, 2012; National Consortium for the Study of Terrorism and Responses to Terrorism, 2009). The history of JI is rather brief. It seems to have been founded in 1992 or 1993 by Abdullah Sungkar (who died of natural causes in 1999) and Abu Bakar Ba'asyir (sentenced in 2011 to 15 years for financing the Bali bombing of 2002), and who were also founders of Pondok Ngruki, a *pesantren* (religious boarding school) in Central Java ("Abdullah Sungkar," n.d.; "Ba'asyir Gets 15 Years in Prison," 2011; International Crisis Group, 2002). JI's goal is to make all of Southeast Asia an Islamic caliphate, or theocracy, with no borders. It uses loose, cell-like groups in four regions that cover most of Southeast Asia, taking advantage of the many islands of Indonesia and even the Philippines for refitting, training, and refuge (International Crisis Group, 2002).

The Indonesian government has aggressively hunted down JI; an Indonesian court has banned them as a "forbidden corporation" and sentenced the two leaders to 15 years in prison (Thompson, 2008). Too, they are on the list of organizations determined by the United States to be a Foreign Terrorist Organization (FTO) (National Coun-terterrrorism Center, 2011). However, before going to prison, Abu Bakar Ba'asyir founded an offshoot of JI, naming it Jamaah Ansharut Tauhid (JAT), with his son Abdul Rohim leading it (Lane, 2012). This group has also been designated as an FTO by the United States, but not by Australia or Indonesia (Lane, 2012). Indonesia reorganized its security organizations in 2010, establishing a National Counterterrorism Agency (BNPT) responsible for pre-vention and counterterrorism efforts, and an elite counterterrorism unit, Detachment or Densus 88 (OSAC, n.d.). In 2012 the last of the Bali bombers, Umar Patek, was sentenced to 20 years in prison (Bland, 2012).

Adding to the pressure on groups like JI and JAT are legislative efforts aimed at reducing the ability of such groups to seek donations to fund operations after al Qaeda funding dried up, with a 2002 Law on Terrorism and a 2010 Money Laundering Law. This has forced terrorists to turn to crime and even computer hacking to fund opera-tions, something the authorities are now monitoring closely (Arnaz, 2012). The Indonesian government has clearly done well in lessening, though not completely erasing, the threat from JI and JAT. Too, while Indonesia remains a secular state, there is a minority group that is attempting to turn it into a theocracy: the radical Islam Defenders Front (FPI). The primary goal of the group is the adoption of Sharia law (Islamic Law) nationwide. It does not care for alcohol, drugs, night clubs, Jews, Buddhists, and Christians, all of which it has attacked, often in the face of the police and authorities; it has around 30,000 members and a leader, Habib Rizieq Shihab (Zenn, 2012). FPI made the news in 2012 when it pressured the government to deny permission for Lady Gaga to perform in Indonesia, even after over 50,000 tickets had been sold ("Lady Gaga 'Devastated,'" 2012).

Finally, it must be emphasized that due to the sheer size of the Southeast Asia area and the fact that there are many islands, getting control of the maritime environment remains a top priority. Indonesia has begun cooperating with the Philippines and Malaysia as well as other countries, based on what was initially a Philippine Coast Watch South program that was envisaged in 2006 and came online in 2008 (Rabasa & Chalk, 2012). It attempts to establish Maritime Domain Awareness or MDA. What this means is that because of the vast oceans and seas and thousands of islands, determined terrorists (including JI), pirates, smugglers, and criminals, no country can get the upper hand without information and intelligence. The program, with the help of Australia and the United States, began in 2011 with around 20 offshore platforms capable of monitoring and interdiction. The new Maritime Research Information Center, Manila, scheduled for completion in 2015, will use this information for strategic threat assessments updated in near real time, and all countries will report information to the Information Fusion Center in Singapore, where many more countries participate (Rabasa & Chalk, 2012; Romero, 2013).

Indonesia remains an interesting place. It successfully held elections in 2014 for both legislative offices and the office of president (Maulia, 2014). What the future holds no one knows, but already ceding Aceh to Islamists may have been the beginning of the end. Perhaps that will not be the case. Indonesia has been dealing with radical groups and terrorist groups for decades and has demonstrated that when groups become violent and dedicated to terror, it will act decisively.

▧ The Philippines

The Philippines or, formally, the Republic of the Philippines, is a vast series of more than 7,100 islands, It is compa-rable in land area to Arizona and is home to more than 103 million people, 80% of whom are Catholic; 5% are Muslim (*CIA World Factbook,* n.d.). The Philippines has had a difficult history of dictatorships, attempted coups, assassinations of politicians, cronyism, and impeachment, and was on *Foreign Policy* magazine's Failed State Index in 2012 (Foreign Policy, 2012). Perhaps one of the reasons is the inability to prevent, or perhaps a tolerance of, human rights violations, pointed out by Human Rights Watch, including extra-judicial killings and enforced disap-pearances (Burgonio, 2012).

The Philippines has been dealing primarily with terrorist groups since gaining its independence in 1946; these are Abu Sayyaf, a violent jihadist group with ties to al Qaeda, two separatist Muslim Groups, the Moro National

Liberation Front (MNLF), the Moro Islamic Liberation Front (MILF), and the Communist Party of the Philippines (CPP) and their military wing, the New Peoples' Army (NPA) (*CIA World Factbook,* n.d.). One scholar, De Castro (2010), has called these, correctly, "Low Intensity Conflicts" or LICs, and he also characterizes them as long; the various governments of the Philippines have tried to eradicate them at times, have held peace talks and/or negotiations and some of the negotiations have lasted for over 40 years (p. 136). The most violent group (no peace talks here) would have to be Abu Sayyaf or **Abu Sayyaf Group** (ASG), which engages in bombings, assassinations, and kidnappings and was founded in 1991 by a former Islamic scholar and a veteran of fighting the Soviets in Afghanistan, Abdurajak Abubakar Janjalani. ASG is on the list as a Designated Foreign Terrorist Organization (National Counterterrorism Center, 2011; Shay, 2009). The goal of ASG is to be an independent Islamic nation in the southern Philippines. It is responsible for the single worst terrorist act in the Philippines, the bombing of a ferry in Manila Bay that resulted in the deaths of 116 people (Shay, 2009). When funding from al Qaeda declined, the group used kidnappings to raise funds and was hunted tirelessly, with U.S. military assistance. The founder and his younger brother were both killed and the group's ranks diminished. It is still out there, though, and in July 2012 it killed plantation workers after operators of a rubber plantation refused to give in to its extortion demands. It also has at a minimum five hostages being held for ransom (Teves, 2012). Most hostages are eventually released or freed by military action, but the saga continues. In 2014, ASG captured two German nationals in the waters off the coast of Malaysia, and they joined two European birdwatchers captured in 2012 (Weiss, 2014).

The Communist Party of the Philippines, National Democratic Front (CPP/NDF) and its armed component the New People's Army or NPA, is the most serious threat due to its size, organization, and goal of replacing the government. It was founded in the late 1960s and is now the longest running Maoist insurgency in the world, over 40 years in existence (Parlade, 2006). It has had violent clashes with the Armed Forces of the Philippines and some success and is attempting to follow a classic guerrilla campaign by consolidating power in villages then extorting or taxing for funding, using propaganda, attacking government and military personnel and sites, and establishing bases (Parlade, 2006). Since its inception it is estimated that over 40,000 people have been killed, with many others wounded and displaced; the total number of current fighters is probably no more than a few thousand (De Castro, 2010; Ploughshares, 2011). The current government of the Philippines, while still actively hunting them down, agreed to peace talks in Oslo, Norway, in the summer of 2012 ("Philippine Government, Rebels Try to Break Impasse," 2012). However, a quick Google search for "attacks by NPA 2012" and "Philippines" comes back with over 373,000 results. A 2014 report providing statistical information on terror attacks for 2013 listed the NPA as number seven in the top ten most active terror groups (National Consortium for the Study of Terrorism and Responses to Terrorism, 2014).

The final two groups are both Muslim, but they are not exactly the same. The **Muslim National Liberation Front** or MNLF was founded in the 1970s, seeking to force concessions from the government with regard to autonomy for islands or provinces that were predominantly Muslim. Waging a war with the government, it finally won concessions in 1996 but not before between 120,000 and 150,000 had died (Lum, 2012). All is not necessarily great, as the Autonomous Region in Muslim Mindanao (AARM), which resulted from the peace deal, complains it does not have a contiguous territory or control over natural resources, and lacks the power to tax (Lum, 2012; Rabasa & Chalk, 2012).

Unhappy with the agreement, a breakaway group formed, the Muslim Islamic Liberation Front, which is more militant and more Islamic and willing to allow Jemaah Islamiya help train members in exchange for a safe haven when needed (De Castro, 2010). While periodically attacking the military for over 15 years, it has been engaged in peace negotiations with the government, in earnest since 2008 ("Phl Gov't, MILF," 2012). It wants a "Bangsamoro Juridical Entity" (BJE) that replaces the AARM, greater autonomy, the power to create its own government, hold elections, and create a judicial system and police, schools, currency, and postal service with resource sharing with the Philippine government (Lum, 2012). An earlier deal fell apart when Christian Catholics living in the area went to court, fearful of the deal, and MNLF remains unsure of what a deal involving the MILF might mean for it (Lum, 2012). Finally, the Philippines was recognized by the Institute for Economics and Peace in 2012 as among the top five countries improving its efforts to achieve peace (Institute for Economics and Peace, 2012).

The bottom line when it comes to the Philippines remains that it is a difficult area to govern with so many islands, coastal plains, and mountainous terrain, but with a great deal of assistance, particularly from the United States in terms of foreign aid, military assistance (on average there are 500 to 600 U.S. military personnel there in some capacity), it has made great strides (Lum, 2012). The Philippines also participates in the Coastal Watch program, which seeks to increase security in the maritime environment (Rabasa & Chalk, 2012).

Burma

The leaders of Burma call the country Myanmar, but not everyone has adopted this. Burma had been ruled by a military junta since 1988 (*CIA World Factbook,* n.d.). The junta allowed elections in 1990 and it lost badly to the opposition party, the National League for Democracy (NLD). The junta then ignored the election results and imposed a 15-year house arrest on the winning party's leader (and Nobel Peace Prize recipient), Aung San Suu Kyi. The Chinese share a border with Burma and did not like the unrest created by the junta. Ruling brutally for years and under pressure from the Chinese to hold elections and promote stability, the junta allowed elections in recent years and moderated its position, so much so that Aung San Suu Kyi has seen her party, the NLD, contest in elections and she has been elected to parliament. The U.S. government and others also removed investment sanctions that had been imposed for 15 years (Kaplan, 2010"; Myanmar's Suu Kyi," 2012).

The foregoing may seem to imply that things are going well in Burma. The country has the potential for improvement but faces enormous challenges even with help from other nations. The junta that ran things into the ground has a history that one would not expect from its democratic name: The State Law and Order Restoration Council. It has a history of viciously battling various ethnic tribes in the mountains for years; while large, over 400,000 in the government's military, the soldiers are unreliable, often underpaid, and there is almost no infrastructure and few formal institutions but they do enjoy a strategic spot between India and China and they are on the Bay of Bengal (*CIA World Factbook,* n.d.; Kaplan 2010). The reason for the fighting is that the various ethnic tribes do not accept the notion of Myanmar or the central government, with tragic results of thousands of burned-out villages and over 500,000 refugees (Kaplan, 2010). Burma, slightly smaller than Texas, and with a mostly Buddhists population, has over 54 million people; abundant natural gas, oil, uranium, coal, and precious stones; the potential for a major deep-water port; and is a natural pathway for pipelines, yet it spends 40 cents per capita on health and $1.10 on education while maintaining one of the largest armies in the world, an expense that consumes 40 percent of the entire government budget (*CIA World Factbook,* n.d.; Kaplan, 2010). Of note also is that in 2007 there were mass demonstrations in response to rises in fuel prices led by prodemocracy proponents and Buddhist monks (called the Saffron Revolution), resulting in a brutal crackdown and thousands of arrests and, to add insult to injury, a tsunami in 2008 left 138,000 dead with thousands injured and homeless (*CIA World Factbook,* n.d.; Kaplan, 2010).

As Kaplan (2010) notes with regard to the future of Burma without a military junta, speculating that it will be replaced with a democratic government:

> Aung San Suu Kyi, as a Nobel Laureate could provide a moral rallying point that even the hill tribes would accept. But the country would be left with no infrastructure, no institutions, and a growing but still frail civil society and NGO community, and with various ethnic groups waiting in the wings that fundamentally distrust the Burmans. (p. 238)

And this from Kaplan (2010), quoting an international negotiator: "There will be no choice but to keep the military in a leading role for a while, because without the military there is nothing in Burma" (p. 238).

What the future holds for Burma is something that no one knows, but in 2012 the junta released numerous political prisoners and began talks with some ethnic tribes, and the United States appointed an ambassador (Clinton, 2012). Perhaps the junta realized the pipeline routes prized by both India and China will go through tribal territories and it may be easier to deal with India and China if the land and the people living there are peaceful (Kaplan, 2010).

Cambodia

The story of Cambodia can break your heart, to paraphrase one author's friend. His name is Don Brewster, a northern California pastor with a tender heart. He and his wife Bridget traveled to Cambodia in 2005 to train pastors from 13 provinces. It was a normal, routine event and while the pastors complained that poverty was extreme in the country, they were optimistic about the future. Returning home, Bridget and Don happened to watch a *Dateline NBC* special on child sex trafficking in Cambodia. Shocked at how appalling it was, they quit their jobs, sold their house, formed a non-profit, and moved to Cambodia to rescue, rehabilitate, educate, and house children (Agape International, n.d.). Don often speaks to our classes when home visiting from Cambodia and when he does he explains that he was stunned with the cavalier attitude that allowed rural poor to sell children to sex traffickers. This is an attractive proposition since the $250 they can get for a daughter will double the annual family income and the daughter is easily replaced by having more children. He did not understand until he realized that the **Khmer Rouge** killed all those with an education, so the education levels and subsequent opportunities are nil, and they killed all monks, priests, and pastors so Cambodians have no moral compass.

This is how that happened. With the world focused elsewhere, in 1975 a ragged band of the Khmer Rouge led by educated-in-France admirer of Chairman Mao, Pol Pot, managed to seize power in Cambodia after a civil war that left over 150,000 dead. The Khmer Rouge, with an army of 700,000 and following the guidance of Pol Pot, immediately closed all factories, schools, hospitals, and universities, abolished all political and civil rights, outlawed money, and forced everyone to the countryside to work on collective farms (Fletcher, 2009; Peace Pledge Union, n.d.). Intellectuals, elites, the upper class, foreigners, anyone speaking a foreign language, Buddhists, and anyone wearing glasses were all executed (Fletcher, 2009; Peace Pledge Union, n.d.). The killing continued until 1979 when Viet Nam grew tired of dealing with attacks from the Khmer Rouge and defeated them but did not destroy them (Fletcher, 2009). They survived in the jungle until 1999 when all their leaders had defected or died. The number of deaths at their hands is impossible to know, with estimates conservatively ranging from one million to two million, but with disease, malnutrition, and executions is likely much higher (Fletcher, 2009). Finally, after years, the UN sponsored a Cambodia International tribunal, which is being held in Cambodia, to prosecute surviving Khmer Rouge leaders, but it is slow going, they all are in their 80s, and yes, ongoing. To date only one individual has been convicted and sentenced (Sato, 2012). As of 2015, the tribunals remained active; two major figures were found guilty in 2014. Though one is 88 years of age and the other is 83, they plan to appeal ("Top Khmer Rouge," 2014).

Cambodia today is slowly improving but life expectancy is well below that of major countries, 50% of the population is under the age of 25, education levels are poor, almost a third live below the poverty line, and outside of the cities there is virtually no infrastructure (*CIA World Factbook,* n.d.). The most recent Human Rights Watch report on Cambodia is not encouraging and notes there is heavy-handed use of the criminal justice system to intimidate political opposition and to limit free speech and demonstrations (Human Rights Watch, 2012).

Japan

Japan has the dubious distinction of being the only country in the world that has been the target of two atomic bombs and a chemical attack. The decision to drop the bombs was made after the stark realization that invading Japan in World War II would likely result in over a million casualties for the United States and other allies with many

© Can Stock Photo Inc. / tanjalagicaimage

▲   This is a photo of Hiroshima, Japan today. It was virtually destroyed by an atomic bomb in 1945 but has been rebuilt.

more millions of Japanese casualties (Stimson, 1947). Japan is an island chain located between the North Pacific Ocean and the Sea of Japan, and it is slightly smaller than California in area. It has a population of slightly more than 127 million people, the vast majority of them Japanese in ethnicity (*CIA World Factbook*, n.d.). Japan clearly has the capacity to build nuclear weapons in months and the material from even tsunami-damaged nuclear power plants to build them, but it has officially renounced ever doing so, although the issue remains debated as it worries about North Korea and China (Kageyama, 2012). The major issues facing Japan today are a slowing economy, an aging, long-lived population (over 83 years of age on average), and a population decline with a drop of 15% by the middle of this century likely (McNicoll, 2005). In fact, with marriage and sex missing as an activity for Japanese and many preferring the single life, Japan may be headed for a slow decline or extinction:

◆   A record 61.4% of unmarried Japanese men aged between 18 and 34 have no girlfriend, up 9.2 percentage points from 2005, the National Institute of Population and Social Security Research in Japan said.

◆   One in four unmarried men and women in their 30s say they have never had sex, and the majority of young women prefer the single life.

◆   An astounding 90% of unmarried young Japanese women said the single life suits them better than marriage. ("No Sex Please," 2011)

Japan faced terrorism early in the 1970s when Fusako Shigenobu, a Marxist-leaning young woman and recent college graduate, decided to move to Beirut, Lebanon, founded the Japanese Red Army in Lebanon (JRA), and began cooperating with Palestinian terrorist groups, managing to remain hidden there for over 30 years (Kyodo News, 2010). While small in number, the group managed to embarrass an unprepared Japanese government and conduct some successful and often deadly terrorist acts. The world first heard about them when some of their members hijacked an airplane and landed it in North Korea in 1970, with some of the perpetrators remaining there ("Background Note: North Korea," 2012; Leheny, 2010). Three members of the group opened fire and killed 24 people in the airport at the Lod Airport in Tel Aviv in May of 1972 (Leheny, 2010; McCurry, 2008). This was followed in 1974 by the storming of the French Embassy in The Hague, which secured the release of fellow JRA members from prison in exchange for the Japanese ambassador and ten other hostages; they were not done and hijacked an airliner in Bangladesh and secured the release of more JRA prisoners, plus received $6 million in ransom money before releasing passengers and crew (McCurry, 2008). This group proved to be a turning point for the Japanese government, which began treating terrorism as a problem that must be solved with international cooperation and not police reaction and negotiation (Leheny, 2010).

Fusako Shigenobu was arrested in Osaka in 2000 and eventually sentenced to 20 years in prison. She has since disbanded the group (McCurry, 2008). She appealed her sentence in 2008 but the appeal was denied in 2010, and she will spend many more years in prison (Kyodo News, 2010; McCurry, 2008). While never large, the JRA's willingness to work with other groups (something Shigenobu denies) made it a deadly threat (McCurry, 2008). The JRA focused the Japanese authorities so exclusively on dealing with leftist or communist groups that they were totally unprepared for the next threat: Aum Shinrikyo (Leheny, 2010).

Aum Shinrikyo was founded in Japan in the 1980s by Chizuo Matsumoto, who was generally viewed as an early failure, not even passing the university entrance exam; he is blind in one eye and has a serious impairment in the other one (Danzig et al., 2011). However, he was charismatic, managed to win and recruit a number of young people who were searching for meaning, and was well-educated and successful. While running an acupuncture clinic, he also added a traditional Chinese medicine pharmacy (Danzig et al., 2011). He joined a strict Buddhist sect but later formed his own, calling it Aum Shinsen no Kai (Aum Mountain Wizards) and began claiming that the world would end in catastrophe or Armageddon, themes used to attract more followers, and he loathed the United States and its materialism (Danzig et al., 2011). The group was soon a full blown Millenarian Cult with overseas chapters and a number of chemists (Leheny, 2010). He later formed a monastic community and required followers to turn everything they owned over to the cult, which changed its name in 1987 to Aum Shinrikyo, which means Aum Teaching of Truth, and Matsumoto changed his name to Shoko Asahara (Danzig et al., 2011). The group grew but became increasingly violent; engaged in politics, which proved embarrassing; and decided to produce and employ biological and chemical weapons to convince people the end was near. The biological weapons were never a success but it managed to manufacture nerve gas. Fearing the police were closing in on them, group members hastily planted the nerve gas sarin (essentially a pesticide that is clear, odorless, tasteless, and deadly if contacted or breathed) in containers on the Tokyo subway line near the Metropolitan Police Department, resulting in 13 deaths and 6,252 injuries, in what can only be described as a horrific attack (Centers for Disease Control, n.d.; Danzig et al., 2011). Following this attack the Japanese Police finally began actively investigating Aum Shinrikyo and determined that members were wealthy, well educated, well trained, and capable of producing large quantities of deadly nerve gas; the group had procured a Russian helicopter to assist in spraying the gas (Danzig et al., 2011; Fletcher, 2012). Asahara was arrested in May of 1995 and his trial took eight years, finally ending in 2004 in a conviction and sentence of death by hanging, with all appeals exhausted in 2006 (Fletcher, 2012). It is designated as a terrorist organization by the United States but not by Japan and remains intact as an organization, eschewing violence though it is under close surveillance (Fletcher, 2012). Over 200 cult members were arrested and convicted in connection with the Sarin attack and 13 are on death row ("Execution of Aum Founder," 2012). Normally Asahara would have been executed by now but police continue to arrest accomplices, some as recently as in the summer of 2012, so the execution has been postponed until after their trials and appeals, which will be lengthy ("Execution of Aum Founder," 2012). The Center for a New American Security produced a report on Aum Shinrikyo (website for report below) that is chilling to read and contains this quote:

> Groups such as Aum expose us to risks uncomfortably analogous to playing Russian roulette. Many chambers in the gun prove to be harmless, but some chambers are loaded. The blank chambers belie the destructive power that the gun can produce when held to the head of a society. Our analysis suggests that the cult's 1995 Tokyo subway attack would have been much more lethal if Aum had not destroyed its purer Sarin when it feared discovery a few months earlier or disseminated the low-purity Sarin more effectively. *Terrorists need time; time will be used for trial and error (tacit knowledge acquisition); trial and error entail risk . . . but Aum found paths to WMD, and other terrorists are likely to do the same.* (Danzig et al., 2011, p. 39; emphasis added)

Aum Shinrikyo awakened the Japanese to the threat of terrorism and they now have credible intelligence and counterterrorism agencies. There have been no major terrorist incidents in Japan following this attack, but it is chilling to realize it could have been much worse. It is clearly evident, based on Aum Shinrikyo's history, that a university trained chemist with access to materials and a laboratory can easily manufacture some very deadly chemical weapons (Danzig et al., 2011).

Summary

This section covered a sweeping number of countries and peoples from a miserable North Korea with a suffering people to the likes of South Korea, a dynamic country, to China and Japan with unique challenges ahead. All have had aspects of terrorism and some, like Japan, illustrate what happens when a nation ignores terrorism or the potential of terrorism. Indonesia, the Philippines, Cambodia, and Burma or Myanamar have all made progress but many challenges remain. The prospect for terrorism, especially Islamic terrorism, remains a threat but for most countries in the region the threat is low level at the present time.

 ## Key Points

- ◆ North Korea is a brutal, rogue country; propped up and protected by China, it will remain so.
- ◆ China must come to terms with an increasingly elderly population and massive migration to cities.
- ◆ Indonesia, the Philippines, and Burma face challenges as well, in part due to geography, poor governance, and natural disasters on a massive scale on an often regular basis, while dealing with various terrorist groups.

KEY TERMS

Abu Sayyaf Group	Japanese Red Army	Muslim National Liberation Front
Aum Shinrikyo	Khmer Rouge	Sarin

DISCUSSION QUESTIONS

1. What is your assessment of terrorism today in China? Japan? Indonesia?

2. How do Indonesia and the Philippines deal with terrorism?

3. Would you invest in Burma?

4. What is the Muslim National Liberation Front and what are its goals?

5. Do you believe Japan now has an effective counterterrorism program?

WEB RESOURCES

2012 Failed States Index: http://www.foreignpolicy.com/failed_states_index_2012_interactive

Agape International Missions: http://agapewebsite.org/. It is an organization based in Cambodia dealing with sex trafficking, a major problem there

CHINAaid.com: http://www.chinaaid.org/. This is the website of the blind pastor from Midland, Texas, who challenged the Chinese government

CIA: The Final Months of the War with Japan: https://www.cia.gov/library/center-for-the-study-of-intelligence/csi-publications/books-and-monographs/the-final-months-of-the-war-with-japan-signals-intelligence-u-s-invasion-planning-and-the-a-bomb-decision/csi9810001.html#rtoc9. This is the CIA site that shows the actual documents made concerning the decision to not invade Japan and opt for the atomic bomb; the decision was made largely due to the huge casualty losses projected.

Coast Watch System report: http://www.rand.org/pubs/occasional_papers/OP372.html. This is the actual RAND Report mentioned with regard to the Coast Watch System

Dateline NBC: Children for Sale: http://www.msnbc.msn.com/id/4038249/ns/dateline_nbc/t/children-sale/#.UAiSbLSe6Sp. This is the series that inspired the Brewsters to act

Falun Dafa Information Center: http://www.faluninfo.net/category/11/

Falun Dafa: http://www.falundafa.org/

Free Tibet: http://www.freetibet.org/about/dalai6

National Security News: Center for a New American Security: http://www.cnas.org/node/6652

Peace Pledge Union: Cambodian genocide: http://www.ppu.org.uk/genocide/g_cambodia.html

U.S. Army Cyber Command: http://www.arcyber.army.mil/. See what the United States is doing about Cyber Warfare

U.S. Department of State: U.S. Relations with Indonesia: http://www.state.gov/r/pa/ei/bgn/2748.htm. A good overview of Indonesia

Uighers: China's Xinjiang Province a "Police State" 3 Years After Riots": http://www.voanews.com/content/uighur-china-xingiang-rights/1364092.html

Voice of the Martyrs: http://www.persecution.com/

World Uygher Congress: http://www.uyghurcongress.org/en/

READING 12

This reading, "Terrorism Risks and Counterterrorism Costs in Post-9/11 Japan," by David Leheny, presented in its original length, reviews Japan's successful solutions to terrorism and evaluates them.

Terrorism Risks and Counterterrorism Costs in Post-9/11 Japan

David Leheny

 ## Introduction

Evidently unaware that such an admission might—were he an Afghani farmer or an Iraqi cab driver—be grounds for water boarding or other 'harsh interrogation methods', Japan's Justice Minister, Hatoyama Kunio (the brother of Hatoyama Yukio, Japan's prime minister as of late 2009), told a stunned audience at the Foreign Correspondents' Club in late October 2007 that a friend of a friend of his was a member of Al Qaeda, who provided advance word of the October 2002 Bali bombing. Hatoyama's point evidently was that, because this particularly helpful terrorist had come in and out of Japan on a number of occasions, Japan really needs to fingerprint and photograph all foreigners entering Japan (though not some categories of special permanent residents, like many *zainichi* Koreans), as a new law now requires. Japanese and foreign journalists alike turned out to be less interested in the increasingly strict procedures for landing at Narita airport than Hatoyama's apparent links to Al Qaeda, so he hurriedly called a news conference, at which he explained that he had never met the terrorist, only that he had heard the story from a friend in a butterfly hobby group. Hatoyama also reassured the reporters that he himself had no advance knowledge of the Bali attack, having heard about the warning from the terrorist (who may not, he added, have been in Al Qaeda

proper, but perhaps just in a sympathetic extremist organization) three or four months after the bombings (*The Economist* 2007).

Hatoyama is hardly the first Japanese official to make an internationally controversial comment to the press, although these have more typically involved exculpatory readings of Japanese wartime history or racially tinged critiques of other nations' populations, such as Governor Ishihara Shintarō 's 2000 description of foreigners as 'sangokujin', an anti-Asian slur (Shipper 2005: 308). Public admissions of connections to terrorists who have aimed specifically to kill citizens of Japan's two military allies, the United States and Australia, have a somewhat less distinguished pedigree. The media attention to the gaffe also obscured the strange way in which Hatoyama sought to make his point. That is, because he himself knew of an admitted terrorist who had come in and out of Japan, all foreigners must be screened on the way in. Rather than encouraging his friend to turn in the terrorist, or to provide information on him or her to the relevant authorities in Tokyo, Jakarta or Canberra, Hatoyama used the case to justify a wide-ranging approach that simultaneously expands the state's access to information that may well be irrelevant to security, while placing much of the cost of the venture on the foreigners who will now have to surrender biometric data to the authorities to be allowed into the country. There was no hint in Hatoyama's discussion

SOURCE: *Japan Forum*, 22(1-2) 2010: 219–237 Copyright © 2010 BAJS. Reprinted with permission.

(which may have been premised on a complete falsehood by his friend, even if Hatoyama has honestly recounted his part in the story) that Japan itself was ever threatened by terrorists, just that terrorists known for targeting other nations might be able to get in and out of Japan with ease. The idea was not that Japan was at that moment a target, but rather that it might become one, and that, before that eventuality, the state would need tighter authority over the comings-and-goings of foreigners who might pose a threat.

Postwar Japan is no stranger to terrorism. During the 1970s and 1980s, it faced continuous challenges by the Japan Red Army (JRA), a leftist organization that staged several audacious attacks inside Japan and even more striking ones—sometimes directed ultimately at Tokyo—outside Japan's borders as well (Steinhoff 1989). In 1995, a millenarian cult released sarin gas on the Tokyo subways, killing twelve (eleven at first, with another dying later of complications from the poisoning), making Japan the first victim of an act of CBRN (chemical, biological, radiological, nuclear) terrorism. More recently, though less dramatically, rightwing militants placed an apparent bomb in the home of a Ministry of Foreign Affairs (MOFA) negotiator often criticized by nationalists as too soft on North Korea. For the most part, the Japanese government has pursued these threats through a combination of policing, surveillance and occasional negotiation. The JRA is by and large a thing of the past; Aum Shinrikyō has been renamed 'Aleph' and has been de-fanged both by legislation and by consistent government monitoring; and rightists tend to focus their venom on targets usually of little interest to Tokyo's authorities, and have been very limited in their use of violence when dealing with more mainstream figures. By virtually any reasonable account, Japan is an extraordinarily safe place, and the risk of terrorism seems minimal.

One would not necessarily know that, however, by the flurry of activity since the 9/11 attacks, largely designed to import knowledge of crisis management, counterterrorism and private security to help Japan prepare for the possibility of a mass-casualty attack. In addition to regular consultations with the United States government, which predated 9/11, and meetings with other G8 members, the Japanese government has sponsored international symposia on risk management for terrorism and private actors have eagerly sought advice from foreign consulting firms on how best to respond to and handle the threat of terrorism. Whatever its past successes in responding to terrorism, Japanese

specialists frequently decry their country's ostensibly tardy response to the threat of transnational, mass-casualty terror, relying in part on the sense that other governments with more experience in developing risk management tools are 'ahead'. In this view, Japan needs to catch up: the government, firms and citizens alike. Japan is now in a world of omnipresent terrorist threats, in which a lack of preparation seems almost like an invitation to attack rather than a credible assessment of low risk. Most interestingly, these discussions normally treat terrorism as some kind of act of God, no more easily predicted than earthquakes, volcanoes or tsunami, rather than as a deliberate act by conscious political actors. Risk management therefore takes a peculiar form, shifting responsibility onto *non-political* rather than political actors and using the demand to reduce risk in part as a justification for the enhancement of the state's coercive and surveillance authority. The state grows even as its responsibility shrinks. As Hook notes in his introductory article, 'managing risk' often involves placing the costs associated with risk onto individuals, an interesting strategy when the political nature of terrorism deeply implicates the state itself.

This article draws on my earlier work on Japanese counterterrorism policy to suggest that there has been something of a reversal of the logic of counterterrorism, a reversal that becomes visible by examining both the construction of risk and the apportionment of its costs. I have argued elsewhere (particularly Leheny 2006) that from the 1970s through the 1990s, terrorism was understood in Japan primarily as a political problem, with Japan's relative safety understood in political and cultural terms, and appropriate responses framed similarly. At this time, the increasingly defined techniques of counterterrorism—police and intelligence techniques spread globally through international institutions and rationalized myths of their efficacy—met with indifference and even resistance in Japan. But the September 11 attacks provoked an increased Japanese commitment to the US-led 'Global War on Terror', a rethinking of Japanese vulnerability and a recalculation of the risks and costs of increased surveillance and engagement. Notably, the suspicion of foreigners evidenced in Hatoyama's comment has not yielded a demand to keep foreigners out of Japan, which would have appealed to romantic notions of Japanese homogeneity and harmony. It instead coexists with political calls to increase the number of foreigners—to respond to problems associated with

Japan's shrinking population—and calls for counterterrorism techniques, in the form of fingerprinting, surveillance, monitoring and intelligence. The costs for this increased scrutiny will fall disproportionately on the foreign population, as well as onto citizens who venture overseas, as they have in most other states. In the Japanese case, this reflects a shift from trusting in something essential and harmonious about Japanese society toward a trust in the efficacy of administrative technique.

This article first traces the postwar development of Japan's counterterrorism policies, locating them as part of a familiar left-right split in Japanese politics that may have left security forces particularly ill-equipped to consider potential threats that did not fit into a Red Army-inspired template. It then touches on the ways in which the costs and risks associated with terrorism have been used politically, particularly in the United States during the 'Global War on Terror'. Third, it explores discourses of risk and responsibility during Japan's moves to support American foreign policy initiatives during the premierships of President George W. Bush and Prime Minister Koizumi *Junichirō*. Finally, it looks at how the public relations materials surrounding the recent changes to entry control procedures represent the allocation of the costs associated with the risk of terrorism. These debates within Japan tend to display the risk of terrorism as being generated by Japan's openness to the world—the ability of the foreign to enter Japan—rather than by Tokyo's activities in the world around it. And they legitimate new policy shifts that permit the government to amass information about foreigners by suggesting that the foreigners themselves are mostly happy to volunteer such information.

 ## Terrorism and Counterterrorism as Japanese Politics

In the midst of an international environment that increasingly criminalized and de-politicized terrorism, Japanese officials had long treated the latter as an ineluctably political phenomenon. As early as 1952, immediately after the end of the US Occupation of Japan, the Liberal Party (LP) hastily enacted the Anti-Subversive Activities Law,

establishing the Public Security Investigation Agency (PSIA) and Public Security Examination Committee (PSEC) within the Ministry of Justice (MOJ). Under the legal authorization of the PSEC, the PSIA was empowered to engage in long-term surveillance of potentially subversive organizations, which in practice meant leftist organizations and parties (Itabashi *et al.* 2002). Although the law was formally invoked only eight times, always with reference to individuals rather than organizations, the expectation that the PSIA would at least monitor the activities of communists, union organizers and other leftists (as opposed to violent rightists and gangsters) helped to enhance the notion that terrorism and counterterrorism were, at least in theory, potentially the continuation of left-right politics by other means.

With various radical groups emerging from the 1960s student movement as factions of a poorly coordinated violent collective known as the JRA, Japan now had an organization potentially worthy of the long-term surveillance and pressure placed by the Japanese police on leftists in general. Espousing a revolutionary agenda that resembled at least the rhetoric of left-wing extremists in Western Europe, JRA leaders aimed also at linking their goals with those of national liberation movements in Palestine and elsewhere, even as their internal practices reflected institutional hierarchies that bore crucial similarities with those of much more conservative Japanese corporate organizations (Steinhoff 1989). With information drawn not only from PSIA-organized surveillance but also informal networks organized around local police boxes (*kōban*), Japanese police sought to turn over JRA members through cajoling, harassment and occasional bullying, seeking actionable intelligence to prevent or forestall potential attacks.

Postwar reforms of Japan's security services, combined with social norms that remained suspicious of overt police coercion, made the left-wing terrorism of the 1970s and 1980s as much a political as a criminal justice matter.[1] The complex constraints on police action were laid bare in the 1972 'Asama Mountain Cottage' incident, when a group of JRA members—themselves perpetrators and survivors of a bloody internal purge—holed up at an inn in Nagano prefecture, taking the cottage owner's wife hostage. Sassa Atsuyuki, then a mid-career National Police Agency (NPA) official, was dispatched to the area, along with members of the elite riot forces from Tokyo's Metropolitan Police Department, with instructions to end the siege by rescuing the hostage, while maintaining good relations with local police and the news media, and without firing a shot. For the NPA, the televised sight of police firing into a house and costing the life of an innocent hostage would have jeopardized its long-lasting efforts to be seen as a friendly police force with no meaningful ties to the fearsome wartime police. Sassa's (1996) controversial account has now served as the basis of a popular film, Harada Masato's *Totsunyu seyo: Asama sansojiken*. However self-serving the notably self-promoting Sassa's depiction of events may be, the frustration he voices with the constitutional and political constraints on violent police action has been relatively widespread. And so the tools deployed by the state against the JRA privileged low-level but persistent pressure on militants, often becoming coercive but lacking the sort of widespread bullying, routinized violence or extrajudicial brutality that have marked counterterrorism programs in other nations.

Many of the JRA's most fearsome attacks occurred overseas, including a 1972 massacre in which one cell opened fire with automatic weapons in Israel's Lod Airport, killing twenty-four people; attacks in Malaysia, the Netherlands and Singapore followed. With virtually no ability to deal with the JRA through police or military tools overseas, the Japanese government negotiated with them when necessary. Most famously, to end a 1977 hijacking of a Japan Air Lines flight that had been forced to land in Dhaka, Bangladesh, the Japanese government paid a $6 million ransom, released some JRA prisoners and secured the release of the passengers and crew. In response to media questions about his decision to capitulate, Prime Minister Fukuda Takeo said 'a human life is heavier than the earth', a humanitarian sentiment that was, according to public opinion polls, highly popular in Japan.

It was, however, controversial overseas. The United States and other major industrialized powers had begun to push for a 'no concessions to terrorists' pledge, under the seemingly logical but empirically unproved assumption that concessions encourage more terrorism. By the 1987 Venice Summit, the Japanese government would be forced to agree with the G7's overall pledge to avoid negotiations with terrorists. This pledge in essence de-politicized international terrorism, removing it from the arena in which

political discussions and negotiations might be used to resolve individual crises and to construct it instead as a law enforcement problem that should be handled through intelligence cooperation, criminalization and prosecution. Other governments had already moved in this direction, though not without controversy, as the UK government found when its refusal to acknowledge Bobby Sands and other Irish Republican Army members as 'political prisoners' (treating them as simple criminals) led to fatal hunger strikes and worldwide opprobrium that arguably provided significant fuel to the Irish nationalist cause. But the international agreements first to ban specific forms of terrorist activity and then to mandate cooperation against it as a form of international crime cemented a notion that terrorism would have to be understood as something non-negotiable and, in essence, apolitical; the cause could not sanctify the methods, and governments would be expected to treat all such threats similarly.

Removing terrorism from the realm of politics proved difficult. With its singular focus on the JRA, Japanese counterterrorism policy in many ways represented a strategy to deal less with a category of political violence than with a specific opponent or set of opponents. Critics have suggested that the attention to leftists, to the exclusion of other potential threats, was an important factor in the NPA's apparent inability to see the growing menace posed in the 1990s by Aum Shinrikyo: a millenarian cult with strong overseas ties and a leadership cadre composed of a suspicious number of chemists. As I have argued with colleagues elsewhere regarding the 1995 sarin gas assault on the Tokyo subways, 'the Aum attack was therefore a massive intelligence failure deeply steeped in politics' (Friman *et al.* 2006: 100). The attack implied to many terrorism specialists that Japan was especially vulnerable to terrorism, though it perhaps should have been interpreted instead as the outcome of intelligence and police capabilities that were ineffective largely because they had been so specifically targeted, and so deeply embedded in political fissures that marked postwar Japanese politics.

It was largely in the aftermath of the Aum attacks that Japanese policy-makers displayed new sensitivity to the potential risks facing Japanese both at home and overseas, including those to which the government could not respond with force or negotiation. Partly for this reason, a Japanese think-tank affiliated with the NPA, the Council for Public Policy (CPP), planned and created a video series designed to instruct Japanese about overseas threats. The first four videos in the series were animated shorts of twenty to thirty minutes apiece; CPP members chose the animated format in part because of its flexibility and popularity and in part because of cost-effectiveness. Two comedies—*Sakusen kōdo Tokyo* and *Sakusen kodo paradaisu* (respectively, 'Battle code: Tokyo' and 'Battle code: paradise')—started the series, addressing a wide array of risks that Japanese might find overseas, from revolutionaries to poorly marked road hazards, all of which might kill careless Japanese who had grown accustomed to their overly protective and safe society. The next two—*Kidnap* (titled in English) and *Tāgettō: Nihon kigyō* (Target: Japanese firm)—were more serious portrayals of specific terrorist attacks that might befall Japanese overseas. Over the course of the four videos (made available through public libraries and some tour companies), the threats grew more and more specifically targeted against Japanese overseas, portrayed primarily as walking bags of poorly protected money, though the message was consistent: Japanese could not rely on their paternalistic government overseas. They would have to exercise *jiko sekinin*, or 'self-responsibility' (Leheny 2006: 137–43).

 ## Risks, Politics and Responsibility

Public, as opposed to personal or private, responsibility for dealing with terrorist threats has been an increasingly hot topic, particularly after the catastrophic costs associated with the 9/11 attacks. The United States government, for example, provided settlements to victims of the attack, provided that they would agree not to sue the airlines for negligence in allowing terrorists on board. New York 'first responders' (emergency fire, security and medical personnel) who perished in the attack were posthumously rewarded with generous payments to their spouses and families; their comrades, many of them made seriously and even fatally ill by exposure to the toxins at Ground Zero, were left largely unprotected, even to the point that left-wing documentary film-maker Michael Moore took several to Cuba for treatment as an indictment of the American healthcare system in his 2007 movie, *Sicko*.

In the face of catastrophic losses that might cripple the American insurance industry, the US Congress hastily passed the Terrorism Risk Insurance Act (TRIA) in 2002, providing a governmental 'backstop' meant to limit private losses in the face of the risk posed by terrorists. One outcome of the law was the development of a private terrorism insurance market, in which firms have evidently been able to map out the potential risks of this otherwise unpredictable sector. Although the American goal has been to ensure that private funds replace public commitments to handle the financial risks of massive terrorist attacks, within three years of the law, economists still argued that the market was not yet sufficiently developed to allow the TRIA to lapse through a built-in sunset clause of the initial legislation. Public commitments are ostensibly designed to structure the construction of a market for the private calculation of and hedging against risk (Hubbard *et al.* 2005). With TRIA as something of a standard-setting piece of legislation, the Japanese government even funded a World Bank comparative study of terrorism insurance, clearly to develop models and practices that might successfully be used in Tokyo.

In some ways, however, the strategies to encourage the private adoption of risk fly in the face of the dominant language of handling terrorism, which views it as behavior so unacceptable that its prevention justifies enhanced coercive authority, widespread public surveillance and even pre-emptive war. That is, the United States and other governments have adopted very nearly a 'zero tolerance' policy for terrorism, which sits uneasily alongside efforts to privatize risk and to shift costs of prevention onto citizens and residents. Writing more generally, and with reference to a non-terrorism related case, economist Michael Power captures some of the intrinsic dilemmas involved:

> Political discourses of 'zero tolerance' sit uneasily with a risk-based ethos. An event such as the demise of Equitable Life [a British mutual life fund that nearly failed in 2000–1], which could be regarded as 'tolerable' from the impersonal point of view of systemic financial risk, was in fact experienced by large numbers of people as a life-changing catastrophe—and reflected in the media as such. People also feel differently about specific risks, e.g. public attitudes to

deaths on the road differ from attitudes to deaths on public transport. All this means that *ex ante* public acceptance of the possibility of failure can never control *ex post* public reaction to actual failure.

(Power 2004: 22)

Because they are generally tailor-made for heavy media exposure and powerful emotional consequences, terrorist attacks lend themselves exactly to this sort of 'zero tolerance' formulation. The 'worst case scenario' logic of terrorism—the possibility of a nuclear attack, release of a deadly virus—would clearly justify state activity that might be deemed illegitimate in any other scenario. By framing terrorism as a uniquely awful phenomenon, one for which the state can have absolutely no tolerance and for which extreme steps may be necessary, officials can break down walls between the necessary and the formerly illegitimate. As Scheppele (2005) notes, the hypothetical terrorist attack—the one for which there can be no tolerance—has increasingly been invoked as a just cause for torture. Faced with cataclysmic requirements, in this logic, the gloves must come off.

 ## Costs and Risks

But this general sense that the gloves must come off implies that the risk of terrorism produces heavier costs of counterterrorism, costs that will largely be borne by those against whom the state uses its unimpeded power. If the state faces the risk of terrorism, risks that it shares with citizens who might be killed, injured or otherwise harmed by an attack, it places many of the costs for responding onto private individuals: citizens, residents, aliens. Inveighing against the overly cautious and bureaucratic constraints on the use of force by police, Tokyo's right-wing, neo-nationalist governor, Ishihara Shintarō, reportedly told the Tokyo Metropolitan Police Department that they should not wait for a terrorist attack to use their guns: 'Don't hesitate—just shoot, and I'll take responsibility' (*Shūkan Posuto* 2004). The likely targets were not difficult to guess, given both the larger climate of discussions about terrorism as well as Ishihara's strong anti-immigrant stance. Sassa Atsuyuki, the frustrated hero of the Asama

mountain cottage incident, had himself indicated that protecting Japan against terrorism would virtually require profiling ('discrimination' [*sabetsu*] was his word), presumably of Arabs or other ethnic groups presumed to be predominantly Muslim (Sassa 2004).

This is, of course, part of a larger political debate about Japan's place in the world and about foreigners' place in Japan. Just as terrorism has merged with other nativist sentiments in the United States, it played into the hands of conservatives who argued that it simply would make sense to authorize more extensive rights for the state to engage in surveillance, coercion and outright violence against potential threats in its midst. After all, facing possible attacks from such enemies as Osama Bin Laden, with casualties well into the thousands, it would be simply irresponsible for the government to do anything other than to ensure preventive and rapidly reactive measures, focused on those most likely to turn worst-case scenarios into reality.

But turning the discussion into one about the risk posed by foreigners within Japan deflects attention from precisely the lessons one might draw from the history of Japan's counterterrorism policies. Japanese efforts in the 1970s and 1980s were marked by more than an absence of force; they were marked, even cruelly, by the wide acknowledgment that the struggle was a political one as much as a criminal one. The NPA and PSIA engaged the JRA relatively ferociously in large part because it had emerged as the hard left of a political fissure that saw many police relatively far on the right. And, even as the police used tough and coercive measures to intimidate leftists, they did so partly in the understanding that the fight was a larger and broader one in which both the left and the right were implicated. Where leftists were punished for their use of violence and intimidated even when peaceful, those on the right were largely ignored even when engaging in brutal violence and terrorism against left-wing journalists, artists and activists. That is, the struggle against the JRA was ineluctably political.

Discussing the post-9/11 threat of terror as political, however, would call into the question the very decisions that had inspired concerns about terrorism in the first place. Bin Laden and other Islamists had rarely referenced Japan in their diatribes, focusing instead on the United States, a small number of European nations and corrupt local and secular governments ruling parts of the Muslim world. But Prime Minister Koizumi's decision to dispatch troops to Iraq in support of the American invasion of 2003 changed that. This had been a controversial decision, largely because it broke with even the malleable limits on Japanese military action overseas, though it earned American support in the form, for example, of an April 2004 decision to list abductions of Japanese in the 1970s and 1980s as one factor in the designation of North Korea as a 'state sponsor of terrorism' (US Department of State 2004). Indeed, many justifications of Koizumi's initiative focused not on the quality of the widely reviled American decision to invade Iraq, but rather on the necessity of maintaining American support in the face of the North Korean threat (e.g. Hara 2004).

Whether one views the decision as a justified engagement to support an ally or as a relatively predictable step toward full remilitarization (as many critics did), it placed Japan in the middle of a political battle between the United States and a wide if poorly coordinated collection of Muslim opponents, a small number of whom demonstrated their willingness to use violence to make their point. The 2004 Madrid bombing and 2005 London subway bombings were major news stories in Japan not only because they represented unusual attacks in other advanced industrial nations but because they served as signals, particularly to critics of Koizumi, of the potential consequences of alliance with the United States.

This was not the script members of the Koizumi cabinet had written; that one had emphasized Japan's responsibility to engage the world constructively, not its likely targeting by religious fanatics. And if police and security officials, along with nationalist hawks like Ishihara, were able to use the new threat to justify steps—enhanced intelligence capabilities, such as a new division in the National Police Agency, or a quicker trigger finger against foreign threats—that they would long have wanted to take anyway, Koizumi and his Chief Cabinet Officer, Fukuda Yasuo, were keen to distance the state from the potential risks now faced by Japanese at home and abroad. Perhaps unsurprisingly, the consequences would be most evident and immediate for Japanese in Iraq.

In April 2004, three young Japanese who had crossed the border from Syria into Iraq were taken hostage by a previously unknown group of militants, the Saraya al-Mujahedeen. Young leftist Imai Noriaki, non-governmental organization worker Takato Nahoko and freelance photographer Kōriyama Sōichirō were shown on televisions around the world with knives to their throats, a particularly devastating image because of the recent beheadings of the journalist Daniel Pearl in Afghanistan and others taken by militant organizations in theatres of combat. The group threatened to kill all three hostages unless the Japanese government withdrew their roughly 550 Ground Self-Defense Force personnel from Japan. Fukuda, the son of the prime minister who had memorably said that 'a human life is heavier than the earth', announced that the government would do no such thing, emphasizing that it had repeatedly warned Japanese not to enter Iraq. To avoid hostage situations, the Japanese would have to take 'self-responsibility'.[2]

This was, of course, not a new term, having been used by the CPP in its overseas safety videos even before the September 11 attacks. But its deployment by a leading politician, made all the more striking by his relationship to a leader who had famously had the state take responsibility for the lives of hostages in an earlier and costly incident, symbolized something new about the way in which the Japanese state would engage the outside world and ask its citizens to pay some of the costs. Little is known publicly about the behind-the-scenes negotiations between Japanese officials, dispatched to Syria, and Sunni clerics who acted as intermediaries to secure the hostages' release. Whether or not there was a ransom paid, the three were in the end more fortunate than Kōda Shōsei, a young tourist who was killed by his captors six months later in Iraq, and Saitō Akihiko, a security guard employed by a British firm, who died while in captivity though evidently from wounds suffered in the firefight that led to his capture.

Crucially, however, public opinion was at best torn on their ordeal. Whereas Prime Minister Fukuda's decision to pay a ransom and release JRA hostages had been a popular one in Japan, as it secured the lives of more than 150 people, public vitriol, largely from the political right, excoriated Imai, Kōriyama and Takato. Journalist Aonuma Yoichiro argued that the three had been illinformed, selfish young people engaged in a bit of careless self-exploration, which they had meaninglessly legitimated through their connection to NGOs (which, unlike Japanese non-profit organizations (NPO), are unlicensed and relatively unregulated). According to Aonuma, writing in the widely read monthly *Bungei Shunjū*, 'One or two people, reinforcing one another's extreme comments at some bar can declare "we're an NGO" [and become one] even if they don't actually do any real overseas activity' (2004: 154). Scribblers on the enormous Japanese Internet message board, Channel 2, referred to the hostages as 'three idiots' (*san-nin no baka*), and conservatives and even many moderates focused not only on the threat to Koizumi's bold foreign policy stance but also on the unseemly and angry way in which the relatives' families, particularly Takato's brother, aggressively demanded governmental capitulation.

Critics on the left returned fire, seizing precisely on the term 'self-responsibility', to suggest that the government had itself taken a distinctively irresponsible stance in its foreign policy, falling in line with a dangerously reckless ally and endangering the reputation and safety of Japanese abroad (see particularly Tachibana 2004). Although public opinion has never embraced the three former hostages, deep ambivalence about the Iraq War has turned into anxiety and even outright opposition, as Prime Minister Abe found even after the withdrawal of Japanese troops from the occupation. Although Koizumi and others had occasionally justified the dispatch of the troops as a principled stand against terrorism, news media reports focused primarily on the strategic aspect of maintaining a strong alliance with the United States. In this debate, Japan absorbed certain costs for keeping itself safe, including its willingness to extend support to the Americans in an unpopular and probably unwise military campaign. Only infrequently did these strategic calculations reflect the idea that citizens themselves might pay part of the cost themselves, by becoming potential targets of future campaigns of scattered intimidation.

✑ Not Yet Closing Time

If officials sought to minimize the state's responsibility overseas, to privatize risk associated with its foreign policy, they were less able to do so at home. Instead, initiatives in the past few years have enhanced the state role while also suggesting that the costs will be borne primarily by the non-Japanese seen to be the source of threat. These choices, however, have been represented as beneficial not only for the Japanese but also for the foreign

targets of state surveillance. The risk is ubiquitous and deadly, but the costs for countering it can be localized and minimized.

With the deaths of Japanese hostages abroad and the feverish rhetoric of Ishihara and other nationalists about the threat at home, one might have expected a wholesale effort to close Japan to immigrants. After all, Japan's increasing foreign population has been the target not just of the far right's invective but also of even mainstream apprehension. As has been amply noted in the literature on Japan's immigration policies, there is a wide—and politically driven—perception that foreigners are predominantly responsible for increasing crime rates in Japan. The argument is demonstrably false, and specialists like Apichai Shipper (2002, 2005) and Ryoko Yamamoto (2004) have called attention both to the political incentives driving the anti-foreigner critique and to the social movements against the labeling. Indeed, particularly at a time that anti-immigrant movements have been on the rise in Europe and North America (though with limited policy effectiveness), it might be surprising that the Japanese government has not gone further to expel foreigners or to prevent their entry.

There are, of course, political costs to anti-foreign initiatives. With an increasingly large and visible foreign presence in Japan, immigrants have not only critics but also defenders and friends, and indeed immigrants themselves have engaged the Japanese NGO community to seek common cause when possible (Shipper 2008). Similarly, anti-foreign sentiments run the risk of antagonizing neighboring governments, whose citizens, many sending remittances home, are among those living and working in Japan.

But it is the economic logic, now overwhelming, that militates against any reduction in Japan's foreign population. With Japan's declining birth rate and long life expectancies, there are now cottage industries not just of demographers assessing the likely shortfalls in Japan's pension system but also of pundits and writers whose doomsday scenarios for Japan seem increasingly persuasive in light of the fact that, if current trends hold for hundreds of years, there will be virtually no Japanese left. Geoffrey McNicoll (2005: 59) points out that by the middle of the twenty-first century, Japan's population will likely drop by about 15 per cent, an alarming decline that is even more catastrophic when considered alongside the very advanced age of many Japanese. Schoppa puts it even more starkly, reflecting on reports in 1997 and 2002 that determined that Japan's population drop was likely not a short-term aberration but rather a longer trend: 'the government will find it impossible to pay for the full costs of the health care and pensions of aging baby boomers if its population of working-age and taxpaying adults shrinks this fast' (2006: 153).

In part for these reasons, it was not altogether surprising to see the previous LDP government taking steps to make it easier for Japan to accept immigrants, even proposing in 2008 that Japan plan to accept ten million foreigners within fifty years. An alliance of eighty LDP members, calling themselves the 'league for promoting exchange with foreigners' (*gaikokujinzai kōryū suishin giin renmei*), and headed by veteran lawmaker Nakagawa Hidenao, recommended the construction of an 'Immigration Agency' (*iminchō*) as well as the creation of employment environments in which foreigners might thrive alongside Japanese (*NB Online* 2008). The proposal was formally conveyed to Prime Minister Fukuda by the LDP's Project Team for 'The Road to a Japanese-Style Immigrant Nation' (*Nihon gata imin kokka e no michi*), embedded in its 'Headquarters for State Strategy' (Sankei 2008). This was not the work of a small number of renegade party members; it instead bore all the hallmarks of a carefully orchestrated party stance on an issue that is probably more intuitively plausible and less controversial now than it ever would have been in the past. But a recognition that Japan needs more immigrants, and a commitment to attract more, should not be taken as a kind of unproblematic shift toward a completely open society. Even the ostensibly more progressive Democratic Party of Japan government has had to pull back on its proposal to extend suffrage in local elections to permanent foreign residents due in part to tensions within the party and with conservative coalition allies (*Straits Times* 2010).

In his 2006 book *Utsukushii kuni e* (Toward a beautiful country), then-presumptive Prime Minister Abe Shinzō argued that Japan needed immigrants: the right kind of immigrants. Very much on the hawkish side of the party, with allies and supporters who might include the most aggressive and voluble critics of non-Japanese, Abe was in some ways taking a chance in devoting a chapter of his book to his simultaneous desires for a new nationalism and for an increase in Japan's migrant population. He tips his hand by praising the soccer players of Arab and North African descent who contributed to France's World Cup victory in 1998; the players had, of course, been not merely athletes playing a game, but patriots doing their duty. But he goes further in an odd section on the Clint Eastwood film *Million Dollar Baby*, not usually seen as a treatise on immigration policy. Abe reflects on the efforts by Clint Eastwood, a boxing club manager, to study Gaelic, which Abe argues—not implausibly—is an effort to engage his Irish-American roots. Citing the critic Matsumoto Ken'ichi, Abe makes a memorable comment about American immigration history: 'Chinese, Koreans, and Hispanics came to the United States believing it to be the "ideal country", but only the Irish immigrants came determined to build the "ideal country"' (2006: 90).

Abe's comment should probably be seen less as an effort to insult Chinese, Koreans and Latinos than as an uncomfortable acceptance of dominant myths, popular among a subset of conservative Irish-Americans, regarding some transcendental meaning of their ancestors' arrival in the United States. Both in the Irish-American version and Abe's version, there may be some troubling racial subtexts regarding who is and who is not an ideal immigrant. But Abe's larger point is about the need for a strong state and for patriots who will support it, and in doing so he moves away from more frequent references to the risks that foreigners pose. Needless to say, his openness to foreigners, and increasingly that of other members of both the LDP and the Democratic Party of Japan (DPJ), cannot be used to suggest that xenophobia has disappeared in Japan any more than the presidency of Barack Obama proves that there is no more racism in the United States. But it suggests a transformation in civic goals, one that recognizes that the risks to Japan of having not enough people are greater and more certain than are the risks associated with allowing them in. Where Japanese safety and internal harmony have seemed to rest with the 'Japaneseness' of the country's people, they now seem to rest in a willingness, including among non-Japanese, to love and to support the Japanese state.

Representing Costs

Hatoyama Kunio's 2007 press conference was a tense one even before he seemed to play a game of 'six degrees of separation' between himself and Osama Bin Laden. In front of an audience of foreign journalists, most of whom travel in and out of Japan with some regularity, he had to defend the necessity of a new immigration policy regime requiring fingerprints, photographs and almost certainly long lines; for these journalists, there was presumably little apprehension that fingerprinting might somehow implicate them in crimes or terrorist plots, but that the procedures themselves would make long trips in and out of Japan even more onerous.

Anticipating some resistance from Japan's foreign residents and from potential tourists, the Japanese government released a five-and-a-half minute video, primarily on the cabinet office's website, providing information in English translation. 'Landing examination procedures for Japan are changing!' has a female Japanese announcer, her voice dubbed into English, explaining the new rules to three initially skeptical foreigners: a white man who appears to be speaking in American-accented English, a black woman who also speaks English with an American accent, and an Asian man whose non-Western accent is left unmolested, although, strangely enough, his English seems somewhat more natural than does that of either of the other actors.[3]

Looking directly into the camera, and apparently asking questions directly to the attractive Japanese female announcer, the white man asks, 'I'd like to know the reason why I have to have my fingerprints taken'. The black woman follows with, 'What are the photographs going to be used for?' and the Asian man says, 'I'd like to hear a more detailed explanation'. The announcer pauses, nods, and responds: 'Yes. Fingerprinting and photographing foreigners not only contributes to the security of Japan but is also for the protection of visiting foreigners so that they can enjoy a safe stay'.

The upbeat music fades out, and a slightly spacier synthesizer accompanies a montage of still shots from famous recent terrorist attacks, with the announcer rattling off the 9/11 attacks, the 2001 bombing in Bali, the 2003 Morocco bombings, the Madrid train bombings of 2004 and finally the 2005 bombings in London. After stressing the shocking nature of mass-casualty terrorism, the announcer continues:

> Based on such international situations [*sic*], the obligation to be fingerprinted and take photographs at the landing examination has been introduced.

> Many foreigners visit Japan every year, and the numbers continue to rise. As a country which has been promoting tourism, Japan's need to protect itself while preserving the safety of visitors from terrorism is extremely important. Keeping this in mind, in order to avoid the entry of terrorists into Japan, it has been decided to impose fingerprinting and photography at immigration.

> Providing personal identification information, such as fingerprints and photographs, confirm that the person applying for entry and the owner of the passport are the same individual, while comparisons with suspicious persons lists can be performed quickly and accurately, assisting in the identification of terrorists and the prevention of terrorism. Also, this information can be used in order to identify illegal entry repeat offenders and is thought to become an effective tool against illegal immigrants and foreign criminals in Japan.

Gesticulating widely and enunciating emphatically, the white man taps his chest and says, 'So, the security of Japan is safety for us'. The black woman smiles and says calmly, 'If so, I do not mind having my fingerprints and photographs taken'.

But the Asian man breaks the mood by saying, 'Hold on. What happens if I refuse to be fingerprinted or have my picture taken?' The announcer pauses, nods, and says, 'If you refuse to be fingerprinted or photographed, you will be denied entry into Japan and ordered to leave'.

Even leaving aside the somewhat troubling racial coding—of reasonable and law-abiding Westerners, of shifty and argumentative Asians—the video is fascinating for its representation of risks and remedies. Foreigners too are threatened, not just Japanese, so they too will benefit from the apprehension of their biometric information and likely pooling of it with similar information held by other gov- ernments. Although the video presents no evidence that these measures actually will work (and such evidence would be hard to find, given the much-lamented difficulty of evaluating the effectiveness of counterterrorism measures), the technical fixes are represented as necessary and productive. After all, a benevolent and responsible state, with no goal other than the safety and security of its citizens and visitors, simply wants to record, store and collect personal information. In essence, the video asks: who, other than a particularly suspicious fellow, would object?

Hatoyama Revisited

Justice Minister Hatoyama's admission to being only two degrees removed from Al Qaeda was the beginning of a busy week for him. Later in the week, in the middle of a Diet subcommittee debate about the need for enhanced intelligence capabilities for Japan, Hatoyama admitted to having provided intelligence to the Pentagon during his early 1970s stint as personal secretary to Prime Minister Tanaka Kakuei. He had apparently been one of the Pentagon's cheaper spies, having earned only a lunch once a month or so to fill them in on events in Tokyo, occasionally ordering eel and sometimes tempura.[4] During a speech in Fukuoka later that week, he announced that '[i]f I were to tell you the truth, the real truth, everyone would be surprised and the media would go crazy. I'm telling you that what's really frightening is that terrorists are hanging around Japan utterly without fear right now' (*Asahi Shimbun*, 4 November 2007).

These remarks may suggest that Hatoyama was having a bit of a 'senior episode', but they also fit well into larger concerns in the discussion of terrorism as a policy problem. In this view, the state needs information and such information is basically apolitical, tied only to security and safety. The problem was not that Japan had allied itself with a country whose military agents were bribing foreign officials for information, but rather that Japan itself had not developed this capability too. And the consequences of the country's lax approach included the many pockets of dangerous militants waiting for the opportunity to strike. Of course, Hatoyama's comments proved controversial precisely because many in his Japanese audience recognized

something troubling in the bald admission of selling off presumably confidential information for a tempura lunch and in his alarmist rhetoric regarding the imminent 9/11-style attack awaiting Japan.

These controversies, however, provide a glimpse of the terrain facing Japanese security officials charged with responding to the threat of terrorism. Some risks have presumably been generated in part by foreign policies and, if the government does not seek to alter these policies, one plausible response is to walk away, by privatizing the risk and limiting the state's exposure to its costs. In other cases, risks provide legitimate opportunities to enhance state authority, and new representational strategies accompany these new initiatives, seeking both to localize costs and to justify them. This is not an indictment of Japan's recent counterterrorism steps; its new security steps *vis-à-vis* foreigners are hardly draconian and they may indeed make people safer, though we will likely never know if they do so. The stockpiled information about foreign residents may also be used purely in beneficial ways and not to scapegoat or corral people for merely overstaying their visas while trying to make ends meet. But the representative strategies surrounding these efforts involve specific ways of considering risk and of allocating costs. They call attention to the difficulties inherent in de-politicizing an ineluctably political problem like terrorism, suggesting that solutions rest in increasing the capacity of a generous and benevolent state, one that can defend even its own aggrandizement, not in rethinking the purposes to which this impersonal organization and its persistent machinery have been directed.

Notes

1. I draw the material in the following paragraphs from Leheny (2006: ch. 5). For other studies of Japanese terrorism and counterterrorism, see Katzenstein (1996, 2002), Hughes (1998, 2007) and Leheny (2010).

2. Hook and Takeda (2007) discuss the concept of 'self-responsibility' more broadly in Japanese politics. Leheny (2006) focuses on it in the context of the Iraq War.

3. The full video can be seen at Japanese Government Internet TV. Available online at http://nettv. gov-online.go.jp/eng/prg/prg1203.html (accessed 22 June 2008).

4. The announcement was reported in numerous sources including Livedoor's news page on 2 November 2007. Available online at http://news.livedoor.com/article/detail/3371859/ (accessed 13 December 2009).

 # References

Abe Shinzō (2006) *Utsukushii kuni e* (Toward a beautiful country), Bungei Shunju.

Aonuma, Yōichirō (2004) 'Iraku no chūshin de ai o sakebu hitotachi' (The people who shriek about love and Iraq), *Bungei Shunjū* June: 148–57.

Asahi Shimbun (2007) 'Hatoyama hōsō, "nihon ni terorisuto ga uroro" to hatsugen' (Justice Minister Hatoyama: there are terrorists hanging around in Japan), 4 November, http://www.asahi.com/politics/update/1104/TKY200711030239.html (accessed 6 November 2007).

The Economist (2007) 'A friend of a friend bombed Bali', 31 October, http://www.economist.com/world/asia/displaystory.cfm?story id=10053265 (accessed 6 November 2007).

Friman, H. Richard, Katzenstein, Peter J., Leheny, David and Okawara, Nobuo (2006) 'Immovable object? Japan's security policy in East Asia', in Peter J. Katzenstein and Shiraishi, Takashi (eds) *Beyond Japan: The Dynamics of East Asian Regionalism*, Cornell University Press.

Hara, Osamu (2004) 'Jieitai iraku hakken wa tai kitachō sen seisaku de mo aru' (The dispatch of the self defense forces to Iraq is also a policy towards North Korea), *Seiron* June: 84–95.

Hook, Glenn D. and Takeda, Hiroko (2007) '"Self-responsibility" and the nature of the Japanese state: risk through the looking glass', *Journal of Japanese Studies* 33(1): 93–123.

Hubbard, R. Glenn, Deal, Bruce and Hess, Peter (2005). 'The economic effects of federal participation in terrorism risk', *Risk Management & Insurance Review* 8(2): 177–209.

Hughes, Christopher W. (1998) 'Japan's Aum Shinrikyo, the changing nature of terrorism, and the post-Cold War security agenda', *Pacifica Review* 10(1): 39–60.

——— (2007) 'Not quite the "Great Britain of the Far East": Japan's security, the US-Japan alliance, and the "war on terror"', *Cambridge Review of International Affairs* 20(2): 325–38.

Itabashi, Isao and Ogawara, Masamichi, with Leheny, David (2002) 'Japan', in Yonah Alexander (ed.) *Combating Terrorism: Strategies of Ten Countries*, University of Michigan Press, pp. 337–73.

Katzenstein, Peter J. (1996) *Cultural Norms and National Security: Police and Military in Postwar Japan*, Cornell University Press.

——— (2002) 'September 11 in comparative perspective: the antiterrorism campaigns of Germany and Japan', *Dialogue-IO* Spring: 45–56.

Leheny, David (2006) *Think Global, Fear Local: Sex, Violence, and Anxiety in Contemporary Japan*, Cornell University Press.

——— (2010) 'Remaking counterterrorism: Japan's preparations for unconventional attacks', in Kay Warren and David Leheny (eds) *Japanese Aid and the Construction of Global Development: Inescapable Solutions*, Abingdon: Routledge, pp. 252–69.

McNicoll, Geoffrey (2005) 'Demographic future of East Asian regional integration', in T. J. Pempel (ed.) *Remapping East Asia: The Construction of a Region*, Cornell University Press, pp. 54–75.

NB (Nikkei Business) Online (2008) 'Jimintō "imin sen-man-nin ukeire" no jitsugensei' (Operationalizing the LDP's 'Ten Million Immigrants' plan), 19 June, http://business.nikkeibp.co.jp/article/topics/20080617/162440/ (accessed 22 June 2008).

Power, Michael (2004) *The Risk Management of Everything: Rethinking the Politics of Uncertainty*, London: Demos.

Sankei (Sankei News Service) (2008) 'Jimintō kokka senryaku honbu, imin 1000 man nin o shushō ni teigen' (LDP Headquarters for State Strategy delivers proposal for 10 million immigrants to the Prime Minister), 20 June, http://sankei.jp.msn.com/politics/policy/080620/plc0806201119003-n1.htm (accessed 22 June 2008).

Sassa, Atsuyuki (1996) *Rengō Sekigun 'Asama Sansō' Jiken* (The Japanese Red Army Asama Mountain Cottage Incident), Bungei Shunjū.

——— (2004) 'Chian o keishi shitekita tsuke o ima harawasareteiru' (We are now paying the price for having neglected public order), *Chūō Kōron* 119(2): 86–91.

Schoppa, Leonard (2006) *Race for the Exits: The Unraveling of Japan's System of Social Protection*, Cornell University Press.

Scheppele, Kim Lane (2005), 'Hypothetical torture in the "war on terrorism"', *Journal of National Security Law & Policy* 1(2): 285–340.

Shipper, Apichai (2002) 'Political construction of foreign workers in Japan', *Critical Asian Studies* 34(1): 41–68.

——— (2005) 'Criminals or victims? The politics of illegal foreigners in Japan', *Journal of Japanese Studies* 31(2): 299–327.

——— (2008) *Fighting for Foreigners: Immigration and its Impact on Japanese Democracy*, Cornell University Press.

Shūkan Posuto (2004) 'Ishihara tō chiji "mayūōwazu utte! Ore ga sekikin o toru"' (Governor Ishihara: 'Don't wait, just shoot! I'll take responsibility'), 16 April: 28–32.

Steinhoff, Patricia G. (1989) 'Hijackers, bombers, and bank robbers: managerial style in the Japanese Red Army', *Journal of Japanese Studies* 48(4): 724–40.

Straits Times (2010) 'Backlash over foreign vote bid', March 22, http://www.straitstimes.com/BreakingNews/Asia/Story/STIStory505015.html (accessed 2 April 2010).

Tachibana, Takashi (2004) 'Koizumi iraku hahei "kurutta shinario"' (An 'insane scenario' for Koizumi's troop dispatch to Iraq), *Gendai* June: 28–41.

United States Department of State Office of the Coordinator for Counterterrorism (2004) *Patterns of Global Terrorism 2003*, Washington: State Department.

Yamamoto, Ryoko (2004) 'Alien attack? The construction of foreign criminality in contemporary Japan', *Japanstudien* 16: 27–57.

David Leheny is the Henry Wendt III '55 Professor of East Asian Studies at Princeton University. His research interests are in Japanese politics and the international relations of the Asia-Pacific. Recent publications include *The Rules of Play: National Identity and the Shaping of Japanese Leisure* (Cornell University Press, 2003), *Think Global, Fear Local: Sex, Violence, and Anxiety in Contemporary Japan* (Cornell University Press, 2006) and (co-edited with Kay Warren) *Japanese Aid and the Construction of Global Development: Inescapable Solutions* (Routledge, 2010). He may be contacted at: dleheny@princeton.edu.

REVIEW QUESTIONS

1. Do Japan's counterterrorism measures seem reasonable?

2. Would you feel comfortable with the rules for entering Japan as a tourist?

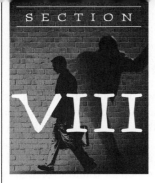

SECTION

VIII

Southwest Asia

Learning Objectives

At the end of the section, students will be able to:

- Describe the major terrorist groups in the region and their associated attacks from the early 1970s to the present.
- Describe the role and importance of Pakistan in promoting religious extremism in the region.
- Describe the evolution and current state of the Tamil Tigers in Sri Lanka.
- Describe and assess the impact of Pakistan's ISI on terrorism in the region.
- Discuss the tensions between India and Pakistan over Kashmir and the impact that conflict has had on terrorism.
- Assess the threat posed by al Qaeda and its affiliates after the 2001 invasion of Afghanistan.

Section Highlights

- Afghanistan and the Taliban
- Pakistan
- India
- Sri Lanka
- Summary
- Readings

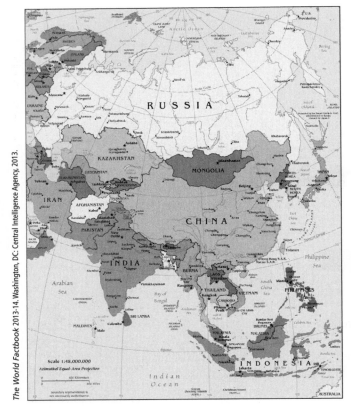

▲　Map of Asia

Introduction

The areas known as South and Central Asia (as distinct from East Asia) sit between Europe and the Middle East to the west and China and Russia to the north and east. The Bureau of South and Central Asian Affairs at the U.S. Department of State includes in this region the countries of Afghanistan, Bangladesh, Bhutan, India, Kazakhstan, Kyrgyzstan, Maldives, Nepal, Pakistan, Sri Lanka, Tajikistan, Turkmenistan, and Uzbekistan. Over the millennia, parts of the region have been ruled by Western European powers, the Ottoman Empire, the Mongols, and other empires seeking control of the vital trade routes between West and East. Many of the countries in Central Asia achieved their independence relatively recently, with the fall of the Soviet Union in the early 1990s.

This section will focus on the nations of **Afghanistan, Pakistan, India**, and **Sri Lanka**. These countries are either currently a focus of U.S. counterterrorism efforts or have had long-standing terrorism experiences that are instructive for modern-day efforts. In particular, this section will highlight the **Taliban** operating in Afghanistan and Pakistan, Islamic fundamentalist groups operating out of the disputed **Kashmir** territory between India and Pakistan, separatist terrorism in India, and the Tamil Tigers separatist group in Sri Lanka.

Afghanistan and the Taliban

Covering 652,230 square miles, Afghanistan is only slightly smaller than Texas and is home to approximately 3.4 million people (*CIA World Factbook*, n.d.). Identifying itself as the Islamic Republic of Afghanistan, it is divided into 34 different provinces, many in rural, inaccessible areas that are difficult to govern and where life is tied to traditions and customs that have not changed in millennia. At least eight different ethnicities are represented, with the two largest being Pashtun (42% of the population) and Tajik (27% of the population[1] (*CIA World Factbook*, n.d.). The country is overwhelmingly Muslim with 80% of the population following the Sunni tradition and 19% identified as Shia. Persian and Dari serve as the two official languages (*CIA World Factbook*, n.d.).

The nation of Afghanistan dates back to 1747 when a tribal leader named Ahmad Shan Durani was able to unite a number of the scattered and fiercely independent Pashtun tribes in the region situated between what are now Pakistan and Iran (*CIA World Factbook*, n.d.). After several failed efforts by the Europeans (especially the British) to control this uncontrollable area, Afghanistan once again became its own state when it gained its

---

[1]The remainder of the population is divided among Hazara (9%), Uzbek (9%), Aimak (4%), Turkmen (3%), Baloch (2%) and other (4%) (*CIA World Factbook*, n.d.).

independence from the "control" of the British in 1919.

In 1978, a communist-led coup won control of the Afghan government. However, gaining control and keeping control were two entirely different prospects. In many ways, the idea of a successful centralized communist government in Afghanistan was doomed from the start (Scheuer, 2004). Still, the then Soviet Union gave it the old college try, invading Afghanistan in an effort to prop up the failing Afghan communists. The result was a disaster for the Soviet Union. With help from Muslims around the world, the Afghan tribes created numerous insurgent guerrilla groups—known as *mujahedeen*—to harass and drive out the Soviets. The mujahedeen believed they were answering a call to jihad against the godless enemies of Islam. These groups were often funded and armed by a Cold War driven

▲   Holy war against communist invaders. Afghan Mujahideen during their *jihad* against occupying Soviet troops

Afghan Mujahideen

United States, which saw a communist-run Afghanistan as the worst possible outcome. After a decade of fighting, millions of dollars spent, and untold lives lost, the Soviets left Afghanistan. Drastically weakened by its long mis-adventure, the Soviet Union fell apart shortly thereafter.

Secure in its success in once again defending the world from communism, the United States also left Afghanistan, leaving chaos, a government vacuum, and many of the mujahedeen in its wake. Not surprisingly, a vicious civil war soon followed with alliances and conflicts among the mujahedeen, as well as tribal militias, ebbing and flowing as territory was held and lost. One of the largest and most effective of the mujahedeen organizations was the Taliban, led by **Mullah Omar.** A Sunni Islamic fundamentalist group founded in the early 1990s, the Taliban sought to unite the tribes of Afghanistan under the rule of a strict interpretation of Islamic law and to remove all secular influence from the country. By 1994, the Taliban controlled almost all of Afghanistan. "By September 1996, the Taliban had captured [the capital city of] Kabul, killed the country's president, and established the Islamic Emirate of Afghanistan. The Taliban's first move was to institute a strict interpretation of Qur'anic instruction and jurisprudence" (National Counterterrorism Center, 2012). While many in Afghanistan welcomed (or at least tolerated) the Taliban as the only group that had managed to bring significant order to the chaos, its strictures were almost a death sentence for women.

The same year the Taliban took control of Kabul, it welcomed back an old friend—bin Laden and his al Qaeda followers who had been expelled from the Sudan. As noted in the previous section, the Taliban provided both a safe haven and an operational base from which al Qaeda launched its war against the United States and the West (see also Scheuer, 2004; Wright, 2006). This turned out to be a dangerous alliance for the Taliban. In response to al Qaeda's destruction of the World Trade Center Towers in New York City and the attack on the Pentagon in Washington, D.C., the United States demanded that the Taliban turn over bin Laden or face dire consequences. Omar refused, and in October 2011 the United States threw its support behind a Taliban-opposition group called the **Northern Alliance**, invaded Afghanistan, and toppled the Taliban government. Ironically, at the time of the 9/11 attacks, the Northern Alliance was on its last legs, having lost its leader to a Taliban assassination only a few days before. Argues one author, had the attacks of 9/11 not taken place, and had the Taliban not sheltered bin Laden, the Northern Alliance likely would not have lasted, giving the Taliban an even firmer hold over the country (Scheuer, 2004).

Still, as of the writing of this book, the Taliban was down but definitely not out. It was replaced by a democratically elected government led by President **Hamid Karzai**. However, the Karzai government had its problems from the beginning. Afghanistan has always been incredibly difficult if not impossible to govern by a centralized administration (Scheuer, 2004) and the Karzai government was seen as corrupt and incompetent with a limited ability to provide essential government services—especially security—to even the more urban parts of the country without significant help and funding from U.S. military and diplomatic forces. In addition, argues former CIA bin Laden unit chief Michael Scheuer, the Karzai government had almost no chance of winning significant support among the Afghan tribal culture. In his book, *Imperial Hubris*, Scheuer describes Karzai as follows:

> A genuinely decent, courageous, and intelligent man, Karzai had nonetheless absented himself from the fight against the Soviets, and also from the one against the Taliban, until he jumped in on the side of the Americans and their overwhelmingly powerful military. With no Islamist credentials and minimal tribal support, the India-educated Karzai was and is a man clearly adept and comfortable hobnobbing with U.S. and British elites, but far less so at chewing sinewy goat taken by hand from a common bowl with an assembly of grimy-fingered Islamist insurgent and tribal leaders and their field commanders. (p. 39)

Scheuer goes on to argue that Karzai's long time in the West and the lack of any fighting or Islamic credentials left his government "missing every component that might have given it a slim chance to survive without long-term propping up by non-Islamic, foreign bayonets" (p. 40). Under the new Afghani constitution, Karzai was barred from running for another term in the 2014 elections. The presidency was won by Ashraf Ghani but the election results were hotly contested and it took eight months—and considerable assistance in the negotiations from the United States—for a government to be formed with Ghani as president and his opponent in the elections—Abdullah Abdullah—holding powers similar to those of a prime minister. The Taliban took the opportunity presented by the uncertainty following the elections to pursue one of its most successful fighting seasons since the war began (Nordland, 2014).

Almost immediately after its initial defeat, the Taliban took shelter in the vast and isolated mountain region that forms the Afghan-Pakistan border and began to regroup. In addition to rebuilding its forces in Afghanistan, it built networks with Taliban groups in Pakistan that both aided in the fight in Afghanistan and waged its own fight against Islamabad (see subsection below). Between 2001 and 2011, it waged an increasing insurgent and terrorist campaign against U.S. forces, the Afghan government, NATO forces in Afghanistan, and any who supported them. One of the strongest examples of groups that overlap the insurgent/terrorist distinction, the Taliban regularly engaged in both strategies. In addition to consistent efforts against U.S. military forces, Taliban targets included the UN assistance mission, NGOs, foreign diplomatic missions, Afghan civilians, and government officials (U.S. Department of State, Country Report, 2011).

Anne Stenersen (2009) suggests a distinction between what she calls the "various layers" of the Afghan Taliban and sets out the distinction as follows:

> The Afghan Taliban leadership (Mullah Omar and his *shura* council, also referred to as the Quetta *shura*) gives general directions and speaks on behalf of the organization, while local commanders in Afghanistan and in the tribal areas of Pakistan carry out militant activities in the Taliban's name, often with a high degree of autonomy. Foreign militant networks such as the "Pakistani Taliban" and al-Qa'ida support the Afghan Taliban insurgency, but they typically have wider agendas and carry out attacks in their own name. (pp. 1–2, fn. 3)

Despite the announced intentions for the United States to pull out its forces and turn security over to the Afghan Army and police forces by 2014, attacks by the Taliban had steadily increased since early in the U.S. invasion. According to the National Consortium for the Study of Terrorism and Responses to Terrorism (START) at the University of Maryland, the monthly number of terrorist attacks increased continuously between 2001 and 2011 (START, June 28, 2011, p.1). As of June 2011, the Afghan Taliban accounted for 57.7% of all attacks in Afghanistan since 1970 and the lethality of its attacks had increased from an average of one victim per suicide attack in 2001 to six victims per suicide attack in 2011. Targets in 2010 included Indian guesthouses, schools, the Kunduz Office of Development Alternatives, a medical mission team, Afghan National Army officers, and the Governor of Kunduz province (who was assassinated)

(U.S. Department of State, Country Reports, 2011). In 2011, in concert with the **Haqqani network** (a Pakistani Taliban group), the Taliban killed 11 people in an attack on the Intercontinental Hotel in Kabul, and, in a separate attack, assassinated the brother of President Karzai. As a popular spot for Western visitors, the hotel was always an attractive target for the Taliban but at the time of the attack, this was enhanced "by the presence of a conference to discuss increasing areas of responsibility for Afghan security forces" (START, June 28, 2011, p. 1).

The Afghan Taliban uses a mixture of guerrilla warfare and terrorist tactics in its attacks and has no compunction about killing both foreign and Afghan civilians. But will, or can, it target foreign civilians outside of Afghanistan as do other Islamist fundamentalist terrorist groups? This is the question Stenersen (2009) examines in her article for the *CTC Sentinel,* and she comes to the conclusion that such a broadening of its geographic area of operation is unlikely. In her article, Stenersen starts by noting that two Afghan Taliban leaders in particular—the brothers Mullah and Mansour Dadullah—are the ones most explicitly linked to threats against Western targets outside of Afghanistan and that that link comes mostly from comments made in a June 2007 video. However, she goes on to note that "in the two years after the video was issued, no firm links have been established between arrested terrorist suspects in Western countries and Dadullah or the Afghan Taliban" (p. 3). In addition, in contrast to the Pakistan Taliban,[2] no members of the Afghanistan Taliban have been directly associated with plots to launch attacks in Western countries. Stenersen suggests two reasons for the Afghan Taliban's failure to pursue Dadullah's threats. First, the Afghan Taliban may not have the capability to do so given that it lacks the broad, multifaceted networks that groups such as al Qaeda can boast. To the extent that Afghan Taliban support networks exist, they are found mostly in Pakistan and the Persian Gulf. Second, Stenersen argues that there is little incentive for the Afghan Taliban to move out of Afghanistan and every incentive to stay put. "Although not stated directly, the Afghan Taliban leadership is probably reluctant to carry out activities that would increase the pressure on its sanctuaries in Pakistan" (2009, p. 4).

 Pakistan

An independent country since 1948, Pakistan is a federal republic with four provinces, one territory, and one capital territory. Covering 796,025 square kilometers, it is approximately twice the size of California (*CIA World Factbook*, n.d.). Of particular strategic value in the region, it encompasses two passes—the Khyber Pass and the Bolan Pass—that served as traditional invasion routes between Central Asia and the Indian subcontinent.

Pakistan has a population of approximately 190.3 million people, making it the world's sixth most populated country (*CIA World Factbook*, n.d.). Just over 44% of the population is Punjabi, while approximately 15% of the population is Pashtun,[3] and another 14% is Sindhi (*CIA World Factbook*, n.d.).[4] Ninety-five percent of the country is Muslim and 75% is identified as following the Sunni faith. Islam is the official religion and Urdu is the official language. Interestingly, according to the *CIA World Factbook*, only 8% of the Pakistani population is identified as speaking Urdu. Forty-eight percent speak Punjabi and 12% speak Sindhi.

The Pakistan constitution provides for a president and a prime minister with a bicameral parliament and a supreme court. In reality, the military has been the most consistent power holder in the nation's history as it has struggled between civilian and military governments. Its legal system combines English common law with personal legal codes for Muslims, Christians, and Hindus (*CIA World Factbook*, n.d.).

Pakistan struggles significantly with both insurgency and terrorism. In 2010, the U.S. State Department noted that the violence "resulted from both political and sectarian conflicts throughout the country, with terrorist

---

[2]The Pakistan Taliban has made explicit threats to carry out attacks in the West and has been linked to a 2008 plot in Barcelona, Spain, to attack the city's transportation system (Stenersen, 2009, p. 3).

[3]The Pakistani Pashtuns reside mostly in the areas bordering Afghanistan where they provided safe-haven and fighters for the Pashtun Afghan Taliban following the U.S. invasion.

[4]The remainder of the Pakistani population is divided among at least four other ethnic groups: Sariaki (8.38%), Muhajir (7.57%), Balochi (3.57%), and other (6.28%) (*CIA World Factbook*, n.d.).

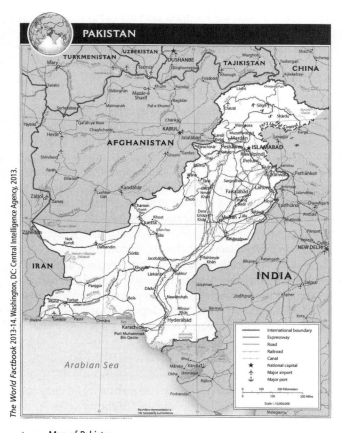

▲   Map of Pakistan

incidents occurring in every province. While government authorities arrested many alleged perpetrators of terrorist violence, few convictions resulted" (U.S. Department of State, Country Reports, 2011). The problem has been exacerbated by the arrival of Afghan refugees and the Afghan Taliban in the most remote areas of North Waziristan (on Pakistan's western border with Afghanistan). This **Federally Administered Tribal Area** (FATA), as well as the Khyber Pakhtunkhwa province, tends to be the central base of operations for most of the violence and the groups responsible. Complicating matters are the continuous difficulties in distinguishing between insurgencies and terrorist groups and the effect that has on developing effective counterterrorism and counterinsurgency policies. A myriad of groups target both noncombatant and military targets. In addition, not all attacks will have claims of responsibility while other attacks will involve multiple claimants. Adding insult to injury, by 2010 al Qaeda and the Taliban had established a strong enough presence in the country that the Pakistani government was unable to challenge either group directly. Instead, the government was left pursuing a policy of trying to "contain" the activity and the groups in "known areas of activity" (U.S. Department of State, Country Reports, 2011).

The Formation of Pakistan: Struggling for an Identity

Pakistan was founded by **Mohammed Ali Jannah** as part of the breakup of Britain's Indian Empire. The mainly Hindu part of the subcontinent formed what is now India, while the majority Muslim northwest corner became the country of Pakistan (Baker, 2011). From the beginning, Pakistan wrestled with whether it was to be a secular or Muslim state. Unfortunately, unlike India's first leader—the secular-minded Jawaharlal Nehru—who governed India for several decades and was able to realize his vision for India, Jannah died the year after Pakistan was founded (Baker, 2011).

Still, despite alternating between weak civilian governments and military dictatorships, Pakistan stayed reasonably secular until 1977. That year, General **Zia ul-Haq** took power and immediately instituted Islamic law and revised the educational curriculum "in an effort to promote nationalism and an Islamic identity" (Baker, 2011). This rapid and all-encompassing move from the secular to the religious might have sparked a rebellion among Pakistan's secular elite but for the Soviet invasion of Afghanistan; Pakistan "rallied in support of its neighbor, out of fear it might be next" (Baker, 2011).

Both before and since General ul-Haq's government, Pakistan struggled between a civilian-led democracy or military-led dictatorships of varying intensity. Yet with the exception of General ul-Haq, both civilian and military governments have remained mostly secular, a somewhat reassuring situation for those who view Pakistan's nuclear weapons with such trepidation. The civilian governments have traditionally been much weaker than the military. In turn, the military has been very good at keeping political parties at odds with each other (Baker, 2011). The military is also very adept at using anti-American sentiment as a diversion from Pakistan's own internal problems and difficult questions regarding the military's willingness or ability to deal with them (Baker, 2011).

The Fight Over Kashmir: The Most Dangerous Place in the World

The main, dominant, even overwhelming theme running through Indian and Pakistani politics—indeed throughout most of the Southwest Asia region—is the dispute between these two countries over the area of Kashmir. As British rule over India came to an end, the agreement among the parties involved was to partition the former imperial territory between the Hindu-majority areas (India) and the Muslim-majority areas (Pakistan). A number of provinces were allowed to choose whether to join the more-secular India or the more religious and Islamic Pakistan. As it happened, the Hindu leader of the Muslim majority provinces of Kashmir and Jammu had closer ties to the British Viceroy, the Earl of Mountbatten, and India's Nehru than to Pakistan and its leader and aligned his province accordingly. Not surprisingly, "the Muslim majority of Kashmir has been unwilling to accede to Indian rule over Kashmir and has been involved in a steady escalation of resistance to India" ever since (Spindlove & Simonsen, 2010, p. 351).

In the years since gaining their independence, India and Pakistan have fought at least three conventional wars and engaged in uncountable other violent confrontations over Kashmir. Numerous terrorist groups on both sides operate in the region, although until recently, the violence stayed within the disputed area or, at most, spilled into either India or Pakistan. Adding to the already engrained hatred between the two countries, Indian military support for Hindu separatists in East Pakistan in 1971 led to the creation of the nation of Bangladesh. "That humiliation informs Pakistan's actions still and its belief that India constitutes an 'existential threat' capable of destabilizing and further dismembering Pakistan" (Baker, 2011). The loss of East Pakistan due to Indian interference also led to Pakistan's nuclear weapons program; it is the first Muslim country to join the "nuclear" club (a status that generates a certain pride and identity for the nation). From the Pakistani perspective, it is also a matter of survival. "Islamabad regards nuclear weapons and their delivery systems as essential to offsetting its conventional inferiority against India and maintaining the South Asian balance of power" (Nuclear Threat Initiative, 2012, p. 1).

In their RAND report in 2009, Fair et al. note Pakistan's stubborn inability or refusal to rethink its regional strategy outside of the conflict with India over Kashmir. Its enmity with India blinds both the civilian and military elite to all other regional considerations if not its own survival. The key manifestation of this blind spot is Pakistan's continued support for numerous insurgent groups in Kashmir (such as Lashkar-e-Tayyiba [LeT], discussed below) even when such support may no longer be in Pakistan's best interests (Fair et al., 2009, p. xvi; Thakur, 2011). As one author succinctly stated: the trouble with Pakistan may be that "it produces more terrorism than can be safely exported" (Thakur, 2011, p. 204)

Pakistan, Proxies, and Regional Policies

Following 9/11, the United States and Pakistan entered into a "partnership" to deal with al Qaeda and the Afghan Taliban. In return for approximately $11 billion in aid between 2001 and 2008, the United States was able to use Pakistan's airspace and cross overland through Pakistan to Afghanistan (Fair et al., 2009). In addition, Pakistan had its army, police, and paramilitary units go after al Qaeda fighters within Pakistan (Fair et al., 2009). But all was not golden in this relationship and what gold there was soon began to melt.

According to a RAND report written in 2009 for the U.S. Air Force, Pakistan had long used insurgent militant groups as proxies—a key part of its approach to its relationship with India and the conflicts over Kashmir (Fair et al., 2009). The report raises concerns that much of that approach is still used with regard to the Afghan Taliban, al Qaeda, and a number of home-grown Islamist militant groups such as the Pakistani Taliban and Lashkar-e-Tayyiba (see below). In addition, Pakistan is reluctant to challenge these groups for fear that they may become even more of a threat to the Pakistani elite and its still secular government than they are now (see also Kronstadt, 2008).

At the same time, as noted by Fair and colleagues,

> many of Pakistan's security and military elites have yet to conclude that militant groups endanger the Pakistani state and therefore pose more harm than good to supreme national interests. This view lingers because many within the security forces still see the militant groups as a useful instrument to keep Afghanistan and India off balance. (2009, p. xvi)

For example, as Baker notes, one reason the military is reluctant to deal with the Haqqani Taliban network in North Waziristan is that "it would risk leaving its eastern flank vulnerable to attack from India" (Baker, 2011). This view persists despite the fact that these groups have now turned their interests to, and are targeting, Pakistani targets in retaliation for the government's cooperation with the United States. Matters are not helped by the government's failure to provide humanitarian aid and basic services to the most rural and isolated parts of the country—a failure highlighted even more by LeT's very visible efforts through the NGO front organization JUD to fill that vacuum.

Adding to the problem is the way in which the alliance was presented to the Pakistani people given the nature of Pakistani politics at the time. Aryn Baker of *Time* magazine explained it well in a 2011 article:

> When the U.S. confronted Pakistan after the terrorist attacks of September 11, 2001, there were no discussions of common goals and shared dreams. There was just a very direct threat: you're either with us or against us. Pakistan had to choose between making an enemy of the U.S. and taking a quick and dirty deal sweetened with the promise of a lot of cash. In the end, Pakistan's cooperation was a transaction that satisfied the urgent needs of the day, brokered by a nervous military dictator, Pervez Musharraf, who failed to explain the value of the U.S. relationship to his people.

But as Baker goes on to note, even without the concerns about al Qaeda and Afghanistan, the United States still needs a stable Pakistan (see also Reidel (2008). The consequences of an unstable nuclear Pakistan riddled with terrorist and insurgent groups and in constant conflict with its non-Muslim and nuclear-powered neighbor could be too nightmarish to contemplate.

Highlights of Terrorist Groups in Pakistan

Lashkar-e-Tayyiba (LeT)

Also known as the Army of the Righteous, **Lashkar-e-Tayyiba (LeT)** is the largest and most proficient of the terrorist groups operating in Kashmir (U.S. Department of State, Foreign Terrorist Organizations, n.d.). Formed in the 1980s as a militant wing of a fundamentalist organization in Afghanistan, its main focus until recently had been to target Indian officials and civilians in Kashmir with the ultimate goal of annexing Kashmir to Pakistan. As indicated above, Pakistan—mostly through its powerful **Inter-Services Intelligence (ISI)** service—traditionally has used radical Islamist groups to promote its own regional agenda, particularly as it applies to India and Kashmir. Indeed, reports "link LeT to . . . ISI which is likely to have facilitated its creation and early activities" (Kronstadt, 2008, p. 4).

If ISI did play a role in creating LeT, it is possible that it has lost much of any control it once had over the group (Kronstadt, 2008; National Counterterrorism Center, 2012; U.S. Department of State, Foreign Terrorist Organizations, n.d.). Since the early part of this century, LeT has expanded its horizons to include Pakistani and Western targets in its sights. As noted by Kronstadt (2008), the group now "seeks not only Islamic rule in all of Kashmir, but it is also a proponent of broader anti-India and anti-Western struggles. . . . LeT is believed to have close links with both Al Qaeda and the Taliban, and over the years it appears to have taken a more expansive, global jihadi perspective" (p. 3).[5]

---

[5]In March 2002, a senior al Qaeda official was arrested at a LeT safe house, suggesting that members of LeT were aiding al Qaeda (U.S. Department of State, Foreign Terrorist Organizations, n.d.).

LeT was designated a Foreign Terrorist Organization by the U.S. Department of State in December 2001 (U.S. Department of State, Foreign Terrorist Organizations, n.d.) and banned—under considerable pressure from the United States—by the Pakistani government in 2002. As a result, the group changed its name to Jamatt-ud-Dawa (JUD). It uses its new name to provide humanitarian assistance and projects to areas of the country where the central government already had limited support. Despite the ban, LeT still operates almost openly in Pakistan, "fueling pervasive doubts that Pakistan's security agencies will honor the promises of cooperation being made by Islamabad's civilian leaders" (Kronstadt, 2008, p. 4).

The actual size of LeT is unknown, but 2010 estimates placed the number at over several thousand (U.S. Department of State, Foreign Terrorist Organizations, n.d.). The group maintains training camps, schools, and medical clinics in Pakistan and has connections and networks throughout Asia (and possibly Western Europe). It also receives funding from expatriate communities in the Middle East and Europe through its "humanitarian" front organizations such as JUD (U.S. Department of State, Foreign Terrorist Organizations, n.d.). LeT (through JUD) is particularly famous for providing vital aid in Kashmir after the October 2005 earthquake, a task the Pakistani government failed at spectacularly (National Counterterrorism Center, 2012).

IN-FOCUS 8.1

Lashkar-e-Tayyiba Attacks on Mumbai, India, on November 26, 2008

On November 26, 2008, Lashkar-e-Tayyiba launched one of its most infamous terrorist attacks on Indian soil to date. Even though other attacks resulted in more casualties, the level of coordination, intelligence, and planning needed to carry out the attacks and the western and global nature of the targets placed LeT in a whole new vein in the eyes of counterterrorism officials and scholars around the world. One hundred eighty-three people were killed and over 300 were injured.[6] Thirty-eight suspects were charged in the aftermath of the attacks, including the lone surviving terrorist, Mohammad Ajmal Amir Kasab (U.S. Department of State, Foreign Terrorist Organizations, n.d.). In May of 2010, Kasab was convicted and sentenced to death for his role in the attacks (Global Terrorism Database, 2012).

A December 2008 Congressional Research Service (CRS) report relates the specific events of that nightmarish evening and the hours that followed:

At approximately 9:30 p.m. local time on the evening of November 26, 2008, a number of well-trained militants came ashore from the Arabian Sea on small boats and attacked numerous high-profile targets in Mumbai, India with automatic weapons and explosives. By the time the episode ended some 62 hours later, about 174 people, including nine terrorists, had been killed and hundreds more injured. Among the multiple sites attacked in the peninsular city known as India's business and entertainment capital were two luxury hotels—the Taj Mahal Palace and the Oberoi-Trident—along with the main railway terminal, a Jewish cultural center, a café frequented by foreigners, a cinema house, and two hospitals. Six American citizens were among the 26 foreigners reported dead. Indian officials have concluded that the attackers numbered only ten, one of whom was captured. Some reports indicate that several other gunmen escaped.

(Continued)

---

[6]The actual number of those killed varies depending on the report. For example, the U.S. Department of State places the number of casualties at 183, the Congressional Research Service at 174, and the Global Terrorism Database at 171. The number of foreigners among the casualties also varies but is generally placed at over 20.

(Continued)

According to reports, the militants arrived in Mumbai from sea on dinghies launched from a larger ship offshore, then fanned out in southern Mumbai in groups of two or three. Each was carrying an assault rifle with 10–12 extra magazines of ammunition, a pistol, several hand grenades, and about 18 pounds of military-grade explosives. They also employed sophisticated technology including global positioning system handsets, satellite phones, Voice over Internet Protocol (VOIP) phone service, and high-resolution satellite photos of the targets. The attackers were said to have demonstrated a keen familiarity with the Taj hotel's layout in particular, suggesting that careful advanced planning had been undertaken.

Indian Prime Minister Shivraj Patil (who resigned in the wake of the attacks) reportedly ordered India's elite National Security Guard commandos deployed 90 minutes after the attacks began, but the mobilized units did not arrive on the scene until the next morning, some ten hours after the initial shooting. The delay likely handed a tactical advantage to the militants. According to a high-ranking Mumbai police official, the militants made no demands and had killed most of their hostages before being engaged by commandos on the morning of November 27. Two full days passed between the time of that engagement and the episode's conclusion when the two hotels were declared cleared of the several remaining gunmen. (Kronstadt, 2008, pp. 1–2)

As noted in this section, India responded by significantly increasing its counterterrorism efforts and strengthening its counterterrorism policies, some would say at the expense of civil liberties. The efforts and the continued emphasis on the threat posed by Pakistan's support for Kashmiri terrorists should come as no surprise. After 2008, Indian leaders and civilians were even more convinced that "Pakistan has long been and remains the main source of India's significant domestic terrorism problems. They continue to blame Islamabad for maintaining an 'infrastructure of terror' that launches attacks inside India" (Kronstadt, 2008, p. 1).

Of most concern to U.S. and Western officials is the expansion of LeT's targets far beyond its original Kashmiri focus. Evidence of its broadened horizons includes the Western nature of the specific sites attacked in Mumbai, a computer found in an LeT safe house containing discussions of at least 320 potential targets in Western Europe and other areas outside of Kashmir and Pakistan, and the involvement with LeT of Americans such as David Headley and others (Thakur, 2011, p. 205).

David Headley was an American citizen who aided LeT in the Mumbai attacks by casing and providing information about various sites to be attacked. As a U.S. citizen, he could move freely about an Indian city without raising any suspicions about links to LeT (or anything else nefarious). In March 2010, Headley pled guilty in U.S. District Court to conspiracy charges for his role in planning the Mumbai attacks as well as plans to attack a Danish newspaper. His plea agreement included an agreement by federal prosecutors not to seek the death penalty or to extradite Headley to India or Denmark (U.S. Department of State, Foreign Terrorist Organizations, n.d.). Headley is not the only U.S. citizen to be linked to LeT. Eleven LeT terrorists were indicted in Virginia in 2003 (National Counterterrorism Center, 2012).

If Pakistan has lost control over LeT, the chances of getting it back are not strong. "Cracking down on the LeT/JUD—especially if it is seen to come from New Delhi, Washington, or other foreign capitals—poses the risk of a serious backlash among Pakistan's religious conservatives who are already vehemently opposed to Islamabad's cooperation with U.S.-led efforts to combat Taliban forces in Afghanistan and western Pakistan" (Kronstadt, 2008, p. 15).

LeT attacks include a January 2001 attack on the Srinagar airport that killed five Indians, and a December 2001 attack on the Indian Parliament building in conjunction with another Pakistani Islamist group called JEM (discussed below). A July 2006 LeT attack in **Mumbai** killed 187 civilians and injured more than 800 others. This attack involved a series of bombs on seven different trains carrying commuters at rush hour. One of those arrested for the attack was LeT's chief operative; at least nine of those arrested were Pakistani nationals (Global Terrorism Database, 2012). Perhaps the most famous LeT attack so far, however, was the November 2008 attacks in Mumbai, India.

Pakistan and Its Taliban Offshoots

Tehrik-e-Taliban Pakistan (TTP). Composed of various militant tribes, **Tehrik-e-Taliban Pakistan** (TTP) was formed in 2007 by Baitullah Mehsud to oppose the Pakistan military in the Federally Administered Tribal Areas. Upon Baitullah's death, leadership fell to Hakimullah Mehsud (National Counterterrorism Center, 2012, p. 116). Attacks as of 2010 included U.S. military bases in Afghanistan, and the U.S. Consulate in Peshawar. The TTP is also suspected of involvement in the death of Prime Minister **Benazir Bhutto** and may have directed a plot to bomb Times Square in New York City in May 2010 (U.S. Department of State, Foreign Terrorist Organizations, n.d.). The group has repeatedly threatened to attack the United States on home soil and the attempted bombing may have been a part of that strategy (National Counterterrorism Center, 2012). Blamed by Islamabad for the majority of terrorist attacks in Pakistan since its founding, many of TTP's recent attacks have been claimed as retaliation for the death of bin Laden (National Counterterrorism Center, 2012).

Jaish-e-Mohammed (JEM). The group known as **Jaish-e-Mohammed** (JEM) was founded by an Islamic extremist by the name of Masood Azhar following his release from prison in India in early 2000. The Pakistan government outlawed JEM in 2002 and its splinter groups in 2003 (National Counterterrorism Center, 2012). The aim of JEM is to unite Kashmir and Pakistan and expel all foreign troops from Afghanistan. One of its more famous attacks was against the legislative assembly building in Srinagar in 2001 that claimed 30 victims. Other attacks that may have included JEM involvement were the Red Mosque uprising in 2007, and a 2009 attack on the Sri Lankan cricket team in Lahore. The main interests of Western counterterrorism authorities in JEM center around the group's joint efforts to work with LeT and other groups in attacks against Western targets. "In June 2008, JEM was working to resolve its differences with other Pakistani extremist groups and began shifting its focus from Kashmir and Afghanistan in order to step up attacks against U.S. and coalition forces" (National Counterterrorism Center, 2012, p. 110). JEM was also involved in the highly publicized kidnapping and beheading of U.S. journalist David Pearl (National Counterterrorism Center, 2012).

Where Does Pakistan Go From Here?

A number of factors within Pakistan could exacerbate instability and radicalization, including never-ending divisions over civilian versus military power, poor health care and high illiteracy, and an inability to leverage its resources to promote and sustain long-term economic growth (Fair et al., 2009, p. xv). In terms of long-term prospects for Pakistan, Fair et al. (2009) argued the following:

> The most likely near-term future is a Pakistan that "muddles" along, neither failing outright nor managing to right its course. Less likely futures include an increasingly theocratic or Islamist state or even a breakup of the state itself. More likely, Pakistan may evolve into a praetorian and authoritarian state tightly under the control of the military and intelligence agencies. All of these options augur more instability inside and outside Pakistan and merit significant efforts to retard their eventuation. (p. xv)

IN-FOCUS 8.2

The Haqqani Network and Pakistan's ISI

A 2012 *New York Times* article reflects on the connection between Pakistan's Inter-Services Intelligence (ISI) service and Islamist terrorist groups by focusing on concerns over ISI's relationship with the Haqqani network. The Haqqani network is a Taliban affiliate whose leaders have operated from Pakistan since early in the Afghan war. The occasion for the article was a June 1, 2012, attack by the Haqqani network on Camp Salerno, an American base in southern Afghanistan near the border with Pakistan. Two Americans were killed. Although tragic, that casualty count was low compared to many attacks in Afghanistan. Of equal concern was the fact that "the attackers [using several suicide bombers with a truck] had penetrated the defenses of a major base to within yards of a dining hall used by hundreds of soldiers." According to the authors of the article, Declan Walsh and Eric Schmitt, "the Salerno attack . . . and others like it have cemented the Haqqani network's standing as the most ominous threat to the fragile American-Pakistani relationship, officials from both countries say." The attack came just weeks after intense Pakistani-American negotiations reopened vital NATO supply lines from Pakistan into Afghanistan. This is hardly the network's first attack, however. According to the *New York Times* article, prior attacks have included the Indian Embassy in Kabul, luxury hotels and restaurants, a NATO facility, and the American Embassy in Kabul.

The concern, and the threat that American officials see to the fragile but still vital relationship with Pakistan, is that many American and Western officials believe that the ISI is supporting the Taliban insurgents—in particular the Haqqani network. This has been a longstanding concern and has been addressed in a number of government and academic analyses (see Kronstadt, 2008; Reidel, 2008; Thakur, 2011) as well as in media pieces such as Walsh and Schmitt's. Pakistani officials admit to ongoing contact with Haqqani officials—a relationship that began in the 1980s when Pakistan was helping to funnel U.S. funds to the group to aid in the fight against the Soviets—but argue that allowing Haqqani leaders to live and openly run businesses in Pakistan does not amount to operational support and approval of actual operations. According to Walsh and Schmitt, some American officials have observed that the Haqqani network is not dependent on ISI, and may be using ISI more than ISI is using it. However, a recent blog in *The Economist* (February 1, 2012) suggests a far more sinister connection based on a leaked NATO report on the Taliban.

Source: Blog, "Pakistan's security state: Reading the Taliban," February 1, 2012. Accessed August 13, 2012, at http://www.economist/blogs/clausewitz/2012/02/pakistans-security-state; Walsh, D., & Schmitt, E. (2012, July 30). New boldness from militants poses risk to U.S.-Pakistan ties. *New York Times*.

 India

India can trace its origins back millennia to when Aryan tribes from the northwest merged with earlier Dravidian inhabitants. European explorers established footholds—both economic and political—starting in the 16th century and by the 19th century, the subcontinent was dominated, and in most areas ruled, by Britain. India remained part of the vast British Empire until after World War II. British exhaustion from the war, combined with peaceful resistance to British rule led by Gandhi and Nehru, led to independence in 1947. The Muslim areas of India—with the exception of Kashmir (discussed above) became Pakistan and the Hindu areas became India.

India is a little over one third the size of the United States, covering 3,287,263 square miles. Its population is divided among three main ethnic classifications: Indo-Aryan (72%), Dravidian (25%), and Other (3%) (*CIA World Factbook*, n.d.). Hindi is the most widely used language although English is assigned "subsidiary official" status and is used for most national, political, and commercial communication (*CIA World Factbook*, n.d.). With approximately 1.2 billion people, India is the second most populous nation in the world after China, with 80.5% of its citizens following the Hindu faith and 13.4% identifying themselves as Muslim (*CIA World Factbook*, n.d.).

As with many members of the British Commonwealth, India has both a president and a prime minister, as well as a bicameral parliament. It is divided into 28 states and seven territories. English common law forms the main basis for India's legal system, with separate personal codes for Muslims, Hindus, and Christians.

According to the U.S. State Department, the

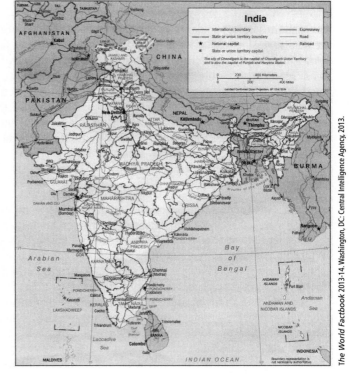

▲   Map of India

The World Factbook 2013-14. Washington, DC: Central Intelligence Agency, 2013.

loss of nearly 1,900 lives in 2010 alone makes India one of the most afflicted countries in terms of terrorism, despite increased and successful counterterrorism policies and strategies (U.S. Department of State, Country Reports, 2011). Kashmir, not surprisingly, remains its main concern, and most of India's recent changes to its counterterrorism policies came as a direct result of the November 2008 Mumbai attack discussed above. For example, two major pieces of legislation passed the Indian Parliament following the attacks. The National Investigating Agency Bill created for the first time in India's history a centralized investigative authority for terrorist organizations and incidents. The bill also bolstered maritime and air security and created additional bases for commando forces. The Unlawful Activities (Prevention) Bill was "meant to facilitate investigations and trials of the accused in terrorism cases. Among other provisions, it . . . doubled (to 180 days) the detention period allowed for suspects and . . . [sought] to restrict the flow of finances that abet terrorist activities" (Kronstadt, 2008, pp. 14–15). One of the key goals of the measures was to increase the extent of the resources India was using to address counterterrorism through its police forces. Prior to the attacks, India had a police to population ratio of 125 per 100,000, which was a "little more than half of the U.N. recommended ratio for peacetime policing" (Kronstadt, 2008, p. 14).

Highlights of Domestic Terrorism in India

Nevertheless, although India's main terrorism problem centers around Kashmir and Pakistan, it does have some significant domestic problems including a religious separatist conflict and a long-standing Marxist-leftist terrorist group.

Naxalites

The **Naxalites** is the popular term for what has been known since 2004 as the Communist Party of India. They follow the Maoist version of that ideology and target police stations, factories, Indian government officials, and multinational corporations (Spindlove & Simonsen, 2010, p. 343). As of 2010, the Naxalites were considered by the Indian prime minister to be the "greatest internal security threat" (U.S. Department of State, Country Reports, 2011). Known for using improvised explosive devices (IEDs) to attack transportation infrastructure as well as Indian security forces, one of their most publicized attacks involved a three-day strike where railways were damaged or destroyed and railway officials attacked. Other targets have included a popular bakery, a cricket stadium, a bus (where 15 civilians and 16 police were killed), a train derailment where 147 were killed, a mosque, and a tourist site (U.S. Department of State, Country Reports, 2011).

Sikhs

In addition to the conflict in Kashmir, India has another domestic separatist problem in its northern Punjab region. The **Sikhs** are a 500-year-old religious group that embodies aspects of both Islam and Hinduism and "emphasizes an inner journey to seek spiritual enlightenment, followed by external behavior to live in peace with the world" (White, 2012). Later leaders of the Sikhs recreated the group as a military organization, beginning in the late 17th century, as a way to defend their faith. This military tradition led to many Sikhs serving with great distinction in the British Army in the latter half of the 19th century ("Origins of Sikhism," 2009).

Any close relations between the Sikhs and the British were permanently damaged in 1919 when a British general—E. H. Dyer—ordered British troops to open fire on Sikhs holding a peaceful protest meeting. Around 400 people were killed and 1,000 wounded ("Origins of Sikhism," 2009). The British immediately promoted and retired Dyer, but "some historians regard the . . . massacre as the event that began the decline of the British Raj, by adding enormous strength to the movement for Indian independence" ("Origins of Sikhism," 2009). Relations had not improved by 1947 when the subcontinent was partitioned between India and Pakistan, but the Sikh community chose to remain with India instead of Pakistan—even though much of the Punjab went to Pakistan—as the lesser of the two evils.

India saw itself as a secular state from the beginning and therefore never could consider giving the Sikhs their own state in the Punjab as this would create a religious state within a secular one. Although India eventually divided the Punjab into three provinces within Sikh control, it was not enough to dissipate much of the anger at what the Sikhs saw as continuing mistreatment and oppression of their religion. "This gave birth to a small violent independence movement in 1977" (White, 2012, p. 245). The violence came to a head in 1984 when the leader of one of the most radicalized groups took refuge with his followers in the **Golden Temple** Complex at Amritsar, the most holy place in the Sikh religion, and the city that was the site of the 1919 massacre.

Indian troops responded to the takeover by entering the complex, killing most of the radicals, and severely damaging the buildings ("Indira Gandhi's Death Remembered," 2009). The response to the actions of the Indian troops was even more tragic. In October 1984, Prime Minister Indira Gandhi—who had ordered the invasion of the temple—was assassinated by two of her Sikh bodyguards. Several days of anti-Sikh rioting followed during which between 2,700 and 3,000 Sikhs were killed (government estimates vary) ("Indira Gandhi's Death Remembered," 2009).

The Golden Temple incident also had international repercussions. In June 1985, a radical Sikh group in Canada, in retaliation for the Golden Temple attack, placed a bomb on Air India Flight 182 bound for New Delhi. The bomb exploded as the plane approached the coast of Ireland, killing all 329 people on board—most of them Canadian Sikhs, and many of them children. The Punjab region is still rocked by violence. Beginning in May 2007, clashes erupted between mainstream Sikhs and groups that had openly supported the Indian Congress Party in that year's elections (Spindlove & Simonsen, 2010, p. 344).

Sri Lanka

A small island off the tip of southern India, Sri Lanka boasts a culture that dates back to the sixth century BCE. In the 14th century CE, a South Indian dynasty established a Tamil kingdom in the northern part of the island—this cultural and geographic split would have long-standing consequences for the island. Alternately ruled by the Portuguese, Dutch, and ultimately the British Empire, what was then known as Ceylon gained its independence from Britain in 1948. The nation's name was changed to Sri Lanka in 1972.

Sri Lanka is only slightly larger than West Virginia, boasting just 65,610 square miles, but holds 22.5 million people (*CIA World Factbook,* n.d.). The vast majority of the population is Sinhalese (73.8%), while only 8.5% are Tamil (either Sri Lankan Tamil or Indian Tamil). Sri Lankan Moors make up approximately 7.2% of the population (*CIA World Factbook,* n.d.). The two national languages are Sinhala and Tamil with Sinhala designated as the official language. The island is a republic with nine provincial designations and a legal system based on a combination of Roman-Dutch civil law, English common law, and Tamil customary law, reflecting the wide variety of settlements and occupations over the centuries. The president serves as both the head of state and the head of the government, with the prime minister serving a mostly ceremonial position. Sri Lanka has a unicameral parliament.

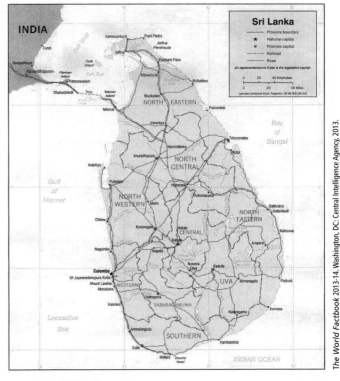

▲ Map of Sri Lanka

The World Factbook 2013-14. Washington, DC: Central Intelligence Agency, 2013.

Terrorism in Sri Lanka: The Tamil Tigers

The **Liberation Tigers of Tamil Elam (LTTE)**, known most often as the Tamil Tigers, was founded in 1976 and evolved from a small terrorist group into one of the longest lasting separatist groups in modern times. Although finally defeated by the Sri Lankan government in 2009 after over 25 years of violent confrontation, the Tamil Tigers still pose a potential threat of resurgence from a large expatriate community and lingering resentments in what was once considered the Tamil homeland. For that reason, and because of the influence the Tamil Tigers had on tactics used by other terrorist groups, LTTE is worth studying in some detail, even though its attacks rarely left the island.

The Tamil Tiger military offensive began in the mid-1970s and developed out of increased concerns regarding Sinhalese discrimination against the minority Tamil population. Originally, power on the island was shared between the two groups. However, by the late 1950s and early 1960s, consistent changes began to emerge both in the Sri Lankan political culture and ultimately in the Sri Lankan constitution that diminished or eliminated many of the rights once enjoyed by the Tamils, including rights to participate in the government, judiciary, or military (Bloom, 2003; Flanigan, 2008). By the 1970s, many of the peaceful Tamil protest groups had begun to turn to violence, and

© Reuters/CORBIS

▲ Tamil Tigers. Women fighters of the Liberation Tigers of Tamil Eelam (LTTE) on their post at a checkpoint.

by the mid-1980s "close to 20 Tamil rebel organizations existed in Sri Lanka, the most powerful of which was the LTTE" (Flanigan, 2008, p. 500). The violence escalated as the government attempted to crack down on the Tamil Tigers and other opposition groups with harsher and harsher measures (Bloom, 2003).

The watershed event for the Tamil Tigers' beginning took place in 1983. In July of that year, 13 Sinhalese soldiers were ambushed by LTTE fighters in the province of Jaffna, which had been under Sri Lankan military occupation for some time. The ambush touched off the worst anti-Tamil riots the island had seen then or since. The violence against Tamil civilians lasted for over three days during which homes were burned, Tamil businesses were destroyed, and looting, rape, and pillaging were prevalent (Bloom, 2003, p. 63). This in turn resulted in increased militancy on the part of the Tamils and increased popular support for armed resistance. LTTE was able to use the riots to increase and consolidate the Tamil resistance movement under its leader, **Vellupillai Prabhakaran** (Bloom 2003; Flanigan, 2008). Anti-Tamil violence was not new to Sri Lanka by this time, but as Bloom (2003) notes, citing a member of the Sri Lankan Parliament, "there was a qualitative difference in the intensity, brutality, and organized nature of the violence in July 1983. No other event is so deeply etched in the collective memories of the victims and the survivors" (p. 63).

At its height, LTTE had approximately 17,000 armed fighters (Flanigan, 2008)—some in units that engaged in traditional guerrilla warfare with Sri Lankan forces while others focused on terrorism against Sri Lankan government and civilian targets. Some of their more famous attacks include the 1991 assassination of former Indian prime minister Rajiv Gandhi (son of Indira Gandhi) while campaigning on behalf of the Indian Congress Party,[7] and the assassination of Sri Lankan president Premadasa in 1993. Unlike most terrorist groups, however, the Tamil Tigers at one point boasted their own intelligence service, air force, and a small naval and amphibious force called the "Sea Tigers" (Flanigan, 2008; U.S. Department of State, Foreign Terrorist Organizations, n.d.).

What the Tamil Tigers were most known for, however, was their cadre of suicide bombers known as the **Black Tigers**. Indeed, the group is considered to have perfected the strategy now being used with such effect by Islamist terrorist groups around the world (U.S. Department of State, Foreign Terrorist Organizations, n.d.). The Black Tiger squad had the advantage of being able to operate not only in Tamil territory but in the Sri Lankan capital of Colombo as well as in India (Flanigan, 2008).

By 2002, the Tamil insurgency was so successful that a cease-fire left the LTTE as the de facto government for the northern and eastern parts of the island. Like Hezbollah in Lebanon, the LTTE filled a vacuum in the Tamil parts of the island left by an inactive or incompetent government presence. Ironically, the Sri Lankan government had been reluctant to send services into these areas for fear of seeming to support the separatists, thus leaving open a perfect opportunity for the LTTE to increase its support among the Tamil population (Flanigan, 2008). (The article by Flanigan [2008] at the end of this section explores the similarities between Hezbollah and the Tamil Tigers in this regard).

But the success of the LTTE as a provincial governing body was not to last. When the cease fire failed and the Tamil Tigers resumed their violent activities, the Sri Lankan government went on an all-out offensive, including removing all civilians from the Tamil areas and declaring all who remained to be combatants and fair game. The offensive was successful. Prabhakaran committed suicide and his second in command was killed in May 2009, leaving the group

---

[7]Rajiv Gandhi became prime minister of India following his mother's assassination by her Sikh bodyguards in 1984. Gandhi incurred LTTE wrath when, as prime minister in 1987, he sent Indian peacekeeping troops into Sri Lanka. His assassination was carried out by a LTTE female suicide bomber; 14 others were also killed in the attack ("1991: Bomb Kills," 1991).

without its most dynamic and unifying leaders. By the end of 2009, those LTTE members who were not dead had fled the island. A 2010 U.S. State Department assessment notes that "in order to address lingering resentment in areas that were formerly held by LTTE combatants, the Sri Lankan government was working to restore civil administration, resettle internally displaced persons, provide immediate infrastructure development, encourage private sector participation, and promote the development of industries" (U.S. Department of State, Country Reports, 2011).

Still, concerns remain that the Tigers have not been completely defanged. Resentments linger despite the Sri Lankan government's laudable efforts, and ethnic hatreds developed over decades, if not centuries, do not dissipate overnight. A significant Tamil expatriate community now exists in India and elsewhere, with a significant fund-raising capability and, ironically, more freedom of movement than the Tigers had in the Tamil provinces. This may increase their ability to create more than a little havoc. According to the U.S. State Department, Tamils may have been responsible for a June 2010 attempted bombing of a train in Tamil Nadu, India. The train was able to avoid derailment and no injuries were sustained (U.S. Department of State, 2011).

Summary

This section focused on the nations of Afghanistan, Pakistan, India, and Sri Lanka. These countries are either currently a focus of U.S. counterterrorism efforts or have had long-standing terrorism experiences that are instructive for modern-day efforts. In particular, this section highlighted the Taliban operating in Afghanistan and Pakistan, Islamic fundamentalist groups operating out of the disputed Kashmir territory between India and Pakistan, separatist terrorism in India, and the Tamil Tigers separatist group in Sri Lanka. Key events such as the November 2008 attack in Mumbai, India, by the Kashmiri terrorist group LeT were highlighted and the future of some of the insurgency movements was assessed. In particular, this section sought to emphasize the difficulty of distinguishing between insurgents and terrorists in this region.

Key Points

- In 1980, the Soviet Union invaded Afghanistan. After a decade of fighting, millions of dollars spent, and untold lives lost, the Soviets left Afghanistan. Drastically weakened by its long misadventure, the Soviet Union fell apart shortly thereafter. A vicious civil war soon followed, with alliances and conflicts among the mujahedeen, as well as tribal militias, ebbing and flowing as territory was held and lost.
- One of the largest and most effective of the mujahedeen organizations was the Taliban led by Mullah Omar. A Sunni Islamic fundamentalist group founded in the early 1990s, the Taliban sought to unite the tribes of Afghanistan under the rule of a strict interpretation of Islamic law and to remove all secular influence from the country. By 1996, the Taliban controlled almost all of Afghanistan.
- That same year, the Taliban government offered sanctuary and an operational base to al Qaeda and Osama bin Laden, from which bin Laden launched numerous attacks against the United States and the West.
- Pakistan struggles significantly with both insurgency and terrorism, particularly in the Federal Administered Tribal Areas (FATA) and the Khyber Pakhtunkhwa province. Complicating matters is the continuous difficulties in distinguishing between insurgencies and terrorist groups and the effect that has on developing effective counterterrorism and counterinsurgency policies.
- The main, dominant, even overwhelming theme running through Indian and Pakistani politics—indeed throughout most of the Southwest Asia region—is the dispute between these two countries over the area of Kashmir.
- In the years since gaining their independence, India and Pakistan have fought at least three conventional wars and engaged in uncountable other violent confrontations over Kashmir. Numerous terrorist groups on both sides operate in the region, although until recently, the violence stayed within the disputed area or, at most, spilled into either India or Pakistan.

◆ Of particular concern is the group Lashkar-e-Tayibba (LeT) which, while traditionally operating out of Kashmir, has recently expanded its targets to include the West inside and outside Asia.

◆ According to the U.S. State Department, the loss of nearly 1,900 lives in 2010 alone makes India one of the most afflicted countries in terms of terrorism, despite increased and successful counterterrorism policies and strategies. Kashmir, not surprisingly remains its main concern, and most of India's recent changes to its counterterrorism policies came as a direct result of the November 2008 Mumbai attack.

◆ Nevertheless, although India's main terrorism problem centers around Kashmir and Pakistan, it does have some significant domestic problems, including a religious separatist conflict with the Sikhs and a long-standing Marxist-leftist terrorist group known as the Naxalites.

◆ The Liberation Tigers of Tamil Elam (LTTE), known most often as the Tamil Tigers, were founded in 1976 and evolved from a small terrorist group into one of the longest lasting separatist groups in modern times. Although finally defeated by the Sri Lankan government in 2009 after over 25 years of violent confrontation, the Tamil Tigers still pose a potential threat of resurgence from a large expatriate community and lingering resentments in what was once considered the Tamil homeland.

KEY TERMS

Afghanistan	Jannah, Mohammed Ali	Northern Alliance
Bhutto, Benazir	Karzai, Hamid	Omar, Mullah
Black Tigers	Kashmir	Pakistan
Federally Administered Tribal Area	Lashkar-e-Tayibba (LeT)	Prabhakaran, Vellupillai
Golden Temple	Liberation Tigers of Tamil Elam (LTTE)	Sikhs
Haqqani network		Sri Lanka
India	Mujahedeen	Taliban
Inter-Services Intelligence (ISI)	Mumbai	Tehrik-e-Taliban Pakistan
Jaish-e-Mohammed (JEM)	Naxalites	Ul-Haq, Zia

DISCUSSION QUESTIONS

1. What led to the creation of the mujahedeen in Afghanistan? What were their goals? Were they successful?

2. Who are the Afghan Taliban? How did they ultimately gain power?

3. Who was/are the Northern Alliance? What is their relationship to the Taliban? What impact did the events of September 11 have on the Northern Alliance?

4. What is Pakistan's defining national security issue? How has that issue affected its relationship with the United States?

5. What is the main concern regarding a conflict between India and Pakistan? How can the United States help mitigate the risk of such a conflict?

6. Describe the attack against Mumbai, India, in 2008. What counterterrorism measures did India take following that attack?

7. What is the ISI? How does it serve the interests of the Pakistani government? How does it conflict with the interests of the Pakistani government?

8. Who are the Sikhs? What was their most famous terrorist attack? What threat do they currently pose to India?

9. Briefly describe the history of the LTTE and assess its current threat.

10. What tactic is LTTE most famous for and how has it influenced other terrorist groups worldwide?

WEB RESOURCES

Central Intelligence Agency: http://www.cia.gov

Global Terrorism Database: http://www.start.umd.edu/gtd

National Counterterrorism Center: http://www.nctc.gov

Nuclear Threat Initiative: http://www.nti.org

U.S. Department of State: http://www.state.gov

READING 13

Reidel addresses the conflict between India and Pakistan over Kashmir. In particular, he examines the role of Pakistan's infamous Inter-Services Intelligence Directorate (ISI) in that conflict. In addition, Reidel takes a broader look at the role of ISI in Pakistan's foreign and domestic policies beyond just the Kashmiri conflict. Reidel's piece was written as part of a 2008 special journal issue addressing the key foreign policy issues faced by an incoming President Obama.

Pakistan and Terror

The Eye of the Storm

Bruce Reidel

Pakistan is the most dangerous country in the world today. All of the nightmares of the twenty-first century come together in Pakistan: nuclear proliferation, drug smuggling, military dictatorship, and above all, international terrorism. Pakistan almost uniquely is both a major victim of terrorism and a major sponsor of terrorism. It has been the scene of horrific acts of terrorist violence, including the murder of Benazir Bhutto in late 2007, and it has been one of the most prolific state sponsors of terror aimed at advancing its national security interests. For the next American president, there is no issue or country more critical to get right, which means developing a policy that will move Pakistan away from being a hothouse of terror.

That goal, however, will be exceedingly difficult to achieve. Over the course of a quarter century, Pakistan has developed an extremely complex nexus of terrorist connections. The Pakistani army and its intelligence service, the Inter Services Intelligence Directorate (known as ISI), created many of the terrorist groups that today flourish in the country and assisted in the growth of terrorist groups founded by others. The Taliban and several Kashmiri terrorist groups provide the best examples. Despite promises to cut off ties to these groups, Pakistan continues to provide them safe haven and in some cases direct support.

Pakistan's army believes these surrogates are critical to its sixty-year-old campaign against India and to securing Pakistan's influence in Afghanistan. Indeed, the two strategies are interrelated as Pakistani officers believe they cannot afford to confront both India and a hostile Afghanistan simultaneously.

Moreover, ISI and Pakistan have a long and intimate relationship with Osama bin Laden and his terrorist apparatus. Before 9/11, these ties included working together with the Taliban and the Kashmiri terror underground, often against India. Since 9/11, Pakistan has become an enemy of bin Laden and al Qaeda, but the environment Pakistan tolerates inside its borders has allowed bin Laden and al Qaeda to continue to operate there. Indeed, today al Qaeda is thriving in Pakistan's borderlands with Afghanistan and using its base in those badlands to conduct terrorist operations around the world.

Pakistan is undergoing a complex political crisis over the issue of the role of the military in government. This crisis diverts critical resources from the job of fighting al Qaeda and allows the nexus of the Taliban-Kashmir-Qaeda to flourish largely undisturbed. Whatever the outcome, the military will remain a critical player in Pakistan and call the shots on Pakistan's support for terrorism.

SOURCE: *The Annals of The American Academy of Political and Social Science*, July, 2008. Copyright © 2008 The American Academy of Political and Social Science.

Ironically, the United States helped create the ISI relationship with terror more than two decades ago. The United States can only kill Frankenstein's creation by fully understanding the nature of the monster and then developing polices with Pakistan to end its involvement with all the elements of the nexus of terror that afflicts the Pakistani host. Coming to grips with Pakistan's obsession with India and with Kashmir is critical to killing the monster. The time may be ripe in 2009 to move.

 # The Origins of the Jihad

The contemporary jihadist terrorist movement has its origins in the war against the Soviet occupation of Afghanistan in the 1980s. For almost a decade the Afghan mujahedin fought the Soviet 40th Red Army in the mountains of Afghanistan. At the beginning of the war, in 1979 and 1980, few observers expected the mujahedin to survive for long. Within the CIA, many expected the Soviets to crush the resistance in a matter of months or a couple of years. Instead, with critical support from the United States and Saudi Arabia, the Soviets were defeated, and the 40th Army retreated back across the Amur River in defeat.

Pakistan was the key to victory, providing the safe haven and support base for the jihad next door. Pakistan's then–military dictator, Zia ul Huq, was prepared to take major risks to assist the Afghan resistance and probably paid the price with his life. Zia was obsessed with the thought that Pakistan might be threatened with destruction by a Soviet-ruled Afghanistan and a Soviet-aligned India. The Pakistani intelligence service was the manager of the war.

The ISI role in the war has been brilliantly related by the head of its Afghan department in the 1980s, Mohammad Yousaf, who lays out in detail how ISI managed the war, including missions deep into the USSR. He argues that Pakistan was betrayed by the United States at the end of the war, when victory was complete over the Soviets. At that point, he argues, the United States abandoned Pakistan by imposing sanctions for its nuclear program, leaving Islamabad to fend for itself in trying to manage the difficult and ugly outcome of the holy war (Yousaf and Adkin 1992).

Many if not most Pakistanis believe the United States used their country in the 1980s to defeat the Soviet Union and then callously betrayed its ally and ignored the consequences of the forces unleashed by the war against the Soviets. In the summer of 1990, just when the war against the Soviets was ending, the United States imposed sanctions on Pakistan for its nuclear weapons program after the CIA determined that the program had reached a sufficient point where it could confirm a weapon had been produced. This action led to the end of U.S. military assistance to Pakistan, including the suspension of delivery of the F16 fighter aircraft already paid for by Islamabad. Many Pakistanis noted that the CIA could have made this judgment at any time in the past decade and only made it when the United States no longer needed Pakistan's help in the Afghan war.

Pakistan was left to deal with the aftermath of the war. More than 2 million Afghanis had taken refuge in Pakistan. This large displaced population bred violence and extremism. A "Kalashnikov culture" developed along the border lands where the refugees and the mujahedin lived, contributing to a growing breakdown in law and order in an area that had never seen strong central government authority.

In Afghanistan, after the Russians withdrew in 1989, the mujahedin tried to march on Kabul but failed to take the city. The ISI told then Prime Minister Benazir Bhutto that Kabul would fall in a matter of days; the CIA told President George H. W. Bush the same (Bhutto 2007). Instead, the fall took almost three years.

The Taliban's rise was the response to the civil war that followed among the mujahedin parties. It arose in the southern Pashtun provinces of the country led by a much-wounded (including losing one eye) veteran of the jihad against the Soviets, Mullah Omar. Omar proclaimed a new holy war to purge the country of the warring parties and to install a pure Islamic government that would restore law and order.

Mullah Omar is a shadowy figure, even by the standards of the jihadist movement. He comes from Oruzugan province in Afghanistan, one of the poorest and most backward parts of a poor and backward country. He is a graduate of one of the most famous madrassas in Pakistan, Darul uloom Haqqania. Omar was outraged by the infighting of the mujahedin and especially by the increasing depravations they exacted on the Afghan people as they became more and more dependent on crime to finance their activities. Apparently the rape of two small boys by one mujahedin leader was the breaking point for Omar. He organized a group of followers to punish the criminals and hung them from the gun barrel of a tank. From that incident the Taliban were born.

Pakistan began providing support to the Taliban from its birth. While still keeping ties open to the other factions, the Pakistanis saw in the Taliban a mechanism to end the civil war and consolidate their influence over Afghanistan through a proxy. The policy began under Prime Minister Benazir Bhutto and continued under her successor, Nawaz Sharif, but the army ran the show. The ISI was confident it could control the Taliban leadership and use the organization to solidify Pakistan's preeminence in the country. Pakistan became the first country in the world to recognize the Taliban government. (Only two countries followed suit: Saudi Arabia and the United Arab Emirates.) An ISI cadre began to advise the Taliban militia, with the Taliban recruiting heavily among the Islamic schools in Pakistan. Increasingly, Pakistani advisors assisted the Taliban military and Pakistani experts handled the logistics to maintain and operate the Taliban's more sophisticated weapons, including tanks and aircraft (Byman 2005).

The Taliban were not the ISI's only creation in the 1990s. In the late 1980s, the Muslim majority in Kashmir, inspired by the Afghan victory, became increasingly resistive, and a major revolt against the Indian occupation broke out. Pakistan moved quickly to both support the insurgency and gain control over it. The ISI, fresh from victory in Afghanistan, used the same tactics and strategy against India that it had used against the Soviet Union and provided training and weapons for the Kashmiris. It gradually created its own Kashmiri groups to do the fighting to ensure loyalty to Islamabad's interests first and foremost. ISI also used Afghanistan as a base for training Kashmiri jihadists to support the insurgency in Kashmir. By training operatives in Afghanistan, ISI sought a measure of deniability from Indian charges that Pakistan was a state sponsor of terrorism.

With Pakistan's help, the war quickly escalated. Thousands died on both sides. Pakistani-sponsored terrorists took the battle outside the province and into India proper, conducting hundreds of horrific attacks against targets throughout India. Pakistani-backed groups, with ISI's help, developed staging posts in Bangladesh and Nepal to help undertake these attacks.[1]

More than once these terrorist operations have raised the risk of provoking a general war between India and Pakistan. After the attack on the Indian parliament in December 2001, a standoff developed that lasted for more than a year. More than a million soldiers deployed, eyeball to eyeball, along the border, with the ever-present threat of a confrontation that could escalate to nuclear war. Despite repeated promises, Pakistan has not broken its links to these Kashmiri groups, which still operate both in Kashmir proper and throughout India.

Enter al Qaeda

In this volatile mix of Afghan and Kashmiri terrorists al Qaeda developed. Osama bin Laden has a long and complex history with Pakistan and the ISI. Bin Laden first arrived in Pakistan in 1980, only a year after the Soviet invasion of Afghanistan. Bin Laden was a rich, twenty-three-year-old volunteer with no experience in military affairs, who began his mission organizing aid from other Saudis. He quickly became a leader in the Arab community in Peshawar, providing funds and equipment to the mujahedin groups.

For the next ten years, bin Laden had a close operational relationship with the ISI. The Pakistani service jealously guarded its position in the Afghan movement, welcoming support from any source but carefully maintaining control of the situation on the ground. It would have been impossible for bin Laden to operate without ISI supervision and constant interaction with its operatives (Lawrence 2005). He was also in contact with the Saudi intelligence service during this period and met on occasion with its leader, Prince Turki bin Faysal. Despite allegations to the contrary, he did not have a link to the CIA, although he was well aware the CIA was funding and arming the mujahedin.

ISI also played a critical role in developing bin Laden's ties with the Taliban when he returned to Afghanistan in 1996. According to the findings of the National Commission on Terrorist Attacks upon the United States (2004), as described in *The 9/11 Commission Report*, ISI set up the first contacts between bin Laden and the Taliban's leader, Mullah Omar, when bin Laden returned in 1996. Part of ISI's motive was to get bin Laden's help with non-Afghan groups they had jointly worked with in the past.

Indeed, bin Laden's interaction with ISI had long gone beyond the Afghan movement. He was also involved in its efforts to create and sponsor Kashmir groups to fight in India. In 1987, the ISI worked with a group of Islamic scholars in Pakistan that included bin Laden and his then-spiritual guide Abdullah Azzam to set up Lashkar-e-Tayyeba, which would become one of the most violent and extreme of the Kashmiri organizations. At first it provided Kashmiri volunteers to fight in Afghanistan, and then it began attacks into Kashmir and ultimately into India's large cities. Bin Laden provided some of the funding to get it started and remained closely connected to the group (Wilson 2007). This relationship with ISI Kashmiri clients would continue after bin Laden returned to Afghanistan in the late 1990s.

On August 7, 1998, bin Laden and his new al Qaeda group carried out their first major operation, attacking the American embassies in Tanzania and Kenya. Within hours of the attack, American intelligence had information that bin Laden was going to be visiting a training camp in Afghanistan to meet with some colleagues. President Bill Clinton ordered a military strike to kill him. Unfortunately, bin Laden changed his plans and was no longer at the target when the missiles struck, missing him by perhaps as little as thirty minutes (*PBS Frontline* 2006).[2] What is revealing is who was at the camp with bin Laden.

The camp was a Kashmiri training facility run by the ISI. The majority of the fighters in the camp were from another Kashmiri group run by ISI, the Harkat ul Mujahedin (HUM). Twenty or so Kashmiris and Pakistani trainers died in the attack (Burke 2003).[3] Despite public protests to the contrary, it is clear that ISI and its Kashmiri cadre were still intimately associated with bin Laden.

In late 1999, the connections between bin Laden, al Qaeda, the Kashmiris, the Taliban, and ISI were even more dramatically illustrated in the hijacking of an Indian airliner from Katmandu in which one passenger was brutally murdered. The hijackers were assisted in gaining access to weapons in the airport by the local ISI station in Katmandu. The hijackers were members of the HUM group and sought the release of one of their leaders, Maulana Massoud Azhar, from jail in India. The flight was diverted to Kandahar, where the Taliban protected the hijackers and negotiated with the Indian authorities. Osama bin Laden was on the ground as well, even hosting

the victory dinner when the hijackers got their demands met. The ISI took Azhar on a victory tour around Pakistan after the ordeal was over to help raise funds for the Kashmiri cause.

Jaswant Singh was then the Indian foreign minister who had the unenviable task of going to Kandahar to deal with the hijackers and get the 155 passengers freed. In his memoirs he describes the hijacking of Indian Air Flight 814 as the "dress rehearsal" for 9/11, since the incident involved many of the same characters and marked their first hijacking of a passenger aircraft. The foreign minister learned that the hijackers had even considered crashing the plane into a target in India if they did not get their demands met (Singh 2007).[4]

Some Pakistanis insist that bin Laden's connection with ISI went beyond support for mutual causes in Afghanistan and Kashmir and extended into meddling in internal Pakistani politics. Former prime minister Benazir Bhutto claimed that ISI recruited bin Laden's help to finance her political enemies and that he provided $10 million to them. Later she claimed that he worked with ISI and others to try to have her assassinated in 1993. She asserted that the mastermind behind the first attack on the World Trade Center in 1993 confirmed this to her when he was captured in Pakistan in 1995 (Bhutto 2007).

The extent of ISI's relationship with bin Laden and his al Qaeda colleagues before 9/11 is impossible to know completely in the absence of cooperation from the ISI itself. Its archives, if they are ever opened, may contain even more evidence of the close partnership between the two.

The United States made a considerable effort in the late 1990s to break up the emerging al Qaeda threat in Afghanistan, using a wide variety of tools. The military option has already been noted. Political tools were used as well. President Clinton raised the issue of Osama bin Laden's safe haven with the Taliban in every meeting he held with Pakistani leaders after the creation of the group, urging and pressing for Islamabad to take action to compel the Taliban to control bin Laden and deliver him to justice one way or another. In March 2000 during his only visit to Islamabad, as Clinton (2004) related, "I told him (Musharraf) I thought terrorism would eventually destroy Pakistan from within if he didn't move against it."

The bilateral track was complemented by a multilateral track. After the East African bombings, the United States secured unanimous United Nations Security Council (UNSC) support for first condemning the Taliban for hosting al Qaeda and then imposing sanctions on their Islamic Emirate of Afghanistan for doing so. UNSC Resolutions 1185, 1214, 1267, 1333, and 1363 called on all countries and especially Pakistan to use their influence with the Taliban. Neither the Taliban nor Pakistan listened to the UN.

 ## After 9/11—The More Things Change, the More They Stay the Same

By coincidence, the head of ISI, Lt. Gen. Mahmoud Ahmed, was in Washington visiting the CIA on the morning of September 11, 2001. His mission from Musharraf was to persuade the Americans to ease off on their policy on the Taliban. No one knew more about the extent of Taliban-Pakistan cooperation than Ahmed. He was immediately summoned to the State Department to meet with Deputy Secretary of State Richard Armitage.

According to Musharraf's account of that meeting, Armitage told Ahmed that Pakistan had to choose whether to be with the United States or against it. Should Pakistan stand by its relationship with the Taliban, according to Musharraf, Armitage threatened to bomb Pakistan back into the stone age. Armitage has denied he threatened to attack Pakistan but has confirmed that he gave the ISI leader an ultimatum to either help America or be seen as an enemy (Musharraf 2006).

Almost immediately Pakistan's policy toward the Taliban changed. The Pakistani ambassador in Washington, Maleeha Lodhi, who had been handpicked for the job by Musharraf, told the press, "We will be urging the Taliban leadership . . . to accede to the demands of the international community . . . and to hand over the person that they are harboring, Osama bin Ladin, so that he is brought to justice" (Hussain 2007, 35). General Ahmed was fired from his post a few days later, apparently for lack of enthusiasm in the new policy toward his former clients.

By September 2001, Pakistani aid to the Taliban was critical to the group. Ahmed Rashid, the leading expert on the Taliban, estimated that up to sixty thousand Pakistani Islamic students had fought in Afghanistan supporting the Taliban, dozens of Pakistani military officers were advising the group, and small units of the elite Special Services Group commandos were engaged in combat operations with the Taliban forces. As he put it, "Pakistan's knowledge of the Taliban's military machine, storage facilities, supply lines and leadership hierarchy was total" (Rashid 2001).

Shortly after the start of military action, Pakistan deserted its Taliban ally. Pakistan's military advisors, pilots, tank crews, and other military personnel in the Taliban army fled the country and went back to Pakistan. Without their allies, the Taliban was even more hopelessly outgunned by the U.S.-backed alliance. The collapse of their resistance came quickly. By the end of the year, the Islamic Emirate was replaced by a government installed by the Northern Alliance. Most of the Taliban fighters simply faded away to their villages in the border region.

Musharraf has explained his decision to abandon the Taliban in his 2006 memoir *In the Line of Fire*. According to his account, Musharraf simulated a war game immediately after Ahmed reported his conversation with Armitage and concluded that Pakistan, even with its nuclear arsenal, could not prevail in a military conflict with the United States. Moreover, Musharraf concluded that the Indians would be the major beneficiary of a Pakistani decision to stand by the Taliban. As he puts it,

> I also analyzed our national interest. First, India had already tried to step in by offering its bases to the U.S. If we did not join the U.S., it would accept India's offer. What would happen then? India would gain a golden opportunity with regard to Kashmir. . . . Second, the security of our strategic assets would be jeopardized. We did not want to lose or damage the military parity that we had achieved with India by becoming a nuclear weapons state. (p. 202)

In short, the decision to reverse a decade of Pakistani policy in Afghanistan was a derivative of the underlying Pakistani concern about India. For this reversal, Pakistan was handsomely rewarded. Musharraf was invited for a state visit to Washington in February 2002 and was promised economic assistance and debt relief. In 2003 Musharraf visited Camp David, and President George W. Bush announced a five-year, $3 billion economic and military assistance package. In 2004 Pakistan was designated a Major Non-NATO Ally, which meant additional technology sharing between the two militaries. In 2005 Pakistan was promised the sale of F16s again to demonstrate the bad old days were truly over. By 2007 more than $10 billion in aid had flowed to Pakistan.

Musharraf was careful to give only selective support to the United States, however, in the war on terrorism. The Taliban apparatus in Pakistan's madrassas was not dismantled, and many Taliban officials continued to operate in Pakistani cities, particularly Quetta, the capital of Baluchistan. By 2004 Taliban officials were openly fund-raising again in Quetta. No major Taliban official has ever been arrested in Pakistan, and Afghan government authorities claim the ISI is again providing direct aid to the Taliban in its operations against NATO and Afghan army forces.

At first Pakistan was also active in arresting al Qaeda operatives, including the alleged mastermind of 9/11, Khalid Shaykh Muhammad, who was captured in March 2003 in Rawalpindi (ironically, the military capital of Pakistan). Musharraf claims to have captured hundreds of Qaeda operatives. It is important to note where they have been caught. Almost all of the captured Qaeda operatives have been apprehended in major Pakistani cities, not in the badlands on the border. Moreover, most have been caught in safe houses belonging to Pakistani-supported groups like Lashkar-e Tayyiba (LeT). Abu Zubayda, the first major operative wrapped up in Pakistan, was in a LeT safe house in Faisalabad (Hussain 2007). Gary Schroen, who led the first CIA team into Afghanistan to fight the Taliban and al Qaeda after 9/11, noted that "since 2002 whenever a raid has been conducted in Pakistan against an al Qaeda safe house, al Qaeda members are found being hosted by militant Pakistanis, primarily from the Lashkar-e Tayyiba group, supporters of the Kashmir insurgency" (Schroen 2005, 361).

The Pakistani army has also engaged in a major military campaign in the badlands, especially in Waziristan. Up to seventy thousand troops have been engaged in counterinsurgency operations there in an on-again, off-again battle. Losses on the side of the army have been heavy, inducing Musharraf to try to arrange peace deals with local tribal leaders who have promised to crack down on al Qaeda but never delivered.

The bottom line is that the top leadership of al Qaeda has not only eluded capture; they have thrived. A sophisticated propaganda machine has developed in the badlands producing dozens of high-quality video and audio tapes for al Qaeda leaders and offering them a public forum for speaking out. More deadly, the badlands have been used to train operatives to carry out operations outside South Asia, particularly in Western Europe. Every major operation in the United Kingdom in the past five years, including the July 7, 2005, underground bombings and the foiled August 2006 plot to blow up simultaneously ten jumbo jets outbound from Heathrow Airport over the Atlantic, have been connected by the British security establishment back to Pakistan and to the al Qaeda leaders hiding there. The German and Danish security authorities have also uncovered plots linked back to Pakistan, with the Danish intelligence chief reporting that the plotters were taking their orders from "a leading al Qaeda person" in Pakistan (Hart and Whitlock 2007).

Musharraf also promised to crack down on the madrassas and end their use by extremists to indoctrinate fanaticism. He said he would end the jihadi culture in Pakistan and halt all cross-border terrorist operations into Kashmir and India. None of this has happened. More than one and a half million students still attend unregulated madrassas in Pakistan. Pakistan's Let militants were involved in the attack on the Indian parliament in December 2001, an attack that diverted Pakistan's military forces away from the Afghan border at the critical moment in the hunt for bin Laden and led them into a year-long standoff with India (Hussain 2007). In July 2006 the police authorities in Mumbai implicated the ISI directly in a bombing campaign on the Mumbai metro system that killed 186 people. Only when the madrassa culture spread directly into Islamabad and threatened to make a mockery of Musharraf's hold on power did he slowly and reluctantly act against their base, the Red Mosque in the capital.

In response to the Pakistani crackdown on their operatives, bin Laden and his deputy Ayman Zawahiri have repeatedly called for Musharraf's overthrow and have plotted to kill him. Two assassination plots have failed. After the attack on the Red Mosque, bin Laden called again for his death and asserted that Musharraf was a pawn of India, Israel, and America. Bin Laden ridiculed Musharraf as a coward who boasts of testing nuclear weapons, but "when the American Foreign Minister Powell came to you, you cowered, bowed and submitted to him like a lowly slave and you permitted the American Crusader forces to use the air, soil and water of Pakistan, the country of Islam, to kill the people of Islam in Afghanistan, then in Waziristan. So woe to you and away with you" (bin Laden 2007).

 ## Lessons Learned

Aside from the liberation of Kabul, America's efforts to persuade Pakistan to take decisive action to control the Taliban, Kashmiri terrorists, and al Qaeda have largely been failures. The record shows that the United States and its global allies have made repeated efforts to encourage Pakistan to act responsibly but have had only limited and temporary success. Even the Pakistani decision in 2001 to break with the Taliban after 9/11 has proven to be short-lived and incomplete. The record also clearly demonstrates that Washington has employed a full range of options in working the Pakistani case, from sanctions to rewards, from unilateralism to multilateralism, and from jawboning to threats. While Armitage's threat seems to have had the most immediate results, it also produced only a temporary and incomplete shift.

The next president should use all of these levers, as appropriate, but he or she must do something much more inventive if we are to break Pakistan's unhealthy relationship with terror. He or she must get into the thicket of the core issue.

Central to Pakistan's policy calculations throughout have been security concerns about India. Pakistani leaders from Zia to Musharraf repeatedly looked at Afghanistan in the context of their struggle with India. Most clearly, Musharraf's own memoir underscores that the decision to align with the United States after 9/11 stemmed from his concerns that India would steal a march on Pakistan and that India would both break up Pakistan and destroy its nuclear arsenal if Pakistan did not move against the Taliban and join the U.S. war on terrorism.

The centrality of India in these episodes reflects its dominant position in Pakistani thinking on virtually every issue. The conflict with India affects all aspects of Pakistan's worldview and its self-image. Even the fundamental issue of population growth and population control is perceived largely in terms of competition with India. Pakistan's population is growing at 2.9 percent annually, one of the highest rates in the world; it has grown from 32 million at the time of partition to 170 million today, far outstripping the economy's efforts to maintain an adequate standard of living for many, let alone reduce poverty. Yet, there is no serious approach to birth control in the country because it is seen as weakening the effort to keep up with the country's larger neighbor. As one scholar has noted,

> Pakistan's political and social elite—the Establishment—have never paid much attention to the population problem. Their overriding concern is a perceived threat from a much larger and faster-growing India. In their eyes, Pakistan is a relatively small state; to some degree the elite have never shaken off the minority complex that was embedded in the Pakistan movement. (Cohen 2004, 234)

The Pakistani army plays the critical role in the formation of the Pakistani worldview and in national security policy formulation. Unlike India, Pakistan places no civilian controls over military activities. Even during the 1990s, when civilian-elected governments ruled the country, the army was in the position of critical decision maker. Under Musharraf, the army has no serious constraints on its policy formulation role. The army has defined the jihadist nexus as a critical force multiplier, necessary to confront India. The result is a "ground reality in which al Qaeda thrives on a vast, deeply entrenched and integrated jihadi infrastructure that straddles the Afghanistan and Pakistan borders. This network includes more than 50 Pakistan-based radical groups who share deep bonds of an Islamic ideology, common political targets—the United States, India and Israel—training facilities and resources" (Behera 2002).

The army has defined the national security agenda since partition in terms of the threat posed by India. It has naturally determined that to deal with the threat, the military must be given a disproportionate slice of the national income. As a consequence, the military gets a very high percentage of the national budget every year. Just as crucially, the army determines the parameters in which diplomacy with India can be pursued. Critical decisions about foreign and security policy are made in the army and only then implemented by the Foreign Ministry.

Former U.S. Director of Central Intelligence George Tenet, who dealt extensively with this issue, concludes in his memoirs that "the Pakistanis always knew more than they were telling us." Tenet noted his belief that "what the Pakistanis really feared was a two-front conflict, with the Indians seeking to reclaim Pakistan and the Taliban mullahs trying to export their radical brand of Islam across the border from Afghanistan. A war with India also posed the grim specter of nuclear confrontation, but from the ruling generals' point of view, the best way to avoid having their nation Talibanized was to keep their enemy close. That meant not cooperating with us in hunting down bin Laden and his organization" (Tenet 2007, 139–40).

To make matters worse, the Pakistani army has a very deep distrust of the United States, which is seen as a fair-weather friend. A common Pakistani joke is that the United States sees Pakistan as a condom, used and then thrown away. An on-again, off-again arms relationship, symbolized by the F16 saga, has left a bitter taste in the officer corps, whose members are convinced that the United States will always abandon Pakistan when its interests no longer require a close relationship.

The implications for the international community of this Pakistani preoccupation with India and the disproportionate role of the army in Pakistani decision making are profound. The United States has properly identified India as an emerging world power and one of the world's most important economies. Washington and New Delhi are drawing increasingly closer, as the two largest democracies in the world. In turn, Pakistan feels even more deeply threatened by the rise of India.

In a best case scenario, the international community, led by the United States, should seek to disarm Pakistan's preoccupation with India by resolving the underlying dispute between the two countries in Kashmir. From the Pakistani perspective, an optimal resolution of Kashmir would lead to the unification of the province with Pakistan, or at least the Muslim-dominated Valley of Kashmir and

the capital of Srinagar. With Kashmir "reunited" with Pakistan, the requirement for nuclear weapons would be reduced, if not removed, and the need for a jihadist option to compel Indian withdrawal from the valley would be gone. This is precisely the outcome that Pakistani leaders have in mind when they urge American leaders to devote diplomatic and political energy to the Kashmir issue.

Of course, it is a completely unrealistic scenario. India has been clear that it will not withdraw from Kashmir. On the contrary, India argues it has already made a major concession by de facto accepting the partition of the state between itself, Pakistan, and China. India is probably prepared to accept the line of control (the ceasefire line of 1948) as the ultimate border with Pakistan but will not accept a fundamental redrawing of borders to put the valley under Pakistan's sovereignty.

But this does not rule out an option that would involve a major effort to resolve the Kashmir problem on a more realistic basis. The basis for such an approach would be to complement the ongoing Indo-Pakistani bilateral dialogue. That dialogue has already produced a series of confidence-building measures between the two countries, reopening transportation links, setting up hotlines between military commands, and holding periodic discussions at the foreign secretary level on all the issues that divide the two. Unfortunately, the dialogue has not seriously addressed the Kashmir issue because of the significant gulf between the two parties and India's refusal to negotiate while still a target of terrorist attacks planned and organized in Pakistan.

The United States has been reluctant to engage more actively in the Kashmir dispute in light of the Indian posture that outside intervention is unwarranted and that Kashmir is a purely bilateral issue. Faced with the likelihood of India's rejection of outside intervention, American diplomacy has put the Kashmir problem in the "too hard" category and left it to simmer. The results are all too predictable. The Kashmir issue periodically boils over, and the United States and the international community have to step in to try to prevent a full-scale war. This was the case during the Kargil crisis in 1999, after the terrorist attack on the Indian parliament in 2001, and again in 2002 when India mobilized its army for war on the Pakistani border.

A unique opportunity for quiet American diplomacy to help advance the Kashmir issue to reach a better, more stable solution may exist in 2009. The U.S. India nuclear deal agreed to during President Bush's July 2005 visit to South Asia should create a more stable and enduring basis for U.S. Indian relations than at any time in history. The deal removes the central obstacle to closer strategic ties between Washington and New Delhi: the nuclear proliferation problem, which has held back the development of their relationship for two decades.

In the new era of U.S. Indian strategic partnership, Washington should be more prepared to press New Delhi to be more flexible on Kashmir. It is clearly in the American interest to try to defuse a lingering conflict that has generated global terrorism and repeatedly threatened to create a full-scale military confrontation on the subcontinent. It is also in India's interest to find a solution to a conflict that has gone on for too long. Since Kargil, India has been more open to an American role in Kashmir because it senses Washington is fundamentally in favor of a resolution on the basis of the status quo, which favors India.

The United States currently has better relations with both India and Pakistan than at any time in the past several decades. The U.S. rapprochement with India, begun by President Clinton and advanced by President Bush, is now supported by an almost unique bipartisan consensus in the American foreign policy establishment and the Congress. At the same time, U.S. Pakistani relations are stronger now than at any time since the Reagan years, and the sanctions that poisoned U.S. Pakistani ties for decades have been removed by legislation supported by both Republicans and Democrats. It is a unique moment.

A Kashmir solution would have to be based around a formula for both making the line of control a permanent and normal international border (perhaps with some minor modifications) and creating a permeable frontier between the two parts of Kashmir so that the Kashmiri people could live more normal lives. A special condominium might be created to allow the two constituencies to work together on issues that are internal to Kashmir, such as transportation, the environment, sports, and tourism.

It is unlikely that the two states will be able to reach such an agreement on their own given the history of mistrust that pervades both sides of the problem. A quiet American effort to promote a solution, led by the next U.S. president, is probably essential to any effort to move the parties toward an agreement.

Resolution of the Kashmiri issue would go a long way to making Pakistan a more normal state and less preoccupied with India. It would also remove a major rationale for the army's disproportionate role in Pakistani national security affairs, thus helping to restore genuine civilian democratic rule in the country. A resolution of the major outstanding issue between Islamabad and New Delhi would reduce the arms race between the two countries and the risk of nuclear conflict. And it would remove the need for Pakistan to find allies, such as the Taliban, LeT, and al Qaeda, to fight asymmetric warfare against India.

Of course, it would not resolve all the tensions between the two neighbors or end the problem of the Taliban in Afghanistan. But more than anything else it would set the stage for a different era in the subcontinent and for more productive interaction between the international community and Pakistan.

The alternative is to let Kashmir simmer and avoid trying to find a means to advance the Indo-Pakistani dialogue. In the long run, this approach is virtually certain to lead to another crisis in the subcontinent. Sooner or later, the two countries will again find themselves on the precipice of war. In a worst-case scenario, a terrorist incident like the July 2006 metro bombings in Mumbai or the hijacking of IA 814 could spark an Indian military response against targets in Pakistan allegedly involved in the planning and orchestration of terrorism. And that could lead to nuclear war.

The next president must adopt a more sophisticated approach to Pakistan and its terror nexus that goes beyond threats and sanctions, beyond commando raids and intelligence cooperation, beyond aid and aircraft sales. It is time to come to grips with what motivates Pakistan's behavior and make peace.

 References

Behera, Navnita Chadha. 2002. Terror trail leads from Kabul to Kashmir. *Asia Times*, May 25.

Bhutto, Benazir. 2007. *Daughter of the East: An autobiography*. London: Simon & Schuster.

bin Laden, Usama. 2007. Come to Jihad: A speech to the people of Pakistan. September. http://www.lauramansfield.com/hh2.rm.

Burke, Jason. 2003. *Al Qaeda: Casting a shadow of terror.* London: IB Taurus.

Byman, Daniel. 2005. *Deadly connections.* Cambridge: Cambridge University Press.

Clinton, Bill. 2004. *My life.* New York: Knopf.

Cohen, Stephen. 2004. *The idea of Pakistan.* Washington, DC: Brookings Institution.

Hart, Spencer, and Craig Whitlock. 2007. Official links German terror plot to Pakistani operatives. *Washington Post,* September 26, p. A16.

Hussain, Zahid. 2007. *Frontline Pakistan: The struggle with militant Islam.* New York: Columbia University Press.

Lawrence, Bruce. 2005. *Messages to the world: The statements of Osama bin Laden.* London: Verso.

Mentschel, Binalakshmi Nepram. 2007. *Is the crescent waxing eastwards? Pakistan's involvement in India's northeast and Bangladesh.* Pakistan Security Research Unit Brief no. 14, June 24, University of Bradford, UK.

Musharraf, Pervez. 2006. *In the line of fire: A memoir.* New York: Free Press.

The National Commission on Terrorist Attacks upon the United States. 2004. *The 9/11 Commission report.* New York: Norton.

PBS Frontline. 2006. Interview with Gary Schroen, the dark side. January 20.

Rashid, Ahmad. 2001. Pakistan, the Taliban and the *US. The Nation,* October 8.

Schroen, Gary. 2005. *First in: An insider's account of how the CIA spearheaded the war on terror in Afghanistan.* New York: Ballantine.

Singh, Jaswant. 2007. *A call to honour: In service of emergent India.* New Delhi, India: Rupa and Co.

Tenet, George. 2007. *At the center of the storm: My years at the CIA.* New York: HarperCollins.

Wilson, John. 2007. Lashkar-e-Tayyeba. Pakistan Security Research Unit Brief no. 12, May 21, University of Bradford, UK.

Yousaf, Mohammad, and Mark Adkin. 1992. *The bear trap: Afghanistan's untold story.* London: Leo Cooper.

Bruce Riedel is a senior fellow in the Saban Center for Middle East Policy at the Brookings Institution. He retired in 2006 after thirty years' service at the Central Intelligence Agency, which included postings overseas. He was a senior advisor on South Asia and the Middle East to the past three presidents of the United States in the staff of the National Security Council at the White House. He was also Deputy Assistant Secretary of Defense for South Asia at the Pentagon and a senior advisor at the North Atlantic Treaty Organization in Brussels.

⊠ Legal Topics

For related research and practice materials, see the following legal topics:

Criminal Law & Procedure Criminal Offenses Crimes Against Persons Terrorism General Overview Criminal Law & Procedure Criminal Offenses Weapons General Overview Transportation Law Air Transportation Air Piracy General Overview

⊠ Notes

1. For a comprehensive look at this network, see Binalakshmi Nepram Mentschel, *Is the Crescent Waxing Eastwards?* (2007).

2. Schroen was CIA station chief in Islamabad when the strike occurred and estimates bin Laden left the camp between thirty minutes to two hours before the missiles struck it (Schroen 2005).

3. Burke interviewed some of the HUM cadre who survived and confirmed the Inter Services Intelligence Directorate (IS') connection (Burke 2003).

4. Jaswant has also discussed the hijack directly with me and his characterization of it as the stage setter for 9/11. See also Zahid Hussain, *Frontline Pakistan* (2007, 62) for more discussion of the nexus behind IC 814's hijacking.

REVIEW QUESTIONS

1. What role did the United States play in creating and supporting the original *mujahedeen* in Afghanistan and Pakistan in the 1980s? Why did the United States take such an approach? What other alternatives might it have pursued to achieve its goals in the region at that time?

2. Reidel argues that Pakistan's ISI has had, and still has (at the time of the article), significant ties to al Qaeda and the Taliban. Are you convinced by his arguments? Why or why not?

3. Is a stable relationship with Pakistan still vital to U.S. interests in Southwest Asia? Why or why not? Does the current relationship with Pakistan help or hinder the U.S. fight against terrorism worldwide?

❖

READING 14

De Silva looks back at the long and tragic 30-year conflict between the government of Sri Lanka and the Tamil Tigers. In May of 2009, the Sri Lankan government defeated the LTTE and now faces the daunting tasks of reconciliation and reconstruction in the majority-Tamil north. De Silva provides recommendations for where the Sri Lankan government should go from here and argues for a significant focus on economic reconstruction.

Post-LTTE Sri Lanka

The Challenge of Reconstruction and Reconciliation

K. M. de Silva

The collapse of the Liberation Tigers of Tamil Eelam (LTTE) in the first half of 2009 took most political observers by surprise, none more so than Western students—including senior politicians and diplomats—of the politics of Sri Lanka. The Rajapaksa government in Sri Lanka skilfully and systematically exploited the defeat of the LTTE to win and retain political support among the Sinhalese population. But, even the government was surprised at the sudden collapse of the LTTE. Like everybody else, it expected the conflict to last much longer.

As the Sri Lankan government of Mahinda Rajapaksa surveys the problems of the island after the Sri Lankan security forces' successful campaign against the LTTE, it would learn very early that it is much more difficult to plan and implement a policy of reconciliation with the Tamil minority than to plan the defeat of the LTTE. For 30 years or more, about a third of the land area of Sri Lanka, and fully two-thirds of the coasts, had been dominated by the LTTE and lay beyond the control of the Sri Lankan state. To bring these areas under the control of the state again is a formidable task, something amounting to a fundamental reconstruction of the territorial structure of the island, the effective reintegration of the northern and eastern provinces with the rest of the country, something that was difficult, if not possible, so long as the LTTE was in control of these areas. These are two of the nine provinces of the island, a structure that

was built by the British in the period from 1833 to 1889. After independence in 1948, the Sri Lanka governments attempted to treat the district, of which there are now 25, rather than the nine provinces, as the largest unit of administration. The provincial structure was revived in the late 1980s partly in response to pressure from the Tamil leadership.

Quite apart from envisaging resistance to the renewal of a possible military challenge posed by any Tamil group, local or overseas, which the government is engaged in doing at present, it would be imperative to deal with the grievances, real or imaginary, of the Tamils, which had given the LTTE so firm a grip on Tamil people, especially those living in the north and east of the island. The northern province is overwhelmingly Tamil, while the Muslims are a substantial minority in the eastern province. They are larger in number, by far, than the Tamils there.

For better or worse, the LTTE proclaimed themselves to be the sole representatives of the Tamil people of Sri Lanka, a position they adhered to for 30 years or more, at much cost to the Tamil people and to the Sri Lankan state. The Tamils paid a huge price for this, both in losing democratic alternatives as well as having to confront the authoritarianism of the LTTE under whom they were compelled to live for several decades. Indeed, there was a systematic destruction of the democratic traditions of the Tamil community, with no compensation for them in the form of an improvement of living standards; on the contrary, there was a decline of living standards. Ever

SOURCE: *India Quarterly*, 66(3), 237–250. © 2010 Indian Council of World Affairs (ICWA).

since the introduction of universal suffrage in Sri Lanka in 1931—or indeed, even earlier than that—the Tamil political leaders had contributed substantially, first, to the construction and then, to the sustenance of Sri Lankan democracy. The LTTE put an end to this and reduced the relationship between the Tamil minority and the Sinhalese majority to a confrontational and military one—a defiant Tamil minority (around 10 per cent of the people) posing a separatist challenge to the Sri Lankan state, with the separatist challenge becoming a military one and the military challenge becoming a terrorist one.

Ever since the LTTE established itself as the predominant force among the Tamils, the relationship between the Tamil minority and the Sinhalese majority (around 70 per cent of the people) was converted to a conflictual encounter in which the economic resources of the island had been drawn into destructive purposes with little or no attention being paid to economic growth or the maintenance of the institutional framework and the infrastructure of the island. The destructive aspects of the decline were more prominent in the areas controlled by the LTTE than in the rest of the island. Nevertheless, the destructive impact of this encounter was not without influence in the Sinhalese areas of the country, admittedly not to the same extent as it was in the areas controlled by the LTTE. One consequence was that instead of any substantial economic growth in the island, one saw a slowing down as economic resources were consumed by the conflict.

For one thing, Sri Lanka which, at one time, was virtually demilitarised, was forced to build up an army, indeed a larger army, per million of the population, than other parts of South Asia, including India and Pakistan. This massive militarisation was a heavy burden on the country but one that was inevitable given the formidable nature of the LTTE challenge to the Sri Lankan state. This was the price the country paid in standing up to and defeating the LTTE. It will be many years before the army could be reduced in size and the Rajapaksa government presently shows no signs of initiating such a reduction in the size of the armed services. In part, fears about a revival of an armed challenge to the Sri Lankan state account for this. The fact, however, is that the prospects of a renewed armed challenge to the Sri Lankan state emerging from the Tamil areas of the country are exaggerated. These are based on a receptivity to fanciful threats being made by sections of the Tamil diaspora in Western Europe and North America. In any event, those threats from the Tamil diaspora could be met more effectively by systematic diplomatic activity rather than continued militarisation. The Sri Lankan state does not need to let down its guard, but it should make a careful and realistic assessment of the potential damage from the aims of fringe groups in the Tamil diaspora. That diaspora amounts to about 800,000 people or about a third or more of the Sri Lankan Tamil population.

The destructive effects of LTTE rule in the Tamil areas of the country involved the diversion of regional and other resources to the maintenance of the LTTE cadres, and for their military and naval campaigns, and included pressure on Tamil families in areas controlled by the LTTE to let children, male and female, join the LTTE forces as soldiers or as child soldiers. There was also the neglect of the regional infrastructure in the areas controlled by the LTTE and the decline of any prospects of economic growth. Nevertheless, the Tamil people were not without pride in the LTTE and the challenge the LTTE posed to the Sri Lankan state. Their violence and destruction was viewed as some sort of compensation for the apparent injustices suffered by the Tamils at the hands of the Sri Lankan state and its officials. The large-scale presence of the Sri Lankan forces—the administration, the police and the defence forces—in Jaffna and the Tamil areas of the country was resented. The LTTE's defiant response to this presence was admired, and regarded as a defence against what were seen as the illegitimate demands and pressures of the state and its officials. The result was that the defeat of the LTTE and the death of Prabhakaran, its leader, and the whole LTTE leadership was treated as an unbearable loss to the Tamil community. Memories of the positive features of the LTTE are likely to remain fresh in the minds of the Tamil people for a long time and, more to the point, would affect the government's attempts at reconciliation—whether this be through policies or gestures. How long these positive memories of the LTTE and its role in the Sri Lankan polity would remain among the Tamils would depend on their receptivity to the Sri Lanka government's reconciliation measures.

The Rajapaksa government has made a start by announcing the appointment of something like a truth and reconciliation commission. How long this commission would take in its deliberations and how it would conduct these deliberations are vitally important factors in any programme of reconciliation and reconstruction. If the South African model of a Desmond Tutu-led reconciliation commission would serve as a point of comparison, much could be achieved. But the ground realities in Sri Lanka would not support any idea of a comparison with the South African situation. There is no equivalent of a Desmond Tutu-like personality to guide the affairs of the Commission; nor is there the likelihood or any credible number of Sinhalese or Tamil personalities to give evidence before such a Commission. For one thing, there is the death of Prabhakaran, the LTTE leader, and the death, also, of the whole LTTE leadership and the absence of an acceptable Tamil leadership, that is to say, acceptable to the Tamil people, that can effectively take over from the LTTE. For another, it is doubtful whether leaders of opposition parties and independent Sinhalese would give evidence before the Commission. Second, a commission of this sort would proceed with great deliberation and this process would be time consuming. And, third, just as time consuming would be the preparation of a report or reports of its deliberations. Fourth, Sri Lanka's recent record in commissions of inquiry does not provide any grounds for optimism about the speedy conclusion of proceedings, or the formulation of proposals and assessment of charges about policies and measures taken.

A more practical alternative would be for a group of advisers of the government to examine the literature on the Sri Lankan conflict that has emerged over the last 30 years or more and having examined this literature, the government could proceed to prepare and implement policies of reconciliation and reconstruction. This process would be quicker and more realistic than waiting for the reports of a truth and reconciliation commission. Another practical alternative would be to invite the Tamil political leadership, emerging from the destruction of the LTTE and representative of other minorities such as the Muslims and the Christians, for discussions on measures of reconciliation and reconstruction. The advantage of these two processes would be that they would be subjected to the ebb and flow of arguments and discussions and evolving political demands. In addition, these measures could work alongside the truth and reconciliation commission, and if properly handled, could reduce the time necessary for the formulation and implementation of an effective policy of reconciliation and reconstruction by such a commission.

The situation the government has to deal with is more complex than a conflict between the Sinhalese and the Tamils. One practical problem that needs to be kept in mind in preparing policies for reconciliation and reconstruction is the fact that Tamil areas in the north of the country and the Tamil and Muslim areas of the east, are among the most underdeveloped parts of the island. This is not a recent or contemporary problem but a historic fact. These areas lack the wealth of resources that the Sinhalese areas of the country have and have had. For example, these areas did not have

the benefits of plantation agriculture that the Sinhalese areas of the country have enjoyed. There was no coffee industry, no tea and no rubber industries in those areas; only some limited extent of land was under coconut production. This situation is not likely to change in the near future, indeed, for years to come. The only possibility is to bring more land under coconut cultivation, if that is at all possible.

Under British rule, the rapidly increasing population in the Tamil areas of the north and east were accommodated in the lower rungs of the administrative structure that evolved in the country, especially in the Sinhalese areas in the southwest of the country. As a result, the Tamils came to dominate the administrative structure of the country for much of the nineteenth century and the early part of the twentieth. There was also the employment in the clerical grades in commercial houses in Colombo, the capital, and elsewhere, that developed under British rule. Sinhalese opposition to the Tamil domination of the public services became a major political factor of discord in the period of the negotiations with the British on the transfer of power in the 1930s and 1940s. In the 1920s and earlier, and indeed till the 1930s and 1940s, the Tamils were in at least 40 per cent of the lower rungs of the public services. Today, this figure is around 4 percent or less. This latter figure needs to be increased and in any policy of reconciliation and reconstruction measures, would need to include such an increase through the equivalent of an affirmative action policy, and on a long-term basis. The reduction of openings for Tamils in state sector employment since the mid-1950s is a grievance of the Tamils.

Any comprehensive analysis of the processes of recovery and reconstruction of the areas of the north and east of Sri Lanka, especially the areas formerly controlled by the LTTE, must begin with the only act of 'ethnic cleansing' in the whole period of conflict between the LTTE and the forces of the Sri Lankan government—the expulsion of the Muslims of the northern province (about 80,000 persons in all) by the LTTE in 1990 and seizure of their property, including houses and other buildings and jewellery and cash, by the LTTE. These Muslims have lived since then as refugees in the south of the country, especially in the Muslim areas of the north-western province but also in the north central province. They also moved to other parts of the Sinhalese areas, especially in the central province, to live among the Muslims there.

Whatever excuses and or reasons the LTTE gave for this exercise in 'ethnic cleansing', the question of compensation for losses suffered never arose. We do not know whether the Muslims formally asked the LTTE for compensation. We do know that the LTTE never paid any compensation. We do not know whether the Muslims expelled by the LTTE from the northern province would wish to go back there and reclaim their houses, shops and other buildings. But now that the LTTE has been defeated, the prospects of the Muslims going back may be more attractive than it would have been if the LTTE remained in power and in control. In any event, there is little prospect that compensation would be available to the Muslims from the resources of the LTTE or from successors of the LTTE. Compensation would have to be negotiated with the Sri Lankan government as part of a wider or overall settlement and responding to the give and take of electoral politics.

There is also the question of the expulsion of Sinhalese from various parts of the northern and eastern provinces. The numbers affected are far smaller than is the number of Muslim victims of LTTE expulsion and, if the people involved are peasants, as they largely are, this problem would be easier to handle for the Sri Lankan government than the problem of the Muslims expelled from the north of Sri Lanka by the LTTE.

Another set of persons who may want to go back to Jaffna are the Sinhalese bakers of the Jaffna area who dominated that trade for decades. A small group, they just moved away in the 1980s and were not expelled. They have expressed a wish to go back to their businesses in Jaffna. In any event, their return is a matter of economic or social forces at work, not merely a matter of concern for those in political power in Jaffna.

 ## Recovery

The 30 years or more of LTTE domination of the northern and eastern provinces were a period of neglect of infrastructure. For instance, the LTTE removed the railway lines in the northern province and used the steel for their own purposes, especially purposes of fortification in the form of bunkers. These railway lines need to be replaced and that replacement will not involve too much time as the track is still available. Only the lines are required and re-laying the lines would not take much time. Fortunately, the Indian government is planning to provide grants for the replacement of these railway lines. Indian assistance is also envisaged for the repair of the port of Kankesanthurai—the most important in the northern province—and the Palali airport in Jaffna.

This would only leave the reconstruction of roads to be undertaken by the Sri Lankan government. The roads of the northern and eastern provinces are not only neglected but also badly in need of repair—whether it be the strategically important A9 linking Jaffna with Kandy and the Sinhalese areas of the country, or the roads in the northern province linking the interior with the coast. This is one of the principal areas of reconstruction and would require grants from the Sri Lanka government; grants would also be necessary for any expansion of the highway system in the northern and eastern provinces. This is an expensive and time-consuming matter but one that is required not only as a gesture of reconciliation but also for welding the road system of the northern and eastern provinces with that in the rest of the country for the wider purposes of national and economic regeneration.

[*Rest of section omitted.*]

 ## Economic Reconstruction

Economic reconstruction is an essential feature of the post-LTTE renewal of the national and regional political systems. The central issue in any policy of economic reconstruction is that the parts of the northern and eastern provinces formerly controlled by the LTTE are some of the most backward areas in terms of economic resources. This is not a recent development but a fact of life in modern times. The British who presided over the transformation of Sri Lanka into a dynamic plantation colony virtually ignored the northern and eastern provinces. The coffee plantations that transformed Sri Lanka's economy in the nineteenth century were all located in the central hills and in Uva. With the collapse of coffee in the 1870s, the renewal of the economy that followed from the 1880s onwards till the 1910s and 1920s came with tea production and with rubber, all of it located in the central hills and the western, southern and northwestern provinces. There was some coconut in the Jaffna peninsula and other parts of the northern and eastern provinces, but the main areas of coconut production were in the western and north-western provinces. The administrative centres of the plantation industry continued to be in Colombo. As in the 1830s to the 1880s, the northern and eastern provinces remained a backwater so far as the plantation economy was concerned.

There was no change in this situation even after independence, except for some significant developments in industry—cement and chemicals—in the early years of independence. These industries were developed by the state in the northern province. By and large, there was little else in the northern province after 1953. Even after the northern and eastern provinces came under the control of the LTTE, there was no industrial development there. Indeed, the LTTE's record as managers of economic development of the areas they controlled was dismal. The records show that they did nothing in the sphere of economic development, nothing that could be considered as a useful contribution to economic growth. The rich rice-producing areas of Mannar in the northern province and the more significant and productive rice-producing areas of the eastern province were neglected. The profits from rice production went largely to the LTTE and those favoured by the LTTE.

Thus, the position after the defeat and collapse of the LTTE is that the northern and eastern provinces are in need of rejuvenation as part of the wider objective of their integration into the national economy. In evolving a set of principles for economic growth in the areas formerly controlled by the LTTE, the Sri Lankan state would need to provide substantial economic resources through the annual budget, for a decade or more, in a spirit of reconciliation, for a successful plan for stabilisation and economic reconstruction. This would be doing something more than the British did for those regions and, more to the point, what the LTTE did.

[*Paragraph omitted.*]

There are two areas in which there is substantial room for development in the northern and eastern provinces— tourism and fisheries.

The beaches of the eastern province have always attracted large numbers of local and foreign tourists. The struggle between the Sri Lankan security forces and the LTTE saw the destruction of many of the hotels in the eastern province and in Trincomalee. The responsibility of the LTTE for this destruction surpassed that of the security forces. These hotels are now being repaired but there is room for larger investments in hotels in the eastern province and the investment would come both from Sri Lankan sources, especially the hotels in Colombo and the western province, and of course, the Tamil diaspora with surplus funds at their disposal, funds which once went to the LTTE.

The northern and eastern provinces have rich fishing resources. They form part of Sri Lanka's fisheries and attract large numbers of people who live on the coasts of the western province and use the period of the southwest monsoon to move to localities in the northern and eastern province when the seas are much calmer there than on the south-western coasts. The island's fishing industry was badly affected by the struggle between the security forces (both the army and the navy) and the LTTE. The fisheries of Jaffna were especially disturbed by both the LTTE and the Sri Lankan security forces. There was, in addition, a special problem of Indian fishermen encroaching on Sri Lankan waters, a problem which was more pronounced in Jaffna than in other parts of the coasts. This is likely to continue.

[*Several paragraphs omitted.*]

The Jaffna peninsula and the northern province, in general, are some of the driest parts of Sri Lanka. Nevertheless, the Tamils of the Jaffna peninsula have developed an efficient system of agriculture dependent on underground sources of water. This agricultural system is efficient largely because of the hard work that goes into these small plots of land, harder by far than in the cultivation of crops in the wetter Sinhalese areas. In the early twentieth century, engineers in Jaffna were debating the merits of exploiting the water resources of the Jaffna lagoon and the Elephant Pass areas for supplying the water needs of Jaffna and the Jaffna peninsula. This scheme was forgotten, partly, because of the expenses involved and partly, because the development of irrigation in other parts of Sri Lanka, culminating in the complex Mahaveli project of the 1970s and 1980s, took priority. These early twentieth century schemes for the supply of water in Jaffna and the Jaffna peninsula could be revived and funded as part of an imaginative part of a policy of reconciliation and reconstruction. It is likely to capture the imagination of the people of Jaffna as a parallel to the impact of the Mahaveli scheme for the Sinhalese population. In recent times, there have been proposals for a diversionary canal to take some of the water from the Mahaveli scheme to parts of the Tamil areas—in particular, the Vanni regions of the northern province. This project is not only very expensive but carries the disadvantage of an excessive evaporation of water in the process of moving from the Mahaveli areas to the northern province. In any event, it would take much longer to bear fruit than an attempt to transfer water from the Jaffna lagoon–Elephant Pass areas to Jaffna and the Jaffna peninsula.

This discussion of irrigation and diversion of water illustrates some of the difficulties that confront the Government of Sri Lanka in any systematic attempt at economic development and reconstruction in the Tamil areas of the country. This process of development and reconstruction is something that goes beyond anything attempted by the British during their rule in the island. It is difficult, expensive and time consuming, but needs to be attempted in any genuine attempt at reconciliation and reconstruction. These schemes could be discussed with the World Bank and the ADB for funding either through grants or soft loans. The model would be the more ambitious funding of the various features of the Mahaveli scheme. Such grants and soft loans were easier to obtain in the 1970s and 1980s. The political objectives of such grants and soft loans—a policy of reconciliation of the Tamils and reconstruction of the northern province if not of the northern and eastern provinces—would make it more likely that such grants would be made as well as soft loans than they would be if this reconstruction of the territories of an aggrieved minority was not at issue.

The Restoration of Democracy in the Tamil Areas of the Country

Reconciliation and reconstruction would lose much of their appeal if they were not accompanied by the rehabilitation of democracy in the Tamil areas of the country. The LTTE is primarily responsible for the systematic destruction of democracy among the Tamils. At the same time, the so-called Tamil moderates cannot be absolved of all blame for this situation in so far as the support they gave the LTTE is remembered, as well as the Tamil moderates being the source of some of the programmes of the LTTE and their political objectives. The separatist claims of the LTTE such as the concept of a Tamil homeland go back to the Federal Party of the 1950s and 1960s. Even when Tamil moderates voted at parliamentary or presidential elections, they tamely permitted the LTTE to lay down the parameters of their programme. For instance, they would not vote if the LTTE declared that they should not vote. Candidates for parliament would either be chosen by the LTTE or in consultation with them; indeed, always with the approval of the LTTE. Such parliamentarians would be guided by the instructions or wishes of the LTTE in their votes in parliament. This subordination of the moderates to the demands of the LTTE is a factor in the decline and collapse of democracy among the Tamils of the northern province. The Tamils in other parts of the island did not succumb to the pressures of the LTTE to the same extent. The Tamil plantation workers were generally free of LTTE pressures, although they were not inconsiderate of LTTE interests.

The virtual collapse of the system of local government in the northern and eastern provinces, in particular, the collapse of the village councils, for which the Sri Lanka government and the LTTE must share the blame, contributed very much to the general decline of democracy in the northern and eastern provinces. The LTTE were not enthusiastic about local government institutions and both the Tamil moderates and the Sri Lankan government accepted this situation as an inevitable consequence of the domination of the LTTE in the areas they controlled. The culpability of the government was seen most prominently in the failure to establish a provincial council for the northern province. The fear was that it would be controlled by the LTTE. A provincial council for the eastern province was established after the Tamils of the eastern province broke away from the LTTE. When elections were held for this council, former LTTE cadres captured power at the expense of the Tamil moderates and the Muslims. The Tamil groups which contested these elections fought as an armed militia. Their inevitable victory was not an example of the operation of democracy so much as an attempt to secure the eastern province as an area free of the LTTE but not free of an armed Tamil militia operating with the consent of the government. Democracy was no part of this whole sequence of events. It was largely a matter of a group of Tamils from the LTTE, breaking away from the LTTE, and working with the Rajapaksa government. Whatever practical advantages there were to the government in its wider struggle with the LTTE, the question of democratic practices did not enter the picture.

The disarming of these Tamil groups now in control of the provincial council of the eastern province is an essential prelude to the democratisation of the eastern province. These armed Tamil groups are used as shock troops for the manipulation of elections in the eastern province. This manipulation is done either for their own purposes—not confined to electoral politics—or on behalf of the Rajapaksa government.

Indeed, if the LTTE was an important factor in the decline and destruction of democracy in the northern and eastern provinces, the government's culpability lies in its tolerance of those armed Tamil groups in the eastern province and elsewhere. The need to protect these erstwhile LTTE cadres from retaliation by the LTTE was used to justify this policy. Now that the LTTE has been demolished, these Tamil groups in control of the eastern province should be compelled to hand back their arms to the police or to the army. The Tamils are a minority of about 20 per cent there.

The fact that these Tamil groups control the provincial council of the eastern province serves to exclude the Muslims who are a dominant minority in that province (as much as 40 per cent of population there) from their legitimate share of power in that province. The fact that these Tamil groups contested the elections to the provincial council as an armed militia was a distortion of the electoral process. That they have not been disarmed yet is striking evidence of the culpability of the government in the continued decline of democracy, and its tolerance of the distortion of democracy in the eastern province. Should these Tamil groups in control of the eastern province be disarmed by the state, it would be necessary to hold a fresh election to the provincial council of the eastern province. This fresh election follows logically from the disarming of the Tamil groups and the Tamil militia there.

The democratisation of the northern and eastern provinces will require the establishment of a provincial council for the northern province. The elections to that provincial council will enable political parties and individuals to vie for positions in that council. And such a council will give the Tamils of the northern province opportunities in fashioning the development of that province; indeed, to become a force in the democratisation of the province.

Along with a provincial council for the northern province, there needs to be the establishment of village councils in both the northern and eastern provinces. This would provide all segments of the population of those provinces—Tamil, Muslim and Sinhalese—opportunities to manage the affairs of those provinces at a local level, which would not only contribute to the democratisation of the country as a whole but would give the local population a voice in the development of their regions. These councils would serve as a check on the bureaucracy, both state and provincial, and a source of information on the needs of the people. They would also be a platform for the maturation of alternative sources of leadership and diminish the power and influence of parliamentarians.

The processes of democratisation are necessarily slow moving. Once these councils are established, they would provide another layer of leadership in Sri Lankan democracy. They could also become centres of parochial interests and obstacles in the wider democratic processes. But that is the price that democratisation involves.

Fortunately, what is required is not the establishment of democracy *de novo*, but the re-establishment of democracy, a return to democratic systems and practices that had been in vogue before the advent of the LTTE. The right to vote had been in existence since 1931 and along with it, parties and associations that appealed to the electorate on the basis of a political platform and organisations that sought places in

their councils on the basis of programmes and promises. For instance, the Tamil Congress had been established in 1944 to win the support of Tamils in all parts of the island in a campaign to fashion a programme of action during the national debates on the transfer of power. The Tamil Congress dominated Tamil politics till around 1955. Their rival, the Federal Party, was established in 1951 and survived till the 1970s as the main party of the Tamils. The Federal Party defeated the Tamil Congress in the mid- 1950s, and like the Tamil Congress, the Federal Party spoke on behalf of Sri Lanka's Tamils. Both were separatist in outlook, the Federal Party much more so than the Tamil Congress. The LTTE's separatism sprang from that of the Federal Party and the Tamil Congress—the Federal Party much more than the Tamil Congress. The Tamil Congress and the Federal Party were driven to political wilderness by the LTTE, with some of their leaders (especially leaders of the Federal Party) being assassinated by the LTTE.

Currently, with the destruction of the LTTE by the Sri Lankan army, and the death of the Tamil Congress leader, G.G. Ponnambalam, the political landscape in the Tamil areas of the north and east of Sri Lanka lacks any significant political party. The national political parties based in Colombo and elsewhere have been unable to establish themselves in the Tamil areas of the island. The present lack of any substantial political parties in the north and east is likely to benefit them. Much more likely is either the emergence of new political parties seeking to build on resources left behind by the Tamil Congress and Federal Party, such as they are, or a revival of either of these parties. Whether they will continue to be purely regional or ethnic parties or search for a national role either on their own or in association with the more prominent Sri Lankan political parties, will form part of the revival of democracy in the Tamil areas of the country. Such a revival will depend on the health of democracy in Sri Lanka as a whole.

The Tamil areas of the country were not the only part affected by the collapse of democracy that followed upon the LTTE's rise to influence. Sri Lankan democracy, as a whole, lost much of its integrity and spirit in the long struggle with the LTTE. A country which had its first election under universal suffrage as early as 1931 continued to have elections even when the LTTE was a prominent political factor in Sri Lanka. But in recent times, national elections have been flawed by malpractices. The exit of the LTTE from the national scene will help in the revival of democracy, both in the Tamil areas which the LTTE dominated and in the broader Sinhalese areas which they did not. The major political parties in the island will have the responsibility of helping in the revival of past standards in electoral politics in the Sinhalese areas of the island, and in the renewal of democracy in the northern and eastern provinces. The latter process will depend heavily on the democratic traditions of the island.

[*A paragraph has been omitted.*]

 ## The Regional Factor

Sri Lanka was caught up in the rivalry between India and China. If India was perturbed at the seemingly increased Chinese influence in Sri Lanka, the relative importance of the assistance given to Sri Lanka in its conflict with the LTTE was the principal cause. When Sri Lanka needed a regular supply of material for the army, India's attitude was ambivalent, at best, while Pakistan, first of all, and China to a much greater degree provided the arms required by the army. The army's victory against the LTTE would have been much more difficult without Pakistan and Chinese, in particular Chinese, arms supplies.

In addition, Chinese supplies and assistance was available without the pressure on Sri Lanka with regard to measures required for the management of Sri Lanka's ethnic conflict—principally legislation and institutions that came from India. In 1986, the LTTE had been routed by the Sri Lanka army under General Ranatunga and an attack on Jaffna—to seal the victory—seemed very likely, when the Indian government intervened on behalf of the LTTE and warned the Sri Lanka government against any attack on Jaffna. The LTTE survived and moved across the Palk straits to Tamil Nadu and elsewhere in India and it took the Sri Lankan governments another 23 years before they could decisively defeat the LTTE again. General Fonseka who fashioned the defeat of the LTTE was a Lt Colonel in the army in 1986.

Had the Indian government not intervened on behalf of the LTTE in 1986, Sri Lanka would have been spared 20 years or more of bitter conflict. In 2009, the LTTE was anxious to get the Indian government to mediate in Sri Lanka again to insist on a halt to the army's campaign against the LTTE and also, on a renewal of peace talks. But on this occasion, the Indian government did not intervene—it only declared an interest in the welfare of the Tamils in Sri Lanka, especially in the welfare of those rendered homeless by the conflict, by which they meant the Tamils of the Jaffna peninsula and the northern province.

After the defeat of the LTTE, the Indian government returned to declaring a concern for the welfare of the Tamils displaced by the conflict. The Indian government announced that they would provide grants for the construction of 50,000 houses being jointly managed by the Indian and Sri Lankan governments. There was pressure from Tamil Nadu to get the Indian government to insist on a system of devolution of power in Sri Lanka that would benefit the Tamils—presumably the Tamils of Jaffna and the northern province. So far, the Indian government has refrained from making this demand.

The Indian government's reluctance on this—so far—is understandable because Sri Lanka's system of devolution based on provinces, introduced in 1987, was modelled on the Indian system of provincial government and moreover, was introduced at the insistence of the Indian government in 1987, and more to the point, it was introduced on behalf of the Tamil minority. The system of district councils introduced in Sri Lanka in 1980 after much discussion and with the support of the Tamil leadership was abandoned in 1987 to make way for the provincial councils which the Indian government advocated, again in response to Tamil pressure. The Tamil leadership preferred provincial councils over district councils. Having secured the provincial councils system in 1987, the Tamils moved to secure the larger objective of an amalgamation of the northern and eastern provinces. This was done by President Jayewardene in 1988 using his special powers as president. Nearly 20 years later, the procedure in the amalgamation process was challenged as unconstitutional by sections of the opposition. The Supreme Court held in favour of this application. The Rajapaksa government accepted the decision with much enthusiasm and moved to create a council for the eastern province. What it did not do was create a council for the northern province. The lack of a provincial council for the northern province remains to be done, a significant gap in the devolutionary process in Sri Lanka.

Confronting the record of Chinese economic assistance in Sri Lanka, the Indian government has followed this example by generous economic assistance of its own, thus engaging in a rivalry with the Chinese. In seeking to influence Sri Lanka in a situation of a rapid expansion of Chinese power and influence in Asia, the Indian government has two obvious disadvantages. First, its concern for the Tamil community in the island makes others in the island suspicious of Indian policies. Second, and perhaps even more disadvantageous, is the pressure from Tamil Nadu, not just the government and the opposition, but also fringe pressure groups who are more enthusiastic than the government and the opposition in Tamil Nadu and much less responsible in utterances and aspirations. In the context of rapidly increasing Chinese interest in Sri Lanka, India would do well to be more expansive in its concerns, to move from being advocates of Tamil interests in the island, and to consider also other minorities, especially the Muslims and, of course, the interests of the Sinhalese majority.

Notes

1. The remainder of the population is divided among Hazara (9%), Uzbek (9%), Aimak (4%), Turkmen (3%), Baloch (2%) and other (4%) (CIA World Factbook, 2012).

2. The Pakistan Taliban has made explicit threats to carry out attacks in the West and has been linked to a 2008 plot in Barcelona, Spain, to attack the city's transportation system (Stenersen, 2009, p. 3).

3. The Pakistani Pashtuns reside mostly in the areas bordering Afghanistan where they provided safe-haven and fighters for the Pashtun Afghan Taliban following the U.S. invasion.

4. The remainder of the Pakistani population is divided among at least four other ethnic groups: Sariaki (8.38%), Muhajir(7.57%), Balochi (3.57%), and other (6.28%) (CIA World Factbook, 2012).

5. In March 2002, a senior al Qaeda official was arrested at a LeT safe house, suggesting that members of LeT were aiding al Qaeda (U.S. Department of State, Foreign Terrorist Organizations, n.d.).

6. Rajiv Gandhi became prime minister of India following his mother's assassination by her Sikh bodyguards in 1984. Gandhi incurred LTTE wrath when, as prime minister in 1987, he sent Indian peacekeeping troops into Sri Lanka. His assassination was carried out by a LTTE female suicide bomber; 14 others were also killed in the attack (BBC News, On This Day, May 21, 1991).

7. The actual number of those killed varies depending on the report. For example, the U.S. Department of State places the number of casualties at 183, the Congressional Research Service at 174, and the Global Terrorism Database at 171. The number of foreigners among the casualties also varies but is generally placed at over 20.

REFERENCES

Baker, A. (2011, May 12). Why We're Stuck With Pakistan. *Time*. Retrieved from http://www.time.com/time/magazine/article/0,9171,2071131,00.html

Origins of Sikhism. (2009). *BBC Religions*. Retrieved from http://www.bbc.co.uk/religion/religions/sikhism/history/history

Bloom, M.M. (2003). Ethnic Conflict, State Terror, and Suicide Bombing in Sri Lanka. *Civil Wars*, 6, 54-84.

Central Intelligence Agency. (2012). *CIA World Factbook*. Retrieved from https://www.cia.gov/library/publications/the-world-factbook/index.html

de Silva, K.M. (2010). Post-LTTE Sri Lanka: The Challenge of Reconstruction and Reconciliation. *India Quarterly: A Journal of International Affairs*, 66, 237-250.

Fair, C.C., Crane, K., Chivvis, S., Puri, S., & Spirtas, M. (2009). Pakistan: Can the United States Secure an Insecure State? Santa Monica, CA: RAND Project Air Force Report.

Flanigan, S.T. (2008). Nonprofit Services Provision by Insurgent Organizations: The Cases of Hizballah and the Tamil Tigers. *Studies in Conflict and Terrorism*, 31, 499-519.

Global Terrorism Database. (2012). Retrieved from http://www.start.umd.edu/gtd/

Harshe, R. (2005). India-Pakistan Conflict Over Kashmir: Peace Through Development Cooperation. *South Asian Survey*, 12, 47-60.

Kronstadt, K.A. (2008). Terrorist Attacks in Mumbai, India and Implications for U.S. Interests. Washington D.C.: CRS Report No. R40087, December 19.

National Counterterrorism Center. (2012). *Counterterrorism Calendar*. Retrieved from http://www.nctc.gov/site/index.html

Nuclear Threat Initiative. (2012). *Country Profiles: Pakistan*. Retrieved from www.nti.org/country-profiles/pakistan

Reidel, B. (2008). Pakistan and Terror: The Eye of the Storm. *Annals of the American Academy of Political and Social Science*, 618, 31.

Scheuer, M. (2004). *Imperial Hubris: Why The West Is Losing The War On Terror*. Washington D.C.: Potomac Books, Inc.

Spindlove, J.R. & Simonsen, C.E. (2010). *Terrorism Today: The Past, The Players, The Future*, (4th ed.) Upper Saddle River, NJ: Prentice Hall.

Stenerson, A. (2009). Are the Afghan Taliban Involved in International Terrorism? *CTC Sentinel*, 2(9), 1-4.

Thakur, R. (2011). Delinking Destiny From Geography: The Changing Balance of India-Pakistan Relations. *India Quarterly: A Journal of International Affairs*, 67, 197-212.

United States Department of State. (2011). Country Reports on Terrorism 2010. Retrieved from http://www.state.gov/j/ct/rls/crt/2010/170258.htm

United States Department of State. (2011). Foreign Terrorist Organizations. Retrieved from http://www.state.gov/j/ct/rls/other/des/123085.htm

White, J.R. (2012). *Terrorism and Homeland Security* (7th ed.) Belmont, CA: Wadsworth-Cengage Learning.

Wright, L. (2006). *The Looming Tower: Al Queda and the Road to 9/11*. New York, NY: Alfred A. Knopf.

REVIEW QUESTIONS

1. What recommendations does the author make with regard to both Sri Lanka and the Tamils? How might these recommendations be applied, if at all, to other separatists conflicts?

2. Can you tell from the article if the author is pro or anti-Tamil? How, if at all, do you think any bias has affected the recommendations being made?

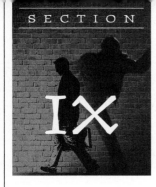

Europe, Turkey, and Russia

Learning Objectives

At the end of this section, students will be able to:

- Describe the major terrorist groups in the region and their associated attacks from the early 1970s to the present.
- Describe the differences between the national/separatist, ideological, and Islamic fundamentalist groups that have operated in Europe over the decades and identify the states where these different types of groups were most prevalent.
- Describe the history of the relations between England and Ireland that eventually led to the "Troubles" in Northern Ireland beginning in 1969.
- Trace the evolution of the conflict in Chechnya from a nationalist movement to one dominated by Islamic fundamentalists.
- Assess how the experience with nationalist/separatist and left-wing ideological groups impacted the ability of the European states to deal with terrorism in the 21st century.
- Assess and analyze the current threat of Islamic fundamentalist groups operating in Western Europe.

Section Highlights

- Nationalist/Separatist Terrorism in Europe
- Left-Wing Groups in Europe
- Sleeping in Europe: Middle-Eastern Terrorists Operating in Western Europe
- Russia and Chechnya

⬛ Introduction

The region referred to by the U.S. Department of State as *Europe and Eurasia* covers all of the Western European countries through to Turkey's border with the Middle East and Southwest Asia, as well as Russia and the countries of the former Soviet Union (variously known as the Balkans or Eastern Europe). This section will highlight the activities of a variety of terrorist groups that operated in Europe from the mid-1960s and early 1970s through the present day. Over the decades—if not, in some cases, centuries—Europe has seen its share of terrorism crossing the spectrum of group types, including: (1) nationalist/separatist groups; (2) left-wing ideological terrorist groups; and (3) Islamic fundamentalists, including those affiliated, however loosely, with al Qaeda.

▲ Map of Europe

The first part of this section will look at some of the world's longest standing nationalist/separatist terrorist groups, including the Provisional Irish Republican Army in Northern Ireland, the **Basque Fatherland and Liberty** (ETA) in Spain, and the **Kurdish Workers Party** (PKK) in Turkey. The second part will look at the legacy of a number of left-wing ideological groups that operated in the region from the mid-1960s into the 1990s, groups such as the **Red Army Faction** (RAF) in Germany, the **Red Brigades** in Italy, **Action Direct** (AD) in France, the **Communist Combatant Cells** (CCC) in Belgium, and **November 17** in Greece. Although many of these groups have died out or pose relatively little danger today, their activities and demise provide important lessons about the existence of, and response to, terrorist groups operating within Western democracies. Finally, the last part of this section will look at the threat of Islamic fundamentalist groups that have been operating in Western Europe and Russia since the early-mid 1990s, sometimes with devastating effects.

There is some debate as to whether Turkey should be addressed as part of treatments of Europe or treatments on the Middle East. Since the days of the Ottoman Empire, Turkey has at one time or another straddled discussions of both regions. It is a member of NATO and has sought membership in the European Community (EC). At the same time, it has moved closer in politics and culture to its Islamic heritage. Today, it certainly sees itself as a major player in Middle East diplomacy and even as a broker between the United States and the West—with which it maintains strong ties—and its Middle East neighbors. Nevertheless, because Part II of this book tries to mirror the bureaus of the U.S. Department of State as an organizing tool, the authors have chosen to leave Turkey in this section for the sake of continuity.

Nationalist/Separatist Terrorism In Europe: Three Case Studies

The United Kingdom, Spain, and Turkey have been home to several well-known terrorist groups of the nationalist/separatist variety—those groups whose stated goal is the overthrow of an occupying power, the establishment of a new state, or, at the very least, increased autonomy or control over the affairs of their own region. What quickly becomes evident is that while the disputes are nationalist in origin, the side one takes in these disputes can also be based on religious affiliation, making the lines between nationalist and religious terrorism blurry at best. Probably nowhere is this seen better than in **the "Troubles"** of Northern Ireland.

The "Troubles": The United Kingdom and Northern Ireland, 1066 to the Present

Conquest and its consequences set the stage for the modern history of both England and Ireland. Starting in the late eighth century, both were subject to conquest and marauding by the Viking tribes—in particular those known as the Danes. Eventually, England was consolidated under the native Saxon rule of Harold Godwinson, but he soon found himself overthrown by William the Conqueror in 1066. From that point on, both England and Ireland found themselves under Norman rule. Indeed, the current British royal family can trace its line back through to that of William.

Conquest, however, be it Viking, Norman, or British, has never set well on the Emerald Isle. Irish loyalty was to tribe and clan rather than sovereign. Once Saint Patrick converted the island to Christianity in the fifth century, that loyalty extended to the Catholic Church and its pope. Never in their history did the Irish see themselves as anything but Irish—different from the British, the Scots, and the Normans in culture, values, language, and belief systems (White, 2012).

Britain and Ireland in the 16th and 17th Centuries

For purposes of this discussion, things began to come to a head between the Irish and the British in the 16th and 17th centuries. In the early 1530s, Henry VIII of England—eager to annul the marriage with his first wife, Catherine of Aragon—broke with the Roman Catholic Church and its pope and founded the Church of England. Not surprisingly, this did not sit well with the Irish who were by now fervently Catholic (Simonsen & Spindlove

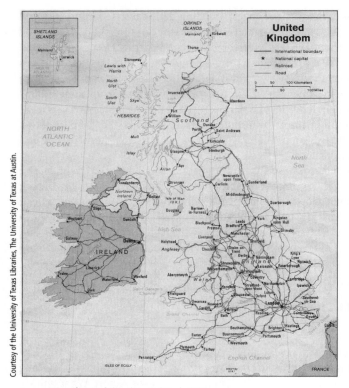

▲   Map of United Kingdom, Political, 1987

2010; White 2012). A rebellion against the English monarchy was the result. The rebellion was less than successful.

Adding insult to injury, Elizabeth I—daughter of Henry VIII and his second wife, Ann Boleyn—established the **Plantation of Ulster** in the late 16th/early 17th century. The Plantation of Ulster allowed English settlers on the Emerald Isle to take over the most prosperous agrarian section of Ireland, the North, and push out the Irish peasants. This move set the tone for a division of territory that ran along religious as well as geographic lines. The British settlers of Northern Ireland followed the Church of England (Protestant) while the displaced peasants—as well as most of those on the remainder of the Isle—were devoutly Catholic.

By the mid-17th century, England was in the midst of a brutal civil war between those who swore allegiance to the Church of England and those who wanted to return Britain to a Catholic nation. English Catholics fled to Ireland for refuge and tried to use Ireland and Irish Catholics as a base for their efforts against the English Protestants. They lost. The two main defeats of the Irish and English Catholics—the battle at Derry in 1689 and the Battle of the Boyne in 1690—are still celebrated today by the Protestants in Northern Ireland, much to the annoyance of the Northern Irish Catholics.

Britain and Ireland in the 18th and 19th Centuries

While the conflicts of the 16th and 17th centuries followed the religious lines of the English Civil War, the conflicts of the 18th and 19th centuries were more strongly characterized by efforts against British rule by both religious groups. By this time, the English settlers of old had been on the Isle for several generations and saw themselves more as Irish than British. It did not help matters that the British elite surrounding England's monarchy saw Ireland as something of a backwater possession. As a result, the religious differences between Irish Protestants and Irish Catholics were put aside (or at least ignored for a time) in favor of getting out from under British rule.

This alliance did not last long, however. Starting in 1845, a devastating series of potato famines occurred across Ireland. Potatoes were the mainstay crop of Ireland, especially among the peasants, who were more often than not Catholic. As a result of the potato famines, the Irish population was decimated as many either emigrated (mostly to the United States) or starved to death. The more prosperous Protestant land owners survived and even increased their wealth. By the end of the 19th century, the Protestants in the North had solidified their hold over the area.

The Split of the Emerald Isle and the Development of the IRA

By the early 20th century, the Irish polity had divided itself into roughly three groups: (1) those who argued for Home Rule whereby Ireland would be granted autonomy with its own parliament under the British monarchy;

(2) the Republicans, who advocated for full independence for all of Ireland including the North; and (3) the Unionists, who wanted to maintain Northern Ireland's union with Britain (White, 2009). The Republicans and the Home Rule advocates were Catholic; the Unionists were Protestant. By the early 20th century, all three had once again taken up arms.

The most famous—if not infamous—of the Republican groups to take up arms was the Irish Republican Army (IRA). The Irish Republican Army had its roots in a group known as the Irish Republican Brotherhood (IRB), which had formed in the 1850s. The IRB was responsible for a disastrous uprising in 1916 known as the Easter Uprising. However, rather than dying out, the IRB developed into the more well-known and long-standing IRA, partly as a consequence of the perceived overreaction by the British to the uprising.

One of the IRA founders—Michael Collins—used the "new" organization to begin a campaign of selective terrorism against British officials and representatives of the British government. The result was the Black and Tan War (1920–1921). The war was fought between a number of Irish armed groups (including the IRA) and a force hastily recruited by the British to deal with the increasing violence.

The Black and Tan War ended with the Treaty of 1921, which granted independence to the main part of the Emerald Isle while keeping Northern Ireland under British rule until a peaceful solution could be worked out. The Treaty—and specifically the provisions regarding Northern Ireland—generated a civil war between the newly constituted Irish Army and the IRA. The IRA rejected any British control of any part of Ireland. In the meantime, Britain's response to what it viewed as an internal Irish conflict was to hunker down, tighten its hold over the North, and stay out of the hostilities as much as possible. As part of this plan, the British created the **Royal Ulster Constabulary** (RUC), consisting mostly of Northern Irish Protestants, to serve as the police force for Northern Ireland.

As a result of the Treaty of 1921 and the civil war that followed, Ireland found itself divided into two entities: Ireland and Northern Ireland. The IRA continued to engage in violence justified by the goal of a fully independent island, but it also began to step up its legitimate political activities through its political party, Sinn Fein. Its increased political activities and its need to maintain a supportive constituency in and out of Ireland made the IRA too moderate for some, leading to the development of the Provisional Irish Republican Army (PIRA) or the "Provos."

The "Troubles": 1969 and Beyond

The **Provos** would be responsible for many of the most tragic and heinous attacks in the conflict over Northern Ireland, especially in the latter part of the 20th century. In fact, their activities in the 1950s and 1960s, in particular their targeting of the RUC, went so far in the terms of sheer violence that they alienated support from even Irish Catholics and ceased operations in 1962. The story of the PIRA might have ended there were it not for the British themselves. In a stunning example of a government being "hoist by its own petard," the British response to events in 1969 actually reinvigorated the Provos.

By the late 1960s, considerable civil rights issues had developed over the imbalance of political and economic opportunities in Northern Ireland. Both areas of opportunity significantly favored Protestants over Catholics. In 1969, the animosity between the two sides erupted into violence. The Northern Irish Government (the vast majority of which was Protestant) then made a fatal calculation by using the RUC and a hated reserve force known as the B-Specials (also staffed by Protestants) to crack down on civil rights workers and demonstrators. Instead of ending the violence, the use of the RUC and the B-Specials exacerbated it.

By August 1969, the British Army had to be called in to control the violence, which hitherto had been an Irish-versus-Irish dispute. In fact, by the time the Army was called in, the violence had escalated so much that even many Irish Catholics welcomed the British Army as the only force that could rein in the RUC and the B-Specials. Instead, misunderstanding the situation entirely, the British saw the situation as a "colonial war" to preserve the British influence in a colony and sided with the Unionists.

The British actions lost them any support they may have had among Irish Catholics and reenergized both the IRA and the Provos. Support for the IRA grew even more in 1972 when a demonstration in Londonderry by Roman Catholic civil rights supporters turned violent as British paratroopers opened fire, killing 13 and injuring 14 others (one of the injured later died). Known as Bloody Sunday, this event precipitated an upsurge in support for the Provos, which advocated violence against the United Kingdom to force it to withdraw from Northern Ireland.

Over the next several decades, the Provos—in various incarnations—conducted bombings, assassinations, kidnappings, punishment beatings, extortion, smuggling, and robbery. Their targets included British government officials and military targets (including the RUC) as well as civilian targets in both Ireland and Britain. They also targeted Catholic civilians suspected of cooperating or sympathizing with the British. Two of their most famous attacks were the bombing of the well-known Harrods department store in London in 1983 and the assassination in 1979 of Lord Louis Mountbatten. Lord Mountbatten—a career Naval officer—had served as the First Lord of the Admiralty and the last Viceroy of British India. He was also a member of the Royal Family (cousin to Queen Elizabeth and uncle to her husband, Prince Philip Mountbatten).

In response to the violence by the Provos, various Unionist (Protestant) paramilitary groups—such as the Ulster Volunteer Force (UVF)—sprung up throughout Northern Ireland. Eventually, the British Army was tasked with trying to control the Unionist terrorist groups while still supporting the by-now hated RUC.

Finally, in 1985, after over 20 years of fighting, the parties agreed to the Anglo-Irish Accord, which attempted to create autonomous governing institutions in Northern Ireland that would include a number of civil rights reforms. Neither side was especially supportive of the agreement and it was not until the **Belfast Agreement** of 1998 that progress began to take shape. The Belfast Agreement established shared governance over Northern Ireland and a revamped criminal justice system that had been plagued with abuses and overzealousness throughout the "Troubles." In addition, the Belfast Agreement resulted in the release of a number of political prisoners, created a Human Rights Commission, and required all paramilitary organizations to disarm. PIRA ceased activities and turned over its weapons in 2005.

The story may not end there. At the time of this writing, slow implementation of the provisions of the Belfast Agreement, as well as hard economic times in Northern Ireland, have resulted in protests and some rioting. At least one new group—the **Real Irish Republican Army** (RIRA)—has emerged claiming to be the successor to the old PIRA and has engaged in some relatively minor attacks in Northern Ireland. Many of the RIRA members are believed to be former PIRA members who left the Provos after the 1997 cease fire (U.S. State Department, 2012). In 2011, the RIRA was linked to at least seven attacks on Northern Irish office buildings, government facilities, banks, and Police Service of Northern Ireland (PSNI) targets using improvised explosive devices (IEDs) (U.S. State Department, 2012). According to a 2012 U.S. State Department report, the RIRA has pledged to continue the use of violent attacks designed to disrupt the peace process and force the removal of British forces from Northern Ireland.

Spain's Basque Region and the ETA

The Basque people are situated in the mountainous region that borders Spain and France. They have their own distinct culture and language and have long viewed themselves as autonomous from Spain and entitled to govern themselves. The Basque population in Spain totals somewhere in the area of three million people (Idoiaga, 2006). The modern-day conflict between the **Basques** and the Spanish government began to take shape in the 1930s, when the then dictator of Spain, Francisco Franco, outlawed the cultural practices of the Basque people, banned the use of their language, and imprisoned or killed numerous Basque civilians. One of the consequences of Franco's actions was the establishment of the Basque Fatherland and Liberty (ETA) group in 1959. The goal of the group was the establishment of a Basque homeland consisting of the Spanish and French Basque regions and based on Marxist governing principles (U.S. Department of State, 2011).

ETA began its campaign of violence in 1968, and since that time, the conflict between ETA and the Spanish government has claimed over 850 lives. ETA targeted Spanish officials, government facilities, journalists, tourism areas, and civilian targets during its campaign. Some of its most deadly attacks in just the first decade of the 21st century included:

♦ A February 2005 attempted assassination of King Juan Carlos and visiting Mexican president Vicente Fox that injured more than 20 people.

♦ A 2006 car bombing at Madrid's international airport.

♦ A July 2009 attack on Civil Guards barracks that wounded more than 60 people, including children. (U.S. Department of State, 2011)

However, as of this writing, ETA has lost much of its support over the last few decades as Spain has given the region considerable gains in autonomy and self-governance. In fact, the Basque Autonomous Community and Navarre "enjoy the highest level of self-governance of any non-state entity with the European Union" (Idioaga, 2006,

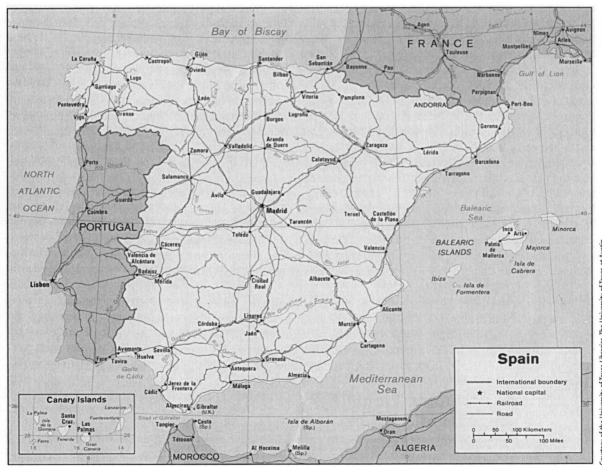

▲ Map of Spain, Political, 1982

p. 3). As a result, the idea of complete independence from Spain simply does not hold the same attraction for much of the region as it once might have. In addition, as of 2013, Spanish and French authorities had arrested approximately 750 ETA members and estimate the remaining active members to number only 100 (U.S. Department of State, 2011). Still, old assumptions regarding ETA's support and capabilities remain. For example, when Madrid's commuter system was attacked in 2004, the early assumptions were that ETA rather than Islamic jihadists were responsible. In October 2011, ETA announced a "definitive cessation" of violent activity. However, "as the group has made and broken several past cease fires, Madrid rejected this . . . announcement and continues to demand that ETA disarm and disband" (U.S. State Department, 2011).

Turkey and the Kurds

The Kurdistan Workers Party (PKK) was started in the Kurdish region of Turkey in 1978 and turned violent beginning in 1984. The **Kurds** are an ethnic group whose geographic origins span the mountainous region crossing Turkey, Iran, Iraq, and Syria. Turkey is home to almost half of the world's 30 million Kurds (Vicks, 2013). Prior to World War I (WWI), this region fell under the auspices of the **Ottoman Empire,** the vast empire that ruled much of the Middle East for centuries and was the precursor of modern-day Turkey. During WWI, the Ottoman Empire fought alongside Germany and Austria-Hungary (the **Central Powers**) against Great Britain, France, Russia, Italy, Japan, and—in the last year of the war—the United States (the **Allied Powers**). When the Central Powers fell in defeat in 1918, the Allied Powers divided the region in ways designed to meet their national interests, often with cavalier disregard for issues of ethnicity and cultural identity. This approach set the stage for many of the region's current conflicts. In doing so, the Allied Powers broke promises to the Kurdish people for their own homeland. Under the secular military that dominated much of Turkish life in the decades that followed, much of the Kurdish culture, identity, and language was vigorously suppressed. By the late 1970s many of the Kurds had had enough.

Founded by Abdullah Ocalan, the PKK originally sought a Marxist-Leninist state for the Kurdish people situated in the region shared by Turkey, Iran, Syria, Armenia, and Iraq. However, according to the U.S. Department of State (2011), "in recent years [the PKK] has spoken more about autonomy within a Turkish state that guarantees Kurdish cultural and linguistic rights." Current estimates place the group's number at about 4,000 to 5,000 with approximately 3,500 located in Northern Iraq (U.S. Department of State, 2011). While the group's targets in the first decade of this century have been predominantly Turkish, the PKK has targeted a variety of groups in the past. These targets included Europeans, Turks, rival Kurdish groups, and supporters of Turkey in Europe and Turkey. Of particular concern to the United States was the group's willingness to target NATO; the PKK attacked a NATO base in 1986 and kidnapped 19 Western tourists in 1999.

In recent years, the PKK—which has now changed its name to the Kongra-Gel (NTC, 2012)—has toned down its left-wing rhetoric in exchange for a slightly more religious bent, a move likely to make it even more of a concern to the secular-minded military, one of the most powerful institutions in Turkey. According to the National Counterterrorism Center (NTC), the activities of Kurdish separatist groups over the decades have resulted in more than 30,000 deaths. Despite an October 2006 ceasefire called by the PKK's leader, Abdullah Ocalan, attacks on Turkish tourist targets continued through 2008 (NTC, 2012). "For decades, the two sides have ground away at each other with a dogged violence that has marred Turkey's otherwise remarkable transformation into a major diplomatic and economic player in the region" (Vick, 2013, n.p.).

Still, as the 21st century moves through its second decade, things seem to be changing. In 2003, Recep Tayyip Erdogan, leader of the Justice and Development Party (AKP), was elected prime minister of Turkey. He was elected to a third term in 2011, and *Time* magazine referred to Erdogan as "the most transformational leader in Turkey since Mustafa Kemal Atatürk founded the modern Turkey in 1923" (Foroohar, 2011, p. 36). Notable about Erdogan is his Islamist yet—for the moment at least—somewhat moderate leanings, which are in stark contrast to the staunchly

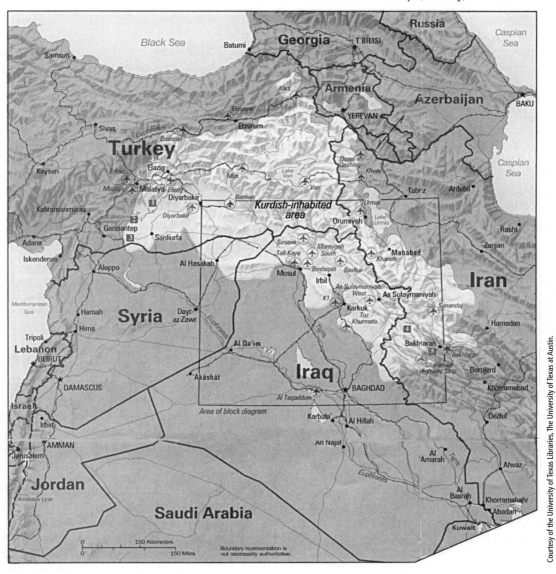

▲   Map of Kurdish Lands, 1992

secular military that still plays a significant role in Turkish political life, despite Erdogan's efforts to limit its power (Foroohar, 2011). A *Time* magazine article noted that:

> Erdogan may be a popular figure who knows how to chest-thump his way to points with a national electorate, but he is also a savvy economic manager and, to some, a reformer who would like Turkey to play a much bigger economic and political role on the global stage. (Foroohar, 2011, p. 38)

As American influence in the region has stumbled, Turkey filled the diplomatic and crisis-soothing vacuum in several key instances.

Perhaps one of Erdogan's most surprising achievements came in April 2013 when, after months of secret negotiations between Erdogan and Ocalan, the jailed leader of the PKK "ordered his fighters out of Turkey and declared that the Kurdish quest for equal rights . . . would henceforth be a purely political struggle" (Vick, 2013, p.1). In addition, according to a 2013 article in *Time* magazine, Kurdish leaders, including Ocalan, say they are no longer interested in a single state but in a loose federation that spans various national borders" (Turgut, 2013).

There are advantages for both sides here. Many of the PKK's more passive supporters have grown tired of the violence in the face of Turkey's growing economic clout and opportunities (Turgut, 2013; Vick, 2013). In addition, the new constitution that Erdogan seeks to establish could accommodate many of the Kurdish goals and aspirations, even if stopping short of full independence. Erdogan, in turn, gets a loyal bloc of voters for future elections and referendums. Ironically, the crisis over ISIS (see Section 6) on Turkey's southern border may further the cause of some autonomy for the Kurds as Turkey begins to see some benefit to a stable Kurdish area as a buffer on its borders with Iraq and Syria.

▧ Left-Wing Groups in Europe

In the latter half of the 20th century, many of the ideological groups operating out of Western Europe grew out of the post-WWII reaction to right-wing ideologies. They were influenced, as well, by a socialist-minded bent in the region dating back to Karl Marx and other left-wing writers prevalent in the Europe of the 19th century. In the 20th century, the emergence of these left-wing groups found inspiration in the growing power of the Soviet Union and the Eastern Bloc. Consequently, the ideology espoused by these groups is referred to as Marxist-Leninist.

Marxism-Leninism refers to the "official ideology of the Soviet Union and the countries of Eastern Europe" once within the Soviet sphere of influence and control. The ideology stems from the teachings of Karl Marx and Vladimir Lenin. At its most basic, Marxism argues that society must be based on

common ownership of the means of production in which the human potential, stunted by the division of labour characteristic of class societies, will be enabled freely to develop its manifold facets. Such a society will obviously have no classes and therefore no need for a state apparatus defined as an instrument of class domination. (Miller, 1991, p. 322)

The ideology is one of reaction against the ills of capitalism and the Industrial Revolution prevalent in the Europe of the 19th century when Marx founded his philosophy. Lenin, as the first leader of the Soviet Union, adopted Marx's teachings and further argued that such a society—what became known as a communist versus a capitalist society—could only be achieved through the revolution of the laboring classes against the elite. In the Russia of 1917, that revolution quickly turned violent and vicious. In this combination of Marx and Lenin, the left-wing ideological groups of Western Europe found their theoretical justification for scrapping the capitalist societies from which they felt alienated and the methodology for doing so.

When the Soviet Union fell in 1989, the entire foundation of the Marxist-Leninist ideology was exposed as an experiment that failed. As a result, most of these groups faded away or died out completely. Still, unlike many of the groups of Southwest Asia or the Middle East, these groups operated—for the most part—in established Western democratic societies and posed a particular challenge to the governments trying to respond to the threat without compromising constitutional or civil liberties. For this reason, they are worth a quick review.

Greece: November 17

According to the National Counterterrorism Center, "Greek domestic terrorism stems from radical leftist and anarchist ideologies that developed as part of the resistance to the military dictatorship that ruled Greece from 1967–1974" (NTC,

2012, p. 102). The most notorious of the groups was November 17 (the Revolutionary Organization of 17 November). Established in 1975, the group was named for a student uprising that had taken place two years earlier.

November 17 sought three goals: (1) an end to the U.S. military presence in Greece; (2) the removal of the Turkish presence in Cyprus; and (3) an end to Greece's ties to the North Atlantic Treaty Organization (NATO) and the European Union (EU). November 17 also dreamed of the establishment in Greece of a Marxist-Leninist state.

The group's initial focus was on assassinations targeting U.S. officials and prominent Greek figures. In fact, one of its first assassinations took place the year it was formed, when CIA Station Chief Richard Welch was assassinated (long before Hezbollah would go after his counterpart in Beirut). Between 1975 and 2011, five U.S. embassy employees were assassinated by November 17 (U.S. Department of State, 2011). In the 1980s, 17N's activities incorporated bombings as well as assassinations, and in the 1990s, its targets expanded to include Turkish diplomats, EU facilities, and foreign firms investing in Greece (U.S. Department of State, 2011).

Also active in the 1970s and 1980s was a group called the Revolutionary Popular Struggle (ELA), which engaged in almost 250 attacks against a wide range of Greek and Western targets (NTC, 2012). "These two groups, as well as several more obscure radical leftist organizations terrorized Greek, U.S., and Western government and commercial interests until the early 2000s" (NTC, 2012, p. 102). The ELA stopped operations in 1994, but the Greek government did not effectively disrupt 17 November until 2002 when the looming 2004 Athens Summer Olympics brought new urgency to the matter. Following a November 2002 attempt against the Athens port of Piraeus, Greek officials arrested 19 members of 17N, including a key leader. Thirteen convictions of 17N members have been upheld in the Greek courts.

Germany: The Red Army Faction

The Red Army Faction (RAF) was founded in Germany in the early 1970s and was often called the Baader Meinhof Gang after two of its founders. The group was known for bombings, assassinations, and kidnappings and had its hand in a number of criminal enterprises used to fund its activities. A Marxist-Leninist group, it targeted both U.S. and German targets as part of its opposition to the ills of capitalism (Spindlove & Simonsen, 2013).

As part of its campaign in the 1970s and 1980s, the RAF assassinated a West German attorney general—Siegfried Buback; kidnapped and killed a key German industrialist—Hanns-Martin Schleyer; attacked and destroyed the officer's mess in the IG Farben building that served to house U.S. Army personnel; and attempted to assassinate U.S. Army General Alexander Haig when he was serving as the head of NATO. Of particular concern with regard to the RAF was its connection to several of the left-wing Palestinian groups (see Section 5). In 1975, members of the RAF participated in the assault on the building belonging to the Organization of the Petroleum Exporting Countries (OPEC) in Vienna, and in 1976 at least two German terrorists were among those who hijacked an Air France flight and held Israeli and Jewish passengers at an old airfield in Uganda until they were rescued by an Israeli commando force. RAF members also trained in weaponry and tactics in Palestinian training camps.

Although the RAF still exists today (Spindlove & Simonsen, 2013), it is no longer the group it was in the 1970s and 1980s and its original ideology has been all but discredited. Its more interesting legacy lies in the capabilities of today's German government to deal with terrorist threats. For example, partly in response to RAF activities, and partly in response to attacks by Palestinian factions against West German targets—including the hijacking of a Lufthansa jet and the 1972 attack on the Israeli athletes during the Munich Olympics—the Germans developed an antiterrorist group known as **Grenzschutzgruppe 9 (GSG 9)**. Today GSG 9 is one of the most respected antiterrorist groups in the world and has been used successfully to prevent and/or respond to terrorist attacks—including its first major effort in 1977 when the group was used to rescue the passengers of a Lufthansa jet being held in Mogadishu, Somalia. While ultimately a failure in terms of achieving its ideological goals, it is responsible for Germany's effective security apparatus, which has foiled a number of Islamist plots over the years.

Italy's Red Brigades

The Red Brigades (RB) were formed in the late 1960s–early 1970s by Renato Curcio and Margherita Cagol. "They believed that a climate of violence would help bring about a revolution in Italy and eventually in all of Europe. Members of the Brigades saw themselves as the vanguard of a world-wide communist revolution" (White, 2012, p. 357). Of all of the left-wing European groups, the Red Brigades came closest in organizational structure to the Marighella model followed by groups in Latin America (White, 2012). The Red Brigades, like most of their European compatriots, targeted domestic as well as Western and NATO targets. Their two most famous attacks involved kidnappings; in one of them, they killed their hostage.

On March 16, 1978, 12 RB gunmen armed with pistols and submachine guns kidnapped renowned Italian politician **Aldo Moro**. He was snatched from his car near a cafe as he was being driven to Parliament. All five of his bodyguards were killed. At the time of the kidnapping, Moro was the leader of Italy's Christian Democratic Party, a five-time prime minister, and the front-runner in the upcoming presidential elections. Just a few days before the kidnapping, Moro had helped avert a government crisis by negotiating a governing coalition between his party and the Communist Party.

The RB claimed responsibility for the kidnapping in a telephone call to a Rome newspaper in which the caller demanded the release of the RB leader and 14 members who were on trial in Turin, Italy, at the time. "For two months, Mr Moro was held at a secret location in Rome allowing him to send letters to his family and politicians—begging the government to negotiate with his captors. The government refused all pleas from family, friends, and the Pope Paul VI to concede to any demands" (BBC, 1978, n.p.). Eight weeks later, Moro's body was found in the trunk of a car in central Rome.

Another of the most famous Red Brigade incidents took place in December 1981 when the group kidnapped Brigadier General **James Dozier**, then the deputy Chief of Staff of NATO's Southern European headquarters in Verona. The Brigade members held the general for 42 days until he was rescued by a specially trained counterterrorism squad. Five terrorists were arrested in the raid and no shots were fired. The rescue of General Dozier signaled the beginning of the group's decline. The captured terrorists gave information about other members, and by 1983, 59 Red Brigade members were on trial for 17 murders, including that of Aldo Moro ("U.S. General Rescued," 1982).

The group briefly reemerged in 1999 when it killed Labour Ministry consultant Massimo d' Antona. A second consultant—Marco Biagi—was killed in 2002. In 2004, Italian officials arrested a number of what they believed were members of this "new" RB on numerous charges, including the murders of Antona and Biagi ("Red Brigades Suspects," 2004).

© Central Press/Hulton Archive/Getty Images

▲ Aldo Moro after his capture by Italy's Red Brigades.

France's Action Direct (AD)

As seen in Section 3, France has a long association both with the term and phenomenon of terrorism. "From a strategic point of view, terror groups have considered France an ideal location from which to strike and then return to hide. It has borders with Spain, Italy, Germany, Switzerland, Belgium, and Luxembourg, plus an efficient transportation system of roads, air, sea, and rail systems" (Simonsen & Spindlove, 2010, p. 163).

France's most famous domestic left-wing group—Action Direct—evolved later than most of the other left-wing groups in Europe, showing up on the scene in the late 1970s and early 1980s and focusing on anti-American and anti-capitalist sentiments. AD also joined with other anti-American causes, such as the PLO's anti-Israel campaign, as well as targeting such symbols as NATO.

Belgium's Communist Combatant Cells (CCC)

Belgium has had experience with both domestic and international terrorism within its borders. In terms of the latter, the PLO's Black September hijacked a Sabena Airlines jet in 1972, and in 1978 the PLO attacked the Iraqi embassy in Brussels, Belgium. In terms of the former, Belgium's left-wing group, the Communist Combatant Cells (CCC), emerged sometime in the 1980s founded by Pierre Carette. The group was probably the shortest lived of the European groups when Carette was arrested in 1988 for a series of 1984 attacks. He was sentenced to life in prison, but was released in 2003. The CCC concentrated its attacks on business, commercial, and NATO targets.

 ## Sleeping in Europe: Middle-East Terror Cells Operating in Western Europe

Today, the threat of the "native" European terrorist groups discussed above has been replaced by the specter of Islamic fundamentalist groups operating in Europe—particularly those with some connection to al Qaeda or al Qaeda affiliates. The idea is that targeting Western civilians in their own backyard will increase pressure on the Western governments, including the United States, to withdraw completely from the Middle East and other Islamic regions. This is not a new idea. Groups from the Middle East and North Africa have long sought to target the West on its home turf as a way to gain significant media attention. Figure 9.1 lists some of the most famous attacks.

What is new is the concern that today's threat comes from groups recruiting from Europe's own immigrant communities rather than from outside groups. In fact, according to a 2010 report, Europol sees Islamic terrorism as the biggest threat to European security (Vidino, 2011). Indeed, as shown in Figure 9.1, several devastating jihadist attacks have been staged in European cities since the early 2000s, including bombings of transportation targets in Madrid and London, the bombing of theaters, trains, and airports in Moscow, and the taking of a school in **Beslan**, Russia.

One of the most frightening aspects of this threat is that many of the terrorists involved are "homegrown." They were young men born and raised in Europe with solid middle-class backgrounds and far more opportunities than many of their counterparts in the Middle East, North Africa, and Southwest Asia. There is often little if any indication of their intent to engage in attacks—including suicide attacks. This not only makes it much harder for law enforcement to become aware of their activities before an attack, but also raises numerous questions about why these young men would turn so violently against the societies they called home.

Vidino (2011) seeks to answer this question and challenges the conventional wisdom that radicalization is a top-down process by high-ranking al Qaeda members. Instead, Vidino argues that the process is one of bottom-up self-radicalization. He also suggests that the operational connections between European radicals and al Qaeda and its affiliates are limited, with most of the support centering around explosives training. In addition, Vidino found that European radicals seek out this training, rather than the other way around. In particular, Vidino focuses his research on the links between "homegrown" terrorism and **al Qaeda in the Lands of the Islamic Maghreb** (AQIM) an al Qaeda affiliate operating out of Algeria and one of the most active of the West European recruiters at the time of his research.

Figure 9.1	Highlights of Terrorist Attacks in Europe by Mid-East Terrorists Groups, 1972–2005

1. 1972—Hostage taking and murder of eleven Israeli athletes at the Summer Olympic Games in Munich.

2. 1975—Attack on the OPEC Headquarters in Vienna; three killed and 70 taken hostage.

3. 1976—Hijacking of Air France flight; plane taken to Entebbe, Uganda, where Israeli or Jewish passengers held until rescued by an Israeli commando force.

4. 1977—Lufthansa flight hijacked by Palestinian supporters of the Red Army Faction; plane held at Mogadishu, Somalia, until hostages rescued by German anti-terror team developed after massacre at the Munich Olympics.

5. 1982—Bombing of the Paris-Toulouse Express.

6. 1983—Bombing of the main rail terminal in Marseille; five people killed.

7. 1984—shooting of demonstrators in St. James Square by gunmen inside the Libyan People's Bureau in London; ten people wounded and one British police officer killed. The shooters held diplomatic immunity and the UK had no other option but to break diplomatic relations with Libya and expel the Libyan diplomats.

8. 1985—Hijacking of **TWA Flight 847** flying from Athens to Rome; one person killed.

9. 1985—Attack on ticket counters at Rome and Vienna airports at Christmas time by Abu Nidal Organization.

10. 1986—Bombing of TWA Flight 840 flying from Rome to Athens; four passengers killed.

11. 1986—Bombing of La Belle Discotheque in Berlin by Libya.

12. 1988—Bombing of **Pan Am Flight 103** by Libya; 270 killed.

13. 2004—Bombings of multiple commuter trains in Madrid; 191 killed, over 1,500 wounded.

14. 2004—Chechen attack on school in Beslan, Russia; over 300 killed, mostly children.

15. 2005—Bombing of bus and metro systems in London; over 50 killed and over 700 injured.

SOURCES: Adapted from Simonson and Spindlove (2010) and Martin (2012)

To set out his thesis about Islamic radicalization in Europe, Vidino divides the evolution of European jihadist networks into four phases: *establishment*, *globalized*, *homegrown*, and *linkage*. Vidino describes the four phases as follows:

The *establishment* phase is the first phase and ran from the late 1980s into the first half of the 1990s. This phase was characterized by the use of Europe as a base of operations by experienced militants from the Afghan war against the Soviets or the conflicts in North Africa (especially in Algeria and Egypt). Most of these jihadists were seeking refuge from repression in their home countries or because they simply were not wanted in their home countries (see also Wright, 2006). "Europe's freedoms, the presence of large diaspora communities, and a lack of attention from local authorities made Europe an ideal logistical base from which militants could continue their activities" (Vidino, 2011, p.1). According to Vidino, the structure of these networks was hierarchal. These networks were autonomous from each other despite a shared ideology. They also tended to refrain from any violent attacks in their host countries.

The second or *globalized* phase began in the second half of the 1990s as a result of two key catalysts. The first was the creation by Osama bin Laden and Ayman Zawahari of a global platform for jihadist activity that they called the World Islamic Front Against Jews and Crusaders. The second catalyst was bin Laden's call to expand the traditional jihadist targets to Western, and particularly U.S., targets, arguing that they were valid targets because the Muslim regimes the jihadists were seeking to topple were dependent on U.S. and Western support. The new global front began generating interactions between and among the formerly autonomous European networks and bin Laden's new focus provided the ideological basis for new activities against these new targets. "The result of these two developments was that, by the end of the 1990s, many of the networks that had been formed throughout Europe had fallen, with varying degrees of allegiance, into the orbit of Bin Laden's project" (Vidino, 2011, p. 2). As a result, al Qaeda needed only to establish a small direct presence in Europe and was instead able to co-opt "already existing networks, particularly the Algerian ones" (Vidino, 2011, p. 2).

The third, or *homegrown*, phase began after 9/11 when America's response significantly disrupted the al Qaeda leadership and its ability to "direct the activities of affiliated networks worldwide, including Europe" (Vidino, 2011, p. 3). This resulted in a hitherto unknown crackdown against jihadist networks in Europe that, while not destroying the networks altogether, forced them to change organization, tactics, and strategy substantially. While these third-phase networks maintained their ideological connection to al Qaeda, they were forced to learn to operate independently. This third phase also revealed a changing demographic. Jihadist networks evolved from older first-generation immigrant jihadists (who often came to Europe with their jihadist membership and credentials already established) to younger second- or third-generation European-born residents who created homegrown networks and clusters and only then went looking for linkages with al Qaeda or other jihadist groups outside of Europe. These homegrown networks demonstrated four similar characteristics: (1) self-radicalization and a lack of recruitment by senior al Qaeda members; (2) a lack of hierarchy and structure; (3) a lack of ties to external organizations; and (4) an increased focus on domestic issues and targets (Vidino, 2011, pp. 4–5).

Vidino's fourth and current phase, *linkage*, is characterized by efforts to establish linkages (but not direct control) between the homegrown networks and al Qaeda or AQIM. These linkages usually involve traveling outside of Europe for explosives training and operational guidance (Vidino, 2011, p. 6).

As part of his description of the *linkage* phase, Vidino (citing Sageman, 2004) argues that the idea of a terrorist recruiter "lurking in mosques, ready to subvert naive and passive worshipers" is more myth than reality (Vidino, 2011, p. 122). While participation in mosques with radical preachers or involvement in Islamic fundamentalist social or other Internet networks may expose already sympathetic and gullible youth to extremist ideology, the reality, says Vidino, is that these individuals become radicalized by themselves though personal experiences and interactions with peers. They create small close-knit networks among friends and family, and *then* approach groups such as al Qaeda or AQIM in an effort to expand their resources and competence (Vidino, 2011, p. 8).

Although al Qaeda–linked or inspired groups in Europe are the main concern, this is by no means the only time groups from the Middle East have operated in Europe. Indeed, as shown in In-Focus 9.1 through 9.4, Western Europe's open societies have long been fertile ground for terrorist groups seeking media-spectacular events against Western targets—indeed providing a far easier target-rich environment than against U.S. targets. For example, the Israeli-Palestinian conflict early on spilled into Europe, at one point almost grounding European tourism to a halt.

IN-FOCUS 9.1

Calling All Travelers—Please Come:
The Devastating Summers of 1985 and 1986

The impact of terrorist attacks can reverberate beyond the tragic loss of life or injury involving innocent bystanders. It can also impact the economic livelihood of citizens and countries alike, sometimes long after the event. Something of this nature can be seen in Europe during the summers of 1985 and 1986. Summer is a time when tourism to Europe is at some of its highest levels for the year as students make use of the region's extensive system of rail travel and youth hostels, and elegant cruise ships ply the ports of the Mediterranean and Baltic coasts. Because of their relatively light security, symbolic nature, likelihood of catching a significant number of innocent victims in a single area, and the inevitable media circus that follows, tourist sites and systems make excellent terrorist targets. In 1985 and 1986, a number of attacks against travel and tourism targets in Western Europe sent the travel industry plummeting.

In June 1985, **TWA Flight 847** was hijacked by Hezbollah as it traveled from Athens to Rome (see In-Focus 9.2).

In October 1985, members of Abu Abbas's **Palestine Liberation Front** (PLF) hijacked the Italian cruise ship *Achille Lauro* out of Alexandria, Egypt, and held the 320 crew members and 80 passengers hostage for two days.

The matter was resolved peacefully when the ship was allowed to dock in Cairo and the Egyptian authorities allowed the terrorists to leave the country for Tunisia. When U.S. officials boarded the ship, however, they discovered that a 69-year-old Jewish American named Leon Klinghoffer had been killed and his body thrown overboard—while still in his wheelchair.

On December 27, 1985, at the height of the winter holiday travel season, terrorists with the Abu Nidal Organization (ANO) simultaneously threw hand grenades and opened fire with submachine guns at the airport ticket counters for El Al airlines at the **Rome and Vienna airports**. Ultimately 18 people were killed and 120 were injured between the two locations. The dead included four of the terrorists—three in Rome and one in Vienna—when law enforcement officials at both airports returned fire. This was the worst—and one of the few—attacks staged against El Al outside of Israel.

On April 2, 1986, a bomb placed under a passenger seat on TWA Flight 840 from Rome to Athens exploded in mid-air. The group claiming responsibility—the Arab Revolutionary Cells—stated that the bombing was in retaliation for the March 1986 U.S. bombing raids against Libyan missile installations. The bomb blew a six foot by three foot hole in the side of the aircraft, and four people, including an infant, were sucked out. In a stunning example of training, competence, and composure, the pilot was able to make an emergency landing in Athens with no further loss of life.

On April 5, 1986, three days after the bombing of TWA 840, a bomb exploded at the **La Belle discotheque**, a popular West Berlin club frequented by American service personnel. Over two hundred people were injured and one U.S. serviceman and one Turkish woman were killed. U.S. intelligence linked the attack to Libya via intercepted messages to its diplomats in East Berlin. The attack was carried out in retaliation for the March 1986 U.S. attack on Libyan missile installations, which in turn was in response to an attack by Libyan forces on a U.S. naval detachment on maneuvers in the Gulf of Sidra. On April 14, 1986, the United States responded with air strikes against Libyan military and intelligence facilities, citing evidence that Tripoli had directly order the La Belle attack (Levitt, 1988). The attack against Pan Am Flight 103, described in In-Focus 9.3, is believed to be in part Libyan retaliation for these events.

IN-FOCUS 9.2

The Hijacking of Trans World Airlines Flight 847

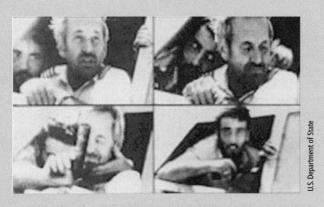

U.S. Department of State

▲ Photo of the pilot of hijacked TWA 847 being interviewed by reporters while a terrorist sits behind him with a very visible gun to his head.

On the morning of June 14, 1985, two gunmen hijacked Trans World Airlines Flight 847 (TWA 847) as it left the Athens airport. The plane carried 153 passengers and crew (129 of them Americans). The flight had originated in Cairo, and was headed toward Rome where it would connect with another TWA flight headed for Boston, Los Angeles, and San Diego. The two hijackers were Lebanese Shiites who demanded the release of more than 700 Lebanese Shiites held by Israel in Atlit prison. The hijackers first forced the crew to fly the plane to Beirut, and over the next two days the plane flew between Beirut and Algiers.

When the hijacked 727 first approached Beirut International Airport around noon on the first day of the hijacking, Beirut airport officials refused the plane permission to land; they had blocked the runway to make landing impossible (Wills, 2003). When the hijackers threatened to blow up the plane, airport officials relented and the plane was allowed to land. After refueling, the pilot—John Testrake—convinced the hijackers that the plane was too heavy to take off with the extra fuel. As a result, 17 women and two children were released (Wills, 2003). The plane then left Beirut for Algiers. Here, U.S. officials convinced the Algerians to let the plane land, believing that it would be easier to control the situation in Algeria rather than Beirut. The plane spent the next five hours on the tarmac in Algiers during which time the Algerians were able to negotiate the release of most of the remaining women and children (Wills, 2003). However, when faced with the refusal of the Algerians to allow the plane to leave, the hijackers again threatened to begin killing passengers. The Algerians relented and the plane headed back to Beirut for the second time.

When the plane returned to Beirut for the second time, about two o'clock in the morning on the 15th, the hijackers demanded a negotiator from Lebanon's Syrian-backed Amal militia led by Nabih Berri. At the time, Berri was also considered the Lebanese "Minister of Justice" (Levitt 1988). When the demand was not met immediately, a 23-year-old U.S. Navy diver named Robert Stethem was shot and killed, his body dumped onto the airport tarmac. At that point, the original two hijackers were joined by a dozen heavily armed members of Amal and Hezbollah. The hijackers from Amal were considered more reasonable and calm relative to the original two hijackers and Amal's leader, Nabih Berri, spoke for the hijackers through the remainder of the crisis.

Once Amal took over the plane, 12 passengers identified by U.S. military identification or perceived as having Jewish-sounding names, were removed from the plane and taken to an unknown location in Beirut (Wills, 2003). Later that morning, the plane again left Beirut for Algeria. This time the plane stayed on the

(Continued)

(Continued)

ground at Algiers for approximately 24 hours while the hijackers and the Algerian government negotiated for the release of more of the passengers. "In all, the Algerians were responsible for the release of sixty-one passengers and the five flight attendants" (Wills, 2003, p. 96).

At 8:00 am, TWA 847 left Algiers and headed back to Beirut, as U.S. officials unsuccessfully tried to figure out a way to get the plane to a location where a rescue mission could be attempted. On June 16, the plane landed at Beirut airport for the last time. Although the runway was blocked as the plane approached, pilot John Testrake radioed the tower: "We have no choice, we have no choice. The hijackers have insisted that we land regardless, even if we have to crash the aircraft." The plane landed in Beirut, and the remaining hostages, mostly American men, were removed from the plane and held in unknown locations throughout the city.

The official and public U.S. response was one of non-negotiation. In reality, over the 17-day crisis, numerous and subtle backchannel contacts with Syria, Iran, Jordan, Israel, and Berri were used to resolve the crisis (Wills, 2003). While President Ronald Reagan also dismissed, at least publicly, any idea of pressuring Israel for release of the prisoners, much of the backchannel conversations and cables involved gaining Israeli assurance that the Atlit prisoners would be released as planned and within a reasonable time after the release of the TWA hostages (Wills, 2003). To Amal and the Lebanese authorities at the time, the hijacking of TWA 847 was "a political opportunity to be exploited through alliance with the hijackers and a partial adoption of their demands, while presenting to the world an image of honest brokers and mediators trying to consider the interests of all 'sides'" (Levitt, 1988, p. 58).

From their side, the Reagan administration's strategy was to "strike the lowest possible profile and hope to keep matters from becoming explosively emotional" (Chaze et al., 1985, p. 19). At the same time, the Reagan administration was haunted by the specter of the Iran hostage crisis that, it is largely believed, brought down the administration of Reagan's predecessor, Jimmy Carter. Had the crisis gone on much longer, it is likely that some military action would have been taken.

One of the aspects the hijacking ended up being most famous for was the nature of the media coverage throughout the 17-day crisis. Over the course of the crisis, the world was treated to a media frenzy at times described by its own ranks as "excessive, irresponsible, sensationalistic, and tasteless," including a memorable interview with pilot John Testrake, taken through the plane's cockpit window with a gunmen clearly brandishing a pistol behind him.

Approximately three years after the hijacking, one of the two original hijackers, Mohammed Ali Hamadei, went on trial in Bonn, Germany, and was convicted of murder, air piracy, hostage-taking, and a number of other charges. Passengers identified Hamadei as the one who brought Stethem to the front of the plane before he was shot. The other hijacker, Hassan Izzaldine, remained at large. At his trial, Hamadei admitted to being one of the two original hijackers, but claimed that Izzaldine was the one who killed Stethem, and that he, Hamadei, tried to stop it. Hamadei was sentenced to life in prison. He was paroled in 2005 after 19 years in prison and was allowed to return to Lebanon (Whitlock, 2005). Unconfirmed and highly conflicting reports on the Internet have him killed in a drone strike in Pakistan in 2010.

IN-FOCUS 9.3

The Destruction of Pan American Airways Flight 103

The cockpit of Pan Am 103 lying on its side in a Scottish field. This became the iconic picture of the attack.

"The destruction of Flight 103 may well have been preventable" (Report of the President's Commission on Aviation Security and Terrorism).

The attack on Pan American World Airways Flight 103 (PA 103) could well be called "the attack that should never have happened." Indeed, all that may have been needed to prevent the destruction of PA 103 was adherence to security procedures already in place. One of those procedures required the strict matching of baggage and passengers at extraordinary risk airports. Any baggage not accompanied by a passenger actually on board the aircraft was required to be hand-searched or not carried on the plane. This rule offered "what should have been foolproof protection against the infiltrated, unaccompanied bag, but this was the rule Pan American World Airways (Pan Am) had dropped at Frankfurt and London in summer 1988" (Wallis, 2001, p. 46). It was an infiltrated, unaccompanied bag that brought down flight 103 on December 21, 1988.

On December 21, 1988, after a 25-minute delay on the tarmac, PA 103 left Heathrow International Airport in London, bound for New York. As was often the case this close to Christmas, the flight was full; 259 passengers and crew members were heading home for the holidays. A number of the passengers were students at Syracuse University returning home from a semester abroad. At 6:56 p.m. London time, London air traffic control handed responsibility for the flight to Scottish Air Traffic Control. Alan Topp, an air traffic controller, confirmed Flight 103 at its cruising altitude of 31,000 feet. Moments later, the green cross that represented Flight 103 on Topp's computer screen vanished, as the aircraft exploded in mid-air.

It was just about 7:00 p.m. when bodies and pieces of the plane began to fall on Lockerbie, Scotland. All 259 people on board the flight and 11 residents of Lockerbie perished that evening. The cockpit, with two of the crew members still inside, landed in farmer James Wilson's field. Another part of the plane landed on the town of Lockerbie, the ignited aviation fuel creating a crater where homes once stood. The remainder of the plane, its passengers, crew, and thousands of pieces of luggage and other personal effects were scattered across 845 square miles of rural Scottish terrain. In a split second, packages meant to be found under the

(Continued)

(Continued)

Christmas tree became evidence in one of the most massive criminal investigations in history. By 8:00 p.m., summoned by John Boyd, the chief constable from Dumfries and Galloway (the jurisdiction that includes the town of Lockerbie), resources from the Royal Air Force and Army were en route to the small town. Their mission would be one of recovery rather than rescue.

Within hours of learning of the attack, experts from a variety of U.S. federal agencies were airborne and headed to Lockerbie, including: the Federal Bureau of Investigation's (FBI's) disaster team (experts in forensic analysis and identification), explosives experts from the FBI and the Bureau of Alcohol, Tobacco, and Firearms, the Federal Aviation Administration (FAA), specialists from the National Transportation Safety Board (NTSB), and engineers from Boeing and engine manufacturer Pratt and Whitney (Emerson & Duffy, 1990; Wallis, 2001). In addition to the law enforcement and other emergency response personnel, within the first 48 hours, family members started to converge in large numbers on the small town of Lockerbie. The diplomatic and airline personnel on the ground were quickly overwhelmed.

The Crime Scene of the Century

Previous FBI efforts to investigate terrorist attacks that took place overseas were stymied either by foreign governments or by the State Department concerned about diplomatic issues (Emerson & Duffy, 1990). In this case, the fact that the plane crashed on the soil of one of our closest allies precluded virtually all diplomatic wrangling over access to the crash site and evidence. Within two days, an unprecedented number of British and U.S. experts were combing the Scottish countryside searching for the smallest of clues and attempting to piece back together the *Maid of the Seas* (the name given to the Boeing 747 designated as Flight 103) (Emerson & Duffy, 1990, p. 43).

The efforts to discover the location of the bomb and how it got there resulted in one of the most amazing criminal investigations in history, involving hundreds of intelligence and law enforcement personnel. Many of the law enforcement personnel, along with a cadre of volunteers, spent months combing a gruesome 845 square miles for evidence that required the most patient and painstaking police work to analyze. The mantra for the searchers was "if it's not a rock and it's not growing, pick it up and put it in a bag" (Emerson & Duffy, 1990, p. 98). Forensic laboratories were set up in Lockerbie and staffed by U.S. and UK experts. In addition, the forensic laboratories at the FBI, and at Fort Halstead in Kent, were at the disposal of the investigators.

Still, as any law enforcement officer will tell, sometimes all the hard work in the world cannot substitute for sheer timing or coincidence. Had the takeoff of Pan Am 103 not been delayed at Heathrow, the explosion would likely have taken place over water, rendering the collection of evidence and any hope of a criminal trial virtually impossible. Yet fate delivered a crime scene. Ultimately, investigators would conclude that the bomb, made of a plastic explosive called Semtex, was placed in a Toshiba cassette-recorder packed in a Samsonite suitcase and placed in the forward cargo hold of Flight 103. The bag was originally placed on an Air Malta flight to Frankfurt and then transferred to Flight 103 from a Pan Am feeder flight running between Frankfurt and London (such transferred bags are known as interline baggage). Thus, the suitcase was not subject to additional searches at Heathrow Airport. More importantly, the bag belonged to none of the passengers from Malta, Frankfurt, or London. Instead, agents working for Libya's intelligence service checked the bag in at Malta and simply let the system send the bag and its destructive contents all the way through to 103's cargo hold.

IN-FOCUS 9.4

When Terrorism Touches a School: The Beslan School Siege

Chechen terrorists representing a violent and often vicious Islamist separatist movement in the region conducted one of the most horrific attacks in modern terrorism history in September of 2004. On September 1, Chechen gunmen (approximately 32) stormed Beslan School No. 1 in Beslan in the Russian republic of North Ossetia, which borders Chechnya. Somewhere around 1,200 children and adults were herded into the school's smaller gymnasium. The floors, walls, and even basketball hoops were then rigged with explosives. The terrorists threatened to detonate the explosives unless their demands were met, including the release of fighters seized during a recent raid in Ingusetia and the removal of Russian troops from Chechnya.

Negotiations between Russian officials and the terrorists started immediately and continued over the next three days with little success. In the meantime, Russian law enforcement quickly lost control of the scene as terrified parents rushed to the scene outside the school, many of them armed and with little patience for negotiations. By the time the hostages had been held in the gymnasium for three days in the sweltering heat with no food or water—forced to drink urine to stave off dehydration—nerves among both hostages and terrorists had worn thin.

On the third day, Russian officials gained permission to send in an ambulance to retrieve some of the dead bodies. We will probably never know for certain what set off the tragic events that followed and what role, if any, the presence of the ambulance personnel played. Conflicting reports have either gunfire going off in the gym or an explosion—the cause of either remains unknown. At one point, either gunfire or a smaller explosion likely jarred one of the explosives on one of the basketball hoops, which then exploded, setting off a chain reaction in the gym. The resulting explosion collapsed the roof of the building, trapping and killing many of the hostages inside. Other reports indicate that some of the terrorists opened fire on the hostages once the shooting began or threw a grenade into a crowd of hostages as they were trying to escape. Meanwhile, back at the front of the school, Russian authorities—upon hearing the shots and explosions—stormed the building. They were quickly overtaken by the hysterical parents who had been surrounding the school since the siege began and started shooting at just about anything that moved. It is unclear how many of the civilian casualties were the result of "friendly fire." In total, more than three hundred people were killed; over half of them children. Russian authorities, who came under intense criticism for the bungled rescue attempt and their inept efforts to control the scene, claim that 31 of the 32 terrorists were killed, but a number of witnesses report that there were more terrorists that escaped.

SOURCE: Drawn from account set forth in Bodansky (2007).

Russia and Chechnya: A Case Study

When the Soviet Union fell in 1989, most of the former Soviet republics either broke away to form their own independent countries or began to agitate for more autonomy from Russia. At the same time, a number of Soviet military officers fighting in Afghanistan with the Soviet Army returned to their former Soviet republics determined to be big fish in a newly discovered, if relatively small, pond. One of these areas was Chechnya, which declared its independence from Russia in 1991. After first ignoring the declaration, by the mid-1990s Moscow decided to deal with the "insurrection" by force and events went downhill from there.

Chechnya sits in the Caucasus Mountains and foothills bordering Georgia and Dagestan with its capital in the city of Grozny. The animosity between Chechnya and Moscow dates back almost two hundred years when Imperial Russia expanded into the region, annexing Chechnya in 1859. The region regained its independence in 1918 with the end of WWI only to lose it again in 1920. In 1944, Stalin brutally dispersed the Chechen people from the region, fearing that they were cooperating with the Nazis. They would not be allowed to return until the 1950s. With the fall of the Soviet Union in 1989, Chechnya once again began to think seriously about independence. By the mid-1990s, a civil war had broken out between various factions and a Moscow-installed government.

While the conflict between Russia and Chechnya is an old one, its current incarnation is not. Bodansky (2007) argues that the conflict that began and escalated in the 1990s was fueled almost entirely by what he calls the "**Islamization**" of the region by Islamic jihadists fully supported by al Qaeda (AQ) and Taliban groups out of Pakistan and Afghanistan. Indeed, in his book *Chechen Jihad*, Bodansky details a number of visits to the region by one of the key AQ leaders, Ayman Zawahari, to advise on recruitment and direct operations. For a time, civil war also erupted between the Islamists and the more secular Chechen leadership, but by the late 1990s, the leadership had caved to the Islamist factions. By 1994, the Chechen leader Dzhokar Dudayev was describing the war as a jihad and calling for worldwide Muslim support. "This call to arms lured the first wave of Afghan and Arab mujahedin to join the ranks of the Chechen forces" (Bodansky, 2007, p. 25). From 1995 to 1996, according to Bodansky, the conflict turned increasingly Islamic as Pakistan's ISI, as well as AQ, funneled mujahedeen in from Afghanistan, Bosnia, and other war zones. In addition, AQ, Pakistan, and other state sponsors of terrorism provided vital and significant training for Chechen Islamists. They also trained in Lebanon's Bekaa Valley under the auspices of Hezbollah. Indeed, Zawahari, second in command of AQ at the time, was substantially involved in the strategic, if not tactical, planning in Chechnya and surrounding regions. By the mid-1990s, the conflict became solely a Muslim cause against the "godless" Russians and their Western supporters with little of the nationalistic character of the original conflict left.

▲ Map of Caucasus and Central Asia

Between 1991 and 2008, START (the National Consortium for the Study of Terrorism and Responses to Terrorism) estimates that there were 3,100 deaths and over 5,100 injuries attributable to terrorist attacks. Ninety percent of these were attributed to Chechen-associated groups or individuals advocating the Chechen (and Islamic jihadist) cause. Almost 17% of the attacks took place in Grozny, while 11.5% of the attacks took place in Moscow, demonstrating the ability and willingness of these groups to target civilians far away from their regional base. Some of the more spectacular attacks in Moscow included:

- A March 2010 attack on a Moscow subway where two female suicide bombers detonated themselves, killing 38 and injuring 95 people (Global Terrorism Database, n.d.).
- An August 2004 attack that resulted in 100 deaths when two planes departing Moscow's airport were destroyed by suicide bombers (Global Terrorism Database, n.d.).
- A February 2004 attack at a Moscow Metro station during morning rush hour with 40 killed and 122 injured (Global Terrorism Database, n.d.).
- An October 2002 attack at a Moscow theater. A three-day standoff occurred at the end of which 41 of the perpetrators and 120 of the hostages were dead (Global Terrorism Database, n.d.).
- The Beslan school attack (START, January 24, 2011).

The attack on the school in Beslan is described in more detail in In-Focus 9.4.

Summary

Europe has been home to a variety of terrorist groups and incidents spanning religious, nationalist/separatist, and ideological motivations. Many of the long-standing nationalist/separatist groups such as ETA in Spain, the PKK in Turkey, and the Provos in Northern Ireland have lost considerable support for their strategies of violence and terrorism as governments have moved more and more to address ethnic and minority grievances in these regions. Still other nationalist, ethnic, and religious groups have sprung up as the breakup of the Eastern Bloc has allowed centuries-old grievances to resurface. In addition, in the latter half of the 20th century, Western Europe was plagued by a number of left-wing Marxist-Leninist groups that sought to overthrow the Western capitalist economies and government structures and replace them with a Soviet-style system. They failed spectacularly in achieving any of their long-term goals, and the experience the Western governments gained in dealing with these ideological groups has proved invaluable in dealing with a more dangerous and current threat, that of Islamic fundamentalist groups using Europe not only as a locale for attacks against Western targets but as a base of operations, planning, and safe-haven. Conventional wisdom suggests that the main players from the terrorists groups in the Middle East (see Sections 5 and 6) are actively recruiting and creating this new breed of terrorists in Europe, but scholars such as Vidino suggest a much more homegrown phenomenon.

Key Points

- The United Kingdom, Spain, and Turkey have been home to several well-known terrorist groups of the nationalist/separatist variety—those groups whose stated goal is the overthrow of an occupying power, the establishment of a new state, or, at the very least, increased autonomy or control over the affairs of their own region.
- Northern Ireland has seen one of the longest standing terrorist conflicts in Europe, between Protestant groups that wish to remain under British rule and Catholics who wish to join with the Republic of Ireland. The conflict between the two groups can be traced as far back as the 16th century, although its current incarnation dates to 1969.

◆ The conflict in Northern Ireland has spawned numerous terrorist groups on both sides of the dispute with the most well-know group being the Provisional Irish Republican Army (PIRA), particularly known for its bombing of Harrods department store in London and the assassination of a member of the Royal Family.

◆ The Basque Fatherland and Liberty group (ETA) began its campaign of violence in 1968, seeking to create an independent Basque state in the mountains on the border of Spain and France. Since that time, the conflict between ETA and the Spanish government has claimed over 850 lives.

◆ As of this writing, ETA has lost much of its support over the last few decades as Spain has given the region considerable gains in autonomy and self-governance.

◆ The Kurdistan Workers Party (PKK) was started in the Kurdish region of Turkey in 1978 and turned violent beginning in 1984. Its goal was to create an independent state in the Kurdish regions spanning Turkey, Iraq, Iran, and Syria.

◆ In the latter half of the 20th century, many of the ideological groups operating out of Western Europe grew out of the post-WWII reaction to right-wing ideologies. The emergence of these left-wing groups found inspiration in the growing power of the Soviet Union and the Eastern Bloc. Consequently, the ideology espoused by these groups is referred to as Marxist-Leninist.

◆ When the Soviet Union fell in 1989, the entire foundation of the Marxist-Leninist ideology was exposed as an experiment that had failed. As a result, most of these groups faded in influence or died out completely. Still, these groups operated—for the most part—in established Western democratic societies and posed a particular challenge to the governments trying to respond to the threats without compromising constitutional or civil liberties.

◆ Today, the threat of the left-wing European terrorist groups has been replaced by the specter of Islamic fundamentalist groups operating in Europe—particularly those with some connection to al Qaeda or al Qaeda affiliates.

◆ One of the most frightening aspects of today's Islamic fundamentalist threat is that many of the terrorists involved in attacks in or stemming from Europe are "homegrown." They are young men born and raised in Europe in solid middle-class backgrounds with far more opportunities than many of their counterparts in the Middle East, North Africa, and Southwest Asia.

◆ While the conflict between Russia and Chechnya is an old one, its current incarnation is not. Bodansky (2007) argues that the conflict that began and escalated in the 1990s was fueled almost entirely by what he calls the "Islamization" of the region by Islamic jihadists fully supported by al Qaeda and Taliban groups out of Pakistan and Afghanistan.

KEY TERMS

Action Direct

Allied Powers

Al Qaeda in the Lands of the Islamic Maghreb

Basque Fatherland and Liberty

Belfast Agreement

Basques

Beslan

Central Powers

Chechnya

Communist Combatant Cells

Dozier, James

Grenzschutzgruppe 9 (GSG 9)

Islamization

Kurdish Worker's Party

Kurds

Marxism-Leninism

Moro, Aldo

November 17

Ottoman Empire

Pan Am Flight 103

Plantation of Ulster

"Provos"

Real Irish Republican Army

Red Army Faction

Red Brigades

Royal Ulster Constabulary

The "Troubles"

TWA Flight 847

DISCUSSION QUESTIONS

1. Describe the main issue between Britain and Northern Ireland. Would you classify this conflict as a religious or nationalist dispute? Support your answer.

2. What event generated the creation of ETA? What were some of its most famous attacks?

3. Why has support in the Basque region for a fully Basque state waned over the last few decades?

4. Using media reports since the Spring of 2013, assess the current state of the PKK and the future likelihood of a peaceful settlement between the PKK and the Turkish government.

5. What led to the creation of November 17? What were its three main goals?

6. Describe the key left-wing groups prevalent in Europe in the latter half of the 20th century. What were their goals? Given that these groups have all but died out, do you think it is still useful to study them? Why or why not?

7. What is currently considered the main terrorist threat in Europe? Do you agree with this assessment? Why or why not?

WEB RESOURCES

British Broadcasting Corporation: http://www.bbc.co.uk

British Library Social Sciences Division: http://www.bl.uk/socialsciences

British Parliament: http://www.parliament.uk

Central Intelligence Agency: http://www.cia.gov

Conflict and Politics in Northern Ireland (University of Ulster): http://www.cain.ulst.ac.uk

Council on Foreign Relations: http://www.cfr.org

Crown Prosecution Service: http://www.cps.gov.uk

Die Spiegel (major German newspaper): http://www.spiegel.de

European Community: http://www.europa.eu

Financial Times (major British newspaper): http://www.ft.com

France at the United Nations: http://www.franceonu.org

French Diplomatie: http://www.diplomatie.gov.fr

Government of Scotland: http://www.scotland.gov.uk

Government of the United Kingdom: http://www.gov.uk

The Guardian (major British newspaper): http://www.theguardian.com

The Independent (major British newspaper): http://www.independent.co.uk

Inter Press Service: http://www.ipsnews.net

Legislation Online (a compilation of legislation across the EC): http://www.legislationonline.org

Max Plank Institute for Foreign and International Criminal Law: http://www.mpicc.de

National Counterterrorism Security Office (UK): http://www.nactso.gov.uk

Northern Ireland Office: http://www.gov.uk/government/organisations/northern-ireland-office

Official Documents of the UK: http://www.official-documents.gov.uk

RAND Corporation: http://www.rand.org

Real Institute Elcano: http://www.realinstitutoekano.org

Royal United Services Institute: http://www.rusi.org

Secret Intelligence Service (UK's MI6): http://www.sis.gov.uk

Security Service (UK's MI5): http://www.mi5.gov.uk

U.S. Department of State: http://www.state.gov

READING 15

In the following article, the author examines evening news coverage in Russia, the United States, and Great Britain in order to compare "the political salience of 'terrorism' in the context of the contemporary 'war on terror'" in all three countries. She suggests that while coverage of terrorism plays an important and emotional role in the national elections in the United States and Russia, it plays a less important and more rational role in the UK. This article provides a good examination of the context of terrorism in the three countries discussed, as well as a good discussion of the media's role in how the terrorist threat is perceived.

Comparing the Politics of Fear

The Role of Terrorism News in Election Campaigns in Russia, the United States and Britain[1]

Sarah Oates
University of Glasgow, UK

 ## Introduction

The spectre of fear can warp, and even come to define, the political agenda of a country. One aspect of the role of fear in politics becomes apparent in the way politicians deploy fear in election campaigns. Various types of fear, ranging from the shock of terrorist attack to nagging anxiety about nuclear war, can serve as useful campaign tools. This article looks specifically at the role of the fear of terrorism in election campaigns in Russia, the United States and Britain. At issue is an attempt to pinpoint whether there is a general way in which terrorism is framed by the news and responded to by the voters—or do countries have a common dialogue between ruler and ruled about terrorism during elections? How does the nightly news talk about terrorism during elections in different countries? How do the citizens of various countries react to the framing of terrorist threat in news during national election campaigns? What lessons could newsmakers learn from the comparison of the coverage of terrorism and the reaction of citizens to this news? This article analyses election news in the 2003/4 election cycle in Russia, the 2004

United States presidential election and the 2005 British parliamentary elections. In addition, the article uses focus groups in the three countries to examine the reactions of citizens to this coverage. The central questions are how the major television channels framed the terrorist threat, how these stories were woven into election coverage, and how voters reacted to this reporting. Evidence from the study suggests that the United States and Russia have a more common experience of the role of terrorism in campaigns. The British media and audience, however, appear to remain more rational and less emotional in discussions about terrorism during a national election.

There have been some useful studies that have compared media in different countries, particularly in times of elections, which offer important comparative analysis and influenced this study.[2] At issue is whether 'campaign language' about terrorism is similar across systems, suggesting dialogue about terrorism with voters has a similar resonance across country boundaries. On the other hand, the exploration of the role of terrorism in election campaigns could suggest that talking about terrorism to the voters is more bounded by national issues, political institutions and media structures than by any sort of global trend. Of particular interest in this study is whether the electoral dialogue about terrorism is more comparable in the two democracies under scrutiny—or whether the lone remaining superpower of the US and the former superpower of Russia have more in common in this regard. In other words, this project is seeking the right questions to ask of an audience when considering how they evaluate and use messages about terrorism and international security in elections. This research is designed to consider, in particular, how to break the traditional notion of the media as what Margaret Thatcher famously called the 'oxygen of publicity' for terrorists. At the same time, there is the suspicion that hard-line politicians find fanning the terrorist threat particularly helpful for gaining or retaining power, while journalists tend to sensationalise the threat in attempts to attract the attention of citizens and consumers.

Hewitt is one of the few authors to highlight the unevenness in coverage of terrorist groups by country. For example, the German media have 'exaggerated the dangers of terrorism and supported government countermeasures wholeheartedly'.[3] In Italy, coverage of terrorism changed significantly in the 1970s, as a tolerance for the members of Red Brigade as like modern 'Robin Hoods' gave way to 'virtually unanimous' condemnation of terrorism in the wake of escalating violence.[4] Hewitt cites bias and unfairness in coverage of terrorists in democratic countries, particularly by the British media in Northern Ireland. He saw the tendency in North America and Britain for the media to ignore the social causes and goals of terrorism.[5] This is a finding echoed by others.[6] Hewitt also discovered that 'terrorist' was not necessarily a negative term for all audiences, being positive for Palestinians in reference to the PLO. Most of the research cited by Hewitt suggests that the level of support respondents in various countries felt for terrorists was much more closely linked to their own proximity to terrorist attacks than to media coverage of terrorism. Although Hewitt wrote this chapter almost a decade before 9/11 and the spate of terrorist attacks in Russia, the point he makes is very salient to the present situation: the public respond more intensely and more emotively when terrorism ceases to be abstract and becomes concrete.

If the traditional view of the relationship between media and terrorism has been the dangerously symbiotic relationship of the two, this study looks at how the public reaction to terrorism is reflected in voting behaviour.[7] This thrusts the study into a large and complex literature regarding the relative balance among party identification, context, the influence of political advertising, the reach of campaign news and other factors. Terrorist threat can pervade this model at every level—for example, it can reinforce or challenge partisan identification if people feel particularly threatened or angry. In addition, the spectre of terrorism will introduce issues and topics into elections, particularly in places such as the United States and Russia where terrorism is a relatively new phenomenon for the public (as opposed to the United Kingdom). Candidates and parties may choose to use a 'fear factor' in their advertising or messages or they may appeal to feelings of nationalism. Candidates and parties may choose to moot particular policies, such as more policing of immigrants or laws limiting hate speech. The voters themselves may seek different messages or react in unexpected ways in the wake of a terrorist attack.

The factor of terrorism in election campaigns has become depressingly relevant in the past decade. The most obvious link between a terrorist attack and elections in recent history was in March 2004, with the deadly train attack in Madrid by Islamic terrorists just three days before Spain's parliamentary elections.[8] Before the tragedy, it had been assumed that the incumbent centre-right party (which initially blamed Basque separatists for the attacks) would consolidate its predominant position. However, the Socialist Party had a surprise victory. In a post-election survey reported in *El País* on 4 April 2004, only about 7 per cent of the respondents felt that the attacks had not affected the Spanish electoral outcome.[9] The effect of terrorism can be considered at every level in the electoral process, from the messages generated by political parties, to the coverage of issues relating to terrorism on the nightly news during the campaign, to how much voters base their decision on concerns about terrorism. The notion of partisan identification generated by Campbell et al.[10] and others (which suggests a relatively weak influence of the media on election results) is still the dominant paradigm in the West. Yet, as the Spanish elections show, terrorism can have a demonstrable effect on elections and this can hold true even if the terrorist act is relatively far removed from election day. Overall, public awareness of terrorism had changed markedly before the elections under study for this article. This was due to 9/11 in the United States and the horrifying series of terrorist attacks in Russia, including the siege of a theatre in Moscow in October 2002 that left more than 170 people dead, and the bombing of the Moscow metro in early 2004.[11] There has been less change in public perception in Britain, where terrorism related to Northern Ireland has been part of the political landscape for decades and the July 2005 London bombings came two months after the 2005 British parliamentary elections.

The study used a parallel methodology to examine the framing of terrorist threat in election campaigns in Russia, the United States and the United Kingdom from 2003 to 2005. Researchers coded the nightly news during the election campaigns on major television channels in the three countries. In Russia, news from the state-run First Channel (Channel 1) and commercial NTV (Channel 4) were collected on weekdays for the month-long campaigns before the December 2003 parliamentary elections and the March 2004 presidential election. In the United States, researchers collected the nightly news on the commercial stations of ABC, CBS, NBC, Fox and CNN from Labor Day 2004 (6 September) to the presidential election on 2 November. In Britain, the project collected the main nightly news on the BBC (public) and ITV (commercial) stations from 7 April to the eve of the parliamentary elections on 4 May.[12] The author and collaborators used a coding frame, listing approximately 100 different categories to label stories in the nightly news. Each news segment was timed and labelled with one or more codes, relating to elections, the economy, terrorism, the military, social issues, entertainment, etc. In addition, time devoted to newsmakers and political parties was tracked. We were then able to define how much of a particular news programme was devoted to specific topics, newsmakers or political parties. This is useful not only for looking at how programmes handle the daily news, but it is particularly helpful for comparing coverage across different channels. In addition, coders recorded a qualitative description of each segment.[13] This analysis, used in assessing Russian election coverage since 1995, generates the percentage of stories that mention specific issues. In all of the countries a team of researchers worked on the coding, and the inter-coder reliability achieved was at least 90 per cent.[14]

The project held focus groups to discuss the impact of terrorism and security issues on vote choice in the three countries under study. In Russia, there were ten focus groups, with eight respondents each, held in Moscow and Ulyanovsk in March and April 2004. Ten US focus groups with an average of 11 participants each were held in December 2005 in Florida, near St Louis and in the Washington DC area. In Russia and the United States, the groups were divided by age (including three groups with college students in the US). In Britain, 17 focus groups were held in the Glasgow area and London in the summer after the May 2005 parliamentary elections. The groups were divided by occupation/class and there was one group for Muslims only. The British groups had an average of seven participants. All of the focus groups were moderated by natives of that country. In the case of both the content analysis and the focus groups, it must be acknowledged that the project is looking at a relatively narrow slice of the entire election/terrorism news phenomenon. By the same token, even a modest amount of material and comments by a few hundred people do suggest some interesting dynamics at play, particularly in the United States and Russia.

 ## The Framing of Terrorist Threat in Russian Elections, 2003–4

Russia's media have become increasingly authoritarian and dominated by the Kremlin, particularly since Vladimir Putin's first election as president in 2000.[15] As such, it is not surprising that the state-run First Channel (Channel 1) provides a biased version of the news that does not question Russia's security policy, particularly the ongoing war in Chechnya. Nor does commercial television (in particular NTV on Channel 4) balance this view effectively.[16] The *Vremya* news programme on Channel 1 is particularly biased during elections, devoting inordinately large coverage to those already in power and friendly to the Kremlin's interests. Those who challenge the Kremlin are either ignored or maligned with unfair reporting, rumour and innuendo.[17] On the other hand, commercial NTV has tended to champion its own political interests and virtually ignored most political parties in its election coverage.[18]

The central themes on Channel 1's *Vremya* during the 2003 parliamentary elections could be described as the efficacy of President Putin; the prominence of top leaders of the pro-government United Russia party and their close political relationship with the president; how the central government fixes problems in the region; and Russia's role in the international sphere. NTV's *Sevodnya* presented somewhat more of the Russian political spectrum and less of Putin, yet the Russian president was still the dominant personality on the newscast. While there was relatively little news on Chechen warfare on *Vremya*, *Sevodnya* still carried some news from the front, although it was only a shadow of the more aggressive war coverage during the 1995 parliamentary campaign on NTV. The most apparent difference was in the choice of which stories to run and how close to the top of the newscast the items appeared. For the presidential election, there was little news to cover as there were no serious opponents to Putin—and the incumbent president did not use either free or paid advertising.[19] Rather, *Vremya* served as a virtual infomercial for Putin.

In looking at the Russian parliamentary campaign in more depth, it is clear that there was little discussion of issues, policies or even ideology.[20] Campaign characteristics were mentioned in 16 per cent of *Vremya*'s stories, compared with 13 per cent for *Sevodnya*. Meanwhile, *Sevodnya* had a heavier emphasis on crime. In addition, *Vremya* had twice as much coverage of the role of the president during the parliamentary campaign. There was more coverage of Chechnya on *Sevodnya*. The commercial news show paid little attention to political parties, with just six mentions of parties over the entire course of the campaign, compared with 38 mentions on *Vremya*. As in earlier years, political parties received a negligible amount of coverage in the broadcast media and there was virtually no discussion of policy even in the relatively more expansive election coverage on state-run Channel 1.[21]

Terrorism was one of the leading topics on Russian parliamentary campaign news, not surprising given both the public interest in the problem in general and the terrorist attack on a train in southern Russia that left more than 40 people dead just two days before the 2003 elections. Major terrorist attacks in Russia also included the seizure of hostages at a Moscow theatre in late 2002 that left at least 170 dead and the deadly Moscow metro bombings in early 2004; and later, beyond the time frame of the research, the

Beslan massacre. Altogether, 9 per cent of the Russian campaign news was devoted to terrorism during the parliamentary campaign. About half of the items (28) on terrorism related to Chechnya and the rest (26) were on other terrorism topics. *Sevodnya* focused more heavily on terrorism as it related to Chechnya, perhaps not surprising in that NTV offered more coverage of the war and Chechen affairs in general. On the other hand, *Vremya* had more coverage (15 items compared with 11 on *Sevodnya*) of terrorism that was not related to Chechnya.

There appeared to be little linkage in the minds of the Russian audience between elections (dull/routine) and terrorism (upsetting/compelling). Yet, the respondents in the focus groups in 2004 did find 'strong' leadership important and their definition of strength was linked to the handling of the Chechen situation and terrorism in general. In this way, although there did not appear to be a primary connection, there was a very important electoral message being sent by Putin and pro-Putin parties in terms of a 'hard line' on Chechnya and terrorists. This stymied any motivation for discussing the Chechen conflict and its consequences in a more meaningful or conciliatory way. Putin's hard stand on terrorism was relevant here, in that many of the participants perceived Putin as a strong, decisive leader, a man they said had promised to 'flush the Chechen terrorists down the toilet'. Many participants made a link between finding Putin 'strong' and 'effective' and feeling that he could deal with Russia's myriad problems, particularly those related to terrorism. In this way, terrorism affected the way in which these respondents analysed the political situation and voted in Russia. The focus-group respondents also felt that the lack of control under democratic regimes—as opposed to the more stringent Russian policy—was responsible for terrorism in both Chechnya and elsewhere. They were frustrated by the apparent inability of the state to control or stop terrorism (either Chechen-related or in the international sphere). They equated this to a lack of state effectiveness in other areas, such as providing employment, pensions or health care. Several times the policies of Soviet dictator Josef Stalin were praised as particularly effective. Worries over security clearly won out over concerns for tolerance. There was a general sense of despair over how to end terrorism, especially as it was so difficult to uncover the real roots of the problem in a multilateral world.

The Framing of Terrorist Threat in the US 2004 Presidential Election

The despair found in the Russian focus groups was shared by many Americans. In contrast to Russia, terrorism and international security were at the heart of the US 2004 presidential election. Ironically, although the US electorate has been relatively safe from terrorist attacks since 9/11, there is evidence that the American psyche and concern about personal security in terms of terrorism have changed fundamentally. Americans continued to be concerned with a personal threat from terrorists throughout the 2004 campaign. As 2004 was the first presidential election in the wake of 9/11 and two major US-led invasions, one would expect a relatively large amount of discussion about terrorism during the campaign. Despite the many differences in the US media system from Russia (i.e. a liberal tradition of freedom of speech, a well-developed civil society and Fourth Estate that can serve as a government watchdog), what is striking about the role of terrorism in election campaigns in the United States and Russia are the similarities rather than the differences.

In an examination of a sample of the news recorded in the 2004 US election campaign (ABC's *World News Tonight with Peter Jennings*, CBS's *Evening News with Dan Rather*, and NBC's *The Nightly News with Tom Brokaw)*, 43 per cent of the news stories were connected to the election. At the same time, terrorism was a frequent buzzword in 22 per cent of these election stories. Terrorism was the second most-talked-about issue (the Iraq War was first). Moreover, terrorism was often mentioned in connection with the war in Iraq. At the same time, many voters reported that the Iraq War and terrorism were key factors affecting their candidate preference.[22] Despite the frequent appearance of terrorism in news stories related to the election, this news on ABC, CBS and NBC failed to provide viewers with substantial information about the candidates' platforms for dealing with the issue. Although 67 per cent of the stories referred to the stands on the terrorist issue by President George W. Bush and Democratic contender John Kerry, such references were vague rather than about specific policy proposals. In the sample, Bush

talked about terrorism *more*, but both candidates tended to talk about it *in the same way*. None of the sampled stories overtly connected terrorism with the 9/11 attacks (although this would have been a natural association for the viewers).

Both Bush and Kerry often commented on terrorism, typically attacking the other's perceived or actual ability to deal with the issue. About half of the news stories in the sample that mentioned terrorism included negative statements from both Bush and Kerry. In several cases, Bush criticised his opponent for being too 'soft' and for the lack of a coherent plan for the 'war on terror': 'John Kerry's the wrong man for the wrong job at the wrong time.'[23] The President's attack was usually followed by a promise by Kerry that he 'will not waver' and 'will hunt down the terrorists wherever they are'. Neither candidate engaged in a discussion of the concrete strategies or methods they planned to use to enhance security. Given that terrorist attacks against Americans dropped in the wake of 9/11, this is not surprising. It was not so much the international security situation and the risk of terrorism itself that had changed (despite some legitimate concerns with copycat attacks or a new allure of global terrorist networks to those disaffected with American foreign policy) as the attitude, awareness and fear on the part of the citizens. Hence, leaders had the choice of either exploiting that fear and claiming they could make the nation 'secure' or they could attempt to reassure citizens that there was no need for a radical shift in internal or external security policy. Both Kerry and Bush chose the first path. One could argue that they were following the advice of their political consultants and that Americans retained a high, if somewhat irrational, fear of terrorist attack. The rhetoric of the campaign, however, only served to underline that irrational fear, making the chance for a reasonable dialogue and debate about American's most useful role in the global sphere increasingly unlikely. Fear-mongering and posturing came to dominate over rational discussion and political debate. In this sense, the American elections came to resemble the Russian elections, even though objectively US citizens had both much less to fear from internal security threats and a much more liberal society.

The focus-group participants did not appear to make particularly rational assessments of the terrorism policy of either Bush or Kerry. This is not surprising, since as noted above neither candidate was putting forth a particularly specific, long-term plan. Bush, of course, was being judged on his actions in the wake of 9/11, including the invasions of Afghanistan and Iraq. It was clear that emotions rather than policy were playing a role in the choice of president—particularly for those who were not committed Republicans or Democrats. That emotion was primarily fear and, given the choice between Bush and Kerry, most found Bush more reassuring on that emotional level. While most focus-group participants were quick to identify Kerry as the candidate who appeared more intelligent (particularly in the debates), there was an overwhelming consensus that Bush was 'stronger'. It is not surprising that Republicans and Bush supporters would feel this way, particularly about an incumbent president during a major terrorist attack and two wars. What was surprising, however, was that those who called themselves Kerry supporters often voiced this opinion, even when they clearly disagreed about the decision of the Bush administration to invade Iraq in 2003. When voters were undecided, they often cited the problems of Kerry's 'weakness' and the appeal of Bush's 'strength' as determinants in their choice. It is particularly interesting in that the general climate of fear, which the focus-group participants discussed and worried over to a large extent, was a new part of the political landscape for younger Americans who could barely remember the Cold War.[24]

Much like their Russian counterparts, the Americans in the focus groups were not too impressed with the campaign coverage in general. They could barely recall any of the paid political advertising, despite the hundreds of millions of dollars spent on publicising the candidates. There was some frustration that the media tended to focus on events relating to terrorism rather than the causes of terrorism, both during the campaign and at other times. They heard frequent references to terrorism, but little that was in-depth or analytical. As a result, many respondents felt that there was little meaningful discussion about terrorism and there was criticism that the media only seemed to 'do a very good job of explaining all the ways that we're not safe'.[25] Overall, there was a feeling of helplessness, dread and sometimes fear that the world was simply a more dangerous place for Americans and that there was little that could be done about it. Like the Russians, they felt disempowered to make meaningful change. Unsurprisingly, however, the American respondents were

less resigned to the trade-off between security and political choice than the Russian respondents. It is important to note, however, that this trade-off could be considered as the same in type, if not in magnitude, in both the US and Russian electoral arenas. Although most respondents denied a direct link between anger over 9/11 and their vote three years later, they would admit to the more diffuse feelings of the importance of 'strength' and hence a preference for Bush. This is a far cry from the 'strong hands' in preference to democracy that is more frequently cited by Russian voters and in opinion polls in the former Communist state, yet there are echoes of this within the 2004 US focus groups as well.

The Framing of Terrorist Threat in the 2005 British Parliamentary Elections

Evidence from the 2005 British parliamentary elections suggests that there can be a more rational discussion and debate about terrorism during a campaign. Granted, much of this evidence is somewhat negative in that there was little discussion of terrorism during the British elections. The most prominent issue relating to international security was Prime Minister Tony Blair's decision to cooperate closely with the Americans in the second invasion of Iraq despite a dearth of British public support for the war. Unsurprisingly, the Labour Party did not highlight this issue in the campaign, focusing instead on the key Labour strength of a strong economy. The other primary factor in the campaign was a lack of realistic competition (a parallel to the Russian case, although not for the lack of a strong political party system). The Conservatives had been unable to find a strong leader or a coherent message since Labour's victory in 1997. While the Liberal Democrats had a promising start with a popular anti-war message, they were hampered by their status as the minority party in Britain, leaving tactical voters uninterested in wasting a vote on a party that could not realistically form a government.

Much like the Russians, findings from the focus groups suggest British voters approached the campaign with a degree of reserve and disinterest. While there was anger at Blair's decision to go to war on shaky evidence for weapons of mass destruction, Labour still proved the most popular party with the voters. The Conservatives pursued a far more negative campaign. They were unable to highlight protest against the war—since the party had supported it at least to the same degree as Labour—but their campaign focused at times on concerns about immigration and the economy. While their complaints about immigration policy were caged in terms of jobs for the British, there was an underlying message of 'British vs. foreigners'. This was embodied in the party slogan 'Are You Thinking What We're Thinking?', ostensibly calling for support for such things as an enhanced police force and more controlled immigration. The British campaign certainly proves that banal slogans are not limited to any particular country. Although many found the campaign slogan of United Russia ('Together with the President!') worryingly free of ideological content, Labour chose the ambiguous, yet upbeat, 'Forward, Not Back'.[26]

It is in the news coverage of the campaign that Britain remains distinct. A preliminary review of the main nightly news on the BBC and commercial ITV in the month-long campaign showed that public television, in particular, was very careful to give time to political parties in segments roughly equal to their success in the last electoral round. In a particularly distinctive feature of British news, both public and commercial television used extensive graphics and studio discussions to discuss policies suggested by political parties. Political leaders and supporters were given opportunities to speak directly for themselves and to respond to allegations during the campaign. Some aspects of campaign 'Americanisation'[27] with a greater focus on media messages over policy statements were evident as viewers were treated to the usual scenes of rallies, walkabouts, school visits, campaign bus tours and a discussion of billboard ads. Britain does not allow paid political advertising on television; hence, the campaign tends to have a quieter and more dignified air than US or Russian campaigns. Parties are given free time instead.[28]

There was ample coverage given to the 2005 campaign on two flagship British news programmes (making it perhaps unsurprising that some British focus-group participants complained of weariness with the campaign). Out of 446 news segments, the campaign was mentioned in 216 of them (48 per cent of the news items) for a total of 10 hours and 23 minutes of items with election coverage aired from 7 April to 4 May.[29] This is comparable to the percentage of news coverage found in the US. Unlike the presence of terrorism items in almost a quarter of the US

election news segments, terrorism was only mentioned in about 10 per cent of all British news segments in the sample (43 out of 445, with 28 on the BBC and 15 on ITV). On the BBC, nine of these segments were considered election news, while on ITV only six segments that mentioned terrorism were also about the elections. Thus, the presence of terrorism issues in the election news in Britain was minimal.

The respondents in 17 focus groups held after the 2005 British elections found little of interest in the election campaign. Much like the Americans, they had a hard time remembering *any* political marketing. Unlike in the US, there was little choice in the election, as it was unlikely that the Conservatives or the Liberal Democrats could unseat Labour. In this sense, British voters were in a similar position to the Russians, with no realistic political alternatives to the current administration. The British focus-group participants saw no particular connection between their vote and terrorism. Terrorism, after all, has been part of the domestic political sphere in Britain for decades. There was anger on the part of many Labour voters about the involvement in the war and some people reported switching their votes because of disaffection with Blair. Given their experience with terrorism (although different from international Islamic terrorism), the British focus-group participants were dubious about the efficacy of a show of 'strength' in eliminating terrorism. They showed sympathy for the American victims of 9/11, but not for the subsequent US invasions. Interestingly, this attitude among the respondents did not noticeably harden after the 7 July 2005 bombings in London (half the focus groups were held after the attacks). This could be related to the restraint in a 'nationalistic' response on the part of the BBC and other British broadcast media (when compared with US coverage of terrorism, for example).[30] On the other hand, it could be related to the fact that most British citizens have been aware of the problems and relative risk of domestic terrorist attacks for decades longer than their US or Russian counterparts. In other words, it no longer had the same resonance for the voters.

 ## Conclusions

The evidence above suggests that terrorism played a varying role in campaigns in recent elections in three countries. Politicians chose to frame things differently—a show of 'strength' was important in the US and Russia, while this was not part of the central campaign rhetoric in Britain. Television news covered the issues differently in the three countries. Terrorism was visible in the campaign news in Russia and the United States, but there was very little in-depth discussion of the issues. In Russia, there was not very much to discuss, as Putin and the main pro-Kremlin party dominated the political sphere and realistic alternatives were essentially absent. In the United States, the economy and the continuing crisis in health care should have been meaningful campaign issues. These problems were covered to a degree, but international security and terrorism were not discussed on the same rational level of comparing and contrasting policy. Terrorism simply did not play a role in the British 2005 elections. Hence, one finds three countries that are involved in the 'war on terror' to varying degrees, but it is relevant in only two of the three countries' national elections. In addition, terrorism plays a role in an emotional, primal way in the United States and Russia rather than as part of a policy discussion.

Could this suggest a sort of 'fear factor' model, in which the political and media systems of some countries lend themselves to an exploitation of concerns about international and domestic security? This would certainly fit into the history of both the United States and Russia, as during the Cold War leaders of both countries used fear of the other to bolster domestic power. Perhaps the answer lies more within this history: that countries that wish to project themselves as major players in the international sphere need to maintain a strong 'back story' of friends and enemies. There is no doubt that both the United States and the Russian Federation have groups that wish to do them violence (although only Russia is engaged in an actual civil war). However, it would appear that this exaggeration of international fear and threat is a very useful tool for leaders to seek or maintain power. It also makes it possible for leaders to limit civil rights, even within democratic societies, through vehicles such as the US Patriot Act. Yet Britain does not currently appear to conform to this model. Do superpowers or ex-superpowers have a particular media dynamic involving patriotism, xenophobia and nationalism? The cases of Russia and the United States would suggest this dynamic is relevant—and would also imply that these media systems and electorates have far more in common than one would expect.

Like Russian voters, it would appear that US voters often acted more as comrades than citizens, motivated by fear and helplessness rather than by a sense of political participation and efficacy. British voters and viewers did not appear to do this. This should motivate us to ask why the British dynamic is so different. One cannot say that terrorism in general is not salient to the British political sphere. Terrorist groups related to Northern Ireland have been a very visible part of the culture and there has been a lively debate about British security measures, such as mandatory identity cards. What is different, however, is the television coverage of global terrorism and Britain's perceived role in the 'war on terror'. It is clear from the focus groups that the British public do not perceive Britain as having a central role in this 'war'. On the other hand, American and Russian citizens seem to link their nation's 'strength' with their own personal security. It does not make much rational sense—which Americans will occasionally comment upon—but the feeling appears to be strong enough to make a significant difference in elections. At issue is the role of television news, especially during campaigns, in this dynamic. It must be said that neither Russian nor US television networks offered a rational review of realistic options for improving international or domestic security. Leaders were not providing particularly useful or realistic statements. However, it should be the role of the media to initiate discussion rather than parrot the press statements of leaders. While this might be difficult in Russia in the current climate, there is no excuse for US journalists to abandon an analytical or watchdog role. It begs the question of the impact of the lack of a real public television sphere in either society. How much can television lead a rational debate—or how much does it merely follow the emotional needs of an audience or the demands of the leaders? In this case, it can transform an election from a democratising institution into an exercise that translates fear into power.

 Notes

1. I acknowledge research support from the ESRC under projects RES 223-25-0028 and RES 228-25- 0048.

2. Including S. Bowler, D. Broughton, T. Donovan and J. Snipp, 'The Informed Electorate? Voter Responsiveness to Campaigns in Britain and Germany', in S. Bowler and D. M. Farrell (eds), *Electoral Strategies and Political Marketing* (New York: St Martin's Press, 1992); Lynda Lee Kaid and Christina Holtz-Bacha (eds), *Political Advertising in Western Democracies: Parties and Candidates on Television* (London: Sage, 1995); P. Norris, *A Virtuous Circle: Political Communication in Postindustrial Societies* (Cambridge: Cambridge University Press, 2000); H. A. Semetko, J. G. Blumer, M. Gurevitch and D. Weaver, *The Formation of Campaign Agendas: A Comparative*

Analysis of Party and Media Roles in Recent American and British Elections (Hillsdale, NJ: Lawrence Erlbaum Associates, 1991).

3. Christopher Hewitt, 'Public's Perspective', in David L. Paletz and Alex P. Schmid (eds), *Terrorism and the Media* (London: Sage, 1992), p. 174.

4. Hewitt, 'Public's Perspective', pp. 174–5.

5. Hewitt, 'Public's Perspective', p. 177.

6. Greg Philo and Mike Berry, *Bad News From Israel* (London: Pluto Press, 2005).

7. Paul Wilkinson, 'The Media and Terrorism: A Reassessment', *Terrorism and Political Violence*, 9(2), 1997, pp. 51–64.

8. Ingrid Van Biezen, 'Terrorism and Democratic Legitimacy: Conflicting Interpretations of the Spanish Elections', *Mediterranean Politics*, 10(1), 2005, pp. 99–108.

9. Cited in Van Biezen, 'Terrorism and Democratic Legitimacy', pp. 104–5.

10. A. Campbell, P. E. Converse, W. E. Miller and D. E. Stokes, *The American Voter* (Chicago: University of Chicago Press, 1960; reprinted 1980).

11. Research for this article took place before the Beslan school attack in September 2004.

12. In both Russia and Britain there is a set election period of approximately one month before each election with a ban on election news starting at midnight on election day. In the United States, there is no formal rule, but the presidential campaign traditionally swings into high gear from right after the Labor Day holiday on the first Monday in September until the eve of the election on the first Tuesday in November. While dates are not fixed for individual elections in Russia and the United Kingdom (although they do follow constitutional patterns), the US presidential election is fixed as the first Tuesday in November every fourth year.

13. For a detailed discussion of the coding as developed for the Russian case, see Sarah Oates, *Television, Elections and Democracy in Russia* (London: RoutledgeCurzon, 2006). For a more detailed discussion of the US coding, see Sarah Oates and Monica Postelnicu, 'Citizen or Comrade?: Terrorist Threat in Election Campaigns in Russia and the US', paper presented at the Annual Meeting of the American Political Science Association, Washington DC, 2005. Information about the New Security Challenges ESRC grant project is available at: http://www.media politics.com/newsecuritychallenges.htm (author's own website).

14. For the US 2004 presidential election, coding was performed by the U-Vote research team at the University of Florida in Gainesville under the direction of Professor Lynda Kaid. In Britain, coding was completed under the direction of the author and Dr Mike Berry. For Russia, the coding was directed by the author, but performed by native Russian speakers (Katya Rogatchevskaia, Andrei Rogatchevski and Boris Rogatchevski).

15. Oates, *Television, Elections and Democracy in Russia*.

16. Oates, *Television, Elections and Democracy in Russia*.

17. See also European Institute for the Media, *Monitoring of the Media Coverage of the 1995 Russian Parliamentary Elections* (Düsseldorf: European Institute for the Media, 1996); European Institute for the Media, *Monitoring of the Media Coverage of the 1996 Russian Presidential Elections* (Düsseldorf: European Institute for the Media, 1996); European Institute for the Media, *Monitoring of the Media Coverage of the 1999 Russian Parliamentary Elections* (Düsseldorf: European Institute for the Media, 2000); European Institute for the Media, *Monitoring of the Media Coverage of the 2000 Russian Presidential Elections* (Düsseldorf: European Institute for the Media, 2000).

18. Sarah Oates, 'Through A Lens Darkly?: Russian Television and Terrorism Coverage in Comparative Perspective', paper prepared for the 'Mass Media in Post-Soviet Russia' international conference, April 2006, University of Surrey, UK. Available at: http://www.media-politics.com/publications.htm (author's own website).

19. Russian law allows for both.

20. The Russian parliamentary elections typically are held every four years, but only the 450 seats of the lower house are still elected by popular vote.

21. For more details, see Oates, 'Through A Lens Darkly?'; Oates, *Television, Elections and Democracy in Russia*; Oates and Postelnicu, 'Citizen or Comrade?'.

22. Pew Research Center for the People and the Press, 'Voters Liked Campaign 2004, but too much "Mud-Slinging"', available at: http://people-press.org/reports/display.php3?ReportID=233 (accessed 20 August 2005).

23. President George W. Bush on NBC, 29 October 2004.

24. It should be noted that there was some debate in the groups as to what was meant by 'terrorism'. Generally, people took it to mean 9/11 and Islamic extremism and the conversation focused on that type of terrorism. However, many respondents correctly pointed out that it was important to define the nature—and scope—of terrorism under discussion because global issues of terrorism, anti-American terrorism and Islamic extremism are not identical.

25. A Washington DC respondent.

26. Pundits were quick to note that this campaign slogan was used by aliens who parodied Bill Clinton and Robert Dole running for president in an episode of the American cartoon comedy show *The Simpsons*.

27. Dennis Kavanagh, 'New Campaign Communications: Consequences for British Political Parties', *Harvard International Journal of Press/Politics*, 1(3), 1996, pp. 60–76.

28. This article does not include an analysis of the 2005 free-time party broadcasts.

29. The election was held on 5 May 2005. The author would like to thank Gordon Ramsay, Hazel King and Murray Leith for their work in coding the British election news.

30. Based on a qualitative analysis by the author of the television coverage of both.

REVIEW QUESTIONS

1. Describe the method used by Oates in this study. Why does the author believe this study is important?

2. Describe the similarities found by Oates regarding the role terrorism media coverage played in the national elections in the United States, the UK, and Russia.

3. Describe the differences found by Oates regarding the role terrorism media coverage played in the national elections in the United States, the UK, and Russia. What explanation does Oates offer for these differences? Can you suggest any others?

4. Choose three countries in which to replicate this study. Justify your choice. Explain how your methodology would differ, if at all, from that used by Oates.

❖

READING 16

In this article, Karyotis traces in detail the evolution of the terrorist group November 17 and the corresponding response by the Greek government. He argues that Greece did not begin to deal effectively with November 17 until terrorism was viewed as a "direct threat to Greek security." As well as providing a useful history of one of the longest-running terrorist groups in Europe, this article shows the interplay between a country's historical and political culture and its response to domestic terrorism.

Securitization of Greek Terrorism and Arrest of the "Revolutionary Organization November 17"

Georgios Karyotis

 ## Introduction

The phenomenon of terrorism in Greece has its roots in the resistance to the military junta, which ruled Greece from 1967 to 1974. In the aftermath of that period, two terrorist organizations first made their appearance: the 'Revolutionary People's Struggle' (ELA) in 1974 and the 'Revolutionary Organization November 17' (17N) in 1975. Since then, approximately 250 groups have claimed responsibility for terrorist acts, but 17N has been the most influential, lethal

and radical group of all, and the main source of violence and terror in Greece.

In 27 years of domestic terrorism, no members of any terrorist group were arrested by the Greek police, leading the US State Department (1990, 2000) to characterize 17N as the 'most dangerous active terrorist organization in Europe', and Greece as 'one of the weakest links in Europe's effort against terrorism'. In June 2002, however, a failed bombing attack at the port of Piraeus led Greek police to their first arrest of a 17N member, and this marked the

SOURCE: *Cooperation and Conflict: Journal of the Nordic International Studies Association*, 42(3): 271–293. Copyright © NISA 2007.

beginning of the end for the group. Within a month, the myth of 17N had dissolved. Two of its hideouts were found containing weapons, files, banners, missiles and bombs, and 19 suspected members of the group were arrested. In February 2003, members of the second major terrorist group, ELA, were also arrested by the police, and Greece finally appeared to be closing a dark chapter of its post-dictatorship history.

My aim in this article is to analyse the reasons for the belated arrest of 17N. I argue that the state's failure to curtail terrorist activity in Greece resulted from the erroneous belief that terrorism was not a direct threat to Greek security. While most European countries had been dealing with terrorism as an important security issue since the mid-1970s, thus adopting strict anti-terrorist laws and increasing cooperation at the European level, terrorism was not perceived as a serious threat, or as a political priority, for the Greek state until the end of the 1990s. As a result, Greece was the only European Union (EU) country in which left-wing terrorist activity remained a serious problem for the authorities.

In exploring the shift towards security in Greek policies on terrorism, I utilize the theory of 'securitization' as developed by the 'Copenhagen School of Security Studies' (CS). Despite its prominence in the literature on security studies, the specific dynamics of securitization remain poorly understood. Adopting a constructivist security approach, I analyse the process through which terrorism was upgraded on the Greek security agenda, as well as the reasons for that move and the consequences. In doing so, I explore political discourses on terrorism as presented in parliamentary discussions and public statements and complement these with personal interviews of members of the Greek political elite.[1] I also examine the state's official response through legislative and policing measures adopted since domestic terrorism first made its presence felt in Greece.

 ## Securitization and the Copenhagen School

Building on, and significantly contributing to, current debates on the concept of security, Barry Buzan, Ole Wæver, Jaap de Wilde and colleagues developed a coherent and comprehensive framework for the study of security. This framework is analysed in the most important book-length publication of the CS: *Security: A New Framework for Analysis*, published in 1998. It is a security framework that is based on two interesting compromises, one conceptual and one methodological, which take into account and aim to overcome the weaknesses of both traditional and new approaches to security.

First, in regard to the polarized debate between traditionalists, who favour a narrow definition of security (e.g. Walt, 1991), and wideners, who call for a more inclusive redefinition of the concept (e.g. Ullman, 1983; Booth, 1991), the CS suggests a middle position. On the one hand, Buzan and colleagues adopt the traditional view that security should be understood as the survival of a referent object in the face of existential threats. On the other hand, in contrast to the traditionalists, they do not want to restrict the discussion of security to military issues nor to the state as the only referent object. To the military sector, they add political, economic, societal and ecological security sectors, recognizing the growing importance of non-military issues. In doing so, they seek to widen and deepen the concept of security without destroying the intellectual coherence of security studies.

Second, and perhaps most importantly, the security framework of the CS is developed from a methodological compromise between Buzan's neo-realist positivism and Wæver's constructivist post-positivism. In his early work, Buzan (1991) adopted the view that security threats were out there, to be observed, measured and analysed. He discussed security on three levels—the sub-state, the state and the international system—but considered the state as the ultimate provider of security and essentially its referent object for all levels. Wæver (1995), on the other hand, rejected the assumption that threats objectively exist and developed a theory to analyse how issues were constructed as existential threats. The two authors explain the resulting compromise between their different metatheoretical positions in their collaborative work:

> Although our philosophical position is in some sense more radically constructivist in holding security to always be a political construction and not something the analyst can describe as it 'really' is, in our purposes we are closer to traditional security studies, which at its best attempted to grasp security constellations and thereby steer them into benign interactions. (Buzan et al., 1998: 35)

The end result is a framework that has security as socially constructed, while attempting to implement an objectivist mode-of-analysis that privileges the role of the state as the primary, but not exclusive, referent of security. As with traditional security studies, and in contrast to some of the new security approaches (for instance,' human security' and critical security studies), the CS scholars 'reject reductionism (giving priority to the individual as the ultimate referent object of security) as an unsound approach to international security' (p. 207). They consider the individual as 'relatively marginal' to understanding international security, which in their view is about 'the relations between collective units and how those are reflected upward into the system' (p. 208).

However, unlike traditional studies, the CS argues that an issue only becomes a security issue when it is presented as such. Thus, they understand security as a 'self-referential' practice, because it is in this practice that an issue becomes a security issue. In that way, they move away from the discussion of what security is, which is essentially a normative question, and instead focus on what security does. They define security as 'the move that takes politics beyond the established rules of the game and frames the issue either as a special kind of politics or as above politics' (Buzan et al., 1998: 23), a process they refer to as 'securitization'.

Any public issue can be located on the spectrum ranging from 'non-politicized' (outside public debate and decision) through 'politicized' (inclusion of an issue in public policy and debate) to 'securitized'. The distinction between these ideal types essentially has to do with the levels of attention governments pay to an issue, and, as attention rises, issues may change their meaning (Baumgartner and Jones, 1993). Kingdon (1995) refers to these times when there is an abrupt shift in how a problem is perceived as 'windows of opportunity' for policy innovation. In turn, changes in perceptions will also influence later policy processes, such as setting relevant goals, identifying appropriate policy responses and giving priority to an issue on the agenda (Hogwood and Gunn, 1984).

An issue is securitized when it is perceived and framed as an existential threat, which underlines its importance and urgency in dealing with it. In such conditions of securitization, the existential nature of the threat gives an actor the 'right to handle the issue through extraordinary means, to break the normal political rules of the game', and to sacrifice other values in the pursuit of security (Buzan et al., 1998: 24). This means that although both are part of politics, securitization can 'be seen as a more extreme version of politicization' (p. 23).

According to the CS, an issue becomes securitized when the term 'security' is mentioned in conjunction with that issue. Wæver (1995: 55) argues that security is not 'a reality prior to language'; it does not exist before it is uttered and can thus be regarded as what in language theory is known as a speech act. 'It is the utterance itself that is the act. By saying the words, something is done (like betting, giving a promise, naming a ship)' (Buzan et al., 1998: 26). The central question that needs to be addressed, then, is '[w]ho can "do" or "speak" security successfully, on what issues, under what conditions, and with what effects' (p. 27).

Securitizing actors are actors who present issues as existential threats to referent objects. Their authority and social power usually derives from their position, which means that although in principal nobody is excluded from becoming a securitizing actor, the field of security is biased in favour of political elites and 'security professionals'. Securitizing actors are different from functional actors, who significantly influence the securitization of an issue by popularizing the security discourse (p. 36). The media are examples of such an actor, and play an important role in any securitization.

According to the CS, presenting something as an existential threat does not itself automatically create securitization. This is what they call a securitization move. An issue is successfully securitized 'only if and when the audience accepts it as such' (p. 25). Uttering security must have a legitimate standing and be accepted by the broader polity for a securitization move to be completed. Consequently, the audience is as important as the securitization actors are. The proof that a securitization move is complete is that 'by means of an argument about the priority and urgency of an existential threat the securitizing actor has managed to break free of procedures he or she would otherwise be bound by' and has persuaded the audience to tolerate violations of rules that would otherwise have to be obeyed (p. 26).

How should one go about researching the process of securitization? The CS argues that '[t]he obvious method is discourse analysis, since . . . [t]he defining criterion of security is textual: a specific rhetorical structure that has to be located in discourse' (p. 176). Security is understood as an inter-subjective and socially constructed practice. Thus, the emphasis is on the security discourse that helps actors make sense of and construct the world. Yet, while discourse analysis is useful in identifying speech acts and the rhetoric of danger, it is not by itself adequate for us fully to understand the process of securitization.

The first limitation of the CS approach derives from its reliance on a single mechanism, speech acts, to explain how an issue is securitized. Although in their collaborative work Buzan et al. differentiate from Wæver's earlier view and acknowledge that it is not only the uttering of the word 'security' that is crucial to the specific nature of the speech act, but also the broader rhetorical performance of which it is a part,[2] they do not develop this aspect of securitization in their analysis, thus reducing the designation of an existential threat to a purely verbal act or linguistic rhetoric. However, as Laclau and Mouffe (1985: 107) point out

> . . . any distinction between what are usually called the linguistic and behavioural aspects of a social practice is either an incorrect distinction or ought to find its place as a differentiation within the social production of meaning, which is structured under the form of discursive totalities.

The importance of non-linguistic elements of securitization is demonstrated by Michael Williams (2003), who argues convincingly that images can also play an important role in securitizing an issue. Similarly, Lene Hansen (2000) argues that the focus on the verbal act of speech cannot adequately analyse security situations characterized by imposed silence, especially in gender relations. Finally, Didier Bigo (2000, 2002) has shown that discourses of danger in societal security issues can also be developed through the implementation of specific security practices, such as bureaucratic procedures (exclusion versus inclusion), profiling of groups (e.g. migrants) and particular security technologies (e.g. visa, identity control and registration). In this article I adopt Laclau and Mouffe's definition of discourse, departing from the CS framework that restricts securitization to a purely linguistic process. The study of the process of securitization of terrorism in Greece thus includes an analysis of parliamentary debates, as one of the main forums where political elites seek to legitimize their policies, as well as visual representations and security practices. Although the latter are not disconnected from political rhetoric and securitizing speech acts, they have in their own right the ability to influence how the audience perceives an issue.

The second limitation which the CS acknowledges is that by using discourse analysis 'we will not find underlying motives, hidden agenda, or such. There might be confidential sources that could reveal intentions and tactics . . . [and thus, discourse analysis] is a poor strategy for finding real motives' (Buzan et al., 1998: 176). Perhaps surprisingly for a constructivist approach, which reflects the ideational turn in security studies, the CS pays little attention to norms that influence the securitizing actor's decision to present an issue as an existential threat, although these might be different from what is presented and discussed in the public debate in order to legitimize a securitizing move. To overcome this limitation, I relied on elite-interviewing in this research in order to come to an understanding of the motives behind the securitization of terrorism in Greece.

A further clarification is required regarding the definitions of politicization and securitization. While securitization clearly signifies heightened anxiety and attention to a perceived existential threat, politicization is too broad a term to describe the importance of an issue in public policy and debate. The term 'latent politicization' is thus introduced to indicate the process when an issue has become part of public debate and policy, which nevertheless is not yet developed and remains peripheral to political discourse and deliberation. Latent politicization is distinguished here from politicization, which indicates not only that an issue is put at the heart of politics but also that an actor manipulates it for political ends.

Using the framework of the CS, the history of domestic terrorism in Greece is analysed in three periods that reflect the changes in the way terrorism was perceived and subsequently dealt with (see Figure 1). Before 1974, terrorism in Greece was non-existent and *non-politicized*. During the first period (1974–89), when Greek leftist terrorist groups first emerged, the political elites failed to recognize the roots, the level and the significance of the terrorist threat (latent politicization). The second period (1989–99) was characterized by intense political debate on terrorism and witnessed the first unsuccessful attempt to securitize the issue (politicization). Finally, the third period (1999 onwards) sealed the securitization of terrorism, which arguably was the catalyst for the arrest on 17 November.

 ## Latent Politicization of Terrorism (1975–89)

During the Greek military junta (1967–74), a number of resistance groups were formed aimed at overthrowing the colonels' regime. These groups contributed to the collapse of the military regime and the restoration of democracy in the summer of 1974, after which most of them were disbanded. However, a small minority of their most extreme members favoured continuation of the struggle, and one of them, Alexandros Giotopoulos, founded the 'Revolutionary Organization November 17' in 1975.[3] The name 'November 17' was chosen after the student uprising in Greece on 17 November 1973 protesting against the military regime. From that time, Giotopoulos was the ideological leader and instructor of 17N, and also the writer of its manifestos, until his arrest in the summer of 2002.

The first act of 17N came on 23 December 1975, when Richard Welch, the United States CIA station chief in Athens, was shot and killed by three unmasked men outside his home. The group had chosen such a high-profile target for its first act in order to attract international publicity and to establish credibility as a revolutionary organization. In a proclamation sent to both the Greek and French press, 17N claimed responsibility for the attack, holding the USA responsible for 'decades of innumerable humiliations, calamities and crimes' inflicted upon the Greek people. However, the proclamation was not published in any newspaper, as the Greek authorities dismissed the possibility that an unknown domestic group could act with such precision and efficiency.

The group struck again a year later, assassinating Evangelos Mallios, a police captain during the military junta. Mallios had been dishonourably discharged from the police force because he had allegedly tortured prisoners during the dictatorship; he was thus presented as a legitimate target. This time the 17N proclamation was published in both the French and the Greek press on 25 December 1976. The feeling that a new movement had been born was originally welcomed among leftist circles in Athens. The assassination of Pandelis Petrou, another former security officer during the military junta, on 16 January 1980, established 17N as a 'revolutionary' group, with what was seen by many as a fair cause which attracted many sympathizers. By targeting the wicked (junta torturers) and the imperialists (Americans) and taking care never to kill innocent bystanders, the group had managed to 'cultivate a Robin Hood image' (Smith, 1999a).

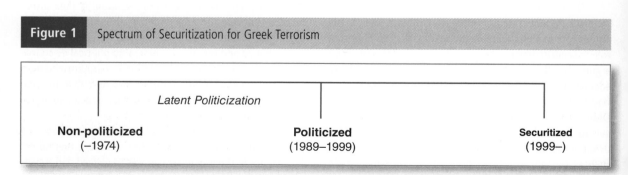

| Figure 1 | Spectrum of Securitization for Greek Terrorism |

Latent Politicization

| **Non-politicized** | **Politicized** | **Securitized** |
| (−1974) | (1989–1999) | (1999–) |

SOURCE: Adapted from Simonson and Spindlove (2010) and Martin (2012)

At a time when most European states were securitizing terrorism and increasing their efforts in dealing with the terrorist threat at both national and European level, the response of the Greek state during this first phase of domestic terrorism was lethargic, inadequate and unplanned. The lack of any coherent strategy in the state's response to terrorism can be attributed to three factors: first, the lack of political consensus on how to define the issue; second, the unwillingness to restrict civil liberties in order to deter the threat; and third, the clear misconceptions over the nature of domestic terrorism.

First, one of the biggest problems was a lack of consensus between the two major political parties, the Panhellenic Socialist Movement (PASOK) and the New Democracy party, on how to define terrorism. The first attempt to deal with the emerging terrorist threat came in 1978, when the New Democracy government introduced an anti-terrorist bill for the first time, called the 'Bill to Combat Terrorism and Protect Democratic Polity' (Law 774/1978). However, the anti-terrorist bill brought about widespread criticism. During the parliamentary debate, all opposition parties characterized it as the first step towards a despotic, undemocratic and tyrannical rule of law. For instance, Andreas Papandreou, the socialist leader of PASOK, argued that Law 774/1978 was 'morally, politically and legally unacceptable'.[4] He concluded that the bill was 'clearly not about terrorists but aimed instead at putting in place the ideological and political conditions to terrorise the Greek population'.[5] As a result, soon after PASOK came to power in 1981, the law was repealed and not replaced.

The second reason terrorism remained low on the political agenda during this first phase was that policymakers were unwilling to implement strict anti-terrorism measures that would curtail civil liberties. Issue definition of a problem is 'influenced by value judgements' (Hogwood and Gunn, 1984: 109), and preserving civil libertarian values in a nation that had suffered the junta's unwarranted use of police brutality was ranked higher than any calls for stricter internal security control. For that reason, although Law 774/1978 was based on Italian and German anti-terrorism legislation, it did not increase police powers in the areas of search, seizure and detaining of suspects. The role of the Greek police compared to that of the police in other

European countries was thus significantly weakened. As a result of the limitations in their power, the police forces and the intelligence services failed to make any substantial progress in identifying the terrorists, missing many opportunities to take advantage of important leads along the way (Kassimeris, 2001).

Third, there were general misconceptions over the origin and seriousness of the threat. In regard to the origin of 17N, conspiracy theories persisted for many years that the group comprised foreigners who wanted to undermine Greece's transition to democracy, worsen its relations with the US and Turkey and isolate Greece politically.[6] According to other speculation, Welch's murder had been the result of internal CIA disagreements related to the succession of the CIA director in the US or to an open CIA–FBI warfare.[7]

As to the seriousness of the terrorist threat, the Greek authorities downplayed its importance throughout this first phase. In 1982 the Greek National Intelligence Agency concluded that '17N is likely to be a "phantom organisation" that possibly does not exist, but is simply a loosely organised group of isolated anarchists that share a common belief in armed struggle' (Papachelas and Telloglou, 2002: 122). Up until the mid-1980s, terrorism had been perceived as an ephemeral phenomenon, attributed to a group of extreme left militants, which had auto-dissolved and would not bother Greece again. To this perception contributed the fact that 17N remained silent for three years, from 1980 to 1983. Thus, in 1983 the PASOK government's Law Minister, George Mangakis, stated 'terrorism in Greece is non-existent'.[8]

Contrary to what the authorities had expected, however, 17N re-emerged, killing US Navy Captain George Tsantes and his driver in November 1983 and wounding US Army Sgt. Robert Judd five months later. Thereafter, 17N continued with sporadic waves of targeted violence, gradually expanding both its operations and its targets. From simple assassinations requiring minimal logistical planning, the group started employing increasingly sophisticated tactics, such as car bombings, rocket attacks and IRA-style improvised mortar bombings. Additionally, 17N diversified its targets by targeting what it called the 'lumpen local capitalist class' that was exploiting the working class and deserved 'punishment by the proletariat'. The group carried out several robberies to fund its operations, and included new assassination targets, such as Greek businessmen, newspaper publishers and judges.

The gradual intensification of terrorist activity in Greece did not alarm the Greek political elites. Policymakers continued to downplay terrorism as a threat to public order, even as terrorists acted with virtual impunity. They considered these terrorist incidents as isolated occurrences of violence that were in no way central to Greek political and social life and in no way comparable to the terrorist problems of Italy, Germany or Spain. Law Minister Mangakis noted in his speech to Parliament:

> What we have in this country is not terrorism but isolated episodes of terrorism like the ones experienced by all nations, even the most peaceful, nonviolent ones, such as Austria and Switzerland. For it is nowadays no longer possible for a country not to have endured some form of political violence.[9]

Such observations signify that the politicization of terrorism during 1974–89 was latent, and terrorism remained a peripheral issue on the political agenda. Even the introduction of the anti-terrorism bill in 1978 had primarily to do with the kidnapping of Aldo Moro and the rising number of terrorist incidents in neighbouring Italy, rather than with a real concern over domestic terrorist activity.[10] Instead, during that period, public debate and policy focused more on other issues and values, such as the promotion of social rights and freedoms and the reconstruction of the economy, which did not allow terrorism to attract any significant amount of attention or resources.

 ## Politicization of Terrorism (1989–99)

A 'window of opportunity' opened to raise terrorism in the Greek policy agenda after the assassination of Pavlos Bakoyannis in September 1989. Bakoyannis was the chief parliamentary spokesperson of New Democracy and the son-in-law of its leader, Konstantinos Mitsotakis, who succeeded Papandreou as Prime Minister in 1990. His landmark killing marked the end of an attitude of tolerance in both the political establishment and the public. First, it changed how the public viewed 17N. Until his assassination, there was a feeling among sectors of the public that the targets of 17N were 'legitimate'.[11] The Bakoyannis killing marked the end of an atypical acquiescence and consent of the public to the acts of 17N.

Second, the assassination brought terrorism to the heart of the political debate because Bakoyannis was the first politician to be targeted by 17N, but also because his murder came at a very unstable political period for Greece, i.e. after successive weak coalition governments. These contributed to an increase in polarization of the party-political debates and to the politicization of terrorism. In particular, New Democracy accused PASOK of being somehow linked to 17N.[12] This allegation was based on the suspicion that because PASOK was the political transformation of the resistance group Panhellenic Liberation Movement (PAK), 17N and PASOK could have been drawn from the same group of people. According to a senior politician, the general belief in the New Democracy camp was that even if PASOK was not behind 17N, there was 'some kind of suppression, hushing-up, non-disclosure and covering up of information, or a least an emotional bond with some people that PASOK might have suspected to be related to terrorism' (Interview, 15 December 2002). Subsequently, the politicization of terrorism did not translate to a constructive political debate on how to deal with the terrorist threat, but instead to competition between the two major parties trying to capitalize on the issue by making short-term political gains.[13]

Meanwhile, the inability of Greece to make any progress in cracking down on domestic terrorism was increasingly threatening Greek–American relations. Repeatedly, the US government issued travel advice to American citizens to avoid Greece because of its terrorist record. In addition, the National Commission on Terrorism (2000) issued a recommendation that the US should impose economic sanctions on Greece until it showed some resolve in cracking down on terrorists. The US government ruled out sanctions, but instead increased its pressure on Greece to deal with the terrorist threat, offering help in identifying the terrorists and in proposing a number of changes to existing Greek legislation. The US also adopted and further fuelled the theories that connected PASOK with 17N, arguing that PASOK was directly or indirectly linked to the terrorists. For instance, an unnamed US official was quoted in 1999 as saying:

> [I]t is logical to assume that people didn't want to look under every rock because of what they might find . . . . If they arrest the leader, for example, and he turns out to be a former best friend of a PASOK leader, that would be embarrassing. (Smith, 1999b)

Similarly, the former CIA Director James Woolsey argued:

> [The US government has] strong reasons to believe that high-ranking members of the Greek government know how to go after this organisation, if they wanted to, but they refuse to act. I won't say anything more, but they know who they are . . . . They are protecting the terrorists.[14]

Such views determined the strategy of the American intelligence services who were involved in the search for 17N. However, according to a former Foreign Minister, Theodore Pangalos, despite its good intentions, 'the involvement of the US government in the fight against terrorism in Greece did not help, but was harmful, because it directed the investigations on PASOK' (Interview, 2 January 2003). The American pressure did contribute to the politicization of terrorism in Greece, however, because it fired up the political debate and forced the authorities to rethink the nature and implications of the terrorist threat.

The First Unsuccessful Securitization Move

As discussed earlier, significantly increased political attention signals the possibility of serious shifts in the framing of an issue and in policy outcomes. In the aftermath of the Bakoyannis murder, the New Democracy government made the first attempt to shift terrorism from normal politics to the security realm. In line with the CS approach, this securitization move entailed the discursive construction of terrorism as an existential threat, as well as the adoption of stricter legal

and policing measures. A new anti-terrorism bill entitled 'Bill for the Protection of Society against Organised Crime' (Law 1916/ 1990) was adopted in 1990, and this included much stricter provisions on terrorism compared to the previous Law 774/1978, which was abolished by PASOK eight years earlier. It significantly increased police powers in the areas of intelligence gathering and detaining suspects without specific charges; it offered protection to judges and their families; it increased the reward offered for police informers; and it prohibited the press from publishing proclamations from the terrorist groups. This was the first time that such draconian measures had been taken in Greece, and these demonstrated a willingness on the part of the government to upgrade terrorism in its priorities.

Parliamentary discussion once again focused on whether the need to fight terrorism justified limitations on civil liberties. According to the government, there could be no freedom when the security of individuals was compromised by terrorists. Vice President of the Government, Athanasios Kanellopoulos, argued that the question over which social good was more important, freedom or security, was 'a pseudo-dilemma . . . In the face of the security threat [that terrorism poses], the freedom of society as a whole will come before the freedom of the individual'.[15] Ioannis Kounenos added that: 'Personal freedom, in a state that is overtaken by fear created by the terrorists, is not freedom. There can be no social or economic stability . . . unless every single citizen of this state feels secure about his or her life'.[16]

Not only did the government present terrorism as an existential threat to the state and society, it also tried to legitimize the new stricter measures adopted, with reference to the policies of other European countries. Eleftherios Papanikolaou referred to the experiences of Italy with the Red Brigades, Germany with the RAF, France with the Action Direct, and cited the adoption of strict laws as the main reason these states succeeded in dismantling the terrorists. Other members of the governing party referred to recommendations of the European Community and the United Nations, as well as to the intensification of European cooperation on terrorism in the TREVI working groups, in order to legitimize the need for stricter antiterrorism provisions in Greece.

On the other hand, the opposition parties rejected the need for the new bill. Their opposition to the proposed Law was structured around two main arguments: first, that the problem with domestic terrorist groups in Greece was significantly different from that of terrorist groups in other European countries, and, second, that it by no means justified limitations to civil liberties. For instance, Ioannis Skoularikis, speaker of PASOK, stated:

> [Greece] does not have a serious problem with terrorism . . . but with a few sporadic, spectacular terrorist acts . . . There is no future for terrorism in Greece because all Greeks are against it . . . The aim of the proposed legislation is to create a climate of fear in order to restrict the rights of the Greek people and subdue any free spirit.[17]

Apart from the opposition parties, large sections of Greek society were also hostile to the new legislation. Public opinion treated the increase in police powers with suspicion and concern over potential infringements and restrictions on civil liberties. However, the clause that provoked the most intense public reaction had to do with the banning of the terrorist communiqués from being published in the Greek press.[18] Until then, terrorist groups in Greece had enjoyed the easiest means of communication with the Greek people (Bakoyannis, 1995; Kassimeris, 1995). Publication of the terrorists' communiqués in the Greek press allowed 17N to have access to the public and advertise their ideas freely. Law 1916/90 was an attempt to put an end to that, but instead turned the media against the government, because they considered this provision as a violation of freedom of speech. Several newspaper editors refused to comply with the legislation and continued to publish 17N communiqués, leading to their arrest and imprisonment. The widespread political opposition, as well as the social reactions to the first serious attempt to upgrade terrorism in the Greek security agenda, indicated that the audience was not ready to accept the government's securitization move and was not willing to accept restrictions on civil liberties. As a result, when PASOK returned to power in 1993, the bill was abolished and not replaced by other legislation.

Overall, during this second phase, terrorism in Greece became an important issue in the political arena.

The politicization of terrorism resulted in more attention being devoted to it, but this did not bring the government any closer to identifying the terrorists, partly because of the narrow focus on the links between PASOK and 17N.[19] Most importantly, during this phase the first securitization move was attempted by the New Democracy government, which involved securitizing speech acts and stricter internal security and legislative measures. However, the securitization move was not successful because the political parties, the media and public opinion could not agree on the existential nature of the terrorist threat. Greek sensitivities as regards civil liberties continued to stand in the way of stricter internal security laws; they put limitations on the role of the police and resulted in ten more years being wasted.[20]

Securitization of Greek Terrorism Post-1999

A significant change of mood became apparent in the political elites in 1999. The Greek government demonstrated a new determination and an unprecedented sense of urgency in dealing with the terrorist threat, which was reflected in the political discourse on terrorism. For the first time, all political parties came to realize the importance of terrorism as a security threat that had to be dealt with immediately and effectively. As a result, after 1999 a process of securitization of terrorism got under way, leading to a gradual reappraisal and re-evaluation of all policies and measures against terrorism.

Speech acts were central to constructing the security discourse. In the public debate, terrorism was presented as a threat to Greek society and national interests. The devastating impact of domestic terrorism on the economy and tourism, as well as on the country's relations with the EU and the US, were some of the issues highlighted. For instance, in 1999 Michalis Chrysohoidis, the Public Order Minister, said (quoted in Smith, 1999a):

> We realise that this has become a huge problem, more serious than perhaps anything else we are currently dealing with . . . I don't think it's too much to say that these terrorist attacks are literally murdering Greece to the point that

counter-terrorism has become the *government's top priority* (emphasis added).

The worst fears of the government were realized on 8 June 2000, when Brigadier Stephen Saunders, a British defence attaché in Greece, was assassinated. 17N claimed responsibility for the attack, arguing that they chose the senior British officer not just because the United Kingdom had taken a leading part in the bombardments of Iraq and Yugoslavia, but also because its policy 'even surpassed the Americans in provocation, cynicism and aggression'.

The timing of the terrorist attack, while the Greek government was trying to persuade the international community of its commitment and ability to eradicate terrorism in Greece, shocked the political establishment. As Chrysohoidis pointed out: 'I believe that apart from the loss of the unlucky victim, this action primarily harms the interests of the country' (BBC News, 2000). At the same time, it brought to an end the conspiracy theories that connected PASOK with 17N. It took place while PASOK was in power and was extremely harmful to Greek interests and PASOK itself. This time the reaction from the major parties was one of solidarity and conviction to find the terrorists, leaving behind the short-term political calculations that had characterized the previous phase.

Because of this new sense of urgency to eradicate terrorism, a new antiterrorism bill was adopted in June 2001. This time it was PASOK that brought the bill to Parliament, which was a significant break from the party's past, considering that it had previously abolished the two laws of 1978 and 1990. The new bill was much stricter than the previous two, yet both major parties supported it, revealing a political consensus that was missing from previous attempts to upgrade terrorism in the security agenda. The bill gave the police greater powers when arresting suspects and also permitted the use of DNA testing to aid in investigations. Collection of personal data, including telephone conversations and videotaping of suspects, was also included in the legislation, along with the legal framework for Greece's first-ever 'witness protection programme' and provisions for granting amnesty to members of terrorist groups who turned state's evidence.

In the parliamentary discussions for the adoption of the law, terrorism was presented as an existential threat by both government and opposition parties. Dora Bakoyanni,

widow of Pavlos Bakoyannis and speaker of New Democracy, stated that it was common sense that:

> ... terrorism has harmed Greek society in its whole; it does not only affect the victims of the attacks but compromises the highest value of all, which is human life. It is common sense that terrorism has been very costly to Greece on a social and national level ... We should let everyone know that the Greek politicians will not accept half-measures but will react to the terrorist threat, although that reaction has already been long overdue.[21]

The minority parties on the left insisted that the bill endangered civil liberties because it legalized the surveillance of citizens and ensured that secret service agents would not be prosecuted for their actions. This was rejected by the majority of the political elites. As Law Minister, Michael Stathopoulos emphatically noted, '*any* restrictions to human rights and freedoms ... are justified in a democratic society, if they are necessary in order to safeguard internal security and public order and to prevent crime' (emphasis added).[22]

The Role of Images

The importance of non-linguistic elements in the construction of the security discourse, discussed earlier in the article, can be seen in the government's communication policy with the public. For the first time, after 1999 the government took initiatives that aimed at sensitizing public opinion on the serious issue of terrorism. These initiatives were based on textual and visual messages designed to remind the public of the consequences, and implications, of terrorist acts.

For example, in a text headed 'One Minute of Silence', and broadcast by all radio and television stations in July 2000, terrorism was depicted as a threat to modern Greece. The text read:

> Terrorism constitutes an insult for the Greeks because of the contempt it displays toward the sanctity of human life, and because it seeks to undermine the social cohesion and political

stability. It is a threat for today's Greece. It is totally alien to Greece's philosophy and logic. It is alien to all of our traditions. The battle against terrorism is a priority. A priority not only for the state but also for the Greek people. It is a commitment undertaken by the government and the society's objective is to continue the effort aimed at uprooting terrorism; *in every possible way*. We owe it to the victims of the terrorists. We owe it to Democracy and its human values. We owe it to Greece. [Emphasis added]

In addition, the government encouraged the formation of an informal group comprising the relatives of the victims of 17N, who until then had remained out of the public eye. The group was formed in December 2001 in order to create a social alliance against terrorism.[23] In particular, the image of Saunders' widow after the attack was one moment that will remain in the mind of anyone who saw it, and it had a political impact. Soon after, images from the terrorist attacks of September 11 further increased public sensitivity to terrorism, as well as public awareness concerning the seriousness and significance of the terrorist threat.

Collectively, securitizing speech acts, as primary and non-verbal messages (e.g. images), as secondary means contributed to the construction of a coherent discourse clearly intended to shift terrorism from normal politics to the security sphere. What was different compared to the first securitization move, however, was that Greek society was ready to accept restrictions on its liberties for the sake of eradicating the terrorist threat. As a result, terrorism became securitized, which justified the adoption of exceptional measures that were previously rejected.

Operational Measures

The securitization of terrorism was the catalyst for dramatic changes at the operational level. First, the Greek authorities sought to establish closer cooperation with French, British, US and other intelligence services in a position to offer information and technical support. In the past, no Public Order Minister would have retained his seat if he openly praised Greek cooperation with other countries' intelligence services, especially American services.[24]

For political and psychological reasons that go back to the dictatorship, such policies were extremely unpopular, even within the security and police forces. The securitization of terrorism changed this. Greece signed new bilateral cooperation agreements on terrorism with many countries, including an agreement with the US in September 2000 and with Turkey in 2001. In addition, the 300-strong anti-terrorist squad established in 1984 was reorganized, with counter-terrorism experts visiting Britain and America for retraining in surveillance techniques and bombing analysis. With these measures, Greece hoped to dispel an inherited mentality of exaggerated mistrust. Furthermore, by cooperating with other countries, Greece wanted to share responsibility for the investigations. As one police colonel pointed out, 'failure of the Greek authorities to capture 17N would also be a failure of our allies helping in the search' (Interview, 5 January 2003).

The contribution of British intelligence after Saunders was murdered was particularly important. The Scotland Yard team was systematic and expert in developing relevant wiretaps and other technical evidence (Buhayer, 2002). In addition, the involvement of the British was not received in the same suspicious and negative light as was the American involvement of the previous years. According to former Defence Minister Gerasimos Arsenis, 'the British pointed our search in the right direction because they did not share the preoccupations of the Americans regarding the links between 17N and PASOK' (Interview, 18 December 2002).

Under the leadership of Chrysohoidis, and with the help of both British and American intelligence, a new round of investigations began. A computerized crime management system was introduced, loaded with all information on 17N in order to compare the files and assist in the cross-linking of information gathered during the history of 17N. Based on the computerized analysis, a report comprising several hundred pages produced in June 2001 offered a systematic overview of the activity of the terrorist group.

In January 2002, the police communicated information that it had the names of 17N members but would not give in to pressures to make arrests until all the required evidence had been gathered (Karakousis, 2002). Two of its members were placed under surveillance and Giotopoulos was identified as the group's leader (Papachelas and

Telloglou, 2002: 218–34). In this climate of persecution, and in an attempt to show that the organization was still invulnerable and active, 17N made the mistake that the police were waiting for in June 2002, with the failed attack at Piraeus. Within 2 months, 19 suspected members of the group had been arrested, including Giotopoulos, the group's leader, and Dimitris Koufodinas, the group's leader of operations (see Kassimeris, 2005). At the trial of the terrorist suspects, which commenced in Athens in March 2003, 15 of the accused were found guilty, while another 4 were acquitted because of lack of evidence. Giotopoulos, the group's leader, received a sentence of 21 life-terms, the heaviest in Greek legal history.[25]

The arrest of 17N signified the complete demystification of the group. The 'phantom organization' was found to comprise personalities most of whom did not match the ideological profiles or the revolutionary personalities that people were expecting to see.[26] In the light of day, the phenomenon of indigenous terrorism in Greece, and 17N in particular, assumed its true dimensions, destroying the myths, fantasies, suspicions and obsessions that had persisted for 27 years.

Reasons for the Securitization of Terrorism

What was remarkable in the securitization of terrorism was that although the terrorist threat had not become any more serious than previously—terrorist incidents were in fact fewer than ever—the political elites were able to come to consensus about how to deal with the issue as an urgent security priority. It is interesting then to explore the reasons for this change, a change that led to the upgrade of terrorism in the security sphere. These can be traced back to changes in Greek security thinking, the impact of European norms and in the momentum gained from international developments and the prospect of hosting the 2004 Olympics.

First, a gradual change in Greek self-perceptions and security thinking took place after Cost as Simitis succeeded ailing Papandreou as leader of PASOK and Prime Minister of Greece in 1996. Simitis initiated a range of modernization programmes that aimed to strengthen the European orientation of Greece, which until then had

been characterized by introversion, opportunism and internal contradictions (Verney, 1990).[27] The focus of the government was on economic reforms that would allow Greece to meet the criteria for entry into the Euro zone, a goal achieved by the end of 1999. At the same time, the new government demonstrated a gradual rethinking of its security priorities and policies.

Ever since domestic terrorism had first made its presence felt in Greece in 1974, the security agenda had been dominated by hostility with neighbouring Turkey, which twice (in 1987 and 1996) brought the two countries to the brink of war over disputes in the Aegean Sea. Foreign policy was shaped with a traditional concept of security in mind, emphasizing the military dimension of politics and supporting unilateral and nationalistic policies. However, Greek–Turkish relations improved dramatically in 1999, following the earthquakes that hit both countries and the resulting mutual empathy and cooperation at various levels that has come to be known as 'earthquake diplomacy'. With the confrontation with Turkey no longer considered a permanently operating factor in the Greek security environment (Lesser et al., 2001), there was room for other issues, such as terrorism, to be upgraded in the Greek security agenda, issues that had previously been overshadowed by the perceived Turkish threat.

Central to this new security thinking was the gradual Europeanization of Greek security policy (Ioakimidis, 2000; Kavakas, 2000). Research under the auspices of the Rand National Security Research Division found that after 1996 Greece became progressively more modern and more European, and increasingly placed virtually all of the country's external policy challenges within a multicultural, European framework (Lesser et al., 2001). As a result, this created what March and Olsen (1989: 160–2) call a 'logic of appropriateness'. More specifically, norms, values and routines embedded within the European institutions gradually became integrated in Greek political life, influencing definitions of political reality and policy outcomes.

Terrorism had been the highest priority in the European internal security agenda during the 1970s and 1980s, but when the Treaty of Maastricht came into force in 1993 the Union shifted its attention to other internal security threats, such as immigration and organized crime (Benyon, 1996). Thus, as Monica den Boer (2003: 1) noted:

'within Europe, it seemed as if the issue of terrorism had temporarily disappeared from the stage'. Yet, the renewed focus on Justice and Home Affairs after the Tampere European Council in 1999 and the September 11 attacks brought terrorism back to the top of the European agenda. In this context, Greece gradually changed its perceptions and policies on terrorism, at the same time as the EU was re-securitizing terrorism and dedicating more resources to dealing with internal security issues.

Schimmelfennig (2000) argues that it is often a rational choice for countries to behave appropriately. The European institutions' greater strength is vested in their ability to define reality for others, so that they internalize the existing order as beneficial to them. The adoption of European norms as regards terrorism in Greece was partly due to a realization that the only way to promote Greek self-interests would be through the EU. Retired Ambassador Byron Theodoropoulos suggested that Greece, by falling into line with the EU counter-terrorist norms and policies was perhaps also expecting some sort of return from the Union in other issues, for instance with regard to Cypriot accession to the EU (Interview, 14 December 2002).

Finally, another main reason that the political elites supported the securitization of terrorism was the prospect of hosting the Olympic Games in Athens in 2004. With the Olympics in mind, both the US and the EU substantially increased the pressures on the Greek government to catch the elusive 17N terrorists. For instance, Wayne Merry (2001), a former US embassy staff member who served in Athens, called for the barring of American athletes from the Athens 2004 Games if the members of the infamous terrorist group were not brought to justice. While in the past American pressures were interpreted as an effort of the US to intervene in Greek politics, the forthcoming Olympics helped the Greek political elites understand that the international concerns regarding terrorism were fully justified and had to be addressed urgently and effectively.

The Greek government did not really fear that 17N would strike during the Olympics. This would be beyond both the group's operational abilities and its ideological platform. However, the political elites were aware that the inability to make progress in dealing with domestic terrorism was damaging the international image of Greece.

For that reason, 'Greece developed a more coherent antiterrorist strategy for the first time, which was structured around the Olympics'.[28] As Pavlos Tsimas noted, the Olympics provided an extra incentive to the Greek authorities to invest more resources in the pursuit of 17N and to cooperate with foreign intelligence services (Interview, 1 October 2002). The belief was that the arrest of the terrorists would contribute to the image of Greece as a secure country and would thus allow it to maximize the economic and political benefits from hosting the Games.

The terrorist attacks of September 11 and the global war on terror further substantiated the feeling of urgency to deal with the terrorist problem at home. It also forced Greek elites to rethink the dangers of international terrorism acting in Greece. Although Palestinian terrorist groups had acted in Greece during the 1970s and 1980s (see Kaminaris, 1999), there was a belief that Greece was not affected by international terrorism, because it was not a Western metropolis and had never had tense relations with the Arab world.[29] September 11 indicated that this could prove to be an illusion. International terrorist groups in need of publicity would possibly view the Athens Olympics as a legitimate and attractive target, which reinforced the ongoing process of securitizing terrorism.[30]

The analysis in this section suggests that norm diffusion played a role in the securitization of terrorism in Greece, but its exact impact is difficult to calculate accurately. Would Greece have securitized terrorism without the influence of European norms and external pressures? Possibly not, because for decades previously it had failed to do so. Essentially, though, regardless of the influence of these norms and pressures, domestic actors supported and constructed the security discourse, through speech acts, visual messages and stricter laws. The changes in Greek security policies and thinking, the pressure from the forthcoming Olympics and the increased anxiety with regard to international terrorism induced Greek political elites to make terrorism the highest priority of the country. As shown, some of these developments have their roots in the mid-1990s, but since 1999 they have begun to materialize in a comprehensive change in the political discourse, a public acceptance of stricter policing and legal measures and a general re-evaluation of all policies on terrorism, all of which contributed to the arrest of 17N.

Conclusions

According to David Fromkin (1975: 687), 'the terrorist's success is almost always the result of misunderstandings or misconceptions of the terrorist strategy'. The analysis in this article reveals that the main reasons the terrorist problem proved to be so resilient in Greece was the failure of the Greek authorities to make a correct diagnosis of the roots, the level and the significance of the terrorist threat. Up until 1999, terrorism had not been perceived as a very serious threat or as a political priority of the Greek state, and as a result there was no coherent counter-terrorist strategy and no systematic and sustained effort to find the perpetrators of the terrorist attacks.

Securitization theory as developed by the CS has been shown to provide a very useful framework through which to explore and understand Greek responses to terrorism. Analysis of the Greek case underlines the importance of perceptions on how issues are dealt with at policy level, which is central to the CS approach and a significant departure from the traditional, realist view that the security agenda is predetermined and closed. In terms of the process of securitization itself, speech acts clearly played a catalytic role in both attempts to securitize terrorism. At the same time, non-verbal images and practices were also influential in the process of securitization and cannot be separated from the securitizing language on which the CS exclusively relies. Finally, this article suggests that identifying the motives behind a securitizing move can contribute to the understanding of the process of securitization and allow for an evaluation of its appropriateness. The Greek experience with indigenous terrorism highlights the potentially positive effects of securitizing an issue, which are not often addressed in the literature. As this article has demonstrated, the securitization of terrorism was the catalyst for the capture of 17N after almost three decades of unpunished terror.

With the arrest of the Revolutionary Organization November 17, Greece seems finally to be closing a dark chapter of its post-dictatorship history. However, this does not mean the total elimination of terrorism in Greece. European experience shows that after a period of time a new generation of terrorists emerge who tend to act in a fragmented and uncontrolled fashion. It is doubtful that another domestic terrorist group will emerge in Greece in the near future with the operational capabilities and scope of 17N, but it is likely that there will be an increase in small intensity terrorist acts such as bombings.[31] Nowadays, the focus of the Greek authorities should inevitably shift to the growing threat from international terrorism, which requires close cooperation within the EU and the global coalition on terrorism. The challenge for the Greek state is to ensure that democracy will continue to be strengthened in a climate of heightened anxiety.

 # Notes

1. A set of 20 personal, semi-structured interviews was carried out in Athens, from April 2002 to October 2003, with members of the Greek political elite. I thank them for their insights and the Public Benefit Institute Alexander Onassis for funding this research. I am also grateful to Roland Dannreuther, David Judge and three anonymous reviewers for their constructive comments and suggestions.

2. The CS notes that 'the security speech-act is not defined by uttering the word *security*. What is essential is the designation of an existential threat requiring emergency action or special measures and the acceptance of that designation by a significant audience' (Buzan et al., 1998: 27).

3. Alexandros Giotopoulos is the son of Dimitris Giotopoulos, a renowned 1930s communist theoretician, and close associate of Leon Trotsky. After his arrest, Giotopoulos denied the accusations and declared that he did not accept the charges, as he did not recognize the system that was making them.

4. Parliamentary Proceedings, *Greek Parliament*, 18 April 1978, p. 2776.

5. Parliamentary Proceedings, *Greek Parliament*, 13 April 1978, p. 2588.

6. For instance, one day after the Welch attack, the first page of the newspaper *Ta Nea* read: 'Great Provocation: Three swarthy men, most likely foreigners, shot dead the CIA chief'.

7. Some had speculated that the CIA was taking in its own laundry. Characteristically, two Greek newspapers had the following front page titles: 'CIA assassinates Richard Welch' and 'Double-agent Welch executed by the CIA' (Kassimeris, 2001: 73). These views were also adopted by some in the US (Kessler, 1994).

8. Parliamentary Proceedings, *Greek Parliament*, 16 May 1983, p. 6429.

9. Parliamentary Proceedings, *Greek Parliament*, 18 May 1983, p. 6452.

10. Parliamentary Proceedings, *Greek Parliament*, 12 April 1978, p. 2483.

11. Personal interview with political commentator and journalist Pavlos Tsimas on 1 October 2002. Although Greek public opinion gradually became more hostile towards 17N and rejected its tactics, even after its arrest 13.1% continued to view its members as ideological revolutionaries (see *To Vima*, 1 September 2002).

12. Similarly, after Bakoyannis's murder, PASOK supporters implied that New Democracy may be behind 17N (Papachelas and Telloglou, 2002: 153–4).

13. For instance, one day after Bakoyannis's assassination, the newspaper *Apogevmatini* carried the front-page title: 'Political Leaders: These are the instigators. PASOK is behind the killers'. On the same day, *Eleftheros Typos* bore the title: 'The PASOK-led 17 November strikes again'.

14. The interview was published in the weekly Greek newspaper *Pontiki* on 8 June 2000.

15. Parliamentary Proceedings, *Greek Parliament*, 12 June 1990, p. 4654.

16. Ibid., p. 4689.

17. Ibid., pp. 4669–70.

18. Most terrorism experts suggest that decreasing the terrorists' access to the mass media will lead to the decline of terrorism (see Crenshaw, 1991).

19. For example, when the French Intelligence agencies identified Giotopoulos and two others as suspects in 1991, they were ignored by both the Greek government and the CIA, who had their own list of suspects connected to PASOK (Papachelas and Telloglou, 2002: 161–2).

20. The Greek case is a characteristic example of the vulnerability of liberal democracies to terrorism (see Chalk, 1996).

21. Parliamentary Proceedings, *Greek Parliament*, 6 June 2001, pp. 9149–50.

22. Ibid., p. 9162.

23. The group adopted the name 'Os Edo' (No More) in a clear reference to the Spanish movement against Basque terrorism called 'Basta Ya'. In a public address, the group stated: 'The terrorists turn our silence into an excuse. In this way, they continue their terrorist activity without any substantial hindrance. The truth is that the merciless killers have managed to harm not only us but the country as well.' See *Kathimerini*, 20 December 2001.

24. Personal interview with Former Public Order Minister Stelios Papathemelis on 25 April 2002. Papathemelis noted that he faced stiff opposition when he tried to enhance cooperation with foreign intelligence agencies in order to identify the terrorists.

25. Vasilis Trikkas, member of the right-wing party Popular Orthodox Rally (LAOS), expressed a minority opinion that leading members of 17N were still free because the prosecuting authorities were 'unable or unwilling to arrest them' (Interview, 18 December 2002).

26. Apart from Giotopoulos, the rest of 17N seemingly had ordinary lives and jobs. Among the 17N terrorists, there was an electrician, a retired printer, a beekeeper, a bus driver and a telephone operator.

27. Greece had participated in TREVI and other European working groups on terrorism since 1981, but was not always willing to cooperate with its European partners. For instance, in 1986 the Greek government refused to comply with an EC decision to impose sanctions on Libya, which was suspected of sponsoring terrorist attacks (Lodge, 1988).

28. Personal interview with Yannis Valinakis, Secretary of International Relations of New Democracy, 20 December 2002.

29. Personal interview, 26 April 2002. The interviewee also noted that the Greek authorities were often tolerant towards Middle-Eastern terrorist groups, expecting in return that Greek interests would not be attacked.

30. The security costs for the Athens Olympics came to a record $1.39 billion, i.e. about the same as the cost of the entire Sydney Olympics in 2000. See *Eleftherotypia*, 10 September 2004.

31. For instance, on 12 January 2007, the left-wing 'Revolutionary Struggle', a spin off group of 17N that emerged in 2003, launched a missile attack on the US Embassy in Athens without causing any casualties.

References

Bakoyannis, Dora (1995) 'Terrorism in Greece', *Mediterranean Quarterly* 6: 17–28.

Baumgartner, Frank R. and Jones, Bryan D. (1993) *Agendas and Instabilities in American Politics*. Chicago, IL: University of Chicago Press.

BBC News (2000) 'Greek Shock at Killing', 8 June. Available at: http://news.bbc.co.uk/1/hi/world/europe/783265.stm; accessed on 15 October 2002.

Benyon, John (1996) 'The Politics of Police Co-operation in the European Union', *International Journal of the Sociology of Law* 24: 353–79.

Bigo, Didier (2000) 'Internal and External Securitizations in Europe', in M. Kelstrup and M. C. Williams (eds) *International Relations Theory and the Politics of European Integration*, pp. 142–68. London: Routledge.

Bigo, Didier (2002) 'Security and Immigration: Toward a Critique of the Governmentality of Unease', *Alternatives* 27: 63–92.

Booth, Ken, ed. (1991) *New Thinking about Strategy and International Security*. London: Harper Collins Academic.

Buhayer, Constantine (2002) 'UK's Role in Boosting Greek Counterterrorism Capabilities', *Jane's Intelligence Review* 14.

Buzan, Barry (1991) *People, States and Fear: An Agenda for International Security Studies in the Post-Cold War Era*, 2nd edn. Boulder, CO: Lynne Rienner.

Buzan, Barry, Wæver, Ole and de Wilde, Jaap (1998) *Security: A New Framework for Analysis*. Boulder, CO: Lynne Rienner.

Chalk, Peter (1996) *West European Terrorism and Counter-Terrorism: The Evolving Dynamic*. London: Macmillan.

Crenshaw, Martha (1991) 'How Terrorism Declines', in Clark McCauley (ed.) *Terrorism Research and Public Policy*, pp. 69–87. London: Frank Cass.

Den Boer, Monica (2003) '9/11 and the Europeanisation of Anti-Terrorism Policy: A Critical Assessment', *Notre Europe Policy Paper* 6.

Fromkin, David (1975) 'The Strategy of Terrorism', *Foreign Affairs* 53: 692–3.

Hansen, Lene (2000) 'The Little Mermaid's Silent Security Dilemma and the Absence of Gender in the Copenhagen School', *Millennium: Journal of International Studies* 29: 285–306.

Hogwood, Brian W. and Gunn, Lewis A. (1984) *Policy Analysis for the Real World*. Oxford: Oxford University Press.

Ioakimidis, Panayotis (2000) 'The Europeanisation of Greece's Foreign Policy: Progress and Problems', in A. Mitsos and E. Mossialos (eds) *Contemporary Greece and Europe*, pp. 359–72. Aldershot: Ashgate.

Kaminaris, Spiros (1999) 'Greece and the Middle East', *Middle East Review of International Affairs Journal* 3: 36–46.

Karakousis, Antonis (2002) 'The Hour of Truth is Near for November 17', *Kathimerini*, English edn., 14 January.

Kassimeris, George (1995) 'Greece: Twenty Years of Political Terrorism', *Terrorism and Political Violence* 7: 74–92.

Kassimeris, George (2001) *Europe's Last Red Terrorists: The Revolutionary Organization 17 November*. London: Hurst & Company.

Kassimeris, George (2005) 'Urban Guerrilla or Revolutionary Fantasist? Dimitris Koufodinas and the Revolutionary Organization 17 November', *Studies in Conflict and Terrorism* 28: 21–31.

Kavakas, Dimitrios (2000) 'Greece', in I. Manners and R. Whitman (eds) *The Foreign Policies of European Union Member States*, pp. 144–61. Manchester: Manchester University Press.

Kessler, Ronald (1994) *Inside the CIA: Revealing the Secrets of the World's Most Powerful Spy Agency*. New York: Pocket Books.

Kingdon, John W. (1995) *Agendas, Alternatives, and Public Policies*, 2nd edn. New York: HarperCollins.

Laclau, Ernesto and Mouffe, Chantal (1985) *Hegemony and Socialist Strategy: Towards a Radical Democratic Politics*. London: Verso.

Lesser, Ian O., Larrabee, Stephen F., Zanini, Michele and Vlachos-Dengler, Katia (2001) *Greece's New Geopolitics*. Santa Monica, CA: RAND.

Lodge, Juliet (1988) 'The European Community and Terrorism: From Principles to Concerted Action', in Juliet Lodge (ed.) *The Threat of Terrorism*, pp. 229–64. Brighton: Wheatsheaf Books.

March, James and Olsen, Johan (1989) *Rediscovering Institutions: The Organizational Basis of Politics*. New York: Free Press.

Merry, Wayne E. (2001) 'Don't Ignore Greek Terrorism', *Christian Science Monitor*, 14 February. Available at http://csmonitor.com/cgi-bin/durableRedirect.pl?/durable/2001/02/14/text/p11s1.html]; accessed on 17 February 2002.

National Commission on Terrorism (2000) *Countering the Changing Threat of International Terrorism*. Available at http://www.gpo.gov/nct/; accessed on 19 February 2002.

Papachelas, Alexis and Telloglou, Tasos (2002) *The 17 November Dossier*. Athens: Estia Publications [in Greek].

Schimmelfennig, Frank (2000) 'International Socialization in the New Europe: Rational Action in an Institutional Environment', *European Journal of International Relations* 6: 109–39.

Smith, Helena (1999a) 'Terrorists Hold Greece Hostage', *Guardian*, 27 May.

Smith, Jeffrey R. (1999b) 'US Presses Greece for Action Against Leftist Terror Group', *Washington Post*, 3 November.

Ullman, Richard (1983) 'Redefining Security', *International Security* 8: 129–53.

US Department of State (1990) *Patterns of Global Terrorism: 1989*. Washington, DC.

US Department of State (2000) *Patterns of Global Terrorism: 1999*. Washington, DC.

Verney, Susannah (1990) 'To Be or Not to Be within the European Community: The Party Debate and Democratic Consolidation in Greece', in Geoffrey Pridham (ed.) *Securing Democracy: Political Parties and Democratic Consolidation in Southern Europe*, pp. 203–23. London and New York: Routledge.

Wæver, Ole (1995) 'Securitization and Desecuritization', in Ronnie D. Lipschutz (ed.) *On Security*, pp. 46–86. New York: Columbia University Press.

Walt, Stephen M. (1991) 'The Renaissance of Security Studies', *International Studies Quarterly* 35: 211–39.

Williams, Michael C. (2003) 'Words, Images, Enemies: Securitization and International Politics', *International Studies Quarterly* 47: 511–31.

Georgios Karyotis is a Lecturer in International Relations at the University of Strathclyde. He holds a PhD from the University of Edinburgh and his main research interests include international security theory, European migration policy, terrorism and South-Eastern Europe politics.

Address: Department of Government, McCance Building, Room 4.43, 16 Richmond Street, Glasgow, G1 1XQ, Scotland, UK. [email: G.Karyotis@strath.ac.uk]

REVIEW QUESTIONS

1. What is "securitization" as used by Karyotis? Why does securitization provide a useful lens with which to study Greece's response to November 17?

2. What methodology does Karyotis use to study the interplay between securitization and November 17? What advantages and disadvantages do you see in his methods?

3. According to Karyotis, what political factors influenced the development and history of November 17?

4. According to Karyotis, what political factors influenced the development and history of the Greek government's response to November 17?

5. What conclusions does Karyotis draw regarding the delay in Greece's response and its ultimate success against the group? Do you agree? Why or why not?

6. Do you think Karyotis's conclusions can be applied to other countries? If so, which ones? Support your answer and selections.

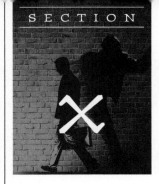

X

The Western Hemisphere, Including Canada, the United States, and South America

Learning Objectives

At the end of this section, students will be able to:

- Describe the major countries that make up the Western Hemisphere and their relationship to terrorism.
- Discuss why it is appropriate to address Canada's slow response to terrorism.
- Describe the impact of lone-wolf attacks by jihadists in the United States, but not Canada.
- Describe the mistakes and lapses that allowed 9/11 to occur.
- Describe the response by the United States to the attacks of 9/11.
- Detail the assistance the United States has provided to Colombia and provide a current assessment of the situation there regarding drugs and the armed groups that threaten peace.

(Continued)

(Continued)

- Provide an explanation for the concern with "Sovereign Citizen" groups and members.
- Detail how al Qaeda is influencing jihadists in the United States.
- Describe the current situation with regard to Shining Path in Peru.
- Detail the rise of the Shia and Sunni Jihadists as well as their beliefs.
- Define the term *jihadist.*
- Detail how Osama bin Laden deliberately provoked the United States and what he then hoped to accomplish.

Introduction

The Western Hemisphere represents a huge land mass and a large number of people, ranging from Canada, the United States, Mexico and Central America, the Caribbean, and South America. The populations are large, with Mexico, Central America, and the Caribbean having over 524 million people, Canada just over 33 million, the United States over 314 million, and South America a population of 389 million (Guardian.co.uk.world.datablog, 2010). The geographic areas of both Canada and the United States alone are quite large, with the United States measuring 9,161,966 square kilometers or 3,537,454 square miles of land and 664,709 square kilometers or 256,645,579 square miles of water (*CIA World Factbook,* n.d.). Canada is slightly larger with 3,855,102 square miles or 9,984,670 square kilometers of land and an additional 891,163 square kilometers or 344, 079 square miles of water (*CIA World Factbook,* n.d.).

This section will focus first on Canada because it presents a textbook example of how deliberately and politically correct Canadians approach difficulties such as terrorism, followed by the United States, Colombia, and Peru.

Canada

Canada is a large country geographically, the second largest in the world (Russia is the largest) with a population that is small relative to the land mass, some 33 million, with about two thirds of English, French, or other European heritage; a small Amerindian population (2%); and a mixed population of primarily Asian, African, and/or Arabian background (26%) (*CIA World Factbook,* n.d.) It has a lengthy, sparsely guarded border with the United States, making it attractive for smuggling in goods and people, including terrorists. Canada remains a country that actually seems surprised that it has terrorists and terrorist supporters in its midst. In fact, on the website of the Canadian Security Intelligence Service is a remarkable paragraph that makes the point that with the exception of the United States, Canada has more terrorist groups than any other country in the world (Canadian Security Intelligence Service, n.d.). The reality is that this diverse, immigrant-friendly, and tolerant country does have very little terrorism in its past. With the possible exception of some violence that surrounded the attempt, later solved by compromise, to make Quebec a separatist province, the only recent attempt was not a success. In an effort to get Canada to stop sending troops to Afghanistan, a group of al Qaeda sympathizers were arrested in 2006 attempting to purchase three tons of ammonium nitrate, the material used in the Oklahoma City bombing; however, the group had been monitored for months and was arrested before it could blow up the Toronto Stock Exchange, a spy agency, and a military base (French, 2010).

However, a successful, remarkable, and significant terror attack illustrates that multiculturalism, tolerance, and diversity can get people killed when it comes to terrorists. Sadly and inexcusably, the 1984 Air India Flight 182 bombing that killed all 329 passengers on a Boeing 747 on June 23, 1985, and a related bombing that killed two baggage

▲   Map of Canada

handlers at Tokyo's Narita airport the same day revealed the cavalier attitude of the authorities to preventing the bombing; the subsequent investigation was tainted with ineptness and political correctness (Milewski, 2007). This was the largest terrorist attack in Canadian history and comprehending the acts of the perpetrators does require some background. Thousands of miles away in India in 1984 events had occurred that resulted in this act of terrorism.

For some background, realize that Canada is home to a large population of Sikhs. They are clearly not assimilated, follow events in India closely, and maintain varied and extensive quarreling organizations that support the founding of an independent state called Khalistan in the Indian state of Punjab. The worst of the organizations was called Babbar Khalsa, had a violent history, and was designated as a terror organization by the European Union, Canada, the United States, and India. In early summer 1984, parts of India were in turmoil. Sikhs in the state of Punjab were viewed by authorities with suspicion and they suspected, correctly, that a large Sikh temple and its founder, Jarnail Singh Bhindranwale, were amassing weapons for another attempt to make part of Punjab a Sikh province. Prime Minister Indira Gandhi ordered an operation resulting in a siege and attack on the Golden Temple at Amritsar, the Sikh's holiest temple, with a result that many found troubling; hundreds died during the siege, including the founder, and many more died in violence in the uprising and riots that followed ("The Bombing of Air India Flight 182," 2006). For her actions in directing the attack on the temple, Prime Minister Gandhi was also assassinated in the fall of the same year at the hands of her Sikh bodyguards (Charlton, 1984).

That was apparently not good enough for Babbar Khalsa and the founder of the organization, Talwinder Singh Parmar, who approached bomb-maker Inderjit Singh Reyat of British Columbia to make the bomb that would bring down Air India Flight 182 and also produce the bomb employed in Japan (Milewski, 2007). Parmar fled the country and was ultimately killed by Indian police in 1992 and the bomb-maker was convicted. Interestingly, two others were tried in 2003 for mass murder and conspiracy in the bombings, Ajaib Singh Bagri and Ripudaman Singh Malik, but they were not convicted; perhaps that was because of the perjury of Inderjit Singh Reyat, who was tried and convicted for that in 2010 with a final appeal denied in 2012 (Woo, 2012).

The entire terrorism act in this case should have been prevented; Parmar was under surveillance, but the investigation was botched and took years to piece together and it was only the attack on the United States in September of 2001 that spurred the Canadians to be a little more serious about terrorism. An in-depth 2010 report on the Air India event conducted by a commission headed by a retired Canadian Supreme Court justice, John Major, characterized it as a case of continued stunning incompetence and sloth with over three pages of line-by-line individual mistakes that could have prevented the bombing, noting that the subsequent investigation was completely botched and the government fled from taking responsibility (Commission of Inquiry Into the Investigation of the Bombing of Air India Flight 182, 2010). The report is massive; in fact, the first volume contains seven chapters, there are five more volumes with thousands of pages, and reading it leads one to conclude it may be the most depressing document in Canadian history. It was so awful in scope that following the release of the report, in 2010, Canadian prime minister Stephen Harper apologized to the families of the Air India bombing victims (Milewski, 2010).

Following the bombing of Air India Flight 182, the Canadian government instituted some changes, such as a Special Emergency Response Team, mandated that unattended baggage would not be allowed on airplanes, and in 1989 developed its national counter-terrorist plan (Charters, 2008). Following the attacks in the United States on September 11, 2001, Canada passed the fairly robust though controversial Anti-terrorism Act after the United States passed the Patriot Act (Canadian Parliament, 2001). The harshest passages were modified in 2007; the section that allowed detention for three days without charges and compelled witnesses to testify was changed ("Canada Rejects Anti-Terror Laws, 2007). The Canadians are suspicious of any rigid counterterrorism legislation that may be seen as "racist" or "un-Canadian" since much of it would necessarily involve minorities, asylum seekers, and refugees (Crepeau, 2011). This quote from a study examining Canada's reaction to terrorism is true:

> Terrorists live in Canada and exploit vulnerable Canadian structures in order to carry out terrorism domestically and abroad. Although the government has taken action in the post-9/11 era, this paper will argue the Government of Canada has not done enough. Canadian complacency and objection to counter-terrorism measures has greatly influenced political inaction. The long borders and coastlines of Canada, and geographic position next to the United States, make Canada an optimum base for terrorist operations and financing. Terrorists who find a hiding place in multicultural society have exploited the openness of Canadian society and the Canadian commitment to liberal democracy. Lax immigration laws have allowed an influx of refugees whose backgrounds the government cannot check, and Canada needs to come to terms with the small minority of immigrants that are in fact terrorists hiding in the midst of society. (Desloges, 2011, p. 1)

What exactly has Canada done to prevent terrorism? A number of accomplishments were made in the past decade. Among changes made were incorporating the Royal Canadian Mounted Police (RCMP), the Office of Critical Infrastructure Protection and Emergency Preparedness, and the new Canadian Border Service Agency (CBSA), together to create a large department called Public Security and Emergency Preparedness Canada (Lerhe, 2009). Much of airport security came under a new agency, Canadian Air Transport Security Authority; additionally, new security centers and a first national security policy, were all funded with a $9.5 billion budget (Lerhe, 2009). A critique written by a think tank dedicated to international engagement and similar issues has assessed these measures though and while there has been progress, all is not well. For example, as Lerhe (2009) notes, many of the agencies do not share information well; as an example, Canada has a Financial Transactions Reports Analysis Centre of Canada that tracks criminal and terrorist money transfers but it does not share information well with other agencies, due to privacy laws. The report does identify other sources of problems such as inadequate funding for coordination and sharing, legal restrictions that are problematic due to the concerns of Canadians, with good reason (Lerhe, 2009). For example, in 2002, a Syrian-born Canadian citizen named Maher Arar was detained returning to Canada via New York. The authorities in New York called the FBI and despite Canadian intervention he was flown to Syria where he was subjected to ten months of torture with the Syrians ultimately concluding that he had no connection to terrorism (ccr.justice.org, n.d.). This outraged Canadians already leery of infringements on civil liberty and ultimately the Canadian government compensated him in the amount of $10 million as well. Lerhe (2009) further notes that there has been inconsistent political leadership, though the latest government in Canada may address this issue, but the one major change advocated remains a secure network that connects all agencies and departments and information-sharing greatly improves.

Canada remains committed to multiculturalism, tolerance, and diversity; has lax though tightening immigration standards; and is a haven for people smuggling. Three recent events, however, may shed light on the future of Canada and terrorism. In September of 2012, after accusing Iran of encouraging Iranians who are now living in Canada to infiltrate communities and disrupt opposition to the Iranian regime, after tiring of Iran's constant threats against Israel, and the worry that a Canadian Embassy in Iran was vulnerable to attacks, the Canadians closed their embassy in Iran and expelled the Iranian diplomats (Campion-Smith, 2012). That was uncharacteristically bold for Canadians. In another move earlier in 2012, Canadians adopted major changes in their immigration policy, including giving the immigration minister the power to deny entry simply based on public policy, limit the humanitarian and compassionate grounds now allowed for appeals, and allow deportation of visitors, refugees, and permanent residents who have committed a crime and served at least six months in jail (Payton, 2012). And finally, in 2012 the Canadians welcomed home a killer who had been held at Guantanamo Bay Naval Base prison facility in Cuba. At the behest of the Obama administration, which wanted a high-risk detainee off its hands, Omar Khadr was released to Canadian authorities even though he had been sentenced to 40 years in prison for killing an American soldier in an ambush; he served a single year in the custody of the United States and will serve the remainder of his 8-year plea bargain in Canada (Levant, 2012). What makes this troubling is that he was not an average teenager when he joined al Qaeda; he was a trained terrorist as was his father, Ahmed, a close associate of Osama bin Laden (Omar returned to Canada after being arrested for terrorism in Pakistan in 1995 at the intervention of then Prime Minister Jean Chrétien). Unlike his father, who would later die in a shootout with Pakistani soldiers, Omar was a killer, designated high risk, and he had vowed to seek revenge (Levant, 2012). No doubt the future will see more of him, but the very fact that Canada would accept someone from al Qaeda points to the very complex, mixed notion of Canadian security.

In summary, Canada remains largely reactive when it comes to terrorism. It is often regarded as allowing the United States to lead in this area, but perhaps that is changing. After Canada closed the Iranian Embassy in Ottawa and expelled the Iranian diplomats, it went a step further, with a Canadian court freezing Iranian property assets in Canada in a bid to assist a U.S. citizen to collect a $13 million judgment against Iran (Zmaneh, 2012). However, perhaps Iran planned to strike back. On April 22, 2013, Canadian officials announced the arrest of two males in Canada on charges of terrorism, plotting to derail a passenger train between Canada and the United States. What makes this so alarming is that the plotters turned out to be directed by **al Qaeda** (a Sunni terrorist organization) from Iran (home to a Shia **jihadist** regime); the Royal Canadian Mounted Police did not assert that it was state-sponsored terrorism, but this does remain a dangerous development that means the threat still is very much real (Clark, 2013).

United States

The United States, a constitutionally governed country, has experienced a large measure of success and for many years felt little direct threat from terrorism or terrorists. This is not to say that it has always been smooth sailing. The United States endured a bitter Civil War (1861–1865), with over 600,000 individuals killed, and a major economic depression in the 1930s. Two major world wars occurred in the 20th century as well as the Korean Conflict, the Viet Nam War, Desert Storm, Operation Enduring Freedom, which was the invasion of Afghanistan, and war with Iraq. Despite adversity, this quote from the CIA sums it up well: the "US remains the world's most powerful nation state" (*CIA World Factbook*, n.d.). The United States has borders with Mexico and Canada, is slightly smaller in size geographically than China, and has slightly more than 313 million people. The majority of the U.S. population is ethnically white, almost 80%, with about 12% of African American descent, around 4.5% Asian, less than 1% Amerindian or Alaska Native, and even fewer than that Pacific Islanders (*CIA World Factbook*, n.d.). A majority of the population

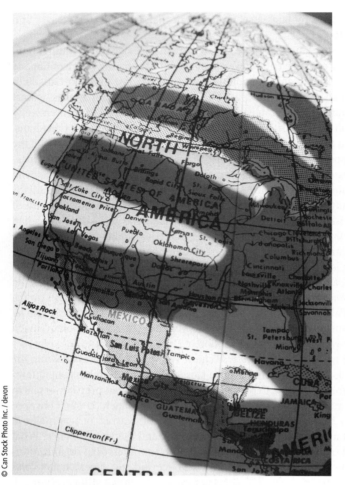

▲ Terrorism is a threat to the Americas. Canada did not take it seriously for some time, while the United States, with the attack on 9/11 had no choice.

is Protestant or Roman Catholic; the Muslim population is less than 1% (*CIA World Factbook,* n.d.).

The United States' experiences with terrorism will be covered by groups and chronologically, beginning with the Ku Klux Klan and the three periods it either thrived or survived; leftwing terrorist groups; the militia/sovereign citizen movement, including the Oklahoma City bombing; single-issue terrorist groups such as the **Animal Liberation Front (ALF)** and the **Earth Liberation Front (ELF)**; and groups that terrorize abortion clinics, although this has been characterized by the federal government as criminal behavior, not terrorism. Finally, the United States has experienced major challenges from jihadists, beginning with the first World Trade Center bombing in 1993 to the present. This remains a major issue today and it is not clear that victory will be achieved any time in the near future, though there have been some successes. Despite warnings from various high-level commissions in the last decade of the 20th century, Americans remained cavalier in their view of the world, confident that would remain untouched by terror. As will be discussed later, that was a mistake.

When the Confederacy lost the Civil War in the United States, initial Reconstruction efforts in the South saw things quickly going back to the status of the pre-war South, with Black Codes adopted by Southern states to minimize the freedom of the former slaves, including limiting voting rights, for example. While things for the South were not quite normal, they were getting better, for whites and those who had been in charge before the war. At this time, the **Ku Klux Klan (KKK)** was founded and it was something of an accident. Bored Confederate veterans, mostly former officers, decided to form a social club in Pulaski, Tennessee, near the Alabama border in December of 1865 (Chalmers, 1981). They decided to name this venture the Ku Klux Klan after the Greek term *kuklos,* for circle, and determined to have hazing rituals for new members, costumes from white sheets, and elaborate names for themselves such as "Grand Wizard" or "Grand Dragon," and soon realized that their costumed riding-around jaunts were terrifying the newly freed slaves (Staff of the KlanWatch Project of the Southern Poverty Law Center, 2011). Of course, there was a history behind this in the South. Often living in small numbers on plantations in remote locations, whites in the South employed curfews and slave patrols to enforce the movement and discipline of slaves. Plantations banded together and formed these mounted slave patrols to provide social control over the slaves. The totality of this effort was incredible, with patrols of a circuit encompassing several large plantations, ensuring that escaped slaves were captured, no slave was off the plantation without a pass, and no slave was being educated (Riechel, 1988). In fact, by about 1750 every Southern colony required slave patrols (Williams & Murphy, 1990). Having experienced the slave patrols of the past, it requires little stretch of the imagination to realize that newly freed slaves were terrorized thoroughly by the rapidly expanding KKK riding through the night. The magnitude of the terror cannot be overstated and the immediate violence was shocking. With the "Mother" Klan in Pulaski serving as a model, slowly other Klan organizations were formed throughout most of the South, though there was little central organization and most were local and forged their own rules. This changed after a secret convention in Nashville, Tennessee, in April of 1867 as a response to the passage of federal laws that replaced the Democratic led governments of ten Southern states and placed them under the command of U.S. Army generals. At the convention, former Confederate general Nathan Bedford Forrest was elected Grand Wizard; he had an empire and soon every Southern state had a Grand Dragon, often a former Confederate general. The Klan performed horrific acts on anyone it deemed to be a threat:

▲   Photo of Early KKK members. Today they appear ridiculous but terrified a large number of minorities across the United States.

> The method of the Klan was violence. It threatened, exiled, flogged, mutilated, shot, stabbed and hanged. It disposed of Negroes who were not respectful, or committed crimes, or belonged to military or political organizations such as the Loyal and Union Leagues. It drove out Northern schoolteachers and Yankee storekeepers and politicians, and "took care of" Negroes who gained land and prospered, or made inflammatory speeches or talked about equal rights. It assaulted carpetbag judges, intimidated juries, and spirited away prisoners. It attacked officials who registered Negroes, who did not give whites priority, or who foreclosed property. (Chalmers, 1981, p. 10)

How bad were things? Covering the period from 1865 to 1869, the PBS television series *American Experience* featured a program on Ulysses S. Grant's election to the presidency and provided a background article titled "The Rise of the Klan" that included some chilling details. For example, prior to the election in 1868, there were over 2,000 murders in Kansas alone, with another 1,000 blacks killed in Louisiana (*The American Experience*, n.d.). How these killings took place is all the more remarkable in that many of those who died were *lynched*, a term that did not even exist before the KKK. Lynching is defined by the National Association for the Advancement of Colored People as when four conditions are met: evidence that a person was killed, the person must have met death illegally, three or more persons must have participated in the killings (to rule out personal vendettas), and this group must have acted under the pretext of protecting justice or tradition. It should be noted that not all of those lynched were minorities, but many were. The Klan continued until 1869, becoming successful but out of control. Fearing the violence was getting out of hand and worried about more federal intervention, Imperial Wizard Forrest disbanded the Klan in

©iStockphoto.com / HultonArchive

January, ordering members to burn all records. The Federal Force Act (later found to be unconstitutional) would eventually completely destroy the Klan (at least for that era), as well as martial law and aggressive prosecutions, but the Klan, in the short period it existed, had succeeded beyond its wildest dreams. Soon the federal troops would leave but the Klan had "kept the freedmen quiet and gotten rid of 'alien' and 'aggressive' leadership among them. It had stopped the incendiarism and generally toned down the Negro, normalized relations, and established law and order" (Chalmers, 1981, p. 19).

It would be almost 80 years before full civil rights were restored for non-whites living in the South. The Klan was dead, but it had ushered in a time of segregation in the land, a doctrine known as "separate but equal," and as noted here this "was half true, everything was separate, but nothing was equal" (Staff of the KlanWatch Project of the Southern Poverty Law Center, 2011, p. 14). The Klan would rise again though, in two more eras with more violence to come. For example, numbers vary but in 1921, the *New York Times* reported that as of that date, according to a study by the Tuskegee Institute, there were 63 cases of lynching in 1921 and a total of 4,096 since 1885 ("63 Lynchings in 1921," 1922). Much of this violence was at the hands of the next generation of the Klan.

The next outbreak of the Ku Klux Klan occurred once again in uncertain times, with a great deal of change and fear creating the climate for the Klan. Following World War I, immigrants fleeing war-torn Europe poured into the United States in huge numbers. There was fear of communists as well, and in this mix was a failed Methodist minister, William J. Simmons, from Alabama. Simmons found that he could sell a dream of an organization that would guard against all the fears. Working full time, Simmons and a recruiter organized over 100,000 Klan members into districts across the country by 1921. The organization became involved politically, especially in the South, and Democratic candidates who wished to win saw fit to be Klan members to win election. Most notable was the late Senator Robert C. Byrd, elected to nine terms from West Virginia, who actually led a filibuster against the Civil Rights Act (Clymer, 2010). The number of Klan members would eventually rise to nearly 3 million and touch almost every state, except Nevada and New Mexico and a few others where the group had little impact, but soon the decline began and Klan membership fell dramatically by 1928. It would continue to decline as America worried about Prohibition and the Depression and fears of immigrants, communists, and Negro rebellion faded.

The final chapter of the Ku Klux Klan has not been written just yet, but it is in draft. Legal challenges, lawsuits, and investigations by the Internal Revenue Service whittled away at the Klan after World War II. It still existed in some places but not as a recognized national force or presence. That would change following moves for civil rights in the United States. The Southern Poverty Law Center calls this the Civil Rights Era Klan, and it came about after the U.S. Supreme Court threw out the doctrine of "separate but equal" in 1954 (Staff of the KlanWatch Project of the Southern Poverty Law Center, 2011). Although not large in numbers during this time period, the Klan resorted to violence, often in the form of bombings. Between 1956 and 1963 there were 130 bombings of churches, homes, and schools (Chalmers, 1981). An example of the violence perpetrated by the Klan was the bombing in May 1964 of the Birmingham, Alabama, 16th Street Baptist Church that resulted in the deaths of four children (Staff of the KlanWatch Project of the Southern Poverty Law Center, 2011). By this time the federal government had grown tired and began aggressively investigating the bombings and murders, including listing a number of Klan groups as targets for infiltration and counterintelligence efforts. The FBI also paid for informants and this greatly assisted the marginalization of the Klan (Chalmers, 1981). Today, the Klan still exists in small pockets across the United States, but it is insignificant as an organization.

Weathering the Klan was a trying experience for America, but the decades of the 1960s and 1970s would prove challenging on a scale that is difficult to imagine today. Looking back more than a decade later, one reporter called the era the "Riot of the Week" period, which reveals how frequently the mostly white, isolated bureaucratic police departments were wholly unprepared for the sheer number and size of riots and protests (Parsons, 1986). Causes of these events ranged across anti-war sentiment, civil rights marches and protests, and simply riots, often instigated by police, with over 100 riots in 1968 alone following the assassination of Dr. Martin Luther King Jr. (Kerner Commission Report, 1969). The Kerner Commission basically blamed poverty, and other national commissions essentially blamed the police. In this time of unrest and social change, a number of groups, mostly left-leaning in their ideology, formed, declared war on America, and began terror campaigns. Two of these now defunct groups will be presented. The first is the **Weather Underground**, a mostly unsuccessful group formed in 1969 by Mark Rudd and others, taking its name from Bob Dylan's "Subterranean Homesick Blues"— "you don't need a weatherman to know which way the wind blows." Though committed to change and wanting to end the Viet Nam War, it proved to be largely ineffective. Two prominent former members, Bill Ayers and Bernardine Dohrn (now married), were indicted but never convicted. Some of the members were convicted and spent time in prison. The group bombed and planned more bombings but proved to be so inept that three of their members were killed when a bomb they were assembling exploded on March 6, 1970 (Rudd, 2011). This effectively ended the group's terrorism.

The next group worthy of a quick examination is the **Black Panthers**, a left-wing African American group founded by Huey Newton and Bobby Seale in 1966. Newton and Seale soon managed to recruit Eldridge Cleaver as a media-savvy sort of chief information officer; he had served time in the California prison system, had written *Soul on Ice*, a best-seller, and had written for the magazine *Ramparts* (Rhodes, 1999). All were from Oakland, California, where the three made headlines patrolling sections of the city that they believed were frequently victimized by police brutality, recruiting others to their cause. The group made national news when it showed up, some members armed, at the state capitol in Sacramento to demonstrate against a proposed new law that would ban allowing individuals to carry loaded firearms in their vehicle or on their person (Rhodes, 1999). This was the Mulford Act and it was aimed specifically at them. The movement spread quickly from city to city but the founders immediately ran into difficulties of their own making. Newton was stopped by a police officer, John Frey. Shots were fired, Frey was killed, and Newton was arrested, convicted, and sent to prison. His conviction would be overturned; he would be released from prison and have two subsequent trials ending in hung juries (Hevesi, 1989). Shortly after that shooting, Cleaver and Bobby Hutton were involved in a shootout with police that left Hutton dead. Cleaver was released on bail of $50,000 and fled to Cuba and later Algeria (Kifner, 1998). Seale was charged with conspiring to riot at the Democratic National Convention in Chicago in 1968 and was charged with ordering the murder of a suspected Federal Bureau of Investigation (FBI) informant who had infiltrated the Black Panthers, though the judge would dismiss charges when the jury was unable to reach a verdict (Cobb, 1982). The Black Panthers would crumble and fall apart by 1974, in large part because of FBI director J. Edgar Hoover's aggressive and illegal counterintelligence program, police raids, infighting, and drugs (Montgomery, 2002). Across the nation, a number of Black Panthers were tried and imprisoned and the group never recovered. The founders had troubled lives; Newton was murdered; Cleaver returned from exile, pleaded guilty, and became a fundamentalist Christian, though he died young; and Seale wrote a cookbook (Kaufman, 1989).

Following their experience with groups like the Black Panthers and the Weather Underground, FBI officials and police officials became alarmed at other single-focus or "special interest" extremism often focused on saving the environment, the earth, or animals. This aggressive approach by groups resorting to terrorism to protect animals or the earth was inspired by a book published in 1962, *Silent Spring* by Rachel Carson. Writing effectively and power-fully while engaging in bad science, she launched a movement that resulted in the banning of many pesticides, including DDT; the predictable result is summed up here: "The legacy of Rachel Carson is that tens of millions of human lives—mostly children in poor, tropical countries—have been traded for the possibility of slightly improved fertility in raptors" (Miller & Conko, 2012).

These emerging threats of single-focus or special interest extremism are deadly, according to testimony by John Lewis, FBI deputy assistant director, before the Senate Judiciary Committee on May 18, 2004 (Lewis, 2004). In his testimony, Lewis characterizes the Animal Liberation Front (ALF) and the Earth Liberation Front (ELF) as emerging domestic threats, though Europe experienced them first. It may be difficult to imagine but these were particularly nasty groups and in the period of 1976 to 2004, the date of his testimony, these two groups were responsible for more than 1,100 criminal acts that resulted in $110 million in damages. Again, to understand these groups, the notion of moral equivalency must be addressed. To these true believers, animals and the earth are equal in rights to human beings. The organization People for the Ethical Treatment of Animals (PETA) has a motto that animals are *not* to be eaten, worn, experimented on, used for entertainment, or abused in any way, which gives you a glimpse of how the ELF and ALF groups feel about anyone abusing animals or harming the Earth, but they take it further. They are leaderless, with small cells of often rather young individuals, united by ideology, who take it upon themselves to commit opportunistic acts such as vandalizing or burning animal testing labs, car dealerships, or ski resorts and then publicizing that on their respective websites, followed up by communiques and interviews from a spokesperson, always sympathetic to their cause but not a participant. Examples of their actions are many but they generally have not harmed individuals. Researchers examined the time frame of 1970 to 2007 in the United States and determined there had been a total of 1,069 criminal acts committed and, based on the global definition of terrorism, they concluded that 17% of these could be classified as terrorism (Carson, LaFree, & Dugan, 2012). ELF has a focus on the Pacific Northwest area of the United States. An example of the types of attacks it engaged in were destroying, using arson, what they thought were labs growing genetically modified poplar trees; they were not, but the damage was done. They then publish photos of the destruction and send out a spokesperson, Craig Rosebraugh, a vegan baker from Portland, Oregon, to issue a communique. Rosebraugh states that he is not a member of ELF, just a sympathizer, and despite investigations, he has never been charged with any crimes related to ELF actions (Murphy, 2001). ALF is also quite ruthless as well. As an example, police blamed ALF for an attack on Long Island that targeted the wife of an executive with Forest Laboratories; her car was vandalized, and a credit card was stolen and used to donate to four charities. ALF also obtained contact information for all of Forest executives' friends and acquaintances, who received messages telling them their friends were harming animals. Spokesperson Jerry Vlasak, a trauma surgeon from Woodland Hills, California, would not acknowledge this incident, but ALF posted a communique regarding the incident on its website (Jones, 2005; Mozingo, 2006). Both ALF and ELF, while still making occasional attacks, have been muted due to relentless investigating by the Federal Bureau of Investigation and other federal, state, and local police agencies, and numerous members are currently serving time in prison. A perusal of the terrorism database at the University of Maryland, Study of Terrorism and Responses to Terrorism (START), does not show any attacks in the United States for 2012 (Study of Terrorism and Responses to Terrorism, 2013). The latest to surrender and make a deal is a Canadian woman, Rebecca Jeanette Rubin, of Vancouver, who turned herself in to federal authorities on November 29, 2012; Rubin was wanted in connection with around 20 acts of arson, committed with both ALF and ELF groups, from 1996 to 2001 in five different states (Coyne, 2012). Rubin is also wanted in California and Colorado for arson.

They have also been relentlessly pursued by legal actions and the passing of multiple federal and state laws, with the bio tech and pharmaceutical industries' encouragement and these industries are even resorting to public relations campaigns (McCullough, 2005). As an example of legislation that ALF deemed draconian was the federal Animal Enterprise Protection Act of 1992; it was replaced in 2006 by the Animal Enterprise Terrorism Act, which lessened some penalties, such as for nonviolent activities, but really began to treat ALF's action as terrorism. While its efforts have lessened, the reality is that ALF has had impact, with industries changing many practices (McCullough, 2005). People have noticed. History will judge ALF but consider this as from a recent post:

> The ELF and ALF could never be the solution to the problems they point to, but neither are they merely incidental to them: radical movements tend to be harbingers of the struggles to come when ossified political systems bury their heads in the sand instead of measuring up to the profound challenges they face and to their own internal contradictions. (Woodhouse, 2012)

In summary, ALF and ELF, while still possible threats, have not demonstrated staying power in this post-9/11 era, dealing with aggressive enforcement and legislation.

As ALF and ELF peaked, the attention of federal authorities began to shift to more right-wing single-interest groups. Already though, scholars are writing dissertations suggesting that groups such as ALF and ELF have been unfairly persecuted. They note that violent anti-abortion individuals and groups, as well as right-wing violent groups, were never targeted as aggressively with investigations and legislation as ALF and ELF were, and they see ALF and ELF members as victims of politics and repression; Shirley (2002) feels that one day they could be seen as sympathetic.

The next group to concern the Federal Bureau of Investigation and other authorities is the **sovereign citizen movement**. Most individuals in the United States are not familiar with the term or the movement, though it encompasses a wide variety of groups whose beliefs are similar. One website that tracks the various groups reports that there may be more than 60 such groups (Seditionist and Sovereignty Movements in the USA, 2014). Sovereign citizens recognize that though they may physically reside in the United States they do not regard themselves as members of the United States but are "sovereign" from the United States (Federal Bureau of Investigation, 2010). They are leaderless, may number as many as 300,000 individuals, and tend to be very small groups instead of large organizations (Southern Poverty Law Center, n.d). They may come together for training, conversation, or discussions of tactics. Many of them do not pay taxes, have no Social Security card, carry no driver's license, and do not obey federal, state, or local laws, policies, and regulations (FBI's Counterterrorism Analysis Section, 2011). They also believe in a redemption theory, that when the U.S. government removed itself from the gold standard it devalued U.S. currency to the point that it is now worthless.

> The Redemption Theory belief leads to their most prevalent method to defraud banks, credit institutions, and the U.S. government: the Redemption Scheme. Sovereign citizens believe that when the U.S. government removed itself from the gold standard, it rendered U.S. currency as a valueless credit note, exchanging one credit document (such as a dollar bill) for another. They assert that the U.S. government now uses citizens as collateral, issuing social security numbers and birth certificates to register people in trade agreements with other countries. (FBI's Counterterrorism Analysis Section, 2011)

This perspective provides their justification for attempts to "rip off" the federal government by filing false income tax returns, using stolen or bogus identification, and committing large amounts of other fraudulent criminal schemes. For an example, a reporter for the *Orange County Register* brings this:

> In August, a South El Monte man pleaded guilty in U.S. District Court to charging about 200 homeowners $15,000 apiece to eliminate their mortgages by filing sovereign citizen-inspired documents, according to the U.S. Justice Department.

> Prosecutors alleged that Ernesto Diaz, 57, earned $2.5 million through this approach. He admitted his process never worked and even cost his brother his house, the Justice Department said. In September, a federal grand jury in Alabama indicted James Timothy Turner, the self-proclaimed president of the sovereign citizen's "Republic for the united States of America," accusing him of paying taxes with fictitious bonds and holding seminars to teach his methods to others, the Justice Department said. (Collins, 2012)

And some members of these groups can be violent. Joe Stack decided a successful kamikaze attack using his private plane on an Internal Revenue Office in a high-rise building in Austin, Texas, was appropriate (Ferran, 2010). In January of 2013 in Louisiana, four deputy sheriffs were shot, with two killed, by a member of the sovereign citizen movement in an unprovoked attack (Galafaro, 2013). Clearly they are on the radar of the FBI, and the Department of Homeland Security worried about the movement and began a focus on it, but received harsh criticism with this result:

Most important, some followers believe they are entitled to use armed force to resist arrest and fight police.

The FBI also is investigating followers for alleged mail fraud and harassment of federal officials through nuisance lawsuits and property liens. Such cases are clogging courts in every state, said Casey Carty, who heads the FBI's sovereign citizen unit. Until recently, federal officials had steered clear of any extensive focus on right-wing extremist groups. In 2009, some members of Congress complained after a Homeland Security Department report warned that such groups might seek to recruit disaffected military veterans returning from Iraq and Afghanistan, as well as others. The report highlighted several groups, including the sovereign citizen movement.

Bowing to the criticism, Homeland Security officials gutted the office that had focused on right-wing extremism. They also canceled planned presentations and shelved a reference guide that the office had produced to inform local police about the movement. (Bennett, 2012)

While there have been a number of violent acts over the years, only one has risen to the level of being horrific. The most active sovereign citizen group in the 1990s was the Patriot Movement, and members of this group, Timothy McVeigh and Terry Nichols, with accomplices Michael and Lori Fortier, perpetrated the bombing of the Alfred P. Murrah Federal Building on April 19, 1995, in Oklahoma City, Oklahoma. This was a deadly act, inspired by a work of fiction, *The Turner Diaries* by Andrew Macdonald, in reality written by William Pierce, a former college professor, racist, and founder of the National Alliance, a minor racist organization in the United States; in that novel the target is the headquarters of the FBI. The toll of the Oklahoma City bombing was enormous, with 168 dead and 490 injured. The nine-story federal building was completely destroyed, two other structures collapsed, 13 more would be condemned, and 86 cars were burned or destroyed in the massive truck bomb explosion. This would be the largest criminal case in U.S. history. FBI agents conducted 28,000 interviews and collected 3.5 tons of evidence and almost one billion pieces of information in the Oklahoma City bombing case ("Lessons Learned," 2006). The Murrah building housed a tempting list of agencies for someone who hated the federal government. Some of the agencies in the Murrah building were: the Bureau of Alcohol, Tobacco, and Firearms; the Drug Enforcement Administration; the Secret Service; the Department of Housing and Urban Development; the Social Security Administration; the U.S. Army and U.S. Marine Corps recruitment offices; the Veterans Administration; the General Accounting Office; the Department of Health and Human Services; the Department of Defense; the U.S. Customs Service; the Department of Agriculture; the Department of Transportation; and the General Services Administration. An office of the Federal Employees Credit Union and the "America's Kids" Child Care Development Center were also housed in the building (Oklahoma Emergency Management, n.d.). And yes, there were children who died and the bombers knew they would (Johnston, 1995). This may have been one the quickest investigations in history as well.

The main perpetrator, McVeigh, had been pulled over by an Oklahoma state trooper just 90 minutes after the bombing for a missing license plate on his vehicle; he was arrested after a concealed weapon was found ("Terror Hits Home," n.d.). After finding a rear axle and its serial number from the Ryder Rental truck used in the bombing, authorities quickly traced it to Junction City, Kansas, where the proprietors made credible composite drawings of the man who had rented the van ("Terror Hits Home," n.d). Agents soon found traces of the explosive material on McVeigh's clothes. He was tried, quickly convicted, and sentenced to death, and Terry Nichols, who helped build the bomb, received a life sentence; Michel Fortier received a 12-year sentence and his wife was granted immunity for her testimony (Madeira, 2012; Romano, 1998). Two hundred forty-two persons witnessed McVeigh's execution on June 11, 2001, ten in person and the rest viewing from Oklahoma City on remote closed circuit television (Madeira, 2012). One wonders how these three military veterans could get involved in something as awful as this. Perhaps it is because of the hateful ideology they adopted as members of a sovereign citizen group. While this movement remains a concern, there have been no other attacks on the scale of the Oklahoma City bombing, but this one scarred Americans and will for decades.

The next developing terrorism concern for America comes from Islamic jihadists. According to the religion of Islam, which means submission, a definition of a jihadist is someone who practices *jihad,* a term meaning "to struggle," but then it becomes confusing. For some the term *jihad* means an inner struggle to lead a godly life. The jihadists who worry the federal authorities and other officials around the globe are those who believe they have an obligation to wage jihad (a struggle using war and violence) on anyone or any country that does not practice Islam, especially the United States, the West, and corrupt Arab regimes (Lahoud, 2010). Hassan al-Banna, the founder of the Muslim Brotherhood, said the following: "It is the nature of Islam to dominate, not be dominated, to impose its law on all nations, and to extend its power to the entire planet" (Wright, 2006, p. 29). Hasan al-Banna meant it and it no doubt contributed to his assassination in 1949, likely by Egyptian intelligence. His ideological successor, Sayyid Qutb, said roughly the same thing: "It is only when the rule of man has been eradicated and Sharia imposed that there will be no compulsion in religion, because there is only one choice: Islam" (Wright, 2006, p. 125).

While that is the view from the Sunni jihadists, a similar view is echoed by the Shias, as covered later in this section. It gets confusing for two reasons. First, those living in the West find this hard to believe. Second, it baffles many because the United States now faces two competing strains of jihad, a Sunni variety and a Shia version and has since 1979. This text will treat each separately and they mostly operate in that manner, though occasionally there are acts of cooperation. For example, it is well known that Iran, a Shia theocracy (the governing of a state by immediate divine guidance or by officials who are regarded as divinely guided), has provided rockets to Hamas, part of the governance in Palestine and a Sunni terrorist organization designated as a Foreign Terrorist Organization by the U.S. Department of State (Karimi, 2012; "Theocracy," n.d.).

Students taught to embrace tolerance, diversity, and multiculturalism will not find those concepts particularly useful when dealing with jihadists of the latter variety. For example, in a speech in 2005 at a conference, the president of Iran, Mahmoud Ahmadinejad, stated that Israel must be wiped off the map (Penketh, 2005). A jihadist does not seek compromise, just one's conversion to Islam or death. Muslims who decide to convert to Christianity or Buddhism, for example, would almost certainly face a difficult, if not deadly, time, unless living in the United States or other Western nation. Are there moderates out there in Islam? Yes, most are, fortunately, but take a look at the former president of Egypt, Mohamed Morsi, a member of the Sunni Muslim Brotherhood (and a 1982 PhD graduate in engineering from the University of Southern California, who then taught at California State University, Northridge, until 1985). Morsi, in 2010, made anti-Semitic (anti-Jewish) videos calling Jews warmongers and the descendants of apes and pigs and has said no woman or non-Muslim individual should be allowed to be president of Egypt based on Islamic Law or Shariah (Kirkpatrick, 2013; Mohamed Morsi, n.d.). Egypt has a peace treaty with Israel and Israel is, of course, concerned. The Muslim Brotherhood assassinated President Anwar Sadat in 1981 for signing that peace treaty, so they are unlikely to honor it for long. Former president Morsi may not be a concern anymore as he was removed in a coup on July 3, 2013, and replaced by a judge, demonstrating that he and the Muslim Brotherhood were too extreme for Egypt (Bradley & Abdellatif, 2013). A new round of elections was held in 2014 and the winner, not surprisingly, was the former general who deposed Morsi, Abdel Fattah el-Sisi; President Sisi also banned the Muslim Brotherhood. Finally, the prime minister of Turkey, Recep Tayyip Erdogan, recently equated Zionism with a crime against humanity (Loiko & McDonnell, 2013). Whether jihadis are Shia or Sunni, if they truly believe what the Koran says literally, they take extreme positions that are dangerous. Again, this is not to suggest that every practicing Muslim embraces this view, and fortunately that is not the case, as the majority reject this, but as two scholars note:

> The point we would like to make is simple. While many Muslims are peace-loving, nonetheless, those who commit acts of violence and terror in the name of God can find ample justification for their actions, based on the teachings of the Qu'ran and the sayings and examples from the prophet Muhammad himself. (Geisler & Saleeb, 2003, p. 319)

The media, academia, and the elite do not share this view of Islam, which is unfortunate, but that does not make it less of a fact. The aforementioned groups are intellectually dishonest and evidence of this is their hypocrisy when it comes to mocking religion. They tolerate and may even participate in bashing Christianity and Judaism, but one will never find them disparaging Islam, because they are afraid of the level of violence and wrath they might incur from members of Islam. Examples of this are many but the following three will be clarifying examples. In 2004, a Dutch filmmaker named Theodore van Gogh, a relative of the famous artist Vincent van Gogh, was shot and stabbed to death on a Tuesday morning in Amsterdam by a Muslim male as a result of making a film criticizing the treatment of women in Islamic society (Frankel, 2004). His collaborator on the film, Ayaan Hirsi Ali, a member of the Dutch Parliament at the time, was forced to go into hiding (Gardels, 2012). Finally, around 150 homes of Christians were burned by mobs on March 11, 2013, in Lahore, Pakistan, after accusations that a Christian had committed blasphemy against the Prophet Muhammad (Steffan, 2013). Is violence a problem with Islam? The authors suggest a reading of the Koran (to use another spelling of the book) and encourage making a decision as an individual, examining such passages as Suras 2, 3, and 4 in particular.

While the United States should have been alerted to the threat posed by the jihadists after the first attack on the World Trade Center in 1993 and the threats and attacks from the Shia jihadists from Iran since 1979, the U.S. posture was one of denial and the use of the criminal justice system to deal with the threats and violence. The perpetrators of the World Trade Center attack in 1993 were prosecuted and the United States moved on. Not that there were not those expressing concern. For example, during the dozen years before the 9/11 attacks there were major efforts by the administrations of Bill Clinton and George W. Bush to improve airline security with reforms that would have made the 9/11 attacks improbable if not impossible. Due to heavy lobbying by the airline industry, including lobbyist Linda Daschle, the wife of then Senate majority leader Tom Daschle, changes were scuttled or watered down to the point of being meaningless. A major commission, the Hart-Rudman Commission, warned shortly before 9/11 that "a direct attack against American citizens on American soil is likely over the next quarter century. The risk is not only death and destruction but also a demoralization that could undermine US global leadership" (Costa, 2001).

The United States has now experienced several attacks, including on two embassies in Africa in 1998, the bombing of the USS *Cole* in 2000, and the attack on September 11, 2001 (Al-Qaida Timeline: Plots and Attacks, n.d.). Two wars have been fought and inconclusively resolved, one directly aimed at al Qaeda and the Taliban sheltering them. The Obama administration seems loathe to address terrorism in a serious manner. It has stated, despite clear evidence, that Major Nidal Nasan, an Army psychiatrist who killed 13 people, is not a terrorist but simply a criminal and characterized the shooting as "workplace violence." Over the past four years, the Obama administration has characterized terrorism as not only that but as "overseas contingency operations" or "man-caused disasters," and generally seems to have a difficult time stating that something is terrorism. Many consider this to be alarming. The most recent example was the bombing of the Boston Marathon in 2013, perpetrated by Chechens who may have been self-radicalized and then encouraged and trained, perhaps in Russia, Chechnya, or Dagestan. Once again, the suspect was read his Miranda rights and treated as a criminal despite compelling evidence that this event was terrorism. The bottom line remains that the United States has a war and the other side, the jihadists, realize it even if some do not.

Iran, the leading Shia jihadist threat to the United States, became a threat when the Ayatollah Khomeini rose to power following the fall of the Shah of Iran, Mohammad Rezā Shāh Pahlavī. The Shah of Iran was deeply unpopular for his brutality, his secret police, the SAVAK (a contraction of the Farsi words for security and information organization), and for his extravagant lifestyle. The fact that he was thrust into power by the Central Intelligence Agency (CIA) of the United States and by Great Britain's Secret Intelligence Service (also called MI6; it's similar to the CIA) in 1953 did not add to his prestige; this coup resulted when the powerful prime minister, Mohammed Mossadegh, nationalized the Anglo-Iranian Oil Company (which would become British Petroleum in 1954), was openly friendly with the Soviet Union, and had reduced the Shah to a mere puppet (Little, 2004; "MI5 or MI6?" n.d.). Things were difficult in Iran during the decade of the 1970s but the Shah proved to be a reliable ally for the United States. However, the Shah instituted major social and economic reforms called the "White Revolution" that angered Iranians. In

1963 he found himself the subject of massive riots and protests by Islamic groups angry about Western modernization, nationalist groups angry over American influence on Iran, and against his dictatorship (Moens, 1991). He responded violently to this threat and exiled Khomeini and survived. Jimmy Carter was elected president in 1976 and the Shah would soon find himself doomed. President Carter championed human rights and encouraged the Shah to institute reforms; the Shah complied and this, seen as weakness on his part, led to large demonstrations calling for his resignation. When the Shah asked for guidance from the Carter administration it was muddled, unclear, and indecisive and soon the country was on the verge of collapse; the Shah once again used massive force to end riots and protests but, worried about the Carter administration, began granting more reforms that just produced more massive violent protests. This resulted in the Carter administration considering having the Shah leave and, fearing a takeover by the military, which President Carter found unthinkable, proposed forming a moderate national government, perhaps under the Shah as a figurehead or perhaps not. Things became serious in December of 1978, when Khomeini began calling for more strikes and protests from exile in Paris (Moens, 1991). In 1979 the Shah decided to appoint a prominent political leader, Shapour Bakhtiar, to form a civilian government and hinted that he might leave for a "vacation." Meanwhile, some in the Carter administration were suggesting that perhaps Khomeini would serve as a figurehead and encourage the formation of a moderate government. It was not to be. On February 1, 1979, Khomeini flew to Tehran and immediately set up an alternative government. The efforts of Bakhtiar failed when the military switched its loyalty to Khomeini, and Bakhtiar left the country (Moens, 1991). But who was Khomeini and what would he do?

Bernard Lewis, perhaps the most noted Middle East historian of our time, was asked the question by the *New York Times* soon after that. Professor Lewis replied that he did not know and reported that, like any university professor, he went to the library to find out. The library at Princeton had one slim volume by Khomeini, called the *Government of the Jurist,* which spelled out the need for, the form of, and the program for establishing an Islamic government. Professor Lewis, now understanding what he was dealing with, wrote an article that the newspaper declined to publish (Lewis, B., 2009).

Immediately after coming to power in Iran, according to a 1987 United Nations Commission on Human Rights report, Khomeini executed approximately 7,000 individuals who were shot, hung, burned, or stoned; they included homosexuals, prostitutes, criminals, and the Shah's officials. Khomeini instituted Sharia law and required women to wear a chador and veil (full-length gown). He allowed a takeover of the American Embassy (American sovereign territory) in Tehran in November 1979 that resulted in hostages held for 444 days. Then there was the attack using Hezbollah, an Iranian-supported terrorist group, on the American Embassy in Beirut, Lebanon, in 1983 that killed 61 and wounded 120, and the bombing of Marine barracks in 1983 that resulted in casualties of nearly 300 (Shahar, n.d.). He followed that up with a disastrous and costly war with Iraq, encouraging young masses of suicide squads for eight years, to no avail, with immense casualties on both sides (Anderson, 1989). Writing in a 2011 article one author recounts the broad brush of Iran's terrorism over the years, including a recent attempt on the life of the Saudi Ambassador to the United States, attacks on a Jewish synagogue in Argentina in the 1990s, and attacks in Saudi Arabia as well (Byman, 2008; Levitt, 2011). Iran has supported al Qaeda and Hamas as well. The latest effort to deal with terrorism by Iran has resulted in sanctions on the country, and the listing of Hezbollah as a designated foreign terrorist organization, as well as members of the Quds Force, a special group within the Revolutionary Guard that often trains, finances, and guides various groups in terrorist attacks around the world (Byman, 2008; Levitt, 2011). The Ayatollah Khomeini died in 1989 but his successor, the Ayatollah Ali Khamenei, has maintained the repression in the country and continued allowing support of the various terrorist groups around the world. Iran has vowed to obtain nuclear weapons and use them, and the prospect of a nuclear Iran is indeed frightening, especially if it has a view of the world that many find troubling. Both Ayatollah Ali Khamenei, the Supreme Leader of Iran, and the former president of Iran, Mahmoud Ahmadinejad, share a belief, similar to that in other religions, that a messianic figure, the Twelfth Imam, hidden by God centuries ago, will return during a time of death and destruction, and many believe a nuclear armed Iran would ensure there would be death and destruction (Petrou, 2011). The Iranian conservative religious Imams do believe this, and Iranian discussions have included whether there should be increased building of hotels as the return of the Twelfth Imam would be a tourist attraction. Former president Carter, concerned with human rights, failed to understand the entire culture of death and destruction part of the Shia jihadists and the Ayatollahs.

Unfortunately, many in the West, and in the United States in particular, find this hard to believe. The Obama administration has engaged in sanctions and threatened war, perhaps. The Iranian regime has suffered setbacks, including deaths by murder or assassination of four Iranian scientists and a devastating computer attack, both of which may have been the result of cooperation between the United States and Israel (Calabresi, Crowley, & Newton-Small, 2013). Recent diplomatic sanctions have damaged the Iranian economy, reducing the purchasing power of its currency by half, exports of oil by half, and producing annual inflation of almost 30% (Calabresi et al., 2013). War is a possibility; President Ahmadinejad's term expired in August 2013, and he was succeeded by Hassan Rouhani, considered a moderate, which may account for the increase in negotiations with Iran; it will be an interesting time for Iran (Petrou, 2011).

The final threat facing the United States is from the Sunni jihadists. Again, as pointed out in the exceptional book by Lawrence Wright, *The Looming Tower: Al-Qaeda and the Road to 9/11,* Sunni jihadists arose primarily in Egypt, outraged that President Sadat signed a peace treaty with Israel. The Muslim Brotherhood renounced violence, and new groups formed from the remnants; many of the members were imprisoned in Egypt and soon realized that violence was their only option for change. In order to really understand what happened, two dates must be examined: September 17, 1978, when Egypt and Israel reached a deal for peace called the Camp David Accords, and March 26, 1979, when the formal peace treaty was signed between Israel and Egypt. This occurred after intense pressure was applied by the United States and President Carter, with much celebration and joy, though with little appreciation for the very real long-term consequences. Too, there was a naive lack of concern about the many people actively seeking to end the Egyptian regime, more motivated than ever. While the peace treaty served to mobilize them to act, and while they surfaced at a similar time to that of the Shia jihadists, they were immediately distracted by the Soviet Union invading Afghanistan and responded by organizing, recruiting, and training Egyptians, Arabs, and sympathizers from all over the world. To explain the rage generated by the peace treaty and the invasion of Afghanistan, this section will present biographies and an examination of the following individuals: Egyptians Sheikh Omar Abdul Rahman (the Blind Sheikh), Ayman Al-Zawahari; Palestinian Abdullah Azzam; and (with Yemen heritage but raised in Saudi Arabia) Osama bin Laden. These are four remarkable, well-educated individuals, working sometimes alone with groups they founded and would lead, sometimes in concert as al Qaeda, to create terrorism that would greatly impact the United States. Understanding who these individuals are, what they wanted, what they accomplished, and how their lives intersected, tells the ongoing story of Sunni jihad. Between them they destroyed at least three embassies, perpetrated 9/11, recruited and trained thousands, killed and injured thousands, and were the proximate cause of the Afghanistan war and the Iraq war.

Omar Abdul Rahman (or Omar Abdel-Rahman) was born in 1938 in Egypt. Juvenile diabetes as a child left him blind but does not seem to have kept him from learning. He reportedly memorized the entire Koran in Braille by age eleven. He attended two universities, earning a doctorate in theology (Holt, 2012). He was a largely undistinguished scholar and cleric but as Egypt fell under Western, secular influences, he became incensed, preaching and organizing for change by leading the Islamic Group, a terror group (Wright, 2006). After being jailed five separate times for issuing **Fatwas** (religious edicts) encouraging violence against secular leaders and regimes, he left for Saudi Arabia where he taught for three years and raised funds for his efforts. He eventually returned to Egypt only to be expelled in 1984 (Holt, 2012). He fled to Afghanistan where he became allied with al Qaeda, and Abdullah Azzam, its founder, as well as Osama bin Laden, the financier of al Qaeda. Upon Azzam's assassination (perhaps by Osama bin Laden), he assumed control of al Qaeda's international arm (Holt, 2012; Wright, 2006). His three passions, according to Mary Weaver, a writer for *The New Yorker* magazine, were defeating the Soviets in Afghanistan, spreading Islam worldwide, and replacing the secular Egyptian government (Weaver, 1993).

Sheikh Rahman initially believed that his preaching and fatwas would mobilize the world for Islam. Imprisoned with Ayman Al-Zawahari for a time in an Egyptian prison, they initially differed on methods, with Al-Zawahari urging violence. Over time Sheikh Rahman would opt for violence to cleanse Egypt of secular rule. After he witnessed the Soviet defeat in Afghanistan, he set his sights on the United States and decided to attempt to destroy it. He entered the United States in July 1990, despite being on a terrorist watch list (Holt, 2012). He immediately began preaching violent jihad in mosques, fund-raising, recruiting, and plotting, and formed al Qaeda in the United States. As one pundit put it, the few jihadists in the United States were the minor league, but now they had a major league player to guide them. Through a series of missteps that resulted in the revocation of his visa, he was nevertheless allowed to return to the United States and attempted to obtain permanent residence status, though that was denied. He then asked for political asylum and managed to remain in the United States while still encouraging violence abroad and plotting against the United States, often sending instructions and sermons on cassette tapes smuggled into Egypt (Weaver, 1993). For example, the Islamic Group in 1992 targeted Egyptian police, foreigners, Christians, and intellectuals, possibly with encouragement from the Blind Sheikh (Wright, 2006).

Aggressively recruiting for al Qaeda in the United States, he managed to plan and execute an attempt to bring down the World Trade Center by bombing it on February 26, 1993. The World Trade Center was badly damaged but not destroyed; the attack killed six and injured 1,042. The operation was financed by Osama bin Laden and the bomber, Ramzi Yousef, was a product of an al Qaeda training camp in Afghanistan (Wright, 2006). Four of Rahman's followers were immediately arrested though Yousef (whose uncle is Khaled Sheikh Mohammed, planner of the 9/11 attacks, also arrested and in Guantanamo Bay Prison) escaped and flew immediately to Pakistan and later to Manila to continue plotting (bin Laden reportedly asked him to figure out a way to assassinate President Clinton) (Holt, 2012; Wright, 2006). Yousef was eventually arrested in Pakistan, after failing to succeed in later plots, was convicted, and is now serving life plus 20 years in a "SuperMax" federal prison in Colorado (Smith, 2013). After a three-month investigation into the bombing , Sheikh Rahman was arrested along with additional members of the terror cell, who were also plotting to assassinate then Egyptian president Hosni Mubarak on an upcoming visit to the United States, and bomb the United Nations building and the Lincoln and Holland Tunnels, and the George Washington Bridge in New York City (Holt, 2012). He was tried in 1995 and convicted of seditious conspiracy and is serving a life sentence at the Butner Federal Medical Center in North Carolina (Holt, 2012; Wright, 2006).

The Blind Sheikh was two years ahead of Abdullah Azzam, al Qaeda's founder, at the university where both received their doctorates, and he served as a mentor to Azzam. When Azzam founded al Qaeda, Omar Abdul Rahman traveled to Afghanistan to work with him and upon his death would travel to the United States and lead al Qaeda there. Ayman Al-Zawahiri, who would become a leader of al Qaeda, was imprisoned with the Blind Sheikh. They led competing organizations that had similar goals.

While Sheikh Rahman remains in prison, he has not been forgotten (Holt, 2012). Former Egyptian president Morsi vowed to seek freedom for the Blind Sheikh, who is in his seventies and reportedly ill. That may not be possible, if relations with Egypt remain tense under President Sisi (Bradley, 2012). Sheikh Rahman will always be remembered for launching the first of many al Qaeda attacks on the United States.

Abdullah Azzam was a Palestinian, born in 1941 in Jenin, Palestinian Territories. He became a member of the Muslim Brotherhood in the 1950s and was strongly influenced by the founder of the Muslim Brotherhood, Hassan al-Banna and his fellow ideologue, Sayyid Qutb (Schnelle, 2012; Wright, 2006). Following Israel's capture of the West Bank in 1967, Azzam fled to Jordan and soon found his way to the al Azhar University in Cairo, and following in the footsteps of his friend, Sheikh Omar Abdul Rahman, the Blind Sheikh, would complete his doctorate in 1973 (Wright, 2006). Living in Jordan, Azzam found employment at a university and his lectures became radical and popular to the point that Jordanian security services began paying attention to him. He then moved to Saudi Arabia to teach at King Abdul Ibn Saud University; it is unclear if he and Osama bin Laden met in Saudi Arabia at this time or later in Pakistan. Azzam made a pilgrimage to Mecca in 1980, meeting an influential member of the Afghan jihad fighting against the Soviets that had invaded Afghanistan the previous year; this was a decisive moment for him. He requested that his university allow him to transfer to a new university in Islamabad, Pakistan, that had ties to his current university. His wish was granted and he moved there in 1981 after the school year ended. He volunteered as an intermediary between the Afghans and Arab volunteers, spending weekends in Peshawar, Pakistan, the center of Afghan resistance to the Soviet invasion, hoping to attract Arab volunteers to fight (Schnelle, 2012). This did not prove to be a successful venture until after he published a book, *The Defense of the Muslim Lands*, in 1984 that increased the number of Arab recruits for the Afghan cause to the point that he quit his university job and with his new friend, Osama bin Laden, formed the Services Bureau or Maktab al-Khadamat to house the now thousands of volunteers, with funding by Saudis, who even paid part of the airfare for recruits to get to Pakistan (Read, 2009; Schnelle, 2012).

Abdullah Azzam was very clear about his desires. According to Lawrence Wright, his slogan was "Jihad and the rifle alone; no negotiations, no conferences, no dialogues," and he wished for a version of Islam that would dominate the world through the force of arms (Wright, 2006, p. 110). He was one of the founders of Hamas, a Palestinian terrorist group that served as an Islamic counter to Yasser Arafat's secular Palestinian Liberation Organization, and founded, along with bin Laden, al Qaeda, the Base. He was reportedly charismatic, courageous, and an effective writer and speaker, and as such generated both goodwill and enemies. He worked tirelessly with Osama bin Laden to recruit individuals to fight the Soviets, often staying with him when in Saudi Arabia (Schnelle, 2012). He used bin Laden to raise funds—some were even provided by bin Laden and the Saudi government. As a Palestinian, he seems to have not been trusted by the Egyptians, particularly Dr. Ayman Al-Zawahari, by then working in hospitals in Pakistan.

Azzam had the Blind Sheikh as a mentor, bin Laden was mentored by Azzam, and Zawahari was an acquaintance, although the latter was jealous, and even the Saudis feared the immense influence of Azzam. He was assassinated on November 24, 1984, by a roadside bomb, along with his two sons (Wright, 2006). His accomplishments include establishing al Qaeda, but perhaps his most significant feat in the area of fund-raising, much of it done in the United States, where he founded 33 branches of bin Laden and Azzam's Services Bureau (which would later become al Qaeda) to support the jihad (Schnelle, 2012). Though Azzam is dead, as noted previously his influence was enormous and continues to have an impact. His writing and speeches mobilized many and he will be remembered primarily as a master propagandist and the founder of al Qaeda; in the end, his rhetoric, writing, and his organizations set the course for much of the deadly terrorism to come (Schnelle, 2012; Wright, 2006).

Ayman Al-Zawahari, an Egyptian, was born in Cairo, Egypt, on June 19, 1951. Very devout, he revered Sayyid Qutb; Al-Zawahari's uncle was Qutb's attorney. Zawahari formed a radical Islamic cell while still in high school, dedicated to making Egypt an Islamic state. While still growing his terrorist cell, he attended university and graduated from medical school in 1974, serving three years in the Egyptian military before opening a clinic. He was one of the first medical doctors to travel to Afghanistan to provide aid to Afghan resistance fighters; there he met Osama bin Laden and ultimately would provide guidance to him, receive financial assistance from him, and end up as second in command of al Qaeda (Wright, 2006).

Al-Zawahari wanted most of all to rid Egypt of its secular government, though he also blamed the United States and the West in general for supporting the Egyptian government with aid. He applauded the takeover of Iran by the Ayatollah Khomeini and wished the same for Egypt. Following Egypt's signing of the peace accord with Israel, his group, now called al-Jihad, began planning to take over the country by killing the leaders, taking over the radio and television building, and expecting this would result in a popular uprising. Unfortunately, one of the members was arrested, leading to a large number of additional arrests, but these somehow missed Al-Zawahari. Unfortunately for the leader of Egypt, Anwar al-Sadat, the officials also missed an al-Jihad cell based within the Egyptian military. A hastily formed plan was developed to assassinate President Sadat. On October 6, 1981, the cell carried out the plan, with a group of four attacking and killing President Sadat with machine guns and grenades (Wright, 2006). This finally resulted in the arrest of Al-Zawahari as he was driven to the airport to go to Pakistan.

Zawahari's trial lasted three years and he was only charged with dealing in arms; he was convicted and sentenced to three years, which he had almost already served. He was released in 1984, an embittered radical (Wright, 2006). Zawahari resumed his medical practice but then, using a false passport, as he was not allowed to leave Egypt, he moved to Saudi Arabia, staying there until moving to Peshawar, Pakistan, in 1986. It is not clear whether Zawahari met Osama bin Laden in Saudi Arabia, but they soon became allies in Pakistan and later in Afghanistan, where Zawahari had traveled frequently and provided medical assistance to the fighters and even to bin Laden. Zawahari was still focused on Egypt and rebuilding his destroyed al-Jihad organization but was intrigued by the global organization, al Qaeda, and the possibilities it offered. Zawahari offered the theoretical underpinnings for jihad, violence, and suicide bombings; bin Laden offered funding and connections (Wright, 2006).

With al Qaeda, Zawahari would become the second in command and assume the role of leader after Osama bin Laden's death on May 2, 2011, but he does have a long history of violent acts, using his own organization, al-Jihad. For example, a faction of Zawahari's al-Jihad had been compromised in the early 1990s. Despite imposing a cell-like structure where no one cell knew any other, the membership director was arrested by Egyptian authorities and had in his possession a computer with the names of eight hundred members. This would ultimately result in the arrest of over one thousand individuals. Zawahari was furious and vowed to strike back. In 1995, on the anniversary of President Sadat's historic visit to Israel, al-Jihad bombed the Egyptian embassy in Islamabad, Pakistan, killing 16 and destroying the embassy. It was a very effective suicide attack that become a model for later al Qaeda attacks (Wright, 2006). He would also plan an attack in Egypt in 1997 that killed 67 foreign tourists (Lake & Seper, 2011). In the end, al-Jihad would merge with al Qaeda. Since the death of bin Laden, he has become an even greater challenge; he lost his wife and son to a U.S. airstrike in 2001 and issued a hit list urging "lone wolf" agents to kill prominent Americans (Lake & Seper, 2011). In his latest role as the leader of al Qaeda, in June of 2013 he urged Sunnis to rush to Syria to fight against the regime, making a general war between Sunnis and Shias in the Middle East a probable event (Williams & Edwards, 2013).

Osama bin Laden was a fortunate son of Mohammed bin Laden, an extremely wealthy, hardworking, uneducated, illiterate construction magnate from Yemen who made his fortune in Saudi Arabia; Osama bin Laden was his 17th son, and although Mohammed bin Laden officially had 54 children and 22 wives, there were other unofficial "marriages" and concubines (Wright, 2006). Osama bin Laden was born on March 10, 1957, though other sources claim it was 1958, to Alia, Mohammed bin Laden's fourth wife, a 14-year-old from Syria (Gatehouse, 2011; Wright, 2006). His father did not seem very involved with his son and Osama traveled very little, coming under the influence of the Saudi Muslim Brotherhood, which he joined in high school. He even attended lectures by Mohammad Qutb, the younger brother of Sayyid Qutb, making Osama bin Laden's outlook of the world and religious ideology similar to those of Azzam and Zawahari (Wright, 2006). When bin Laden's father was killed in a plane crash, he inherited a considerable amount of money (Gatehouse, 2011). He attended King Abdul Aziz University but left with one year of study remaining, already a polygamist with four wives (Wright, 2006).

Following the invasion of Afghanistan in 1979, Osama bin Laden seems to have traveled there, possibly working for Saudi Intelligence, to aid Afghans fighting the Soviets (Gatehouse, 2011). It is certain that he was there by 1983 or 1984 as he set up the Services Office (Maktab al-Khadamat) along with his mentor, Abdul Azzam (Gatehouse, 2011; Wright, 2006). In 1986 bin Laden created his own training camp for Arab fighters, called al-Masadah or the Lion's Den; their clashes with the Soviets produced uneven results. Eventually, with Abdullah Azzam, he would form al Qaeda and live to see the Soviets defeated.

Returning to Saudi Arabia following the defeat of the Soviets, he was reportedly outraged that the United States was allowing U.S. troops to base in Saudi Arabia to prepare for the first Gulf War. This view was no doubt influenced by Ayman Al-Zawahari, with whom he worked closely in Afghanistan, and he soon adopted and espoused a vision of global jihad, aimed at toppling the United States next. His violent rhetoric became too much for the Saudi government to tolerate and he was strongly encouraged in 1991 to leave the country; he went to Sudan, the home of an extreme Islamic regime. Here he worked openly as a businessman, largely staying out of the limelight. It was Al-Zawahari's bombing of the Egyptian embassy in 1995 with his al-Jihad group that got them expelled from Sudan in 1996 (Gatehouse, 2011).

Al Qaeda, bin Laden, and Al-Zawahari were then welcomed to Afghanistan by the Taliban (the term means "student" and refers to the Islamic fundamentalists who fought against the Soviets) and their leader, Mullah Omar. From Afghanistan, al Qaeda would soon stage one of several major attacks, but not before bin Laden would issue a fatwa, or his Declaration of War Against Americans Occupying the Land of the Two Holy Places, on August 23, 1996. Only then would he act. After years of planning and surveillance, largely conducted by Ali Mohamed, a member of both al Qaeda and, for a time, the U.S. military, al Qaeda was ready to act. On August 7, 1998, two U.S. embassies were bombed on the same day with deadly results, one in Kenya and one in Tanzania. These actions of al Qaeda were condemned around the world, including by Islam, when investigations proved al Qaeda was behind the attacks. Bin Laden hoped the bombings would lure the United States into Afghanistan. While al Qaeda did carry out the attacks, it had help, as a federal judge determined in 2011, who stated that Iran and Hezbollah aided and abetted al Qaeda in the embassy bombings (Thiessen, 2011; Wright, 2006).

Following the success of the embassy bombings, the United States was slow to react. Al Qaeda, on the other hand, was not. As early as 1999, plans were being made for the attack on the U.S. mainland in 2001, with an attack also planned for 2000 on a U.S. naval vessel. The FBI and the CIA knew something was coming but bureaucratic infighting, incompetence, and a refusal to cooperate and share information doomed their efforts. For example, John Patrick O'Neill, an FBI agent based in New York City, was convinced that al Qaeda had sleeper agents in the United States as early as 2000, but the FBI hierarchy ignored this as a manageable threat (Wright, 2006).

Al Qaeda's next attempt at a major event was thwarted by inexperience and just plain bad luck. On January 3, 2000, the USS *The Sullivans,* an American destroyer, was refueling in Aden, Yemen; it was a tempting, vulnerable target but the attack failed when the al Qaeda operatives overloaded the boat they planned to use and it sank (Miniter, 2003; Wright, 2006). This failed attempt and the lack of information-sharing following it is an example of the ineptitude that existed at the time. The national security advisor to President Clinton, Richard Clarke, was aware in the summer of 2000 that something was up; a CIA informant told intelligence officials that there would likely be an attack on a U.S. naval vessel in the eastern Mediterranean, but Clarke was not aware of the January attempt on the USS *The Sullivans* and he also did not know that the Navy was using Aden, Yemen, as a refueling station (Miniter, 2003).

The USS *Cole* was a formidable ship, a billion-dollar guided-missile destroyer with advanced stealth technology. But the ship's sentries ship carried unloaded shotguns. On October 12, 2000, in the port of Aden, Yemen, as refueling the *Cole* was just getting under way, two men in a small fishing boat laden with C-4, a plastic explosive used by the U.S. military, fashioned into a shaped charge design to direct the blast inward, sailed to amidships and detonated the bomb (Miniter, 2003; Wright, 2006). This resulted in the deaths of 17 sailors, injuries to 39 others, and a 40' by 40' hole in the ship's hull. Due to courageous efforts by the crew, the ship was saved though severely damaged and would require 14 months and $250 million to repair (Miniter, 2003; Nagle, 2002; Wright, 2006).

The FBI immediately began an investigation but it did not go well. Leading the FBI's on-scene investigation in Yemen was John O'Neill, the lead investigator for the Nairobi, Kenya, embassy bombing by al Qaeda. O'Neill immediately crossed swords with the U.S. ambassador to Yemen, Barbara Bodine, over the size of the team, her rules, and her interference. She hampered the investigation and eventually had O'Neill recalled before he and the team made sufficient in-roads to determine whether it was an al Qaeda attack. The CIA was ahead of the FBI and proved it was, and the FBI later agreed with that assessment. It was an al Qaeda attack exploited by bin Laden; he wrote poetry glorifying the event and it was used as a recruiting tool (Miniter, 2003; Wright, 2006).

President Clinton's term was winding down, he was making a final effort for an Israeli/Palestinian peace deal, and there was no a response to the bombing of the USS *Cole* by bin Laden and al Qaeda (Miniter, 2003; Wright, 2006). As Lawrence Wright notes in *The Looming Tower:*

> Bin Laden was angry and disappointed. He hoped to lure America into the same trap the Soviets had fallen into: Afghanistan. His strategy was to continually attack until the U.S. forces invaded; then the Mujahadeen would swarm upon them and bleed them until the entire American empire fell from its wounds. It had happened to Great Britain and to the Soviet Union. He was certain it would happen to America. The declaration of war, the strike on the American embassies and now the bombing of the *Cole* had been inadequate, however, to provoke a massive retaliation. He would have to create an irresistible outrage (2006, p. 375).

In the end, bin Laden would simply have to wait for a new president of the United States. As Richard Miniter (2003) explains in *Losing bin Laden:*

> In the last days of his administration, Clinton decided not to fire a parting shot at bin Laden. The terrorists of the world were left with another lesson: even American warships could now be attacked with impunity. The world's sole superpower would not dare to strike them. During the Clinton Administration, fifty-nine Americans were killed by bin Laden's operations. And while almost fifty terrorists had been tracked and captured, and dozens of plots had been foiled and six terror cells smashed, the administration had waged no real war against its overt enemy. Instead, the administration reacted in fits and starts—half-measures that frustrated those who knew what needed to be done. What was needed was a full-fledged war on terror to kill bin Laden and destroy al Qaeda. But that would be left for the next administration, the administration of George W. Bush, which took on an overt war against the terrorists. (p. 229)

Bin Laden and al Qaeda now had the planning well under way for directly attacking the United States. Fortunately for al Qaeda, the CIA and the FBI were not sharing information. Two future hijackers were in the United States, according to the CIA, but the CIA neglected to mention it to the FBI. And it continued to get worse, as an infamous "Gorelick Memo" written by Jamie S. Gorelick, the second-in-command at the Clinton Justice Department, placed limits on communication between law enforcement officials and prosecutors investigating counterterrorism cases, despite opposition by prominent U.S. attorney Mary Jo White (Editorial: "Memos Show," 2004).

The summer of 2001 proved to be very busy for bin Laden and al Qaeda as well as for the dedicated, frustrated, and isolated agents of various U.S. agencies who sensed an attack was coming. For al Qaeda, June would see it formally absorbing the Egyptian terror organization al-Jihad, with bin Laden in charge and Zawahari second in command. Bin Laden had an immediate mission for the Egyptians, which was to kill a hostile, famous warlord in Afghanistan: Ahmed Shah Massoud. Massoud had repeatedly attempted to warn the world that al Qaeda was in Afghanistan and a dangerous threat, including telling American officials that bin Laden intended to carry out an attack on a much larger scale in the United States than the embassy bombings (Wright, 2006). Massoud would be killed just prior to 9/11. While bin Laden continued to plan the upcoming September attack, the FBI and CIA held numerous meetings and planning sessions in New York and Washington, D.C., in an attempt to predict what was coming and where it would occur. There was little progress or cooperation. If the CIA had shared its information with the FBI, and the FBI had shared its information with its own agents, 9/11 would not have happened. In addition, the National Security Agency had intercepts and intelligence but failed to share them. Lawrence Wright's compelling account of this period, *The Looming Tower,* details just how awful the blunders were, and the 9/11 Commission Report sheds even greater light on this. The point here is that despite ample warnings by multiple sources, 9/11 happened. It is most depressing reading the 9/11 Commission Report and the assertion that an FBI supervisor knew that an attack involving an aircraft was imminent:

> There was substantial disagreement between Minneapolis agents and FBI headquarters as to what Mouss-aoui was planning to do. In one conversation between a Minneapolis supervisor and a headquarters agent, the latter complained that Minneapolis's FISA request was couched in a manner intended to get people "spun up." The supervisor replied that was precisely his intent. He said he was "trying to keep someone from taking a plane and crashing into the World Trade Center." (9/11 Commission Report, 2004)

Americans were not told that hijacking was no longer the goal of jihadists, but that they sought death and destruction using a hijacked airplane. As a result, four airplanes were hijacked; three planes reached their intended targets, and one was forced down by now alarmed passengers in a field in Pennsylvania. In the end, more than 2,600 people died in the World Trade Center attacks, 125 died at the Pentagon, and 256 died on the four planes (The 9/11 Commission, 2004).

WASHINGTON — President Obama said Tuesday that he will delay a planned withdrawal of U.S. troops from Afghanistan, and maintain a force of 9,800 in the war-torn nation through the end of this year.

SOURCE: http://www.usatoday.com/story/news/2015/03/24/obama-afghanistan-ashraf-ghani-troop-withdrawal/70368016/

This would be the zenith of Osama bin Laden's career as a terrorist. There would be other major attacks, in Bali in 2002, in Spain in 2004, and in London in 2005, and scores of others but none approached the level of the 9/11 attacks ("Al-Qaeda Attacks," 2011). His wish was granted, the United States quickly attacked Afghanistan and moved on to attack Saddam Hussein and Iraq. The outcome of both wars remains to be seen. American forces have largely left Iraq and violence continues; most American and other NATO forces left Afghanistan in 2014 and uncertainty is the only sure thing in that theater. The cost for the United States was high but the United States was not destroyed as bin Laden hoped. According to an article written in the fall of 2011, the cost in terms of lives lost by American military in the Iraq and Afghanistan wars was over 6,000 (the 2013 figure is about 6,600) with approximately 43,000 wounded; the economic cost of the 9/11 attack was $170 billion, and a $50 billion tab annually for homeland security and two wars bring the total to more than $4 trillion, 20 times more than the monetary cost of the 9/11 attack (defense.gov, 2013; Yerger, 2011). In the end, Osama bin Laden would not live to see the finality of his actions; he was killed in a daring raid by U.S. Navy Seals on May 2, 2011, in Abbottabad, Pakistan (Knickerbocker, 2011). The harsh interrogation techniques that some characterized as torture, utilized during the administration of George W. Bush, did aid in finding and finishing bin Laden, though the Obama administration publicly denies this (Gans, 2012). It is an impressive record for just four individuals: Abdullah Azzam forged an ideology and an organization; both Sheikh Omar Abdul Rahman and Ayman Al-Zawahari, though often focused on Egypt, would change and join Osama bin Laden and make terrorism a reality for much of the world. They managed to lead movements that would cost the United States and other nations dearly, though in the end, their real goal will not be accomplished, that of defeating the United States. They were involved in assassinating the president of Egypt, destroying three embassies, and attacking the World Trade Center twice. In addition they attacked a U.S. warship and nearly destroyed it, made other successful attacks in Bali, Spain, and London, and left a legacy for others to carry on. Two are now dead, one is in prison, and the other is hiding. Thousands have benefited from their guidance and training.

So, where does this leave the United States going forward? From President Obama's perspective, the War on Terror may now recede into the background, as he concluded in a 2013 speech at the National Defense University. He declared that counterterrorism and intelligence gathered and shared, along with the arrest and prosecution of terrorists, is now the position of the United States; he also defended his policy of using drone strikes to kill terrorists, including American citizens, such as Anwar Awlaki, and vowed to find a way to close Guantanamo Bay prison (Obama, 2013). Andrew C. McCarthy's take on the speech was that it attempted to deflect from various scandals:

> Because the president is embroiled in not one but three scandals (and counting), involving his dereliction of duty in connection with the Benghazi massacre, as well as his administration's serial abuses of prosecutorial and regulatory power (siccing the Justice department on the press and the IRS on the Tea Party) and once again he maintains Islamic extremism is the cause of violence and that this is a lie. (McCarthy, 2013)

In summary, the United States seems to be returning to the glory days of pre-9/11. It remains to be seen what the future will bring. The president seemed optimistic in his speech; he maintained that:

- ◆ The core of al Qaeda in Afghanistan and Pakistan is on a path to defeat.
- ◆ Unrest in the Arab world has also allowed extremists to gain a foothold in countries like Libya and Syria.
- ◆ Finally, the United States faces a real threat from radicalized individuals (Obama, 2013).

The United States will likely be dealing with terrorism for many years to come. The Sunni and Shia jihadists have not gone away completely.

▲   Map of Colombia

Colombia

Colombia has dealt with terrorism for more than five decades. Two left-wing groups, the **Revolutionary Armed Forces of Colombia** or **FARC**, and the **National Liberation Army** or **ELN**, were the major players in Columbia's terrorism. A third right-wing group also contributed to the violence: The **United Self-Defense Forces of Colombia** or **AUC**, which formed to counter both the FARC and ELN and protect landowners and large agricultural interests. All three groups are on the U.S. State Department's list of designated terrorist organizations (*CIA World Factbook,* n.d.; Hanson, 2008, 2009). The toll has been incredible, with thousands of Colombians dead and over 5.3 million internally displaced as a result of the continual violence (Fitzpatrick & Norby, 2013).

▲   Photo of a FARC radio tower. This illustrates that they were opportunistic and controlled large amounts of territory and even had their own crude by effective communications systems.

Colombia is the northern-most country in South America, with a majority population of mestizos (meaning a mix of European and South American Indian), whites, and mulattos it is overwhelmingly Catholic. It is the only country in South America with coastlines on the Northern Pacific and the Caribbean Sea; it has coastal lowlands, highlands, and three ranges of the Andes Mountains, which complicate road construction and travel (*CIA World Factbook,* n.d.). An occasional volcanic eruption or earthquake and periodic drought make it an interesting place. It is slightly smaller than twice the size of Texas (*CIA World Factbook,* n.d.; Rochlin, 2011). Agriculture is the main economic activity in Colombia and until 1991 coffee was the major export ("Country Guides: Colombia," n.d.). Oil is now Colombia's major export, and due to terrorist activities, exploration for oil has been impossible until the present. More than 70% of the country is still unexplored for oil (Hussain, 2013). Colombia has just completed a major oil pipeline, thanks in part to financing from Canada, a competitor, and oil production has surged 80% in the last seven years (Hussain, 2013).

Colombia now has a stable government, and a United States-Colombia Free Trade Agreement was ratified in 2011 and implemented in 2012, providing hope for a brighter future (Beittel, 2012; *CIA World Factbook,* n.d.). Colombia does, however, have a major problem and that is largely because of the ease of growing illicit but very profitable coca, opium poppy, and cannabis; Colombia meets almost the entire U.S. demand for cocaine; much of the heroin found in the United States is from Colombia. The combination of drugs, money, and a cause can make a very profitable terrorist organization.

In Colombia's case, though, it is much more complicated than that. During the 1940s and 1950s Colombia underwent a period of turmoil called La Violencia (the violence) that resulted in the deaths of 200,000, as liberals (agrarian and landowning elites) fought conservatives (agro-exporters and merchants), political uncertainty, a dictatorship, and a coup, resulting in a weak federal government (Rochlin, 2011). Colombia has seen so much violence throughout its history that it even has an academic discipline that studies it: violentology (Rochlin, 2011). This unrest led to the founding of the Revolutionary Armed Forces of Colombia or FARC, and the National Liberation Army (ELN); both are committed, deadly, violent terrorist organizations that embrace a Marxist ideology (Hanson, 2009). To explain: Marxists view capitalism as the problem. In a capitalist system, such as that in the United States or Colombia, Marxists see private ownership of the means of production, distribution, and exchange of wealth as evil (this would include land, especially in a country like Colombia where 1.15% of the population owns 52% of the land; Fitzpatrick & Norby, 2013). As a result of this, there is intense competition resulting in the exploitation of the proletariat (working class) by the bourgeoisie (owners and controllers of the means of production). According to Karl Marx, this continual competition requires that minimal wages be paid for labor; the bourgeoisie becomes richer and the proletariat becomes poorer and more miserable. The conditions of poverty and exploitation produce all forms of social problems. A classless, free society, free of economic exploitation, will emerge only when the proletariat ultimately destroys the economic system in violent revolution. Again, the capitalist system breeds social inequities that lead to social pathologies, including deviance and crime. Marx also points out that there is an interrelationship between deviant and non-deviant aspects of society. Deviance serves many purposes in support of the existing society. Without deviance and crime, police, judges, and law professors would have no jobs (Platt, 1982). What both FARC and ELN wanted was a communist nation, or at least that is what they desired in the beginning.

FARC and ELN have similar ideologies and initially had noble ambitions of achieving land reform, a more equitable distribution of wealth, and social justice. Those ambitions were sacrificed as they both turned to violence and fought the government for control of areas of the country. This violence and ability to wage war on the government was funded by massive drug exports and kidnappings on a scale that is almost beyond belief. They are very different organizations and their history and current status will be presented next.

FARC began as a peasant-based movement, adopting a Soviet-style doctrine and a strict military structure. It was founded in 1965 by Jacobo Arenas of the Colombia Communist Party and Manuel "Supershot" Marulanda, a Liberal Party guerrilla (Rochlin, 2011). FARC evolved over the years from a small group in the 1960s to a much more robust force in the late 1970s that enjoyed wealth and influence in much of the interior of the country. It is somewhat ironic that a communist organization seemed to forget its ideology and became a capitalist criminal organization, using kidnapping, drug trafficking, and extortion, and boasting of their foreign investments (Fitzpatrick & Norby, 2013; Rochlin, 2011). In 2008, Cuban leader Fidel Castro, a long supporter of FARC, unexpectedly complained that FARC at the time retained about 700 people it had kidnapped (Rochlin, 2011). FARC's annual income from all of this was reported to be around $500 million (Hanson, 2009). By 1982 FARC believed it was strong enough to go on the offensive against the government and opened 48 different fronts (Rochlin, 2011). With the election of President Clinton, the United States, which had been a firm supporter of the Colombian government's attempts to destroy or minimize FARC, seemed to turn away and ignore Colombia and the terrorists. This changed in 1998 when Colombian president Andrés Pastrana ceded a territory to the FARC the size of Switzerland in exchange for negotiations. This only emboldened FARC, its numbers swelled to over 20,000, and it became more heavily involved in narcotics trafficking, kidnapping, and extortion. Examples are many but here are just of few of their deeds:

- the November 2005 kidnapping of 60 people, many of whom are being held hostage by FARC until the government decides to release hundreds of FARC's comrades serving prison sentences
- the February 2002 hijacking of a domestic commercial flight and the kidnapping of a Colombian senator on board
- the February 2002 kidnapping of a Colombian presidential candidate, Ingrid Betancourt, who was traveling in guerrilla territory. Betancourt is the most prominent member of a group of hostages held by FARC
- the October 2001 kidnapping and assassination of a former Colombian minister of culture
- the March 1999 murder of three U.S. missionaries working in Colombia, which resulted in a U.S. indictment of FARC and six of its members in April 2002 (Hanson, 2009)

FARC proved so alarming that the United States developed and implemented, with the Colombian government, Plan Colombia, a multi-year aid effort to upgrade, train, and equip the Colombian military, and to increase institutional and social programs, though these would only get about a fourth of the projected $7.5 billion, which eventually reached $9 billion) (Fitzpatrick & Norby, 2013; Rochlin, 2011). Much of the training, intelligence, and interdiction would be conducted by private military contractors, something of a revolution at the time, and with little oversight or transparency. The plan initially allowed for a maximum of 400 contractors from the United States, though this was raised to 600 in 2004 (Rochlin, 2011). Note that it limited the number of contractors who were U.S. citizens. An American company with a contract to work on Plan Colombia could hire any number of mercenaries or specialists from another country to get the individuals needed (Rochlin, 2011). Fortunately, a new and more effective administration soon took over in Colombia. President Uribe aggressively went after FARC, using support from Plan Colombia, and significantly weakened them. He served two terms and became the most popular president in recent memory.

FARC has had some reversals, so it may be willing to demobilize. In 2008 in a raid into Ecuador, Colombian forces, probably with an assist from the United States, attacked an encampment of FARC rebels and killed their long-time spokesman, Raul Reyes (Romero, 2009), and dozens of others. In addition to killing Reyes, Columbian forces found a treasure trove of information. Reyes had three laptops, external hard drives, material for an autobiography, extensive files and records, and many details of meetings and strategy, in violation of FARC's strict operational security policies (Crandall, 2011). The group's founder and leader, Manuel "Sureshot" Marulanda, died of a heart attack the same year; top military leader Jorge Briceño (also known as Mono Jojoy) was killed in a massive raid in 2010 and his successor, Alfonso Cano, was killed in November of 2011 ("Colombia's Armed Groups," 2013). President Santos has decided to negotiate an end to the conflict with the weakened FARC, with cease fires observed and broken, but talks are continue in Cuba (Otis, 2013).

The ELN was founded by university students, oil workers, and Catholic priests who followed liberation theology, a socially conscious form of Roman Catholicism that combines Christian and Marxist teachings. One of the early recruits, and later a leader, was a former priest born in Spain, Manuel Pérez (Insight Crime, n.d.; Ruiz, 2001). After the deaths of some founding members, ELN is now headed primarily by Nicolas "Gabino" Rodriguez; now 65 years of age, he is considered the "most enduring rebel" (McDougall, 2009); Molinsky, 2013). ELN's ideology is similar to FARC's but it never attracted the same number of followers. ELN initially avoided the drug trade and focused on extortion, protection money, and racketeering, targeting the mining and energy companies in the Magdalena area of northeast Colombia, near the Venezuela border (McDougall, 2009). It should be noted that in its early years, the ELN did not engage in drug trafficking, but this soon changed and the once quasi-religious group became a less ideological group and more of a criminal one (Insight Crime, n.d.). At one point its numbers shrank to less than a hundred members after multiple defeats at the hands of the Colombian military in 1973. However, new oil fields and pipelines in the Aruca area it controlled saw it gain in numbers to around 5,000 by 1998, with an income of around $150 million a year (McDougall, 2009).

For some perspective on the exploits of ELN, a review of some of its activities is needed:

◆ October, 1998, in Machuca, ELN blew up an oil pipeline and the blast killed 67 villagers
◆ April 12, 1999, in Bucamaranga, a city in the northeast of the country, five well-dressed men boarded an Avianca plane bound for Bogota with 41 passengers and crew. Shortly after take-off, they put on ski masks and produced pistols. They forced the plane to land in a jungle air strip where they were met by an armed force of the ELN. Some hostages were then released quickly, particularly the elderly and children, but many languished
◆ On a Sunday morning, in the wealthy suburb of Santiago de Cali, 30 ELN commandos disguised as soldiers entered the church Iglesia La Maria during mass, announced there was a bomb, and evacuated the worshipers (soon to be hostages) to waiting trucks. The 140 hostages were the largest abduction in Colombia's history (Ruiz, 2001).

The ELN did negotiate a truce with the FARC in 2009 and has had on-again off-again negotiations with the federal government. The Santos administration has ignored the ELN's wishes to join current peace negotiations and views the ELN as less of a threat than FARC (Molinski, 2013). The predictable response from the ELN has been violence. It has killed two police officers in separate ambushes, twice blown up the Caño Limón pipeline of California-based Occidental Petroleum, and in January 2013, abducted five employees of Canada's Braeval Mining Corporation, including a Canadian and two Peruvians ("Fear of Missing Out," 2013; Molinski, 2013).

The final group presented is the right-wing paramilitary organization, the United Self-Defense Forces of Colombia, or AUC. AUC was founded by Carlos Castaño. His father was killed by FARC when his family could not pay the ransom following his abduction (McDougall, 2009). He and his three brothers attempted to work with the Colombian military but soon decided they could do better on their own. They received training from an Israeli mercenary while serving in Pablo Escobar's Medellin drug cartel. They deserted and formed a vigilante group that would eventually absorb other paramilitary groups to become the AUC. The AUC provided protection from FARC and ELN, initially with the blessing of the government and support of businessmen and wealthy landowners, ranchers, and mining and petroleum companies unhappy with the government's efforts to protect them (Trent, 2012). The AUC did well with protecting these interests but it also engaged in the drug trade, tapped oil pipelines, and expanded into the gambling and construction industries, with 70% of its earnings eventually coming from cocaine (McDougall, 2009; Trent, 2012). Its numbers grew to over 30,000, but so did the atrocities and complaints as it fought FARC and ELN, often for control of areas and the lucrative drug trade. The AUC massacred and assassinated supporters of FARC and ELN (Beittel, 2012). Much of its income came from extortion. An example of this is Chiquita Brands International, Inc., based in Cincinnati, Ohio. Chiquita was told to cease making protection payments to the AUC. In a plea deal that resulted in a $25 million fine and no prison time, executives acknowledged that they made payments of approximately $2 million to the AUC from 2000 until February 2004, a year after their own attorneys and the U.S. Justice Department told them to cease (Meyer, 2007; Meyer & Kraul, 2007). In another case, Drummond Coal of Birmingham, Alabama, was accused of, but denied during Congressional hearings, providing vehicles, equipment, and supplies to the AUC (Meyer & Kraul, 2007).

While hope remains that FARC (which released the last of its hostages in April of 2012, some held for over a dozen years) and eventually the ELN will reach a negotiated peace deal, it is worth noting that President Uribe did conduct similar negotiations with the AUC. On July 15, 2003, the AUC and Colombian authorities agreed to demobilize by 2005 (Beittel, 2012). The process, established by the Justice and Peace Law, has been heavily criticized for not adequately punishing members of the AUC and not providing equitable reparations for victims, but it did reduce the levels of violence. Not taking part in this was the founder, Carlos Castaño, killed at a ranch in the department of Antioquia in 2004 in a power struggle for control of the AUC; nine individuals (including his brother Vincent, now missing and presumed dead) were convicted and sentenced for his murder in 2011 ("Carlos Castaño Murder," 2011). The demobilization began in 2004 and ended in 2006, with 31,000 members of the AUC demobilized and 17,000 weapons turned in, although some failed to demobilize and instead organized criminal groups, or Bacrims, now engaged in the drug trade (Beittel, 2012). Some of the Bacrim groups are Los Urabeños, Los Ratrojos, the Popular Revolutionary Anti-Terrorist Army of Colombia (ERPAC), Los Machos, Los Paisas, and the Aguilas Negras, and all engage in murders, massacres, threats, extortion, and acts of sexual violence (Beittel, 2012). One group calls itself the Anti-land Restitution Army, which is against efforts to finally provide justice for those displaced from their lands (Beittel, 2012).

In summary, Colombia remains a violent and dangerous country, despite major improvements. Some of the challenges are:

- Deal with corruption, with 11,000 politicians and businessmen at all levels of government involved in the 2006 "parapolitics" scandal, accused of using paramilitaries for political support and benefits, and another in 2008 involving the military dressing up murdered civilians as rebels to increase the body count (Beittel, 2012; Trent, 2012).
- Deal with the Bacrims, new groups of paramilitaries that are basically criminal gangs, although government success with destroying FARC provides hope (Fitzpatrick & Norby, 2013).
- Negotiate and find a peace deal with FARC and eventually ELN (Molinski, 2013).
- Deal justly with displaced persons, compensating for land lost during the internal conflicts with FARC, ELN, and the AUC. Implement the 2011 Victims' and Land Restitution Law as well as provide economic reparations. The cost of this will exceed $30.5 billion over 10 years (Beittel, 2012). This will be difficult as FARC, as owners of much stolen land, opposes the law, large landowners (however they acquired land)

oppose it, and the Bacrims oppose it (Fitzpatrick & Norby, 2013). Making matters more difficult, some farmers who have been provided land as restitution have now sold it to large corporations, much to the chagrin of the government (Muñoz, 2013).

- ◆ Monitor activity of groups, such as Hezbollah, that have had a presence in South America thanks to the late Venezuelan president Hugo Chàvez. For example, U.S. and Colombian authorities cooperated in 2008 to arrest a major drug ring in Colombia that supplied drugs and money to Hezbollah (Kraul & Rotella, 2008).

Colombia on the whole is in much better shape than it was a decade or two ago. Challenges remain but there have been vast improvements in security and level of violence. By 2010, Colombia's poverty rate was down to 34% from a high of 50% in 2002. Economic growth has been steady, reaching nearly 6% in 2011, and foreign direct investment doubled over the last six years (Beittel, 2012).

Peru

Peru is situated in western South America, on the South Pacific Ocean, between Chile and Ecuador; it is slightly smaller than Alaska (*CIA World Factbook,* n.d.). Geographically challenging, Peru has three very different areas: the coast, the Andes, and the jungle of the eastern slopes of the Andes, making travel difficult because of the terrain. This produced isolation for many communities, poor road networks, and little infrastructure (Ambrose, 2005). Despite these disadvantages, Peru's population of 29 million has achieved one of the highest economic growth rates in the region (Wilkinson, 2011). Peruvians are ethnically diverse with Amerindians making up 45% of the population, followed by mestizo (37%), white (15%), and black, Japanese, Chinese, and other (3%). The population is over 80% Catholic; 15% are evangelical Protestant (*CIA World Factbook,* n.d.).

Peru suffered under military dictatorships for years. This mismanagement resulted in a failed economy, nationalized mines, expropriated land from the wealthy, and an increasingly brutal insurgency. In 1990 Peru elected as its president Alberto Fujimori, a virtually unknown economist ("Alberto Fujimori, n.d.; *CIA World Factbook,* n.d.). Fujimori was the son of Japanese immigrants, a graduate of a Peruvian university who earned a graduate degree in mathematics from the University of Wisconsin ("Alberto Fujimori," n.d.). His election brought a decade of economic progress and a reduction in the amount of terrorism conducted by the Marxist and left-leaning Shining Path or *Sendero Luminoso,* and the smaller Cuban-inspired **Túpac Amaru Revolutionary Movement** *(Movimiento Revolucionario Túpac Amaru)* or **MRTA** (*CIA World Factbook,* n.d.; Gregory, 2009). It proved to be a very difficult time, however, largely because Fujimori became increasingly authoritarian, unpopular, and an essentially lawless figure; he is currently in prison having been convicted of numerous crimes, including ordering massacres and authorizing kidnappings (*CIA World Factbook,* n.d.; Kraul & Leon, 2013). While it was clear that much of the violence that resulted in almost 70,000 deaths occurred because of terrorism by Shining Path (around 54%) and the MRTA, many of the deaths, estimated to be 23,000, were at the hands of the government and government-sanctioned death squads (Kraul & Leon, 2013; Truth and Reconciliation Commission, 2003; Wells, 2012).

Reading some of the accounts of Peru during Fujimori's time as head of the government almost seems much like watching *The Sopranos* but it is really a sad and brutal soap opera. Fujimori suspended the Peruvian congress and judiciary and instituted martial law. His wife, Susana Higuchi, a civil engineer, disagreed with his actions and ran against his reelection bid only to lose; they later divorced and she also claimed he had her tortured ("Alberto Fujimori," n.d.; Wilkinson, 2011). Unable to complete his last term as president because of scandals, corruption, and massive protests, and as a dual citizen of Japan, he fled there and even ran unsuccessfully for office ("Alberto Fujimori," n.d.). He planned another run for the presidency of Peru, and moved back to Chile only to be arrested by Chilean authorities and be extradited to Peru where he was tried and convicted for the fifth time and is currently serving a 25-year sentence ("Alberto Fujimori," n.d.; Bridges, 2005). Elections in Peru have since occurred on a regular basis and the country is now a stable democracy, though not without challenges. In the most recent election, Fujimori's daughter Keiko, running on a pro-business platform and promising to pardon her father, lost the election to the incumbent, Ollanta Humala, who has said he will not pardon her father (Kraul & Leon, 2013).

To understand Peru and the problems it faced, one has to not only know Alberto Fujimori and the draconian actions he took in essentially declaring war and never negotiating, but also the two groups that he clearly believed forced his hand (Nieto, 2011). Indeed, it is astounding that due to the use of fear, intimidation, and patronage, few spoke out though everyone knew that intimidation, arrests, and murders were occurring, in large part at the hands of the government (Burt, 2006). In fact, one activist, when asked why she did not speak out, explained that *"Quien habla es terrorista"* ("Anyone who speaks out [in protest] is [considered to be] a terrorist") (Burt, 2006). Those groups that Fujimori and his government waged war on are the Shining Path, founded by Abimael Guzmán, a college professor, and the Túpac Amaru Revolutionary Movement or MRTA, founded by three college

▲   Map of Peru

radicals, Nestor Cerpa Cartolini, Victor Polay Campos, and Miguel Rincon (Ambrose, 2005; Baer, 2003). The latter group is named after Túpac Amaru, a famous Incan warrior who battled the invading Spaniards unsuccessfully (Baer, 2003). The two groups were very different in terms of lethality, with Shining Path the most brutal and violent, but their ideologies were similar. This text will first cover the Shining Path group (a designated foreign terrorist organization by the United States) and conclude with the MRTA, no longer a designated terrorist organization (Gregory, 2009). Included in discussions of each will be the government's response to their violence.

The Shining Path group was organized in the late 1960s by Abimael Guzmán. Guzmán moved to the city of Ayacucho to teach at the San Cristobal de Huamanga University in 1962 (Ambrose, 2005). He was already aware of the plight of the peasants, their poverty, oppression, and their difficult subsistence lifestyle. He had witnessed this initially as a college student conducting a census following an earthquake. Guzmán found squalid homes, no indoor plumbing, and polluted streams with contaminated water the peasants drank (Ambrose, 2005). Guzmán soon adopted a Marxist position and vowed to found a movement to destroy the Peruvian state and form a utopian Marxist state run by the peasants (Ambrose, 2005; Gregory, 2009). Using a traditional guerrilla warfare model, he began recruiting on two fronts, university students and peasants, but to reach the latter, he had to learn the local language, Quechan, which further isolated the people of the region (Ambrose, 2005).

The Shining Path built, trained, and recruited but did not engage in violence until 1980. Guzmán was patient. He nurtured, propagandized, armed, trained, and raised money for all this with an alliance with coca farmers, helping them process their crops and then traffic the coca. The Peruvian government's attempt to eradicate the crops added to his and Shining Path's appeal in rural areas (Anderson, 1989; J. Smith, 1989). Guzmán also did not neglect women, using his position as a college professor to recruit female Indian students from the university's education department, who were training to be teachers (Nash, 1992). They would help him spread the teachings of Shining

Path. Many would go on to become leaders of armed bands and 8 of the 19 members of Shining Path Central Committee's leadership were women; according to Shining Path literature, 40% of Shining Path members were women (Ambrose, 2005; Nash, 1992). They were also regarded as more ruthless than the men; in one account, witnesses reported that a man had been shot in the arm by Shining Path and was crying. A woman from Shining Path ended that by slitting his throat with a knife (Nash, 1992). Peruvians were stunned in 1992 when one of the country's prima ballerinas, Maritza Garrido Lecca, turned out to be a member of the Central Committee (Nash, 1992). A police videotape that showed her with raised fists shouting "Communism will take over the world" added to their discomfort (Nash, 1992).

It is shocking that with support from never more than 10,000 idealistic intellectuals and peasants, Shining Path controlled nearly one third of the country after ten years of violence (Ambrose, 2005; Gregory, 2009). They also in all likelihood were responsible for 37,000 deaths (Truth and Reconciliation Commission, 2003). Perhaps it was the brutal nature of their efforts that gave them success for a time. Guzmán once said to his followers that 10% of the Peruvian population would have to be assassinated for Shining Path to take power and that he was prepared to cross a "river of blood" to do it (P. Joshua, 2008). Shining Path even strung dead dogs from lamp posts in the capital of Peru to signal the brutality to come (P. Joshua, 2008). Specific examples of several of Shining Path's actions will be highlighted next:

- On May 18, 1989, Shining Path rebels killed four people and exploded bombs across Peru to mark the ninth anniversary of their first armed raids. This brought the number killed in the past three days to 31, with attacks on university and secondary school professors who resisted their indoctrination ("Maoist Shining Path Rebels," 1989).
- On January 1, 1988, at midnight, Shining Path knocked out power lines in the mountains carrying power to Lima and 24 states, leaving 750 miles of the coast without power ("Peru Greets," 1988).
- During a visit from Pope John Paul II to Peru, the Shining Path welcomed him by destroying power stations and blacking out parts of Lima for 30 minutes. It followed up by emblazoning a hill north of the city with a hammer and sickle emblem (Chavez, 1985).
- One of the most deadly attacks, called the Tarata Bombing, was on July 16, 1992. It occurred in a fashionable district in Lima called the Miraflores Ward. Two trucks, each packed with 1,000 kilograms of explosives (more than a ton of explosives), were detonated, killing 25 people, wounding 200, and damaging or destroying 183 homes along with 400 businesses and 63 parked cars ("40 Killed," 1992).

The sheer number of attacks by Shining Path is staggering, over 4,000, and at times it seemed as if at least one occurred daily. They ran the gauntlet from bombings of schools and shopping malls to ambushes and targeted assassinations (Ambrose, 2005). The government, especially under President Fujimori's no-holds-barred approach, finally defeated it, though Shining Path has not been eradicated. The beginning of the end for Shining Path was the capture of Abimael Guzmán on September 12, 1992, along with eight others, and his computer containing Shining Path's membership roster, organizational structure, and locations and details of weapons caches (Ambrose, 2005). Since Peru was under martial law, a military court sentenced him to life in prison in 1992 for "treason against the fatherland," but a federal appeals court later overturned the military verdict. A later trial resulted in a mistrial, but he was finally found guilty and sentenced to life in prison in 2006 (Shining Path, 2013).

Shining Path continued to worry Peruvians for some time to come, but it was never the major threat it once was. In 2013, the trial of the last remaining leader of the group was held for "Comrade Artemio," the nickname of Florentino Cerron Cardoso ("Peru Snares Elusive Shining Path Leader," 2003; Stone, 2013). He was fined $183 million and sentenced to life in prison ("Court Hands Shining Path Leader Life Sentence," 2013). Shining Path remains as an organization but it seems evident that it is now a criminal enterprise engaged in drug trafficking, no doubt in concert with Mexican drug cartels. One source calling Shining Path a "family clan" that is driven by money

(Hearn, 2012). The group recently shot down a military helicopter and killed several soldiers attempting to dislodge it (Neuman, 2012).

Shining Path was largely defeated but at a terrible cost and massive abuse by the government. Thousands were tortured and jailed, often being falsely accused, and some sources claim the government murdered or forced disappearances of 23,000 people, mostly carried out by the armed forces (Wells, 2012). Few in the government were ever prosecuted though former president Fujimori remains in prison, as does his security chief, Vladimiro Montesinos ("At Home in Peru's Nastiest Cell-Block," 2007; "Shining Path," n.d.). Shining Path has attempted to come back as a political party, with lawyers for Shining Path's imprisoned founder, Abimael Guzmán, forming the Movement for Amnesty and Fundamental Rights or Movadef (Wells, 2012). What alarmed many in Peru was that the group managed to get enough signatures to apply for status as a political party, demonstrating that many of the younger generations don't know or don't care what happened in the conflict (Neuman, 2012). Election officials rejected its efforts in January 2012, ruling that it "adhered to anti-democratic principles" and other technicalities (Neuman, 2012). In response, President Ollanta Humala's government is drafting a law to make joining Shining Path illegal and proposing prison for anyone who "denies or minimizes" terrorist acts during the Shining Path conflict (Wells, 2012). It is also instructive that during the trial of "Comrade Artemio" in 2013, the prosecution claimed that he was behind the founding of Movadef and funded it to continue Shining Path ("Artimeo' have founded the Movadef," 2013). Many years must pass before the scars created by the Shining Path and the government's brutal response will heal. Peru is now regarded a reliable regional leader against terrorism.

The final group examined is the Túpac Amaru Revolutionary Movement or MRTA, founded by Nestor Cerpa Cartolini, Victor Polay Campos, and Miguel Rincon in the early 1980s (Baer, 2003; Gregory, 2009). Never cooperating with the Shining Path, it was much smaller in terms of numbers while pursuing the same goal, a Marxist society for Peru. It rarely challenged the military, usually attacked Peru's wealthy elites and their holdings, and generally sought to avoid violence (Baer, 2003). Examples of MRTA terrorism are much fewer and less serious than Shining Path's. In one of its first acts, members robbed a bank (Nieto, 2011). The one major battle with the Peruvian military occurred in April of 1989; now called the Battle at Los Molinos, it was a defeat for the MRTA and resulted in the deaths of 50 of its members and six in the Peruvian military (Truth and Reconciliation Commission, 2003). The defeat cost not only lives but also demonstrated MTRA's military weakness. The Shining Path soon moved into this area (Nieto, 2011; Truth and Reconciliation Commission, 2003). For the most part MTRA evaded the military and only two more times directly attacked government forces. In 1989 it killed the defense minister, General Enrique Lopez Albuhar, and shortly after that kidnapped a mining executive, David Ballon Vera, who was found starved to death with signs that he had been tortured (Schemo, 1997a). In 1990, almost a year after its major defeat, it mustered sufficient force to attack a police outpost; that lasted two hours with two members of MRTA killed (Nieto, 2011).

Victor Polay was arrested in June 1992, devastating the organization as he was one of the founders, and soon he was sentenced to life in prison (Baer, 2003). Another leader, Miguel Rincon, was captured with an American, Lori Berenson, soon after, as they planned to attack the Peruvian congress in 1995 (Nieto, 2011). Berenson, a former Massachusetts Institute of Technology student, had dropped out of school and joined the MRTA. She was sentenced to life in prison, with a subsequent trial reducing that to 20 years. She is now out of prison on parole after serving 15 years in Peru and is not yet allowed to leave Peru (Tran, 2010). Following these setbacks, the organization struck back in a spectacular fashion that gained headlines worldwide for months (Baer, 2003; Gregory, 2009). On December 17, 1996, a Christmas party was under way at the Japanese ambassador's residence in Lima. Around 450 guests, including ambassadors and diplomats, were enjoying the evening when their waiters produced automatic weapons and a hole was blown in the back wall (Fennel & Chauvin, 1997). MRTA's demands were extensive:

- ◆ Freedom for all imprisoned MRTA members
- ◆ Transfer of hostages and prisoners to a jungle location

- ◆ Payment of a "war tax"
- ◆ An economic program for Peruvian poor
- ◆ All hostages would be shot unless demands were met (Escobar, 1996)

The residence was occupied for 126 days, with the MRTA releasing, in exchange for various things, the majority of the hostages. Hours after the initial incident it released 80 women. The following day it released the ambassadors of Canada, Greece, and Germany, ordering them to negotiate the release of an estimated 500 imprisoned MRTA members (Fennel & Chauvin, 1997). Over the next weeks and months, the remaining hostages were Peruvian military and government officials, with very few MRTA prisoners released (Schemo, 1997b). Finally, on April 22, 1997, government anti-terror experts launched an attack on the residence. Using sophisticated listening devices to track the captors' presence, 140 troops, utilizing five tunnels dug under the residence, surprised and killed all of the captors, with one hostage dying in the attack (Fennel & Chauvin, 1997; Schemo, 1997b).

This was the end of the MRTA as an organization. It would never recover, though it had had the one shining moment with the hostage-taking event that caused President Fujimori and the government a "massive loss of face," but in the end he prevailed, again (Fennel & Chauvin, 1997). The MRTA had an impact, though a small one. The final tally for its violence puts it at less than 2% of those killed in the internal conflict (Truth and Reconciliation Commission, 2003). Finally, Peru experienced a very difficult time largely due to the decision by Fujimori to adopt a no-negotiation strategy and a win-at-any-cost war against Shining Path and the MRTA. Was it worth it? That is the question that only time can answer. Millions if not billions of dollars that could have improved infrastructure and built schools and roads were instead spent on an internal war. The reputation of the Peruvian military and the government was severely damaged. At the same time, the war ended subsistence lifestyles for millions of poor living in rural areas, as they migrated to the coast to escape the conflict (Nieto, 2011). One final note of irony on Peru is irresistible. Imagine a special prison in Peru situated on the Callao naval base near Lima. Confined in this prison are three inmates in subterranean cells that look out on a small exercise yard. The inmates are allowed to use the exercise yard alone periodically, glared at by the other inmates who shout insults. You do not have to imagine this is reality. The first inmate is Victor Polay, one of the founders of the Túpac Amaru Revolutionary Movement. The second inmate is Abimael Guzmán, the leader and founder of the Shining Path. The third inmate is former resident Fujimori's security minister, the one who had this prison built to house terrorists. He is Vladimiro Montesinos, also known as "Rasputinos," convicted, as was his boss, former president Fujimori, of massacres and kidnapping ("At Home in Peru's Nastiest Cell-Block," 2007). Former president Fujimori, because of his age and health, is in a different prison and that fact sparked outrage recently when a video was leaked of him speaking on a cell phone, discussing his garden (Vigo, 2013).

⬔ Summary

This section presented an overview of terrorism and responses to it in Canada, the United States, Colombia, and Peru. These countries took very different responses to terrorism. Canada, not feeling threatened by terrorism, was slow to act and took years to come up with reasonable policies, procedures, and organizations to deal with the problem. It did finally accomplish a reasoned and balance approach, in keeping with its tolerant attitude.

The United States, on the other hand, went from almost no coherent policy on terrorism, other than treating it as a crime, to an extremely aggressive approach. It launched two wars and passed draconian legislation such as the USA PATRIOT Act, or the "Uniting and Strengthening America by Providing Appropriate Tools Required to Intercept and Obstruct Terrorism (USA PATRIOT) Act of 2001. The Patriot Act was renewed in 2005. The United States, under both President Clinton and President Bush, engaged in the practice of rendition, that is, shipping suspected terrorists to countries that are more willing to go beyond "enhanced interrogation" and engage in torture, often getting it wrong as was the case mentioned earlier of Maher Arar, a Canadian citizen ("The Story of Maher Arar," n.d.; Wright, 2006). The practice is also called extraordinary rendition and can be defined as:

> Extraordinary rendition, or irregular rendition, is a policy where individuals known to be members or affiliates of terrorist organizations are seized and covertly transferred to a third country detention facility for debriefing. The process is extrajudicial, done in secret, and typically not carried out exclusively by U.S. personnel. (Scheuer, 2007)

The United States may still be doing this, though the Obama administration has denied it, but it does use predator drones to kill terrorists, including U.S. citizens, which has been the subject of much criticism (International Human Rights and Conflict Resolution Clinic Stanford Law Global Justice Clinic NYU School of Law, 2012). Clearly, the Obama administration seems to have no trouble spying on U.S. citizens and any number of other individuals and governments around the world, using Section 215 of the Patriot Act and asking the Foreign Intelligence Surveillance Act (FISA) Court to permit it (FISA was passed in 2008 and amended in 2012; Saavage, Wyatt, & Baker, 2013). It may be working, as there have been attacks on the United States by terrorists but none on a large scale.

Colombia adopted the approach of both fighting the left-wing, Marxist terrorist organizations aggressively, even condoning right-wing para-militaries that attacked them, sometimes in concert with the Colombian Armed Forces. It also benefited greatly from support from the United States and Plan Colombia. Colombia still faces challenges but on the whole enters the decade with a brighter future. Again, it must deal with corruption, as it seems to face a major corruption case every two or three years. It needs to continue to be very aggressive with the Bacrims, ending their violent criminal behavior. Colombia must reach an acceptable peace deal with FARC and the ELN and continue to implement the restitution law and provide economic reparations.

Peru, the final country, is a special case. From the beginning it chose to deal with the Shining Path and the MRTA as criminals or terrorists that would be defeated. Negotiations were never contemplated. No assistance from others was requested. Shining Path and the MRTA perished. Alberto Fujimori's administration followed this approach; in the end it proved to be effective but the cost was high. The Peruvian military has been demonized for the abuses it heaped on suspected terrorists or their supporters. Perhaps just as damaging was the knowledge that journalists, activists, and many others were aware of the excesses of the military and their "hit squads" but remained silent out of fear. Peru must remain vigilant as several of the causes that attracted many to join the Shining Path or MRTA, such as poverty, land redistribution, inequity, lack of infrastructure, and isolation, still have not been substantially addressed.

▧ Key Points

- Canada has never taken terrorism seriously and was very slow to respond and change after 9/11.
- After almost a decade, Canada has reorganized and has a plan but still does not view terrorism as a pressing threat.
- The Canadian response to the Air India Flight 182 was a failure that hopefully will never be repeated.
- The United States has seen terrorism from domestic organizations, such as the Ku Klux Klan, and three left-wing terror groups, such as the Weather Underground, and now must deal with jihadist terror.
- The United States fought two wars in response to 9/11 and continues to battle jihadists.
- Colombia seems to be finally coming to terms with terrorism and hopefully will reach a deal with FARC to end conflict and terror and move forward.
- Peru has made inroads against Shining Path but must remain vigilant, as Shining Path is still a potential threat.

KEY TERMS

Al Qaeda

Animal Liberation Front (ALF)

Black Panthers

Earth Liberation Front (ELF)

Fatwa

Ku Klux Klan (KKK)

Jihadists

National Liberation Army (ELN)

Revolutionary Armed Forces of
Colombia (FARC)

Sovereign citizen movement

Túpac Amaru Revolutionary
Movement (MRTA)

United Self-Defense Forces of
Colombia (AUC)

Weather Underground

DISCUSSION QUESTIONS

1. Was 9/11 preventable?

2. What should the United States do about Iran?

3. The United States seems to be returning to the past, treating terrorism as a manageable criminal problem. Is that appropriate—yes or no? Support your answer.

4. Israel is an ally of the United States. What should the United States do if Israel decides it must attack Iran and destroy Iran's nuclear facility?

5. Should a president of the United States have the authority to authorize drone strikes to kill U.S. citizens?

6. Is extraordinary rendition ever justified?

WEB RESOURCES

9/11 Commission Report: Executive Summary: http://govinfo.library.unt.edu/911/report/911Report_Exec.pdf

9/11 Commission Report: http://www.9-11commission.gov/report/911Report.pdf

Canadian Security Intelligence Service: http://www.csis-scrs.gc.ca/prrts/trrrsm/index-eng.asp

Carson, Rachel, Indictment of: http://www.21stcenturysciencetech.com/articles/summ02/Carson.html

Drone Strikes: http://livingunderdrones.org/wp-content/uploads/2012/10/Stanford-NYU-LIVING-UNDER-DRONES.pdf. This is an interesting study that details the results of many drone strikes and questions the utility. Well worth reading.

FARC, History of: http://www.farc-ep.co/

FBI "Vault"; Black Panther Party: http://vault.fbi.gov/Black%20Panther%20Party%20

Final report of the Commission of Inquiry into the investigation of the bombing of Air India flight tabled (a parliamentary term used in Canada when a report is brought to the House of Commons or released to the public): http://www.cba.org/cba/submissions/pdf/07-25-eng.pdf

Ku Klux Klan: http://www.kkk.com/

MI5 or MI6? https://www.mi5.gov.uk/careers/working-at-mi5/working-with-mi6-and-gchq/mi5-or-mi6.aspx

National Alliance website: http://www.natvan.com/

Peru: http://www.cverdad.org.pe/ingles/ifinal/conclusiones.php. This is an excellent summary of the period in Peru from 1980 until 2000.

Pew Survey on Muslims: http://www.pewforum.org/uploadedFiles/Topics/Religious_Affiliation/Muslim/worlds-muslims-religion-politics-society-full-report.pdf

Rise of the Ku Klux Klan: http://www.pbs.org/wgbh/americanexperience/features/general-article/grant-kkk/. Great site for a history of the Ku Klux Klan

Sadat, Anwar, Brief Biography: http://www.ibiblio.org/sullivan/bios/Sadat-bio.html

The Third Jihad: Radical Islam's Vision for America: A documentary by a Muslim moderate worried that Muslims are violent: http://www.youtube.com/watch?v=4XUub1no1qw

Turner Diaries; they are considered racist and offensive: http://www.jrbooksonline.com/PDF_Books/TurnerDiaries.pdf

READING 17

The following journal article is presented in its entirety. It is a look back at an ethnographer sitting down with racists. Read the article and answer the questions at the end of the section.

An Ethnographer Looks at Neo-Nazi and Klan Groups

The Racist Mind Revisited

Raphael S. Ezekiel
Harvard School of Public Health

Americans today often learn about Nazis and the Ku Klux Klan through television clips of rallies or marches by men uniformed in camouflage garb with swastika armbands or in robes. These images often carry commentary implying that the racist people are particularly dangerous because they are so different from the viewer, being consumed by irrationality. The racists and their leaders are driven by hatred, it is suggested, and one can scarcely imagine where they come from or how to impede them.

Over 4 years, the author of *The Racist Mind: Portraits of American Neo-Nazis and Klansmen* (Ezekiel, 1995) met about once a week with the young members of a neo-Nazi group in Detroit, periodically holding semistructured interviews with the members and the somewhat older group leader. Over 3 years, he interviewed at length national and middling leaders in the neo-Nazi and Klan movement and attended and observed movement gatherings such as the Aryan Nations national conclaves in Idaho, the regional Klan assembly at Stone Mountain, Georgia, and cross burnings in Michigan. At these gatherings, the writer talked with participants, listened to their conversations with one another, and listened alongside the participants to the speeches of the movement leaders. The resulting volume describes leaders, followers, and gatherings and employs lengthy quotations from transcripts to buttress reflections on the White racist movement and the meaning of membership in the militant groups. The first half of this article reviews those findings; the second half, beginning with the discussion of Becoming a Neo-Nazi, considerably extends the book's reflections, particularly suggesting steps that would make youth less susceptible to recruitment by racist organizers. It will close with comments on our own social responsibilities.

The methodological core of this work was candor. I was open with my respondents about my identity: That I was a Jew, a leftist, and a university professor. I was direct about my agenda: That I believe (as I do) that most people build for themselves lives that make sense to themselves, and that in my work, I go to people whose lives seem strange to others and ask them to relate to me in their own words the sense of their lives. I told them

SOURCE: *American Behavioral Scientist*, Vol. 46 No. 1, September 2002 51–71. © 2002 Sage Publications

that I would be using their own words to let others see their meaning, adding my own thoughts. Most people whom I approached were cooperative. (It is relevant that my skin is Caucasian pale. Although I told them that I was a Jew, I did not resemble their rather medieval image of a Jew.)

 ## Movement Size

The militant White racist movement is far from monolithic; it is a loose confederation of small groups made coherent by the organizing work of major leaders and united by common ideology. The Southern Poverty Law Center (2000a) reported that in 1999, there were 36 Klan organizations (with a total of 138 chapters), 21 neo-Nazi organizations (130 chapters), and 10 racist skinhead organizations (40 chapters). The Klan was once the anchor of the movement and kept its distance from the Nazi organizations as they emerged, but since the 1980s, the two sets of organizations have more or less merged in what some have called the Nazification of the Klan. Concepts and symbols are mixed indiscriminately among the various groups.

Membership estimates were available in 1994 from reliable monitoring organizations: The Center for Democratic Renewal (D. Levitas, personal communication, autumn 1994), the Southern Poverty Law Center (D. Welch, personal communication, autumn 1994), and the American Jewish Committee (K. Stern, personal communication, autumn 1994) estimated hard-core membership in the militant White racist movement at 23,000 to 25,000. Of these, 5,500 to 6,000 belonged to one or another of the Klans; 3,500 were skinheads; and 500 to 1,000 were in Nazi groups or in groups close to the Nazis. The remainder of the hard core was less easily identified; the Center for Democratic Renewal referred to them as the Christian Patriot Movement; they were to be found in politically active churches of the Christian Identity sect and in rural groups scattered across the country. The monitoring groups estimated that 150,000 sympathizers bought movement literature, sent contributions to movement groups, or attended rallies, whereas another 450,000

people who did not actually purchase movement literature did read it. The movement is small but has had impact beyond its size because of a reputation based on a history of violence.

 ## Movement Ideology

The movement's ideology emerged as one interviewed leaders, listened to their speeches, and read movement newspapers and pamphlets. Two thoughts are the core of this movement: That "race" is real, and those in the movement are God's elect. Race is seen in 19th-century terms: race as a biological category with absolute boundaries, each race having a different essence—just as a rock is a rock and a tree is a tree, a White is a White and a Black is a Black.

Whites are civilization builders who have created our modern world, both its technology and its art. People of color are civilization destroyers, most characteristically showing their essences in the social pathologies of inner-city populations. The races have separate origins, as explained in the theology of Christian Identity (a major influence throughout the White racist movement):[1] Whites are the creation of God, who is White; people of color, whom they refer to as "the mud races," have originated in the mating of Whites with animals. Along with White people and people of color, the world includes a third and very dangerous species, the Jews, who have resulted from the mating of Eve with the Serpent. Whites are actual humans and the children of God; Jews are not human but the children of Satan; and people of color are semihuman. God has created the world, which the humans—Whites—are to rule; the people of color, like cattle in the field, should not be hated but tolerated and set to work to meet the needs of the humans. Satan has created the Jews to destroy the Whites and seize the world for Satan. These two forces, the army of God and the army of Satan—the Whites and the Jews—are to struggle with each other until one is destroyed. The Israelites of the Old Testament were early Aryans, unrelated to modern Jews; Jesus thus was an Aryan, not a Jew (Ezekiel, 1995, p. xxvi; Zeskind, 1986).

Most White people, the ideology states, are uninformed about the real nature of things and believe the soothing and ill-intentioned lies of the "Jewcontrolled" media. The Jews have made great strides in their war against God, convincing Whites that they must give up their position of dominance in America and cede privilege and power to the African Americans, the Latinos, the Asians, the feminists, the gays, and immigrants. Through their domination of the media, churches, schools, major corporations, and government, the Jews and their White dupes have succeeded in reducing drastically the power of White people; the White race now faces its extinction. Only the members of the movement have grasped this truth and are working to awaken and mobilize White people to defend themselves before the White race has been destroyed. The movement is a defense organization. Members of the movement are very special people, a minority that is not afraid of the truth, is loyal to God, and is willing to fight the workings of Satan and his Jews. Movement members are, in fact, the Chosen People.

The Jews work to destroy the White race through contamination; they call on Whites to engage in race mixing with Blacks and other races, which will produce a hybrid variety that has lost the White essence and thus the God-given strength and virtue of the pure Whites. The call to self-destruction is carried by the media, the churches, and the government, all controlled by the Jews and their flunkies. Only the White racist movement understands the real meaning of the ongoing changes in society and in the culture, understands that a single hidden aim lies behind those seemingly unrelated changes, and that a single hidden force plans them and brings them about. The Jews, in turn, recognize that only the White racist movement stands between them and success, and so the Jews use their organizations, the media, the churches, and the federal government to attack the White racist movement. To the extent the Jews succeed, Whites are taught to lose their race pride and movement leaders are sent to prison.

The struggle between God's agents and Satan's agents is a war of annihilation; only one side will survive. Any measure is justifiable in this war for survival. If innocent people die, it is unfortunate but a given in a war of survival.

All this is heard repeatedly in leadership presentations, and its apocalyptic energy animates the larger movement gatherings. But one wonders how much of the detail is salient for members on a day-by-day basis. If one listens at length to ordinary members, one hears pieces of this God-and-Devil story. But what comes through as central in the members' thinking is that Whites are losing ground, the world is changing, and the member may not do well in the world. The Whites are losing, and the member is losing. These are people who are scared and who draw important comfort from being members of a group.

Official ideology in the movement speaks extensively about the characteristics of the target Others, and this is what the general public assumes is the core of the movement. But, as I shall show in this article, the members teach us that the actual emotional center of the group is thoughts and feelings about the Self. The group is valued most for what it can do for the member's sense of himself.

 ## Gender

I say "sense of himself" because this is a men's movement. Some women are around but always in quite traditional supportive roles. They are the girlfriends or wives of members, and at gatherings they can serve the food they have cooked. In my 7 years around the movement, I heard many speeches, but never one by a woman. I never saw a woman in a leadership role. Women were servants and nurturers.

Those were the roles open to real women. The men also had their sexualized fantasies about special women who were imagined as trophies. The drawings in their publications, like their chatter, suggested a junior high school mindset: scantily clad women holding AK-47s below their jutting breasts were "saving their love for real men."

 ## Homophobia

Fear of homosexual rape was evident; strong Black men who "wanted to rape White women" might also commit anal rape on the White adolescent members. At a less problematic level, there was buy-in to traditional straight American male attitudes about homosexuality: Gay men and lesbian women were perverts. Contemporary shifts in mainstream attitudes toward understanding were seen as the results of the Jewish campaign to undermine the strength of the White race; through their control of the media, the Jews were able to make it seem as though gay men and lesbian women were ordinary people who should be accepted rather than people who were violating God's design. At movement rallies, people were led repeatedly in the chant, "Praise God for AIDS!"

 ## Targets

At the ideological level—in the writings and speeches of leaders—the contemporary Klan has joined the neo-Nazis in identifying the Jews as the prime source of evil. Leadership speeches throughout the movement present "the Jew" as the central enemy, with African Americans, Latinos, and Asians as the rather dumb members of "the mud races" who are pawns of the Jews, as are many brainwashed Whites. The leadership ranks gay men and lesbian women with Jews in the enemies list.

Among the rank and file, the picture is more traditional. Most followers whom I have met exhibited intense prejudice against African Americans that tended to reflect the general prejudice of their families and neighborhoods. Followers could repeat the party line about the Jews, but my strong impression from interviews and from watching socialization into the Detroit group was that new members arrived with strong antipathy toward Blacks but little interest in Jews. They came in hating Blacks and liking the idea that the movement represented Whites in a struggle against Blacks; after entry, they had to be taught who the Jews are and why they should hate them.[2]

 ## Leaders

There is no White racist movement without its leaders. There are people who are resentful, people who are needy, people who are adventurers, but by themselves they are not a movement. The leaders, themselves a particular kind of adventurer, combine charisma, ideology, and organizational capacity to create White racist groups and, from the groups, the movement.

Racist leaders rise through their own talents. The life stories they tell in the interviews speak of an initial time of puzzlement and casting about, a point of enlightenment, a discovered capacity to draw followers, and a determined struggle to bring the truth forward. Some have college degrees, but in essence all are self-educated with the certainty and the blind spots this entails. The utter certainty is a great deal of their power. The march to prominence has taken place in a context of competition; the racist organizations are in a constant competition with one another, and a leader gains importance as his capacity to attract members and media attention grows. The power to attract members comes from the leader's certainty and his capacity with words and body to be the living expression of the resentment and anger of the listeners. Moreover, he can make his listeners feel that they are part of something that is happening, that these are not empty words.

In many ways, the leader is operating in a vacuum: Middle-class politicians and clergy do not speak to this audience. The audience surmises, accurately, that the Establishment does not see them. The good life seen in advertisements will not be coming to them; their spokespeople are not on the talk shows; their futures will have little wealth and less glory. They do not feel respected.

The leader works with this raw material. The leader radically differs from the media's depiction of him. He is not irrational, and his primary motivation is not race hatred. He is rational and, in many cases, intelligent. He has a flaw: Within his self-education, he has rejected mainstream explanations of the social world and sees himself as one of those original thinkers who is at first scorned but later will be proven correct; this enables him to ignore pieces of personal experience that might disconfirm his ideas.

Because his own life course and thinking are fairly unbound by mainstream assumptions and he is basically self-defined, conspiracy theories are congenial. Within his self-education, he has rejected a sense of the world as tediously complicated, as a result of manifold complex interacting forces. He pictures himself as atomistic and self-determined, and it is logical then that he can believe great effects are caused by tiny groups of hidden men through hidden instrumentalities.

In most cases, the leader is not extremely racist. Racism is comfortable for him, but not his passion. At core, he is a political organizer. His motive is power. Racism is his tool. He feels most alive when he senses himself influencing men, affecting them.

The interviews with national leaders were lengthy—2 or 3 hours, repeated two, three, or more times. As we sat in interview, the respondent would take calls on the phone from lieutenants and would also speak to them in person. From those interactions and from the interviews, a pattern emerged. The leader, usually, is a man who is clever, who is shallow, and who does not respect people. He thinks almost all people are dumb and easily misled. He thinks almost all people will act for cold self-interest and will cheat others whenever they think they will not be caught. His disrespect includes his followers. He respects only those, friend or foe, who have power. His followers are people to be manipulated, not to be led to better self-knowledge. He loves, in abstract words, those whom he feels are disadvantaged. He loves, to this listener's perception, an idea of himself.

As I recall the stories the leaders told me and the things they said to their followers, everything that comes to mind is masculine: The actors in the stories are masculine; the stories are about combat, domination, and subjugation; the stories are not about nurturance or about cooperative effort that adds new elements, not about creativity or about tenderness. In a very fundamental way, the world of the leaders and the followers is an only-masculine world, a world impoverished of half the range of human feeling and thought—like the Army, like prison.

 Followers

The first years of research were with a Detroit neo-Nazi group that I will call the Death's Head Strike Group. The 1995 book tells how I made the contact with the group and gives portrayals and reflections in depth. The conception of followers that grew from this one group of followers was not contradicted as I met others at movement gatherings.

The members of the Death's Head Strike Group were all male, other than the cell leader's very young lover, who soon left, and several other women who later chanced by briefly. The group was small, with a nucleus of 7 to 10 members and 10 or 15 others in a looser connection. Still another 10 or 15 friends could be mobilized for a specific action. Members were young, ranging from 16 to 30, with a median age of 19. They had come to the group in batches based on friendship clusters. The majority had

come from one of three distinct Detroit neighborhoods. People in two of these neighborhoods were extremely poor; the neighborhoods had once been White but now only two or three White families to a block remained, the other families being African American. The third neighborhood was half White and half Black, with families ranging from working-class downward—a struggling but not destitute neighborhood.

Almost every Strike Group member (18 of 20 who were interviewed) had lost a parent when young; usually the loss was of a father (16 of 18). Most of the losses (15 of 18) were due to divorce or separation. The other 3 were due to death or to causes that had never been revealed to the child. The median age at time of loss was 7. The fathers had been working men. After leaving, they maintained no contact with the child or the family and did not send money to the home. Stepfathers or transient boyfriends of the mother tended to be cold, rough, and abusive. Several members had spent portions of their childhood in foster homes.

A few members spoke spontaneously of parental alcoholism or violence; a few others were responsive when I asked questions. Seven members reported alcoholism, six reported family violence. Seven members spontaneously mentioned serving time at detention centers, jails, or prisons. I suspect, from the stories involving street fights and the hints about drug use and pilfering, that there was more undisclosed penal time.

There was little money in the homes. Most of the mothers worked as cooks or waitresses in small eating places or drew disability payments. Most of the members had no jobs and no prospect of work. A couple had steady work at low wages; a few found occasional work in the neighborhood, for example, tearing down a shack for someone. Industrial employment in Detroit was shrinking rapidly, and the prospects for these young men were very poor, especially because they had little work experience or education. They had left school early. The school history of 16 members is known. Six had quit school in the 9th grade, 3 in the 10th, and 4 in the 11th. The 3 who had graduated from high school had each taken a semester or two at a community college.

These young men were living in startling social isolation. The impact of parental loss and poverty depends on the sort of parenting by the remaining parent and on the quality of other social supports. Aside from their mothers, about whom little is known, social supports were minimal. Ties to siblings tended to be weak or nonexistent, and only one member spoke of someone from the extended family who had played a role in his life. None ever mentioned a teacher who had been important to him, or a coach or scout leader, or anyone from a church, the neighborhood, or a social agency.

The members had grown up in neighborhoods in which they had to fight a lot. This would have not been easy because most of them were fairly slim, rather slight. They depended on fighting real hard, once something broke out. They were wiry and tough, but preoccupied with their thinness.

They were not good physical specimens. A surprising number had been born with a childhood disease or deficiency, such as being born a blue baby or "born with half a liver." There are a lot of hospital stories in the interviews.

Very early in the interviewing, I sensed an underlying theme of fear. At an unspoken but deep level, the members seemed to feel extremely vulnerable, that their lives might be snuffed out at any time like a match flame in the wind. This makes the appeal of the Nazi symbols understandable. When I asked them what they knew about Nazism, they referred to late-night movies on television. If you are afraid that you will disappear, how appealing are the symbols of a force that was hard, ruthless, even willing to murder to achieve its goals.

None of the members could establish a long-term intimate relationship. Several had caused pregnancies, but neither they nor the cell leader was able to establish a fathering role. Eventually, two of the members did become seriously involved with women, one of them fathering a child; each of these young men drifted from the group as he became involved. One of them soon was living with his woman friend and held down both a full-time job and a part-time job, where he began to have friendly bonds with some of his African American coworkers; he abandoned the Strike Group.

The group did not have conventional meetings; rather, the members hung out at the leader's apartment. Periodically, they were transported in the windowless rear of a rented Ryder van to some outlying town where they would put on a rally for a few minutes until counter demonstrators drove them away. I think this action also assuaged their fears.

If you are not quite sure that you are alive, how reassuring to stand shoulder to shoulder with your comrades, withstanding for a few minutes the taunts, threats, and hurled snowballs, chunks of ice, and flash-light batteries of the counter-demonstrators. When the police quickly shepherded them safely back to their van, they rode back to Detroit, and, that afternoon and for weeks thereafter, rejoiced in rather inflated memories of their courage, feasting their eyes repeatedly on newspaper photos.

✒ Becoming a Neo-Nazi

How does a working-class kid in Detroit become a neo-Nazi? What are the factors that make adolescents vulnerable to recruitment by racist organizers? Figure 1 is a schematized representation of factors suggested by the interviews. Social factors, listed on the left, intersect with personal and family psycho-dynamics. A range of alternative outcomes, shown on the right, may follow.

The tally of social factors begins by noting the *presence of a racist group.* Where there are no groups and where there is no effort at recruitment, recruitment is probably unlikely. Every young person who has been recruited stands for hundreds of others who just as readily might have been recruited if there had been an organization on the scene.

Many of the Detroit youths I met with had written away for membership cards in the Klan when they were in junior high school. This was a mail transaction that gave them a card to carry in their back pockets, with whatever boost that gave their egos as they moved about in the racially mixed schools in which they were the minority. They could not recall how they had found out about this mail-order opportunity. They later heard about the Detroit neo-Nazi group because it got a lot of publicity on Detroit television. Occasional placards shown on the video clips included the group's phone number, so they could call its leader. After meeting with him a few times, they would start coming around regularly. When one joined, a couple of friends usually followed.

The growing number of White racist Web sites on the Internet make racist propaganda widely accessible. Monitoring organizations fear that this will aid recruitment (Southern Poverty Law Center, 2000b, 2001a; Weitzman, 1998, 2000).

| **Figure 1** | Schematized Representation of Suggested Factors Influencing a Youth Becoming a White Racist Activist |

Presence of
Racist Group

Social Dislocation

Economic Pressure (?)

Social Isolation

Racist Ideology   Family
                                   Dynamics,
                                   Personal

Macho Ideology   Psychodynamics

Absence of
Democratic Ideology

Absence of
Cross-Cutting Loyalties

Differential Outcomes,
such as:

ordinary coping

numbness

malaise

alcoholism

chronic anger

individual
violence

racist activism

Data are absent, and some are skeptical (Southern Poverty Law Center, 2001b).

The diagram's second social factor is *social dislocation*. Widespread changes in American society mean that previous status hierarchies are disrupted or threatened. (Note parallels to developments in Weimar Germany [Kershaw, 1998].) Most members of the American White racist movement believe that they, as White men, are members of an endangered species. Very little about their futures can be taken for granted. Many cues tell them that old values, which they have assumed would benefit them for life, are challenged by new values. Real social change is involved here, as well as exaggerated perceptions of change and endangerment.

White Americans have made only an awkward accommodation to the increased political strength of African Americans. The work of Howard Schuman and his associates addresses the complexity of this issue: Looking at surveys of probability samples over decades, they found White Americans verbally endorsing some egalitarian values, while steadfastly opposing concrete steps that would implement those values (Schuman, Steeb, Bobo, & Krysan, 1997).

In early versions of this diagram, I listed *economic pressure* as a social factor. People that I met seemed to come from families with incomes below the median and sometimes well below the median. I am no longer confident that my impressions justify a claim linking economic factors to membership in racist groups. Several lines of research challenge this assumption. First, Leonard Zeskind (personal communication, autumn 1994), whom I consider the most astute political observer of the White racist movement, believes that the movement is a representative cross-section of American society. Second, James Aho (1990), in a careful study of the movement in Idaho, found educational levels that did not suggest economic pressures. (Interestingly, he noted that respondents in the more extreme portion of the sample "seem either to be college graduates or high school dropouts") (p. 141).

Finally, a strong set of new studies from Yale casts doubt on a linkage between economic status and racist group membership or racist crime. The most impressive of the Yale studies (Green, Strolovitch, & Wong, 1998) demonstrated that the number of bias-crime incidents in New York City neighborhoods between 1987 and 1995 was not related to neighborhood economic status (unemployment rate, poverty rate, or median income) but to turf patterns: Racially motivated crime rose when there was a rise in non-White migration into neighborhoods in which Whites had for a long time enjoyed a large majority. A second study (Green, Abelson, & Garnett, 1999) examined responses from a probability sample of North Carolinians about political and economic matters. Elegant procedures permitted the inclusion of an identified subsample composed of members of White supremacist groups and of hate crime perpetrators. (Unfortunately, this subsample is small.) Although the two populations differ predictably in political views, the subsample's more negative view of the eco- nomic condition and prospects of their communities differs from the general population's assessment to a degree that is only "small to moderate" (p. 447). A third study (Green, Glaser, & Rich, 1998) readdresses, with more sophisticated techniques, historical data on lynchings and economic changes and found "little robust support" for a frustration-aggression hypothesis. Interestingly, the ensuing discussion highlights a more nuanced conception of the linkage of economics and racist activity: The authors point to historical periods in which propagandists from the political, business, or labor communities mobilized racial hostility by identifying a racial group as the cause of economic problems—an analysis that parallels this article's characterization of the White racist leaders as people who are fundamentally political beings.

Despite the power of the Yale studies, this issue may not be conclusively settled. We are dealing with groups with secret memberships, and the historical record in Weimar Germany dramatically commands attention (Kershaw, 1998).

The third social factor is *social isolation*, which has been discussed at length above. The importance of social support has been widely studied; see, for example, the discussion of social support in Cohen and Herbert's 1996 review of health psychology. The young men in the Strike Group may or may not have been close to their mothers, but the lack of other meaningful adults meant unusual vulnerability. The racist group offered comradeship, authority figures, and a home to young people who lived in what might be termed *spiritual poverty*.

The fourth and fifth social factors are *racist ideology* and *macho ideology*. These help determine the direction that the conversion process takes. The ideology of racism, which was passed on to the new member by the group leader and reinforced in conversations with other members, gives the new recruit a continuing sense that there is an important reason for the group to exist. This is more than a casual friendship group. The movement makes its claim, in the ideology, to a turf and declares its role as defending that turf. The members struck me as people who felt rather orphaned, and the racist ideology permits the member to construct in his mind a new family, the mythologized White race. In the member's conversations, we hear the fantasy that someday this great White family will realize what he has done for them and then they will embrace him.

Macho ideology is a familiar presence in authoritarian movements (Adorno, Frenkel-Brunswik, Levinson, & Sanford, 1950; Smith, 1965; Stone, Lederer, & Christie, 1993). This is an ideology of pseudomasculinity, an ideology that glorifies toughness and fears tenderness or nurturance as weakness. This stance buttresses the ego of shaky male individuals; it can be especially important to adolescents, and in this case, we are speaking of particularly fearful adolescents.

The listing of social factors concludes by noting absences: the *absence of democratic ideology* as a real part of the mental life of the youths, and the *absence of cross-cutting loyalties* that might make exclusivist appeals uncomfortable. Research relevant to the latter appears in Urban and Miller (1998) and Marcus-Newhall, Miller, Holz, and Brewer (1993). The two absences become apparent when one asks how it can be that these young people do not experience revulsion when presented with the authoritarian and racist worldviews that are the center of the recruitment process. What is not there that one might expect? The absences, as we shall see, have direct implications for prevention of recruitment.

The diagram assumes that the impact of the social forces depends on the particular characteristics of the individual's psyche and the dynamics in the individual's family. Psychodynamics are examined by Dunbar (2000), Dunbar, Krop, and Sullaway (2000), Hopf (1993), Staub (1989), and Sullivan and Transue (1996). Staub covered historical and social issues as well as psychodynamics. Dunbar, et al. (2000) compared men convicted of racist homicide to men convicted of nonracist homicide.

Hopf's (1993) review of qualitative and clinical work on authoritarians and their families yields psychological portraits that fit the neo-Nazi youth of the Strike Group to a startling degree. On pages 128 to 130, she reconstructs Ackerman and Jahoda's 1950 study. Ackerman and Jahoda interviewed psycho-analysts at length about anti-Semitic non-Jewish patients in their caseloads. The following characteristics were described as universal for these anti-Semitic patients:

1. A vague feeling of fear, linked to an inner picture of the world around them that appears to be hostile, evil, and difficult to master;

2. A shaky self-image, identity problems, and fluctuations between overestimation of self and self-derogation;

3. Difficulties in interpersonal relationships manifested in part in a high degree of isolation and hidden in part behind functioning facades. "But at best such disguises deceive the outer world and sometimes the self; they never lead to the establishment of warm, human relationships" (Ackerman & Jahoda, 1950, p. 33);

4. The tendency to conform and fear of attracting attention;

5. Problems in coping with reality; often there are weak bonds not only to other persons but also to external objects (content of work, occupation in leisure time, and so forth);

6. Problems in the development of an autonomous set of ethics.

Hopf's review leads one back to the qualitative chapters of *The Authoritarian Personality* (Adorno et al., 1950). Indeed, the entire Stone et al. volume and Smith's 1965 data argue for readdressing the central concepts of the Berkeley research (also see Smith's 1997 review). Note, on the other hand, a strong, recent dissent by Martin (2001), who argued that even in the qualitative sections, the members of the Berkeley group were fatally naive in their methodology, fell into systematic error because they had reified scale positions as existing human types, and consistently misinterpreted data in a self-serving fashion.

In either case, the psychological has consequence. The diagram proposes that in the presence of the stipulated social factors, some people of particular personal and family psychological patterns will enter into a period of activism in the White racist movement. The diagram proposes as well that small differences in the social and individual inputs will result in quite a range of possible outcomes. One can well imagine people who might become lonely cranks, or drunkards, or even quite ordinarily competent adults. The devil, as always, is in the details.

Prevention

Personal Connection

What would make individual White adolescents less vulnerable to the recruitment efforts of neo-Nazi and Klan organizers? Recall the social isolation of the Detroit youths. During those repeated interactions, I became aware that warmth was increasing between us, despite my identifying myself as a Jew and a progressive.

I had my own personal issues. On a pivotal afternoon very early in the project, I was driving to Detroit to continue our conversations and thinking about the life of one of the young men. I had been getting a sense of what his life had been and what its onward trajectory was likely to be. What could be done, I asked myself, that would help him have a more competent sense of himself, that would encourage him to take a firmer grasp on his life—to begin to understand that his life mattered and that it could be directed in a hopeful way? I pondered and abruptly shook myself: "What am I doing, worrying about a Nazi?" I thought about it. And then from my gut came the reply: "He is also a kid. It cannot be wrong to be concerned about a kid."

The neo-Nazi youths were reacting to me as well. Their greetings, their remarks, and their bearing showed that I was becoming a person who mattered to them. That made sense. I would sit and speak with an individual for a long time, talking with him about his life, taking his life seriously, looking into his eyes as we spoke. This happened again and again. Probably no one had acted that way with these youths for a long time.

This sort of interaction may be a critical ingredient in programs that address the needs of disadvantaged kids. Every child needs an adult who *sees* him or her: an adult who does not disappear, who shows by attention and action over time that he or she takes the child seriously and that the child matters.

This is not an elegant formula. It is labor-intensive and lacks multiplier effects. But it may be fundamental. In addition to macro-economic changes, perhaps we need direct personal action if we are to reduce the amount of youth violence, of teenage pregnancy, of youthful gang activity, of racist activity, and of all the other ways that disadvantaged youths hurt themselves and others. We perhaps should tithe ourselves—a tithing of time for children in need of relationships. See in this connection the discussions of mentoring in Freedman (1993) and in Tierney and Grossman (2000).

Community

Complementary ways in which we could address the social isolation of youths such as these tie to the word *community*. Research with the Peace Corps and experience in teaching have convinced me of the power of context: Given a meaningful challenge that is difficult but not insuperable, within an artful combination of structure and freedom, young people can mature and become competent to a degree that would not be predicted from a simple examination of their past (Ezekiel, 1969; Smith, 1966). These lessons were in my mind as I interacted with the neo-Nazi youth and asked myself about alternative scenarios that could have been played out in their lives. They were neither mindless nor hate-filled. They were poorly educated and fearful at the core. What they wanted

profoundly was to have close relationships and to feel that their lives mattered.

The Nazi group offered them this feeling to a degree and for a while, but other endeavors could probably have done this as well or better. I thought I probably could have led many of them away from their leader (because I was a warmer person and cared more for them) to some other group. But what other group would fit their needs? The best fit, I felt, would be a radical environmental group such as Earth First. Such a group would have given them adventure and a chance to shock the Establishment, while doing something of intrinsic value and enjoying camaraderie. Earth First's goals would have jibed with the anticorporate bias of the youths and with their romanticism about the outdoors. Working in that organization would not have particularly affected the racial prejudice these kids harbored, but it would have met their need to act out in a shocking and socially relevant fashion while gaining group affection. The youths probably would have remained racist—but so would their peers who did not wear swastikas. The point is that these young people seemed to have no intrinsic need to act out their racism but did have needs to have companionship, to be shocking, and to feel that their lives had meaning.

Community organizations could build much more broadly on the similar hungers of great numbers of kids, who could learn community in contexts of challenge. Our culture tends to relegate young people to roles that are neither meaningful nor honorific. What is the significance of being overindulged or socialized to see oneself primarily as a consumer? What scope do we offer an adolescent who wishes to prove his or her significance in the world? Churches, synagogues, mosques, neighborhood organizations, scouting organizations, and political or ethnic organizations could begin to build youth groups in which there was serious challenge; there are plenty of hard and meaningful tasks to be taken on. A critical need, again, would be for adult leadership that would not fade out.

Schools, Democracy, Antiracist Education

Schools, like community groups, may play a role. The youths I met had first become involved in racist activity in junior high school. Their prior (and subsequent) schooling had not led them to harbor a concept of community. The classroom had seldom been shaped as a community in which class members had felt mutual responsibility for one another. On the contrary, the classroom probably had reflected the desperation and the atomization of the society outside the school.

Equally, the schools had left no feel for democracy. The youths had no positive association to the word, which seemed to them a meaningless term used by adults for hypocritical purposes. School had afforded little chance for real impact on decisions that mattered, opportunities to learn in action the meaning of the word democracy. Both community and democracy can be taught through experience in the classroom, when schools consider these goals part of the curriculum and invest energy in building related skills.

For the neo-Nazi youths, the teaching in school of multiculturalism had been another adult exercise in hypocrisy. Black History Month was an annual annoyance. It is easy for an adult-led discussion to seem like sermonizing. I would suggest that education about racism should begin with respect for the constructs and emotions that the students bring with them into the classroom. The students have ideas and emotions about race that are the product of their own lives. They have heard their parents, their neighbors, and their friends, and they have had their own experiences. To ignore their emotions and constructs around race is to ignore the sense that they make of their own experiences.[3]

Teaching about racism, I want to suggest, is a subtopic of teaching about identity. Perhaps the first step is to help the student think through his or her own sense of identity and to look for its roots. What was the life of your grandparents and what does that life tell you about yourself? What are the legends or myths in your family—why is it special to be a Kelly or a Krueger? What has been the experience of religion that you have heard about from your parents, and what has been your own experience—how have these helped to make you the person you are? Young people, regardless of race or ethnicity, should be helped to see where their own sense of identity comes from and how it affects their own lives. And to see its many different facets. Only then can the student begin to acknowledge that other people also have a sense of identity, and that it also had multiple roots. And also plays a role in their lives.

How Do We Resemble Them?

How Do We Differ?

It is fashionable to repeat Walt Kelly's *Pogo*: "We have met the enemy, and they are us." And it is worth noting that the neo-Nazis are not totally alien to White Americans. A social attitude does not exist in the mind as an isolated single entity. Real attitudes, or orientations, are laid down throughout life in layer after layer. If you visited South Africa and spoke with older White South Africans, you would expect to find their minds affected by having grown up in a society that was intensely racist. White Americans grow up in a society in which race has been and is profoundly important.

If I grow up living next to a cement factory and inhale cement dust every day, cement dust becomes part of my body. If I am White and grow up in a society in which race matters, I inhale racism, and racism becomes part of my mind and spirit. (I do not presume to speak here for the experience of people of color.) There will always be layers of myself that harbor racist thoughts and racist attitudes. This is not to say that those must remain the dominant parts of my mind and spirit. It is to say that it is mistaken to presume that I have no traces of racism in me.

The task is to get acquainted with those layers of oneself—to learn to recognize them and not be frightened by them. It is not a disgrace to have absorbed some racism. It is a disgrace not to know it and to let those parts of ourselves go unchecked.

I overcome those layers of myself by getting acquainted with them and by adding additional layers that are not racist. How do I do that? By action: I try to behave in a nonracist fashion or an antiracist fashion in the external world and absorb this experience as another layer of myself.

There is perhaps a parallel in clinical work. In therapy, I may learn to recognize the parts of me that were shaped by early experience with my parents (or, rather, what my childish mind thought was early experience with my parents). I learn to understand that in some circumstances—for example, in a disagreement with a superior—pieces of those early attitudes are likely to get activated. I can learn to think about this before going into the boss's office, and prepare myself not to be blindsided by infantile parts of myself that are not relevant to the situation at hand. And, over time, I can add layers of nondefensive experience to my psyche.

If, then, those of us who are White have grown up in the same society as the racists and have absorbed some of the same cement dust, are we the same as them? The organized White racist movement rests on the following four axioms: that race is real, that White is best, that the language of human interactions is power, and that society's surface conceals conspiracy. We European Americans have layers of ourselves that also hold the first two of those axioms. We may have been taught in school that race is merely a social construct and that White superiority is a myth. But that teaching runs up against what we are taught by our lives, every day. Race does matter in America. And White ends upon top.

We can learn to not be captive to the layers of ourselves that are racist. "Am I racist?" is not the question. The question is, To what degree am I racist in what situations? And the more important question, What are the concrete effects of my actions (or inactions)? In the 1970s, my interviews with African American families in the Detroit inner-city included interactions with a woman named Ruby and her children (Ezekiel, 1984). Ruby lived on 12th Street (as it was then named), and her children had cornflakes with water for their daily breakfast. Ruby and her children were real; the contrasts between their lives and mine were painful. I learned to ask myself, when people spoke on the radio or at the university about a program or a policy, how it would affect Ruby's children: Would it help them to have milk with their cornflakes, or would they keep on eating cornflakes with water?

 ## The Guilt of the Organized Racists

Between Reconstruction and 1945, 3,000 to 5,000 African American men and women were tortured and killed by lynch mobs. The Ku Klux Klan was a dominant force behind those killings. Local officials were often themselves Klansmen but, in any case, did not obstruct the Klan. During the civil rights struggles of the 1950s and 1960s, Klansmen instigated mob assaults on Freedom Riders and the like, and carried out bombings and murders of activists—or of little girls in a church—under cover of darkness. Organized White racism has a long and bloody history. Its goal has been to preserve White domination in America. Its primary weapon has been terror.

The Klan and neo-Nazi groups hold a different position today. Town and county officials are much less likely to be secret members and are much less likely to be cowed by overt White racist demands. To this writer's perception, White racism remains a major strand in American culture, but as a political force, it has expressed itself more often as a covert message within mainstream politics. Major presidential candidates have not hesitated to win support by suggesting that too much is being done for the undeserving poor, a code word for African Americans. Demonization of the poor has proceeded apace, buttressed by an almost unspoken assumption (a statistically inaccurate assumption) that most of the poor are non-Whites. This demonization may have served as a distracting cover for a reapportionment of wealth from middle-income families upwards (Collins & Yeskel, 2000).

Probably the greatest effect of White racism today is its capacity to slow institutional change. Policies that help institutional racism to continue to flourish do much more to hurt minority people than do hate crimes. High infant mortality rates in the inner cities and policies that let them continue are more dangerous than the Klan.

Actual hate crimes, for the most part, are committed by people who are not members of organized White racist groups (Dunbar, 2000; Dunbar et al., 2000). Deep racial distrust and antipathy mark our culture and would exist without the dramatic statements and demonstrations of the White racist groups. But the statements and the rallies of those groups increase the temperature, and the advocacy of specific steps pinpoints actions that perpetrators can take.

The leaders and the lieutenants of those groups are morally responsible to a nontrivial degree for racial violence in the United States (a responsibility they gladly claim in private conversation). Indeed, the future for which they avowedly work is one in which racial violence increases until the long predicted race war erupts and White America wins back its God-ordained dominance. The followers in the groups, the willing actors in the theater produced by the leaders, share in that moral responsibility. And of course, where leaders or followers have committed crimes, they are fully responsible.

Racism, Hate Crimes, and Responsibility

I chose to talk with members and leaders of White racist organizations as part of a broader project of understanding White racism in America. I have gained the impression, since publication of *The Racist Mind*, that more and more of the general public and the educated public are letting the task of talking about hate crimes displace from the agenda the task of thinking about racism.

Perhaps this is not surprising. You and I do not commit racial assaults, and no one we know does. It is interesting and unthreatening to imagine the world of those other people, whoever they may be, who engage in racist violence. And how nice if that form of contemplation can also be the only price White people have to pay for living in an unjust society. So that the more indignant and outraged I can be about the evils of the Klan and neo-Nazis, the more virtuous I can feel. And the more virtuous I feel about their misdeeds, the less I need to listen to tiresome critics who talk about racism and the need for institutional changes.

If I were to think about the true and continuing effects of racism, I would have to think about the ongoing social order, in which I am a part and for which I have responsibility. All of us are ready to say that Klan murders are evil. But what are we ready to do, today, about the continuing racially based maldistribution of health, of wealth, and of hope?

Notes

1. A recent report suggested that the influence of Christian identity in the movement is in decline, being replaced, especially among young members, by racial Odinism (Southern Poverty Law Center, 2001c).

2. In interviews, speeches, and conversations, nothing was said about Catholics or the Catholic Church.

3. The teacher must proceed, I think, with some humility. It is not unreasonable for a White American kid to have absorbed some racism. That young person is growing up in a fairly racist society—that is, a society in which race strongly affects life chances (health, longevity, income, wealth). He or she hears on all sides conversations in which race is an emotionally charged subject. He or she lives, often, in a neighborhood that is segregated by race. This young person learns over and over that race matters in America. To preach to an adolescent that race does not matter, or that we should act as though it does not matter, rightly invites skepticism. Teachers need to wrestle in their own minds and guts with these issues before trying to educate others. Are teachers ready to be honest with children about the actual state of our society and to talk honestly about the steps that may need to happen for the society to be less racist? This may require talking about economics, the great unspeakable in our culture. Teachers may need to spend time in protected settings, working through their own understanding; they may also need to do a fair amount of reading. Simple preaching is not going to accomplish the task (Ezekiel, 1998, 1999a, 1999b, 1999c).

References

Ackerman, N. W., & Jahoda, M. (1950). *Anti-Semitism and emotional disorder: A psychoanalytic interpretation*. New York: Harper-Collins.

Adorno, T. W., Frenkel-Brunswik, E., Levinson, D., & Sanford, R. N. (1950). *The authoritarian personality*. New York: HarperCollins.

Aho, J. A. (1990). *The politics of righteousness: Idaho Christian patriotism*. Seattle: University of Washington Press.

Cohen, S., & Herbert, T. (1996). Health psychology: Psychological factors and physical disease from the perspective of human psychoneuroimmunology. *Annual Review of Psychology, 47*, 113–142.

Collins, C., & Yeskel, F. (2000). *Economic apartheid in America: A primer on economic inequality and insecurity*. New York: New Press.

Dunbar, E. (2000). *Toward a profile of violent hate crime offenders: Behavioral and ideological signifiers of bias motivated criminality*. Manuscript submitted for publication.

Dunbar, E., Krop, H., & Sullaway, M. (2000). *Behavioral, psychometric, and diagnostic characteristics of bias-motivated homicide offenders*. Manuscript submitted for publication.

Ezekiel, R. (1969). Setting and the emergence of competence during adult socialization: Working at home vs. working "out there." *Merrill-Palmer Quarterly of Behavior and Development, 15*(4), 389–396.

Ezekiel, R. (1984). *Voices from the corner: Poverty and racism in the inner city*. Philadelphia: Temple University Press.

Ezekiel, R. (1995). *The racist mind: Portraits of American neo-Nazis and Klansmen*. New York: Viking Penguin.

Ezekiel, R. (1998). *Anti-bias education with 5th- and 6th-graders*. Unpublished paper for the Museum of Tolerance, Simon Wiesenthal Center.

Ezekiel, R. (1999a). *Teaching tolerance: Today, tomorrow, five years from now*. Unpublished paper for the Museum of Tolerance, Simon Wiesenthal Center.

Ezekiel, R. (1999b). *Teaching tolerance: Self-examination*. Unpublished paper for the Museum of Tolerance, Simon Wiesenthal Center.

Ezekiel, R. (1999c). *Teaching tolerance: Learning from history*. Unpublished paper for the Museum of Tolerance, Simon Wiesenthal Center.

Freedman, M. (1993). *The kindness of strangers: Adult mentors, urban youth, and the new voluntarism*. Cambridge, United Kingdom: Cambridge University Press.

Green, D., Abelson, R., & Garnett, M. (1999). The distinctive political views of hate-crime perpetrators and white supremacists. In D. Prentice & D. Miller (Eds.), *Cultural divides: Understanding and overcoming group conflict* (pp. 429–464). New York: Russell Sage.

Green, D., Glaser, J., & Rich, A. (1998). From lynching to gay bashing: The elusive connection between economic conditions and hate crime. *Journal of Personality and Social Psychology, 75*, 82–92.

Green, D., Strolovitch, D., & Wong, J. (1998). Defended neighborhoods, integration, and racially motivated crime. *American Journal of Sociology, 104*, 372–403.

Hopf, C. (1993). Authoritarians and their families: Qualitative studies on the origins of authoritarian dispositions. In W. Stone, G. Lederer, & R. Christie (Eds.), *Strength and weakness: The authoritarian personality today* (pp. 119–143). New York: Springer-Verlag.

Kershaw, I. (1998). *Hitler: 1889–1936: Hubris*. New York: Norton.

Marcus-Newhall, A., Miller, N., Holz, R., & Brewer, M. B. (1993). Cross-cutting category membership with role assignment: A means of reducing intergroup bias. *British Journal of Social Psychology, 32*, 125–146.

Martin, J. (2001). The authoritarian personality, 50 years later: What lessons are there for political psychology? *Political Psychology, 22*, 1–26.

Schuman, H., Steeb, C., Bobo, L., & Krysan, M. (1997). *Racial attitudes in America: Trends and interpretations* (Rev. ed.). Cambridge, MA: Harvard University Press.

Smith, M. B. (1965). An analysis of two measures of "authoritarianism" among Peace Corps teachers. *Journal of Personality, 33*(44), 513–535.

Smith, M. B. (1966). Explorations in competence: A study of Peace Corps teachers in Ghana. *American Psychologist, 21*, 555–566.

Smith, M. B. (1997). The authoritarian personality: A re-review 40 years later. *Political Psychology, 18*, 159–163.

Southern Poverty Law Center. (2000a, winter). Active hate groups in the U.S. in 1999. *Intelligence Report, 97*, 30–35.

Southern Poverty Law Center. (2000b, winter). Hate groups on the Internet. *Intelligence Report, 97*, 36–39.

Southern Poverty Law Center. (2001a, spring). Active hate sites in the Internet in the year 2000. *Intelligence Report, 101*, 40–43.

Southern Poverty Law Center. (2001b, spring). Cyberhate revisited. *Intelligence Report, 101*, 44–45.

Southern Poverty Law Center. (2001c, spring). The new Romantics. *Intelligence Report, 101*, 56–57.

Staub, E. (1989). *The roots of evil: The origins of genocide and other group violence*. Cambridge, United Kingdom: Cambridge University Press.

Stone, W., Lederer, G., & Christie, R. (1993). *Strength and weakness: The authoritarian personality today*. New York: Springer-Verlag.

Sullivan, J., & Transue, J. (1996). The psychological underpinnings of democracy: A selective review of research in political tolerance, interpersonal trust, and social capital. *Annual Review of Psychology, 47*, 625–650.

Tierney, J., & Grossman, J. (2000). *Making a difference: An impact study of Big Brothers/Big Sisters*. Philadelphia: Public/Private Ventures.

Urban, L., & Miller, N. (1998). A theoretical analysis of crossed categorization effects: A meta-analysis. *Journal of Personality and Social Psychology, 74*, 894–908.

Weitzman, M. (1998, October). *The inverted image: Anti-Semitism and anti-Catholicism on the Internet*. Paper presented at the fifth biennial Conference on Christianity and the Holocaust, Princeton, NJ.

Weitzman, M. (2000). Workshop 4 on education: Use and misuse of the Internet. In *Proceedings: The Stockholm International Forum on the Holocaust*, pp. 250–253.

Zeskind, L. (1986). *The "Christian Identity" movement: Analyzing its theological rationalization for racist and anti-Semitic violence*. Atlanta, GA: Center for Democratic Renewal.

REVIEW QUESTIONS

1. Do you believe that the author captured the movement's ideology accurately?

2. The author has some suggestions to remediate and redirect youth headed in a racist direction. Do you believe they are valid and potentially good solutions?

3. What is the role of women in these groups?

READING 18

"Killer Drones: The 'Silver Bullet' of Democratic Warfare?" by Frank Sauer and Niklas Schörnig is an article that argues that weaponized drones are inexpensive, precise, and potentially dangerous for any democracy as their use could undermine democratic values. While not condemning the use of drones as weapons, the authors suggest much more thought should occur prior to a massive drone arms race. Read the article and consider what a policy for a country, such as the United States, Great Britain, or Canada, would entail. Issues to consider are: who controls the drones? Should it be the military or the intelligence agencies? Who decides on appropriate targets and locations?

Killer Drones
The "Silver Bullet" of Democratic Warfare?

Frank Sauer
Bundeswehr University Munich, Germany

Niklas Schörnig
Peace Research Institute Frankfurt, Germany

 ## Introduction

Unmanned aerial vehicles (UAVs), commonly known as 'drones', and unmanned systems in general represent perhaps the most important contemporary development in conventional military armaments.[1] These systems offer numerous advantages to the military, especially in relation to dirty, dull, or dangerous tasks (US Department of Defense, 2007: 19). Machines can operate in hazardous environments; they require no minimum hygienic standards; they do not need training; and they can be sent from the factory straight to the frontline, sometimes even with the memory of a destroyed predecessor. Given sufficient power supply, they do not tire. And, lastly, the use of unmanned systems in dangerous situations such as forward reconnaissance, bomb disposal, or the suppression of enemy air defenses means that human soldiers can be given the best possible force protection—namely, not being exposed to the enemy in the first place. From a military perspective,

it thus seems obvious why the demand for unmanned systems rose tremendously over the last decade—and looks set to continue to do so.[2]

Today, it is commonly believed that 40 or more countries are developing military unmanned systems (Singer, 2009: 241). Much of the data behind such a claim is hearsay and quite tough to verify. However, the most recent edition of *Military Balance*, published annually by the International Institute for Strategic Studies, can serve as a guidepost and a first handle on numbers. Here, 34 countries are listed as holding either medium- or heavy-sized UAVs (IISS, 2011: 24–6). The list of UAV holders reveals a peculiarity: two-thirds of these countries are democratic states.

While the 'drone hype' is commonly said to be driven mainly by the United States and Israel, it seems that democracies in general have been the first to jump on the unmanned bandwagon. Democracy indices—such as Polity IV, on which we draw here—need to be taken with an

SOURCE: Security Dialogue August 2012 vol. 43 no. 4 363–380.

appropriate grain of salt, yet they can be helpful in system-izing and corroborating this first impression. With Polity IV ranging from 10 (strongly democratic) to –10 (strongly autocratic), 24 of those 34 countries listed in *Military Balance* turn out to have a polity score of 6 or higher—in other words, they can safely be called democratic. So, why are democracies in the driving seat of this development?

Some might point to a fairly obvious answer: because they can be. The financial and technological resources required for pursuing drone warfare are most readily available to wealthy states, the majority of which are democracies. Adhering to the 'technological imperative' (for a concise summary, see Reppy, 1990: 102–3), so this theoretical argument goes, they employ these superior resources to build the best technology and arm their militaries with it, simply because it is possible for them to do so. Yet, no country's defense budget, not even that of the USA, is limitless. Consequently, political decisions about how to allocate resources have to be made, all while maintaining the technological edge. So, for instance, why did the US Army abandon its 'Comanche' advanced helicopter project in 2004, thereby swapping a fast and stealthy hi-tech helicopter for what were then technologically inferior, slow, and non-stealthy drones (Fulghum and Wall, 2004)? The simplistic argument according to which technology is the sole driver, while common place and not entirely implausible, obviously carries only so far.

We are not the first scholars to raise the question of why democracies are so intrigued by unmanned weapon systems and particularly drones at the moment, but a review of the burgeoning literature on military use of unmanned systems shows that the issue has hitherto only been dealt with cursorily and in passing. There is currently no general, systematic, and theory-driven study seeking to probe the peculiar nexus of democracy and the use of unmanned systems.[3] We thus aim to address this lacuna by providing a critical exploration of the reasons why it is democracies that are spearheading the development of military unmanned systems, as well as the consequences of this situation. In this regard, we will not deal with specific procurement and employment practices by specific democratic states in specific cases, but rather try to paint a picture with somewhat broader brush strokes. This is the first goal we are aiming for with this article. In addition, we wish to contribute to the growing body of theoretical literature that takes a skeptical stance towards what is known as 'democratic peace theory'. The reasoning behind our adoption of the theoretical framework of the democratic peace

only to critically question it is that, if we are to understand the peculiar nexus between democracy and unmanned systems, a perspective that retains 'democracy' at the center of the analysis is required. By fleshing out the 'antinomies' of democratic peace theory (Müller, 2004), we contribute an intrinsic, first-order critique, one that will be more substantiated than an outright rejection of the entire concept up front.

Our antinomic reading of democratic peace theory ties in with the recent critical turn of the so-called democratic distinctiveness programme that emerged from the democratic peace debate (Geis and Wagner, 2011).[4] Instead of naïvely taking a supposed democratic peacefulness at face value, this entails further questioning the precise ways in which democracies are distinct from other regime types, as we discuss in detail in this article's second section. Such a theoretical framework enables us to systematically account for the 'dark side' of democratic distinctiveness by identifying the specific *inherent ambivalences* in democracies that are responsible for their *aggressive behavior*—behavior that is out of tune with what the conventional, positive bias of classical democratic peace theory gives reason to expect. More precisely, we argue in the third section that the *same* specific interests and norms that are conventionally taken to be pivotal for democratic peacefulness—the need to reduce costs, the short-term satisfaction of particular 'risk-transfer rules' for avoiding casualties, and the upkeep of a specific set of normative values—constitute the special appeal of unmanned systems to democracies. In turn, we demonstrate in the fourth section that by relying on these systems to satisfy said interests and norms, democracies will end up *thwarting* the latter in the long run—*inter alia* by rendering themselves only *more* war-prone. However, despite its skeptical stance, the article ends on an optimistic note. As we subscribe to the idea that free speech and deliberation are constitutive features of democracies, we believe that critically self-reflecting the mid- and long-term effects of robotic warfare could lead to more responsible behavior in the future.

 From Democratic Peace to The Liberal Study of International Conflict

Until the early 1980s, conventional wisdom in political science held that democracies were not different from states with other regime types when dealing with matters of peace and war. Then, Michael Doyle (1983) found 'proof' that democracies—in contrast to any other group of states, however defined—had virtually not fought each other since at least 1815,[5] thus starting what has since been termed a '"democratic turn" in peace and conflict research' (Geis and Wagner, 2011: 1556). Doyle's 'separate peace' was deemed so robust in regard to different definitions of war and democracy, as well as the various statistical methods applied, that Jack Levy (1988: 662) called it 'as close as anything we have to an empirical law in international relations'.

Much 'democratic peace' research, especially in the theory's early years, has been criticized for its trivial optimism, one that takes democratization to be a panacea for world peace. Equipped with notions of linear causality and an alleged seal of approval from established science, the theory's supposed practical relevance found fertile ground among political elites from the 1990s onwards, beginning with Bill Clinton's doctrine of democratic enlargement and subsequently peaking in the neoconservative verve to spread democracy by force during the George W. Bush administration (as criticized, for example, in Russett, 2005). Yet, more skeptical scholars had long been objecting that the self-congratulatory picture of democracy should be thoroughly questioned and that even the pivotal analytic concept of 'democracy' itself is both value-laden and historically contingent (see, for example, Oren, 1995; Hobson, 2009).

So, while some scholars settled with the empirical finding without giving the theoretical basis much further thought (see the critique by Ray, 1998: 39), others felt inspired to dig deeper. As Anna Geis and Wolfgang Wagner (2011: 1555; see also Hasenclever and Wagner, 2004: 465) point out, 'many of these studies are inspired by Immanuel Kant's famous essay on "Perpetual Peace"', and they have by now led to a more critical and broader 'liberal study of international conflict'. Instead of focusing on the interdemocratic peace as such, this 'democratic distinctiveness programme' brings together research on such diverse issues as specific democratic compliance with international law (Morrow, 2007; Slaughter, 1995), the special interest and capability of democracies in establishing and maintaining international institutions (Hasenclever and Weiffen, 2006), why democracies tend to win the wars they fight (Reiter and Stam, 2002), and how they fight these wars (Mandel, 2004; Shaw, 2005; Watts, 2008).

One concrete tie-in into this strand of research is the concept of 'antinomies' of the democratic peace (Müller,

2004). This approach highlights the ambivalence of democratic behavior, particularly in relation to the empirical observation that democracies tend to keep the peace among themselves while at the same time behaving in a strikingly *belligerent* fashion towards non-democracies.[6] Focusing on democratic aggressiveness, scholars coined the notion of 'democratic wars' (Geis et al., 2006)—that is, wars that are typical for democracies and consistent with specific norms, such as a 'humanitarian intervention' in a non-democracy to end human suffering (see also Freedman, 2006/7). In this line of thought, the gap between the ambivalent empirical findings of the democratic peace theory is closed (Geis et al., 2012), yet liberalism's plain progressivism towards general peacefulness is fundamentally questioned in turn (Rengger, 2006: 133). The antinomy concept helps to expose how classic democratic peace theory systematically turns a blind eye to some unwelcome conclusions deriving from most basic assumptions and empirical observations. More precisely, since, as proposed by Harald Müller (2004: 516n6), an antinomy 'is understood as a law-like proposition from which a secondary proposition and its very opposite can be deduced' (see also Müller and Evangelista, 2008: 2), democratic *peace* theory can be said to suffer from an optimistic bias when deducing democratic behavior only *one* way. After all, this simply ignores the potentially wider variety of behavior derivable from the *same* assumptions, in particular the belligerent behavior democracies display towards non-democracies. Consequently, the antinomic approach acknowledges that democracies are distinct but harbor *inherent tendencies* for both non-violent *and* violent behavior, *both* democracy-specific.

From Specific Institutions and Norms to Democratic Distinctiveness

Why is it plausible to assume that democracies are distinct? With the first preliminary article of his famous tractate *Perpetual Peace*, Immanuel Kant ([1795] 1957: 12–13) provided an essential starting point for research on democratic distinctiveness in the 1980s. Here, he argued that when those who decide about war or peace are obliged to fight and bear the costs,

> they would be very cautious in commencing such a poor game, decreeing for themselves all the calamities of war. Among the latter would be:

having to fight, having to pay the costs of war from their own resources, having painfully to repair the devastation war leaves behind, and, to fill up the measure of evils, load themselves with a heavy national debt that would embitter peace itself and that can never be liquidated on account of constant wars in the future.

Kant refers to the weighing of both material and immaterial costs. Some scholars focus on the monetary side, arguing that democracies limit military expenditure in peacetime in favor of non-military investments (Müller and Becker, 2008: 105). The fortune of democratic state leaders is said to hinge on the production of enough public good to ensure their re-election, which is why military spending is less useful to them, creating, in this line of thought, a specifically democratic incentive to limit respective costs during peacetime (Fordham and Walker, 2005: 144). In addition, excessive waste of taxpayers' money is also said to be distinctively constrained by institutional checks such as parliaments, free media, or bureaucracies auditing and evaluating procurement processes (Fordham and Walker, 2005: 142–5). Systematic quantitative studies corroborate the notion that higher degrees of democracy result in lowered defense spending during times of peace (see, for example, Fordham and Walker, 2005; Yildirim and Sezgin, 2005). Now, some nevertheless deem all these theoretically derived and statistically tested claims at best counterintuitive in the light of, to cite the prime example, the history of defense spending in the USA. Yet, our argument here does not require us to definitively establish that democracies *really* engage in less military spending during peacetime. In fact, for the sake of the argument we can assume this for the remainder of the article, because our aim is to question not the validity of this distinctive feature but rather that of an ensuing short-circuited interpretation, the conclusion that has customarily been drawn from it in the democratic peace literature. This warrants some elaboration.

An article by Julide Yildirim and Selami Sezgin (2005: 99) is cited as an example of how statistical results about the effect of democratization on military spending yield dubious conclusions when interpreted solely and uncritically through the prism of a taken-for-granted democratic peace paradigm, culminating as it does in the generalizing conjecture that 'worldwide attempts trying to increase the level of democracy may result in a more peaceful world, reducing military expenditures and hence wars'. A second example article by Benjamin Fordham and Thomas Walker (2005: 141, 154–5) also provides strong empirical support for the 'demilitarizing effect of democracy' in an analysis of 'a wide range of states since 1816', but in their conclusion Fordham and Walker reflect more critically on caveats such as a state's possible turn to an imperialistic foreign policy. To systematically follow this trail further is precisely what our reading of democratic peace theory is about. Antinomically scrutinizing the distinctive democratic inclination to save costs means picking up on Fordham and Walker's merely tentative qualms by arguing systematically that—all things being equal—the *opposite* of Yildirim and Sezgin's conclusion might very well be the case: namely, more aggressive behavior of democracies towards non-democracies. We will elaborate on this further in the third and fourth sections of this article.

We now turn to the second relevant category of costs of war, the non-monetary ones, especially having to fight and risk one's own life and limb. The conventional reading of Kant by authors such as Ernst-Otto Czempiel has been to suggest that the citizen's decision is practically predetermined owing to the potential costs of war being perceived as prohibitively high on all accounts. War would in fact vanish *if only* 'the consent of the citizens is required to decide whether or not war should be declared' (Czempiel, 1995: 9).

However, a careful reading of Kant's statement ('they would be very cautious') suggests nothing but an actual—indeterminate—weighing of pros and cons. Given a genuine choice, it seems the possibility of democracies engaging in wars of aggression is not entirely ruled out – as long as the costs are considered acceptable (Schweller, 1992: 241; Fearon, 1995: 386).

In this regard, Martin Shaw (2005) reveals the importance of the successful application of 'risk-transfer rules'— with minimization of casualties to troops in democracies being the most important of these. In democracies, the possibility of satisfying this rule under both national and international public scrutiny heavily influences the decision for or against a military engagement. Shaw's reasoning fits well with the US example, in which casualty avoidance among military forces has become a mission goal in itself, pointing to the possibility of democracies waging wars of aggression if a 'zero casualty doctrine' can be implemented (see also the case of Israel in Levy, 2011: 79).

One might say that *every* military is interested in low casualty rates in order to ensure that it remains able to fight another day. However, it is plausible to assume that democracies are indeed distinct owing to a particularly low tolerance for casualties for two reasons, one utilitarian and one normative.

The utilitarian argument suggests that decisionmakers in democracies fear losses among their own more than authoritarian leaders because rising numbers of casualties in a conflict will have adverse effects on public support for the military mission (Mueller, 1973; Gartner and Segura, 1998). More precisely, pertinent research suggests that the relevance of casualties for public opinion differs according to the type of conflict and is inversely related to the national interest understood to be at stake. In 'wars of necessity' (Freedman, 2006/7) for self-defense and national survival, any democratic population is willing to tolerate high casualties among its own troops. Yet, the tolerance for casualties is comparably lower in so-called wars of choice (Freedman, 2006/7), such as humanitarian missions (Larson, 1996). Democratic publics are thus casualty-phobic insofar as additional casualties create more disapproval if the public perceives them as being unnecessary or in vain (Gelpi et al., 2006). This is of crucial relevance with regard to unmanned systems and the types of wars democracies fight, as we will argue in the third section of this article.

The normative argument relates to the 'externalization hypothesis' (Risse-Kappen, 1995: 499–500) of democratic peace theory, which states that shared liberal norms and principles make democracies appreciate one another and keep the peace among themselves. Figuring as the normative foundation of most modern democracies, the founding idea of liberal thinking is that every individual human being is the bearer of innate, indefeasible *rights*—for example, the rights to equality before the law, physical integrity, personal freedom, and so on (see, for example, Owen, 1994). This is an ideal-type sketch, but research does in fact support the notion that such shared norms reduce conflict initiations (Peterson and Graham, 2011) and that democratic institutions—or, more precisely, specific democratic features such as party competition—reduce human rights abuses (Bueno de Mesquita et al., 2005). This suggests that, in addition to the utilitarian argument, democracies should by and large be more casualty-averse because they distinctively value the life of human individuals.

But, democratic distinctiveness regarding norms goes beyond casualty aversion. Conventional wisdom holds that the rule of law is an essential feature of modern-type democracies (Helmke and Rosenbluth, 2009). Domestically, democracies generally regard the rule of law and its limits on the use of force as paramount. Violence is regarded as a means of last resort and subject to the test of appropriateness and judicial oversight. Democracies also tend to stay wedded to these accustomed principles in the realm of foreign policymaking (see, for example, Dixon, 1994; Weart, 1998; Slaughter, 1995). They are, for instance, more eager to implement international law by incorporating pertinent rules into their military handbooks (Simmons, 1998: 84). Of course, exceptions to the rule exist. And 'creating' the 'legal' conditions for waterboarding quickly comes to mind as an example of antinomically engendered processes. But, to not get ahead of ourselves, the important point to hang on to is that research suggests that democracies demonstrate higher levels of compliance with the laws of war once they have committed themselves by treaty ratification (see, for example, Morrow, 2007).

Now, understanding human rights to be universal, democracies by and large respect them in relation to their own military personnel, enemy soldiers, and the adversary's civilian population. They distinctively subscribe to three crucial notions of the law of armed conflict: discrimination between military and civilian targets, proportionality of violent means applied, and attributable responsibility for actions in war. Again, pointing to examples of heinous misconduct by democracies such as occurred, for instance, during the Vietnam War seems the quick way to dismiss all this lock, stock, and barrel. However, Stephen Watts (2008) argues that even in Vietnam the USA acted significantly more carefully in relation to the civilian population than the Soviet Union did in Afghanistan under similar conditions. To qualify this: democratic restraint only holds as long as winning and minimizing one's own casualties is assured (Downes, 2006). This normative hierarchy, placing the lives of one's own soldiers over those of the adversary's civilian population, is consistent with findings by other scholars (Mandel, 2004; Shaw, 2005; Geis et al., 2010) and corroborated by insights regarding compliance: when confronted with severe violations of the rules of war by an opponent, democracies tend to succumb to temptation and react in a tit-for-tat fashion (Morrow, 2007). Most importantly, antinomically speaking the desire to enforce and export universal human rights or to militarily uphold international law might even be a distinctively democratic reason to engage in wars of choice with non-democracies in the first place. So, all in all, during conflicts with non-democracies liberal norms remain present—but they can then engender behaviors and actions that are unexpected from the vantage point of classic democratic peace theory, thus pointing to the antinomic character of democratic distinctiveness.

To sum up our argument so far, assuming democracies to be distinctively set up in terms of institutions and their subscription to specific liberal norms and values seems generally valid. Yet, this does *not* necessarily render them immune to aggressive behavior—and unmanned systems play a very problematic role in this regard.

From Liberal Weapon of Choice to Silver Bullet

The types of wars fought by democracies have changed over the last decades. In contemporary asymmetric conflicts and wars of choice, clear criteria for progress or 'victory' are hard to establish and mission objectives may change significantly over time. Accordingly, decisionmakers face the possibility of a sudden shift in the public's mood when casualties rise without palpable progress. This might even lead into a 'casualty trap' (Schörnig, 2009)—a stalemate situation where military operations are ceased to avoid additional casualties, thereby forestalling mission accomplishment.

Bearing the arguments set out in this article's second section in mind, we can discern three risk-transfer paths that are currently followed by democratic decisionmakers in response to the looming casualty trap. Relying on private military companies (PMCs) rather than regular service personnel to circumvent public scrutiny is one (Avant and Sigelman, 2010: 259; see also Schooner, 2008), while relying on a combination of air power and locals to limit the exposure of one's own troops is another way of dealing with democracy-specific casualty-sensitivity. However, both have produced mixed results; the preferred solution is the third one—namely, the replacement of 'labor' (soldiers) by 'capital' (technology). Advanced cruise missiles and potent conventional warheads might do (Sapolsky and Shapiro, 1996). Yet, those come into conflict with liberal norms and the law of armed conflict. Excessive firepower is actually a disservice. In order to minimize civilian casualties, *less*

firepower has to be applied, but in a precise fashion, thereby affecting the militarily relevant target only (Shaw, 2005: 87–8) and 'influencing', not annihilating, the opponent (Mandel, 2004: 71).

Clearly, then, the liberal weapon of choice should make it possible to minimize civilian casualties and heed the laws of armed conflict while avoiding friendly casualties by substituting capital for labor. Yet, this seems irreconcilable with limiting expenditures. It now begins to become clear why unmanned-systems technology seems so attractive.

First, unmanned systems are considered cheaper than their inhabited counterparts. Hence, they cater to the distinct democratic interest in limiting military expenditure during peacetime. Numerous reasons are usually pointed out for this, all seemingly valid at first sight. Obviously, there is no need for expensive life-support systems. Also, training a drone pilot is cheaper than training a fighter jock.[7] Their salaries are lower, too. In addition, maintenance of unmanned systems is also said to be cheaper, as their airframes are not as complex as those of manned planes.

Second, and most importantly, unmanned systems offer themselves in the light of democracies' distinct casualty aversion because they are said to protect troops not merely by distancing but by removing them from the battlefield.[8] With 'friendlies' at less or no risk at all, decisionmakers have less to fear in terms of a public-opinion backlash. 'Reaper' drones, for example, can simply loiter over a potential target without risking the life of a human pilot. They can also risk a closer approach for better situational awareness in the event of doubt. Explosive-ordnance-disposal systems render the issue even more obvious—no wonder their use has risen exponentially over the last few years and that they are being 'fetishized' as 'life-savers' (Roderick, 2010).

Third, at first glance unmanned systems seemingly also allow for heeding the norms and laws of war. In combination with its precision-guided munitions, the real-time intelligence, surveillance and reconnaissance (ISR) from the Reaper's unblinking eyes—supposedly—allows for striking with enough precision to minimize or even avoid civilian casualties and unnecessary damage. By replacing firepower with precision and the capability to wait hours for the optimal moment to engage, so the argument goes, the least force necessary in accordance with the law of armed conflict's norms of discrimination and proportionality can be applied. Not having to worry about a pilot's life is an advantage again: during the air war over former Yugoslavia, NATO fighters had to maintain a minimum height to avoid Serbian anti-aircraft fire, thereby reducing the accuracy of bombings. As drones can take bigger risks, flying lower will increase accuracy.

Unmanned systems seem a perfect fit for democratic warfare through their appeal to the utilitarian and normative characteristics of democracies. Because they are ascribed the unique capability of satisfying the rule of 'risk-transfer war', respecting the laws of armed conflict *and* limiting expenditure at the same time, they are even more than the weapon of choice: they seemingly provide a 'silver bullet' for democratic decisionmakers. Yet, with current killer drones as only a first stepping stone in what Robert Mandel (2004) has termed 'the quest for bloodless war', democracies currently fuel two trends.

The first of these is weaponization. UAVs, for example, started out as single-purpose observation drones, but have since become both communication relays and multi-sensor ISR and weapon platforms. Since the sensor (formerly the UAV) and the shooter (formerly a manned airplane, an artillery unit, etc.) no longer have to be coordinated but are now two-in-one, unmanned combat air vehicles (UCAVs) reduce the sensor-to-shooter gap from hours to minutes or seconds, increasing efficiency and thus providing an extremely valuable capability from a military point of view. Only a limited number of countries currently operate weaponized systems. However, the trend is already well underway, and there is little to no opposition to it. New, small-yield missiles will make weaponized UAVs appear even more suitable for 'precision warfare' in the future.

The second trend is autonomy—that is, 'the capacity to operate in the real-world environment without any form of external control', including, eventually, independent intelligent decisionmaking (Lin et al., 2008: 105; see also Sparrow, 2007: 65–6). Starting out as mere remote-controlled devices, modern UAVs are capable of performing a number of tasks on their own for extended periods of time. Following a preprogrammed route is routine. Even complex tasks like take-off, landing, or responding to emergencies like damage or even partial wing loss are safely handled. Consequently, UAV operators are changing their role from 'pilots' in-the-loop to mere supervisors on-the-loop, splitting their attention between several

airborne drones. And, even though current unmanned systems lack strong artificial intelligence, there is an unfaltering trend towards greater autonomy. Some fully autonomous weapon systems already exist, such as automatic close-in weapon systems for terminal defense against missiles or artillery shells (like the US 'Phalanx') and fixed border sentries in South Korea and Israel (Lin et al., 2008: 13–14, 18–19; Marchant et al., 2011: 276–7). More are likely to follow, as we will argue below.

To sum up our argument so far, killer drones seemingly lend themselves as the silver bullet of democratic warfare, explaining the distinct democratic eagerness to employ them. Yet, we are going to claim that many of the characteristics ascribed to drones are not holding up under closer scrutiny. More importantly, in the long run democracies may be disregarding numerous problematic normative consequences while striving for more, weaponized, and eventually autonomous systems.

From Silver Bullet to Boomerang

In general, proponents of unmanned systems and robots expect them to reduce human suffering and death in war in the short run. As Singer (2009: 431) points out, some might even hope that robots will 'finally end our species' bent toward war' in the long run, although Singer himself is sceptical of this claim. Indeed, critics deem this long-term hope utopian and point to a history of similar, dashed hopes following the invention of machine guns, dynamite or nuclear weapons. Critics are also more cautious regarding the short-term benefits and suspect them to be Janus-faced at best and possibly outweighed later on. The rapidly growing body of literature on the ethical, legal, political, and military problems of unmanned systems reflects this general divide.[9]

Rather than repeating the already long list of issues under debate or merely adding more bits and pieces to it, this article contributes a substantiated and systematic critical perspective. By drawing on the existing literature with the two trends of weaponization and autonomization in mind, we critically examine unmanned systems in the light of their long-term compatibility with the utilitarian and normative democracy-specific reasons we established above as being crucial for using them in the first place. These entail cost reduction, casualty avoidance, and the heeding of legal norms and liberal values. As it turns out, the silver bullet harbors the danger of coming back as a boomerang.

Costs

In terms of material costs, the first problem to consider is proliferating technology. Like many modern military applications today, unmanned systems have been made possible through the 'spin in' of civilian technology into military hardware (Sparrow, 2009: 28; see also Oudes and Zwijnenburg, 2011: 20–1). For example, basic drone technology—that is, airframes, propulsion, telemetry, control software, in short everything but sophisticated military components such as sensors, weaponry, etc.—can be borrowed from the civilian sphere. The downside, of course, is that drone technology is comparably easy to obtain and widespread (accordingly, even non-state entities such as Hamas supposedly use drones, as do a number of PMCs; see Krishnan, 2011: 62; Singer, 2011: 79). In addition to their distinctive interest in unmanned systems, democratic states possess a comparative advantage in terms of research and development (R&D), since they can rely on competitive civilian markets to bring up innovative solutions, generating a broader range of technological choice for both consumers and the military (Müller and Becker, 2008: 103). However, it is easier for authoritarian states to channel massive resources into refining specific systems once the basic designs have been re-engineered, as the Cold War experience suggests (see, for example, Evangelista, 1988: 22–49). China, for instance, based its first handful of UAV concepts on copies of existing designs, and shortly afterwards, as of 2010, could already display around 25 different UAV models, including weaponized ones, at arms shows (Minnick, 2010). So, democratic states spearhead innovation by investing significant resources in R&D, but 'in both technology and war . . . there is no such thing as a permanent first mover advantage' (Singer, 2011: 79). In trying to quickly reap the alleged benefits of using drones and in order to enhance force protection for currently deployed troops, NATO states have fielded equipment that—by former standards—was not sufficiently mature. As a result, many systems have been lost owing to malfunction (the US

drone that crashed in Iran in 2011 is a case in point), increasing the risk of unrecovered wrecks fueling a technological transfer and drone technology being sold on the international black market. So, while innovators bear significantly higher costs, followers can adopt a pick-and-choose approach, invest in (copies of) proven concepts, rely on technology transfer, or substitute individual features with cheaper civilian technology. The problem of proliferation is hence exacerbated by the fact that the comparably *costly* efforts by democracies to 'keep the edge' are paving the way for aspirants who can simply trail the development with indirect help (Von Kospoth, 2009).

So far, the assessment suggests that democratic efforts to reduce the monetary costs are doomed to failure owing to the dynamic interaction of innovation and replication. Irony is added by the fact that while drones are favored by democracies because they are commonly considered cheaper for a number of seemingly plausible reasons that we laid out earlier, the verdict on this promise is not actually in yet. For example, most US Air Force operating costs, including those related to drones, are either unknown or misreported (Wheeler, 2011), and a 'single Predator or Reaper requires as many as 170 personnel to launch, command, recover and repair, plus handle the imagery it gathers', as David Axe (2011) notes. According to Axe, the Pentagon learned that, counter intuitively, '"unmanned" aircraft actually require lots of manpower', and manpower is, after all, a major driver of costs.

Casualties

Unmanned systems are understood to minimize casualties among one's own troops. According to their proponents, this is particularly helpful in asymmetric conflicts between states and non-state actors. However, the use of unmanned systems, particularly weaponized ones, may also aggravate the downside of such conflict settings. Operating them abroad may invite guerilla warfare or even terrorist attacks as a response to their overwhelming conventional superiority. Paul W. Kahn (2002: 6) concludes that 'the asymmetrical capacities of Western—and particularly U.S. forces—themselves create the conditions for increasing use of terrorism'. Given that democratic states tend to value the lives of their civilians even more highly than those of their soldiers (Mandel, 2004: 11), attempts to reduce casualties on the battlefield abroad might backfire. We will return to this train of thought in the next subsection.

Proponents of precision drone strikes also hope that the constant threat of 'bolt from the blue' attacks will frustrate opponents into surrender, with the civilian population remaining unharmed at the same time. Yet, the drone strikes in the Afghan–Pakistan border region are not corroborating such a view. Estimates of the numbers of civilian casualties, even while differing wildly (Rogers, 2010: 13–15; see also Ahmad, 2011; Woods, 2011), suggest that the hastily fielded technology is less discriminate and proportionate than was hoped. Against this background, some critics ask whether relying on technological supremacy and drone attacks is helpful in these asymmetric engagements. It might prove counterproductive by creating a 'siege mentality', with 'public anger' ultimately solidifying the power of the extremists, thus protracting the conflict rather than bringing it to a swifter and less bloody end (Kilcullen and McDonald Exum, 2009: 19–20; see also Sullins, 2011: 164–5; Oudes and Zwijnenburg, 2011).[10]

A more general argument in the debate over the rise of drones and military robotics is that their sheer existence lowers the threshold for military engagement for democracies. As argued above, the political risk of casualties provides a major restraint to democratic leaders in their decision to commit troops for so-called wars of choice. But, the more technology allows for removing soldiers from the battlefield, lifting the Clausewitzian 'fog of war' and creating a general asymmetric advantage, the less likely losses among one's own troops become and the lower the threshold to engage with military means. This already applies to unarmed UAVs, as they provide troops with instant and risk-free high-quality intelligence, surveillance, and reconnaissance. Now, to avoid a misunderstanding: States are of course obliged to provide their troops with the best protection available. However, implementing this option also leads to the antinomic effect of political decisionmakers supporting military missions they would not have supported under different, more costly circumstances. The growing numbers of weaponized drones and robots only make this slope more slippery. The Obama administration's argument for not having to ask the Congress to authorize the Libya campaign under the War Powers Act is a case in point. White House legal counsels argued that the military engagement was limited, conducted without the involvement of US ground forces, and thus free of any risk of friendly casualties (Savage and Landler, 2011; Saletan, 2011).[11]

Finally, and recalling the problem of proliferation, if both major and regional powers continue to build up capacities, global and regional robot arms races seem a likely consequence. A whole body of arms control literature raises the serious conjecture that accelerating arms races will have a destabilizing effect on state relations and increase the risk of military conflict—with, again, more rather than fewer casualties looming.

Laws and Norms

As argued in the second section of this article, democracies by and large have a special inclination towards heeding the norms and laws of armed conflict. Now, according to Paul Kahn, a soldier's right to kill his or her opponents depends on the condition of mutual risk. In this line of thought, the quest for bloodless war presents a 'deep challenge' to the *morality* of warfare as such (Kahn, 2002: 3). Snipers during World Wars I and II, for example, tended to be executed on the spot for violating this condition of mutual risk. Snipers in a war zone risk getting caught. Soldiers piloting weaponized drones from the other side of the globe do not. This riskless setup is not only in stark contrast to military notions of honor and valor (the US Air Force, for instance, wonders what a drone pilot could possibly be awarded a medal for). Rather, one might even conclude that answering this development with improvised explosive devices and suicide bombings—both sometimes regarded as cowardly ways of fighting—is only a consequent reaction to the sheer impossibility of a 'fair' fight.

Weaponized drones raise tricky *legal* issues as well, for example with respect to limiting the use of force to distinct combat zones in which—and *only* in which—the law of armed conflict would permit the killing of enemies (O'Connell, 2010). And, to hark back to the flip side—namely, the aforementioned aspect of inviting attacks against the 'homeland': Is the off-duty drone pilot (in civilian clothing) on route to her house a legitimate target?

More generally speaking, what are the consequences of the 'drone stare' distancing soldiers from the battlefield physically and emotionally, thus quite literally 'dehumanizing' the conduct of combat (Wall and Monahan, 2011; see also Oudes and Zwijnenburg, 2011: 21–3)?

Kahn (2002: 4) concludes that riskless warfare should not even be regarded as war but rather as 'police enforcement'—with much stricter rules and regulations than the law of armed conflict applying. Yet, it is unclear whether proponents of armed unmanned systems are prepared to accept this or whether they would rather stick to targeted killings executed by drones and cubicle warriors (for a critical discussion, see United Nations, 2010: 24–5).

Considering in more detail, lastly, the ongoing trend towards greater autonomy in mobile weaponized systems, it becomes clear that autonomy begets autonomy. Processes and decisions in automated war will become so swift that residual human interference means an unacceptable military disadvantage. With humans moving from in-the-loop to on-the-loop today and owing to the ever-increasing pace of the decisionmaking process out-of-the-loop, the incentive to procure and field more and more autonomous armed systems will be overwhelming, making it only a matter of time before autonomous weapon systems—armed robots—arrive on the battlefield. Their key characteristic is that they will be making decisions about life and death as they autonomously decide on what or whom to engage. This raises a whole new set of ethical and legal questions, in particular regarding fundamental rules of discrimination, proportionality, and responsibility.

For one thing, computer and artificial intelligence experts disagree over whether robots could ever discriminate sufficiently between combatants and non-combatants. Proponents argue that autonomous robots in the not-so-distant future, implanted with an 'ethical governor', might even be able to act in a more 'humane' fashion than human soldiers in specific battlefield scenarios (Arkin, 2009; Canning, 2009; Marchant et al., 2011: 280; for a critical perspective, see Wallach and Allen, 2009). 'Literally selfless robots who do not prioritize their own continued existence over obeying their ethical programming' (Lin et al., 2008: 52), with a capability for speedy judgment unimpaired by stress, fatigue, or limited cognitive abilities, would enhance compliance with the law of armed conflict, thereby reducing or eliminating the likelihood of massacres and atrocities. Critics reply that law is by definition subject to debate and controversy, and therefore virtually impossible to put into hard software code (Sharkey, 2008). Defining a 'civilian' is already a huge task, especially in the context of irregular warfare. Reliably distinguishing a civilian in the haste of battle is an even bigger challenge (Marchant et al., 2011: 282–3). Acquiring the required tacit knowledge, let alone a gut feeling, is something currently imaginable robots will simply not be capable of.

Similar questions arise for the rule of proportionality. Human soldiers are in a constant gray area when weighing military ends and means and endangering civilians or civilian infrastructure to comply with international law. It seems impossible to put this fuzziness into binary code (Sharkey, 2009; 2011: 45–6; Sparrow, 2011: 99; Asaro, 2011: 116–7). Proponents of robotic warfare certainly have a point when arguing that in the course of history humans have made myriads of errors, mis-judgments, or conscious decisions leading to senseless loss of innocent lives, and that robots, on the whole, should be judged in the light of that history rather than against theoretical standards (Arkin, 2009). This, however, raises the adjoining question of responsibility.

A human acting against the law of armed conflict can principally be held responsible for his or her actions, either by a court martial or by civilian authorities. Opponents of robot warfare therefore ask: 'Who is to be held responsible for the lethal mishaps of a robot?' (Sharkey, 2008: 88). Robert Sparrow (2007: 67) argues 'that it is a fundamental condition of fighting a just war that someone may be held [morally and legally] responsible for the death of enemies killed in it'. Since robots cannot fulfill this condition, fielding them in war is unethical, he concludes. Others (Lin et al., 2008: 64–6, 73–4) retort that a 'slave morality' programming circumvents this problem by reducing robots to mere instruments, with the commanding officer as the sole responsible authority. But, how can a military robot be autonomous and enslaved at the same time? As it is the very essence of autonomy to (learn and) make individual decisions in contrast to preprogrammed ones, a weaponized autonomous robot may act in ways unforeseeable to its designer or military commander. In fact, this superior 'warfighting creativity' is the idea behind constructing an autonomous military robot in the first place. The problem thus persists, because prosecuting either designer or officer would then be unjust, while penalizing the robot would be meaningless (Sparrow, 2007: 69–73)—leaving a fundamental challenge for international law.

Conclusion

In this article, we explored the peculiar nexus of democracy and the use of unmanned weapons systems in a systematic and theory-driven fashion. By critically refracting democratic peace theory through the prism of the democratic distinctiveness programme, we demonstrated that an antinomic reading of democratic peace is capable of yielding deeper insights into the procurement decisions of democratic states as well as the consequences of the currently ensuing drone hype.

Stating that democracies are characterized by a set of distinctive interests and norms, we argued that this setup causes killer drones and armed robots to appear as a silver bullet for political and military decisionmakers. These systems are seemingly cheaper and supposedly help states to heed the provisions of the law of armed conflict. They are considered especially suitable for the casualty-averse risk-transfer war that democracies prefer. However, we further argued that this supposed silver bullet might well come back as a boomerang. By fielding more weaponized and autonomous systems, in the long run democracies will not only be burdened with the mounting costs of an arms race but will also be rendered more war-prone (in relation to non-democracies), all while employing weapons that are at best dubious from the perspective of morality and the laws of armed conflict.

Unmanned systems are not a silver bullet. But, they are not *necessarily* a condemnable weapon of evil either. They *can* unarguably protect humans in various ways, they *might* eventually turn out to be cheaper in some respects, and they do not violate international law *per se*. Nevertheless, before democracies rush on, more deliberation on the problematic consequences pointed to in this article is needed. And here we believe democracies to be distinct as well. Even if democratic decisionmakers focus on the short-term benefits of robotic weapons, the political arena is open, and our article suggests that nothing is overdetermined. In the light of democracies' dark side, it might thus be up to nongovernmental organizations and civil society to recognize the need for a broad, sober, and informed discussion on the pros and cons of unmanned systems—including debate on the option of arms control to curb detrimental effects on civilians and peace at large.[12] In short, it is high time for democracies to bite the (silver) bullet and face the implications of their obsession with killer drones.

 Acknowledgements

For their comments, the authors also thank the members of the international relations research colloquia at Bundeswehr University and Ludwig-Maximilians-University in Munich. In addition, they especially thank Anna Geis, Eva Herschinger, Carlo Masala, Kimo Quaintance, and Donald Riznik for their helpful reviews of the initial manuscript, as well as Henning Halbe, Sören Kieserling and Annabel Schmitz for their research assistance. Lastly, the authors are indebted to the anonymous reviewers as well as the editors of *Security Dialogue* for their helpful comments and suggestions. Both authors are members of the International Committee for Robot Arms Control (ICRAC). The authors gratefully acknowledge the suggestions made by participants at the ICRAC workshop on 'Arms Control for Robots', held in Berlin on 20–22 September 2010.

Notes

1. In the past, pilotless aircraft used for target practice were sometimes called 'drones'. Today, as in this article, the term 'drone' is used as shorthand for tele-operated 'unmanned aerial vehicles'. Some drones are already capable of executing certain tasks autonomously. Nevertheless, we follow Lin et al. (2008: 4, 100–5) in reserving the terms 'robot' or 'robotic (weapon) system' for *at least* semi- and especially fully-autonomous systems. More generally, we use the terms 'unmanned system' or 'unmanned weapon system'.

2. To give but one example, the current 'Aircraft Procurement Plan' of the US Department of Defense plans for a slight decrease of the total aviation force until 2021. Yet, the number of aircraft in the unmanned-system category 'will more than *triple*' in the same period (US Department of Defense, 2011: 14, emphasis added).

3. Schörnig (2010) contains elements of such an approach, albeit in an earlier and less systematic take on the issue.

4. The notion 'democratic distinctiveness programme' goes back to John Owen (2004).

5. Earlier work on the democratic peace by Babst (1972) had not been noted by most scholars. In addition, Small and Singer (1976) had found no statistical proof for Babst's claim.

6. There has been a 'monadic revival' in connection with this, however, arguing that democracies are more peaceful *on average* when compared with similar autocracies (Russett and Oneal, 2001; MacMillan, 2003).

7. We owe this point to an anonymous reviewer.

8. It is worth noting in this respect, however, that accidental killings of US forces by US drones in 'friendly fire' incidents are not unheard of.

9. For a short but comprehensive overview of issues, see Lin et al. (2008). Exemplarily see also the books by Arkin (2009) and Wallach and Allen (2009), as well as the contributions by Singer (2010), Lin (2010), Arkin (2010), Strawser (2010) and Sharkey (2010) in the special issue of the *Journal of Military Ethics* on ethics and emerging military technologies (Lucas, 2010). Most recently, see Lin et al. (2012), Marchant et al. (2011), and Oudes and Zwijnenburg (2011). For in-depth discussions, see the interviews in Dabringer (2011).

10. The military's general fondness for drones notwithstanding, this critique even left a mark in the Afghanistan war for a short time, when General Stanley McChrystal curbed the number of drone strikes to reduce civilian casualties—directives that were largely revoked by his successor General David Petraeus.

11. A possible 'upside' to a lower threshold to war comes to mind—namely, meeting the international community's 'responsibility to protect' via unarmed systems—though this is a discussion we cannot go into here. Suffice it to say that the option to wage war at less cost and without risk lends itself to more democratic aggression towards non-democracies in general.

12. See, for example, the website of the International Committee for Robot Arms Control (ICRAC) at http:// www.icrac. net/ (accessed 18 July 2012).

✎ References

Ahmad MI (2011) The magical realism of body counts. *Aljazeera*, 13 June. Available at: http://english. aljazeera.net/indepth/opinion/2011/06/2011613931606455.html (accessed 20 October 2011).

Arkin RC (2009) *Governing Lethal Behavior in Autonomous Robots*. London: Chapman and Hall/CRC. Arkin RC (2010) The case for ethical autonomy in unmanned systems. *Journal of Military Ethics* 9(4): 332–341.

Asaro P (2011) Military robots and just war theory. In: Dabringer G (ed.) *Ethical and Legal Aspects of Unmanned Systems*. Vienna: Institut für Religion und Frieden, 103–119.

Avant D and Sigelman L (2010) Private security and democracy: Lessons from the US in Iraq. *Security Studies* 19(2): 230–265.

Axe D (2011) US drones trump China theatrics. Available at: http://the-diplomat.com/2011/02/07/us-drones-trump-china-theatrics/4/ (accessed 20 October 2011).

Babst DV (1972) A force for peace. *Industrial Research* 14: 55–58.

Bueno de Mesquita B, Cherif F, Downs GW and Smith A (2005) Thinking inside the box: A closer look at democracy and human rights. *International Studies Quarterly* 49(3): 439–458.

Canning JS (2009) 'You've just been disarmed. Have a nice day!' *IEEE Technology and Society Magazine* 28(1): 13–15.

Czempiel EO (1995) *Are Democracies Peaceful? Not Quite Yet*. PRIF Report 37. Frankfurt: Peace Research Institute Frankfurt.

Dabringer G (ed.) (2011) *Ethical and Legal Aspects of Unmanned Systems*, Vienna: Institut für Religion und Frieden.

Dixon WJ (1994) Democracy and peaceful settlement of international conflict. *American Political Science Review* 88(1): 14–32.

Downes AB (2006) Desperate times, desperate measures: The causes of civilian victimization in war. *International Security* 30(4): 152–195.

Doyle MW (1983) Kant, liberal legacies, and foreign affairs (I). *Philosophy & Public Affairs* 12(3): 205–235.

Evangelista M (1988) *Innovation and the Arms Race: How the United States and the Soviet Union Develop New Military Technologies*. Ithaca, NY: Cornell University Press.

Fearon JD (1995) Rationalist explanations for war. *International Organization* 49(3): 379–411.

Fordham BO and Walker TC (2005) Kantian liberalism, regime type, and military resource allocation: Do democracies spend less? *International Studies Quarterly* 49(1): 141–157.

Freedman L (2006/7) Iraq, liberal wars and illiberal containment. *Survival* 48(4): 51–65.

Fulghum DA and Wall R (2004) Comanche helicopter program killed. Available at: http://www.aviationweek.com/aw/generic/story_generic.jsp?channel=awst&id=news/03014wna.xml (accessed 20 October 2011).

Gartner SS and Segura GM (1998) War, casualties, and public opinion. *Journal of Conflict Resolution* 42(3): 278–300.

Geis A, Brock L and Müller H (eds) (2006) *Democratic Wars: Looking at the Dark Side of Democratic Peace*. Basingstoke: Palgrave Macmillan.

Geis A and Wagner W (2011) How far is it from Königsberg to Kandahar? Democratic peace and democratic violence in international relations. *Review of International Studies* 37(4): 1555–1577.

Geis A, Müller H and Schörnig N (2010) Liberale Demokratien und Krieg [Liberal democracies and war]. *Zeitschrift für Internationale Beziehungen* 17(2): 171–201.

Geis A, Müller H and Schörnig N (2012) *The Janus Face of Liberal Democracies: Militant 'Forces for Good'*. Cambridge: Cambridge University Press (forthcoming).

Gelpi CF, Feaver, PD and Reifler J (2006) Success matters: Casualty sensitivity and the war in Iraq. *International Security* 30(3): 7–46.

Hasenclever A and Wagner W (2004) From the analysis of a separate democratic peace to the liberal study of international conflict. *International Politics* 41(4): 465–471.

Hasenclever A and Weiffen B (2006) International institutions are the key: A new perspective on the democratic peace. *Review of International Studies* 32(4): 563–585.

Helmke G and Rosenbluth F (2009) Regimes and the rule of law: Judicial independence in comparative perspective. *Annual Review of Political Sciences* 12: 345–366.

Hobson C (2009) Beyond the end of history: The need for a 'radical historicisation' of democracy in international relations. *Millennium* 37(3): 631–657.

International Institute for Strategic Studies (IISS) (2011) Unmanned aerial vehicles: Emerging lessons and technologies. In: IISS (ed.) *The Military Balance 2011.* London: IISS, 20–26.

Kahn PW (2002) The paradox of riskless warfare. *Philosophy & Public Policy Quarterly* 22(3): 2–8. Kant I ([1795] 1957) *Perpetual Peace.* New York: Bobbs-Merrill.

Kilcullen D and McDonald Exum A (2009) Death from above, outrage down below. *New York Times,* 16 May. Available at: http://www.nytimes.com/2009/05/17/opinion/17exum.html (accessed 20 October 2011).

Krishnan A (2011) Ethical and legal challenges. In: Dabringer G (ed.) *Ethical and Legal Aspects of Unmanned Systems.* Vienna: Institut für Religion and Frieden, 53–69.

Larson EV (1996) *Casualties and Consensus.* Santa Monica, CA: RAND.

Levy J (1988) Domestic politics and war. *Journal of Interdisciplinary History* 18(4): 653–673.

Levy Y (2011) How casualty sensitivity affects civilian control: The Israeli experience. *International Studies Perspectives* 12(1): 68–88.

Lin P (2010) Ethical blowback from emerging technologies. *Journal of Military Ethics* 9(4): 313–331.

Lin P, Bekey G and Abney K (2008) *Autonomous Military Robotics: Risk, Ethics, and Design.* San Luis Obispo, CA: California Polytechnic State University.

Lin P, Bekey G and Abney K (eds) (2012) *Robot Ethics: The Ethical and Social Implications of Robotics.* Cambridge, MA: MIT Press.

Lucas, G Jr. (2010) Special issue on ethics and emerging military technologies. *Journal of Military Ethics* 9(4): 289–431.

MacMillan J (2003) Beyond the separate democratic peace. *Journal of Peace Research* 40(2): 233–241.

Mandel R (2004) *Security, Strategy, and the Quest for Bloodless War.* Boulder, CO: Lynne Rienner.

Marchant GE et al. (2011) International governance of autonomous military robots. *The Columbia Science and Technology Law Review* 12: 272–315.

Minnick W (2010) Zhuhai Airshow goes unmanned. Available at: http://china-defense.blogspot.de/2010/11/chinese-uavs-at-airshow-china-2010-in.html (accessed 1 May 2012).

Morrow JD (2007) When do states follow the laws of war? *American Political Science Review* 101(3): 559–572.

Mueller JE (1973) *War, Presidents and Public Opinion.* New York: John Wiley.

Müller H (2004) The antinomy of democratic peace. *International Politics* 41(4): 494–520.

Müller H and Becker U (2008) Technology, nuclear arms control, and democracy. In: Evangelista M, Müller H and Schörnig N (eds) *Democracy and Security: Preferences, Norms and Policy-Making.* London: Routledge, 102–119.

Müller H and Evangelista M (2008) Introduction. In: Evangelista M, Müller H and Schörnig N (eds) *Democracy and Security: Preferences, Norms and Policy-Making.* London: Routledge, 1–13.

O'Connell ME (2010) Lawful use of combat drones. Presentation at the hearing 'Rise of the drones II: Examining the legality of unmanned targeting', House of Representatives Subcommittee on National Security and Foreign Affairs, US Congress, 28 April 2010.

Oren I (1995) The subjectivity of the 'democratic' peace. *International Security* 20(2): 174–184.

Oudes C and Zwijnenburg W (2011) *Does Unmanned Make Unacceptable? Exploring the Debate on Using Drones and Robots in Warfare.* Utrecht: IKV Pax Christi.

Owen JM (1994) How liberalism produces democratic peace. *International Security* 19(2): 87–125.

Owen JM (2004) Democratic peace research: Whence and whither? *International Politics* 41(4): 605–617.

Peterson TM and Graham L (2011) Shared human rights norms and military conflict. *Journal of Conflict Resolution* 55(2): 248–273.

Ray JL (1998) Does democracy cause peace? *American Review of Political Sciences* 1: 27–46.

Reiter D and Stam AC (2002) *Democracies at War.* Princeton, NJ: Princeton University Press.

Rengger N (2006) On democratic war theory. In: Brock L, Geis A and Müller H (eds) *Democratic Wars: Looking at the Dark Side of Democratic Peace.* Basingstoke: Palgrave Macmillan, 123–142.

Reppy J (1990) The technological imperative in strategic thought. *Journal of Peace Research* 27(1): 101–106.

Risse-Kappen T (1995) Democratic peace: Warlike democracies. *European Journal of International Relations* 1(4): 491–517.

Roderick I (2010) Considering the fetish value of EOD robots. *International Journal of Cultural Studies* 13(3): 235–253.

Rogers C (2010) *Civilians in Armed Conflict: Civilian Harm and Conflict in Northwest Pakistan.* Washington, DC: Campaign for Innocent Victims in Conflict.

Russett B (2005) Bushwhacking the democratic peace. *International Studies Perspectives* 6(4): 395–408.

Russett B and Oneal JR (2001) *Triangulating Peace: Democracy, Interdependence, and International Organizations.* New York: W.W. Norton.

Saletan W (2011) Koh is my god pilot: Can the president wage a drone war without congressional approval? The Obama administration says yes. *Slate,* 30 June. Available at: http://www.slate.com/articles/health_and_science/human_nature/2011/06/koh_is_my_god_pilot.html (accessed 20 October 2011).

Sapolsky HM and Shapiro J (1996) Casualties, technology, and America's future wars. *Parameters* 26(2): 119–127.

Savage C and Landler M (2011) White House defends continuing U.S. role in Libya operation. *New York Times,* 15 June. Available at: http://www.nytimes.com/2011/06/16/us/politics/16powers.html?_r=2&src=me&ref=us (accessed 20 October 2011).

Schooner SL (2008) Why contractor fatalities matter. *Parameters* 38(3): 78–91.

Schörnig N (2009) In der Opferfalle. Die Bundesregierung und die zunehmenden Gefallenen der Bundeswehr in Afghanistan [Facing the Afghan casualty trap. Rising public awareness puts German government under pressure]. HSFK-Standpunkt 2/2009. Frankfurt: Peace Research Institute Frankfurt.

Schörnig N (2010) Robot warriors: Why the Western investment into military robots might backfire. PRIF Report No. 100, Frankfurt: Peace Research Institute Frankfurt.

Schweller RL (1992) Domestic structure and preventive war. *World Politics* 44(2): 235–269.

Sharkey N (2008) Grounds for discrimination: Autonomous robot weapons. *RUSI Defence Systems* 11(2): 86–89.

Sharkey N (2009) Death strikes from the sky. *IEEE Technology and Society Magazine* 28(1): 16–19.

Sharkey N (2010) Saying 'no!' to lethal autonomous targeting. *Journal of Military Ethics* 9(4): 369–383.

Sharkey N (2011) Moral and legal aspects of military robots. In: Dabringer G (ed.) *Ethical and Legal Aspects of Unmanned Systems.* Vienna: Institut für Religion und Frieden, 43–51.

Shaw M (2005) *The New Western Way of War.* Cambridge: Polity.

Simmons BA (1998) Compliance with international agreements. *Annual Review of Political Sciences* 1: 75–93.

Singer PW (2009) *Wired for War.* New York: Penguin.

Singer PW (2010) The ethics of killer applications: Why is it so hard to talk about morality when it comes to new military technology? *Journal of Military Ethics* 9(4): 299–312.

Singer PW (2011) The future of war. In: Dabringer G (ed.) *Ethical and Legal Aspects of Unmanned Systems.* Vienna: Institut für Religion und Frieden, 71–85.

Slaughter AM (1995) International law in a world of liberal states. *European Journal of International Law* 6(1): 503–538.

Small M and Singer D (1976) The war-proneness of democratic regimes, 1816–1980. *The Jerusalem Journal of International Relations* 1(4): 50–69.

Sparrow R (2007) Killer robots. *Journal of Applied Philosophy* 24(1): 62–77.

Sparrow R (2009) Predators or plowshares? Arms control of robotic weapons. *IEEE Technology and Society Magazine* 28(1): 25–29.

Sparrow R (2011) The ethical challenges of military robots. In: Dabringer G (ed.) *Ethical and Legal Aspects of Unmanned Systems.* Vienna: Institut für Religion und Frieden, 87–102.

Strawser BJ (2010) Moral predators: The duty to employ uninhabited aerial vehicles. *Journal of Military Ethics* 9(4): 342–368.

Sullins JP (2011) Aspects of telerobotic systems. In: Dabringer G (ed.) *Ethical and Legal Aspects of Unmanned Systems.* Vienna: Institut für Religion und Frieden, 157–167.

United Nations (2010) Report of the Special Rapporteur on extrajudicial, summary or arbitrary executions, A/HRC/14/24/Add.6, New York.

US Department of Defense (2007) *Unmanned Systems Roadmap 2007–2032.* Washington, DC: US Department of Defense.

US Department of Defense (2011) *Aircraft Procurement Plan: Fiscal Years (FY) 2012–2041.* Washington, DC: US Department of Defense.

Von Kospoth N (2009) China's leap in unmanned aircraft development. Available at: http://www.defpro.com/daily/details/424/ (accessed 20 October 2011).

Wall T and Monahan T (2011) Surveillance and violence from afar. *Theoretical Criminology* 15(3): 239–254.

Wallach W and Allen C (2009) *Moral Machines.* Oxford: Oxford University Press.

Watts S (2008) Air war and restraint: The role of public opinion and democracy. In: Evangelista M, Müller H and Schörnig N (eds) *Democracy and Security: Preferences, Norms and Policy-Making.* London: Routledge, 53–71.

Weart SR (1998) *Never at War.* New Haven, CT: Yale University Press.

Wheeler W (2011) Flying blind: Most USAF aircraft operating costs are unknown. Available at: http://www.defencetalk.com/reports/AirForceOSCosts.pdf (accessed 20 October 2011).

Woods C (2011) Drone war exposed. Available at: http://www.thebureauinvestigates.com/2011/08/10/most-complete-picture-yet-of-cia-drone-strikes/ (accessed 20 October 2011).

Yildirim J and Sezgin S (2005) Democracy and military expenditure. *Transition Studies Review* 12(1): 93–100.

Frank Sauer is a Lecturer and Research Associate at the Political Science Department of Bundeswehr University, Munich, and a doctoral candidate at the Goethe University Frankfurt. In his research, he focuses on nuclear weapons, especially their (non-)use and proliferation. He has also published on a range of other topics, including the transformation of armed forces and German foreign policy. Email: frank.sauer@unibw.de.

Niklas Schörnig is Senior Research Fellow at the Peace Research Institute Frankfurt (PRIF) and Visiting Lecturer at the Goethe University Frankfurt. His research focuses on, *inter alia*, the transformation of Western armies after the end of the Cold War, US and European defense-industry issues, arms dynamics and arms control, liberal and realist international relations theory, and Australian foreign policy. Email: schoernig@hsfk.de.

REVIEW QUESTIONS

1. The authors of this article worry that there is a rush to use drones. What is your opinion given the overview they present? It might also be helpful to read this as well before you form an opinion: http://www.livingunderdrones.org/ (you can download the report for free from the site).

2. What do you believe should be the U.S. policy on drones being used to kill foreign terrorists and terrorists who are U.S. citizens that the federal government has decided to kill?

3. Should any administration be able to use a drone to kill someone in the United States? A serial killer? A foreign terrorist? A U.S. citizen driving on the interstate who is deemed a terrorist?

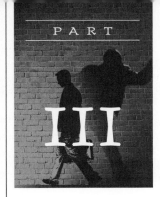

III

Homeland Security

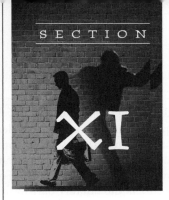

Homeland Security

Before and After 9/11

At the end of this section, students will be able to:

- Describe the major features of organizations involved in Homeland Security before 9/11.
- Discuss the thinking that led to the creation of Homeland Security and ultimately the Department of Homeland Security.
- What did the 9/11 Commission Report recommend?
- What does the new organization the Department of Homeland Security do?
- Discuss how successful databases for tracking track terrorists have been.
- Discuss how the bulk of funding for Homeland Security has been spent.
- Describe the shortcomings of current oversight of the intelligence community.

Introduction

The United States seems to develop a predictable response when it comes to defending itself. After assisting in the decisive defeat of Germany and its allies in World War I, the U.S. military quickly demobilized. Despite clear warning signs of another war in Europe, the United States, still recovering from the Great Depression, ignored the ominous storm clouds during the 1930s and did not begin to manufacture weapons, mobilize, and train armies. And war did come and once again the United States found itself in a world war, from 1941 to 1945. Following the defeat of Germany and Japan, the United States once again rapidly shrank the military only to find itself embarrassed in Korea in 1950 when the fledging Republic of South Korea was attacked. Only quick reinforcements, from the United Nations and allies, allowed the United States to fight to a draw, and a cease fire, which is still in effect.

A similar situation, a pendulum effect, occurs when it comes to dealing with terrorism. The United States adopted a cavalier approach in the 1990s to combating terrorism. Terrorism was random, occurred mostly overseas, was infrequent, and if it happened an agency would respond, depending on where and what kind of event transpired, an investigation would begin, prosecutions would follow, and justice would be served; the pendulum was all the way to the left. When a major attack occurred, such as the 9/11 attack, the pendulum swung all the way to the right; the United States attacked two countries and went to war, developed what many regarded as draconian laws that infringed on civil liberty, such as the Patriot Act and now, more than 13 years later, the United States seems to be swinging back to the 1990s approach to terrorism. To begin this section, a review of the 1990s homeland security posture is provided, but first, a prescient warning that was not heeded:

> To deter attack against the homeland in the 21st century, the United States requires a new triad of prevention, protection, and response. Failure to prevent mass-casualty attacks against the American homeland will jeopardize not only American lives but U.S. foreign policy writ large. It would undermine support for U.S. international leadership and for many of our personal freedoms, as well. Indeed, the abrupt undermining of U.S. power and prestige is the worst thing that could happen to the structure of global peace in the next quarter century and nothing is more likely to produce it than devastating attacks on American soil. (U.S. Commission on National Security/21st Century, 2001)

The chilling quote above comes from what is known as the Hart-Rudman Commission, taken from its final report released on February 15, 2001. Many are to blame for 9/11, but a good place to begin assigning blame is Congress. Congress knew things were not working well with the federal agencies sworn to protect Americans. As is often the case in Washington when faced with difficult problems, a commission is created that studies the issues and makes recommendations that then die. Congress asked the Hart-Rudman Commission to consider such things as security and intelligence for the coming decade and the commission did. Unfortunately, Congress failed to act. Congress also felt some pressure to do something about airline security, especially after TWA Flight 800 went down off the coast of Long Island in 1996, at first attributed to a bomb. Things were not particularly good when it came to airline security, and the Department of Transportation inspectors found they could breach airport security with impunity, routinely smuggling firearms, hand grenades, and bomb components through every security checkpoint at every airport visited (Yeoman & Hogan, 2002). Two commissions dealt with this and made sensible recommendations, but lobbyists, working closely with the Federal Aviation Administration, defeated every recommendation. One of the commissions was chaired by then-Vice President Al Gore but not even he could overcome the intense lobbying by the airline industry, including Linda Daschle, lobbying for the Air Transport Association; she was also the wife of then Senate majority leader Tom Daschle (Yeoman & Hogan, 2002).

Certainly President Bill Clinton also shares some blame. Following the embassy bombings in Africa in 1998, Sudan was facing sanctions for allowing Osama bin Laden to mastermind those attacks from Sudan. Lawrence

Wright, in his book *The Looming Tower*, asserts that the Sudanese minister of state for defense, Major General Elfatih Erwa, met with the American ambassador to Sudan, Timothy Carney, and offered to hand bin Laden over to the United States. His offer was declined (Wright, 2011).

This section will cover the concept of homeland security in detail. The notion of homeland security and the organization that currently exists, the **Department of Homeland Security (DHS),** was introduced in Section 1. This section will begin by describing homeland security before 9/11 and the important changes following 9/11. The current state of affairs with regard to homeland security will be presented next, along with updates on potential threats the United States is aware of. The United States was woefully unprepared prior to the attacks on 9/11. Clearly the United States has a Department of Defense and a strong military, which is sometimes used proactively, but its focus is external, not internal. In fact, there is a public law often called the Posse Comitatus Act that prohibits use of the military to assist with law enforcement in the United States (18 USC § 1385—Use of Army and Air Force as Posse Comitatus, n.d.). This section will focus on homeland security organizations other than the military, pre-9/11 and after 9/11, and emphasize the DHS and the intelligence agencies that assist it under the **Office of the Director of National Intelligence (ODNI).**

Michael Chertoff, the secretary for the Department of Homeland Security from 2004 to 2009, points out that before 9/11, using the 1990s approach, most of our concerns were with limited terrorist attacks in the United States, while most of the authorities' focus was on overseas bombings and airplane hijackings (Chertoff, 2011). These attacks, often on embassies or military personnel, were all "handled" by the existing criminal justice agencies, using the Foreign Intelligence Surveillance Act, Title III of the Omnibus Crime Control and Safe Streets Act, and related statutes and regulations. As covered in the previous section, strict, perhaps overly strict, interpretations of the rules for sharing information existed. For example:

> Exchange of information collected by foreign and domestic agencies was determined by a strict set of rules that was (perhaps somewhat incorrectly) interpreted as forbidding pure "intelligence" information from being collected for law enforcement purposes, and—conversely—made it difficult to share criminal justice-derived information with other agencies. When terrorists were apprehended either in the United States or abroad, they were accorded the treatment of any other criminal defendant, including receiving warnings about the right to silence, and a full-blown criminal jury trial. (Chertoff, 2011)

He also points out that it became abundantly clear after 9/11, especially after the exhaustive 9/11 Commission Report, how inadequate the architecture was for dealing with attacks and threats. Some of the shortcomings were obvious and needed quick action. Among them, the inability to coordinate and integrate information, with electronic surveillance governed by rules that did not foresee mobile phones and the Internet, and simply prosecuting offenders or terrorists would not do. As 9/11 demonstrated, the enormous devastation that modern terrorism produces meant a shift to interdiction and prevention.

For background, realize that the federal government had a limited role when it came to policing, especially before 1900, and after 1900 it never created any centralized police agencies, but spread law enforcement or policing functions among about 70 different federal agencies, many with narrow jurisdictions. The Tenth Amendment to the United States Constitution left "police powers" (law enforcement, order maintenance, and disaster relief) to the states and local governments (Travis & Langworthy, 2008). Under this model, states and local governments make up most of the police agencies and personnel in the United States. There are nearly 1.1 million local and state police personnel compared to 105,000 various types of federal police (Burns, 2013). This number does not include private police or security personnel, which, based on studies, place that number at over 4 million, and many of these are quite sophisticated. The federal government can assist only if a governor of a state requests assistance in an emergency or natural disaster beyond the capabilities of the state. A similar situation existed when it came to intelligence agencies, with the director of the Central Intelligence Agency (CIA) supposedly supervising the 16 other very parochial intelligence agencies, with little authority over them and their often very narrow responsibilities, such as the National Reconnaissance Agency (NRO). If an intelligence agency needs to see what is occurring in hostile territory or difficult terrain, the NRO will task one of the many satellites at its disposal to find out.

When a crisis developed requiring policing or more intelligence, the federal government would react and create an agency to deal with the problem or authorize an agency to create a police force, investigation arm, or an intelligence agency. This meant that the development of policing at the federal level, as David Johnson (1981) notes, was "haphazard," with federal agencies creating a police force only when it became abundantly clear one was needed. For example, the Justice Department asked Congress for authorization to hire investigators in 1907, but Congress refused. Ignoring Congress, Attorney General Charles Bonaparte created the Bureau of Investigation, President Theodore Roosevelt then ordered the transfer of eight Secret Service agents to the new bureau, and Bonaparte appointed William Burns, the head of a famous private security firm, to head what would soon become the Federal Bureau of Investigation (Travis & Langworthy, 2008). Similar situations happened when it came to the intelligence field. For example, following the Korean War, each branch of the U.S. military had an intelligence service that produced costly, duplicative, and often conflicting intelligence. The solution, of course, was to study the problem, under President Dwight Eisenhower, and then, under President John Kennedy, create the Defense Intelligence Agency (DIA) (Defense Intelligence Agency, n.d.). The DIA now employs approximately 16,500 military personnel and civilians engaged in producing foreign military intelligence (Defense Intelligence Agency, n.d.). Of course, all branches of the military still have their own intelligence organizations today.

The homeland security posture of the federal government before 9/11 was one of dispersed roles and responsibilities. For example, pre 9/11, the **Federal Emergency Management Agency (FEMA)** essentially was homeland security. FEMA was established by President Jimmy Carter in 1979 to consolidate over 100 programs and agencies addressing various types of national emergencies, and it had a decent reputation after a bumpy beginning (Carafano & Weitz, 2005). Under President Clinton, the FEMA director was elevated to Cabinet rank and the director, James Witt, hired experienced emergency management personnel from many of the states, streamlined procedures, and even managed a healthy increase in its budget. It was generally regarded as a successful agency at the time, became responsible for the Federal Response Plan, and was assigned the role of lead federal agency to respond to a terrorist attack (Carafano & Weitz, 2005). This would change, of course, when it later found itself as a small player in the giant Department of Homeland Security.

To best illustrate the problem of the pre-9/11 world at the federal level, no better example exists than that presented by Lawrence Wright in his book, *The Looming Tower*, covered in the previous section. The CIA, the FBI, and the National Security Agency all had information that if shared would have prevented 9/11. Before 9/11, federal agencies that were supposed to protect people were fragmented, and they did not coordinate or cooperate well with each other. In other words, at the federal level there would be a reaction that would succeed in dealing with a threat, after the fact, but there was simply no coordinated effort aimed at preventing a threat in the first place. The Oklahoma City bombing, for example, serves as another reminder of how a cooperative effort by local, state, and federal authorities yielded decisive and quick results reacting to terrorism in the pre-9/11 world.

The Oklahoma City bombing was primarily designed to destroy the Alfred P. Murrah Federal Building and it did a credible job. Since it was a federal building, President Clinton declared a national emergency as it was a property for which the federal government had "preeminent" or "primary responsibility" and this placed FEMA in charge of disaster relief, able to call on other federal agencies for assistance (Winthrop, 1997). While the FBI was investigating the case and quickly made arrests, FEMA appointed a Federal Coordinating Officer and activated seven of the possible 12 support functions, including communications, public works and engineering, information and planning, mass care, resource support, health and medical services, and urban search and rescue, drawing from the various federal agencies that had capabilities in these areas (Winthrop, 1997). To give perspective on the number of actual agencies involved in this, the timeline of events will be presented:

- Immediately following the blast, local and state search and rescue units and civilian volunteers entered the building
- The Oklahoma City Fire Department set up an Incident Command System to handle search and rescue and the massive influx of resources and took responsibility for search/rescue and recovery operations
- The Oklahoma City Police Department set up a perimeter and provided security
- The FBI began the criminal investigation
- The State Emergency Operations Center was up and running 23 minutes after the blast and included public safety, health, and the American Red Cross (ARC)
- President Clinton declared it a disaster, making ARC the lead agency for food, shelter, first aid, relief supplies, and welfare
- FEMA sent in 665 rescue personnel, primarily search and rescue personnel, from across the nation
- By 3:30 p.m. a compassion center was activated for the families of victims (Office for Victims of Crime, 2000)
- The FEMA director arrived late in the afternoon and a federal coordinator was appointed (The Oklahoma Department of Civil Emergency Management, n.d.)

This effort was successful, and while mistakes were made due to unforeseen acts, such as the massive swarm of reporters and journalists that overwhelmed Oklahoma City authorities, it was a quick investigation, search, rescue, and recovery, typical of a response in the pre-9/11 world.

On September 11, 2001, Sunni jihadists from al Qaeda attacked the United States, with two planes striking the World Trade Center in New York; another plane slammed into the Pentagon near Washington, D.C. not realizing the rules of the game had changed. Death awaited them and their kidnappers planned on becoming martyrs (9/11 Commission Report, 2004). A fourth plane, United Flight 93 out of Newark, New Jersey, was late in taking off. Some passengers and crew called friends and family on their cell phones when it became apparent they had been hijacked. From these calls they learned of the earlier hijackings and attacks on the World Trade Center and quickly concluded they must act, though unarmed. They rushed the cockpit and from voice recordings investigators now know the hijackers then crashed the plane onto a field in Pennsylvania, killing all on board (9/11 Commission Report, 2004). The nation was shocked and President George W. Bush moved quickly and aggressively responded. On September 17, just days after the attack on America, President Bush made plans for war in Afghanistan and Iraq (Kessler, 2003). He declared a "War on Terror" in a speech on September 20, while announcing at the same time a new Office of Homeland Security (Bush, 2001). This was followed by the passage by Congress of $40 billion in funds for rebuilding New York City and the Pentagon (Bush, 2001). The focus of the administration's efforts quickly became clear. The act would not be treated as a criminal attack and prosecuted, but would be the opening salvo in a war or wars fought to a conclusion (Owens, 2009). It would also be the beginning of a disastrous overreach by the federal government. What one scholar calls an "overreaction" produced bureaucracies that cannot be sustained, especially considering that the United States has "12,883 miles of coastline, legally admits 177 million foreigners each year and shares 5,550 miles of border with its top trading partner" and most of what DHS does is not homeland security (Friedman, 2011, p. 78). Of course, the same scholar also recognizes that when things like a 9/11 attack occur, a government must do "something" quickly to ease fears and he characterizes this as "security theater" and that is expensive, as there are bureaucracies that want to continue to stress the threat to increase their funding and personnel (Friedman, 2011).

Attorney General John Ashcroft, told by President Bush to never let this happen again, wanted a strict act on deterring and preventing terrorism by intercepting communications. Although he did not get everything, much of what he asked for was passed by Congress. The intrusive Patriot Act was passed on October 26, 2001, as the Uniting and Strengthening America by Providing Appropriate Tools Required to Intercept and Obstruct Terrorism (USA PATRIOT) Act (American Civil Liberties Union, n.d.). The Patriot Act was problematic for many, including a number of scholars who called it "draconian," as it included in Section 215 the ability to gather "any tangible thing" pertinent to a terrorism investigation, and Section 206 permitted roving wiretaps that allowed the government to obtain intelligence surveillance orders that identify neither the person nor the facility (American Civil Liberties Union, n.d.) (Owens, 2009). In addition it had a number of provisions dealing with financing of terrorism and a provision that allows the use of national security letters (NSLs) that grant the government the ability to obtain the communication, financial, and credit records of anyone deemed relevant to a terrorism investigation even if that person is not suspected of unlawful behavior (American Civil Liberties Union, n.d.). The Act has been amended several times; the last time was in 2011 and it was basically extended as is until 2015.

More actions followed. President Bush asked Congress to approve 30,000 airport security staff in the Aviation and Transportation Security Act and make them private contractors. Congress insisted they be federal employees under the new **Transportation Security Agency (TSA)**. One of the authors had a regional TSA head speak to some of his college students. This regional chief recounted how he was hired and soon arrived at a major airport to find that he was the only employee, there were no offices, and no one knew what to do with him. Congress also insisted on forming a cabinet-level Department of Homeland Security (DHS) versus the president's vision of an Office of Homeland Security and this was signed into law in 2002, creating the 15th cabinet-level department (advocated by Senator Joseph Lieberman following a recommendation from the Hart-Rudman Commission (U.S. Commission on National Security/21st Century, 2001; Friedman, 2011; Schimmel, 2012). As one report notes:

> The typical federal response to a crisis is to reorganize and most government reorganizations follow a classic pattern: various agencies, units and functions are merged into a new or expanded agency of department, with a Cabinet secretary, director or administrator vested with formal chain-of-command authority of a hierarchical organization. (The Partnership for Public Service, 2011, p. 7)

War came to Afghanistan on October 7, 2001, with the United States and the North American Treaty Organization (NATO) launching Operation Enduring Freedom (Taddeo, 2010). After intense lobbying of Congress by Bush administration personnel throughout much of 2002, the administration obtained a Congressional authorization for the use of military force in Iraq ("U.S. Congress, 'Authorization,'" 2002). Secretary of State Colin Powell then argued before the United Nations Security Council that Iraq was a hostile regime that might share its weapons of mass destruction (which almost everyone believed Iraq possessed, including almost all Democratic members of Congress) with terrorists, managed to get a similar United Nations resolution passed in the Security Council (see FIgure 13.2). War with Iraq began on March 19, 2003, with a massive missile strike on Baghdad followed by an invasion with 130,000 troops (Long, 2003).

With the foregoing for background, the stage is set for an examination of post-9/11 homeland security. After a brief period of time in which the Office of Homeland Security operated from the White House, the DHS is now a massive agency that by 2013 had mushroomed into an organization of gigantic proportions with more than 240,000 personnel (Department of Homeland Security, n.d.). The organization's history, challenges, successes, and failures will be covered. First, though, two definitions:

- *Antiterrorism:* passive measures designed to reduce the likelihood of terrorist attack. Activities related to antiterrorism are undertaken by the private sector, or government facilities not directly involved in the counterterrorism fight.
- *Counterterrorism:* those active measures undertaken by law enforcement, the intelligence community, militaries, and diplomats which are designed to hunt down and neutralize terrorist groups. This is almost always within the jurisdiction of governments.

Conducting threat assessments, planning, and employing security measures are passive actions to reduce the likelihood of an attack. Counterterrorism is proactive and the DHS employs a number of strategies designed to prevent terrorism (Johnson, 2013, p. 9).

The DHS began as a merger of 22 domestic and law enforcement agencies involving 180,000 employees with a mandate to increase internal defenses (The Partnership for Public Service, 2011). From the beginning it suffered from multiple issues, including inadequate planning (the agency was supposed to be up and running 60 days after the enactment of the law establishing it), disorganization, resistance from merged agencies, multiple unions, different payroll and procurement systems, and many of these problems are not completely solved today (The Partnership for Public Service, 2011). Part of the problem is its size; it is the third largest federal agency, with the Department of Defense the largest and the Department of Veterans Affairs the second largest. Eight percent of the federal work force works for the DHS, with a quarter of those working at DHS employed by the Transportation Security Agency (TSA) (Congressional Budget Office, 2012). Making things infinitely worse, and failing to follow recommendations of the 9/11 Commission, Congress opted for one primary oversight committee in the House and one in the Senate but allowed nearly 95 other committees and subcommittees to exercise jurisdiction and oversight as well, leading to excessive hearings that are often redundant and a distraction from the mission (Thompson, 2010).

An obvious example of how bad things were at DHS was the FEMA response to Hurricane Katrina. Director Ridge wanted to replace the head of FEMA and open a regional homeland security office in New Orleans prior to Katrina but was overruled by the White House. The blundering response to Hurricane Katrina was no doubt not a surprise to DHS; FEMA was not clear on what its mission was at the time of Katrina in an agency focused on terrorism, and when Katrina struck New Orleans, one of four positions was vacant at the agency (The Partnership for Public Service, 2011). Other problems soon arose. Various agencies now in the DHS had "regions" across the country. None of the "regions" of one agency matched those of another. Still other problems required immediate attention, such as having to decide how to deal with 15 basic and special pay systems, 10 different hiring methods, eight overtime pay rates, seven different payroll and benefit systems, 19 different performance management systems, and 17 unions (The Partnership for Public Service, 2011). Following the brief stay of Director Ridge, barely two years, the DHS was headed by Michael Chertoff for four years and in 2009 Janet Napolitano became secretary, leaving in 2013; her successor is Secretary Jeh Johnson. A report by the Partnership for Public Service notes that chaos reigned throughout the Ridge years at DHS, Chertoff was able to impose discipline and reorganize DHS during his tenure, and there were high attrition rates at DHS, TSA, and FEMA (The Partnership for Public Service, 2011).

Today the DHS is heavily criticized and considering the amount of funds it has received with dubious results it may well deserve it. To date, the authors are not aware of any terrorist threats deterred by the DHS. Certainly, Secretary Napolitano claimed "the system worked" after the foiled "underwear bomber" attempt on Christmas 2009, but the bomber was caught by alert passengers, not the DHS (Applebaum, 2011). Yes, there have been attempts at attacking the United States but these have been detected and prevented by others, such as the **National Security Agency (NSA)** (King, 2013). The vast majority of funding for DHS, now at $60 billion a year, has gone to pay the salaries of employees, focus on aviation security, with little proof of effectiveness, and deter attacks on major sporting events, such as the National Football League's Super Bowl, with some emphasis on rail and port security. Secretary Napolitano resigned in 2013 to assume a post running the University of California system. At the time there were 15 vacant positions at the top of DHS (Kenny, 2013). DHS and especially the TSA and FEMA routinely are among the worst employers or agencies to work for when their employees are surveyed. In 2011, DHS was ranked 31 out of 33 large agencies and FEMA ranked 231 out of 240, with TSA near the bottom, 232 out of 240 agencies, even worse (Stier, 2012). DHS has five major missions. Each one will be presented, along with the actions by the agency to accomplish the tasks associated with the mission and an evaluation of the current situation.

The first mission of the DHS is preventing terrorism and enhancing security. DHS does this by the use of fusion centers, 77 altogether (DHS, n.d.). These fusion centers are supposed to bring together various law enforcement and intelligence personnel and are designed to produce useful intelligence. However, a bipartisan 2012 report by the Senate Homeland Security and Governmental Affairs Permanent Subcommittee on Investigations found that they produce intelligence of uneven quality, some of it shoddy, and have no useful intelligence despite spending between $289 million and $1.4 billion since 2003 (note they do not even know the specific amount spent) (Clark, 2012b).

The DHS supports this mission by grant funding to various cities and urban areas. However, this has been heavily criticized as well. In a decade DHS has spent nearly $35 billion on various grants to prevent terrorism, with few appreciable results. In a 2012 report by Tom Coburn, a Republican senator from Oklahoma, that some consider disturbing, one finds such gems as reports that Fargo, North Dakota, received more than $8 million in homeland security grants. Fargo, North Dakota, has averaged fewer than 2 homicides a year since 2005 but believed it needed a "new $256,643 armored truck, complete with a rotating [gun] turret" using homeland security funds among other items (Coburn, 2012). Other outrages are:

> The decision by officials in Michigan to purchase 13 sno-cone machines and the $45 million that was spent by officials in Cook County, Illinois on a failed video surveillance network have already garnered national attention as examples of dubious spending. Both were defended or promoted by DHS. (Coburn, 2012, p. 4)

The report also notes that the DHS Inspector General in 2012 stated that FEMA had no system in place to monitor grants to determine whether they were spent on anything related to preventing terrorism (Coburn, 2012). Troubling as well is the militarization of police. When grants funding community-oriented policing dried up and were replaced by grants emphasizing homeland security, police responded by requesting aid, particularly when it came to establishing and arming Special Weapons and Tactics (SWAT) teams and obtaining armored personnel carriers (Balko, 2013). Now everyone has a SWAT team, including the Department of Education and by 2005, 80% of cities with a population between 25,000 and 50,000 have SWAT teams, many armed by DHS grants (Balko, 2013). In addition, one has to wonder why the DHS has ordered 1.6 billion rounds of ammunition, enough ammo for the U.S. Army to wage war for 20 years, and no doubt contributed to the ammo shortage (Benko, 2013). Congress needs to take a serious look at the DHS and reform it or end it. In an article calling for abolishing the DHS, Charles Kenny (2013) notes that the odds of an American being killed in a terrorist attack here or abroad is 1 in 20 million, and DHS is simply not worth it.

What about the TSA? It in particular deserves praise, scorn, and criticism. It has spent nearly $60 billion thus far and on the positive side it has accomplished the following:

- It finally, after mistakes, has acceptable explosive detection technology for baggage screening
- In 2013, the TSA had less revealing and more effective full body scanners
- Screener training has improved
- Aircraft now have reinforced cockpit doors
- Pilots are armed
- It is now testing an expedited screening program after several earlier models proved unsuccessful, and maintains a no-fly list and another called "selects" for those requiring extra screening, but this is still a troubled effort. The dead Boston Bomber, Tamerian Tsarnaev, was on a watch list and not detected—another mistake.
- It now matches all travelers with a Secure Flight database after earlier failures
- TSA now screens all air cargo domestically and was working on international cargo, as of 2013, though it took almost 11 years to do that

TSA is working on a risk-based approach to screening, meaning that if one is willing to provide the TSA a great deal of information, such as date of birth, birthplace, employer, overseas travel, passport number, agree to a background check, be finger printed, be interviewed, and pay a fee, the passenger would get expedited screening. Additionally, TSA is working on implementing new protocols for children, the elderly, and wounded veterans (Clayton, 2013; Ott, 2011).

TSA, nevertheless, has critics. A congressional joint study by the House Committees on Transportation and Infrastructure and Oversight and Government Reform in 2011 concluded that flying is no safer than it was before 9/11, TSA has grown to over 65,000 personnel (a 400% increase), and has procured contracts to hire and train more than 137,000 staff ($17,500 per hire) (Kenny, J. , 2012). Senator Rand Paul, a Republican from Kentucky and not a fan of TSA, attempted to get rid of it in 2012 through legislation but the bill died in committee (Paul, 2012). For further evidence that TSA could do a better job, in August 2012 a stranded jet-skier climbed across a fence at New York's John F. Kennedy International Airport, walked past a $100 million system of alarms, motion detectors, and closed circuit television cameras, and managed to get into Delta Airlines Terminal Three where an employee wondered why he was wearing a wetsuit (Sweetman, Mackenzie, & Eshel, 2012). And finally, despite supposedly tough economic times, the TSA also ordered new uniforms for all, to the tune of $50 million in February 2013 (Harrington, 2013)

TSA is not only responsible for aviation security, but for all other transportation security, though aviation has been the primary focus. When it comes to port security (maritime security) and rail security, Congress did not specify in much detail how this was to be accomplished. The rail industry has worked with TSA to meet its standards and the TSA audits rail systems periodically. Recently 16 mass transit and rail agencies were commended for highest security levels (Federal Transit Administration, 2013).

Regarding port security, TSA has proposed and implemented a number of steps, including a Transportation Workers Identification Credential (TWIC) for port and vessel transportation workers that costs the individual $129 and is good for five years (Sadler, 2013). Maritime security is also carried out by employing Maritime Domain Awareness, meaning the U.S. Coast Guard (USCG) and Customs and Border Protection (CBP) (DHS entities) and the National Oceanographic and Atmospheric Administration (a Department of Commerce agency) obtain information on vessels, ownership, the maritime cargo supply, and the marine environment. Vessel surveillance relies on an automatic identification system that provides real-time data on vessels, their identity, speed, route, tonnage, and destination (Altiok, 2009). The CBP requires 24-hour notice by means of a sea automated manifest system for cargo container bound for the United States before it is loaded and 24-hour notice prior to arrival for bulk materials (Altiok, 2009). The Container Security Initiative was supposed to ensure that eventually all cargo would be scanned by radiation technology or nonintrusive imaging technology but thus far only about 5% of containers are screened (Wolf, 2013). Why is this concerning? It is not unreasonable to expect a foreign state or terrorist group working with a foreign state to attempt to ship and detonate a nuclear weapon in a U.S. port with disastrous results, especially economic. In a *New York Times* editorial written by three members of Congress, they expressed outrage at the failure of DHS:

> In June of 2012, several members of Congress expressed outrage at the failure of DHS to meet the deadline for 100 per cent screening and cited a RAND report that such an attack could cost between $45 billion and $1 trillion dollars in global trade losses as well as tens of thousands of deaths. (Nadler, Markey, & Thompson, 2012)

DHS is also responsible for securing our borders and for immigration. Again, this is an area that DHS has difficulty dealing with in any meaningful way because Congress fails to legislate solutions. DHS has no idea how many people are in the country illegally, whether from entering illegally or staying on an expired visa. Despite some discussions following 9/11, there is no national identification card. When it comes to border security, Congress and the Government Accountability Office have been asking for years how secure the borders are and get poor or changing answers. An editorial in the *Arizona Republic* in July of 2013 stated that DHS "doesn't know or doesn't want us know how effective its border-security efforts are" using a metric of "number of apprehensions" to say they are doing an effective job (Editorial Board, 2013). That is, as a Government Accountability Office report noted in 2013, a measure of activity, not program results (GAO Report, 2013).

The next major mission area the DHS is responsible for is safeguarding and securing cyberspace. DHS efforts in this area are only beginning, and much of this is not its fault, but that of Congress, where legislation on **cybersecurity** has stalled due to privacy and liability concerns. The situation is complex when it comes to cybersecurity and the federal government. Recently the Office of Management and Budget, by federal law responsible for oversight of government information systems, transferred several of these responsibilities to the DHS according to a Government Accountability Office (GAO) report (Government Accountability Office, 2013). In the same report the GAO notes there is not an integrated national strategy when it comes to cybersecurity and recommends there be an integrated strategy that includes milestones, performance measures, and a clear definition of roles and responsibilities; it further notes that DHS has not developed a capability for predictive analysis of cyber threats (Government Accountability Office, 2013). So where is the DHS when it comes to cybersecurity? DHS is currently working on this following an Executive Order signed by President Obama in February of 2013.

The Executive Order calls for the DHS to lead a voluntary public/private partnership on cybersecurity (DHS is also responsible for securing all of the .gov domains). Specifically it demands the DHS "provide strategic guidance, promote a national unity of effort and coordinate the overall federal effort to promote the security and resilience of the nation's critical infrastructure" (Obama, 2013). It also tasks the National Institute of Standards and Technology to develop standards, methodologies, procedures, and processes to enable uniform cybersecurity. The idea behind this is that vital private and public organizations should all receive classified threat information on cyber-crime and at the same time would agree to share information about cyber-attacks on their organizations. Making this all the more difficult is the fact that the U.S. military now has a U.S. Cyber Command whose mission seems to duplicate the DHS mission (Sternstein, 2013). Prior to the order on cybersecurity, in Presidential Policy Directive 20, a classified document that was signed in 2012, President Obama authorizes cybersecurity attacks, and the military has been drafting cyber rules of engagement while the Defense Advanced Research Project Agency has been working on cyber warfare tools. Congress, of course, is unhappy (Schwartz, 2012). One member complained:

> "No executive order can possibly do what needs to be done to protect our networks and our nation. It also cannot take the place of legislation. Strengthening cybersecurity must be collaborative and bipartisan."
> Rep. Mac Thornberry, R-Texas, vice chairman of the House Armed Services Committee. (Sternstein, 2013)

The DHS has now established an Integrated Task Force to figure it all out.

The final mission of the DHS is that carried out largely by FEMA, which remains a troubled agency, as noted earlier, with poor rankings when it comes to being a desirable place to work. FEMA has improved since its inadequate response to damage by Hurricane Katrina. The DHS and FEMA have operated under a National Preparedness Goal that resulted in a National Preparedness System and a National Preparedness Report to uniformly assess risks and threats and to prepare responses. FEMA has robust grant and training programs, and encourages exercises on a regular basis. Noteworthy is the effort made by FEMA to beef up Urban Search and Rescue Teams. Today there are more than 300 of these nationwide, only 55% of which even existed before 9/11; FEMA's encouragement and the many grant programs have been useful (Manning, 2013). Clearly the response to Hurricane Sandy in 2012 was much better, considering the size of the storm, in which an estimated 375,000 homes were damaged or destroyed. FEMA sent Incident Management Assistance Teams to three states even before the storm, 68 disaster recovery centers were set up, and FEMA passed out a million liters of water and ready-to-eat meals (Manuel, 2013). Ninety-seven people died as a result of Hurricane Sandy, and there were $50 billion in damages; debris removal continued well into 2013.

Before summarizing and commenting on the DHS, the authors offer some insight for football fans that most are undoubtedly not aware of. The DHS protects fans with unprecedented security measures, especially those attending the NFL's Super Bowl. The first Super Bowl after 9/11 was designated a **National Security Special Event**, meaning that the Secret Service, a DHS agency, assumed the role of mandated lead agency (Schimmel, 2012). This placed the assets of the U.S. government at the disposal of the Secret Service, ensuring security. The Super Bowl also becomes a placed where other federal agencies, with Secret Service approval, can employ their new technologies, such as testing facial recognition software, for example, or in the case of the U.S. Immigration and Customs Enforcement, screening 30,000 applications from volunteers wanting to work at the Super Bowl, to determine if they are in the country illegally (Schimmel, 2012).

In summary, the DHS and many of its organizations have major issues and need reform and better oversight. Some believe Congress should close the DHS down. Much of this issue rests with Congress failing to police itself and to provide a single, consistent point of guidance from one committee in the House of Representatives and one in the Senate. Absent that, DHS and its agencies will continue to flounder, grow, and engage in mission creep, meaning doing things that hardly seem part of the mission. For example, the TSA has now developed a program called Visible Intermodal Prevention and Response squads or VIPR intended to provide security screening at train stations, national political conventions, or sporting events, such as the Indianapolis 500 racing event (Nixon, 2013). There is some question whether this is even legal. Finally, we should mention SPOT and FAST (not your dogs). This first refers to a TSA program called Screening Passengers by Observation Technique (SPOT). This is a behavioral profiling program used by TSA. Does it work? From January of 2006 to November of 2009, approximately 232,000 people were identified using this technique and were given additional screening. The result was a total of 1,710 arrests, most for outstanding criminal warrants (Weinburger, 2010). Future Attribute Screening Technique or FAST is the next attempt. It employs barely scientific means that attempt to measure the body temperature, blink rate, and body movements of people walking through a screening machine and, based on those, provide a measure of malintent. It is clearly not ready for prime-time and like SPOT, likely a waste of money, much like the DHS (Weinburger, 2010).

The other part of protecting the homeland is the intelligence community. Prior to 9/11, the intelligence community was "stove piped," meaning that it was an agency that worked for a particular military service and/or had a very specific charter or purpose. The head of the Central Intelligence Agency did serve as the primary adviser on intelligence to the president, supposedly coordinating the other intelligence agencies and their budget submissions, but really had very little influence on the other agencies, which are:

Air Force Intelligence	Drug Enforcement Administration
Army Intelligence	Federal Bureau of Investigation
Central Intelligence Agency	Marine Corps Intelligence
Coast Guard Intelligence	National Geospatial-Intelligence Agency
Defense Intelligence Agency	National Reconnaissance Office
Department of Energy	National Security Agency
Department of State	Navy Intelligence
Department of the Treasury	

Pre-9/11, most intelligence agencies collected information or intelligence by many different means and then analyzed it. The means included human intelligence, signals, imagery, measures and signals intelligence, and open source intelligence (Federal Bureau of Investigation, n.d.). Most intelligence agencies were created because, again, Congress or a governmental agency decided it needed something. For example, the National Security Agency was created within the Department of Defense in 1952 to handle the increasingly demanding need for unifying signal intelligence among the military intelligence services (National Security Agency, n.d.). The 9/11

Commission spent a great deal of time analyzing the intelligence community in its lengthy report. Congress recognized, drawing on the Hart-Rudman recommendations and those from the 9/11 Commission, a need to reform the intelligence community to better advise the president, the National Security Council, and the Homeland Security Council. The Intelligence Reform and Terrorism Prevention Act (IRTPA) of 2004 did not provide for the same type of hierarchical organization as DHS. It created the Office of the Director of National Intelligence (ODNI) "to oversee and help coordinate the work of the 16 separate intelligence agencies that traditionally operated independently and did not share information" (The Partnership for Public Service, 2011, p. 19). It did provide for a **director of national intelligence (DNI),** who now serves as the president's chief intelligence officer, replacing the head of the Central Intelligence Agency as the president's chief intelligence adviser. Executive Order 12333 was also changed to state that the intelligence community falls under the leadership of the ODNI (Rettig, 2013). However, the law was ambiguous, giving the DNI oversight responsibility for getting all of the intelligence agencies to begin cooperating and working together, but not the authority to do so (The Partnership for Public Service, 2011). The law also created a **National Counterterrorism Center** to integrate and analyze threats, which does at least get personnel from many different agencies together in one location, including many outside the intelligence community, though they seem to have missed the Christmas Bomber of 2009 and the Boston Bombers (Clark, 2012b).

The ODNI began with 11 employees and rapidly expanded to around 1,750 and frankly there have been problems with the structure, with four different individuals in charge of the organization in the short period it has existed, not counting an acting director (Clar, 2012b). The intelligence community has been beset by leaks, forcing the current DNI, James Clapper, to add a question on polygraph exams about unauthorized disclosure of classified information (Clark, 2012b). Although the ODNI needs structural and organizational changes, some things have been successful. ODNI implemented a joint duty concept where senior executives and senior professionals would be required to spend some time working at different intelligence agencies, hoping to change the culture, reduce insular behavior, and encourage collaboration and cooperation (The Partnership for Public Service, 2011). Too, the current DNI, in 2010, managed to wrest some of the national intelligence budget from the Pentagon, $53 billion beginning in 2013 (Nakashima, 2010). The ODNI now runs the intelligence community using an approach intended to reduce redundancies among the various intelligence agencies by using 17 national intelligence managers designated by a region or domain, such as cybersecurity or finance (Clark, 2012b).

Clearly more needs to be done; President Obama called the DNI job "the most thankless job in Washington" (Clark, 2012b). Despite the legislation and changes to the executive order, the DNI has only partial control of budgets, is restricted to creating and disseminating integrated intelligence products, and lacks executive authority (Rettig, 2013). The ODNI and the DNI have no direct authority over any of the intelligence agencies, and many intelligence agencies work for cabinet-level officials, such as the Defense Intelligence Agency (Rettig, 2013). What must be done is to elevate the DNI to a cabinet level, grant the position specific budgetary and management power, and integrate foreign and domestic intelligence, something that has still not been done (Partnership for Public Service, 2011; Rettig, 2013).

The United States requires first-rate intelligence, but at the moment is not getting it. The director of the CIA, David Petraeus, resigned in 2012 after the FBI informed DNI James Clapper it was investigating Petraeus for having an extramarital affair; Clapper called Petraeus and suggested he resign (Entous & Gorman, 2012). Clapper was criticized for poor intelligence that resulted in the deaths of four individuals in Benghazi, Libya, on September 11, 2012, an event not fully explained and then covered up by President Obama, Ambassador to the United Nations Susan Rice, and Secretary of State Hillary Clinton, who all blamed it on a film trailer (Hanson, 2013). Clapper also lied to Congress; when asked directly during a Senate hearing if the NSA collected data on Americans, he replied, "No sir" (Miller, 2013). It turns out the NSA does and it is a massive effort collecting phone records, videos, e-mails, and documents, and the American Civil Liberties Union is suing (Nakashima & Wilson, 2013).

Summary

The United States needs some homeland security and robust intelligence agencies. With the DHS, Congress most likely went overboard and created a monster agency that will take years to manage properly and clearly much of what it does is not even related to homeland security. Congress still exercises excessive oversight with nearly one hundred committees often providing conflicting guidance and priorities. This must be addressed but Congress is the body that must do it and at the time of this writing, there is gridlock and no plan to fix it.

Congress did not go far enough when it came to dealing effectively with changing the intelligence community. The efforts were tepid and ineffectual, and the result has placed the ODNI and DNI in the nearly impossible position of having responsibility with little true authority. Peter Drucker, the now deceased management guru, once said that some jobs are "widow makers," meaning some jobs are so inherently difficult they will defeat the best and brightest (Wartzman, 2008). As structured, even with a supportive president, the ODNI and DNI will struggle.

Key Points

- Congress no doubt went overboard creating such a large agency as the DHS. It does many things that have little relationship to Homeland Security.
- Congressional oversight, using more than 100 committees and subcommittees, is unworkable.
- TSA has a number of critics but efforts to kill it or diminish it have so far failed.
- The reforms in the intelligence community did not go far enough and may have even made things worse.

KEY TERMS

Cybersecurity

Department of Homeland Security (DHS)

Director of national intelligence (DNI)

Federal Emergency Management Agency (FEMA)

National Counterterrorism Center

National Security Agency (NSA)

National Security Special Event

Office of the Director of National Intelligence (ODNI)

Transportation Security Agency (TSA)

DISCUSSION QUESTIONS

1. Was 9/11 preventable? Could it happen today? Why or why not?

2. The United States seems to be returning to the past, treating terrorism as a manageable criminal problem. Is that appropriate? Yes or no? Support your answer, considering the odds of dying in a terrorist attack are 1 in 20 million.

3. Should the NSA, with approval from a court, be able to keep every telephone conversation you have had in the last year, as well as every text message sent and received?

4. Should there even be a DHS?

5. How would you resolve the issue of the DNI lacking authority over the intelligence community?

WEB RESOURCES

Central Intelligence Agency: https://www.cia.gov/index.html

FEMA has a number of courses that anyone can complete for free. You learn something, get a certificate, and it goes on your résumé: https://training.fema.gov/IS/crslist.aspx

National Counter Terrorism Center: http://www.nctc.gov/

Various intelligence agencies: http://www.dni.gov/index.php/intelligence-community/members-of-the-ic; http://www.fbi.gov/about-us/intelligence/disciplines

READING 19

The following journal article is presented in its entirety. It is a look at the relationship between the various homeland security entities, the National Football League (NFL), contractors, and technologies that are emerging. Read the article and answer the questions at the end of the article.

Protecting the NFL/Militarizing the Homeland

Citizen Soldiers and Urban Resilience in Post-9/11 America

Kimberly S. Schimmel
Kent State University, USA

Every citizen must be a soldier. This was the case with the Greeks and the Romans, and must be that of every free state. (Thomas Jefferson, Third President of the United States)

In a provocative post-9/11 blog post, legal scholar Anthony Sebok (2003) proposes the term 'citizen soldier' to explain the symbolic ramifications of the Support Antiterrorism by Fostering Effective Technologies (or SAFETY) Act, passed by the US Congress in 2002 to ensure that lack of insurance and concerns over liability did not impede the development of anti-terrorism technologies. Arguing that the new legal protections offered civilian companies creating 'qualified anti-terrorism technology' effectively turns the US into a battlefield, civilian companies into government defense contractors, and all US citizens into soldiers[1] by implication, Sebok suggests a troubling new conflation between military and civilian discourses and practices. While the concept of the citizen soldier is not itself new, as the opening quote acknowledges, its (re)emergence in the post-9/11 landscape hints at a different set of obligations and sacrifices expected of those occupying the role—that is, of all of us.[2]

In this article I examine emergent security efforts of the National Football League (NFL), its stadiums and its fans, suggesting that these efforts—which are encouraged and enforced by the US government and US military— render the NFL a uniquely militarized sport association and the annual Super Bowl game a uniquely militarized sport mega-event. This approach extends an ongoing research inquiry into urbanization and militarization and builds directly upon two prior studies:[3] the first (Schimmel, 2006) linking stadium development and urban growth ideology with the militarization of US urban society; and the second (Schimmel, 2011) examining US news media (print and online) discourses of stadium and event security at the Super Bowl. In the present article, I focus on intersections between the NFL's security practices and the US Department of Homeland Security's (DHS) counter-terrorism agenda, including new policies and legal structures that support pre-emption, protection, and preparation activities that manage and mitigate the effects of terrorist attacks. As I will argue, the intensifying and mutually supportive relationship between the NFL and the various forces employed to keep it safe is both unique in US sports and implies the citizen-soldiering of football fans in

SOURCE: *International Review for the Sociology of Sport* 47(3) 338–357. © The Author(s) 2012

somewhat unexpected ways. Coaffee's recent work on urban resilience in the UK (Coaffee, 2009; Coaffee and Wood, 2006) contends that in the present historical moment, 'security is becoming more civic, urban, domestic, and personal' (Coaffee, 2009: 9). How, if at all, does resilience play out in the context of anti-terrorism policy in post-9/11 America? How, if at all, does US professional football articulate with that policy? I elaborate on the notion of urban resilience below and also address the question, why the NFL?

Urban Resilience

In his 2003 book *Terrorism, Risk and the City: The Making of a Contemporary Urban Landscape*, Jon Coaffee analyzes governmental security strategy in the UK, focusing on the decision in the 1970s to create a heavily fortified 'ring of steel' around parts of Belfast and London to keep them safe from terrorist attacks. In an updated version of the book in 2009, now centered on the global city and subtitled *Towards Urban Resilience*, Coaffee expands his analytic scope to include the changes wrought by the 9/11 attacks in both the UK and (to lesser extent) the US. Noting that 11 September made the rings of steel approach to 'counter-terrorist tactics appear inadequate' and that security policy began to shift to more anticipatory and preparatory measures (2009: 84), Coaffee engages with and problematizes the gradually emergent discourse of urban resilience (or what Medd and Marvin [2005] term the governance of preparedness) in global anti-terrorist strategies and policies. He notes that while the term resilience first emerged in research on the stressors faced by ecological systems, it has been applied over the past decade to a broad range of phenomena, including human social systems, economic recovery, disaster planning, and now security (Coaffee, 2009).

Defined by the UK Resilience website as 'the ability to detect, prevent and if necessary handle disruptive challenges [including] challenges arising from the possibility of a terrorist attack' (Coaffee, 2006: 504), a resilience approach translates to anti-terrorist policies that are 'increasingly anticipatory and pre-emptive as preparation for the inevitable attack' (Coaffee, 2009: 9). The specific trends that accelerated post-9/11 and are positioned by Coaffee and Wood (2006) within an urban resilience framework include:

1) the growth of electronic surveillance within public and semi-public urban spaces, in particular automated software-driven systems. (2006: 507);

2) the increased popularity of physical or symbolic notions of the boundary and territorial closure. (e.g. gated communities, defended airports etc.; 2006: 508);

3) the increasing sophistication and cost of security and contingency planning undertaken by organizations and different levels of government. (2006: 508); and

4) the linking of resilience and security strategies to competition for footloose global capital. (2006: 58)

Coaffee and Wood conclude by suggesting that the concept of security has 'come home', or has been 'reterritorialized' such that the influence of national and international security discourses and procedures is effectively scaled down to urban communities and ordinary citizens. Contradictorily, however, this reterritorialization co-exists in government policies with the positioning of terrorist threats as 'everywhere and nowhere' (Coaffee and Wood, 2006: 514), which risks generating a permanently fearful—and thus more easily controllable—populace (Coaffee, 2009).

In this perspective of the city as 'battlespace' (see Graham, 2009), what is the role of sport in general and football in particular? I have argued elsewhere (Schimmel, 2006, 2011) that the ongoing militarization of urban space, hastened by the 9/11 attacks, is made visible in specific sport practices, discourses and policies and seems more palatable to Americans in a sport landscape than in other seemingly 'terrorist-ready' contexts (see also Warren, 2004). In part, sport represents an example of the 'scaled-down' or more 'domestic' and 'personal' contexts (to borrow Coaffee's terminology) in which national and international security strategies play out in the 21st century. As such, the four trends noted above can be traced in the specific policies and practices regarding NFL security. However, my aim in this article is not simply to apply Coaffee's broad arguments about urban resilience to the case of US football. Rather, I will argue that the security strategies of the US government and those of the NFL are mutually beneficial in ways that help secure the NFL's position and profitability and, more importantly, help implement and secure consensus for the US Department of Homeland Security's 'war on terror'. This relationship is intentional, is intensifying, is mutually supportive, and was institutionalized in 2009 through the National Infrastructure Protection Plan and the SAFETY Act noted in my opening paragraph.

In the following section I briefly describe the National Football League and its premier event (the Super Bowl) in its contemporary urban context; this material will be familiar to some readers but not others. I then detail anti-terrorist strategies undertaken by the NFL post-9/11 and explore the NFL's 'exceptionalism' in the context of urban resilience policy in the US. I conclude by a return to the notion of the citizen soldier and her/his enlistment in the US war on terrorism.

The NFI and the Super Bowl in Urban Context

From the first 'paid' football players in Pittsburgh, Pennsylvania in the 1890s, through its early growth phase in the single-industry towns of Ohio, to the placement of contemporary franchises, professional football has always been an urban game. The NFL was founded in 1920 in Canton, Ohio, a city located in the middle of Ohio's fiercest football rivalries and home of football's best team at the time (profootballhof.com, 2011). The enduring success of the League is due to cooperation among team owners and their adoption of the business model established in the 1870s by Major League Baseball. The model contained three elements that became taken-for-granted features of the American professional sports industry: cartelization, monopoly, and monopsony. These business practices have established Major League Baseball, The National Football League, the National Hockey League and the National Basketball Association as some of the most powerful firms in the history of the United States.

Cartelization, which is illegal in the US in businesses other than sports,[4] refers to the general practice of owners of business firms acting together to make decisions about the production and distribution of their products (in this case sports). Sport franchise owners have a remarkably complex set of rules designed to restrict business competition for athletic labor and divide geographical markets for individual franchises. Even though each team is a separate business entity, the owners look out for the collective interest of the league. For example, all leagues have rules for revenue sharing among teams, and owners vote on the placement, ownership, and number of franchises in the league. In addition, television and radio broadcasts, admission to games, and the sale of team-related merchandise are all subject to league regulations. In short, monopoly practices manipulate the distribution of professional sport to consumers. Monsopsony practices, on the other hand, manipulate the cost of acquiring sport labor. For example, league rules specify the procedures for drafting new players and binding them to contracts, thereby assuring bidding wars will not break out for athletic talent. Eventually, other sport owners for exactly the same reason embraced cartelization and economic concentration as did Major League Baseball: the creation of artificial scarcity in the marketplace increased consumer demand. Teams literally could not survive outside of a prestige league; they either merged into it or folded. For example, in the 1920s football had two leagues and 58 teams, but by the 1950s mergers, failures, and exclusions reduced the number of leagues to one and the NFL had control of the entire US market.[5]

Currently the NFL consists of 32 franchises (all except the Green Bay Packers are privately owned) located in 31 different cities across the US. The NFL has placed franchises in 24 of the 30 largest 'designated market areas' and in every region of the country (citydata.com). Because of the limits imposed by the League's business practices, however, there are more cities that want to host a team and that have the population base to support it than there are teams to go around. The NFL's artificial scarcity, combined with a history of franchise relocation encourages bidding wars between cities wishing to obtain or retain teams. Success or failure at these efforts is seen to symbolize a city's emerging or declining urban status and business climate. The NFL is thus linked to the larger growth discourse that accompanies the US model of urban development, where a fragmented and pluralistic urban system, market-oriented politics, and a deep commitment to localism are, in fact, the national urban policy. Here in the 'capital of capitalism' (see Kantor, 2010), fierce competition for capital investment and jobs produces winner and loser cities (and sections of cities) and glaring social inequities.

The NFL's urban linkage goes far beyond the symbolic, however, to include material dimensions such as the reconstitution of urban space and the use of public funds for the purpose of sport-related infrastructure development and mega-projects such as stadia. NFL stadia are large (ranging from 63,000 to 92,000 in seating capacity), the vast majority are publicly owned, and they are expensive (all are taxpayer supported). New stadium construction becomes a central feature of the bidding wars for franchises because it is usually necessary if a city is to retain its 'home' franchise amidst owner's threats to leave, or is to entice one in another city to relocate. Also, in some cities, new stadiums are necessary before those cities can host the NFL's premiere event, the Super Bowl championship game (discussed below). The public's return on investment is touted to include numerous material benefits, including employment growth and revenue creation that address social inequities and benefit all city residents. While over two decades of social science research refutes these claims the discourse connecting sport-related infrastructure development to urban growth and regeneration remains dominant and public investment in major sport facility construction continues a space. Between 1990 and 2009, over US$22 billion of public money was spent to build sport stadiums and arenas and to subsides real estate developments in the immediate areas surrounding them (Brown et al., 2004; Coakley, 2009; Delaney and Eckstein, 2007).

The NFL's Super Bowl is the most watched, single-game sporting event in the US. Played annually since 1967, it has become one of the most highly watched television shows of the year, with Super Bowl 2010 drawing the largest TV audience in US history (Newyorktimes.com, 2010). Competition related to Super Bowl occurs on at least three levels: franchises compete to qualify for the game; television networks compete for the rights to broadcast it; and cities compete for the opportunity to host it. Unlike championship games and series in other US professional sports in which the championship game is located in the cities whose teams qualify, NFL owners 'award' their championship game to a city with a 'Super Bowl ready' stadium based upon a competitive bidding process.

The Super Bowl is preceded by two weeks of massive media hype that includes not only analysis of players and teams, but also special features about the sport of football more generally, and about the city and stadium that will host the event. More spectacle than mere game (see for example, Butterworth, 2008; Falcous and Silk, 2005; Real, 1974), the Super Bowl build-up involves numerous NFL entertainment venues and a 600,000 square foot theme

park called the 'NFL Experience', leading up to the production of 'Super Bowl Sunday' with its pre-game and halftime shows created to dramatize the connection between the National Football League and the 'American' way of life. Patriotic themes and messages are an established Super Bowl tradition, as is the involvement of the US military in Super Bowl rituals. From fighter jet 'flyovers', to the presentation of the US flag by military Color Guard, the marching of military bands, the singing of military choirs, and 'live look ins' to US military bases at home and abroad, the NFL and the US military have shared more than 40 years of Super Bowl history (Carden, 2009).

Since the terrorist attacks in the United States on 11 September 2001, however, military representation at—and involvement in—the Super Bowl has taken on new dimensions.[6] New relationships are emerging, and established relationships are intensifying, between the US's premier sport event, the NFL that controls it, and the various forces employed to keep it safe. Super Bowl stadiums, as well as symbolizing a city's urban status and late-capitalist generation 'successes', are now also positioned as 'terrorist targets' in need of protection. New stadiums are thus developed based not only on franchise owners' demands for profit and control, but also on the NFL's increasingly security-focused requirements for hosting a Super Bowl[7] (McCourt, 2010). These requirements include both the physical structure that houses the game and the urban spaces and communities in which the stadiums are located,[8] which are increasingly viewed as terrain in which military tactics and weaponry are necessary to controls crowds and prevent and respond to terrorists attacks (Graham, 2004; Schimmel, 2006; Warren, 2004). In the post-9/11 period, the expanding powers of the US government combined with ever-increasing partnerships between the NFL, military forces, law enforcement, and private security entrepreneurs created increasingly militarized domestic urban terrains.[9]

 ## Forging Linkages/ Expanding Powers: The US Government, Military Contractors, and the NFI

Just six weeks after the 9/11 attacks, the US Congress passed legislation that gave the President sweeping new powers of search and surveillance and expanded terrorism laws to include 'domestic terrorism'. The 325-page Uniting and Strengthening America by Providing Appropriate Tools Required to Intercept and Obstruct Terrorism (USA Patriot) Act of 2001 was passed on 26 September and was followed by numerous executive orders, regulations, policies and practices. Little more than one year after that, President George W. Bush signed the Homeland Security Act of 2002 creating the nation's 15th cabinet-level Department of Homeland Security (DHS). The 'most extensive reorganization of the federal government in the past fifty years' immediately consolidated 16 Federal Offices and for the first time in US history established a single federal department whose priority mission is to 'prevent terrorist attacks within the United States' (US Department of Homeland Security, 2002).

Over the course of the decade, the NFL and US government counter-terrorism agencies and private contractors linked up in mutually beneficial ways, developing the security-related practices that in 2009 were formally endorsed by the US Department of Homeland Security. Of most relevance to this article is the number of public spaces, arenas and events that were deemed 'target-worthy' by the US government following 9/11, with the Super Bowl receiving special emphasis.[10] The first Super Bowl after 9/11 was the first sporting event and only the 12th event overall to be designated as a 'National Security Special Event' (NSSE). Previous NSSE's included Presidential inaugurations, Democratic and Republican National Conventions, and United Nations Assemblies (US Congressional Research Service, 2007). Once designated, the Secret Service (the federal agency responsible for protecting the President) assumes the mandated role as the lead agency in providing protection and 'strengthening existing partnerships with federal, state, and local law enforcement' (US Department of Homeland Security, 2003).

Consider the security-related activities that were introduced at Super Bowl events or that were associated with the NFL regular season between 2001 and 2010 (see Table 1).

Without question, the extra security demands of hosting an NFL team, and especially the Super Bowl game, have been leveraged by all levels of US government and police to forge linkages across various agencies and to expand capabilities of tracking and surveillance. Through Super Bowl security planning, local police gain access to Pentagon-level private security contractors and have the opportunity to obtain specialized tactical equipment, training, and other tangible resources at a reduced cost (Parker, 2007). At the federal level, in addition to extending the reach of the Department of Homeland Security, agencies such as the Immigration and Naturalization Services have used the cover of the Super Bowl event to search for undocumented workers. Transportation and Safety Administration (the agency overseeing US airport security) has used the Super Bowl to expand its security methods into civilian settings. The Federal Bureau of Investigation has used the event to check its 'most wanted' list. The linkages between government, security entrepreneurs, and the NFL expands yearly to include multiple partners across multiple scales; planning for Super Bowl security currently begins two years prior to the event and can involve up to 70 agencies (Heinze, 2006).

This security expansion at NFL events justifies the accelerated, intensifying militarization of US urban space in the post-9/11 era. The ongoing claim by the NFL and multiple government sources that 'everything changed' on 9/11 reminds us constantly of the violence of the attacks and of our continued vulnerability. Fear and uncertainly regarding when 'they' are going to hit us next becomes the new normal. The Super Bowl, the cities and stadiums that host it, and we, as football fans and residents of urban communities, are portrayed as being under constant threat. In this discourse, both the city and the event are, in military parlance, 'target rich environments'. As Grey and Wyly (2007) summarize, numerous capital, political, and legal resources 'were invested in the ideological construction of a suddenly vulnerable American Homeland' (p. 329). In US cities, they argue, increasing aspects of 'everyday life and death now take place in the certainty of uncertainty in an endless American war on terror' (p. 330). I am extending this argument to suggest that since 'everything changed on September 11' and since the deployment of state and corporate power is the 'necessary' (and thus uncontestable) response to protecting our 'freedom', there are few public outcries to the fact that massive military build-up now accompanies the Super Bowl and that it is an extraordinary incursion into urban civic life—it alters traffic patterns, restricts movement throughout the city and commerce in NFL 'clean zones' that extend a mile out from the stadium, and subjects citizens to military operational and security procedures that they do not encounter anywhere else, including at US airports (the NFL even prohibits 'running' in NFL zones).[11]

I suggest that the convergence of NFL interests and the US national domestic security agenda was formally

Table 1	Security activities related to NFL events, 2001–2010

Year and location	Security-related activity
2001 Super Bowl, Tampa, FL	FaceTrak™ biometric technology digitally scans the faces of all 100,000 fans and workers (without their knowledge) entering the stadium and matches them against a database, sorting them according to their criminal histories (Trigaux, 2001).
2002 Super Bowl, New Orleans, LA	By White House designation, the Super Bowl becomes a National Special Security Event or NSSE (US Congressional Research Service, 2007).
2003 Super Bowl, San Diego, CA	The Federal Immigration and Naturalization Service launches an 'Operation Game Day' dragnet in the city (located 15 miles from the Mexican border) in which 69 foreign-born security guards and taxi drivers are arrested (O'Driscoll, 2003).
2004 Super Bowl, Houston, TX	The NFL hires a private security firm specializing in crowd management to screen everyone entering the stadium (Vaishnav, 2004).
2005 All NFL Games	The NFL institutes a new rule requiring 'pat downs' of every fan entering an NFL stadium and announces a 'Fan Code of Conduct' under which unruly patrons can be ejected or denied entrance without refund to stadiums and parking lots (McCarthy, 2008).
2005 Super Bowl, Jacksonville, FL	Local law enforcement teams-up with Pentagon security contractor GTSI and its InteGuard Alliance partners to install 100 VPN (Virtual Private Network) encrypted video cameras throughout the stadium and into the city, which are designed to 'expand' and 'stay for decades' (McEachern, 2005).
2006 Super Bowl, Detroit, MI	The Federal Bureau of Investigation sends 250 personnel to the city where local police are assisted by an additional 50 federal, state, and local agencies. SWAT teams, 'aided by digital maps covering every inch' of the stadium, are 'ready to respond on a moment's notice'. The US asks Canadian officials to restrict private plane travel near the stadium, which is located one half mile from Canada (NewsEdge Corporation, 2006).
2007 Super Bowl, Miami, FL	An Israeli airport security company, New Age Solutions, is contracted to expand its Behavior Pattern Recognition (BPR) training from Miami International Airport to the custodial staff at the Super Bowl, making Miami the first place in the US where civilian employees are trained in BPR methods to 'look for bad people, not just bad things' (Machlis, 2007).
2008 Super Bowl, Glendale, AZ	Northrop Grumman Corp. is contracted to provide hazardous duty robots (Hulme, 2008). The US military provides PC-12 surveillance aircraft, Blackhawk helicopters, and Cessna Citation Interceptors to patrol the area. The North American Aerospace Defense Command (NORAD) sends the 162nd Fighter Wing F-16 Fighting Falcons to fly a series of defense deterrence missions over the cities of Tucson, Nogales, and Phoenix 'to demonstrate NORAD's quick-response capability' (borderfirereport.net, 2008).
2009 Super Bowl, Tampa, FL	US Immigration and Customs Enforcement, looking for people who are in the country illegally, screens up to 30,000 applications of individuals hoping to volunteer or work at Super Bowl facilities (Zink, 2008). The Transportation and Safety Administration (TSA) links with 70 local police officers that have been trained in the TSA's behavior detection methods (Catalanello, 2009). The city of Tampa expands its use of E*SPONDER and Microsoft Surface technology, originally purchased in 2005 through a US Department of Homeland Security grant, to integrate all aspects of Super Bowl security (Microsoft.com, 2009).

(Continued)

| Table 1 | (Continued) |

Year and location	Security-related activity
2009 US Supreme Court	In January the US Supreme Court refuses to hear the appeal of an NFL fan who argues that the NFL's 'pat downs' violate his US Constitutional 4th Amendment right against unreasonable searches (Varian, 2009).
2010 Super Bowl, Miami, FL	In the lead-up to the game, local police perform 498 biometric searches in high crime areas of the city with newly purchased Printrak Mobile Automated Identification System's handheld devices designed to capture biometric information and search it against the police department's criminal database. During Super Bowl security operations, officers perform 111 biometric searches, identify 58 people who had previous arrest records, and arrest 25 individuals (Morphotrak.com, 2010).
2010 The US Pentagon	In June, the US Pentagon announces it will adopt the NFL's 'instant reply' technology to monitor battlefields in Iraq and Afghanistan (homelandsecuritynewswire.com, 2010).

institutionalized with the adoption of the National Infrastructure Protection Plan (NIPP) embedding resilience strategies into domestic counter-terrorism policy and with the exemption, granted in 2009, of the NFL from terrorism-related lawsuits. In the next section, I present an overview of the NIPP. The purpose here is not to suggest that the US has adopted a resilience approach to all domestic counter-terrorism policy, nor is it to present an in-depth analysis of the range of measures included in a resilience agenda. Rather, the purpose is to trace the emergence of resilience discourse in one example of US counter-terrorism policy, show its increased emphasis over time, and locate the NFL's exceptionalism within the policy framework.

 # Critical Infrastructure Protection and Sport Stadiums: Emphasizing Resilience in Security Policy

Within the critical infrastructure and key resources protection area, national priorities must include preventing catastrophic loss of life and managing cascading, disruptive impacts on the US and global economies across multiple threat scenarios. Achieving this goal requires a strategy that appropriately balances resiliency—a

traditional American strength in adverse times—with focused, risk-informed prevention, protection, and preparedness activities so that we can manage and reduce the most serious risks we face. (NIPP, 2009: i) (Michael Chertoff, 2nd Secretary of the US Department of Homeland Security under George W. Bush, 2005–2009)

The practices employed by the NFL and its US counter-terrorism partners—increasing electronic surveillance of urban space, increasing sophistication of security planning, territorial closure, linking security strategy to the competition for mobile capital—are all evidence of the resilience-oriented trends outlined by Coaffee and Wood (2006) and occurred at the same time in which the US government first embraced, and then emphasized, resilience in formal counter-terrorism policy. Resilience, as Coaffee (2009) and others (see for example, Jacobs, 2005; Medd and Marvin, 2005) explain, is a transdisciplinary concept that integrates natural, physical, and socio-political features. Resilience against terrorism has become a salient concept for policy-makers in recent years as security policy shifts from building up more solid walls of protection (see Davis, 1995) toward actions that minimize the consequences and emphasize the 'bounce-backability' (see Coaffee and Wood, 2006: 508) of a terrorist attack. This shift is evident in the US, for example, in national policy devoted to protecting the nation's Critical Infrastructure and Key Resources (CIKR).

The Homeland Security Act of 2002 that created the DHS also gave it wide-ranging responsibilities for leading and coordinating the nation's CIKR protection, originally defined as targets whose destruction 'could create local disaster or profoundly damage our nation's morale or confidence' and 'high profile events [. . .] strongly coupled to our national symbols or national morale' (US Congressional Research Service, 2004: 8). Included in the government's list of CIKR are sports stadiums and arenas, namely all 112 that host the nation's professional sports league franchises, including the NFL (US Department of Homeland Security, 2010b). The NIPP provides the overarching approach for integrating the nations' CIKR protection initiatives in a single effort.

The first NIPP (2006) was developed in response to Homeland Security Presidential Directive 7, which defined responsibilities for the DHS and certain other federal agencies that represent 18 different industry sectors responsible for various CIKR, such as national monuments, nuclear reactors, and government facilities. Sports stadiums are located within the DHS's commercial facilities sector (NIPP, 2009). Homeland Security Presidential Directive 7 did not include an explicit emphasis on resiliency. Thus, the dominant focus of the 2006 NIPP was on asset protection, defined by the DHS as 'actions that deter the threat, mitigate vulnerabilities, or minimize the consequences associated with an attack or disaster' rather than on asset resiliency, which is defined as '[. . .] the ability to resist, absorb, recover from, or successfully adapt to adversity [. . .]'. Over time, some of the stakeholders 'believed that the concept of continuity and resilience in and of itself, was not articulated and addressed as clearly as needed for their purposes' (US Government Accountability Office, 2010).

Policy emphasis on resilience increased in 2009 with the second NIPP document and through the recommendations of the National Infrastructure Advisory Council, a group with 30 Presidentially appointed members from private industry, academia, and state and local government. The 175-page 2009 NIPP (which remains the most current version), subtitled 'Partnering to enhance protection and resilience', emphasizes resilience with the same level of importance as protection. This did not, however, stop the criticisms of some stakeholders, who felt that resiliency was still under-emphasized. Eventually members of Congress asked the US Government Accountability Office (GAO) to compare the 2007 and 2009 NIPP policy statements with respect to their emphasis on resilience. The GAO found that the 2006 NIPP used resilience or resiliency-related terms 93 times while the 2009 NIPP used those terms 183 times, almost twice as often. In addition, the 2009 NIPP inserts the term resilience in some chapter titles and refers to resiliency alongside protection in the introductory section of the document. According to DHS officials interviewed for the study, the changes in the 2009 NIPP were not representative of a major shift in domestic security policy, but rather were intended 'to increase attention to and awareness about resiliency as it applies within individual sectors' (US Government Accountability Office, 2010: 23). In September 2009, the National Infrastructure Advisory Council published its final report and recommendations to the President. Titled 'Critical Infrastructure Resilience', the report has links to 40 other studies, committee reports and recommendations that have the term resilient or resiliency in their titles (National Infrastructure Advisory Council, 2009).

As Coaffee and O'Hare (2008) have explored, building resilience into infrastructure security fuses risk-management policy-making agendas across a range of scales and stakeholders. I am suggesting that in the US, the inclusion of sports stadiums and arenas in the NIPP created a *unique and unprecedented* institutionalized relationship between sports league owners, facility managers, and the DHS providing structures through which resilience strategies and protocols can be explored (and funded). Thus, 'old' strategies for protecting sport venues and events can be vetted by the DHS and new ways of managing the risks of contemporary forms of terrorist attack can be developed and deployed. US domestic securitization strategies, which now embrace resilience more than ever before, extend through sport in ways not seen (and not possible) in a pre-9/11 context. The first DHS assessment of large stadiums occurred in 2005 through an online Vulnerability Self-Assessment Tool (Hall, 2008). Post-NIPP stadium security strategies emphasize collaboration of multiple partners, emergency planning and the exercise of 'what if' scenarios (see Elmer and Opel, 2006). In addition to the protection strategies discussed above, the resilience framework in the NIPP has led, for example, to funding from the DHS, through the commercial facilities sector of Infrastructure Protection, to initiate the development of a modeling tool for stadium evacuation in the event of terrorist attack or other disaster. The project, publically announced in March 2010, is a collaboration between REGAL (a company contracted by the Secret Service), and the University of Mississippi's National Center for Sport Safety and Security (US Department of Homeland Security, 2010a). As another example, in April 2010 the DHS hosted a two-day sports league security conference and tabletop exercise that was attended by over 200 representatives from US sport leagues, private security companies, academia, law enforcement, and every level of government. The DHS emphasized that the tabletop exercise challenged participants to develop ways to work together and to 'share best practices' because 'high-profile sporting and related events [. . .] can also be high-target opportunities for terrorists and other criminals' (US Department of Homeland Security, 2010b).

The stadiums that host NFL franchises and events make up a small portion of the nation's sport stadiums and arenas contained in the NIPP and the number of fans attending NFL events is the smallest among the US professional sport leagues: the NFL drew 17 million fans in the 2009 season, compared to 21 million each for the National Hockey League and the National Basketball League, and 73 million for Major League Baseball (US Department of Homeland Security, 2010b). The NFL stadium with the largest occupancy capacity, FedEx Field, seats just under 92,000 people, and the second largest, New Meadowlands Stadium, seats 82,000. There are 20 National Collegiate Athletic Association (NCAA) stadiums and arenas that each hold over 90,000 spectators (Kosk, 2010). Consider also that as many as 400,000 stock car racing fans may attend a single NASCAR event. Yet, as I have argued, the NFL and its stadiums and events enjoy a uniquely privileged position within US domestic security policy. But the NFL's exceptional sport status within US counter-terrorism policy goes beyond support for security actions per se, to include legal protection in the event that those actions fail to deter a terrorist attack. The NFL's ability to 'bounce-back' from a terrorist event is aided by DHS and by the same legal structures that protect military contractors in the battlefield. This protection is of obvious financial importance to the League, for as the NFL's Chief of Security (and former FBI official) stated, 'An attack from a terrorist organization could put us out of business' (Frank, 2009). Coaffee and Wood (2006) suggest that contemporary

securitization is more personal than ever before. As I show below, in post-9/11 US all citizens are enlisted into the war on terrorism—in ways that are not completely voluntary.

 ## Citizen-Soldiering the Homeland

> The National Football League [has] won exemption for lawsuits under a post-9/11 law that prohibits them from being sued if terrorists attack a site they are protecting. The protection extends only to companies' services and equipment that the Homeland Security Department has approved as being effective in anti-terrorism [. . .] and whose products have Homeland Security's highest reliability rating. (Frank, 2009: 3A)

Shortly after 9/11 numerous lawsuits were filed against airports, airplane manufacturers, security companies and government agencies (e.g. Port Authority of New York/New Jersey) for failure to protect the US citizenry. The cases were allowed to proceed after a federal judge ruled that the use of an airplane as a suicide weapon was potentially foreseeable (andyfrain.com, 2010). As a result, even companies whose products were not part of the immediate post-9/11 litigation feared the possibility of future suits from victims, with unlimited liability costs, if another terrorist attack were to occur. Insurance costs for all terrorism-related 'potentially foreseeable' risks became incredibly expensive and the federal government grew concerned that the massive 'liability could stifle the entrepreneurial spirit for developing technologies and products that disrupt attacks and enable an effective response' (NIPP, 2009: 89). Congress responded by passing the Support Anti-terrorism by Fostering Effective Technologies (or SAFETY) Act of 2002.

The SAFETY Act, which was a part of the larger Homeland Security Act (discussed above), had the specific purpose of ensuring that the threat of liability does not deter 'potential sellers from developing, commercializing, and deploying' anti-terrorism technologies (NIPP, 2009: 89). The SAFETY Act incentivizes the deployment of new and innovative anti-terrorism technologies by providing liability protections to companies whose products or services have successfully passed DHS review and are awarded

Qualified Anti-Terrorism Technology (QATT) status. There are two levels of protection available: the first is a QATT 'Designation', which limits the amount of liability insurance the seller must maintain; the second and highest is a 'Certification', which effectively eliminates the seller's liability against victim's claims arising from acts of terrorism. A Certification provides the seller with a complete defense in litigation related to the 'performance of the technology in preventing, detecting, or deterring terrorist acts or deployment to recover from one' (safetyact.gov). Certification status means that the product is among those 'approved for Homeland security' and placed on the Approved Products List for Homeland Security on the SAFETY Act website.

In March 2009, the NFL was awarded the DHS's Certification through the year 2016 for its 'NFL Best Practices for Stadium Security', a nine-page document outlining the League's guidelines for stadium and event security and operations. Details contained in the NFL Guidelines are not available to the public (QATT specifics are exempt from the US Freedom of Information Act); however, numerous media outlets reported that the DHS awarded the NFL's practices of digital surveillance, spectator searches, the enforcement of barricaded zones, threat assessments, and the hiring, vetting, and training of personnel (see for example, Frank, 2009; Renieris, 2009). In a *USA Today* report, a Homeland Security spokeswoman commented on the success of the SAFETY Act protection in leading to the 'wider deployment of anti-terrorism technologies and services' (Frank, 2009: 3A). Other companies receiving the DHS Certification include aviation giant Boeing Corp., for its strengthened flight deck doors on aircraft, and IBM for software that can more accurately verify names and identities. The protection provided to the sellers of certified anti-terrorism technologies such as the NFL is linked to previously established legal doctrines.

Government Contractor Defense

According to the regulations codified in the SAFETY Act, a Certification entitles the seller to assert the Government Contractor Defense (GCD), which 'immunizes sellers' against claims arising from acts of terrorism (US Federal Register, 2006: 33149). The GCD, spanning more than 50 years of jurisprudence, is the primary defense historically used by military contractors or 'private military firms'. In laymen's terms, the GCD provides that a US soldier or

soldier's family cannot sue the Pentagon's suppliers for injury or death the soldier incurs while on duty, even if caused by defective equipment (e.g. a gun fails to fire, a gas mask leaks). The foundation of contemporary GCD is the US Supreme Court case *Boyle v United Technologies Corp., 487 US 500* (1988) involving the estate of a marine helicopter pilot who drowned when his escape hatch failed to allow him to escape his downed aircraft. In deciding for the helicopter manufacturer, the US Supreme Court set forth a three-pronged test against which all future immunity claims would be made: that the United States approved of the manufacturer's specifications; that the equipment conformed to those specifications; and, that the supplier warned the United States about the dangers in the use of the equipment (see Radowsky, 2005: 14).

Radowsky's (2005) analysis delineates the numerous ways in which the *Boyle* decision has been used to expand the GCD far beyond its original military context and design defect origins (her analysis does not, however, include the SAFETY Act exemption). Part of this expansion, Radowsky illustrates, has been 'horizontal', extending beyond military contractors and procurement contracts to shield items provided by sub-contractors and nonmilitary contractors. Through the SAFETY Act, the DHS extends the GCD to sport and to civilian companies and cases where products or services are provided under government approval. Boyle's three-pronged test becomes part of the vetting process used to determine which products and services receive a Qualified Anti-Terrorism Technology Designation and which are awarded DHS Certification with sovereign immunity. With respect to the NFL, its DHS Certification means that if a terrorist action occurs at an NFL event, or within NFL zones of the city, victims and victims' estates cannot sue the League to compensate for death or injury, even if the NFL's security practices are shown to be faulty. In practical terms, the SAFETY Act exemption means that the NFL is legally viewed as the same as a Pentagon supplier, and we, the citizens, are the soldiers at war.

'If You See Something, Say Something'™

I argue that citizen soldiers in the US are now being recruited by the DHS to participate in a public awareness campaign to report 'suspicious activity' to state and local law enforcement agencies. The 'If you see something, say something'™ campaign was launched in July 2010 in conjunction with the National Suspicious Activities Reporting Initiative (NSI), but received its widest advertisement when DHS Secretary Janet Napolitano personally announced it at the 2011 Super Bowl in Arlington, Texas, encouraging individual citizens to play an 'active role in keeping the country safe'. In addition, the 2011 Super Bowl campaign included print and video advertisements and a training video for 'NFL employees to ensure that both employees and fans have the tools they need to identify and report suspicious activities and threats' (US Department of Homeland Security, 2011a). Licensed to the DHS, the campaign was originally implemented by New York City's Metropolitan Transportation Authority but is now a part of the NSI effort to 'develop, evaluate, and implement common processes and policies for gathering, documenting, processing, analyzing, and sharing information about terrorism-related suspicious activities' (US Department of Homeland Security, 2011b).

In their discussion of the reterritorialization of security, Coaffee and Wood (2006) argue that security discourses, procedures, and even material examples of national security deployments are telescoping down to smaller scales. Homeland security defense against global threats now extend 'right down to personal safety as one continuous spectrum of security' (Coaffee and Wood, 2006: 515). Crime prevention, anti-social behavior measures, and security all become merged together in a wide range of policy agendas, accompanied by the rhetoric that we are living in a dangerous and uncertain world. In this context, anti-terror initiatives have become the new normal in US cities, reshaping the spaces and interstices of daily life (see Wekerle and Jackson, 2005). Moving beyond Coaffee and Wood, I suggest that the DHS's 'If you see something, say something'™ campaign enlists ordinary citizens into the war on terror underscoring the concept that 'an alert public plays a critical role in keeping our nation safe' (US Department of Homeland Security, 2011b). We are all citizen soldiers, and the war is everywhere.

 ## Conclusion

The citizen soldier concept is an ideal abstraction, used especially in the American Revolution based on the notion that citizens have an obligation to arm themselves to defend their communities against foreign invaders and domestic tyrants. In his examination of cities as battlespaces (a term he borrows from Blackmore, 2005), Stephen Graham (2009) invokes the concept of the citizen soldier in his discussion of the ways in which military urbanism gains legitimacy by its seamless extension into popular, urban, and material culture. US weapons systems and Predator drones, for example, cross over into the latest warfare consumer video games and US military vehicles cross over into civilian Sports Utility Vehicles in a circuit of cultural militarization linking domestic cities to occupied ones. Within the new military urbanism, Graham argues, the everyday sites and spaces of the city are becoming the main battlespaces both at home and abroad. The battlespace concept is distinct from notions of war taking place in battlefields, which have time and geographical limits. Battlespace, on the other hand, is limitless, 'nothing lies outside of it', it is the 'boundless and unending process of militarization where everything becomes the site of permanent war' (Graham, 2009: 389). It is this concept, according to Graham, which lies at the heart of contemporary efforts to urbanize and militarize security doctrine, and it works by collapsing conventional military-civilian binaries. Security doctrine now blurs the operational and legal separations between for example, war and peace and global and local scales.

A number of sport studies scholars have examined the ways in which the NFL is linked to manufacturing consent for the 'war on terror' through its militarized cultural spectacles and media discourses associated with it (see for example, Butterworth, 2008; Falcous and Silk, 2005; King, 2008; Schimmel, 2011; Silk and Falcous, 2005). What I have suggested here goes beyond that. I can be a fan of the NFL's brand of professional football, consume its products and watch its competitions through the media, and yet, I retain some degree of agency enabling me to resist its ideological messages, subvert their dominate meanings, or withhold my consent for the military operations the League endorses. At the present historical moment, however, I cannot attend an NFL event or be present in the urban location that the NFL is 'protecting', without in effect, being on active duty in the US war on terror. In the event of a terrorist attack, my physical presence appears to legally equate me with a soldier at war. However, as Sebok (2003) provocatively points out, US military soldiers who are injured in the line of duty benefit from veteran pensions and government-provided medical care. A citizen soldier who is injured in a terrorist attack has no right to financial compensation (under conditions of SAFETY Act Certification). The NFL goes to great lengths to assure fans that they are in the 'safest place on earth' when they attend a game (Schimmel, 2011). In fact, however, and as I have argued in this article, it is not fans that are protected, it is the NFL.

Postscript

The 10th anniversary of the 9/11 attacks fell on a Sunday, the day the NFL opened its 2011 season. A night game was scheduled at FedEx Field between the New York Giants and the Washington (DC) Redskins. Throughout most of 2010 and 2011, the NFL was embroiled in a labor dispute (settled in July 2011) resulting in an ownership 'lock out' of the players that threatened the entire 2011 NFL football season. According to ESPN Sports television (Outside the Lines, 8 Sunday 2011) the players planned to argue to the courts that the 2011 season must begin on time because the 'nation needs' the NFL to play on 11 September 2011.

Acknowledgements

Special appreciation is expressed to C. Lee Harrington for comments made on earlier drafts of this article and to Madeline B. Davis for research assistance.

Funding

This research received no specific grant from any funding agency in the public, commercial, or not-for-profit sectors.

⊠ Notes

1. The acceptance, popularization, and glorification of the military in the guise of the US soldier has changed over time. From the late 1960s through the 1990s oppositional voices problematized the relationship more so than is currently the case. I am indebted to an anonymous reviewer for this insight.

2. The authorial position for this article is that of a US citizen.

3. Schimmel (2006) included an exploratory examination of the urban militarization involving the 2005 and 2006 Super Bowls. Schimmel (2011) is a media analysis of the security discourse surrounding all the Super Bowls in the post-9/11 era based upon data from internet posts and newspaper articles from US news media sources published between 2000 and 2010. The present study draws upon these prior studies, and in addition includes new data from legal documents and US national and local government agencies' policy documents, reports, and commissioned studies.

4. The US Congress, based upon its Constitutional power to regulate interstate commerce, passed the Sherman Antitrust Act in 1890 to prohibit monopolies of any kind (The Lenix Information Project, 2011). In 1922, the US Supreme Court formally granted Major League Baseball an exemption from the Sherman Act ruling that 'the business [of] giving exhibitions of baseball' did not constitute interstate commerce. Since then there have been numerous US Congressional hearings considering the sports exemption, but none have dismantled it. The other US professional sports leagues benefit from the precedent set in the baseball case, though there some situations in which the Sherman Act has been applied to them differently (US Senate Committee on the Judiciary, 2011).

5. Portions of the preceding text were adapted from Schimmel (2001).

6. For a description of the ways in which the NFL and FOX Television packaged the first Super Bowl after the 9/11 attacks, including a discussion of the Super Bowl's 'delimited construction of citizenship in contemporary America' and its affirmation of 'civil religion' see Butterworth (2008).

7. For example, McCourt (2010) reports on the development of the newly opened New Meadowlands Stadium, site of the Super Bowl in 2014, which was designed to 'focus on deterrence as a secured, hard target' and includes a command and control center developed in partnership with the NFL and the US Department of Homeland Security's Command, Control and Interoperability Division. For a broader discussion of the 'security legacies' left by sports mega-events, see Giulianotti and Klauser (2010).

8. For example, the NFL requires 19,000 'quality hotel' rooms within an hour's drive of the stadium for use by NFL-related groups and another 25,000 hotel rooms within an hour's drive for use by the general public; the establishment of a one-mile 'clean zone' around the stadium and pertinent downtown areas where unauthorized merchandise vending is prohibited; the cities provide the NFL with cost-free venues for entertainment and parties hosted by the NFL; that the cities provide 600,000 square feet of space for the NFL's temporary theme park 'The NFL Experience'; and that cities must give up control of parking lots (car parks) around the stadium.

9. For a broader discussion of the intensified depth and mutuality of the sport–war nexus in the post-9/11 era, see King (2008).

10. It is important to note that in the period between 2000 and 2010 there were no credible, specific threats against NFL stadiums or the Super Bowl event reported in US print and online media. However in 2008 and again in 2009, the FBI and DHS released a threat assessment bulletin that warned operators of transit systems, stadiums, hotels and entertainment complexes to be on heighten alert for attack. As reported by the Associated Press, the bulletin made specific reference to an al-Qaida training manual that lists 'blasting and destroying the places of amusement, immorality and sin . . . and attacking vital economic centers'. However, media reports also quoted DHS officials as stating they had, 'no information regarding the timing, location or target' of any attacks (see Hays and Barrett, 2009). A widely reported threat of a dirty bomb attack against NFL stadiums in 2008, turned out to be a hoax.

11. The use of facial recognition technology (FRT) has resulted in a fair amount of public discourse. After its utilization in the 2001 Super Bowl, the RAND Center commissioned a report on the issues of privacy rights and the countervailing benefits of FRT to national security. The author of that report presented a number of policy recommendations to maximize FRT utility while minimizing threats to privacy (see Woodward, 2001). In the time period immediately following the 9/11 attacks, justifications for the use of FRT grew immensely (see Lyon, 2003).

 # References

Andyfrain.com (2010) *Andy Frain services receives Certification by the Department of Homeland Security under SAFETY Act.* Available at: http://.www.andyfrain.com/safetyact.php (accessed 25 October 2010).

Blackmore T (2005) *War X: Human Extensions in Battlespace.* Toronto: University of Toronto Press.

Borderfire.net (2008) *CBP, National Guard announces Super Bowl restrictions*, 1 February. Available at: http://www.borderfire.net/index3php (accessed 2 February 2008).

Brown M, Nagell M, McEvoy C and Rascher D (2004) Revenue and wealth maximization in the National Football League: The impact of stadia. *Sport Management Quarterly* 13(4): 227–236.

Butterworth M (2008) Fox Sports, Super Bowl XLII, and the affirmation of American civil religion. *Journal of Sport and Social Issues* 32(3): 318–323.

Carden M J (2009) NFL, military continue Super Bowl traditions. *American Foreign Press Service*, 29 January. Available at: http://www.af.mil/story=123133219 (accessed 18 April 2010).

Catalanello R (2009) Airport-style safety measure in the mix. *St. Petersburg Times*, 24 January, p. 6B.

Coaffee J (2003) *Terrorism, Risk and the City: The Making of a Contemporary Urban Landscape.* Burlington, VT: Ashgate.

Coaffee J (2006) From counter-terrorism to resilience. *European Legacy— Journal of the International Society for the Study of European Ideas (ISSEI)* 11(4): 389–403.

Coaffee J (2009) *Terrorism, Risk and the Global City: Towards Urban Resilience.* Burlington, VT: Ashgate.

Coaffee J and O'Hare P (2008) Urban resilience and national security: The role for planners. *Proceedings of the Institute of Civil Engineers: Urban Design and Planning* 161: DP4, 171–182.

Coaffee J and Wood DM (2006) Security is coming home: Rethinking scale and constructing resilience in global urban response to terrorism risk. *International Relations* 20(4): 503–517.

Coakley J (2009) *Sports in Society: Issues and Controversies.* New York: McGraw-Hill.

Davis M (1995) Fortress Los Angeles: The militarization of urban space. In: Kasinitz P (ed.) *Metropolis—Centre and Symbol of Our Times.* London: Macmillan, 335–368.

Delaney K and Eckstein R (2007) *Public Dollars, Private Stadiums: The Battle over Building Sports Stadiums.* Piscataway, NJ: Rutgers University.

Elmer G and Opel A (2006) Surviving the inevitable future: Presumption in the age of faulty intelligence. *Cultural Studies* 20(4/5): 774–792.

Falcous M and Silk M (2005) Manufacturing consent: Mediated sporting spectacle and the cultural politics of the 'War on Terror'. *International Journal of Media and Cultural Politics* 9(1): 59–65.

Frank T (2009) NFL exempt from terrorism lawsuits; Post-9/11 law aims to foster better technology. *USA Today*, 10 March, p. 3A.

Giulianotti R and Klauser F (2010) Security governance and sport mega-events: Toward an interdisciplinary research agenda. *Journal of Sport and Social Issues* 34(1): 49–61.

Graham S (2004) Introduction: Cities, welfare, and the states of emergency. In: Graham S (ed.) *Cities, War and Terrorism: Towards and Urban Geopolitics.* Oxford: Blackwell, 1–26.

Graham S (2009) Cities as battlespace: The new military urbanism. *City* 13(4): 384–402.

Grey M and Wyly E (2009) The terror city hypothesis. In: Gregory D (ed.) *Violent Geographies.* London: Routledge, 329–348.

Hall S (2008) Sport venue security: Planning and preparedness for terrorist-related incidents. *The SMART Journal* 4(2): 6–15.

Hays T and Barrett (2009) *Stadiums, hotels warned to watch for terrorists.* Available at: http:// www.huffingtonpost.com/2009/09/22/stadiums-hotels-warned-to_0_n_294654.html (accessed 23 September 2009).

Heinze C (2006) *Super security.* Available at: http://www.emergencymgmt.com/story.php?id= 102351 (accessed 2 February 2007).

Homelandsecuritynewswire.com (2010) *US military adopts NFL's instant reply technology*, 4 June. Available at: http://www.homelandsecuritynewswire.com/us-military-adopts-nfls-instant-reply-technology (accessed 19 October 2010).

Hulme G (2008) *Super Bowl. Robo security*, 2 February. Available at: http://www.informationweek.com/blog/main/archives/2008/02/super_bowl_html (accessed 3 February 2008).

Jacobs B (2005) Urban vulnerability: Public management in a changing world. *Journal of Contingencies and Crisis Management* 13(2): 39–43.

Kantor P (2010) City futures: Politics, economic crisis, and the American model of urban development. *Urban Research & Practice* 3(1): 1–11.

King S (2008) Offensive lines: Sport-state synergy in an era of perpetual war. *Cultural Studies<>Critical Methodologies* 8(4): 527–539.

Kosk N (2010) *Get in on the game*, 7 October. Available at: http://www.securityinfowatch.com/ printer/1317918 (accessed 18 November 2010).

Lyon D (2003) *Surveillance after September 11*. Malden, MA: Polity Press.

Machlis D (2007) US gets Israeli security for Super Bowl. *The Jerusalem Post*, 4 February. Available at: http://www.sport.jpost.com (accessed 21 February 2007).

McCarthy M (2008) NFL policy cracks down on unruly fans. *USA Today*, 2 August, p. 1A.

McCourt M (2010) A 'touchdown' for security, 1 August. Available at: http://www.securitymagazine.com/copyright/BNP (accessed 18 October 2010).

McEachern C (2005) Security scores at Super Bowl—The city hosting this year's game turns to GTSI. *VARbusiness.com*, 24 January, p. 45.

Medd W and Marvin S (2005) From the politics of urgency to the governance of preparedness: A research agenda on urban vulnerability. *Contingencies and Crisis Management* 13(2): 44–49.

Microsoft.com (2009) *Tampa authorities deploy latest technology to Super Bowl security*, 28 January. Available at: http://www.Microsoft.com/industry/government/news/e_sponder_super_bowl.mspx (assessed 29 September 2009).

Morphotrak.com (2010) *Printrak mobile AFIS solution: Proven and tested*, 9 January. Available at: http://www.morphotrak.com (accessed 30 January 2010).

National Infrastructure Advisory Council (2009) *Critical Infrastructure Resilience Final Report and Recommendations*. Available at: http://www.dhs.gov/xlibrary/assests/niac_critical_infra-structure_resilience.pdf (accessed 1 March 2011).

National Infrastructure Protection Plan (NIPP) (2009) *National Infrastructure Protection Plan: Partnering to Enhance Protection and Resilience*. US Department of Homeland Security. Available at: http://www.dhs.gov/xlibrary/assets/NIPP_Plan.pdf (accessed 6 February 2011).

NewsEdge Corporation (2006) *Supersizing bowl security for Super Bowl XL*, 13 January. Available at: http://www.securityinfowatch.com (accessed 1 January 2006).

Newyorktimes.com (2010) *Super Bowl dethrones M*A*S*H as the most-watched television show in US history*, 8 February 2010.

O'Driscoll P (2003) High visibility of police keeps fans feeling safe. *USA Today*, 27 January, p. 7C.

Parker R (2007) A local perspective on major event police planning in a post-September 11 environment: Super Bowl XLI, Miami, Florida. *Police Chief Magazine* 74(9), 2 September. Available at: http://www.policechiefmagazine.org (accessed 22 September 2009).

Profootballhof.com (2011) General history: Chronology (1869–1939). Available at: http://www. profootballhof.co/hostory/general/chronology/1869-1930.aspx (assessed 14 May 2011).

Radowsky K (2005) The government contractor defense & its impact on litigation against military contractors. *The Military Law Taskforce*, 20 December. Available at: http://www.nlgmltf.org/pdfs/History_Litigation_Mil_Contractors.pdf (accessed 12 May 2011).

Real M (1974) The Super Bowl: Mythic spectacle. In: Newcomb H (ed.) *Television: The Critical View*. New York: Oxford University Press.

Renieris E (2009) NFL = No financial liability? Post-9/11 law gives NFL exemption for terrorism lawsuits. *Vanderbilt Journal of Entertainment and Technology Law-JETBlog*, 16 March. Available at: http://www.jetl.wordpress.com/2009/03/16/bfl-no-financial-liability-post-911- law-gives-nfl-exemption-from-terroris-lawsuits (accessed 10 January 2010).

Schimmel KS (2001) Take me out to the ballgame: The transformation of production-consumption relations in professional team sport. In: Harrington CL and Bielby DD (eds) *Popular Culture: Production and Consumption*. Oxford: Blackwell, 36–52.

Schimmel KS (2006) Deep play: Sports mega-events and urban social conditions in the USA. *Sociological Review* 54(s2): 160–174.

Schimmel KS (2011) From 'violencecomplacent' to 'terrorist-ready': Post-9/11 framing of the US Super Bowl. *Urban Studies* 48(15): 3277–3291.

Sebok AJ (2003) *The SAFETY Act of 2002: Does its decision to protect antiterrorism technologies from tort lawsuits make sense?* Available at: http://www.writ.news.findlaw.com/ sebok/20031208.html (accessed 25 October 2010).

Silk M and Falcous M (2005) One day in September/A week in February: Mobilizing American (sporting) nationalism. *Sociology of Sport Journal* 22: 447–471.

The Lenix Information Project (2011) *The Sherman Antitrust Act*. Available at: http://www.linfo. org/sherman.html (assessed 1 May 2011).

Trigaux R (2001) Cameras scanned fans for criminals. *St. Petersburg Times*, 31 January, p. 1A.

US Congressional Research Service (2004) *Critical Infrastructure and Key Assets: Definition and Identification*. CRS Report for Congress, Order Code RL32631, 1 October. Available at: http://www.fas.org/spg/crs/RL32631.pdf (accessed 30 April 2011).

US Congressional Research Service (2007) *National Special Security Events*. CRS Report for Congress, Order Code RS22754, 6 November.

US Department of Homeland Security (2002) *National Strategy for Homeland Security*. Washington, DC. Available at: http://www.whitehouse.gov/homeland/book/nat_strat_his.pdf (assessed 30 April 2011).

US Department of Homeland Security (2003) *National Special Security Events Fact Sheet*, 9 July. Available at: http://www.dhs.gov (accessed 14 July 2006).

US Department of Homeland Security (2010a) *Science & technology bloggers' roundtable: SportEvac transcript*, 16 March. Available at: http://www.dhs.gov/ynews/gallery/gc_1270044985681.shtm (accessed 12 April 2011).

US Department of Homeland Security (2010b) DHS Hosts sports leagues conference and table-top exercise. *The Blog @ Homeland Security*, 28 April. Available at: http://www.blog.dhs.gov/2010/04/dhs-hosts-sports-leagues-conference-and.html (accessed 12 April 2011).

US Department of Homeland Security (2011a) *Secretary Napolitano announces, 'If you see something, say something,' campaign at Super Bowl XLV*. Office of the Press Secretary, 31 January. Available at: http://www.dhs.gov/ynews/releases/pr_1296509083464.shtm (accessed 12 April 2011).

US Department of Homeland Security (2011b) *If you see something, say something campaign™: Report suspicious activity to local law enforcement or call 911*. 29 March. Available at: http://www.dhs.gov/files/reportincidents/see-something-say-something.shtm (accessed 14 April 2011).

US Federal Register (2006) *Rules and Regulations 71(110): 33147–33168*, 8 June. Available at: http://www.safetyact.gov (accessed 30 April 2011).

US Government Accountability Office (2010) *Update to National Infrastructure Protection Plan Includes Increased Emphasis on Risk Management and Resilience: GAO-10-296*. Available at: http://www.gao.gov/news.items/d10296.pdf (accessed 1 May 2011).

US Senate Committee on the Judiciary (2011) *Professional Sports and the Federal Antitrust Act*. Available at: http://judiciary.senate.gov/about/history/SportsAntitrust.cfm (accessed 1 May 2011).

Vaishnav A (2004) Super Bowl XXXVIII; Going into this with eyes wide open security measures will be stepped up. *The Boston Globe*, 27 January, p. D10.

Varian B (2009) Fans can expect to be frisked. *St. Petersburg Times*, 22 January, p. 1B.

Warren R (2004) City streets—The war zones of globalization: Democracy and military operations on urban terrain in the early twenty-first century. In: Graham S (ed.) *Cities, War and Terrorism: Towards and Urban Geopolitics*. Oxford: Blackwell, 214–230.

Wekerle GR and Jackson PSB (2005) Urbanizing the security agenda: Anti-terrorism, urban sprawl and social movements. *City* 9(1): 33–49.

Woodward JD Jr (2001) *Super Bowl Surveillance: Facing up to Biometrics*. Available at: http://www.rand.org/pubs/issue_papers/IP209.html (accessed 20 September 2009).

Zink J (2008) Huddle promises safe Super Bowl, 10 December. Available at: http://www.Tampabay.com/news/publicsafety/security-huddle-promises-safe-bowl/933452 (accessed 9 September 2009).

REVIEW QUESTIONS

1. Do you believe that the author accurately describes the relationship between the National Football League and the Department of Homeland Security?

2. Do you believe the security solutions presented are valid and potentially good?

3. Discuss in a paragraph how the NFL, contractors, and the U.S. government counter-terrorism agencies linked up in a mutually beneficial way.

4. Does the NFL enjoy special privileges within U.S. domestic security policy? If yes, how?

5. Are you surprised after reading this article at the extent of collaboration between the NFL, homeland security agencies, and contractors? Yes or no? Explain your answer.

6. Are you a citizen soldier if you attend an NFL game?

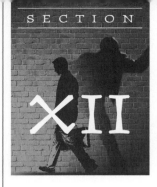

The Special Case of Weapons of Mass Destruction (WMD) Terrorism

Learning Objectives

At the end of this section, students will be able to:

- Describe the three main obstacles terrorists must be able to overcome in order to carry out a significant WMD attack.
- Discuss the relationship between expertise, resources, motivation, and a WMD attack.
- Analyze the different WMD agents based on five common characteristics.
- Describe and evaluate the differences in the threats posed by biological, chemical, and nuclear weapons.

Introduction

Planning for terrorism, especially a Weapons of Mass Destruction (WMD) attack, is difficult, made all the more so by a fear factor—a function of the difference between **risk perception** and **risk assessment**. Risk perception refers to the subjective, intuitive feelings individuals hold as to their own personal safety (Slovic, 1987, p. 280). Risk assessment involves the use of quantitative, technologically advanced analysis of actual risks from disasters or crises (Slovic et al., 2004, p. 311). The unpredictability of a terrorist attack makes risk assessment difficult. This limits what can be communicated to the public, which in turn leaves private citizens to form their own perceptions of risk. In terms of terrorism, these perceptions are often out of proportion to the actual risk of an attack (Slovic, 2004). The public's fear of terrorism, whether warranted or not, can force governments to "take actions that reassure people, even if such actions are not justified on technical grounds (i.e. even if they do not really reduce the threat but only appear to do so)" (Sunstein, 2003, pp. 131–132). This in turn distracts policy makers from focusing on the need to educate the public on the actual risks in advance of a crisis (Slovic, 2004, p. 989).

The challenges of uncertainty and fear are exacerbated when dealing with attacks that fall within the WMD rubric—attacks using chemical, biological, radiological, or nuclear weapons. It is exceedingly difficult to judge or estimate the potential harm of such attacks without knowing the precise agent used, where and when it was delivered, and in what form. Planning for a bioterrorism attack, in particular, is further complicated by the fact that officials may not even know an attack has taken place until two to three weeks after the release of the agent. Senator Daniel K. Akaka echoed these concerns during a 2001 committee hearing on decontaminating public buildings following the anthrax attack:

> Bioterrorism is different from other forms of terrorism. A bioterrorist attack will not be preceded by a large explosion. First responders will be physicians and nurses from our local hospitals and emergency rooms who may not realize that there has been an attack for days or weeks. (U.S. House of Representatives, November 8, 2001)

In addition, policy makers cannot know for certain how the public will react. In the event of a catastrophic WMD attack, to what extent will mass hysteria and panic set in? Research in this area suggests that the public may not comply with quarantine or evacuation orders and, in fact, the imposition of a quarantine may make matters worse by generating panic where there was none (Preston, 2007, p. 4).

This uncertainty makes the formation of policies to address terrorist attacks, especially WMD attacks, problematic. It is difficult for policy makers to formulate comprehensive and cohesive policies on prevention and response when they are uncertain as to what, precisely, they are trying to prevent or respond to. The consequences of creating ineffective or ill-designed policies can be harsh. If the policies designed to prevent or respond to terrorist attacks fail, the results are most often calculated in terms of fatalities, injuries, and property damage.

Because of the potentially catastrophic consequences of a WMD attack, this section is designed to provide a brief introduction to the nature and threat of WMD agents. This section starts with a discussion of the three main obstacles—**acquisition, weaponization**, and **delivery**—that terrorists must overcome in order to carry out a significant WMD terrorist attack. The discussion is followed by a discussion of the basic nature of biological, chemical, and **nuclear agents**. It should be noted that these discussions are written from the perspective of a social scientist and are meant to be introductory. For further information regarding the science underlying each of these types of agents, students should refer to the appropriate scientific disciplines. The list of web resources included at the end of this section also provides some useful starting points.

 ## Acquisition, Weaponization, and Delivery: The Perfect WMD Storm

In understanding and assessing the threat of a terrorist attack on U.S. soil using a weapon of mass destruction, or WMD, one first needs to understand the basic requirements for a "successful" WMD attack. There are three: acquisition, weaponization, and delivery. All three of these elements are required regardless of the type of WMD agent being used.

First, a terrorist group must acquire the agent or agents they are considering using. The difficulty of this task depends on the nature of the agent. **Biological agents** may be easier for a non-state actor to acquire than **fissile material** for a nuclear weapon, due to issues of availability, cost, and security obstacles.

Second, a terrorist group must weaponize the agent they have acquired. Weaponization refers to the process of making the actual agent into a form that can cause the most casualties. For example, many of the agents available for use in a bioterrorism attack (such as anthrax) occur in nature and simply are not that dangerous if one comes into contact with them. However, when in 2001 natural anthrax spores were converted into a fine, highly sophisticated powder, the danger presented by the anthrax increased exponentially.

Third, a terrorist or terrorist group must develop an accurate and widespread delivery system so that their chosen weapon has the widest impact (translation: casualties) possible. For example, Aum Shinrikyo—the group responsible for the March 1995 sarin gas attack against the Tokyo subway system—attempted at least nine prior bioterrorism attacks, including spraying liquid anthrax slurry from vans in downtown Tokyo and attempts to target U.S. naval bases with botulinum. Only five days before the sarin gas attack, the Aum cult attempted to disperse a biological agent using briefcases with vents and battery-powered fans (Preston, 2007, p. 233). In the end, they lacked the expertise to deploy a truly mass-casualty delivery system and settled for a very unsophisticated delivery system (plastic bags of the gas punctured by umbrellas) that caused far fewer casualties than the potential for sarin gas suggests.

Although the world has seen efforts to mount a WMD attack by a terrorist group (see text box for the two most famous), the casualties involved were minimal compared to the potential casualties from some of these weapons. So far, getting all three factors—acquisition, weaponization, and delivery—to come together at the same time by a single terrorist group—the WMD perfect storm, if you will—has not been an easy endeavor. In general, the more deadly the attack being sought, the more money, expertise, and resources are needed to make things come together, with expertise often being the key missing ingredient. In addition, beyond acquisition, weaponization, and delivery, there is a fourth factor: that of the group's motivation.

Until recently, most terrorism scholars believed that terrorists would not cross this terrible line. This especially applied to those groups pursuing a nationalistic or ethnic autonomy agenda, who were dependent on support among the larger populations who would turn against on them if such weapons were used. With the advent of groups pursuing a dominantly religious agenda, where they are ordered by divine authority to create as many casualties as possible as a goal in and of itself, that motivational line may more likely be crossed (Falkenrath et al., 2000). Still, the most deadly terrorist attack to date involved not a weapon of mass destruction, but several good old-fashioned passenger jets loaded with good old-fashioned jet fuel. Today's terrorist groups may simply see conventional weapons as capable of meeting their goals without the increased obstacles involved with WMD. However, the threat of a WMD attack cannot and should not be ignored as the consequences of such an attack could be truly horrifying.

 ## The Agents of WMD

Every potential WMD agent, whether it be chemical, biological, radioactive, or nuclear, can be described using five characteristics: lethality (toxicity), speed of action, specificity, controllability, and residual effects. The level of lethality/toxicity refers to the extent to which those coming in contact with the agent are likely to die. For example, there is no cure for the Ebola virus, so its level of lethality is considered high. Speed of action refers to how quickly victims will come in contact with and react to the agent. Specificity refers to whether or not the agent is specific to a particular animal or plant species, or humans, and the extent, if any, to which the agent can jump between species. For instance, the H1N1 influenza virus began life as a bird virus and only later mutated into a form that could jump to humans. Indeed, many recent policy recommendations for dealing with outbreaks of infectious diseases (deliberate or otherwise) include better coordination between the veterinary and human public health communities (Tucker & Kadlec, 2001). The remaining two characteristics, controllability and residual effects, refer to how well the terrorists can control the area over which the agent is spread (and not get caught in it themselves) as well as how long the effects of the agent will last over what amount of territory and with what strength. The remainder of this section will describe each type of WMD in more detail and conclude with a review of the debate over the nature and extent of the WMD threat.

Bacteria, Viruses, and Toxins . . . Oh My!

To date in the United States, there has been only one significant terrorist attack by a non-state actor using a biological agent—the anthrax attacks of September and October 2001. (In-Focus 12.1. describes these attacks and their aftermath). Biological weapons are derived from living organisms or infectious material derived from them. They are dependent on their ability to multiply within a living host in order to do any damage. In general, biological agents are considered more toxic, slower acting, more specific to certain species, less controllable, and longer lasting than their chemical cousins.

IN-FOCUS 12.1

The Anthrax Attacks

Anthrax is an acute infectious disease caused by spores known as *Bacillus anthracis*. These spores are commonly found in the soil and can lay dormant for years. Anthrax can be absorbed through the skin (cutaneous anthrax), ingested by eating contaminated food (gastrointestinal anthrax), or inhaled through the lungs (inhalational anthrax). Whether or not contact with anthrax spores results in an infection, and whether or not that infection is fatal, depends on the nature of the contact and the nature of the spores at the time of contact (U.S. Government Accountability Office, 2003).

In September and October 2001, letters laced with a powdered form of anthrax were mailed to the offices of members of the U.S. Congress and to the offices of ABC News, CBS News, and the *New York Post* in New York City, and American Media, Inc. (publisher of the tabloids *National Inquirer* and the *Sun*) in Boca Raton, Florida. Five people died and 22 were infected. The first letter was postmarked on September 18. By the time the first case had been diagnosed on October 4—in a tabloid photographer from Palm Beach, Florida, who died the next day—there were already seven people who had been infected but not diagnosed (Gursky, Inglesby, & O'Toole, 2008). In addition, anthrax spores, likely due to cross-contaminated letters, were found in the mailrooms of several federal buildings, including the Central Intelligence Agency (CIA), a U.S. House of Representatives office building, the U.S. Supreme Court, the White House, and the Walter Reed Army Institute of Research (Gursky et al., 2008, p. 394).

The Centers for Disease Control and Prevention (CDC) identified six epicenters from the initial exposure spread (U.S. Government Accountability Office 2003). The epicenters involved four different states plus the District of Columbia. Several post office facilities were shut down, including the main processing facility in Washington, D.C. Purging the buildings of the contamination required considerable resources in finances and personnel. Ironically, anthrax-laced letters were only found at two of the six epicenters. The remainder of the contamination came from cross-contamination of locations, equipment, and general mail. The mailings and contamination raised the level of risk perception and fear throughout the country. People were afraid to open mail, and numerous false alarms were called into the Federal Bureau of Investigation (FBI) regarding envelopes with a powdery substance or unknown, suspicious packages.

The strain of the anthrax used in the attacks was a highly sophisticated version of the bacteria that had been manufactured at the U.S. Army's main biodefense research facility at Fort Detrick, Maryland. Despite an early identification of the source of the anthrax, however, the case remained open for almost seven years. Finally, in late July 2008, a scientist named Bruce Ivins, who was working on the anthrax vaccine at Fort Detrick, committed suicide. His death came shortly before he was to be arrested on five counts of capital murder. At a press conference on August 6, 2008, the FBI attributed Ivins's motivation to his concern that the vaccine program he was working on was going to be shut down. The case was officially closed by the FBI that same day.

Although the anthrax infections were limited to the East Coast of the United States, the Government Accountability Office (GAO) noted that the attacks had national implications. "Because mail processed at contaminated postal facilities could be cross-contaminated and end up anywhere in the country, the local incidents generated concern about white powders found in locations beyond the epicenters and created a demand throughout the nation for public health resources at the local, state, and federal levels" (U.S. Government Accountability Office, 2003, p. 10). As a result, the anthrax attacks became an inextricable part of the discussions and investigations regarding homeland security and WMD terrorism.

Biological agents come in five types: **bacteria, viruses, toxins,** rickettsiae, and fungi. This section will address the first three as they hold many of the most well-known pathogens. Bacteria are actual living organisms that are present in nature and capable of living outside the human body. Some of the more well-known bacteria include typhoid, brucellosis, plague, tularemia, and anthrax. Viruses, while considered living organisms, cannot live outside a host (plant, animal, or human) and are dependent on their ability to replicate inside a host for their long-term survival. Viruses work as a collection of genetic codes inside a protein that fool a body's own cells into replicating the virus's DNA rather than the DNA of healthy cells. If a virus cannot find a non-immunized host in which to work its damage, it will die out. It was this dependency that allowed Donald Henderson and the World Health Organization (WHO) to eliminate smallpox in the early 1970s using a ring vaccination system. In addition to smallpox, the virus category includes

▲ Photos of the anthrax-laced letters sent through the mail to media outlets and congressional buildings in September and October of 2001.

hemorrhagic fevers such as Ebola and Marburg as well as the influenza strains (the flu). In contrast to bacteria and viruses, toxins are not considered living organisms. Instead, they are non-living chemical substances produced as by-products by living organisms, such as the venom produced by rattlesnakes. Botulism and ricin (sometimes used in assassinations) are also part of the toxin family.

Each of these types of biological agents can be characterized by the five factors discussed above. The Centers for Disease Control and Prevention (CDC)—the main government agency tasked with protecting the U.S. population from infectious diseases—has further divided biological agents into various categories depending on the lethality, ease of contagion, and level of preparedness and response required. The website for the CDC—www.cdc.gov—lists the various biological pathogen categories and the agents that fall into each one. The seriousness of a pathogen does not depend on what type of biological agent it is. For example, Category A—which houses the most dangerous pathogens—includes the viruses Ebola, Marburg, and smallpox as well as the bacteria for anthrax and plague.

Two key issues concern biological weapons experts at the time of this writing. The first is about emerging infections completely unrelated to any terrorist act, especially given the adaptability of bacteria and viruses. The growing and widespread use of antibiotics has in many cases encouraged these pathogens to mutate into resistant forms. The technology available for new vaccines and antibiotics is barely able to keep up with a pathogen's ability to adapt, and experts worry that at some point we could lose that race (Tucker & Kadlec, 2001). Adding to the problem is globalization whereby people can move across the globe faster and more frequently than we ever anticipated, bringing native pathogens to a new destination where populations may have far less immunity. The article by Tucker and Kadlec (2001), found in the readings for this section, addresses this issue in more detail.

The second concern, closely related to the first, is that of bioengineered weapons. The field of **bioengineering** has expanded by leaps and bounds in recent years. The vast majority of bioengineering efforts focus on designing new cures for serious diseases and making full use of new technologies to develop effective vaccines and treatments for a variety of health concerns. However, this same technology, in the hands of terrorists or rogue states, could allow such groups to more easily conduct a biological terrorist attack using bioengineering to design strains of diseases that are resistant to current treatments or that combine various pathogens to create more lethal and harder to treat agents. The advantage to bioengineered weapons is that they take relatively fewer resources and people to manufacture. On the other hand, creating such weapons will require considerably more expertise than most terrorist groups have access to, at least for the moment.

Chemical Agents

Unlike biological agents, **chemical agents** do not occur in nature; they must be manufactured. Most chemical agents fit into one of four categories: choking and incapacitating agents (such as chlorine or phosgene), blood agents (such as

IN-FOCUS 12.2

Terror on the Subway

On March 20, 1995, a new terrorism line was crossed—an attack by a non-state actor against a civilian target using a chemical agent in circumstances that left no doubt that it was a terrorist attack (Falkenrath, 2001, p. 161). The attack was carried out by a millennium cult known as Aum Shinrikyo (Aum) whose goal was to bring about an Armageddon that only the cult members would survive. Aum members followed the teachings of their guru, known as the Asahara. They drank his blood as a path to enlightenment and believed that only the Asahara's followers would survive the end of the world. At the time of the attack, the cult had 40,000 members worldwide and an estimated 1 billion dollars in assets (Pangi, 2008). Members included experts in physics, engineering, and the computer sciences, as well as high ranking government officials.

On March 20, 1995, at 7:30 a.m., Aum members boarded five different trains of the Tokyo subway system at the height of the morning rush hour. The Aum members were carrying umbrellas and plastic bags filled with sarin gas, an odorless and colorless nerve agent that can be lethal in sufficient dosages. Shortly after boarding the trains, the Aum members punctured the bags with the umbrella tips, releasing the gas. The cult members then quietly exited the trains. All five trains converged on the Kasumegaseki station where most of Tokyo's government offices are located. At that point, chaos ensued.

> Subway stations were forced to evacuate passengers en masse, many choking, vomiting, and blinded by the chemicals. They fled up the stairways and collapsed in the streets while fire, police, and emergency medical responders, most unprotected, ran down the stairs to assist victims. The scene was immediately broadcast over television and radio. Images of confusion and chaos dominated the nine o'clock news and provided Tokyo and the world with its first glimpse of an act of terrorism with a weapon of mass destruction. (Pangi, 2008, p. 433)

Fortunately, the gas was not in its purest, most lethal form. As a result, the gas emitted an odor that alerted passengers and first responders (Pangi, 2008). Even so, 12 people were killed and 5,500 were injured (Preston, 2007).

The attack on the Tokyo subway generated considerable debate both inside and outside Japan. Was the subway attack a fluke or a harbinger of worse to come? Supporting the latter viewpoint was the fact that this was not Aum's first attempt to conduct a catastrophic WMD attack. The cult possessed the financial resources to amass production laboratories, equipment, highly educated technical personnel, and even a Russian helicopter. Attempts to launch a WMD attack, prior to the attack on the subway, included spraying liquid anthrax slurry from vans in downtown Tokyo and attempts to target U.S. naval bases with botulinum. Only five days before the sarin gas attack, the Aum cult attempted to disperse a biological agent using briefcases with vents and battery-powered fans (Preston, 2007, p. 233). In fact, the only piece of the puzzle missing in all the cult's efforts was specific expertise on bioagents and weaponization. As noted by Preston, "it is important we not discount how great a difference the input of even one former bioweaponeer" would make (p. 234).

(Continued)

(Continued)

Despite its huge financial investments, however, the cult failed to pull off nine out of ten of its WMD attacks. As noted in the report of the Gilmore Commission,

Aum's attempts to develop a variety of lethal agents or devices, indicates that, despite Aum's considerable resources and the superior technical expertise and state-of-the-art equipment and facilities at its disposal, the group could not effect a truly successful WMD attack. The lesson of Aum is that any non-state entity faces organizational and significant technological difficulties and other hurdles in attempting to weaponize and deliver chemical and biological weapons, arguably providing a refutation of the suggestion voiced with increasing frequency about the ease with which such weapons can be made and used. (Gilmore, 2000, p. xi)

While no American citizens were harmed in the attack, there are several reasons why this attack led U.S. policy makers to take notice. First, the attack demonstrated that previous assumptions that terrorists would avoid using such weapons were no longer valid. The attack on the Tokyo subway "demonstrated that terrorist use of advanced weapons of mass destruction was no longer a theoretical possibility" (Falkenrath, 2001, p. 161). Second, the attack also highlighted the threat posed by the large numbers of biological and chemical experts left unemployed after the fall of the Soviet Union. Finally, the timing of the Tokyo attack—only one month before the bombing of the federal building in Oklahoma City in April 1995—gave policy makers the opportunity to add chemical and biological terrorism concerns to the debates over the response to Oklahoma City.

▲   Response to Tokyo Subway Attack, March 1995

cyanide and cyanogen chloride), blister agents (such as the mustard gas used in WWI), and nerve agents (such as sarin gas or VX gas). The nerve agent sarin was used in the 1995 attack on the Tokyo subway system described in In-Focus 12.2.

Chemical agents are viewed as less effective than biological agents. For one thing, they are not contagious. Victims have to come into direct contact with the agent, which makes a significant and effective delivery system vital for any kind of high-casualty attack. Chemical agents also tend to dissipate faster than biological agents given such factors as weather and prevailing wind patterns. Once the contamination dissipates, or is cleaned up, there is little further threat to the targeted population.

Nevertheless, chemical agents pose a serious threat partly because of several obstacles faced by law enforcement agencies in tracking the threat. First, chemical agents tend to be the easiest of the weapons of mass destruction to acquire; many can be ordered on the Internet with just a credit card. In addition, many chemicals are considered "dual use" or **precursor chemicals**. This means that any given chemical agent can have a perfectly legitimate and even beneficial use on its own and will not be highly regulated. It's when two or more chemicals are combined in a particular fashion that they can become deadly. For this reason, prevention efforts that rely on tracking purchases and possession of certain agents have had limited success.

Nuclear Physics 101

A nuclear attack by a non-state actor (terrorist group) is what homeland security and emergency management experts call a "low-probability, high-consequence" event. The chances of a terrorist group getting access to, or improvising its own, nuclear weapon are slim, and the more deadly and powerful the weapon the lower the probability of an attack. Still, if one does occur the results could be devastating. As a recent *Frontline* episode made quite clear, we are not well prepared to mitigate the effects of even a small nuclear attack ("Nuclear Aftershocks," 2012).

In order to launch a terrorist attack using a nuclear weapon, a terrorist group has essentially two options: (1) to steal or buy an existing nuclear weapon (particularly the warhead); or (2) steal or otherwise acquire nuclear material and build its own bomb (Hecker, 2006). To understand the nature and likelihood of either of these options, it is first necessary to understand a little of how nuclear weapons work. So we will start this topic with a short version of Nuclear Physics 101.

The first thing one must understand about nuclear terrorism is the difference between a nuclear weapon and a radioactive dispersal devise (RDD), sometimes called a "dirty bomb." A nuclear weapon uses a conventional explosive to create a super-critical mass of fissionable nuclear material. Once started, this super-critical mass is capable of a self-sustaining chain reaction. It is this material and the chain reaction that gives a nuclear bomb its incredible destructive capability far beyond the magnitude of conventional explosives. An RDD, in contrast, uses conventional explosives to spread radioactive material over as wide an area as possible (the wider the area to be affected, the more sophisticated the bomb will need to be). However, since there is no chain reaction or nuclear yield, the explosion itself can cause more damage than the dispersed material. In many cases, it might be the fear of contamination from the radioactive material—whether realistic or not—that causes the most upheaval.

Nuclear weapons are based on fissile material, which itself is a type of radioactive isotope. Radioactive isotopes are drawn from elements having the same number of protons but different numbers of neutrons. For example, uranium isotopes—the most commonly used material—have 92 protons but can differ in their number of neutrons; Uranium-235 (U-235) has 92 protons and 143 neutrons while Uranium-239 (U-239) has 92 protons and 146 neutrons. When the neutrons in an element collide, they break into smaller nuclei in a process called fission which releases nuclear energy. Isotopes that can sustain this chain reaction long enough to cause a nuclear explosion are called fissile material.

Two of the most commonly used fissile materials in nuclear weapons are U-235 and Plutonium 239 (PL-239). **Uranium** occurs in nature and can be mined. However, when first pulled out of the ground the ore contains only 7% uranium. In order to create a strong enough chain reaction to be useful, the uranium needs to be enriched. The enrichment process enhances or increases the amount of uranium relative to other elements in the material. At 10% uranium, one would need 4,000 kg to make a basic bomb. If the uranium is enriched to 20% it becomes known as highly enriched uranium (HEU). The United States uses HEU at 90% for its weapons grade uranium, only 25 kg of which are needed for a bomb. Unlike uranium, **plutonium** does not occur in nature but is manufactured. The facilities, money, expertise, and resources needed to produce plutonium are probably far outside the capability range of non-state actors. The theft of plutonium, however, could pose a danger, as only 8 kg of plutonium are needed to make a bomb.

Acquiring the needed nuclear material is only one obstacle a terrorist would have to overcome in order to carry out a nuclear attack. The next step would be to build the bomb itself. There are generally two types of devices relevant to this discussion: a gun-assembly device and an implosion-type device. The gun-assembly device can use only highly enriched uranium. A gun-assembly device uses an explosion to combine in a horizontal fashion the necessary material. This device is considered to be the smaller of the two devises and the easier of the two to construct. The implosion-type device is a spherical device that creates the needed energy by imploding the nuclear material rather than breaking it apart. This device is larger than the gun-assembly device and considered more difficult to construct. However, it is a little more versatile as it can be built using either uranium or plutonium. The yield from either device can be considerable depending on the nature of the material used. Indeed, both these devices were used in the attacks against Hiroshima and Nagasaki at the end of World War II.

There are also several different potential delivery methods depending on whether one is attempting a surface or non-surface explosion (Falkenrath et al., 2006). On the surface—the most likely delivery method—the device could be delivered in a small truck or large SUV. With regard to a non-surface explosion, the likelihood would be a small boat, an air detonation via a small plane, or a sub-surface detonation set off in a subway (in that order of likelihood). A high-altitude delivery (such as via a missile) also needs to be considered, but that type of attack is most probably limited to a military delivery only (Falkenrath et al., 2000; Hecker, 2006).

Since the last actual detonation of a nuclear device in an urban environment took place over 65 years ago, experts are not certain as to the effects and impact on today's cities. Certainly, few if any survivors will be found at or close to ground zero. Farther away from ground zero, however, the picture becomes less clear. Questions abound as to the effects modern skyscrapers and transportation infrastructure would have on such variables as blast damage, thermal impact, and the spread of radiation. We do know, of course, that the residual radioactive environment would have a significant impact on immediate rescue and recovery operations, let alone the long-term impact at the target zone and the surrounding environment.

Summary

To date, there have only been two major WMD attacks by non-state actor groups on civilian targets (see In-Focus 12.1 and 12.2). Despite the fear, constant media coverage, resources expended in response, and the tragic deaths involved, these attacks were not even close to the low-probability, high-consequence WMD attack that terrorism experts fear. The attacks did, however, reveal significant gaps in our ability to respond to such an event. The effectiveness of any response depends first on understanding the nature of the threat. This section has described the nature of the agents potentially involved in a WMD attack and has attempted to assess the threat posed by the use of biological, chemical, or nuclear agents by a non-state actor terrorist group. While much of this is educated speculation, the consensus is that the more lethal and high-casualty the attack, the more resources and expertise would be required to mount it. Ultimately, given the casualty counts available to terrorist groups using conventional weapons, the use of WMD may simply not be worth the effort.

Key Points

- The challenges of creating policies for preventing and responding to terrorist attacks are exacerbated when dealing with attacks that fall within the WMD rubric—attacks using chemical, biological, radiological, or nuclear weapons. It is exceedingly difficult to judge or estimate the potential harm of such attacks without knowing the precise agent used, where and when it was delivered, and in what form.
- In order to carry out a WMD attack, three requirements must be met: acquisition of the agent, weaponization of the agent, and delivery of the agent.
- In general, the more deadly the attack being sought, the more money, expertise, and resources are needed, with expertise often being the key missing ingredient.
- Every potential WMD agent, whether it be chemical, biological, radioactive, or nuclear, can be described using five characteristics: lethality (toxicity), speed of action, specificity, controllability, and residual effects.

◆ Biological weapons are derived from living organisms or the infectious material derived from them. They are dependent on their ability to multiply within a living host in order to do any damage. In general, biological agents are considered more toxic, slower acting, more specific to certain species, less controllable, and longer lasting than their chemical cousins.

◆ Planning for a bioterrorism attack, in particular, is further complicated by the fact that officials may not even know an attack has taken place until two to three weeks after the release of the agent.

◆ Chemical agents do not occur in nature; they must be manufactured. Most chemical agents can be divided into four categories: choking and incapacitating agents, blood agents, blister agents, and nerve agents. Chemical agents are viewed as less effective than biological agents.

◆ A nuclear attack by a non-state actor (terrorist group) is what homeland security and emergency management experts call a low-probability, high-consequence event. The chances of a terrorist group getting access to, or improvising its own nuclear weapon, are slim and the more deadly and powerful the weapon the lower the probability of an attack. Still, if one does occur, the results could be devastating.

◆ In order to launch a terrorist attack using a nuclear weapon, a terrorist group has essentially two options: (1) to steal or buy an existing nuclear weapon (particularly the warhead); or (2) steal or otherwise acquire nuclear material and build its own bomb. Experts view the latter as the more likely of the two.

KEY TERMS

Acquisition	Chemical agents	Risk assessment
Bacteria	Delivery	Risk perception
Bioengineering	Fissile material	Toxins
Biological agents	Nuclear agents	Uranium
Centers for Disease Control and Prevention (CDC)	Plutonium	Viruses
	Precursor chemicals	Weaponization

DISCUSSION QUESTIONS

1. Why is planning for a response to terrorist attacks so difficult? What is it about a WMD attack that increases the difficulty?

2. What is the difference between risk assessment and risk perception?

3. What are the four main types of agents in the WMD rubric? Briefly describe each one.

4. What three obstacles must a terrorist group overcome to launch a WMD attack? Briefly describe each obstacle. Which one do you think is the most difficult to overcome? Why?

5. Describe each of the five main characteristics by which the CDC evaluates a WMD agent.

6. What are the two main threats today with regard to biological pathogens?

7. Why are chemical agents considered less effective than biological agents?

8. What is the most likely method by which a non-state actor could acquire a nuclear weapon?

9. Of the three types of WMD agents (biological, chemical, or nuclear) which one do you believe poses the biggest risk concerning a terrorist attack. Why?

WEB RESOURCES

Centers for Disease Control and Prevention: http://www.cdc.gov

Center for Strategic and International Studies: http://www.csis.org

Department of Homeland Security: http://www.dhs.gov

Federation of American Scientists: www.fas.org

Johns Hopkins Center for Public Health Preparedness: http://www.jhsph.edu/research/centers-and-institutes

National Counterproliferation Center: http://www.counterwmd.gov

National Institute of Allergy and Infectious Diseases: http://www.niaid.nih.gov

Nuclear Threat Initiative: http://www.nti.org

RAND Corporation: http://www.rand.org

UN Office for Disarmament Affairs: http://www.un.org/disarmament/WMD

READING 20

There is considerable debate within the counterterrorism community as to the nature of the risk posed by a terrorist use of a nuclear weapon. On one side stands the "not-if-but-when" crowd who view this type of attack as almost inevitable. Others argue that the risk of a terrorist nuclear attack is greatly exaggerated and that compared to the risk of death from car accidents or even major natural disasters, the risk of being killed or injured in a terrorist nuclear attack is infinitesimal. Still, there is no arguing that a nuclear attack is the epitome of a low-probability, high-consequence attack. As such, Hecker argues that not only is it vital to keep nuclear material out of terrorist hands, but that governments have underestimated the difficulties of that task. He ends his article with a number of policy prescriptions.

Toward a Comprehensive Safeguards System

Keeping Fissile Materials
Out of Terrorists' Hands

Siegfried S. Hecker

What is the likeliest route for terrorists to acquire a nuclear weapon? Theft or diversion of an intact weapon from a nuclear state is the most direct, but not the most probable. A recent National Academies report (National Research Council 2002) and several other extensive studies (Falkenrath, Newman, and Thayer 1998; Ferguson et al. 2004) stress the importance of protecting nuclear weapons but conclude that improvised nuclear devices (IND) built from stolen or diverted fissile materials, either plutonium or highly enriched uranium (HEU), pose a greater threat. The general consensus of nuclear weapons experts is that terrorists would face significant but not insurmountable challenges to build a primitive but devastating nuclear device and that it would most likely be delivered to the intended target by truck, boat, or light airplane.

Fortunately, the technologies and materials required to enrich uranium or construct reactors to produce plutonium are considered beyond the reach of even the most

sophisticated terrorist groups today. Moreover, procurement and construction activities are not easily carried out clandestinely, although recent revelations of the sophistication of A. Q. Khan's proliferation ring raise concerns. So the good news is that terrorists are unlikely to make weapons-usable HEU or plutonium from scratch. The bad news is that they can potentially steal it or buy it because of huge amounts available worldwide, with some being inadequately secured. Keeping these materials out of the hands of terrorists is a much greater challenge than securing nuclear weapons.

The challenge of preventing theft or diversion of fissile materials is now widely recognized. Presidents Bush and Putin addressed the matter in their 2005 Bratislava accord (Bush and Putin 2005). The G-8 leaders pledged cooperation to this end in their Gleneagles Statement on Nonproliferation that same year (G-8 Statement 2005), and it is also addressed by UN Security Council Resolution 1540 on Nonproliferation (United Nations Security Council

SOURCE: *Annals of the American Academy of Political and Social Sciences*, Vol. 607, September 2006, pp. 1121–132.

2004). Moreover, several comprehensive treatises detail this challenge and offer potential solutions (Falkenrath, Newman, and Thayer 1998; Albright and O'Neill 1999; Ferguson et al. 2004; Allison 2004, 2005; Bunn and Weir 2005; Perkovich et al. 2005; National Research Council 2006a, 2006b). Allison (2005) captured the essence of these studies with his challenge to governments around the world to keep fissile materials just as secure as treasures in the Kremlin Armory and gold in Fort Knox.

Securing all fissile material around the world, however, is considerably more challenging than locking it up to a "gold standard." The first section of this article describes those challenges, which are both technical and political. Plutonium and HEU are used in weapons, research, power reactors, and some industrial applications in forms that can be turned into weapons-usable materials with routine chemical processing. Such materials are processed, shaped, transported, stored, and used, and some inevitably wind up in waste streams. After exploring why securing fissile material is more difficult than is generally appreciated, I then go on to assess the components of a comprehensive safeguard system, addressing both the general and specific challenges posed by the current threat environment.

 ## Five Characteristics of Fissile Materials

In this section, I present five reasons why securing fissile material is more difficult than generally appreciated. The characteristics of nuclear material must be understood to establish a comprehensive safeguards system.

Existing Inventories of Fissile Material Are Far Larger Than the Amount Required for a Nuclear Bomb

Most states with nuclear weapons have stopped producing weapons-grade plutonium and HEU; in fact, the United States and Russia are reducing their inventories because they exceed current weapons requirements. The Institute of Science and Security (ISIS) reports that approximately 1.9 million kilograms of HEU and 1.83 million kilograms of plutonium exist worldwide (Albright 2005). Approximately 1.4 million kilograms of

plutonium are found in highly radioactive spent fuel and would not be very attractive to terrorists. The remaining 2.3 million kilograms of weapons-usable fissile material, however, must be protected. But to truly prevent a nuclear terrorist attack, we must be able to account for a few tens of kilograms out of more than 2 million available worldwide.

> *Most states with nuclear weapons have stopped producing weapons-grade plutonium and HEU [highly enriched uranium]; in fact, the United States and Russia are reducing their inventories because they exceed current weapons requirements.*

The results of a recent study on plutonium in the United States underscore the problem of numbers (U.S. Department of Energy (DOE) 1994). The United States produced or acquired 111,400 kilograms of plutonium since 1943. In 1994, the total inventory was 99,500 kilograms. Although there are explanations for the "missing" 11,900 kilograms, the uncertainties between physical inventories and accounting are many times the amount required for a bomb.[1] This same study shows that even in the United States, where nuclear safeguard technologies and methodologies were first developed and applied, the accounting system alone cannot come close to ensuring that significant amounts of nuclear materials are not missing. The investigation also illustrates that confidence in the security of nuclear materials should rest not on the numbers but, instead, must rely on the integrity of the nuclear safeguards system.

Fissile Materials Exist in Every Imaginable Form

These materials are not like gold bricks at Fort Knox. Plutonium and uranium are highly reactive metals that oxidize rapidly, especially in humid conditions or in the presence of hydrogen. Furthermore, plutonium is constantly created and destroyed during reactor operation and transmutes into other elements over time. For weapons applications, plutonium and uranium are used in metallic form—often alloyed with other chemical elements. For reactor applications, they are used in metallic or ceramic (principally oxide) forms. To make weapons or reactor fuel elements, they are processed using industrial processes such as dissolution in acids or salts; gasification; melting and casting; powder processing; electrochemical processing; shaping, machining, welding, or pressing; and waste processing and storage. For plutonium, all such operations are conducted in specially designed laboratories to prevent exposure to airborne plutonium. It is no surprise that operating losses and inventory differences are large when tons of plutonium or HEU are processed. Moreover, large-scale processing without adequate control and accounting leads to the potential of plant operators covertly diverting small but significant quantities of these materials.

Fissile Materials Exist in Many Locations, Not Just in a Few Storage Vaults

Plutonium and HEU exist in enrichment and fuel fabrication facilities, reactors, reprocessing plants, and storage facilities. The materials are typically well secured in weapons. Historically, however, security for nuclear research reactors and facilities has not been adequate. In states that reprocess spent fuel, plutonium also exists in reprocessing plants and mixed-oxide fuel fabrication plants. And of course, these materials exist in transport, all of which are not always secure.

Not only do these materials exist in many locations within one country, they exist in multiple countries. In addition to states with nuclear weapons programs, they exist in countries that have reprocessing plants or use mixed-oxide fuel. The greatest concern, however, is the use of HEU in research reactors around the world. In the United States, the Atoms for Peace program supported building such reactors in more than forty countries. The Soviets had a similar export program. The security environment in many countries was inadequate to protect fresh HEU fuel. Today, roughly 120 research reactors in forty countries still use HEU.

Fissile Materials Are Difficult to Measure and Handle

Safeguard systems must be able to measure fissile materials accurately. Monitoring and accounting of plutonium is hampered because plutonium must be handled in glove boxes or other ventilated enclosures and stored in airtight containers because of its radiotoxicity. Masses for inventories are measured by weighing, destructive assay methods employing wet chemistry, and nondestructive assay methods such as calorimetry (measuring heat content that is related to isotope concentrations) or neutron- and gamma-ray-based radiation measurements. The extraordinary scientific complexity of plutonium metal presents additional challenges. Plutonium exists in seven different crystal structures with varying densities. Adding a few atomic percent gallium or aluminum to pure plutonium will change its density by as much as 25 percent, complicating mass determinations (Hecker 2001). Oxidizing plutonium metal to plutonium dioxide (typical for storage and reactor applications) drops the density of pure plutonium by nearly a factor of two.

Gamma ray detectors are used to make nondestructive measurements of isotopic composition in plutonium and uranium. Chemical analysis is often required to ascertain the precise chemical composition, which is especially important for plutonium because it changes composition with time by transmutation. These measurements and analytical capabilities are not available in many locations that house plutonium or HEU.

Military Secrecy Hampers Safeguards and Transparency

In the early years, information regarding both bomb and reactor materials was classified. The Atoms for Peace program declassified much of this information. But some details about plutonium chemistry and isotopic compositions were kept secret until the Energy Department released its plutonium study in 1994 (a similar study on HEU was never published). Russia and China still keep isotopic and chemical compositions of weapon nuclear materials secret, and most locations and amounts remain out of the public domain.

Implementation of a comprehensive safeguards system is imperative to protect weapons-usable materials worldwide.

Although secrecy is necessary to protect a state's nuclear weapons program, excessive secrecy and, in particular, compartmentalization impede implementation of a rigorous safeguards system. Communication may be limited among sites that produce, use, and dispose of these materials. It can impede accounting, the establishment of systemwide inventories, and the sharing of best practices. Responsible government officials cannot assess systemic vulnerabilities due to a lack of transparency. Likewise, little information allows other nations to judge the adequacy of each others' nuclear materials security. Excessive secrecy also precludes states from sharing crucial information about the chemical and isotopic composition of fissile materials stockpiles, which makes attribution in case of theft or a detonation more difficult.

For these five reasons, simply locking up all of the materials is not a feasible course of action. Many states do not even know what "all" is. Implementation of a comprehensive safeguards system is imperative to protect weapons-usable materials worldwide.

 ## Toward a Comprehensive Safeguards System

Each state that possesses weapons-usable fissile materials must provide for their physical protection, control, and accounting—the three pillars of a rigorous, comprehensive safeguards system. Such a system (nuclear materials protection, control, and accounting, or MPC&A) was first developed in the United States forty years ago. It became the model for the International Atomic Energy Agency (IAEA) international safeguards system. Uneven and incomplete application of domestic and international safeguards contributes to inadequate fissile materials security worldwide today.

The international nuclear safeguards system is designed to assure the international community that states party to the Nuclear Non-Proliferation Treaty (NPT) and similar agreements honor their commitments not to proliferate nuclear weapons. The traditional system attempts to verify nondiversion of declared nuclear materials; it focuses on correctness of a state's declaration. The strengthened safeguards system, which includes the Additional Protocol developed after the Gulf War, expands verification to provide credible assurance of the absence of undeclared nuclear materials; it focuses on completeness of declaration (Goldschmidt 1999).

Although international safeguards are necessary to prevent diversion of nuclear materials by a state, they are not sufficient to prevent theft of weapons usable material by determined individuals or groups. Pellaud (1997) pointed out that IAEA safeguards agreements with more than 130 states cover some 900-plus facilities and locations but only 20,000 kilograms of HEU and 500,000 kilograms of plutonium (including fifty tons of separated plutonium) compared to the roughly 1.9 million kilograms that exist worldwide. Nuclear materials in military programs are not subject to international safeguards. The United States entered into voluntary IAEA safeguards agreements in 1977, but these exclude facilities with direct national security significance. India, Pakistan, and Israel never signed the NPT, and North Korea withdrew.

Adequate security, therefore, depends on rigorous application of domestic safeguards in addition to the international safeguards that may apply. The U.S. domestic safeguard system is designed to protect nuclear materials against external threats such as terrorists and against insider threats. The principal safeguard against external threats is physical protection. The more insidious insider threat also requires additional rigorous internal controls and accounting.

The Soviet Union focused on physical protection (guns, guards, and high fences) along with stringent personnel screening. Its nuclear materials security record was excellent because the Soviet police state with its omnipresent KGB and a system of grave consequences deterred the insider threat as well. However, with the social, political, and economic upheaval that followed the dissolution of the Soviet Union, its past practices become Russia's liability. Physical protection alone is no longer adequate.

Modem safeguard systems combine physical protection with MPC&A. Physical protection consists of measures to protect nuclear material or facilities (and their transportation) against sabotage and theft. Nuclear facilities that require physical protection include all research, development, production, and storage sites; nuclear reactors; fuel cycle facilities; and spent fuel storage and disposal facilities. These measures include guards, fences, and exclusion areas around facilities, in addition to perimeter and interior intrusion detection systems. Measures also include limited access and egress to facilities, buildings, and rooms. Technologies employed include systems such as microwave, electric field, and infrared systems on the perimeter and ultrasound, infrared, and motion detection closed circuit television on the interior. Finally, neutron, gamma ray, and metal detectors at points of egress add an important element of defense.

MPC&A are designed to offer accurate nuclear materials inventory information, control nuclear materials to deter and prevent loss or misuse, provide timely and localized detection of unauthorized removal of materials, and ensure in near real time that all nuclear materials are accounted for and that theft or diversion has not occurred. Proper material control limits the handling of nuclear materials only to authorized and properly identified personnel and ensures that two people are present during nuclear material transactions. It helps track nuclear material from one site to another, from facility to facility, and from room to room. It ensures that there are a limited number of entries and exits, and alarms alert authorities to potential theft or diversion. It identifies nuclear material for tracking purposes.

Modem material accounting also employs statistical and computer-based measures to maintain knowledge of quantities of nuclear material present in each area of a facility. It relies on inventories and material balances to verify the presence of material or to detect a loss. In the United States, the Nuclear Materials Management and Safeguards System (NMMSS) implemented in 1976 contains current and historical data on inventories and transactions involving source and special nuclear materials within the United States and on all exports and imports. It tracks all transactions, including domestic and foreign transfers, operating losses, inventory differences, and burn up (transmutation and fission). Reconciliation of facility books with NMMSS also ensures that control indicators are furnished to those who perform oversight responsibilities and that anomalies are identified.

I provide this level of detail to demonstrate the complexity of securing nuclear materials. Effective MPC&A systems must be integrated with operational and safety practices. In the United States, it remains a challenge to provide adequate protection against changing terrorist threats. In view of the DOE plutonium report, it is not possible to guarantee that kilogram quantities of plutonium are not missing. We must rely on the integrity of the MPC&A system and its application for our confidence that such materials are not outside of state control. U.S. facilities operators must account for every gram of these materials in virtual real time. To declare any of it as an "inventory difference" or "waste" requires rigorous justification and verification.

It is imperative that each state with nuclear facilities implement its own rigorous, comprehensive safeguards system to prevent theft or diversion of weapons-usable materials. Although both countries have made progress in recent years, Russia and China have much work to do to achieve a modern safeguards system. Little is known about Pakistan and India. States that currently employ such systems and the IAEA should expand significantly their efforts to provide technical assistance to these nations. The G-8 should reprioritize its nuclear security financial assistance to help states develop their own rigorous MPC&A systems. These efforts will also help states meet their counterterrorism obligations under UN Security Council Resolution 1540. In addition, the international safeguards system should be strengthened by universal adoption of the Additional Protocol and greater access for IAEA inspectors, along with stricter enforcement by the UN Security Council.

Each state must also develop a complete registry of weapons-usable plutonium and HEU along the lines of the DOE plutonium study. The IAEA already has registry requirements for states that hold safeguarded materials, but as pointed out above, that constitutes only a fraction of the total worldwide. Such registry studies (both public and classified) will help identify historical anomalies and potential vulnerabilities in nuclear material inventories.

 ## Other Vulnerabilities

Since rigorous safeguard systems have not been in place since the advent of nuclear materials, and since many countries still fall short today, nuclear materials could already be in the wrong hands or at least outside state-controlled systems. Fortunately, there are few known incidents of theft to date. The IAEA (2006) illicit nuclear trafficking database shows 196 incidents involving nuclear materials from 1993 to 2004. Only eighteen involved fissile materials, three with kilogram quantities of HEU and three with gram quantities of plutonium.

Each state should enhance its internal detection and tracking capabilities for illicit trafficking of nuclear materials and enhance its border and port security. These efforts should be aided by international efforts such as the DOE Second Line of Defense program, which has helped to install radiation detectors (and train personnel) at airports, seaports, and border crossings in Russia and other states. It is imperative that each state identify past weaknesses and anomalies in fissile material inventories.

Efforts to interdict potential shipments of nuclear materials such as the Proliferation Security Initiative should be strengthened. Increased intelligence sharing is important. Cooperative sting operations may flush out material outside state-controlled systems. Enhanced emergency response capabilities will help manage the consequences of an attack and potentially help disable suspected terrorist devices. Finally, forensics and attribution will be important, both for response and for preventing repeat attacks.

Over the longer term, we must also guard against "mining" of low-grade materials, such as nuclear waste, spent fuel, and lost or abandoned materials. We must also pay much greater attention to safeguarding alternate nuclear materials such as neptunium and americium, which have been produced in multiple-ton quantities and may eventually become a terrorist bomb threat (Albright 2005). We must safeguard any process or nuclear material that is easier to obtain and less costly than building an enrichment plant or a reactor. In addition, the commercial nuclear industry must redouble its safeguards efforts as nuclear power expands worldwide.

Why have terrorists not yet crossed the nuclear threshold? Perhaps it is the lack of access to weapons-usable fissile material. But nuclear attacks may also present an unacceptable level of risk and uncertainty to terrorists—not only risk of injury or death in preparing the mission, but potential failure of the mission. For example, even nuclear-capable states still experience criticality accidents that kill nuclear workers because of misjudgments in material handling. And terrorists are much more certain of success using chemical explosives, with which they have much greater familiarity.

Materials for dirty bombs include roughly a dozen radioisotopes that are ubiquitous in international use as radiation sources for medicine, industry, and agriculture—and readily available to determined terrorists.

Terrorists have also not yet crossed the radiological dispersal bomb (dirty bomb) threshold. A dirty bomb will disperse radioactive materials but not cause a nuclear detonation and mushroom cloud. Materials for dirty bombs include roughly a dozen radioisotopes that are ubiquitous in international use as radiation sources for medicine, industry, and agriculture—and readily available to determined terrorists. A dirty bomb would not kill many people, but it would cause enormous psychological trauma and economic disruption (Ferguson et al. 2004). Regardless of whether or not terrorists are just about to cross the nuclear bomb threshold, we must assume that some of them eventually will. The best preventive measure is to keep the weapons-usable material out of their hands.

Today's Greatest Threats

To deal with today's urgent threats, it is important to consider specifically tailored solutions in addition to the generic recommendations made above. To that end, this section briefly describes what I see as the greatest threats in the current security environment. These six threats represent the highest probability of theft or diversion of several tens of kilograms of weapons-usable plutonium or HEU and these materials getting into the hands of terrorists. Once armed with such materials, terrorists will be able to build an improvised nuclear explosive device and detonate it virtually anywhere in the world.

1. Pakistan heads the list. It has all technical prerequisites: HEU and plutonium; enrichment, reactor, and reprocessing facilities; a complete infrastructure for nuclear technologies and nuclear weapons; largely unknown, but questionable, nuclear materials security; and missiles and other delivery systems. It views itself as threatened by a nuclear India. It has a history of political instability; the presence of fundamental Islamic terrorists in the country and in the region; uncertain loyalties of civilian (including scientific) and military officials; and it is home to A. Q. Khan, the world's most notorious nuclear black marketer. Helping Pakistan secure its nuclear materials during these challenging times is made difficult by the precarious position of its leadership and the

anti-American sentiments of much of its populace. Yet such cooperation is imperative.

2. North Korea is a threat because it has withdrawn from the NPT and has separated roughly forty to fifty kilograms of plutonium (Hecker 2004). Although it is unlikely that this material will be stolen, we cannot dismiss the possibility that plutonium (especially if more is accumulated) may be exported to terrorist groups. This is most likely to occur when North Korea perceives the existence of the regime or its nation as terminally threatened. I believe that the North Korean nuclear threat is solvable, but the slow pace of the six-party talks demonstrates how difficult that is. Preventing the export of plutonium must be highest priority.

3. HEU-fueled research reactors around the world are still operating in about forty countries, many with inadequate safeguards. Fresh fuel for these reactors takes little processing to convert to weapons-usable HEU. These reactors have constituted a grave terrorist threat for three decades. Much has been done to close reactors or retrofit them with low-enriched uranium. The DOE Global Threat Reduction Initiative has increased the pace of these efforts during the past two years. So long as any HEU exists in inadequately safeguarded facilities, however, it presents an unacceptable risk. The solution is an accelerated U.S.-Russian led effort to take back all HEU, backed by G-8 financing.

4. The Russian nuclear complex was most vulnerable in the early and mid-1990s. We are fortunate that nothing really terrible happened in the Russian nuclear complex. Credit goes to the loyalty of Russian nuclear workers and to the Nunn-Lugar Cooperative Threat Reduction program. Over the past five years, the Russian government has enhanced physical security at its sites and reduced economic hardship for its nuclear stewards. But the Russian complex remains excessively large, and the amount of weapons-usable materials is staggering. Unfortunately, cooperative efforts have yielded significant improvements in control and accounting in only a limited number of facilities. To my knowledge, Russia has neither a baseline inventory of fissile materials produced nor a

reconciliation of what exists today with what has been produced and used. There is apparently no incentive to pursue one. Enhanced physical protection and reemergence of strong security services provide only temporary protection. It is time for Russia to make the commitment to and investment in a comprehensive, modern MPC&A system for all of its facilities. The United States can help, but only if Russia takes the lead.

5. Kazakhstan returned Soviet nuclear weapons to Russia under the Nunn-Lugar program, but it did not return all weapons-usable material. Project Sapphire brought nearly six hundred kilograms of HEU from Kazakhstan to the United States in 1994, but there are still HEU-fueled reactors and additional quantities of HEU in Kazakhstan (Albright 2005). It inherited a Soviet BN-350 fast reactor along with several tons of lightly irradiated plutonium. It also inherited the huge former Soviet nuclear test site at Semipalatinsk. U.S.-Kazakh cooperation has enhanced security of reactor installations and the BN-350 fuel. The security of the test site, however, has declined dramatically since the days of Soviet ownership, raising concerns about vulnerable materials that may have been left behind by the Soviets. In addition, the apparent decision to keep the spent BN-350 fuel in Kazakhstan creates significant risks if the Kazakh regime were to take on Iranian-style leadership.

6. Iran is last on this short list because it apparently does not yet have weapons-usable materials. It is clearly determined to get them, however, and when it does, it will move to second place. The only apparent solution in Iran is to prevent it from making weapons usable material. It is imperative that Russia and China support current European and U.S. efforts to prevent the completion of enrichment capabilities and the development of other worrisome nuclear technologies.

This short list illustrates the extreme urgency of the threat posed by loose fissile material. But it also emphasizes the need for tailored nonproliferation strategies. Others may propose a different list with different priorities;

indeed, my long list also includes China, India, and Israel, as well as the additional incremental risk from increased commercial nuclear power, nuclear wastes, and the alternate nuclear materials mentioned above.

I agree with the 2005 Gleneagles communique, which concludes that "the proliferation of weapons of mass destruction (WMD) and their delivery means, together with international terrorism, remain the pre-eminent threats to international peace and security." But the key element, keeping weapons-usable materials out of terrorists' hands, is much more difficult than is generally appreciated. A greater sense of urgency is required—not only on part of the United States but on the part of states that have more benign views of the risks of nuclear terrorism and believe that nuclear proliferation is an American problem. Quite the contrary, loose fissile material must be the top security priority of every nation.

Siegfried Hecker is a visiting professor at the Center for International Security and Cooperation (CISAC) at Stanford University and an emeritus director of Los Alamos National Laboratory. A metallurgist, he focuses on plutonium research, stockpile stewardship, nuclear threat reduction, global nonproliferation, and counterterrorism. He is actively involved with the U.S. National Academies, serving on the Council of the National Academy of Engineering, as chair of the Committee on Counterterrorism Challenges for Russia and the United States, and as a member of the National Academy of Sciences Committee on International Security and Arms Control Nonproliferation Panel.

 Note

1. The "missing" 11,900 kilograms were explained as follows: 3,400 kg expended in wartime and tests, 2,800 kg declared as inventory differences, 3,400 kg as waste (normal operating losses), 1,200 kg as fission and transmutation, 400 kg as decay and other removals; 100 kg in U.S. civilian industry; 700 kg exported to foreign countries; and a 100 kg rounding difference along with classified transactions. Inventory differences are defined as the difference between the quantities of material in accounting records compared to those determined in physical inventories. They were previously identified as "material unaccounted for," which included operating losses. Today, the operating losses are counted separately, and inventory differences result primarily from statistical measurement uncertainties; recording, reporting, and rounding errors;

uncertainties of the amount of material held up in the processing plant; measurement uncertainties because of wide variations of material that contain fissile materials (material matrix) during processing; uncertainties associated with waste; and unmeasured materials associated with accidental spills or releases of materials. Waste (normal operating losses) is defined as intentional removals from the inventory as waste because they are technically or economically unrecoverable. Examples include discharges to cribs, tanks, settling ponds, or disposal facilities (burial sites).

 References

Albright, D. 2005. *Global stocks of nuclear explosive materials: Summary tables and charts.* July 12, updated September 7. Washington, DC: Institute for Science and International Security.

Albright, D., and K. O'Neill. 1999. *The challenges of fissile materials control.* Washington, DC: Institute of Science and International Security.

Allison, G. 2004. How to stop nuclear terror. *Foreign Affairs* 83: 64–74.

———. 2005. *Nuclear terrorism: The ultimate preventable catastrophe.* New York: Owl Books, Henry Holt and Co. LLC.

Bunn, M., and A. Weir. 2005. *Securing the bomb 2005: A new global imperative.* Cambridge, MA/Washington, DC: Project on Managing the Atom, Harvard University/Nuclear Threat Initiative.

Bush, G. W., and V. V.: Putin. 2005. Joint statement on nuclear security issued in conjunction with the summit in Slovakia, Feb. 24, 2005. Washington, DC: The White House, Office of the Press Secretary.

Falkenrath, R. A., R. D. Newman, and B. A. Thayer. 1998. *Americas Achilles' heel: Nuclear, biological, and chemical terrorism and covert attack.* Cambridge, MA: MIT Press.

Ferguson, C., W. C. Potter, A. Sands, L. S. Spector, and F. L. Wehling. 2004. *The four faces of nuclear terrorism.* Monterey, CA: Monterey Institute—Center For National Security Studies.

G-8 Statement on Nonproliferation, Gleneagles, Scotland, 2005. http://www.fco.gov.uk/Files/kfile/PostG8_Gleneagles_Counter Proliferation.pdf.

Goldschmidt, P. 1999. The IAEA safeguards system moves into the 21st century. *Supplement to the IAEA Bulletin* 4L:Sl-S20.

Hecker, S. S. 2001. The complex world of plutonium science. *MRS Bulletin* 26:672-78.

———. 2004. *Visit to the Yongbyon Nuclear Scientific Research Center in North Korea.* United States Senate Committee on Foreign Relations Hearing, 108th Cong., 2nd sess., January 21, 2004. http://www.senate.gov/-foreign/hearings/2004/hrg04012la.html.

International Atomic Energy Agency (IAEA). 2006. *Illicit nuclear trafficking: Facts & figures.* February 25. Vienna, Austria: IAEA.

National Research Council. 2002. *Making our nation safer: The role of science and technology in countering terrorism.* Washington, DC: National Academies Press.

———. 2006a. *Protection, control, and accounting of nuclear materials: International challenges and national programs.* Washington, DC: National Academies Press.

____. 2006b. *Strengthening long-term nuclear security: Protecting weapon-usable material in Russia.* Washington, DC: National Academies Press.

Pellaud, B. 1997. IAEA safeguards: Experiences and challenges. Presented at the International Atomic Energy Agency Symposium on International Safeguards, Vienna, Austria, October. http://f40.iaea.org/worldatom/Periodicals/Bulletin/Bull394/pellaud.html.

Perkovich, G., J. T. Mathews, J. Cirincione, R. Gottemoeller, and J. Wolfsthal. 2005. *Universal compliance: A new strategy for nuclear security.* Washington, DC: Carnegie Endowment for International Peace.

United Nations Security Council. 2004. *Resolution 1540 on Nonproliferation of Weapons of Mass Destruction.* Adopted April 28. New York: UN.

U.S. Department of Energy. 1994. *Plutonium: The first 50 years.* Washington, DC: U.S. Department of Energy.

REVIEW QUESTIONS

1. What are the different ways non-state actor groups could acquire a nuclear weapon? Which one does Hecker believe is the most likely?

2. What problems does Hecker identify in tracking and accounting for fissile material?

3. What evidence does Hecker provide for his concerns? Are you convinced?

4. Identify two of the policy prescriptions that Hecker proposes for dealing with the threat of nuclear terrorism. Which of the two do you believe would be the most effective? Why?

READING 21

National security has long been thought the purview of law enforcement, intelligence, diplomatic, and military agencies. Tucker and Kadlec argue that national security efforts should also include public health, agricultural, and zoonotic experts and officials. They also argue that concerns about bioterrorism should not outweigh concerns about naturally emerging infectious diseases and that the most current threat stems from newly emerging infections that are resistant to current antibiotics. The authors conclude their article by arguing that improving the public health system to deal with these naturally emerging threats also improves our ability to deal with the bioterrorist threat.

Infectious Disease and National Security

Jonathan B. Tucker and Robert P. Kadlec

 In Brief

With the global AIDS pandemic and the emergence or re-emergence of other deadly pathogens, infectious disease increasingly poses a major threat to U.S. national security. Because emerging infections and bioterrorism are two sides of the same coin, an improved ability to detect and contain such outbreaks will require better

SOURCE: Strategic Review, Spring 2001, pages 12–20.

communication and coordination among the medical, veterinary, public health, defense, law enforcement, and intelligence communities.

During the 1960s and 1970s, powerful antibiotic drugs and vaccines appeared to have banished the major infectious diseases from the United States, leading to complacency and the neglect of programs for disease surveillance and prevention. Over the past two decades, however, infectious disease has returned to the United States with a vengeance. Since 1980, the U.S. death rate from AIDS and other infectious diseases has increased by about 4.8% per year, compared with an annual decrease of 2.3% for the 15 years before 1980.[1]

According to an unclassified report by the National Intelligence Council, "New and reemerging infectious diseases will pose a rising global health threat and will complicate U.S. and global security over the next 20 years. These diseases will endanger U.S. citizens at home and abroad, threaten U.S. armed forces deployed overseas, and exacerbate social and political instability in key countries and regions in which the United States has significant interests."[2] To give but two examples, AIDS is undermining the economies and civil societies of many countries in sub-Saharan Africa, and the deterioration of the Russian public health system, as reflected in declining life expectancy and birth rates, is sapping that country's social and economic vitality.[3]

The United States is currently poorly organized to address the health and security challenges posed by the global threat of infectious diseases. An examination of the 1999 outbreak of West Nile encephalitis in New York City supports this conclusion and gives rise to several policy recommendations.

The Return of Infectious Disease

Worldwide since 1973, 20 well-known diseases, including tuberculosis, malaria, and cholera, have re-emerged in more virulent or drug-resistant forms or have spread geographically. Over the same period, scientists have identified at least 30 previously unknown diseases for which no cures exist.[4] Examples of emerging infections include the deadly outbreaks of Ebola hemorrhagic fever in Africa, the worldwide AIDS pandemic, Lyme disease, hepatitis C, bovine spongiform encephalopathy (BSE or "mad cow disease"), Nipah virus, and new strains of influenza. AIDS was not recognized until the 1980s but now infects some 36 million people worldwide and kills 3 million annually.

Several factors have contributed to the problem of emerging infections. The inappropriate use or overuse of antibiotic drugs for treating humans and livestock has fostered the evolution of resistant strains of tuberculosis and other bacterial diseases, even as development of new generations of antibiotics has lagged.[5] Climate change and ecosystem disturbances, such as clearing rainforests for economic gain or human settlements, have altered the geographical distribution of disease vectors such as rodents, monkeys, and mosquitoes, increasing their contact with humans. Rapid population growth and rural-urban migration have given rise to "megacities," enabling diseases that once remained isolated in rural areas to spread to large urban populations. Finally, the rising volume of tourism, trade, and imported agricultural products associated with economic globalization has created new opportunities for the introduction of disease vectors and pathogens.

The first line of defense against the importation of infectious disease is at the nation's borders. U.S. customs and public health officials attempt to stop infected travelers and agricultural pests at airports and seaports by means of immigration screening, food insperion, quarantine, and fumigation. Yet because most U.S. cities are within a 36-hour commercial flight of any part of the world, or less than the incubation period of many infectious diseases, infected individuals may not be visibly ill when they cross a U.S. border. The risk of disease importations is greatest in major hubs of global commerce such as New York City, Los Angeles, and Miami.

> The United States is currently poorly organized to address the health and security challenges posed by the global threat of infectious diseases

A parallel concern is the potential for the deliberate introduction of disease agents by terrorists. This issue first came to the attention of U.S. policy makers after a Japanese cult called Aum Shinrikyo released the chemical nerve agent sarin on the Tokyo subway in March 1995, killing 12 people and injuring scores more. A subsequent investigation revealed that on at least nine occasions in 1990 and 1993, the cult had disseminated a biological agent (anthrax or botulinum toxin) in an attempt to inflict mass casualties, but had failed for technical reasons.[6] Although it is unlikely that a small terrorist group would have the technical and financial resources to carry out a major bioterrorist attack, the necessary know-how, cultures, and dissemination equipment might be provided by a state sponsor, or by recruiting scientists and engineers formerly employed in a state-level biowarfare program.[7]

Defense analysts also worry about the possible use of biological agents by hostile states as a means of "asymmetric warfare" David and Goliath strategics in which small countries would seek to circumvent or blunt the conventional military supremacy of the United States and its ability to intervene in regional conflicts.[8]" Such strategies might involve the use of disease agents to attack troops or civilians, destroy U.S. crops or livestock, or contaminate the nation's food supply. Biological attacks could be carried out on a scale large enough to hamper or deter U.S. intervention abroad, yet without crossing the mass-casualty threshold that could credibly trigger nuclear retaliation. Even in the face of U.S. deterrent threats, a rogue state or terrorist group that believed it could carry out an attack without attribution might be tempted to do so, particularly in the heat of crisis or war.

A covert biological attack could be mistaken for a natural outbreak of disease, particularly if it involved the use of an indigenous pathogen delivered by food or water contamination rather than by a clearly artificial means of dissemination. In 1984, for example, members of the Oregon-based Rajneeshee cult used salmonella bacteria to contaminate 10 restaurant salad bars in a trial run of a scheme to manipulate a local election by making large numbers of voters too sick to go to the polls. After 751 people fell ill with food poisoning, public health investigators concluded that the outbreak had resulted from natural sources. The true cause did not emerge until a year later, when a member of the cult confessed to the crime.[9]

The West Nile Outbreak

Rapidly detecting and containing a major epidemic—whether the result of a natural emerging infection or a terrorist attack—would pose significant challenges to the U.S. public health system as it is currently configured. City and state health authorities would be on the front lines, backed up by the medical detectives and virus hunters at the Centers for Disease Control and Prevention (CDC) and other federal agencies. The 1999 epidemic of West Nile encephalitis in New York City revealed some serious gaps in the current system that must be remedied if the nation is to be prepared for future outbreaks.

The first manifestation of the epidemic appeared in early July 1999, when common birds such as sparrows, robins, and crows began to die in unusual numbers in northern Queens and the South Bronx. The sick birds were unable to fly and had trouble balancing, symptoms of neurological damage. One month later, humans began to be stricken with an illness characterized by encephalitis, or inflammation of the brain, although a connection with the bird die-off was not suspected at the time.

In late August, city health investigators identified a total of eight patients with severe encephalitis at hospitals in northern Queens. Normally, only nine cases of the disease were reported citywide in an entire year.[10] The patients were all elderly and had spent time outdoors in the evening hours, engaged in activities such as gardening or smoking.[11] Recognizing the possibility of an epidemic, health department officials called doctors at 70 hospitals around the city and identified 30 additional cases.[12] Patient blood samples, shipped in early September to the CDC's Division of Vector-Borne Infectious Diseases in Fort Collins, CO, tested positive for St. Louis encephalitis, a viral disease endemic to the U.S. Southeast that can be transmitted to humans through the bite of a mosquito that has previously fed on the blood of an infected bird. The diagnosis was unexpected: in the previous 5 years, only nine cases of St. Louis encephalitis had been reported in New York State, and the disease had never before been seen in New York City.[13] In the absence of a vaccine or treatment, the only way to contain the epidemic was through public education and mosquito control.

Over Labor Day weekend, several exotic birds in an outdoor cage at the Bronx Zoo died; necropsies showed brain lesions indicating possible encephalitis. The zoo pathologist suspected a link between the bird die-off and the human disease outbreak, but when she offered to collaborate with the researchers and send them hard tissue specimens for analysis, they flatly refused. Increasingly concerned and frustrated, she called a veterinary pathologist she knew at the U.S. Army

Medical Research Institute of Infectious Diseases (USAMRIID) at Fort Detrick, MD. Although USAMRIID does not normally respond to civilian requests, the personal contact agreed to analyze the samples. In the interim, the New York State virology laboratory failed to confirm the CDC's diagnosis of St. Louis encephalitis as the cause of the human epidemic. In mid-September, scientists attending a conference in Albany on the encephalitis outbreak suggested sending patient samples to a virology lab at the University of California, Irvine, for a second opinion.[14]

Within days of each other, researchers at the California lab, USAMRTTD, and the CDC independently determined that the human and bird deaths were linked not to St. Louis encephalitis virus but to a close relative called West Nile virus. On September 27, the CDC admitted officially that its initial identification had been incorrect.[15] Like St. Louis encephalitis, West Nile virus is transmitted to birds and humans by mosquitoes. First isolated in 1937 in the West Nile district of Uganda, the virus has been reported throughout East Africa and western India, as well as in Australia, Egypt, Israel, South Africa, and Eastern Europe.[16] Infection usually produces a mild illness involving fever, headache, sore throat, backache, fatigue, rash, nausea, diarrhea, and breathing problems.[17] Serious neurological injury and death resulted in 5 to 10% of cases, mainly children, elderly people, and individuals with a weakened immune system.

> The 1999 epidemic of West Nile encephalitis in New York City revealed some serious gaps in the current system that must be remedied if the nation is to be prepared for future outbreaks

By the end of 1999, the West Nile epidemic had spread into the tri-state area of New York, New Jersey, and Connecticut, sickening 62 people and killing seven of them, as well as thousands of wild birds throughout the region and several horses on Long Island.[18] The strain of the virus responsible for the U.S. outbreak was nearly identical to one that had been isolated in Israel in 1998.[19] Epidemiologists speculated that the virus could have found its way to New York City by several possible routes, including the migration of infected birds, the travel of infected persons from the Middle East, the illegal importation of birds or other domestic pets, or the unintentional introduction of virus-infected mosquitoes on an aircraft bound for John F. Kennedy International Airport.[20]

The three-week delay in the diagnosis of West Nile virus revealed some important gaps in the U.S. public health system. CDC scientists investigating the outbreak had suffered from "tunnel vision" by screening only for encephalitis viruses commonly found in the United States and neglecting those linked to foreign outbreaks or possibly developed for biological terrorism. The CDC also failed to recognize the link between the human and animal outbreaks. Community newspapers in northern Queens had reported bird die-offs as early as late June 1999, or five weeks before the human outbreak began. If the veterinary investigation had begun earlier and been pursued more aggressively, it is possible that the human epidemic could have been mitigated or even averted.[21]

The West Nile incident suggests that rapid identification and containment of an emerging infection calls for greater agility than current institutional arrangement permits. Throughout the outbreak investigation, communication among the 18 participating local, state, and federal agencies was complex and difficult, and was achieved primarily through conference calls lasting several hours.[22] Most of the cooperation between animal health and public health agencies resulted from informal relationships rather than official coordinating mechanisms. Moreover, the various diagnostic laboratories lacked a consistent approach and methodology, making it difficult to compare results. This experience indicates the need for better information sharing among agencies, as well as a common database for disease surveillance and laboratory tracking.

The next emerging or re-remerging disease introduced into the United States may be far more deadly. In the worst-case scenario, an emerging pathogen would have the attributes of the 1918 strain of influenza virus, which was highly contagious through the air and uncharacteristically lethal to young, healthy individuals. Known as Spanish Flu, this disease caused a global pandemic that claimed more than 20 million lives in less than two years.[23] The speed at which the U.S. public health system identifies and contains such an outbreak could mean the difference between life and death for a large number of Americans.

Infectious Disease Surveillance

In the event of an emerging infection or covert bioterrorist attack, medical and public health practitioners would provide the country's first line of defense.[24] An outbreak would most likely be detected when the initial victims sought treatment at emergency rooms and doctors' offices. Rapid containment at this stage through isolation, vaccination, and drug therapy could save lives and, in the case of a contagious agent, prevent further spread. If health care providers are to be the sentinels of a future epidemic, however, they must possess the necessary knowledge and professional awareness. Timely diagnosis will be delayed if general practitioners and emergency-room physicians are unfamiliar with the signs and symptoms of emerging infections or bioterrorist agents.

Infectious diseases surveillance often involves requiring physicians to report certain diseases or "syndromes" (clusters of symptoms) to local health departments. Even if such reporting is mandatory, however, it is often incomplete. Emergency room doctors may be too busy to comply or may not know to whom to report. For this reason, it is essential to establish simple reporting mechanisms and clear communication channels among medical practitioners and city and state health departments. To respond effectively whenever physicians call, day or night, health departments will need more expertise in infectious diseases, epidemiology, and information technology; expanded diagnostic laboratory capacity and electronic communication links; and good working relationships with emergency management agencies and local law enforcement.[25]

A complementary approach to epidemiological surveillance, known as "data-mining," involves the monitoring and analysis of a variety of public health indicators that can provide early warning of an unusual outbreak of infectious disease. The New York City Office of Emergency Management has created a monitoring system that collects the following information on a regular basis:

- Admissions to hospitals through the emergency room for influenza-like illness, and fever of unknown origin;
- Infectious disease admissions to hospital intensive-care units;
- Volume of calls to Emergency Medical Services (EMS) for patients with severe respiratory or gastrointestinal symptoms;
- Pharmacy sales of cold and cough medicines, antibiotics, and anti-diarrheal medication;
- Cases of influenza-like illness in nursing homes;
- Numbers of city employees calling in sick to public schools, fire departments, and police departments and
- Infectious disease-related deaths, and deaths of people under 50 who are otherwise healthy (from the Medical Examiner's offices).

Although most of the data collection is now performed manually, the city plans to automate it on a health alert network or a web site so that a variety of public health indicators can he monitored on a continuous basis.[26]

Problems of Coordination

Today, the U.S. response to a serious epidemic of infectious disease arising from a natural emerging infection or an act of bioterrorism would be seriously constrained by poor communication and coordination among the diverse array of federal, state, and local agencies responsible for public health, animal health, law enforcement, and intelligence collection. Improving interagency communication and coordination will require overcoming some formidable obstacles, including fragmented jurisdiction over the issue and cultural gaps among agencies.

Many emerging infectious diseases and biowarfare agents are "zoonotic," meaning that they originate in animals but can infect humans. Given the considerable overlap of animal and human pathogens, animals can serve as useful sentinels for outbreaks of zoonotic diseases. Sheep, for example, are far more sensitive to anthrax infection than humans. Nevertheless, the West Nile investigation exposed a major gap between the veterinary and public health communities. Although the key to identifying the causative agent lay in merging information from the parallel investigations of the bird and human outbreaks, problems of interagency communication and coordination delayed the correct diagnosis for three weeks.[27]

Most of the cooperation between animal health and public health agencies in the West Nile incident resulted from informal relationships rather than official coordinating mechanisms

Why the disconnect? The expert communities that address health issues related to people, domesticated animals, and wildlife are separated organizationally, geographically, and jurisdictionally, but infectious diseases do not respect these artificial boundaries. State and local veterinary agencies and the U.S. Department of Agriculture focus on the health of domestic pets, horse livestock, and other economically important species. Particularly low priority and funding accrue to the health of wildlife, particularly nonendangered species such as crows and rats, which are the responsibility of Parks Departments and animal control officers. Monitoring the health of zoo animals is another "gray area" with no clear leadership. Only six zoos in the United States employ full-time pathologists.[28]

The dual threats of emerging infections and bioterrorism also pose major conceptual and technical challenges for the U.S. intelligence community. During the West Nile investigation, the belated diagnosis of the virus raised red flags with CIA analysts because of an eerie coincidence. In April 1999, Michael Ramadan, a self-declared Iraqi defector who claimed to have worked for twenty years as a body-double for Saddam Hussein, published a memoir in England titled *In the Shadow of Saddam*. He asserted that in 1997 the Iraqi leader had ordered the development of a highly virulent strain of West Nile virus as a bioterrorist weapon.[29] Additional concern was raised by the fact that during the 1980s, the CDC had shipped an Israeli strain of West Nile virus to a microbiologist in Basra, Iraq, ostensibly for public health research.[30] Nevertheless, further analysis by U.S. law enforcement, public health, and intelligence experts turned up no evidence that the Iraqis had developed West Nile virus as a biological weapon.[31]

This incident demonstrated the difficulty of distinguishing a natural outbreak from a deliberate attack. One problem facing the CIA was that monitoring bioterrorist threats requires a good technical understanding of infectious diseases and epidemiology, fields with which national security experts are generally unfamiliar. Trained epidemiologists would have recognized immediately that the West Nile virus was a poor candidate as a bioterrorist weapon because of its relatively low virulence and reliance on mosquitoes for transmission to humans. The putative terrorists would have had to import virus-infected mosquitoes, release them, and wait for them to spread the disease—an unlikely scenario.

Although information exchanges between U.S. public health and intelligence specialists have occurred on an ad hoc basis, efforts to institutionalize this process face some major obstacles. One problem is that public health and national security agencies have very different organizational cultures. Infectious disease experts and epidemiologists view themselves as members of the international scientific community, publish their research results and interact openly with colleagues from politically sensitive countries. Few scientists at the CDC, the National Institutes of Health, or the Department of Agriculture are cleared for classified information or have access to encrypted phone and fax lines. In contrast, the national security community protects operational security and intelligence sources and methods by imposing restrictions on the flow of information through security clearances, compartmentalization, and "need to know"-practices that are anathema to public health specialists.

Because of these cultural differences, a sense of mutual distrust exists between the two communities. For national security agencies, working closely with medical doctors and epidemiologists poses a risk to the security and effectiveness of their operations. Public health practitioners, for their part, are concerned that cooperation with defense, law enforcement, or intelligence agencies could taint their public image. During an investigation of a disease outbreak, local officials may be willing to speak openly to scientists pursuing a public health mission, but they will be much less forthcoming if the investigators are perceived to be surrogates for law enforcement agencies that could pursue legal prosecutions. At the international level, the political sensitivities are even greater. Any link between the CDC and the U.S. intelligence community—however innocent—would arouse intense suspicions abroad that could undermine the ability of public health experts to investigate natural disease outbreaks in foreign countries.[32]

Policy Recommendations

New policies are needed to overcome the current institutional obstacles and improve the nation's preparedness for rapidly identifying and containing outbreaks of disease associated with emerging infections or bioterrorism. We must:

- *Strengthen global epidemiological surveillance.* The United States requires a "defense in depth" against emerging infections that includes improved surveillance and reporting at all levels, from local to international. Clearly, it would be preferable to identify and contain a dangerous epidemic while it is still outside U.S. borders. To this end, the United States should expand technical and financial assistance to the World Health Organization in its ambitious but underfunded effort to build a global network of disease-reporting stations, diagnostic laboratories, and epidemiological response teams.

- *Assess strategies for* disease *surveillance.* The pilot surveillance system being developed by New York City should be carefully evaluated and, if the results are promising, implemented nationwide with technical and financial support from the CDC. To facilitate prompt and accurate data reporting, the CDC should also expand its current efforts to establish electronic disease-reporting networks at the state and local levels. At present, half of the emergency rooms in the United States are not connected to the Internet.[33]

- *Conduct awareness training of primary care providers.* State and local public health departments should make concerted efforts to reach out to emergency room doctors, family practitioners, general practitioners, physicians' assistants, and nurse-practitioners, who would be the "first observers" in the event of an unusual outbreak of disease. This outreach effort will need to be multifaceted, involving both physician education about what unusual signs and symptoms to look for and how to report them to the appropriate public health authorities. Such training programs should be subsidized by federal grants to medical and veterinary schools, professional societies, and certification organizations, and might be made a formal requirement for obtaining a medical license.

- *Bridge the gap between the public health and animal health communities.* Because outbreaks of zoonotic disease in animals can provide advance warning of an impending human epidemic, it is essential to establish a surveillance network for livestock and wild animals in which veterinarians observe and report unusual patterns of animal disease to state and local public health departments. Epidemiologists also need to gain a better understanding of the complex relationships between human and animal health. To some extent, the West Nile outbreak has been a catalyst for greater interdisciplinary cooperation among veterinarians, physicians, ecologists, and wildlife biologists.[34]

The dual threats of emerging infections and bioterrorism also pose major conceptual and technical challenges for the U.S. intelligence community.

- *Bridge the gap between federal public health agencies and the U.S. intelligence community.* Although the CIA has recruited bioscientists for its analytical staff, this solution is not optimal because infectious disease experts need to interact freely with colleagues from other countries if they are to remain current and well informed. A better approach would be to provide for institutionalized exchanges of people and training between public health and intelligence agencies. Rotating scientists from the CDC and USAMRIID into the intelligence community for temporary details of six months to a year would provide a continuous infusion of expertise, although discretion would be needed to avoid undermining the CDC's effectiveness in the international arena. Another option would be to create an intra-governmental coordinating body of experts from public health and intelligence agencies that would meet periodically to review unusual outbreaks of infectious disease when some suspicion of covert biowarfare or bioterrorism exists.

- *Strengthen basic scientific research on bioterrorist threat agents.* Knowledge of the pathophysiology, virulence factors, immunology, and genomic structure of disease agents is vital for the development of diagnostic tests, therapeutics, and vaccines. Yet such knowledge is limited for the roughly two dozen classical biological warfare agents and is almost nonexistent for the more than 100 microbial pathogens of potential bioterrorist or biowarfare concern.[35] Another need is to expand research on zoonotic diseases. Unlike countries such as Canada, the United States lacks an animal disease laboratory capable of studying deadly zoonotic infections under maximum-containment (Rio safety Level 4) conditions. In early 2000, the U.S. Congress denied funding to build a BL-4 laboratory at the U.S. Department of Agriculture's Plum Island Animal Disease Center on Long Island, NY, because of concerns about its proximity to New York City.[36] As an alternative, a BL-4 laboratory for the study of zoonotic agents should be established in a less populated area.

- *Enhance the role and capabilities of the U.S. Army Medical Research Institute of Infectious Diseases (USAMRIID) at Fort Detrick, MD.* As the Department of Defense's premier laboratory for the medical aspects of biological warfare defense, USAMRIID houses a BL-4 laboratory for research on contagious and incurable viruses that infect humans. Because of its military mission, the Army laboratory has unique expertise in the diagnosis, pathology, and therapeutics of biowarfare agents that complements the CDC's strengths in the area of emerging infections. USAMRIID is at a disadvantage, however, in that it is subordinated to the Army budget and personnel system. Because USAMRIID is a national resource, it should receive funding and personnel levels commensurate with that role.

- *Create incentives for vaccine and antibiotic development and production by the private sector.* A logical next step in the creation of a national infrastructure for meeting the challenges of emerging diseases and bioterrorism is to improve the country's development and production base for vaccines and anti-microbial drugs. As highlighted by the limited supply of seasonal influenza and tetanus vaccines, the number of commercial pharmaceutical companies producing vaccines for public health use has steadily declined. Moreover, because the pharmaceutical industries do optimize efficiencies by producing medications in batches in response to market demand, companies have a limited capacity to "surge" the production of a drug in the event of an unexpected rise in demand or a major medical emergency. To reverse these negative trends, the U.S. government should encourage the pharmaceutical industry to reinvest in antibiotic and vaccine development and production by means of tax incentives and a limited safe-harbor from legal liability.

In conclusion, the natural emergence of a deadly and contagious infectious disease like the Spanish Flu of 1918, or the deliberate release of a pathogen as an act of bioterrorism or asymmetric warfare, could limit in a major loss of life and social disruption. Prudent investments in medical education, public health surveillance, basic science research, improved intra-agency coordination, and expansion of the nation's pharmaceutical production capacity offer the best forms of insurance against such a catastrophe.

 Notes

1. National Intelligence Council, *The Global Infectious Desease Threat and Its Implications fur the United States* NIE 99–17D, January 2000 [www.cia.gov/cia/publications/nie/report/nic99–17d.html].

2. Ibid.

3. Michael Wines, "An Ailing Russia Lives a Tough Life That's Getting Shorter," *New York Times,* December 3, 2000, p. A1.

4. National Intelligence Council, The Global Infectious Disease Threat.

5. Denise Grady, "Drug-Resistant Bacteria Still on the Rise," *New York Times,* December 28, 2000, p. A17.

6. David E. Kaplan, "Aum Shinrikyo (1995)," in Jonathan B. Tucker, ed., *Toxic Terror: Assessing Terrorist Use of Chemical and Biological Weapons* (Cambridge, MA: MIT Press, 2000), pp. 207–226.

7. Jonathan B. Tucker, "Chemical and Biological Terrorism: How Real a Threat?" *Current History,* Vol. 99, no. 636 (April 2000), pp. 147–153.

8. Jonathan R. Tucker, "Asymmetric Warfare," *Forum for Applied Research and Public Policy,* Vol. 14, no. 2 (Summer 1999), pp. 32–38.

9. Seth Carus, "The Rajneeshees (1984)," in Tucker, ed., *Toxic Terror,* pp. 115–137.

10.   Marrelle Layton, M.D., M.P.H., "Outbreak Surveillance and Management at the State and Local Level: Current, Realities," presentation at the Second National Symposium on Medical and Public Health Response to Bioterrorism, Washington, DC, November 28, 2000, preliminary transcript.

11.   U.S. Centers for Disease Control and Prevention, "Outbreak of West Nile-Like Encephalitis—New York, 1999," *Morbidity and Mortality Weekly Report,* Vol. 48, no. 38 (October 1, 1999), pp. 845–849.

12.   Jennifer Steinhauer and Judith Miller, "In New York Outbreak, Glimpse of Gaps in Biological Defenses," *New York Times,* October 11, 1999.

13.   Minority Staff; Senate Government Affairs Committee, "Expect the Unexpected: The West Nile Virus Wake-Up Call," Report to Senator Joseph I. Lieberman, Ranking Member, July 24, 2000 [www.senate.gov/-lieberman/newsite/mos.cfm].

14.   US. General Accounting Office, "West Nile Outbreak: Lessons for Public Health Preparedness," Report No. GAO/HEHS-00-180, September 2000, p. 51.

15.   David Barstow, "With New Virus, Experts Suspect More Died of Encephalitis," *New York Times,* September 27, 1999.

16.   Lawrence K. Altman, "The Doctor's World: Encephalitis Outbreak Teaches an Old Lesson," *New York Times,* September 28, 1999.

17.   Andrew C. Revkin, "Clues to an Alien Virus: Scientists Begin to Crack the Mysteries of West Nile," *New York Times,* August 8, 2000.

18.   Jennifer Steinhauer, "West Nile Virus Data Cheers, and Puzzles, Health Officials," *New York Times,* September 25, 2000.

19.   R. S. Lanciotti, J. T. Roehrig, V. Deubel, J. Smith, et al., "Origin of the West Nile Virus Responsible for an Outbreak of Encephalitis in the Northeastern United States," *Science,* Vol. 286, no. 5448 (December 17, 1999), pp. 2333-2337. See also, Lawrence K. Altman, "Scientists Say Virus in Encephalitis Outbreak Is Like an Israeli Strain," *New York Times,* December 5, 1999.

20.   Revkin, "Clues to an Alien Virus."

21.   Layton, "Outbreak Surveillance and Management."

22.   Steinbauer and Miller, "In New York Outbreak, Glimpse of Gaps in Biological Defenses."

23.   Gina Kolata, *Flu: The Story of the Great Influenza Outbreak of 1918* (New York: Farrar, Straus & Giroux, 1999), p. 7.

24.   Andrew C. Revkin, "Mosquito Virus Exposes a Hole in the Safety Net," *New York Times,* October 4, 1999.

25.   Layton, "Outbreak Surveillance and Management."

26.   Dr. Jerome M. Hauer, former director, Mayor's Office of Emergency Management, New York City, interview with Tucker, December 5, 2000.

27.   U.S. General Accounting Office, "West Nile Outbreak," p. 40.

28.   Tracy McNamara, presentation at Workshop on Agro-Terrorism, Cornell University, November 13, 2000.

29.   Richard Preston, "West Nile Mystery," *The New Yorker;* October 18 and 25, 1999, pp. 90–108.

30.   Jonathan B. Tucker, "Lessons of Iraq's Biological Warfare Programme," *Arms Control/Contemporary & Security Policy,* Vol. 14, no. 3 (December 1993), p. 238.

31.   Vernon Loeb, "CIA Finds No Sign N.Y. Virus Was an Attack," *Washington Post,* October 12, 1999, p. A2.

32.   Christopher F. Chyba, *Biological Terrorism, Emerging Diseases and National Security* (New York: Project on World Security, Rockefeller Brothers Fund, 1998), pp. 24–25.

33.   Tara O'Toole, Johns Hopkins Center for Civilian Biodefense Studies, personal communication to Tucker, January 22, 2001.

34.   Kirk Johnson, "West Nile Side Effect: A Wealth of Data on Wildlife Death," *New York Times,* November 27, 2000

35.   Deputy Group Chief, CIA Non-Proliferation Center, personal communication to Kadlec, September 27, 2000.

36.   Judith Miller, "Long Island Lab May Do Studies of Bioterrorism," *New York Times,* September 22, 1999, pp. Al, B6; John Rather, "Congressman Opposes Disease Center Upgrade," *New York Times* [Long Island Weekly], January 30, 2000, p. 5.

REVIEW QUESTIONS

1.   What is the relationship between infectious diseases and terrorism? Why do the authors view infectious disease as a national security threat?

2.   What are the various threats from infectious pathogens described by the authors? Which ones do they feel pose the greatest threat?

3.   What is the relationship between infectious diseases found in humans and those found in livestock and other animals? Why should diseases in animals be of concern to counterterrorism officials?

4.   Identify two public health policy changes recommended by the authors. Which of the two do you believe would be the most effective? Why?

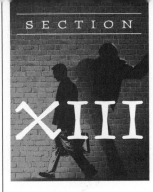

XIII

Law and Terrorism

Domestic and International Legal Regimes

At the end of this section, students will be able to:

- Describe the main types of legal regimes available to deal with the prevention of and response to terrorist attacks.
- Describe the key legal provisions discussed in this section.
- Discuss the difficulties of defining terrorism in different legal regimes.
- Discuss and evaluate the differences between civil and criminal approaches to dealing with terrorism.
- Analyze the effectiveness of terrorism laws in balancing short-term and long-term response and prevention needs.
- Assess the current role of the United Nations in responding to international terrorism.

Section Highlights

- What's in a Legal Regime?
- Definitions and the Law
- U.S. Domestic Law on Terrorism and Homeland Security
- International Law and Terrorism: Efforts of the United Nations

 Introduction

This section examines the legal regimes dealing with terrorism and homeland security in both the domestic (U.S.) and international arenas. For purposes of this section, *legal regimes refers to the laws, legal structures, and binding authorities that define the possible actions governments can take in response to the issues of terrorism and homeland security.*

Legal regimes in this arena can be particularly challenging to both establish and implement because they are often formed or modified in reaction to specific terrorist attacks that have resulted in significant loss of life. These legal regimes represent attempts at creating practical, rational, and theoretical solutions to very emotional problems, often impacting strongly held ideological, cultural, or religious beliefs. Their creation or modification may be reactive in the short term, but the legal regimes developed also need to provide long-term abilities to both prevent and respond to terrorist attacks across a wide variety of situations.

> This policy challenge of balancing corrective action based on lessons of recent incidents with the need to continue ongoing fundamental prevention and preparedness activities is evident in the development of homeland security; it continues to impact homeland security and emergency management law, policy, and practice. (Bentley, 2012, p. 23)[1]

Homeland security has emerged as one of the most interdisciplinary fields of study and analysis. The legal regimes that define government and individual action mirror the field's vast diversity. For that reason, it is impossible to cover all of terrorism and homeland security law in a single book, let alone a single section. Coverage choices are necessary and unavoidable. After an introduction to the content of legal regimes as used in this section and a brief discussion of definitional issues, this section, therefore, will highlight examples of the legal regimes established by the U.S. federal government. There are hundreds more examples that could have been added and thousands more across state and federal statutory and regulatory provisions. This section can provide only a taste of what is out there. A list of web resources at the end of the section is provided for further and more detailed study. Finally, this section will conclude with a brief look at international legal efforts through the auspices of the United Nations.

 What's in a Legal Regime?

A legal regime can combine a variety of different types of authorities granted to governments. In the United States, these can include the U.S. Constitution, federal legislation, the **Code of Federal Regulations (CFR)**, case law

---

[1]The reactive nature of policy arenas such as terrorism and homeland security is often studied through the theoretical frameworks of agenda setting and focusing events. For overviews and research in these areas, see Birkland (1997, 2001); Cobb and Elder (1983); and Kingdon (1995).

handed down by the federal court system (including the U.S. Supreme Court), executive orders, and—in the area of national security—a variety of **presidential directives** that may or may not have the same force of law as executive orders (U.S. General Accounting Office, 1992). In addition, because almost all disasters, including terrorist attacks, are local, state provisions play a significant role in terrorism and homeland security law in the United States.

Executive orders are especially prevalent in the areas of terrorism and homeland security because of the national security emphasis these orders take. Ultimate responsibility for national security typically rests with the president and the executive branch. Executive orders and directives have been issued under a variety of names and for a variety of purposes and do not require legislative approval (although the president may still be dependent on congressional funding for implementation). In the terrorism and homeland security area, several different types of orders have been issued by different presidential administrations, including:

- Presidential Decision Directives (PDDs)
- National Security Action Memoranda
- National Security Directives (NSDs)
- National Security Decision Directives (NSDDs)
- National Security Presidential Directives (NSPDs), and
- Homeland Security Presidential Directives (HSPDs) (Bentley, 2012, p. 35)

One of the most significant challenges in developing laws to deal with terrorism and homeland security concerns in a democracy is to balance the need for security with respect for constitutional and civil liberties. In addition, the laws and rules established must be realistic given the needs of everyday life in an open society. This balancing act was noted quite powerfully by Secretary of Homeland Security Jeh Johnson—a former member of the Judiciary Committee of the American Bar Association (ABA)—in remarks to the ABA's 2014 Annual Convention:

> In the name of homeland security, I can build you a perfectly safe city, but it will be a prison. I can guarantee you a perfectly safe, risk-free commercial flight, but every passenger will be strip-searched and not permitted any food, luggage, or freedom of movement during the ride. We can build more walls, install more invasive screening devices, ask more intrusive questions, expect more answers, and make everybody suspicious of those different from themselves. But we should not do this at the cost who we are as a Nation of people who respect the law, cherish privacy and freedom, celebrate diversity, and who are not afraid. This is our greatest strength as a Nation. (Johnson, 2014)

In addition to issuing executive orders, the president has the power to enter into **treaties** with foreign governments with the consent of the U.S. Senate. Treaties to which multiple countries are signatories are often referred to as **conventions**. There are a number of international conventions under the auspices of the United Nations dealing with international terrorism issues to which the United States is a signatory. In addition, congressional legislation may be needed to implement provisions of a convention to which the United States has agreed. This type of legislation is called implementing legislation.

 ## Definitions and the Law

Defining the term *terrorism* is one the most difficult challenges in international law. There is no single all-encompassing definition that crosses all boundaries and borders. Many consider the effort to find one to be a wild-goose chase that needlessly wastes time and effort that could be spent on more narrow, focused efforts in areas where compromise and agreement among the relevant parties is more likely, or at least possible.

Why is the term so hard to define? As Hoffman (2006) notes in his seminal text on terrorism, one of the key reasons the term is so difficult to define is because its meaning has changed so much over the centuries, usually defined by whomever was in power at the time. It has also become a strictly pejorative term, so applying the term to one group or another is often a politically charged issue. For example, many nations that are now members of the UN gained their independence through revolutions against other UN members. Some of these revolutions involved the use of terrorist-type attacks and tactics (see previous sections) and those involved were certainly called terrorists by the governments they revolted against. It is no wonder, then, that some of these nations are reluctant to apply the term to current nationalist or ethnic groups involved in violent conflict.

But while the difficulty of defining terrorism in the international arena may be understandable, one would think that in a single country with no such conflicts, a single legal definition would not only be simpler to achieve, it would be considered vital. How does one develop a legal regime against a phenomenon without being able to define it? No such luck. The United States also lacks a single definition of terrorism. Instead, there exists in U.S. law a wide variety of statutory and regulatory definitions relevant to different government departments and agencies and reflecting the institutions' particular mission (Hoffman, 2006). In-Focus 13.1 highlights a few of these definitions. A number of these definitions are also found, or referenced, in the U.S. statutes discussed below.

IN-FOCUS 13.1

U.S. Legal Definitions of Terrorism

Part of the law requiring the _U.S. Department of State_ to publish an annual report on terrorism around the world.

Section 2656f(a) of Title 22 of the United States Code:

1. the term "international terrorism" means terrorism involving citizens or the territory of more than one country;

2. the term "terrorism" means premeditated, politically motivated violence perpetrated against non-combatant targets by subnational groups or clandestine agents.

Part of the U.S. Criminal Code:

Section 2331(1) of Title 18 of the United States Code:
[T]he term "international terrorism" means activities that . . . involve violent acts or acts dangerous to human life that are a violation of the criminal laws of the United States or of any State, or that would be a criminal violation if committed within the jurisdiction of the United States or of any State; [and] appear to be intended . . . to intimidate or coerce a civilian population; . . . to influence the policy of a government by intimidation or coercion; or . . . to affect the conduct of a government by mass destruction, assassination, or kidnapping; and [which] occur primarily outside the territorial jurisdiction of the United States, or transcend

national boundaries in terms of the means by which they are accomplished, the persons they appear intended to intimidate or coerce, or the locale in which their perpetrators operate or seek asylum.

Definition Used by the Federal Bureau of Investigation:

28 C.F.R. 0.85(l)

The unlawful use of force or violence against persons or property to intimidate or coerce a Government, the civilian population, or any segment thereof, in furtherance of political or social objectives (See also National Institutes of Justice, 2014)

Definition Found in the USA PATRIOT Act:

Section 2331 of Title 18 of the United States Code as amended by Section 803 of the USA PATRIOT Act

Activities that (A) involve acts dangerous to human life that are a violation of the criminal laws of the U.S. or of any state, that (B) appear to be intended (i) to intimidate or coerce a civilian population, (ii) to influence the policy of a government by intimidation or coercion, or (iii) to affect the conduct of a government by mass destruction, assassination, or kidnapping, and (C) occur primarily within the territorial jurisdiction of the U.S.

Definition Used by the Department of Homeland Security

Homeland Security Act of 2002, PL 107-296, 116 Stat. 2141

A. Any activity that involves an act that—

i. is dangerous to human life or potentially destructive of critical infrastructure or key resources; and

ii. is a violation of the criminal laws of the United States or of any State or other subdivision of the United States; and

B. appears to be intended—

iii. to intimidate or coerce a civilian population;

iv. to influence the policy of a government by intimidation or coercion; or

v. to affect the conduct of a government by mass destruction, assassination, or kidnapping. (see also Hoffman, 2006, p. 31)

Definition Used by the Department of Defense

The unlawful use of violence or threat of violence, often motivated by religious, political, or other ideological beliefs, to instill fear and coerce governments or societies in pursuit of goals that are usually political (U.S. Department of Defense, 2014).

U.S. Domestic Law on Terrorism and Homeland Security

U.S. federal law dealing with terrorism and homeland security covers a wide variety of civil and criminal provisions relevant to numerous government functions and agencies. This discussion will take a closer look at the statutes and provisions dealing with general homeland security issues, **critical infrastructure protection**, the detention of enemy combatants, emergency management and disaster response, law enforcement and civil liberties, public health preparedness and response, and transportation security. A list of some of the key provisions in these areas, in chronological order, can be seen in In-Focus 13.2.

IN-FOCUS 13.2

Key U.S. Anti-Terrorism and Related Statutes, 1961–2007

1961	Amendment to Section 902 of the Federal Aviation Act of 1958 to include air piracy as a federal crime (PL 87-197).
1974	Anti-Hijacking Act (PL 92-366).
1978	Foreign Intelligence Surveillance Act (FISA) (PL 95-11).
1980	Classified Information Procedures Act (PL 96-456).
1984	Destruction of an Energy Facility Act (PL 98-743)
1985	International Security and Development Corporation Act (dealing with amendments to airline terrorism provisions) (PL 99-83).
1986	Omnibus Diplomatic Security and Antiterrorism Act (PL 99-399).
1994	Air Piracy Act (PL 103-272)
1996	Antiterrorism and Effective Death Penalty Act (PL 104-132)
2000	Disaster Mitigation Act (PL 106-390)
2001	USA PATRIOT ACT (PL 107-56)
2002	Homeland Security Act (PL 107-296)
2002	Public Health Security and Bioterrorism Preparedness and Response Act (PL 107-188)
2004	Intelligence Reform and Terrorism Prevention Act (PL 108-458)
2005	Detainee Treatment Act (PL 109-148)
2006	Military Commissions Act (PL 109-366)
2007	Implementing Recommendations of the 9/11 Commission Act (PL 110-53)
2007	Protect America Act (PL 110-55)

The above timeline is a combination of timelines created in Abrams (2008) and Bentley (2012).

Homeland Security

The actual term *homeland security* was not even present in the lexicon until after the events of 9/11. Following the attacks, the term took on a life of its own. In the 15 years since that fateful day, the term has come to refer to a field of study, a professional endeavor of amazing complexity, a key term for grant applications, and an entirely new cabinet-level federal department—the **Department of Homeland Security** (DHS).

In fact, the tragic events of September and October 2001 resulted in more domestic terrorism policy changes—in the form of legislation, executive orders, and federal regulations—than in the entire decade prior to the attacks (Rubin, 2004). By the end of 2002, the attacks had resulted in ten pieces of legislation, two executive orders, one Presidential Homeland Security Decision Directive (HSDD), and the creation of DHS, the largest reorganization of the federal government since the creation of the Department of Defense (DoD) in 1949 (Rubin, 2004).[2]

The department actually started life as the Office of Homeland Security established by Executive Order No. 13228 issued on October 11, 2001, exactly one month after the attacks on the World Trade Center and the Pentagon. It was headed by an Assistant to the President for Homeland Security with a mission to "develop and coordinate the implementation of a comprehensive national strategy to secure the United States from terrorist threats or attacks" (Nemeth, 2010, p. 65). Concerns about sufficient authority, budgetary control, and the ability to coordinate over 22 different agencies and bureaus led to the office being quickly overshadowed and replaced by a department.

DHS was established with the passage of the **Homeland Security Act of 2002**. The act sets forth the mission of the department to "prevent terrorist attacks within the United States, reduce the vulnerability of the United States to terrorism, and minimize the damage, and assist in the recovery, from terrorist attacks that do occur within the United States" (Bentley, 2012, p. 29). The department is headed by a secretary who serves on the Cabinet.

In addition to setting out the mission of the new department, the Homeland Security Act laid out a blueprint for the key areas of concern for homeland security, including borders, information and infrastructure, WMD threats (see Section 12), and emergency preparedness and response (Nemeth, 2010). The act has been revised several times since its passage, including several reorganizations of DHS along different functional lines.

The same year as the passage of the Homeland Security Act, President Bush issued a number of Homeland Security Presidential Directives (HSPD). Among them was HSPD-4: National Strategy to Combat Weapons of Mass Destruction (WMD).[3] This directive came on the immediate heels of the anthrax attacks against media and congressional offices in Washington, D.C., and Florida. Although the directive dealt with all aspects of the WMD threat, it did note that: "Our approach to defend against biological threats has long been based on our approach to chemical threats, despite the fundamental differences between these weapons. The United States is developing a new approach to provide us and our friends and allies with an effective defense against biological weapons" (HSPD-4, 2002; see also Ryan & Glarum, 2008).

The Role of the Military in Homeland Security Matters

The Homeland Security Act also reaffirmed the **Posse Comitatus Act of 1878**, which forbids the use of military forces for the purposes of domestic law enforcement. However, that prohibition has never been absolute. For instance, while the National Guard and the Coast Guard fall under the jurisdiction of the armed forces and are considered active duty military in wartime, in peacetime they fall under the authority of state governors and DHS, respectively, and as such are exempt from the Posse Comitatus restrictions. Thus they have been able to be used in order-keeping roles following disasters, riots, and disturbances such as those in Los Angeles following the verdicts in the Rodney King case, and following the attacks on 9/11.

Because the United States has faced an increased threat of mass casualty attacks at home, especially involving the use of WMD agents, "there has been a growing debate among politicians and government policy-makers and scholars over what role, if any, the U.S. military should take in domestic operations" (Nemeth, 2010, p. 72). This broadening of the debate has been aided by the creation of the **U.S. Northern Command (USNORTHCOM)** in 2002

---

[2]Many consider the establishment of the Department of Defense (DoD) as dating back to the National Security Act of 1947. In fact, the 1947 act created the National Military Establishment (NME), which consisted of the Department of War and the Department of the Navy headed by a new Secretary of Defense. The Department of the Air Force was also created in the same act. In 1949, the National Security Act was amended to place all service secretaries (Army, Navy, and Air Force) under the authority of the Secretary of Defense. The NME was renamed the DoD at that time.

[3]WMD refers to terrorist attacks using chemical, biological, radiological, or nuclear agents or weapons. It is also sometimes referred to as CBRN terrorism.

to "provide command and control of Department of Defense (DoD) homeland defense efforts and to coordinate defense support of civil authorities" (Nemeth, 2010, p. 72, citing United States Northern Command, "About USNORTHCOM," July 24, 2009 (available at http://www.northcom.mil/about/index.html). Since its creation, the command has assisted in a variety of operations on U.S. soil including Hurricane Katrina, wildfires in the western states, and counter-drug operations (Nemeth, 2010). The willingness to broaden the use of the military in domestic homeland security matters is also aided by the widespread understanding that a significant amount of expertise and experience in dealing with WMD lies with military units such as NORTHCOM and the medical research and biodefense facilities at Fort Detrick, Maryland.

Critical Infrastructure Protection

In April 2005, Kenneth Falkenrath—a renowned expert in nuclear proliferation and terrorism—noted that a basic thrust of al Qaeda's strategy on September 11 was to identify a "commonplace system in our midst that we relied upon every day and attack in such a way that they could achieve catastrophic secondary effects" (United States Senate, 2005, p. 18). This is essentially the idea behind critical infrastructure—the network of facilities that provide such basic items as power, water, food, and transportation and that allow a modern society to function on a day-to-day basis.

Since the attacks of 9/11, the lists of sectors considered to be critical infrastructure has grown tremendously. The website of the Department of Homeland Security lists 16, including chemical facilities, commercial facilities, communications, critical manufacturing, dams, defense industrial base, emergency services, energy, financial services, food and agriculture, government facilities, health care and public health, information technology, nuclear reactors, materials, and waste; transportation services, and water and wastewater sectors. Here are several examples of some of the laws and other provisions in this area, especially those passed since 2001.

Executive Order No. 13231: On Critical Infrastructure Protection in the Information Age, October 16, 2001

Issued within a month of the 9/11 attacks and within days of the anthrax mailings, Executive Order No. 13231 on Critical Infrastructure Protection in the Information Age created the President's Critical Infrastructure Protection Board to "coordinate and have cognizance of Federal efforts and programs that relate to the protection of information systems" key to critical infrastructure protection. These included the information systems essential to telecommunications, energy, financial services, manufacturing, transportation, health care, and emergency services sectors.

Homeland Security Presidential Directive-9: Defense of United States Agriculture and Food, January 30, 2004

President Bush issued HSPD-9 to "establish a national policy to defend the agriculture and food system against terrorist attacks, major disasters, and other emergencies" (Ryan & Glarum, 2008, p. 216). Recognizing the need to consider food and agriculture as much a part of critical infrastructure as the more traditional energy and transportation sectors, the directive established as the policy of the United States under the coordination of the Secretary of Homeland Security to "protect the agriculture and food system from terrorist attacks, major disasters, and other emergencies by:

(a) identifying and prioritizing sector-critical infrastructure and key resources for establishing protection requirements;

(b) developing awareness and early warning capabilities to recognize threats;

(c) mitigating vulnerabilities at critical production and processing nodes;

(d) enhancing screening procedures for domestic and imported products; and

(e) enhancing response and recovery procedures (HSPD-9, 2004; see also Ryan & Glarum, 2010, p. 216).

The Food Safety Modernization Act (FSMA) (PL 111-353), July 2011

The website of the U.S. Food and Drug Administration (FDA) calls the FSMA "the most sweeping reform of our food safety laws in more than 70 years" (U.S. Food and Drug Administration, 2014). The act is designed to prevent both intentional and unintentional contamination of the U.S. food supply using enhanced regulatory authorities to focus on prevention rather than relying on response after contamination has taken place (Painter, 2013). The FSMA also requires the secretaries of Health and Human Services (HHS) and Agriculture to develop a National Agriculture and Food Defense Strategy, including an implementation plan and research agenda, as well as a number of other mandatory reporting requirements. As of 2014, the national strategy had not been published (Painter, 2013; U.S. Food and Drug Administration, 2014).

Chemical Facility Anti-Terrorism Standards (CFATS), (PL 109-295)

As part of the Department of Homeland Security Appropriations Act of 2007, DHS was authorized to regulate chemical facility security. In implementing that authorization, DHS issued the **Chemical Facility Anti-Terrorism Standards (CFATS)** in the Code of Federal Regulations ranging from sections 17688 to 17745. These regulations provide a process whereby facilities submit information and security plans to DHS, DHS then reviews and approves the plans, and then inspects their implementation (Shea, 2014).

According to a recent Congressional Research Service study, as of August 2014, DHS had authorized 1,838 submitted plans, conducted authorization inspections of 1,348 of those plans, and approved 970. Although progress is being made, this is still well below DHS's projected timelines for encompassing all eligible chemical plants in the United States (Shea, 2014, p. 12).

Detention of Enemy Combatants

"In the post 9-11 world, international humanitarian law, military law, and U.S. criminal law intersect in how the U.S. government handles detention of suspected terrorist associates" (Bentley, 2012, p. 37). Nowhere is this more evident than with the issue of how the United States handles "enemy combatants." The term refers to those members or associates of the Taliban or al Qaeda—whether U.S. or non-U.S. citizens—caught in Afghanistan and Pakistan and most frequently held at the U.S. Naval facility in Guantanamo Bay, Cuba.

Some of the key issues surrounding the treatment of these detainees include interrogation procedures, the right of the detainees to know the charges and evidence against them (including classified evidence), the right to counsel and speedy trial, and the "application of U.S. constitutional and statutory procedures for trying the detainees or releasing them to other countries" (Bentley, 2012, p. 38).

▲ Photo From Abu Ghraih Prison in Iraq

The procedures and policies used to deal with the detainees were based on a series of statutes such as the Detainee Treatment Act of 2005, the Authorization for the Use of Military Force of 2001, and the Military Commissions Act of 2006. In-Focus 13.3 provides a copy of the introduction to a Congressional Research Service Report (Elsea, 2014) that summarizes the incredibly complex legal issues involved in these statutory schemes.

IN-FOCUS 13.3

The Military Commissions Act of 2009: Overview and Legal Issues

Introduction

The use of military commissions to try suspected terrorists has been the focus of intense debate (as well as significant litigation) since President Bush in November 2001 issued his original Military Order (M.O.) authorizing such trials.[1] The M.O. specified that persons subject to it would have no recourse to the U.S. court system to appeal a verdict or obtain any other sort of relief, but the Supreme Court essentially invalidated that provision in its 2004 opinion, *Rasul v. Bush*.[2] In response, Congress enacted the Detainee Treatment Act of 2005 (DTA).[3] The DTA did not authorize military commissions, but amended Title 28, *U.S. Code* to revoke all judicial jurisdiction over habeas claims by persons detained as "enemy combatants," and it created jurisdiction in the Court of Appeals for the District of Columbia Circuit to hear appeals of final decisions of military commissions.

The Supreme Court, after finding that Congress's efforts to strip it of jurisdiction did not apply to a case already pending before the Court, *Hamdan v. Rumsfeld*,[4] invalidated the military commission system established by presidential order. The Court held that although Congress had in general authorized the use of military commissions, such commissions were required to follow procedural rules as similar as possible to courts-martial proceedings, as required by the Uniform Code of Military Justice (UCMJ).[5] In response, Congress promptly passed the Military Commissions Act of 2006 (MCA 2006)[6] to authorize military commissions and establish procedural rules that were modeled after, but departed from in some significant ways, the UCMJ. The MCA 2006 also amended the Detainee Treatment Act in order to strip the judiciary of habeas jurisdiction in all cases brought by detainees, including pending cases,[7] but the Supreme Court held that provision to be an unconstitutional suspension of the Writ of Habeas Corpus.[8]

President Bush reconstituted the military commissions under the MCA 2006 by issuing Executive Order 13425.[9] The Department of Defense (DOD) issued regulations for the conduct of military commissions pursuant to the MCA 2006[10] and restarted the military commission proceedings, which resulted in three convictions under the Bush administration. One detainee, David Matthew Hicks of Australia, was convicted of material support to terrorism pursuant to a plea agreement in 2007.[11] In 2008, Salim Hamdan was found guilty of one count of providing material support for terrorism and sentenced to 66 months' imprisonment, but credited with five years' time served.[12] Both men are now free from detention. Ali Hamza Ahmad Suliman al Bahlul of Yemen was found guilty of multiple counts of conspiracy and solicitation to commit certain war crimes and of providing material support for terrorism in connection with his role as Al Qaeda's "propaganda chief."[13] He refused representation and boycotted most of his trial, and was subsequently sentenced to life imprisonment. The latter two convictions were reversed on appeal by the U.S. Court of Appeals for the D.C. Circuit.[14] The government sought and was granted a rehearing *en banc* in the *Bahlul* case to appeal the decisions.

No challenge to military commissions under the MCA 2006 reached the Supreme Court. President Obama halted the proceedings upon taking office in January 2009 in order to review whether to continue their use. The President issued an Executive Order requiring that the Guantánamo detention facility be closed no later than a year from the date of the Order.[15] The Order required specified officials to review all Guantánamo detentions to assess whether the detainee should continue to be held by the United States, transferred or released to another country, or be prosecuted by the United States for criminal offenses.[16] The Secretary of

Defense was also required to take steps to ensure that all proceedings before military commissions and the United States Court of Military Commission Review were halted, although some pretrial proceedings continued to take place. One case was moved to a federal district court.[17]

In May 2009, the Obama administration announced that it was considering restarting the military commission system with some changes to the procedural rules.[18] DOD informed Congress about modifications to the Manual for Military Commissions, to take effect July 14, 2009.[19] The Senate passed the Military Commissions Act of 2009 (MCA 2009) as part of the Department of Defense Authorization Act (NDAA) for FY2010, S. 1391, to provide some reforms the administration supported and to make other amendments to the Military Commissions Act, as described below. The bill that emerged from conference (H.R. 2647) contained some, but not all, of the proposals submitted by the Obama administration, and was enacted October 28, 2009, P.L. 111-84.

President Obama's Detention Policy Task Force issued a preliminary report July 20, 2009, reaffirming that the White House considers military commissions to be an appropriate forum for trying some cases involving suspected violations of the laws of the war, although federal criminal court would be the preferred forum for trials of detainees.[20] The disposition of each case was assigned to a team composed of Department of Justice (DOJ) and Department of Defense (DOD) personnel, including prosecutors from the Office of Military Commissions. Appended to the report was a set of criteria to govern the disposition of cases involving Guantánamo detainees. This protocol identified three broad categories of factors to be taken into consideration:

- Strength of interest, namely, the nature and gravity of offenses or underlying conduct; identity of victims; location of offense; location and context in which the individual was apprehended; and the conduct of the investigation.
- Efficiency, namely, protection of intelligence source and methods; venue; number of defendants; foreign policy concerns; legal or evidentiary problems; efficiency and resource concerns.
- Other prosecution considerations, namely, the extent to which the forum and offenses that can be tried there permit a full presentation of the wrongful conduct, and the available sentence upon conviction.

Federal prosecutors are to evaluate their cases under "traditional principles of federal prosecution."

On November 13, 2009, Attorney General Holder announced his decision to transfer the five "9/11 conspirators," who include Khalid Sheikh Mohammed, Walid Muhammed Salih Mubarak Bin Attash, Ramzi Bin Al Shibh, Ali Abdul-Aziz Ali, and Mustafa Ahmed Al Hawsawi, to the Southern District of New York to stand trial.[21] Five other detainees to be tried by military commission included Omar Khadr, a Canadian citizen captured as a teenager and charged before a military commission for allegedly throwing a hand grenade that killed a U.S. medic in Afghanistan;[22] Abd al-Rahim al-Nashiri, whose military commission charges related to the October 2000 attack on the USS *Cole* were previously withdrawn in February 2009; Ahmed Mohammed Ahmed Haza al Darbi, accused of participating in an Al Qaeda plot to blow up oil tankers in the Straits of Hormuz;[23] and two other detainees about whom no further information was given.[24]

As the deadline for closing the detention facility at Guantánamo passed unmet, the Obama administration reportedly completed its assessment, determining that about 50 of the detainees held there would continue to be held without trial, that around 40 detainees would be prosecuted in military commission or federal court, and that the remaining 110 detainees would be released once a suitable country has agreed to take each of

(Continued)

(Continued)

them.[25] However, the transfer of 30 detainees of Yemeni nationality was stymied because an Al Qaeda affiliate in Yemen is suspected to have been behind the attempt to blow up a civilian airliner on Christmas Day 2009.[26]

Notes

1.   Detention, Treatment, and Trial of Certain Non-Citizens in the War Against Terrorism §1(a), 66 Fed. Reg. 57,833 (November 16, 2001) (hereinafter "M.O."). President Bush subsequently determined that 20 of the detainees at the U.S. Naval Station in Guantánamo Bay held in connection with the conflict were subject to the M.O., and 10 were eventually charged for trial before military commissions. *See* Press Release, Department of Defense, President Determines Enemy Combatants Subject to His Military Order (July 3, 2003), *available at* http://www.defense.gov/releases/release.aspx?releaseid=5511. According to the Defense Department, that determination is effectively "a grant of [military] jurisdiction over the person." *See* John Mintz, *6 Could Be Facing Military Tribunals*, WASH. POST, July 4, 2003, at A1. In 2004, nine additional detainees were determined to be eligible. *See* Press Release, Department of Defense, Presidential Military Order Applied to Nine More Combatants (July 7, 2004), *available at* http://www.defenselink.mil/releases/release.aspx?releaseid=7525. In November 2005, five more detainees were charged. *See* Press Release, Department of Defense, Military Commission Charges Approved (November 7, 2005), *available at* http://www.defense.gov/releases/release.aspx?releaseid=9052.

2.   *Rasul v. Bush*, 542 U.S. 466 (2004). Persons subject to the M.O. were described as not privileged to "seek any remedy or maintain any proceeding, directly or indirectly" in federal or state court, the court of any foreign nation, or any international tribunal. M.O. at §7(b). However, the Bush administration shortly thereafter indicated that defendants were not intended to be precluded from petitioning a federal court for a writ of habeas. *See* Alberto R. Gonzales, *Martial Justice, Full and Fair*, NY TIMES (op-ed), November 30, 2001. The government did not rely on the M.O. as the legal basis for asserting detainees had no right to pursue writs of habeas corpus, but the Court's opinion served as a warning that military commission verdicts would be subject to collateral review. For a summary of *Rasul* and related cases, see CRS Report R41156, *Judicial Activity Concerning Enemy Combatant Detainees: Major Court Rulings*, by Jennifer K. Elsea and Michael John Garcia.

3.   Title 10 of P.L. 109-148 and Title 14 of P.L. 109-163. The two versions of the Detainee Treatment Act (DTA) were identical as enacted, but subsequent amendments have resulted in some differences in the text.

4   *Hamdan v. Rumsfeld*, 548 U.S. 557 (2006), *rev'g* 415 F.3d 33 (D.C. Cir. 2005).

5.   10 U.S.C. §801 *et seq.* Military commissions were said to be authorized pursuant to 10 U.S.C. §§821 and 836.

6.   P.L. 109-366, 120 Stat. 2600, codified at chapter 47A of Title 10, *U.S. Code* (2006).

7.   P.L. 109-366 §7.

8.   *Boumediene v. Bush*, 533 U.S. 723 (2008). For an analysis of the case, see CRS Report R41156, *Judicial Activity Concerning Enemy Combatant Detainees: Major Court Rulings*, by Jennifer K. Elsea and Michael John Garcia.

9.   Exec. Ord. No. 13425, 72 Fed. Reg. 7737 (February 14, 2007).

10.   Department of Defense, The Manual for Military Commissions ["M.M.C. 2007"], January 18, 2007, *available at* http://www.defenselink.mil/news/MANUAL%20FOR%20MILITARY%20COMMISSIONS%202007%20signed.pdf.

11.   Press release, Department of Defense, Detainee Convicted of Terrorism Charge at Guantánamo Trial" (March 30, 2007), *available at* http://www.defenselink.mil/releases/release.aspx?releaseid=10678. Hicks was sentenced to seven years' confinement. As part of his pretrial agreement, his sentence was limited to nine months' confinement to be served in Australia, with six years and three months suspended.

12.   Press release, Department of Defense, Detainee Transfer Announced (November 28, 2008), *available at* http://www.defenselink.mil/releases/release.aspx?releaseid=12372.

13.   Press release, Department of Defense, Detainee Sentenced to life in Prison (November 3, 2008), *available at* http://www.defenselink.mil/releases/release.aspx?releaseid=12331.

14.   *Hamdan v. United States*, 696 F.3d 1238 (D.C. Cir. 2012); *Al Bahlul v. United States*, 2013 WL 297726 (D.C. Cir. January 25, 2013) (percuriam).

15. Exec. Ord. 13492, Review and Disposition of Individuals Detained at the Guantánamo Bay Naval Base and Closure of Detention Facilities, 74 Fed. Reg. 4,897 (January 22, 2009).

16. *Id.* at §4.

17. Press Release, Department of Justice, Ahmed Ghailani Transferred from Guantánamo Bay to New York for Prosecution on Terror Charges (June 9, 2009), *available at* http://www.justice.gov/opa/pr/2009/June/09-ag-563.html. Ghailani was ultimately convicted and sentenced to life in prison. *See* Benjamin Weiser, *Ex-Detainee Gets Life Sentence in Embassy Blasts*, N.Y. TIMES, January 26, 2011, at A18. For more information, see CRS Report R41156, *Judicial Activity Concerning Enemy Combatant Detainees: Major Court Rulings*, by Jennifer K. Elsea and Michael John Garcia.

18. Peter Finn, *Obama Set to Revive Military Commissions*, WASH. POST, May 9, 2009.

19. Letter from Robert M. Gates, Secretary of Defense, to Senator Carl Levin, May 15, 2009.

20. Memorandum from the Detention Policy Task Force to the Attorney General and the Secretary of Defense, July 20, 2009, http://www.scotusblog.com/wp-content/uploads/2009/07/law-of-war-prosecution-prelim-report-7-20-09.pdf.

21. Press Release, U.S. Department of Justice, "Departments of Justice and Defense Announce Forum Decisions for Ten Guantánamo Detainees," November 13, 2009, *available at* http://www.justice.gov/opa/pr/2009/November/09-ag-1224.html.

22. Khadr pleaded guilty in 2010 and was sentenced to 40 years in prison. He will serve eight years pursuant to his plea agreement, and has been transferred to Canada to serve the remaining portion of his sentence.

23. Al Darbi pleaded guilty in February 2014 under an agreement that provides a sentence of between 13 and 15 years. *See* Charlie Savage, *Guantánamo Detainee Pleads Guilty in 2002 Attack on Tanker Off Yemen*, NY TIMES, February 20, 2014.

24. One of these may have been Majid Shoukat Khan, who has pleaded guilty to conspiracy and other crimes in connection with the August 2003 bombing of the J.W. Marriot hotel in Indonesia and an attempted assassination of former Pakistani president Pervez Musharraf. The other may have been Noor Uthman Muhammed, who pleaded guilty to conspiracy and providing material support for terrorism in connection with service at the Khalden terrorist training camp in Afghanistan. He was sentenced in February 2011 to 14 years' imprisonment, but his plea agreement provided for only 34 months.

25. *See* Charlie Savage, *Detainees Will Still Be Held, but Not Tried, Official Says*, NY TIMES, January 22, 2010.

26. *Id.*

Congressional Research Service. The Military Commissions Act of 2009 (MCA 2009): Overview and Legal Issues, Jennifer K. Elsea, Legislative Attorney, March 7, 2014.

In addition to the statutory provisions, a number of now famous U.S. Supreme Court cases have dealt with detainee and enemy combatant issues. Figure 13.1 shows a list of some of these key cases.

Figure 13.1	Key U.S. Supreme Court Cases Dealing With Detention of Enemy Combatants and Related Issues

2004	*Rasul v. Bush*	542 U.S. 466 (2004)
2004	*Hamdi v. Rumsfeld*	542 U.S. 507 (2004)
2004	*Rumsfeld v. Padilla*	542 U.S. 426 (2004)
2006	*Hamdan v. Rumsfeld*	548 U.S. 557 (2006)
2008	*Boudhediene v. Bush*	553 U.S. 723 (2008)

For excellent, if detailed, discussions of these cases, see Gurule and Corn (2011) and Dycus, Banks, and Raven-Hansen (2007).

Emergency Management and Disaster Response

Emergencies are local. Even the response to the most significant of mass casualty terrorist attacks will involve—and possibly overwhelm—local and state first responders before federal resources get involved. For that reason, most legal authority for issues such as quarantine and isolation, mandatory evacuations, and command and control are situated in state statutory and regulatory law, with federal authorities able to step in if a state cannot or fails to respond adequately. Even when National Guard forces are used in response to a disaster, as they often are, they do so under the authority of the state's governor. In addition, though there are general national guidelines and mandates for response and recovery plans, the plans themselves are established at the state and local levels. Requests for federal aid (especially military) and for federal disaster declarations must be made by the governor to the president. The White House cannot issue disaster or emergency declarations absent such a request.

Still, the federal government is by no means uninvolved in emergency response and preparedness to terrorist attacks. The more significant and deadly the attack, the more likely and extensive federal involvement will be. This is especially true for attacks using weapons of mass destruction such as chemical, biological, nuclear, or radioactive agents. Here are a few examples of federal provisions in this area.

Disaster Relief Act of 1974 (PL 93-288)

The Disaster Relief Act was the first law to formalize federal disaster relief to both states and individuals and families in one place. It also created the authority for the president to issue emergency declarations. Most importantly, the law placed a new emphasis on a multi-hazard, or all-hazards, approach to emergency management; the idea is that the same capacities, response, and recovery plans can be applied to all different types of disasters whether natural or man-made. Prior to 1974, "emergency management was fragmented and preoccupied with confronting individual disasters or specific types of disasters as if each disaster were unique or as if each category or type of disaster had its own independent set of response needs" (Sylves, 2008, p. 55). The all-hazards approach was generally accepted until the events of 2001, after which a debate has ensued over whether or not the response to terrorist attacks, especially those involving Weapons of Mass Destruction, requires specialized expertise, training, and planning not sufficiently covered by the all-hazards approach.

Executive Order No. 12148: Federal Emergency Management (July 20, 1979)

Under this order, President Jimmy Carter placed all of the functions and authority of the 1974 Act under the director of the new Federal Emergency Management Agency (FEMA). The order also moved emergency functions from several different federal agencies into the new organization and delegated to FEMA's director the authority to establish "federal policies for all civil defense and civil emergency planning, management, mitigation, and assistance functions of executive branch agencies" including responses to major terrorist events (Sylves, 2008, p. 57).

Robert T. Stafford Disaster Relief and Emergency Act of 1988 (PL 100-707) (November 23, 1988)

The Stafford Act significantly amended the 1974 Act to establish the framework for federal disaster assistance, including requirements for requesting assistance, the type of assistance available, and the role of the Federal Emergency Management Agency (FEMA) (Bentley, 2012). FEMA has the responsibility for implementing the provisions of the Stafford Act through regulations published in the Code of Federal Regulations. According to Sylves, "the effects of the Stafford Act were so profound for emergency managers that this law came to demarcate the beginning of modern-era national disaster management" (2008, p. 60).

In 2000, the Stafford Act was amended by the Disaster Mitigation Act (PL 106-390) to require all states to include long- and short-term hazard mitigation plans in their emergency response strategies, specifically requiring attention to both natural and man-made disasters (Bentley, 2012; Sylves, 2008).

Presidential Decision Declaration 69: U.S. Policy on Counterterrorism (June 21, 1995)

In this order, President Bill Clinton attempted to clarify the relationship between counterterrorism response and emergency management by designating lead agencies for crisis management (Federal Bureau of Investigation) and consequence management (FEMA) in response to dealing with a terrorist attack. Crisis management deals with the tasks of investigating the attack, tracking down and punishing (where possible) the perpetrators, and preventing future attacks, especially in the immediate aftermath of the incident. Consequence management deals with the immediate first response to the attack, the mitigation of further casualties, and recovery tasks—the traditional emergency response issues. According to Richard Falkenrath (2001), a researcher on emergency preparedness issues and WMD terrorism, the order signaled an increased White House concern with the issue but did little to increase budgets or transagency coordination. In addition, the distinction between consequence and crisis management has come under fire in recent years. Opponents argue that the distinction is confusing and fails to recognize the overlap between the two missions. In addition, the fact that consequence management falls within DHS, while crisis management does not, creates more coordination problems than it solves.

Defense Against Weapons of Mass Destruction Act of 1996 (PL 104-201) (September 23, 1996)

Known as the Nunn-Lugar-Domenici amendment after the three Senators responsible for its passage, this act responded to concerns following the April 1995 Oklahoma City bombing and the March 1995 sarin gas attacks in Tokyo, regarding the poor state of capabilities in the United States to deal with WMD attacks, especially at the state level. The Amendment to the fiscal year (FY) 1997 defense authorization bill directed the secretary of defense to "carry out a program to provide civilian personnel of Federal, State, and local agencies with training and expert advice regarding emergency responses to a use or threatened use of a weapon of mass destruction or related materials" (Falkenrath, 2001, p. 162). Since that time, numerous state and local agencies have applied to the program for funding for equipment and training. First situated in the DoD, the program was eventually transferred, not to FEMA, as many had assumed, but to the Department of Justice. With the founding of DHS, the coordination of state and local response was moved to the new department, but military experts still provide a substantial amount of the actual training and advice to civilian first responders.

Law Enforcement and Civil Liberties

A vast array of criminal laws that deal with terrorism can be found throughout the U.S. Penal Code. Since 2001, a much greater emphasis has been placed on criminalizing any material, financial, physical, or other involvement in any organizations even tangentially related to groups on the list of Foreign Terrorist Organizations. These organizations now include a number of "charitable" organizations that U.S. and international authorities believe have been established by terrorist groups, especially in migrant and immigrant communities, to raise funds for both legal and illegal activities. Other laws have been changed to make it easier for federal authorities to investigate and prosecute those suspected of terrorist activities. Three examples are discussed here: the USA PATRIOT Act, the Real-ID program, and the Antiterrorism and Effective Death Penalty Act of 1996.

Antiterrorism and Effective Death Penalty Act of 1996 (PL 104-132) (April 24, 1996)

This act was passed in response to several incidents, including the 1993 bombing of the World Trade Center in New York City and the 1995 bombing of the Murrah Federal Building in Oklahoma City (see Section 10). The law deals with a wide-ranging array of issues, some specific to terrorism, others more broadly related to criminal procedure issues such as procedures to reform habeas corpus proceedings to lessen the amount of time death row inmates can

PEKKA SAKKI/AFP/Getty Images

▲   Signing of the 2006 USA PATRIOT Act

spend appealing their sentences. Areas addressed in the 1996 act include victim compensation, jurisdiction for victims to sue sovereign states accused of terrorist attacks in U.S. courts (a direct response to Libya's downing of Pan Am flight 103 in 1988), prohibitions on international fundraising for terrorist groups, nuclear, biological, and chemical weapons prohibitions, increased penalties for certain terrorist-related offenses, and the implementation of the convention on tagging and detecting plastic explosives.

USA PATRIOT Act (PL 107-56) (October 26, 2001)

Using one of the more effective acronyms in U.S. legislative history, the full name of the act is: *Uniting and Strengthening America by Providing Appropriate Tools Required to Intercept and Obstruct Terrorism.* In-Focus 13.4 shows the different areas of criminal and civil law addressed in the act. The Patriot Act amended numerous criminal and civil provisions in the U.S. code, including the Electronic Communications Privacy Act, the Computer Fraud and Abuse Act, the Foreign Intelligence and Surveillance Act, the Family Educational Rights and Privacy Act, the Money Laundering Control Act of 1986, the Bank Secrecy Act, the Right to Financial Privacy Act, the Fair Credit Reporting Act, the Immigration and Nationality Act, the Victims of Crime Act of 1984, and the Telemarketing and Consumer Fraud and Abuse Prevention Act. Spurred by the events of 9/11, the almost 350-page bill passed within a month of that tragic day. "Given the intensity of the times, it is not surprising that a bill of this import found formal approval in so short a span. The time influenced the aggressive nature of the bill" (Nemeth, 2010, p. 72).

Although spurred by the attacks on 9/11, several of the law enforcement provisions addressed concerns that had been raised by the law enforcement community for a number of years prior. For example, the provisions regarding wiretap warrants for phones and other electronic communications updated provisions that had been based on assumptions that suspects were using landlines, assumptions that had not been true for close to a decade. Previous wiretap and warrant laws had also not taken into consideration the explosion of e-mail, text messaging, social networks, and other Internet-based communications. The act also expanded the definition of terrorist organizations to those supporting groups engaged in acts of violence, and significantly increased reporting requirements for banking and other financial institutions (Bentley, 2012). One of the many controversial provisions was one allowing the use of "sneak and peak" warrants where law enforcement authorities in certain circumstances can execute a search warrant of a suspect's residence with no notification to the suspect either before or after the execution of the search (Nemeth, 2010).

The Patriot Act has been amended several times since its passage and was reauthorized in 2006 despite significant controversy and concerns that the act was unconstitutional in many aspects and violated individual civil liberties in the name of increased security. Critics also accused federal agencies, particularly DHS and the FBI, of abusing their power by claiming concerns or suspicions of terrorism where they were not warranted in order to trigger the provisions of the act. DHS responded by noting concrete examples where the act has led to the apprehension and prosecution of terrorists (Nemeth, 2010).

IN-FOCUS 13.4

The USA Patriot Act of 2001: Table of Contents

TITLE I: Enhancing Domestic Security Against Terrorism

TITLE II: Enhanced Surveillance Procedures

TITLE III: International Money Laundering Abatement and Antiterrorism Financing Act

TITLE IV: Protecting the Border

TITLE V: Removing Obstacles to Investigating Terrorism

TITLE VI: Providing for Victims of Terrorism, Public Safety Officers, and Their Families

TITLE VII: Increased Information Sharing for Critical Infrastructure Protection

TITLE VIII: Strengthening the Criminal Laws on Terrorism

TITLE IX: Improved Intelligence

TITLE X: Miscellaneous

Real ID Program

A much more specialized, but also controversial consequence of 9/11, was the **Real ID** program. This program attempts to set standard federal guidelines for the issuance of driver's licenses in an effort to ensure their validity and security. It was spurred in part by fears of terrorist sleeper cells hiding in U.S. suburbia, but also by broader concerns over U.S. immigration policy that had been a significant political issue long before 2001 and continues to plague us today. Proponents of the program see it as a way of controlling illegal immigration. Opponents of the program see it as a blank check to discriminate against legal and illegal immigrants. The main controversy regarding the program centers around criticisms that it creates a national identify card—an idea that has long been anathema in American culture. DHS responds to this criticism by noting that Real ID only sets minimum standards for licenses and that states maintain the responsibility for designing and issuing their own unique licenses or other identification (see http://www.dhs.gov/real-id-public-faqs; see also Nemeth, 2010, p. 75).

Public Health and Preparedness

As early as the 1995 sarin gas attack against a Tokyo subway, public health officials, relevant congressional committees, and an array of interest groups and scientific experts have taken up the issue of improving public health preparedness and response to a terrorist attack. Much of the concern in this area has centered around the threat of bioterrorism, which was exacerbated considerably by what was considered a lackluster and flawed response to the anthrax attacks of September and October 2001. However, many of the measures considered and implemented also addressed the ability of federal, state, and local emergency responders to prepare for and respond to the medical needs of victims of any large-casualty terrorist attack whether from conventional or non-conventional agents or methods. In addition, many medical and infectious disease experts argue that the more serious public health threat is that stemming from naturally occurring infectious diseases—especially those resistant to current antibiotics—and that improving our ability to deal with naturally occurring outbreaks also improves our ability to deal with a bioterrorist attack on civilians, as well as threats to our food and water supplies. Examples of measures in this area include a mix of statutory and presidential directive efforts.

Public Health Service Act of 1944 (PL 78-410) (July 1, 1944)

This act has been amended a number of times in its 70-year history, including recently, in March 2013, by the Pandemic and All-Hazards Preparedness Reauthorization Act (PL 113-5). A public health emergency may be declared under section 319 of the act. Such a declaration triggers a number of federal resources and authorities. However, even in the absence of a public health emergency declaration, "public health officials, both state and federal, may exercise principal health authorities to control communicable diseases" where necessary, including the authority to order quarantine or isolation, impose travel restrictions, or require inoculations or medical examinations (Ryan & Glarum, 2008, p. 220). In addition, under section 361 (42 U.S.C. section 264) of the act, the Centers for Disease Control and Prevention's (CDC) Division of Global Migration and Quarantine is empowered "to detain, medically examine, or conditionally release persons suspected of carrying certain communicable diseases" (Ryan & Glarum, 2008, p. 221).

Public Health Security and Bioterrorism Preparedness and Response Act of 2002 (PL 107-188) (June 12, 2002)

Key provisions in this act deal with the ability of U.S. authorities to control the possession, storage, and spread of dangerous pathogens. Under the act, the Department of Health and Human Services (HHS) is required to maintain a list of select agents that pose a threat to public health and safety. All facilities in possession of any agents on the list must register with the department (Ryan & Glarum, 2008). Background checks are required for anyone "deemed to have a legitimate need for access to select agents" and all institutions possessing the agents are required to maintain comprehensive security plans and risk assessments (Ryan & Glarum, 2008, p. 215).

Homeland Security Presidential Directive-18: Medical Countermeasures Against Weapons of Mass Destruction, (January 31, 2007)

This directive is aimed at the need to develop and stockpile medical countermeasures necessary to treat victims of chemical, biological, radiological, nuclear, and explosive agents. It also includes provisions for research and development into new countermeasures. "Although it is not feasible to develop countermeasures against every possible biological threat, the directive aimed at tackling some of the more important ones" (Ryan & Glarum, 2008, p. 217).

Homeland Security Presidential Directive-21: Public Health and Medical Preparedness (October 18, 2007)

This comprehensive directive addresses the ability of the U.S. public health system to provide mass casualty care following any significant terrorist attack, whether related to weapons of mass destruction or not. For example, at the time of the directive, hospital surge capacity for large urban hospitals was estimated to hover at approximately 50 beds. This meant that beyond an unexpected increase of 50 patients, most hospitals would be overwhelmed and unable to take on additional patients, especially the serious trauma cases that a mass casualty attack would entail. The surge capacity for rural hospitals was even less.

The directive addressed a variety of public health system issues including biosurveillance, countermeasures stockpiling and distribution, mass casualty care, community resilience, risk awareness, and education and training (Ryan & Glarum, 2008). Provisions regarding exceptions to medical privacy rules during a public health emergency were also addressed. Finally, the directive sought to encourage "the engagement of the private sector, academic, and other non-government entities in preparedness and response efforts" and highlighted the "important roles of individuals, families, and communities" (Ryan & Glarum, 2008, pp. 217–218).

Project BioShield Act of 2004 (PL 108-276) (July 21, 2004)

Debates over legislative proposals to address the development of new vaccines began in 2003. These proposals eventually led to the Project BioShield Act of 2004 (BioShield). BioShield was designed to provide market incentives to pharmaceutical and other biotechnology companies to engage in vaccine research and to ensure a market for new vaccines by providing for government purchase. It also provided for expedited funding and approval procedures.

The general need for a BioShield program was not hotly contested. The need was well summarized in the April 2003 testimony of Dr. Michael Friedman of the Pharmaceutical Research and Manufacturers of America:

> The President's proposal speaks primarily to the early and to the later stages and the lengthy high risk and costly process of bringing new medicines into the market. It does not, however, speak to the time-consuming and resource-intensive middle portion of that process which is largely our responsibility. Further, research into biothreat countermeasures presents challenges beyond those ordinarily encountered in non-biodefense R&D. These include scientific challenges, economic challenges, and legal challenges. (U.S. House of Representatives, 2003)

Friedman goes on to discuss that some products will be distributed with fewer or less-clinical trials than usual, opening up companies to huge liability and a dearth of insurance companies willing to cover the risk. In addition, the need for rapid development of products will require considerable communication and coordination among companies, opening them up to antitrust liability (according to their legal counsel), and countermeasure research will divert resources from other serious diseases such as diabetes.

The economic concerns of a limited vaccine market were not the only issues raised by community participants as the details of BioShield were debated. Infectious disease specialists such as those represented by the Infectious Disease Society of America (IDSA), while supportive of BioShield, cautioned against an overemphasis on bioterrorism, noting that the most pressing issue with infectious diseases was the growing resistance of bacteria to current antibiotics (U.S. House of Representatives, 2003, p. 103; U.S. House of Representatives, 2001, p. 97). In addition, participants from industry, interest groups, and academia cautioned against a government reliance solely on vaccines to deal with a bioterrorism attack to the exclusion of other forms of treatment.

By October 2004, oversight hearings were addressing proposed legislation called BioShield II to deal with deficiencies in its implementation. While government agencies and congressional supporters lauded DHS efforts to spur research and development into vaccines and other treatments and promised further funding, other members of Congress, members of the pharmaceutical industry, and biotechnology experts were less enthusiastic. Concerns regarding BioShield centered around the slow pace with which BioShield was implemented and the awarding of contracts to newer rather than established biotechnology and pharmaceutical firms, the need for the process to be streamlined with fewer agencies involved, and more transparency as to the requirements for government funding (U.S. House of Representatives, 2005a, p. 72; U.S. House of Representatives, 2005b). Still, as of 2013, an additional $2.8 billion in advanced funding was authorized for BioShield through FY 2018 (Painter, 2013, p. 15).

Implementation of Public Health Laws

Throughout this section, we have provided examples of the many laws, executive orders, and other provisions dealing with numerous terrorism and homeland security issues. The existence of these provisions, however, does not always guarantee compliance or effective implementation. A good example of this dilemma can be found in the research conducted by Jacobsen, Wasserman, Botoseneanu, Silverstein, and Wu (20112) into the implementation of public health laws by state and local public health and emergency management officials. Their research was spurred by evidence of inadequate responses to recent disasters:

> Time and time again, perhaps most starkly in the case of Hurricane Katrina, we have witnessed the consequences—at times deadly—of responses to public health emergencies that were delayed or inadequate because of ambiguity over fundamental issues about who is responsible for what. (Jacobsen et al., 2012, p. 298)

The researchers compared the **objective legal environment**—specific legislation, regulations, and judicial decisions—with the **perceived legal environment**—the understanding by practitioners of how the objective legal environment should be interpreted—and found a "clear disconnect between the objective legal conditions (federal and state specific) and what the respondents perceived them to be, especially at the local level" (2012, p. 305). In addition, local officials questioned the adequacy of public health preparedness laws because of a lack of testing in actual large-scale events and because of concerns that the population might not actually comply with such requirements as quarantine or isolation (Jacobsen et al., 2012, p. 308).

Transportation

When one thinks of transportation security, the main focus is often aviation. This is not surprising given the history of terrorist attacks against U.S. aviation and passengers. The attacks of 9/11 may have been the most devastating in U.S. aviation history, and the first of that nature on U.S. soil, but they were hardly the only ones. Others, some discussed in previous sections of this text, include the hijacking of Trans World Airlines (TWA) Flight 847 in 1985, the bombings of TWA and Pan American Airways (Pan Am) flights in 1986, and the destruction of Pan Am Flight 103 over Lockerbie, Scotland, in 1988.

Legislation, federal regulations, and executive orders regarding aviation security date back to the late 1960s and early 1970s, with the first metal detectors required at airports starting in 1972. The most recent legislation prior to the 9/11 attacks was the Aviation Security Improvement Act of 1990 (PL 101-604), which was a direct result of the Pan Am 103 disaster and the recommendations of the presidential commission formed in response. The concerns addressed in the 1990 act were wide-ranging, including improvements in hiring and training of airport screeners, full screening of all checked baggage, improved security of the tarmac and air operations areas to restrict access to airplanes, screening of cargo and air freight, an increased role for aviation security within the structure of the Federal Aviation Administration (FAA), and increased international cooperation. The commission formed to investigate the Pan Am 103 disaster also echoed previous recommendations that responsibility for aviation security be removed from the jurisdictions of the airlines and placed with the federal government as a law enforcement function.

Sadly, many of the concerns of 1990 remained concerns on September 10, 2001. In fact, other than the changes in cockpit security as a direct result of the nature of the 9/11 attacks, none of the aspects of the aviation security efforts following 9/11 were new. There was just a new urgency to them.

Aviation and Transportation Security Act (ATSA) of 2001 (PL 107-71) (November 19, 2001)

The changes made to the aviation industry in the ATSA were wide-ranging and significant, under both the act itself and its implementing regulations. First, aviation security was placed under a newly formed federal agency—the Transportation Security Agency (TSA)—ultimately under DHS. Airport screeners were now federal employees with mandated background checks and longer hour requirements for training. Changes were made to the security layout of airports, and access to the boarding areas was limited to ticketed passengers. New restrictions were placed on the carrying of any objects that could be used as a weapon on to a plane, including the box-cutter type devices that were used to hijack the four planes on 9/11. New technologies for passenger screening were put in place—although not without considerable controversy regarding privacy issues. Prior to 9/11, explosives screening focused on carry-on luggage and selective screening of checked luggage for international flights. The ATSA mandated 100% screening of all checked baggage on both domestic and international flights to and from the U.S. (Painter, 2013). In addition, under the Implementing the 9/11 Commission Recommendations Act of 2007 (PL 110-53), TSA was required to physically screen all cargo placed on passenger flights. "While TSA has met the requirement for cargo screening domestically . . . additional work is needed to implement similar measures for U.S.-bound international flights" (Painter, 2013, pp. 40–41).

Even given the tragedy and loss of life of the 9/11 attacks, the implementation of improvements in aviation security, while substantial, have been less than ideal. With no successful attacks in the United States since 2001, passengers and the airlines have once again become concerned with ease and economic viability of air travel as much as with security. A 2013 Congressional Research Service report noted that over a decade later, the following issues remain:

◆ Effectively screening passengers, baggage, and cargo for explosive threats;
◆ Developing effective risk-based methods for screening passengers and others with access to aircraft and sensitive areas;
◆ Exploiting available intelligence information and watch lists to identify individuals that pose potential threats to civil aviation;
◆ Developing effective strategies for addressing aircraft vulnerabilities to shoulder-fired missiles and other standoff weapons; and
◆ Addressing the potential security implications of unmanned aircraft operations in domestic airspace (Painter, 2013, p. 40)

Other Areas of Transportation Interest

Aviation security is not the only concern of transportation experts and officials. Long the stepchildren of the transportation security industry compared to aviation, other sectors received significantly increased statutory and regulatory attention after 2001. The TSA now regulates security not only of aviation, but also for the maritime, mass transit (including rail), and commercial trucking industries (in coordination with the Department of Transportation's Federal Motor Carrier Safety Administration). Of particular concern, given the threat of the use of weapons of mass destruction, was the transportation of hazardous chemicals that could be either stolen and released by terrorists or released by accident. Significant media attention was given to transportation routes for hazardous chemicals, including nuclear waste material and chemicals by train and truck through populated areas. New rules were established for hazardous materials licenses and any commercial drivers wishing to carry a hazardous materials endorsement on their license must also pass a TSA background check.

 International Law and Terrorism: Efforts of the United Nations

International efforts against terrorism take place between nations (bilateral treaties), among nations that share a geographical region (such as the European Union or the Organization of American States), and worldwide (among members of the United Nations). UN bureaus and organizations such as the International Civil Aviation Organization (ICAO) and the International Maritime Organization (IMO) address security and terrorism issues relevant to their particular missions. This section will discuss different types of treaties and recent efforts of the UN Security Council against transnational terrorism, with a focus on recent research by Galicki (2005) and Heupel (2007).

A Comparison of Types of Conventions

The first international treaty against terrorism was the Convention for the Prevention and Punishment of Terrorism of 1937 (Geneva Convention of 1937). "Although the Geneva Convention of 1937 . . . unfortunately, has never entered into force, one cannot underestimate its importance as the first comprehensive and multilateral antiterrorist convention" (Galicki, 2005, p. 744).

Since then, as Galicki notes in her 2005 study, conventions against terrorism can be divided into categories related to scope and geography. They can be either **comprehensive conventions** or **sectoral conventions**, and either regional or universal. A number of treaties can also combine these characteristics. For example, a treaty can be comprehensive and universal (such as the 1937 Convention) or comprehensive and regional (such as the Inter-American Convention Against Terrorism of 2002). A treaty can also be universal and sectoral—involving all members of the UN but focusing on a specific topic or type of attack (such as the Convention for the Suppression of Unlawful Seizure of Aircraft of 1970) (Galicki, 2005, p. 744).

Galicki (2005) reviewed international terrorism–related treaties across all four categories dating back to the 1937 convention and found that "among universal antiterrorist conventions, there is an overwhelming majority of those of sectoral nature, whereas in the case of **regional conventions,** comprehensive treaties prevail" (p. 744). In other words, the more countries involved in a convention (universal), the more narrow the focus of the topic (sectoral rather than comprehensive). This makes particular sense given the difficulty of defining the term "terrorism" in international agreements. An agreed upon definition is more likely to be found among countries that share a regional focus and regional concerns. In universal treaties, that issue is sidestepped somewhat by defining the act by the particular topic the treaty is designed to address. The following are some examples:

- "unlawful acts against the safety of civil aviation"
- "crimes against internationally protected persons, including diplomatic agents"
- "taking hostages"
- "theft, robbery, or any other unlawful taking of nuclear material or of credible threat thereof"
- "unlawful acts against the safety of fixed platforms located on the continental shelf" (Galicki, 2005, p. 745)

Galicki also notes the increased attention paid to the intersection of transnational organized crime and terrorism through the UN's Palermo Convention of 2000. While the Palermo Convention does not specifically mention terrorism in its list of offenses, many of the offenses listed, such as extortion, money laundering, and participating in an organized criminal group, are conducted by terrorist groups as part of their operations and fund-raising activities.

The UN Security Council (SC): Efforts Against State Versus Transnational Terrorism

State-sponsored terrorism—where a specific state conducts a terrorist attack against citizens of its own state or others—has been on the decline, according to Heupel (2007), while **transnational terrorism** has been on the rise, forcing the **United Nations Security Council** (SC) to adjust its approach to responding to acts of international terrorism. The SC is one of the few international bodies that has the authority to impose binding counterterrorism obligations on most nation-states. Therefore, even though its impact has been mixed, as noted by Heupel (2007), its response to international terrorism is worth examining.

Over the decades, the SC has taken action, through its resolution and sanctions process, against several state sponsors of terror. These have included SC Resolutions 731 and 748, in 1992, as well as SC Resolution 883 in 1993, against Libya for the downing of Pan American Airways and Air France airliners, and SC Resolution and SC Resolution 1054 in 1996 against Sudan for failing to turn over terrorist suspects in several major attacks and continuing to harbor and support multiple terrorist groups, including bin Laden and al Qaeda (Heupel, 2007).

Transnational terrorism presents its own set of problems that differ in many respects from the more traditional issues dealt with by the SC. Transnational terrorism relies less on direct one-to-one support from a particular state or set of states. It "commonly uses weak or failing states as safe havens and receives support from various non-state actors" (Heupel, 2007, p. 478). Since the organizational structure of transnational groups such as al Qaeda cross numerous state lines, dealing with the phenomena is much more difficult. Still, by examining both binding obligations passed by the SC and strategies implemented to encourage compliance, Heupel argues that the SC has adapted to the shift from state-sponsored to transnational terrorism.

> Over time, the SC appears to have realized that the specific features of transnational terrorism require new strategies. It therefore turned to imposing sanctions against terrorists and their non-state associates and committed all states to implementing these sanctions. Furthermore, the SC obligated all states to strengthen their domestic capacities to prevent and suppress terrorism. Finally, the SC recognized that most states needed implementation assistance and therefore, in an unprecedented way, applied managerial compliance strategies. (Heupel, 2007, p. 479)

Several SC Resolutions were issued before 2001 against the Taliban and al Qaeda, including Resolutions 1189, 1193, 1214, 1267, and 1333 (Heupel, 2007). While a laudable attempt to get at al Qaeda, these resolutions were similar to the previous ones against Libya and Sudan in that they were targeted against a single state, Afghanistan, and the group its government harbored and supported.

The sea change came with SC Resolution 1373, passed on September 28, 2001. This resolution, rather than being directed against a single government, required member states to take broad, far-reaching steps to combat terrorism (United Nations, 2001). Areas addressed in the resolution include financing terrorism, support—passive or active—of any persons involved in terrorist acts, providing safe havens to terrorist groups, assistance in criminal investigations and prosecutions of terrorist acts, preventing the free movement of terrorist groups by improving border controls, and increased cooperation against terrorism by accession to existing international protocols and conventions (United Nations, 2001). A SC Committee on Counterterrorism was also established. A copy of Resolution 1373 can be seen in Figure 13.2. As noted by Heupel, "Resolution 1373 . . . clearly acknowledges the distinct features of transnational terrorism—its use of developed and developing states around the globe, its cross-border network structure—and intends to provide answers to this distinct threat" (2007, p. 489).

Summary

This section examines the legal regimes dealing with terrorism and homeland security in both the domestic (U.S.) and international arenas. Legal regimes in this arena can be particularly challenging to both establish and implement because they are often formed or modified in reaction to specific terrorists, yet at the same time must provide long-term abilities to both prevent and respond to terrorist attacks across a wide variety of situations. This section focused on legal regimes based on U.S. federal law. Even across U.S. law, definitions of terrorism may vary depending on the mission of the relevant government agency or department. Federal laws dealing with terrorism and homeland security issues existed long before the attacks on the World Trade Center and the Pentagon in 2001, but exploded in frequency and coverage since. The U.S. legal regimes dealing with terrorism and homeland security now include statutory, regulatory, and executive provisions covering such areas as Homeland Security, Critical Infrastructure Protection, Detention of Enemy Combatants, Emergency Management and Disaster Response, Law Enforcement and Civil Liberties, Public Health and Preparedness, and Transportation. This section highlighted some examples of provisions in each of these areas.

International efforts against specific types of terrorism, specific states harboring terrorist groups, and transnational terrorism are also in place. Galicki's (2007) research compared various types of conventions established under the auspices of the United Nations and found that conventions dealing with either a specific target of terrorism (sectoral) or a limited geographic region (regional) were the most prevalent and effective, but that there is room for more comprehensive and **universal conventions** as well. Heupel's (2005) research found that the United

Figure 13.2	United Nations Security Resolution 1373 (2001): On Threats to International Peace and Security Caused by Terrorist Acts

Adopted by the Security Council at its 4385th meeting, on 28 September 2001

The Security Council,

Reaffirming its resolutions 1269 (1999) of 19 October 1999 and 1368 (2001) of 12 September 2001,

Reaffirming also its unequivocal condemnation of the terrorist attacks which took place in New York, Washington, D.C. and Pennsylvania on 11 September 2001, and expressing its determination to prevent all such acts,

Reaffirming further that such acts, like any act of international terrorism, constitute a threat to international peace and security,

Reaffirming the inherent right of individual or collective self-defence as recognized by the Charter of the United Nations as reiterated in resolution 1368 (2001),

Reaffirming the need to combat by all means, in accordance with the Charter of the United Nations, threats to international peace and security caused by terrorist acts,

Deeply concerned by the increase, in various regions of the world, of acts of terrorism motivated by intolerance or extremism,

Calling on States to work together urgently to prevent and suppress terrorist acts, including through increased cooperation and full implementation of the relevant international conventions relating to terrorism,

Recognizing the need for States to complement international cooperation by taking additional measures to prevent and suppress, in their territories through all lawful means, the financing and preparation of any acts of terrorism,

Reaffirming the principle established by the General Assembly in its declaration of October 1970 (resolution 2625 (XXV)) and reiterated by the Security Council in its resolution 1189 (1998) of 13 August 1998, namely that every State has the duty to refrain from organizing, instigating, assisting or participating in terrorist acts in another State or acquiescing in organized activities within its territory directed towards the commission of such acts,

Acting under Chapter VII of the Charter of the United Nations,

1. *Decides* that all States shall:

    (a) Prevent and suppress the financing of terrorist acts;

    (b) Criminalize the wilful provision or collection, by any means, directly or indirectly, of funds by their nationals or in their territories with the intention that the funds should be used, or in the knowledge that they are to be used, in order to carry out terrorist acts;

    (c) Freeze without delay funds and other financial assets or economic resources of persons who commit, or attempt to commit, terrorist acts or participate in or facilitate the commission of terrorist acts; of entities owned or controlled directly or indirectly by such persons; and of persons and entities acting on behalf of, or at the direction of such persons and entities, including funds derived or generated from property owned or controlled directly or indirectly by such persons and associated persons and entities;

(d) Prohibit their nationals or any persons and entities within their territories from making any funds, financial assets or economic resources or financial or other related services available, directly or indirectly, for the benefit of persons who commit or attempt to commit or facilitate or participate in the commission of terrorist acts, of entities owned or controlled, directly or indirectly, by such persons and of persons and entities acting on behalf of or at the direction of such persons;

2. *Decides also* that all States shall:

(a) Refrain from providing any form of support, active or passive, to entities or persons involved in terrorist acts, including by suppressing recruitment of members of terrorist groups and eliminating the supply of weapons to terrorists;

(b) Take the necessary steps to prevent the commission of terrorist acts, including by provision of early warning to other States by exchange of information;

(c) Deny safe haven to those who finance, plan, support, or commit terrorist acts, or provide safe havens;

(d) Prevent those who finance, plan, facilitate or commit terrorist acts from using their respective territories for those purposes against other States or their citizens;

(e) Ensure that any person who participates in the financing, planning, preparation or perpetration of terrorist acts or in supporting terrorist acts is brought to justice and ensure that, in addition to any other measures against them, such terrorist acts are established as serious criminal offences in domestic laws and regulations and that the punishment duly reflects the seriousness of such terrorist acts;

(f) Afford one another the greatest measure of assistance in connection with criminal investigations or criminal proceedings relating to the financing or support of terrorist acts, including assistance in obtaining evidence in their possession necessary for the proceedings;

(g) Prevent the movement of terrorists or terrorist groups by effective border controls and controls on issuance of identity papers and travel documents, and through measures for preventing counterfeiting, forgery or fraudulent use of identity papers and travel documents;

3. *Calls* upon all States to:

(a) Find ways of intensifying and accelerating the exchange of operational information, especially regarding actions or movements of terrorist persons or networks; forged or falsified travel documents; traffic in arms, explosives or sensitive materials; use of communications technologies by terrorist groups; and the threat posed by the possession of weapons of mass destruction by terrorist groups;

(b) Exchange information in accordance with international and domestic law and cooperate on administrative and judicial matters to prevent the commission of terrorist acts;

(c) Cooperate, particularly through bilateral and multilateral arrangements and agreements, to prevent and suppress terrorist attacks and take action against perpetrators of such acts;

(d) Become parties as soon as possible to the relevant international conventions and protocols relating to terrorism, including the International Convention for the Suppression of the Financing of Terrorism of 9 December 1999;

(Continued)

Figure 13.2 (Continued)

(e) Increase cooperation and fully implement the relevant international conventions and protocols relating to terrorism and Security Council resolutions 1269 (1999) and 1368 (2001);

(f) Take appropriate measures in conformity with the relevant provisions of national and international law, including international standards of human rights, before granting refugee status, for the purpose of ensuring that the asylum-seeker has not planned, facilitated or participated in the commission of terrorist acts;

(g) Ensure, in conformity with international law, that refugee status is not abused by the perpetrators, organizers or facilitators of terrorist acts, and that claims of political motivation are not recognized as grounds for refusing requests for the extradition of alleged terrorists;

4. *Notes* with concern the close connection between international terrorism and transnational organized crime, illicit drugs, money-laundering, illegal arms-trafficking, and illegal movement of nuclear, chemical, biological and other potentially deadly materials, and in this regard *emphasizes* the need to enhance coordination of efforts on national, subregional, regional and international levels in order to strengthen a global response to this serious challenge and threat to international security;

5. *Declares* that acts, methods, and practices of terrorism are contrary to the purposes and principles of the United Nations and that knowingly financing, planning and inciting terrorist acts are also contrary to the purposes and principles of the United Nations;

6. *Decides* to establish, in accordance with rule 28 of its provisional rules of procedure, a Committee of the Security Council, consisting of all the members of the Council, to monitor implementation of this resolution, with the assistance of appropriate expertise, and *calls upon* all States to report to the Committee, no later than 90 days from the date of adoption of this resolution and thereafter according to a timetable to be proposed by the Committee, on the steps they have taken to implement this resolution;

7. *Directs* the Committee to delineate its tasks, submit a work programme within 30 days of the adoption of this resolution, and to consider the support it requires, in consultation with the Secretary-General;

8. *Expresses* its determination to take all necessary steps in order to ensure the full implementation of this resolution, in accordance with its responsibilities under the Charter;

9. *Decides* to remain seized of this matter.

Nations Security Council has also learned to adapt its sanctions authority to deal with transnational as well as state-sponsored terrorism.

▨ Key Points

- Legal regimes in this arena can be particularly challenging to both establish and implement because they are often formed or modified in reaction to specific terrorist attacks, yet at the same time must provide long-term abilities to both prevent and respond to terrorist attacks across a wide variety of situations.
- Legal regimes dealing with terrorism and homeland security can be found in state and federal statutes and regulations, executive orders, case law, and international conventions and treaties.

◆ Legal definitions of terrorism in the United States differ depending on the government institution using the term and its mission or purpose.

◆ The U.S. domestic legal regime includes both civil and criminal provisions.

◆ A significant number of federal homeland security functions are now housed in the Department of Homeland Security, the largest reorganization of the federal government since 1949.

◆ The sectors now considered part of critical infrastructure protection have expanded considerably since 2001 and now include 16 different sectors or industries.

◆ Laws and regulations dealing with emergency management and public health preparedness have swung between an emphasis on an all-hazards approach and a more type-specific approach as the United States struggles with how to deal with mass-casualty attacks on U.S. soil.

◆ Since 2001, a much greater emphasis has been placed on criminalizing any material, financial, physical, or other involvement in any organizations related to groups on the list of Foreign Terrorist Organizations. Other laws have been changed to make it easier for federal authorities to investigate and prosecute those suspected of terrorist activities.

◆ United Nations conventions on international terrorism can be either comprehensive, sectoral, regional, or universal and can combine characteristics of more than one type. Conventions dealing with either a specific target of terrorism (sectoral) or a limited geographic region (regional) were the most prevalent and effective, but there is room for more comprehensive and universal conventions as well.

◆ The United Nations Security Council has also learned to adapt its sanctions authority to deal with transnational as well as state-sponsored terrorism.

KEY TERMS

Chemical Facility Anti-Terrorism
  Standards (CFATS)

Code of Federal Regulations (CFR)

Comprehensive conventions

Conventions

Critical infrastructure protection

Department of Homeland Security

Enemy combatants

Executive orders

Homeland Security Act of 2002

Objective legal environment

Perceived legal environment

Posse Comitatus Act of 1878

Presidential directives

Real ID

Regional conventions

Sectoral conventions

State-sponsored terrorism

Transnational terrorism

Treaties

United Nations Security Council

Universal conventions

U.S. Northern Command
  (USNORTHCOM)

DISCUSSION QUESTIONS

1. Why are legal regimes dealing with terrorism and homeland security so challenging to establish?

2. What are some of the ways U.S. law defines terrorism? Why would a single government need multiple definitions of terrorism? Do you think it is possible to develop a single federal definition? Why or why not?

3. What are some of the areas of law that fall under the rubric of terrorism and homeland security? Of the ones discussed in this section, pick three you feel are the most important. Support your answer.

4. What do you feel is the most challenging aspect of developing domestic legal regimes to deal with terrorism and homeland security? Support your answer.

5. What do you feel is the most challenging aspect of developing international legal regimes to deal with terrorism and homeland security? Support your answer.

6. What are the four types of conventions described in this section?

7. Develop a title and three provisions for an international convention dealing with an issue or issues related to terrorism. Identify your convention based on the four types above. Justify the need for your convention.

8. Review the provisions of Security Council Resolution 1373. What amendments would you make to the resolution and why?

WEB RESOURCES

Congress.gov (official legislative website): http://www.congress.gov

Federation of American Scientists (collection of presidential orders and directives): http://www.fas.org/irp/offdocs

International Law Association: http://www.ila-hq.org

Legal Information Institute: http://www.law.cornell.edu

United Nations: http://www.un.org

United Nations Actions to Counter Terrorism: http://www.un.org/en/terrorism

United Nations Security Council: http://www.un.org/en/sc

University of Maryland (collection of CRS reports on terrorism and homeland security): http://www.law.maryland.edu/marshall/crsreports

U.S. Department of Homeland Security: http://www.dhs.gov

U.S. Department of Justice (website highlighting the provisions of the USA Patriot Act): http://www.justice.gov/archive/ll/highlights.htm

U.S. Department of State, Office of Legal Affairs: http://www.state.gov/s/l

U.S. House Judiciary Committee: http://www.judiciary.house.gov

U.S. Senate Judiciary Committee: http://www.judiciary.senate.gov

READING 22

Among the most controversial areas in the USA PATRIOT Act are those surrounding privacy in this new information age. The concerns have been compounded by a spate of domestic and international spying scandals that have hit the media in recent years. Garson examines some of the implications for information privacy and security evident in the renewal of the Patriot Act in 2006 and the Real ID program.

Securing the Virtual State

Recent Developments in Privacy and Security

G. David Garson
North Carolina State University, Raleigh

As has been widely reported, information security rose to first place priority after 9/11. The FY 2005 Office of Management and Budget (OMB) report on agency compliance with the Federal Information Security Management Act of 2002 found that the percentage of systems certified and accredited as meeting security standards had risen to 85%, an increase over 77% in FY 2004 (Dizard, 2006b). Such improvements in information technology (IT) security are attributed to efforts mobilized by the Y2K problem of 2000 and terrorist attacks on September 11, 2001, and to the subsequent emphasis given to security by the Chief Information Officers (CIO) Council and in IT budgets in the 2001 to 2006 period. With investment in a presumably more information-secure state, these questions may be asked: What have we bought with our money? Is it the same kind of state we have appreciated for so many years, or are we changing it into something different from what we have known?

There is no doubt that we have mandated IT security at unprecedented levels. As of September 30, 2005, the Federal Acquisition Regulations Council has required federal agencies to incorporate IT security in all purchases, including conforming to agency-specific security requirements, and to consult security specialists when making IT purchases (Miller, 2005; Vijayan, 2005). This implemented the Federal Information Security Management Act of 2002 (part of the E-Government Act of 2002), which called for heightened security throughout system life cycles. Similarly, at the state level, the Department of Homeland Security's (DHS) FY 2006 grant application kit for competition for $3.9 billion in IT funds urged states to conform to new 2005 national standards for the National Information Exchange Model, an XML-based model for information sharing. Although cybersecurity planning had been encouraged previously, for FY 2006 for the first time, states and localities seeking grant funds were called on to develop and implement a comprehensive cybersecurity plan and establish an information security officer as a 24-7 point of contact. Using the carrot of the $2.5 billion State Homeland Security Grant Program and the $862 million Urban Area Security Initiative, DHS set forth numerous additional specific security guidelines for states and localities (Lipowicz, 2006b).

SOURCE: Garson, G. D. (2006). Securing the virtual state: Recent developments in privacy and security. *Social Science Computer Review, 24,* 489–496.

Renewal of the PATRIOT Act, 2006

The legislation providing the overarching framework for security is, of course, the PATRIOT Act, which was renewed recently. In mid-December 2005, the Senate rejected the reauthorization of the PATRIOT Act by a vote of 52 to 47, with opposition coming largely on privacy rights grounds. A week later, a 5-week extension was granted, and then on February 2, 2006, another 5-week extension was approved by Congress. Finally, on March 2, 2006, the Senate passed a modified version of the renewal act by a vote of 89 to 10. Fourteen of 16 PATRIOT Act provisions were made permanent, with 2 others (including roving wiretaps) given a renewal for 4 years. In spite of the widespread general attention to civil liberties issues, few Americans or even social scientists have given attention to the details of the renewal.

Particularly controversial was Section 605 of the House version of the PATRIOT Act renewal, which created a permanent national police force under the Secret Service with the power to make arrests without warrant for any offense against the United States committed in their presence or for any felony under the laws of the United States. Although the Secret Service had already exercised such powers in the District of Columbia, Section 605 nationalized the scope of a secret police force with powers of warrantless arrest. The renewal of the PATRIOT Act made Section 605 permanent.

Other controversial aspects of PATRIOT Act renewal included:

- "Sneak and peek" searches, which are physical searches of homes and offices without notice, before or after, that privacy was being invaded. At issue was whether such searches would have to be followed by notice 7 or 30 days after the search. The renewal bill continued sneak and peak searches and provided for 30-day notification, but law enforcement appeals could extend notification indefinitely.

- Roving wiretaps of selected premises were provided. At issue was whether enforcement authorities would have to determine that a specific target was probably present before surveillance was initiated. The renewal bill extended roving wiretaps for 4 more years.

- PATRIOT Act Section 215 authorized the FBI to take out secret court orders forcing businesses and organizations to disclose personal records (e.g., library records, Internet service provider records, bank records). At issue was whether the FBI should be required to show a connection between the records and suspected terrorism. The renewal bill imposed no such requirement but renews Section 215 for only 4 years.

- Section 215 also criminalized the disclosure by librarians, Internet service providers, or others of searches of their records. At issue was whether organizations should have a right to inform their members, clients, or customers that their privacy had been violated. The renewal as passed extended the gag order for 1 year, after which it gave subpoena recipients in terrorist investigations the right to challenge in court the requirement that they refrain from telling anyone. As few libraries or other subpoena recipients would be willing to undertake an expensive court fight with the government, which would claim secrecy was required by national security, critics called this concession to civil liberties merely "cosmetic."

- Between passage of the PATRIOT Act in 2001 and 2005, more than 30,000 National Security Letters (NSLs) were issued by the FBI, forcing businesses and organizations to turn over records. At issue was whether the current requirement of relevance needed to be tightened to restrict wholesale use of NSLs. One proposal was to require NSLs to be approved by the FISA Court or a federal judge. The renewal continued NSLs but did stipulate that libraries could not be recipients of NSLs unless they ran Internet servers and that the FBI could not demand the name of a lawyer consulted about a NSL, as had been the practice. The renewal law made explicit that judicial review of NSLs is allowable.

- Also at issue was a proposal to make disclosure of an NSL a crime with a 5-year jail term if there was intent to obstruct an investigation (in the view of enforcement authorities, not in the view of the organization). The renewal bill retained NSL disclosure penalties. The renewal law also provided that nondisclosure orders were no longer to attach to NSL requests on an automatic basis.

- At issue was whether organizations would have the right to appeal Section 215 and NSL records search requests and, if so, whether the appeal process would be pro forma or meaningful. The renewal bill did provide for individuals receiving Section 215 orders to seek judicial review. The renewal bill supported the right of NSL recipients to seek legal counsel but did not spell out any appeal rights apart from appealing gag orders.
- The PATRIOT Act employs a vague and expansive definition of terrorism, to include a broad range of political activities. Domestic terrorism is defined to include acts "dangerous to human life" in violation of criminal law, if the act involves intimidation or coercion of a civilian population, influence of government policy by intimidation or coercion, or affecting the conduct of government through mass destruction, assassination, or kidnapping. The ACLU (2002) has noted that Greenpeace, Operation Rescue, Vieques Island, and World Trade Organization protesters and the Environmental Liberation Front had all engaged in activities that could subject them to being investigated as engaging in domestic terrorism. The renewal act did not alter the definition of terrorism.
- The PATRIOT Act allowed information gathered in warrantless searches to be used in criminal trials, where such information would otherwise be inadmissible as having been illegally obtained. The renewal act left this intact.

In summary, in the PATRIOT Act renewal, the Bush administration pushed for permanent expansion of relatively unrestricted police powers in the name of antiterrorism but faced opposition from Democrats, joined by libertarian Republicans, opposed to conferring further police state powers on the president ("Congress Extends," 2006). In the end, however, the PATRIOT Act was renewed with expansive executive powers for domestic spying, with most areas of controversy resolved in favor of executive power at the expense of civil liberties.

 ## The Domestic Spying Affair of 2005 to 2006

When security powers are increased, as with the PATRIOT Act renewal, public trust rests on the credibility of officials with respect to the use of their powers. This credibility was challenged when, just before Christmas 2005, the *New York Times* revealed that after the 9/11 terrorist attacks in 2001, President Bush had authorized the National Security Administration (NSA) to intercept telephone calls and e-mail traffic without the benefit of court-issued warrants. The Bush administration defended its action on the grounds of homeland security and the need to pursue its antiterrorist efforts. As authority, Bush cited his presidential powers as commander in chief, his inherent powers to act to protect the American people, and Congress's post-9/11 resolution authorizing him to wage war on terrorists. The administration at first claimed that spying was limited to calls or e-mails in which at least one end of the communication was in another country but later admitted that domestic-to-domestic traffic had also been surveilled, allegedly "by accident." The president's attorney general, NSA advisor, and other administration officials rallied behind the president (Lichtblau, 2006).

The issue of domestic spying immediately became front-page news, with some Democrats calling for creation of an independent counsel to investigate possible violation of the Fourth Amendment's prohibition against unreasonable searches and seizures—the constitutional provision on which is the basis for the requirement for court authorization of warrants and wiretaps. The warrantless electronic surveillance program of the Bush administration had bypassed court approvals, even after-the-fact ones, even by the secretive Foreign Intelligence Surveillance Court (FISC, established in the 1970s under the Foreign Intelligence Surveillance Act). The PATRIOT Act had designated FISC as the only court authorized to issue surveillance orders in investigations of terrorism, and it was thought to approve almost all requests submitted to it.

Domestic spying seemed to be part of a general policy, not an aberration, because also in December 2005, the ACLU had released new FBI records (released as a result of a lawsuit over FBI activities in relation to the 2004 national political conventions) showing ongoing monitoring of antiwar, civil rights, environmental, and other activist groups. "Our government is spying on Americans— unapologetically, unnecessarily and with no regard for the Constitution," the ACLU stated (Fisher, 2005).

Although Congressional leaders, including Democrats, had been briefed about the domestic spying program, the briefings had bound them to secrecy, precluding even obtaining legal opinions regarding the constitutionality of the program. When the issue became public, various Senators made known that they had sent Vice President Cheney letters raising legal issues, but with no response. The no-discussions policy was further emphasized when, immediately after the domestic spying affair became public, the Bush Justice Department moved to investigate and prosecute the whistle-blowers who had leaked the story.

 ## Privacy Protection in the Age of Homeland Security

On the administrative side, the DHS is a major recipient of increased IT security powers and funds. Special issues of privacy have surrounded the DHS since its inception. Under criticism from civil libertarians fearing the escalation of intrusions into individual privacy based on the PATRIOT Act and other legislation, Congress created a DHS Privacy Office at the same time that it created the DHS itself in 2002. Although the Privacy Office won praise for its reports critical of privacy protections under the Transportation Security Department's Computer-Assisted Passenger Prescreening System, civil liberties advocates have continued to charge that the DHS Privacy Office has too little authority to compel cooperation in investigating privacy complaints, relying instead on voluntary compliance. Moreover, because the Privacy Office must clear reports through the DHS secretary, civil libertarians fear the possibility of political stifling of critical reports in the future. In 2005, Rep. Bennie Thompson (D-Miss.) introduced the "Power Act" to give subpoena powers to the DHS Privacy Office and otherwise enhance its independence (Lipowicz, 2005a), but such reforms have gone nowhere.

Perhaps in response to criticism regarding its information security policies, the Bush administration in 2005 added privacy to the quarterly President's Management Agenda scorecard, which rates agencies from red *(worst)* to yellow to green *(best)* on a variety of criteria. As of the end of 2005, only 9 of 26 agencies were found to conduct privacy impact assessments for at least 90% of their major IT systems (Mosquera, 2005). Moreover, in the past, privacy impact assessments have focused on procedural rather than substantive privacy, leaving civil libertarians unsatisfied.

In February 2005, the OMB issued a memorandum requiring all federal departments and agencies to designate "the senior official who has the overall agency-wide responsibility for information privacy issues" (OMB, 2005). The OMB noted that consistent with the Paperwork Reduction Act of 1995, the agency's chief information officer could perform this role. The controversy, of course, was that the CIO also was responsible for information access, security, and other matters that could conflict with privacy goals and thus suffered conflict of interest. The OMB memo fell short of requiring separate, let alone independent, privacy officers in each department. Moreover, a 1998 memorandum (OMB, 1998) had earlier required designation of departmental privacy officers and the conducting of privacy reviews, and the issuing of the 2005 memo was a tacit admission that there were shortcomings in implementation of the Privacy Act of 1974, the Computer Matching and Privacy Protection Act of 1988, the Paperwork Reduction Act of 1995, and the Commerce Department's Principles for Providing and Using Personal Information ("Privacy Principles"), published by the Information Infrastructure Task Force in June 1995.

 ## Administrative Problems of the DHS

A prime mandate of the DHS is data integration. Toward this goal, in 2002, in the aftermath of the 9/11 terrorist attack on New York and Washington, D.C., the Joint Regional Information Exchange System (JRIES) was formed, eventually falling under the aegis of the DHS. Its purpose was to create a national antiterrorism information-sharing network linking police intelligence units and serving as a vehicle for sharing daily antiterrorism information among federal, state, and local officials. However, in 2005 negotiations over JRIES broke down, with major city police intelligence directors terminating full integration with the Homeland Security Information Network (HSIN) because DHS wished to include nonlaw-enforcement officials (e.g., state homeland security advisors) in the network, although police wished sensitive information to be restricted to law enforcement officials, if only for legal reasons (Lipowicz, 2005c). Currently, police share much information with HSIN but have been seeking funding for their own separate network for sensitive information.

Community officials have complained that security intelligence via HSIN has been slow to come, and when it does come, it lacks useful detail (McKay, 2005, p. 34). Partly in response to such criticisms, the DHS reorganized in 2005, under the direction of Michael Chertoff, the new DHS secretary, creating a new Office of Intelligence and Analysis and revising the Homeland Security Advisory System. Chertoff made better sharing of information with state and local government one of his six top priorities. HSIN data sharing comes via the HSIN-Secret network, which Chertoff has acknowledged already is accessible by "tens of thousands of users" (DHS, 2006). That is, inherent in the DHS information systems design is the integration of massive personal databases, in part drawing on information from the PATRIOT Act and other unprecedented increases in police powers, then allowing a precleared, extremely large group of users to access it. The question is, Is abuse inevitable under such a design? Unfortunately, the cloak of secrecy may keep the answer to this question hidden for some time.

In early 2006, the inspector general for the DHS reported a variety of shortcomings with DHS's web security. Though DHS monitored computers on its web, monitoring did not necessarily result in management action. One problem was volume, with 65 million security alerts in a 3-month study period. A greater problem was that DHS's automated security tools could not identify specific workstations which generated the messages. Also, the security alert messages, such as "detect.misuse.porn" did not necessarily indicate what they purported (e.g., non-pornography traffic could generate the pornography warnings, which were 10% of all security warnings) but instead were triggered by use of such words as *oral* (Lipowicz, 2005c). The inspector general's audit also found DHS lacked security accreditations and certifications, compliance with which it expected of others. The security and other DHS IT failures were blamed by the inspector general

in part on the fact that the CIO reports in a relatively low position in the DHS hierarchy, with only 50 of 5,000 IT professionals reporting to him (Chabrow, 2005).

In addition to problems with data integration and web security, the IT track record of DHS in managing its own affairs is also problematic. In January 2006, the DHS cancelled its contract with BearingPoint, Inc., for the Emerge2 comprehensive management system (Electronically Managing Enterprise Resources for Government Effectiveness and Efficiency 2), in spite of having invested $9 million in the $229 million contract (Temin, 2006). The abandonment of Emerge2 was a tacit admission of failure in its one-size-fits-all enterprise software approach to financial systems. DHS instead opted to allow its diverse constituent agencies to choose from among an array of existing public and private sector financial services providers, hoping later to be able to pool the data from these different providers in a way that still supports data warehousing for purposes of central reporting and data analysis. The Emerge2 name was continued, but now with what DHS optimistically called a heterogenous "centers of excellence" strategy (Dizard, 2006a). DHS claimed the $9 million was not wasted but was well spent as it resulted in clarifying requirements and training plans and a web portal to be used by DHS financial managers. Although some hailed DHS's action for catching a probable megafailure early on, both the Emerge2 program manager and the DHS chief financial officer quit in late 2005 or early 2006, leaving the task of creating a DHS financial management system to others. At this writing, DHS has yet to prove that it is up to the IT mandates under which it was created half a decade ago.

 ## Toward a National ID System

The United States is moving rapidly toward a de facto national ID system, something once associated with autocratic regimes such as apartheid-era South Africa. In this, the United States is not at all alone. Belgium, for instance, was the first European Union country to issue national ID cards to its citizens, with 1 million cards in circulation in 2005, expected to rise to more than 8 million by 2009. The cards are used for both public and private sector authentication, including online tax transactions, document requests, library loans, and access to local services such as swimming pools and private sector transactions such as ticket purchases (Government Technology News Staff, 2005).

Bush- and Republican-sponsored, the Real ID Act of 2005 passed mostly along party lines and largely without debate in Congress or much attention from the media. It was embedded in an emergency military appropriations bill for Iraq to make it hard for opponents to vote against it. Its stated purpose is "to establish and rapidly implement regulations for State driver's license and identification document security standards." The Real ID Act compels states to design their driver's licenses by 2008 to comply with federal antiterrorist standards. Federal employees would then reject licenses or identity cards that do not comply, which could curb Americans' access to airplanes, national parks, federal courthouses, Social Security, even bank accounts. It will take effect in 2008.

Under the Real ID Act, DHS is given authority to specify standards, which could include biometrics (fingerprints, iris scans), radio frequency identification tags, and DNA information. It will include basic identity information and a digital photo. State departments of motor vehicles will have to require much more identification than in the past. (Another unfunded mandate to the states.) Departments of motor vehicles will require most license applicants to show a photo ID, a birth certificate, proof of their Social Security number and a document showing their full name and address. All of the documents then would have to be checked against federal databases.

The Real ID Act probably will involve creation of a massive national database on individuals, though there is still a chance the Real ID cards will only be checked locally on an individual basis (e.g., Does the cardholder have the fingerprints the card says he or she has?). Whether there will be such a database and who will have access to it, including commercial access, has not been determined.

 ## Summary

Although e-government is often advertised in terms of advancing democracy and civic participation, the dark side is that it may also involve threats to democratic values such as personal privacy. When e-government is part of a large effort to secure the state, using IT as a major part of security strategy, earnest efforts may focus on technical

achievements that run roughshod over such values, creating a democratic backlash. This happened in South Korea's attempt to implement an electronic education system and electronic national IDs, precipitating fierce popular opposition led by the Korean Teachers' Union and others who denounced e-government initiatives as threats to privacy and who were able to deal e-government implementation a major setback as a result (Jho, 2005). Thus far in the United States, there has been no similar popular reaction against trends outlined in this article. However, social science theory suggests that in the long run, the success of any institutional effort, including the building of a secure virtual state, rests on its conformity with the fundamental values of the culture in which it is found. Presumably, privacy and democracy remain core values of American political culture. This fact provides a cautionary flag to those who think that the pendulum that has brought information security to the highest priority may not in the predictable future swing back toward other priorities of our democratic culture.

 ## References

ACLU. (2002). *How the USA Patriot Act redefines "Domestic Terrorism."* Retrieved, March 8, 2006, from http://www.aclu.org/natsec/emergpowers/144441eg20021206.html

Chabrow, E. (2005). Who's in charge around here? *Government Enterprise/Information Week.* Retrieved March 19, 2006, from http://www.governmententerprise.com/showArticle.jhtml>articleID=60401377

Congress extends Patriot Act 5 weeks. (2006). *CNN.com.* Retrieved February 6, 2006, from http://www.cnn.com/2006/POLITICS/02/02/patriot.act.ap/

Department of Homeland Security. (2006). *Press release: Remarks by the Secretary of Homeland Security Michael Chertoff 2006 Bureau of Justice Assistance, U.S. Department of Justice and SEARCH Symposium on Justice and Public Safety Information Sharing.* Retrieved March 29, 2006, from http://www.dhs.gov/dhspublic/display?content=5485

Dizard, W. P., III. (2006a). DHS moves quickly to redirect Emerge2. *Government Computer News, 25*(2). Retrieved February 14, 2006, from http://www.gcn.com/25_2/news/380431.html

Dizard, W.P., III. (2006b). OMB sees improvement in agencies' IT security. *Government Computer News.* Retrieved March 2, 2006, from http://www.gcn.com/voll_nol/daily-updates/383491.html

Fisher, W. (2005). Congress to probe domestic spying. Inter Press Service News Agency. Retrieved January 2, 2006, from http://www.ipsnews.net/news.asp?idnews=31530

Government Technology News Staff. (2005). Belgium's e-government electronic identity card program. *Government Technology.*

Retrieved February 17, 2006, from http://www.govtech.net/mag azine/channel_story.php/97003

Jho, W. (2005). Challenges for e-governance: Protests from civil society on the protection of privacy in e-government in Korea. *International Review of Administrative Sciences, 71(1),* 151–166.

Lichtblau, E. (2006). Bush defends spy program and denies misleading public. *New York Times.* Retrieved January 2, 2006, from http://www.nytimes.com/2006/01/02/politics/02spy. html

Lipowicz, A. (2005a). DHS chief privacy officer Kelly steps down. *Government Computer News.* Retrieved September 30, 2005, from http://www.gcn.com/voll_nol/daily-updates/37145–1. html

Lipowicz, A. (2005b). DHS plans to beef up cybersecurity. *Government Computer News.* Retrieved September 30, 2005, from http://www.gcn.com/voll_nol/daily-updates/37146–1.html

Lipowicz, A. (2005c). JRIES homeland security network falls victim to policy dispute. *Government Computer News.* Retrieved October 8, 2005, from http://www.gcn.com/voll_nol/daily-updates/37223–1 .html

McKay, J. (2005, November). The security shuffle. *Government Technology,* pp. 32-40.

Miller, J. (2005). IT security requirements now part of the FAR. *Government Computer News.* Retrieved October 4, 2005, from http://www.gcn.com/voll_nol/daily-updates/37162-1.html

Mosquera, M. (2005). GAO: Many agencies' financial management systems inadequate. *Government Computer News.* Retrieved September 26, 2005, from http://www.gcn.com/voll_nol/daily-updates/37080-1 .html

Office of Management and Budget. (1998). *Privacy and personal information in federal records* (Memorandum C99-05). Retrieved November 30, 2005, from http://www.whitehouse.gov/omb/ memoranda/m99-05-a.html

Office of Management and Budget. (2005). *Designation of senior agency officials for privacy* (Memorandum CM-05-08). Retrieved November 30, 2005, from http://www.whitehouse.gov/omb/ memoranda/fy2005/m05-08.pdf

Temin, T. R. (2006). When to fold 'em. *Government Computer News* 25(3). Retrieved December 14, 2005, from http://www.gcn.com/25_3/ commentary/3813 8-1.html

Vijayan, J. (2005, October 10). Feds make security a priority in IT purchases. *Computer World,* p. 7.

G. David Garson is a professor in the Public Administration Program of North Carolina State University. He is editor of the *Social Science Computer Review,* and his most recent book is *Public Information Technology and E-Governance: Managing the Virtual State* (2006). He may be contacted at david_garson@ncsu. edu.

REVIEW QUESTIONS

1. What are some of the most controversial provisions of the Patriot Act regarding information security and privacy? Which one concerns you the most? Why?

2. What is Section 605? Why is it controversial?

3. What is the mission of the DHS Privacy Office? What have been the main criticisms leveled at the office since its inception?

4. What is the REAL ID program? Do you think the United States should have a national identity card? Why or why not? Does the REAL ID program serve this need? Support your answer.

5. Based on Garson's article, do you feel that individuals still maintain privacy in their online activities? To what extent? Support your answer.

❖

READING 23

This section of our text explores some examples of laws and legal structures that can be used to deal with terrorism. But how effective are domestic and international legal regimes in actually protecting democracies from terrorism? Choi examines this question by looking at 131 countries from 1984 to 2004 in an effort to delineate the causal relationship between the rule of law and the existence or non-existence of terrorist conflict in a county.

Fighting Terrorism Through the Rule of Law?

Seung-Whan Choi[1]

The question of how best to combat terrorism is one of today's most highly debated topics among academics, policy makers, and politicians alike. In particular, deterring terrorism is considered an important foreign policy objective for governments. The Bush administration and its defenders championed the advancement of democracy in terrorism-prone countries as a practical foreign policy goal (Gause 2005).[1] Currently, the Obama administration promotes the idea that the development of an appropriate legal framework and the preservation of political freedom and social justice are a winning strategy for democratic countries in response to potential terrorist threats (Hinnen 2009, April 28). Existing scientific studies, however, present contradictory causal arguments about the effect of democratic governance on reducing terrorism. A majority of studies claim that, because democracies promote high levels of civil liberties such as freedom of association and legal rights for accused criminals (e.g., terror suspects), they are more likely to be vulnerable to potential terrorist attacks (e.g., Eubank and Weinberg 1994, 2001). In contrast, a relatively small number of studies maintain that, because democracies encourage political participation and nonviolent resolution of conflicts, their chance of experiencing terrorist incidents is subsequently diminished (e.g., Eyerman 1998). As these unresolved and ongoing debates demonstrate, current scholarship fails to offer a concrete answer to the question of whether democracies attract more terrorist attacks than nondemocracies.

Treating the "rule of law" as one of the most fundamental characteristics of liberal democratic societies, I conceptualize it as the synthesis of effective and impartial judicial systems and ordinary citizens' recognition of the law as legitimate. I present a causal explanation, which posits that because ordinary citizens can peacefully resolve grievances through democratic rule of law systems, they lack the feelings of hopelessness and desperation that motivate terrorist action. Consequently, legitimately held rule of law systems serve to insulate democracies from terrorist attacks. Built on negative binomial regression, population-averaged negative binomial regression, and rare event logit models of 131 countries during the period from 1984 to 2004, I find that all things being equal, democratic societies that maintain a strong rule of law experience notably fewer domestic and international terrorist incidents than those societies with a relatively weak rule of law.

 ## The Dampening Effect of the Democratic Rule of Law on Domestic and International Terrorist Incidents

Before exploring the link between the rule of law and terrorism, these two concepts must be clarified, as their definitions remain controversial. Terrorism is a particularly difficult concept to define because of its value-laden nature: one country's terrorist may be another country's freedom fighter. For analytical clarity, I follow the definition of terrorism of LaFree and Ackerman (2009, 348) as "the threatened or actual use of illegal force, directed against civilian targets, by non state actors, in order to attain a political goal, through fear, coercion or intimidation."[2] Domestic terrorism includes incidents such as the Oklahoma City bombing since they arise only against domestic targets of the terrorists' home country. International terrorism, then, is a situation in which a terrorist incident in country A involves perpetrators, victims,

[1]Department of Political Science (M/C 276), University of Illinois at Chicago, Chicago, IL, USA

SOURCE: Choi, S.W. (2010). Fighting terrorism through the rule of law? *Journal of Conflict Resolution,* 54, 940–966.

institutions, governments, or citizens of country B (Enders and Sandler 2006, 7; Dugan 2010). These subtle differences are important, as they provide clarity to a discussion of terrorism in general.

Like the concept of terrorism, the rule of law is also "subject to various definitional and normative disputes" (O'Donnell 2004, 34). For analytical parsimony, I limit myself to two fundamental components that should be present in most democratic societies with a high-quality rule of law: (1) fair, impartial, and effective judicial systems and (2) a nonarbitrary basis according to which laws and the legal system as a whole can be viewed as legitimate.[3] As legal scholar Joseph Raz (1977, 198–201) argues, fair and impartial judicial systems require at least an independent judiciary branch with fair-minded judges, prosecutors and lawyers, as well as strong and stable law enforcement or police (for a similar view, see Fuller 1969). Institutionalizing an independent judiciary system reflects a strong commitment by government to the basic principle that all people are equal before the law and those people deserve the opportunity to have their grievances and disputes heard and settled in court.

Only when fair and independent judicial bodies have been institutionalized are citizens able to have trust and confidence in legal norms, procedures, courts, and the police. When this is the case, citizens are more likely to consult established laws and legal procedures to reconcile political and personal differences rather than turn to physical violence as a means of dispute resolution. Indeed, it is only when citizens believe in the likelihood of a fair and impartial legal ruling in court that citizens are willing to turn to domestic justice systems. Undoubtedly, such a high level of citizen trust in the legal institutions of the state brings a beneficial degree of order to the political and social relations of a society (Hardin 2001; O'Donnell 2004).

The Linkage between the Rule of Law and Domestic Terrorism

The above discussion leads to the inference that ordinary citizens have incentives to use political violence against other citizens, political figures, institutions, or the government under three conditions: (1) when they hold grievances, (2) when they find no peaceful means of resolving these grievances, exacerbating feelings of hopelessness, and desperation, and (3) when they view terrorist action as a legitimate and viable last resort to vent their anger and frustration. The lynchpin of this line of reasoning is that as long as ordinary citizens have access to a peaceful mechanism for conflict resolution, they are less likely to contemplate terrorist violence as a practical option to settle disputes. Along this line, I argue that since liberal democracies promote a high-quality rule of law system, which serves as an effective conflict resolution mechanism, they are likely to experience fewer activities of domestic terrorism.

As a fundamental building block of democratic societies, a high-quality legal system "serves to protect people against anarchy as well as from [the] arbitrary exercise of power by public officials and allows people to plan their daily affairs with confidence" (Wilson 2006, 153). Since liberal democratic judicial systems ensure independent adjudication of legal rules, they create a fair chance for the interests at stake in each case to be properly heard in efficient but inexpensive legal outlets. Thus, in the presence of an effective, independent judicial system in liberal democratic societies, ordinary citizens do not need to resort to illegal terrorist measures to resolve their complaints and grievances. Eyerman (1998, 154) makes a similar observation: since democracies "increase the expected return of legal activity and offer multiple channels of non-violent expression without the threat of government retaliation," they assuage potentially growing bitterness and dissatisfaction that may turn ordinary people into terrorists (see also Frey and Luechinger 2003). In contrast, where sound judicial systems are lacking, dissatisfied people are likely to embrace the principle of retributive justice and become more likely to initiate terrorist attacks.

Furthermore, since democratic citizens are socialized to trust in the fairness and impartiality of the legal system in times of disputes, they subscribe to established laws as a means to settle political grievances. From this perspective, engaging in violence would be self-defeating behavior ultimately undermining a legal institution seen as important and necessary. Furthermore, because democratic citizens see these institutions as both fair and legitimate, citizens will tend to subscribe to the established legal order, even if they disagree with individual legal statutes and rulings. Democratic citizens trust that legal adjudication produces a right and fair result, even if it is not the result they might have wanted.[4]

It is then not hard to imagine why ordinary people in democratic countries would be less likely to become perpetrators of domestic terrorism than those in nondemocratic countries, where the legal system is suited mainly for the rich and powerful: a nonarbitrary creation of law and a dispassionate legal system that metes out appropriate punishment make extralegal violence untenable and/or undesirable. Because citizens who live in countries without the rule of law view their own governments as illegitimate, public policy decisions as arbitrary, and peaceful participation futile, they are more likely to resort to attacks against domestic targets (or to support terrorist groups that do so). It is important to note that, in fighting domestic terrorism, law-abiding citizens in democratic societies are no less important than the actual presence of an independent judiciary with fair-minded judges and law enforcement officials. As we have seen, judicial institutions alone cannot produce a high-quality rule of law. Other factors within a society, especially the citizenry, must be actively involved. Exclusive reliance on legal authority is less likely to create and maintain safe and healthy communities if democratic citizens do not willingly cooperate with judicial institutions to resolve grievances and if democratic citizens do not serve as watchful eyes and ears against illegal activities of domestic terrorism (Hogg and Brown 1998; Hardin 2001). Alex P. Schmid (2005, 28), Senior Crime Prevention Officer of the United Nations, presents a compelling argument relating to this point: "where the rule of law is firmly in place, it ensures the responsiveness of government to the people as it enables enhanced critical civil participation. The more citizens are stakeholders in the political process, the less likely it is that some of them form a terrorist organization. In this sense, it can be argued that *the rule of law has a preventive effect on the rise of terrorism*" (emphasis added). In sum, ordinary people within democracies can resolve grievances through rule of law systems, which they have trust in, thereby mitigating the likelihood that they will commit terrorist acts, and resulting in less politically motivated violence.

The Linkage between the Rule of Law and International Terrorism

It appears that existing studies of international terrorism suffer from two common misperceptions. First, many studies put forward religious and ideological motives as the main causes of international terrorism (e.g., Reich 1990).[5] Typically, the terrorist activities of Al-Qaeda and the Taliban are seen as examples of organizations that advance their religious and ideological agenda. However, these studies overlook the fundamental question of what causes ordinary people to become terrorists in the first place. Religion and ideology, by themselves, do not necessarily drive ordinary citizens to resort to terrorist violence. When ordinary citizens with grievances lack peaceful outlets of conflict resolution, they tend to join radicalized terrorist groups that justify their violent actions through the selective use of religion and ideology. Second, some students of terrorism tend to misperceive the nature of international terrorist incidents. The international aspect of terrorism does not necessarily require the involvement of notorious international terrorist organizations such as Al-Qaeda. As noted earlier, as long as the origin of victims, targets, or perpetrators in political violence can be traced back to at least two different countries, this violence is regarded as international terrorism.

I argue that ordinary people have incentives to terrorize foreigners and foreign facilities when two conditions are met: (1) when they hold grievances against foreigners who violate political and legal rights of local citizens and (2) when these local people, due to poor-quality rule of law in the home country, do not believe in the effectiveness of pursuing justice peacefully. Students of terrorism often fail to observe the fact that when local people have grievances against Western foreigners, they have little chance of resolving them through the legal authority due to an omnipotent presence of foreign power or an unequal international treaty in which foreigners' crimes are immune from the domestic jurisdiction. This impotence of domestic justice systems makes local people feel helpless and desperate. Consequently, disgruntled local people turn to terrorist violence as a last resort.

There are several examples that illustrate how disgruntlement among locals later transforms into violence at the hands of terrorists. In May 2006, several Iraqis abducted two U.S. soldiers at a checkpoint and they were subsequently murdered. The Iraqis learned that the two soldiers raped and killed fourteen-year-old Abeer al-Janabi and committed the murder of her mother, father, and six-year-old sister in their home south of Baghdad (Robertson and Kakan 2009, May 8). On January 12, 2009, several Pakistanis, who were displeased with America's political support for Israel, terrorized the U.S. consulate in Karachi rather than seek peaceful channels of conflict resolution (see http://chinaconfidential.blogspot.com/2009/01/pakistani-students-storm-us-consulate.html). These two examples show how distressed local people are inclined to make use of terrorist violence against foreigners when they do not have an adequate rule of law system to hear their grievances.

There are four main archetypal narratives that can better illustrate and explain the causal mechanisms underlying the relationship between the rule of law and international terrorism. The first causal mechanism involves situations where ordinary citizens within their own country feel hopeless and desperate against foreigners who abuse fellow citizens' legal rights at home or abroad and who exploit the home country's political and economic interests. When foreigners are not subject to domestic legal jurisdiction, or when they are unfairly protected by the home country's justice systems, residents of the home country are likely to take justice into their own hands through locally coordinated terrorist attacks against foreigners and foreign facilities. An example is the insurgency of Iraqi civilians against armed privately contracted soldiers who operate not only with virtual immunity from Iraqi law but also from the laws of their own countries (see Broder and Risen 2007, September 20).

The second causal mechanism is an extension of the first, where hopeless citizens become international terrorists as a strategy to advance their domestic agendas. In this instance, discontented citizens who are frustrated with a low-quality rule of law at home, go abroad to carry out their attacks against foreign targets of the host country. These attacks represent an attempt to rectify foreign exploitation of their home country or to undermine Western support for brutal regimes (e.g., in the Middle East). This is done either because foreign targets are more vulnerable to attack or because there is some strategic advantage in putting the attack on an international stage rather than on a domestic one. In the former case, foreign targets may be more subject to attack due to easy access (Enders and Sandler 2006). In the latter case, the purpose of international terrorist attacks is to evoke domestic opposition

in the host country, demanding the end of the foreign presence (Pape 2005; Wade and Reiter 2007). The suicide car bombing of the UN headquarters in Baghdad on August 19, 2003, is an illustrative example. The followers of the late Abu Musab al-Zarqawi, a Jordanian militant Islamist, intentionally targeted the United Nations and killed at least eight Iraqis and fourteen foreigners including Sérgio Vieira de Mello, a Brazilian UN diplomat (Enders and Sandler 2006).

The third causal mechanism of international terrorism is one in which discontented people are angry at politically influential foreign targets operating within their own country. However, in these situations, citizens possess no readily available means to retaliate against those foreigners or their well-guarded foreign facilities. In such cases, feelings of powerlessness among disgruntled citizens may lead them to elicit the support of international terrorists because they see it as the best strategy to redress their frustrations and grudges (Tessler and Robbins 2007). These circumstances provide ideal opportunities for international terrorist groups to build inroads with the disaffected locals, giving these groups easy access to material resources, safe havens, and better channels through which to execute militant operations against foreign targets in the host country (i.e., foreign terrorist attacks on some other foreign target). For example, many Iraqis welcome and support Al Qaeda operatives from other countries, like Pakistan, to fight against U.S. forces.

The fourth causal mechanism involves situations where citizens have grievances against their own government but have no avenue for redress because corrupt domestic justice systems take the government's side. This breakdown of the basic perceptions necessary for the rule of law to materialize allows for the possibility of "mob rule" and lawlessness. However, the citizens themselves are often too weak to revolt, which makes them likely to turn to outside sources such as international terrorist groups to take action on their behalf (i.e., foreign terrorist attacks on a domestic target). Possessing global financial resources and disciplined members operating in autonomous terrorist cells, international terrorists are capable of luring local people who feel alienated and disadvantaged, using them to help push forward and carry out their own terrorist plots. For example, disgruntled Pakistani tribesmen joined together with foreign Al Qaeda members, Uzbek militants, and Taliban fighters to initiate terrorist attacks against the people and places attached to the Musharraf government (Masood 2008, January 18).

To recap, the rule of law reinforces a political system's legitimacy by protecting the rights of citizens and foreigners and by providing the means for them to settle grievances in nonviolent ways. It thus acts as a cornerstone of liberal democracies, making it unnecessary for ordinary people to rely on terrorist violence as a last resort to resolve disputes. An independent judiciary with fair-minded judges and police officers, who enforce the letter of the law, creates a nonviolent environment in which the public recognizes established laws as a legitimate channel to settle disputes peacefully. Thus, the combined impact of impartial judicial systems and ordinary citizens' recognition of the law as legitimate is likely to reduce all types of terrorism in democratic countries.[6] This leads to the following hypothesis:

> *Hypothesis 1*: The democratic rule of law has a dampening effect on domestic and international terrorism: fair and impartial judicial systems along with the public's recognition of the law as legitimate will discourage any type of terrorist acts.

Conclusion

With a focus on either domestic or international terrorism, many existing studies report mixed findings when the question of whether democracy encourages or reduces terrorist events is addressed. This study provides a novel probe into determining if adherence to the rule of law tends to reduce the incidence of domestic and international terrorist attacks. I have argued for the rule of law as an essential feature of liberal democratic governments and hypothesized that the synergistic effect of fair and impartial judicial systems and legitimate nonarbitrary law discourages ordinary citizens from resorting to politically motivated violence over peaceful resolution of conflict. In short, democratic rule of law systems offer ordinary citizens nonviolent ways of settling grievances, so they are more likely to overcome feelings of desperation, which could otherwise lead to seeking terrorist violence as a last resort. The empirical results show that a strong rule of law tradition produces

a dampening effect on political violence, regardless of the type of terrorism.

I believe that scholars and policy makers should pay closer attention to the *immediate précipitants* of terrorism. Examining the enabling conditions of terrorism such as government constraints, political competition, and failed states often overlooks its root causes. Only when we move beyond our present penchant for *permissive factors* will we begin to provide a blueprint for how best to challenge terrorism (Crenshaw 1981). I hope that my focus upon the rule of law variable that has the potential, when rule of law is absent, to directly motivate ordinary people to turn to terrorist activity is a move in this direction. In particular, my rule of law argument and findings are relevant to the literature of international terrorism in which religion and ideology are often considered the main causes of international terrorist incidents (e.g., Reich 1990). I strongly emphasize that local people do not necessarily resort to terrorist violence against foreigners or foreign facilities simply due to their religious and ideological beliefs. When local people with grievances are recruited to radical religious or ideological groups, they become international terrorists to vent their frustration and rage. People who live in countries lacking the rule of law have a much smaller chance of resolving grievances peacefully and are more likely to turn to international terrorist violence. In this context, the underlying cause of international terrorism should be traced back to a poor quality rule of law, not religion or ideology.[7] Simply put, a weak rule of law tradition provides international terrorist recruiters with some of their most effective recruitment materials, thereby leading to a rise of international terrorism.[8]

The empirical findings I report in this study have important implications for future policy making. Although some democratic countries have been tempted to bend the rule of law in response to growing threats of terrorism, it is essential to continue to cherish and uphold the tradition of a high-quality legal system to deter disgruntled people from turning into domestic terrorists. Furthermore, although the thrust of American foreign policy has been toward democratization, the empirical findings presented here suggest that the cultivation of the rule of law is more likely to reduce the occurrence of terrorist events. For these reasons, emphasis should not be put on simple democratic procedures like electoral systems and legislative debate, but on more civic education, and judicial and legal training. In fact, the empirical findings suggest that the significance of the role of the U.S. Agency for International Development (USAID) and similar agencies in the European Union are critical, since one of their primary missions is to fund independent judiciaries in developing countries (see USAID 2008). These kinds of agencies should work together to find more financial resources and help establish a sound rule of law tradition in terrorism-prone countries.

Contrary to current policy practice, it is not rational to push for quick elections as a deterrent to terrorism over the institutionalization of the rule of law system. Indeed, terrorists know that they can neither come to power nor wield political influence through free elections, therefore it makes little sense to espouse free elections as the highest priority in deterring terrorism in nascent democracies (Gause 2005). For example, Bosnia was able to hold free elections and showed strong political participation within a year of the Dayton Peace Accords, but the elections were swiftly followed by violent ethnic conflict and terrorist attacks. The domestic judicial system was essentially too corrupt to deter political violence, and many ordinary Bosnians chose to use terrorist tactics in resolving their grievances rather than leave their fate to untrustworthy legal services (Zakaria 2004). As Schmid (2005, 28) points out, "in order to help countries in transition to reach a higher level of law enforcement, judges, prosecutors, lawyers, and police need to be instructed or trained to bring national practices in line with recognized international standards . . . on advancing the development of an interdependent judiciary and promoting more just legal systems." If legal practices are not used in the service of fairness but rather to entrench sharp inequalities, supporting terrorism may come to be seen as a viable alternative for victims.

 Notes

1. Ironically, the Bush administration and its defenders simultaneously argued in favor of some undemocratic practices,

such as secret wiretapping, as effective tools in challenging terrorist activities against Americans.

2. This definition appears to be among the most comprehensive (see Dugan 2010, 9).

3. A comprehensive legal definition of the rule of law can be found in Schmid and Boland (2001, xi).

4. It should be noted that the democratic rule of law cannot simply mean the legal coercion of citizens but must also imply some public understanding and acknowledgment of the legitimacy inherent in the tendency of law's operations. For the purposes of this study, I contend that this is achieved when the coercive element of law is somehow linked up with, and justified by, a notion of public autonomy relevant to our intuitions concerning democracy. In this sense, an effective and legitimate rule of law is obtained when there is a middle ground between, or coexistence of, pure voluntary adherence to law and pure submission to legal coercion. In this middle ground, laws are seen as legitimate, entailing a recognition of, if not a voluntary subscription to, laws as sovereign in political and social proceedings, insofar as laws are produced in some nonarbitrary fashion, and are adjudicated and processed by fair and impartial judicial systems (see Rawls 1971/1999; Hobbes 1985). In short, the democratic rule of law means that citizens voluntarily subscribe to the legal system, even when being coerced to act in accord to individual laws.

5. On the contrary, Pape (2005) offers strategic logic as a cause of suicide terrorism, not religion or ideology.

6. One may argue that democratic citizens and elected officials may be heavily implicated in terrorism abroad, especially in nondemocratic countries regarded as illegitimate and dangerous. Although this possibility appears to be at odds with my line of reasoning, it is exceptional because only a few major powers like the United States could carry out such "preventive or extreme measures" (see Barber 2003; Hobsbawm 2007). More importantly, this possibility should be referred to as state or state-sponsored terrorism, which is beyond my conceptualization of terrorism since I only focus upon those terrorist acts committed by individuals or subnational groups.

7. Of course, my country-year level analysis lacks a direct test of which factor is the main cause of international terrorism, the rule of law or religion. This would require individual level data across country. Because no such data are currently available, I instead conduct a preliminary test with country-year data. Since Islamic terrorism is the center of attention, I create a variable identifying Islamic countries based on the criterion of the *CIA World Factbook* (2008): a country is considered Islamic if a plurality of its population practices Islam as their religion. Appendix 2

shows the results where the significance of the rule of law is confirmed: it is clearly a contributing factor in reducing the likelihood of terrorism even if the Islam variable is accounted for in the estimation.

8. Of course, some exceptional cases are the international terrorist attacks perpetrated by groups whose members were born and bred in strong rule of law states (e.g., Earth Liberation Front).

 Appendix 1

Table A1	List of Sample Countries			
Albania	Czech Republic	Indonesia	Mozambique	Somalia
Algeria	Czechoslovakia	Iran	Myanmar (Burma)	South Africa
Angola	Denmark	Iraq	Namibia	Spain
Argentina	Dominican Republic	Ireland	Netherlands	Sri Lanka
Armenia	Ecuador	Israel	New Zealand	Sudan
Australia	Egypt	Italy	Nicaragua	Sweden
Austria	El Salvador	Ivory Coast	Niger	Switzerland
Azerbaijan	Estonia	Jamaica	Nigeria	Syria
Bahrain	Ethiopia	Japan	Norway	Taiwan
Bangladesh	Finland	Jordan	Pakistan	Tanzania
Belgium	France	Kazakhstan	Panama	Thailand
Bolivia	Gabon	Kenya	Papua New Guinea	Togo
Botswana	Gambia	Korea North	Paraguay	Trinidad & Tobago
Brazil	Germany	Korea South	Peru	Tunisia
Bulgaria	Germany East	Kuwait	Philippines	Turkey
Burkina Faso	Germany West	Latvia	Poland	Uganda
Cameroon	Ghana	Liberia	Portugal	Ukraine
Canada	Greece	Libya	Qatar	United Arab Emirates
Chile	Guatemala	Lithuania	Romania	United Kingdom
China	Guinea	Madagascar	Russia	United States
Colombia	Guinea-Bissau	Malawi	Saudi Arabia	Uruguay
Congo Brazzaville	Guyana	Malaysia	Senegal	Venezuela
Costa Rica	Haiti	Mali	Sierra Leone	Yemen
Croatia	Honduras	Mexico	Singapore	Yemen South
Cuba	Hungary	Moldova	Slovak Republic	Yugoslavia
Cyprus	India	Morocco	Slovenia	Zambia

Acknowledgment

The author would like to thank Laura Dugan, Hyeran Jo, Nolan McCarty, Gregory Miller, Gary LaFree, Quan Li, Shali Luo Josh Pakter, Jeffrey Pickering, Matthew Powers, Abraham Singer, and Brandon Valeriano for their helpful comments at the various stages of this project.

 Declaration of Conflicting Interests

The author declared no potential conflicts of interests with respect to the authorship and/or publication of this article.

 Funding

The author received no financial support for the research and/or authorship of this article.

 References

Abadie, Alberto. 2004. Poverty, political freedom, and the roots of terrorism. NBER Working Paper no. W10859. http://ssrn.com/abstract=611366 (accessed October 9, 2009).

Barber, Benjamin R. 2003. *Fear's empire: War, terrorism, and democracy*. New York, NY: W.W. Norton & Company.

Broder, John M., and James Risen. 2007. Armed guards in Iraq occupy a legal limbo. *New York Times*, September 20.

Chenoweth, Erica. 2010. Democratic competition and terrorist activity. *Journal of Politics* 72 (1): 16–30.

CIA World Factbook. 2008. *Central intelligence agency: Books*. https://www.cia.gov/library/ publications/the-world-factbook/ (accessed October 9, 2009).

Crain, Nicole V., and W. Mark Crain. 2006. Terrorized economies. *Public Choice* 128 (1/2): 317–49.

Crenshaw, Martha. 1981. The causes of terrorism. *Comparative Politics* 13 (4): 379–99.

Davis, Kevin E. 2004. What can the rule of law variable tell us about rule of law reforms? *Michigan Journal of International Law* 26 (1): 141–61.

Drakos, Konstatinos, and Andreas Gofas. 2006. The devil you know but are afraid to face. *Journal of Conflict Resolution* 50 (5): 714–35.

Dugan, Laura. 2010. The making of the global terrorism database and what we have learned about the life cycles of terrorist organizations. Unpublished paper.

Enders, Walter, and Todd Sandler. 2006. *The political economy of terrorism*. Cambridge: Cambridge University Press.

Engene, Jan Oskar. 2004. *Terrorism in Western Europe: Explaining trends since 1950*. Northhampton, MA: Edward Elgar.

Eubank, William, and Leonard Weinberg. 1994. Does democracy encourage terrorism? *Terrorism and Political Violence* 6 (4): 417–43.

———. 2001. Terrorism and democracy: Perpetrators and victims. Terrorism and Political Violence 13 (1): 155–64.

Eyerman, Joe. 1998. Terrorism and democratic states: Soft targets or accessible systems. *International Interactions* 24 (2): 151–70.

Frey, Bruno S., and Simon Luechinger. 2003. How to fight terrorism: Alternatives to deterrence. *Defence and Peace Economics* 14 (4): 237–49.

———. 2005. Measuring terrorism. In *Law and the state: A political economy approach, ed.* Alain Marciano and Jean-Michel Josselin, 142–81. Cheltenham, UK: Edward Elgar.

Frey, Bruno S., Simon Luechinger, and Alois Stutzer. 2007. Calculating tragedy: Assessing the costs of terrorism. *Journal of Economic Surveys* 21 (1): 1–24.

Fuller, Lon L. 1969. The morality of law. New Haven, CT: Yale University Press.

Gause, F. Gregory, III. 2005. Can democracy stop terrorism? *Foreign Affairs* 84 (5): 62–76.

Gleditsch, Kristian. 2002. Expanded trade and GDP data. *Journal of Conflict Resolution* 46 (5): 712–24.

Gleditsch, Nils, Peter Wallensteen, Mikael Eriksson, Margareta Sollenberg, and Harvard Strand. 2002. Armed conflict 1946–2001. *Journal of Peace Research* 39 (5): 615–37.

Global Terrorism Database. 2009. Personal email exchange with Erin Miller, Global Terrorism Database staff, February 3.

Greene, William H. 2003. *Econometric analysis*. 5th ed. Upper Saddle River, NJ: Prentice Hall.

Gujarati, Damodar N. 2003. *Basic econometrics*. 4th ed. New York, NY: McGraw-Hill.

Hardin, Russell. 2001. Law and social order. *Philosophical Issues* 11: 61–85.

Hinnen, Todd. 2009. *Prepared remarks to the Washington Institute for Near East Policy*. http://www.washingtoninstitute.org/html/pdf/hinnen.pdf (accessed April 28, 2009).

Hobbes, Thomas. 1985. *Leviathan*. London: Penguin Books.

Hobsbawm, Eric. 2007. *Globalisation, democracy, and terrorism*. London: Brown Book Group.

Hogg, Russell, and David Brown. 1998. *Rethinking law & order*. Annandale, NSW: Pluto Press.

Iqbal, Zaryab, and Christopher Zorn. 2006. Sic semper tyrannis? Power, repression, and assassination since the Second World War. *Journal of Politics* 68 (3): 489–501.

IslamOnline.net & News Agencies. 2009. *Singapore Muslims fight extremism*. http://www. islamonline.net/ (accessed May 16, 2009).

Ivanova, Kate, and Todd Sandler. 2007. CBRN attack perpetrators: An empirical study. *Foreign Policy Analysis* 3 (4): 273–94.

Kahn, Joseph, and Tim Weiner. 2002. World leaders rethinking strategy on aid to poor. *New York Times.* Section A(1), 3, March 18.

King, Gary, and Langche Zeng. 2001. Explaining rare events in international relations. *International Organization* 55 (3): 693–715.

Krueger, Alan, and Jitka Maleckova. 2002. Does poverty cause terrorism? The economics and the education of suicide bombers. *The New Republic*, June 24.

LaFree, Gary, and Gary Ackerman. 2009. The empirical study of terrorism: Social and legal research. *Annual Review of Law and Social Science* 5: 347–74.

LaFree, Gary, and Laura Dugan. 2007. Introducing the global terrorism database. *Terrorism and Political Violence* 19 (2): 181–204.

LaFree, Gary, Laura Dugan, and Susan Fahey. 2007. Global terrorism and failed states. In *Peace and conflict*, ed. J. Joseph Hewitt, Jonathan Wilkenfeld, and Ted Robert Gurr. Boulder: Paradigm Publishers.

LaFree, Gary, Sue-Ming Yang, and Martha Crenshaw. 2009. Trajectories of terrorism: Attack patterns of foreign groups that have targeted the United States, 1970–2004. *Criminology & Public Policy* 8 (3): 445–73.

Lee, Joonghoon. 2008. Exploring global terrorism data. *ACM Crossroads* 15 (2): 7–16.

Li, Quan. 2005. Does democracy promote or reduce transnational terrorist incidents? *Journal of Conflict Resolution* 49 (2): 278–97.

Li, Quan, and Drew Schaub. 2004. Economic globalization and transnational terrorist incidents: A pooled time series analysis. *Journal of Conflict Resolution* 48 (2): 230–58.

Marshall, Monty, and Keith Jaggers. 2007. *POLITY IV project: Political regime characteristics and transitions, 1800–2006.* Dataset Users' Manual. Severn, MD: Center for Systemic Peace.

Masood, Salman. 2008. 12 killed in suicide bombing at Shiite mosque in Pakistan. *New York Times*, January 18.

Mickolus, Edward F. 1982. *International terrorism: Attributes of terrorist events, 1968–1977.* (ITERATE 2). Ann Arbor, MI: Inter-University Consortium for Political and Social Research.

Mickolus, Edward F., Todd Sandler, Jean M. Murdock, and Peter Flemming. 2006. *International terrorism: Attributes of terrorist events, 1968–2005* (ITERATE 5). Dunn Loring, VA: Vineyard Software.

O'Donnell, Guillermo. 2004. Why the rule of law matters. *Journal of Democracy* 15 (4): 32–46.

Pape, Robert A. 2005. *Dying to win.* New York, NY: Random House.

Piazza, James A. 2008. Incubators of terror. *International Studies Quarterly* 52 (3): 469–88.

Political Instability Task Force. 2007. *Internal wars and failures of governance, 1955–2006.* http://globalpolicy.gmu.edu/pitf/pitfpset.htm (accessed October 9, 2009).

Rawls, John. 1971/1999. *A theory of justice, revised edition.* Cambridge: Harvard University Press.

Raz, Joseph. 1977. The rule of law and its virtue. *Law Quarterly Review* 93: 195–211.

Reich, Walter. Ed. 1990. *Origins of terrorism.* Cambridge, UK: Cambridge University Press.

Robertson, Campbell, and Atheer Kakan. 2009. Iraqis seek death penalty for American. *New York Times*, May 8. http://www.nytimes.com/2009/05/09/world/middleeast/09green.html?_r=1&scp=1&sq=Abeer+al-Janabi&st=nyt (accessed October 9, 2009).

Ross, Jeffrey Ian. 1993. Structural causes of oppositional political terrorism: Towards a causal model. *Journal of Peace Research* 30 (3): 317–29.

Rotberg, Robert I. 2002. Failed states in a world of terror. *Foreign Affairs* 81 (4): 127–40.

Schmid, Alex P. 1992. Terrorism and democracy. Terrorism and Political Violence 4 (4): 14–25.

_____. 2004. Statistics on terrorism: The challenge of measuring trends in global terrorism. *Forum on Crime and Society* 4 (1–2): 49–69.

_____. 2005. Terrorism and human rights: A perspective from the United Nations. *Terrorism and Political Violence* 17 (1–2): 25–35.

Schmid, Alex P., and Etihne Boland. Eds. 2001. *The rule of law in the global village: Issues of sovereignty and universality.* Milan: ISPAC.

Tessler, Mark, and Michael D. H. Robbins. 2007. What leads some ordinary Arab men and women to approve of terrorist acts against the United States? *Journal of Conflict Resolution* 51 (2): 305–28.

Tomz, Michael, Gary King, and Langche Zeng. 1999. *RELOGIT: Rare events logistic regression, Versionl 1.1.* Cambridge, MA: Harvard University.

USAID. 2008. *This is USAID.* http://www.usaid.gov/about_usaid/ (accessed October 9, 2009).

U.S. Census Bureau, Population Division. 2008. International database (IDB) International Data Base (IDB). http://www.census.gov/ipc/www/idb/index.php (accessed October 9, 2009).

Van Belle, Douglas A. 2000. *Press freedom and global politics.* Westport, CT: Praeger.

Vanhanen, Tatu. 2000. A new dataset for measuring democracy, 1810–1998. *Journal of Peace Research* 37 (2): 251–65.

Wade, Sara Jackson, and Dan Reiter. 2007. Does democracy matter? Regime type and suicide terrorism. *Journal of Conflict Resolution* 51 (2): 329–48.

Weinberg, Leonard, and William Eubank. 1998. Terrorism and democracy: What recent events disclose. *Terrorism and Political Violence* 10 (1): 108–18.

Wilkinson, Paul. 2001. *Terrorism versus democracy.* Portland, OR: Frank Cass.

Wilson, Jeremy M. 2006. Law and order in an emerging democracy. *Annals of the American Academy of Political and Social Science* 605 (1): 152–77.

World Bank. 2008. *Worldwide governance indicators, 1996–2007.* Washington, DC: World Bank.

Zakaria, Fareed. 2004. Islam, democracy, and constitutional liberalism. *Political Science Quarterly* 119 (1): 1–20.

1. How does Choi define the rule of law?

2. What links does Choi identify between the rule of law and domestic terrorism?

3. What links does Choi identify between the rule of law and international terrorism?

4. What is Choi's hypothesis?

5. What are his conclusions? Do you agree? Why or why not?

6. What do you believe are the implications of Choi's findings for U.S. domestic and foreign counterterrorism policy?

SECTION

XIV

The Future
of Terrorism

Learning Objectives

At the end of this section, students will be able to:

- Describe major issues facing the world in the next decade and the impact on terrorism.
- Describe the major threats that will face the United States in the next decade with regard to terrorism.
- Discuss what al Qaeda looks like today as a threat, and what its future looks like?
- Discuss what the FBI/DNI believes is the next big threat.
- Discuss how much longer al Qaeda and the current religious-based wave of terrorism will last.
- Discuss the notion that we will have a "fifth wave" of terrorism and where its impact will be the greatest.
- Discuss the tension between the National Security Agency and citizens' privacy. When and where should a person who is an American citizen expect privacy?
- Describe Remote Biometric Identification and suggest means to control it.

 Introduction

The *future* refers to what is to come or events that will occur. Predicting the future for the coming decades will certainly be difficult and predicting the future of terrorism even more so. Turning to experts on the future of the world, though, seems a good beginning. For example, the World Future Society (n.d.) has been publishing annual forecasts for over 40 years and it has a good track record. Taking a look at some of the latest predictions and applying them to terrorism may be useful. For example, one forecast is that much of Asia and Africa will suffer from massive dust bowls from overgrazing, soil erosion, and deforestation; China's economy will stop growing and shrink; the United Nations estimates that by 2025, 2.8 billion people will live in water-stressed environments and, finally, water

▲ Drought remains an issue in many parts of the world, suggesting that conflict may result from shortages. This is a major dam in South Africa.

tables are now falling in countries that have half of the world's population. What does this mean for the future of terrorism? It means that many countries and regions already dealing with strife and terrorism will face unprecedented challenges, as the current Director of National Intelligence (DNI), James Clapper, noted in prepared remarks before the Senate Select Committee on Intelligence in 2013. While certainly worried about other more pressing concerns, DNI Clapper (2013) stated:

> In addition, some non-state terrorists or extremists will almost certainly target vulnerable water infrastructure to achieve their objectives and continue to use water-related grievances as recruitment and fund raising tools.

and

> Terrorists, militants, and international crime organizations can use declining local food security to promote their own legitimacy and undermine government authority. Growing food insecurity in weakly governed countries could lead to political violence and provide opportunities for existing insurgent groups to capitalize on poor conditions, exploit international food aid, and discredit governments for their inability to address basic needs.

Given that worrisome scenario, this section will cover the following topics in an attempt to lay out the future of terrorism.

◆ Al Qaeda: what is its current state and what kind of a threat does it pose to the United States?
◆ Current threat priorities from the DNI and the intelligence community, including how much longer the "War on Terror" will last
◆ The Middle East, where it is no longer an Arab Spring but perhaps an Arab Winter, lasting a decade
◆ The United States, Europe, and Islam
◆ The NSA as an agency and the technologies, methods, and resources it and other federal agencies use that may save us while impairing our privacy and freedom, including the use of drones
◆ Will there be a "fifth wave of terrorism" in the future?

The Threat Today

Al Qaeda presents a threat to the United States but at the moment perhaps not an imminent one, as there are so many other opportunities and events occurring in the Middle East and North Africa. In testimony before the Foreign Affairs Subcommittee on Terrorism, Nonproliferation, and Trade in the U.S. House of Representatives, a researcher from the RAND Corporation, Seth G. Jones (2013), noted that al Qaeda and its affiliates around the world are interested in establishing Islamic states or regions in the Middle East or Northern Africa. There remains the core or central al Qaeda headed by Ayman al-Zawahari; he would like to see continents conquered as Islamic states though Egypt is his first priority. There are a number of affiliated (these swear allegiance to al Qaeda) and allied groups of al Qaeda around the world that communicate with al Qaeda, and inspired groups, meaning groups that are inspired to act by al Qaeda examples or rhetoric. None of these groups poses a threat to the United States currently and Jones suggests the United States proceed with a light footprint when dealing with al Qaeda, meaning no huge armies but Special Forces and perhaps Central Intelligence Agency assets, support of local governments, and trying to challenge their ideology as un-Islamic. How long will this war on al Qaeda last? In testimony before Congress, Michael Sheehan, the Assistant Secretary of Defense for Special Operations and Low Intensity Conflict, stated it would be another 10 to 20 years. That is consistent with a "wave of terrorism" lasting about 40 years according to David Rapoport's (2006) four waves of modern terrorism thesis covered in an earlier section.

The current threats and priorities, according to DNI Clapper and the intelligence community, are many and complex. In his testimony he actually addresses the first global threat as the one in the Cyber Domain, meaning there is a concern about our critical infrastructure being damaged or controlled by outsiders or terrorists in a cyber attack, hacking into the Supervisory Control and Data Acquisition (SCADA) systems that control all of our infrastructure, including our air transportation system (Villeneuve, 2009) and, yes, there is a government report that the Federal Aviation Administration is vulnerable to cyber attacks. This is a very real and frightening concern, and the Department of Homeland Security (DHS) and the Federal Bureau of Investigation (FBI) are finally getting serious about this, with the FBI elevating the cyber threat to the number three national security priority, after only counterterrorism and counterintelligence. The FBI (2010), in addition to making cyber threats a higher priority, also has a program called InfraGard. InfraGard is a public/private partnership for all organizations and representatives of all sectors of our infrastructure that convenes meetings monthly in most major cities where threat information is exchanged and channels exist to report breaches of security, data, and threats. DHS even created a new position, a deputy under secretary for cybersecurity, and hired Phyllis Schneck, with a very solid background in private industry, to fill it (Napolitano, 2013); this demonstrates that DHS recognizes the threat and is taking concrete action. Cyber espionage, usually committed by Russia and China, remains a concern. These two, among others, have penetrated government, business, academic, and private sector organizations. In testimony before Congress, Joseph M Demarest (2013), assistant director of the FBI's Cyber Division, worried about foreign intelligence services, terrorist groups, organized criminal enterprises, and hacktivists (those who hack into sites for social and political goals). *The Wall Street Journal* recently had an article suggesting that over the next two decades, machines will do all the driving of the more than 253 million long-haul trucks on the road (Berman, 2013) but imagine what happens when someone hacks into the system that controls just a few of them. The Affordable Care Act, known commonly as ObamaCare, will create a massive database of personal information (Hurtubise, 2013). The Center for Medicare and Medicaid Services chief, Marilyn Tavenner, whose department will oversee what is termed the data hub, admitted she has not attended any briefings on cyber security by the FBI or DHS.

The second area of concern for the DNI in his recent testimony was al Qaeda and terrorism, and much of his assessment mirrors that already covered; he does express concern about homegrown violent extremists in the United States but places the likelihood of attacks in the near term at fewer than ten per year. The DNI also expresses concern about two particularly nasty jihadist groups, Al-Shabbab from Somalia and Boko Haram from Nigeria. Thus far they have acted only regionally but are very much anti–United States, anti-Christian. and anti-West. It should also be noted that the United States and others have still not satisfactorily figured out how to stop suicide bombings. In Iraq, for example, there were more vehicle and suicide bombings by al Qaeda in 2012 than in 2011. This is an area that must be addressed as the sophistication continues to grow, with the latest worry being explosives concealed in breast implants set off by injecting a fluid via a needle (Goldhill, 2013).

The third major area addressed by DNI Clapper is that of weapons of mass destruction and he focuses primarily on Iran and its attempts to construct nuclear weapons, assessing that Iran has made progress in pursuing weapons-grade uranium and that it would use missile technology to deliver the weapons. There is a brief mention of North Korea and its intention to conduct an additional nuclear test in the near term, and concern about chemical weapons use in Syria. To sum up, the DNI regards al Qaeda and other terrorist groups as worrisome but not an immediate threat to the homeland, but cyber threats are rising almost exponentially, and the United States does have to keep an eye on Iran and North Korea.

The Middle East

The next focus is on the Middle East, which currently is a dangerous area. In December of 2010, a desperate street vendor in Tunisia set himself ablaze, frustrated by repeated mistreatment by local officials. This triggered spontaneous protests so massive that Tunisian dictator Zine El Abidine Ben Ali fled into exile while other protests in Libya and Egypt soon toppled those governments. Things remain unsettled in Tunisia, Libya, and Egypt. As part of the Arab Spring, the situation in Syria has evolved into a massive civil war that began in March of 2011, with over 150,000 people killed, millions fleeing the fighting (Salaheddin & Karam, 2013), and chemical weapons being employed (Hjelmgaard, 2013), and there is no end in sight. Syria could be interesting to watch as Hezbollah, a Shia terrorist group from Lebanon that is supported by Iran, is actively fighting in Syria alongside supporters of Bashar Assad, against al Qaeda and others, mostly Sunnis. One of the authors recently taught an online course on terrorism. One student had been based in Baghdad. He was a former Marine with combat experience, now working for a private security firm. When asked what he thought would happen in the Middle East, he predicted a war between the Sunnis (Saudi Arabia and others) and the Shias (Iran). He could be correct.

Why was there an Arab Spring? Most researchers and scholars point out that the unrest resulted from socioeconomic inequities, the perception of official corruption, a large youth population, and modern social communication means. Of all those factors perhaps the most important is the large numbers of youths in the Middle East and North Africa who are educated but unemployed. In an article just after the first of the year in 2011, one journalist noted that:

> Sixty percent of the regions' people are under 30, twice the rate of North America, found a study from the Pew Forum on Religion and Public Life. And with the unemployment rate at 10 percent or more, North Africa and the Middle East also have the highest regional rates of joblessness in the world. For the region's young people, it's four times that. (Knickmeyer, 2011)

These, Knickmeyer termed the Arab World's Youth Army. Thus far they have done a credible job of forcing out the dictators of Libya, Egypt, and Tunisia. When secular Egyptians realized that then-President Morsi, a duly elected leader, was about to turn Egypt into a Sunni version of the Iranian theocracy, they managed to force the Egyptian Army to act again and arrest and replace him. Clearly the United States and the Obama administration misunderstood or ignored the nature of the Muslim Brotherhood. The Muslim Brotherhood was covered extensively in an earlier section but as a reminder, it has been around since its founding by Hasan al-Banna in 1928 in Egypt, with a goal of making that country an Islamic state; it has a history of violence though it prefers quiet political organizing. Michael Totten (2013) interviewed Eric Trager, an expert on Egypt, and Trager stated:

> The United States has done a very poor job managing perceptions in Egypt. The administration assumed if it wasn't critical about Morsi's behavior domestically, they'd win his cooperation on foreign policy. The problem is that Morsi was only willing to cooperate with us on foreign policy in the short run. The Muslim Brotherhood wants to consolidate power in Egypt and then create a global Islamic state. It's a key part of their ideology and their rhetoric. They talk about it with me. They can't be our partners.

The same expert also makes it clear why the West and Americans should care about Egypt:

> For the simple reason that Egypt is a lynchpin of American foreign policy in the Middle East. It's important for counterterrorism, for maintaining the peace treaty with Israel, ensuring overflight rights so our planes can deliver goods to the Persian Gulf, to check Iran's interests, and ensure passage through the Suez Canal.

The Middle East will remain a precarious region for some time and how it all ends is unknown. Egypt will have a military government for some time, though a new constitution and elections have been promised. In what must be one of the great ironies of all time, an Egyptian court released former president Mubarak from jail (Bradley & El-Ghobashy, 2013; Kirkpatrick & Colwell, 2013) where he had been held for two years facing various charges. While democratically elected though unpopular, President Morsi will remain jailed. Israel watches this closely, realizing that it does need a stable Egypt to counter Iran, as does the United States. The lack of seriousness with the current approach of apparently embracing the Muslim Brotherhood over the Egyptian military gives one pause. The Egyptian military saved Egypt from becoming a Sunni Iran. As an *Investor's Business Daily* editorial noted:

> With a lethal combination of arrogance and naivete, the U.S. has mishandled a crisis of its own making in the Mideast's most important Arab power. Egypt might now turn back toward Moscow. . . . "Like any politician at this level, I've got a healthy ego," the president said shortly before being elected in 2008. Unfortunately, the Obama ego has been downright deadly for America globally. (Editorial Board, 2013)

Summarizing, no one knows what will happen in Libya, Tunisia, Syria, or even Egypt and the rest of the Middle East. It is clear that the Muslim Brotherhood, with support from perhaps as many as 30% of Egyptians, elected one of its own and immediately began to change and head on a path to quickly becoming an Islamic state. The majority of Egyptians rejected that, demanding a more balanced, more secular government. Perhaps that is the model for other countries there.

The United States, Europe, and Islam

The United States has problems with a sluggish economy and several enemies around the world, but is the greatest enemy within the United States itself? Will America see Shariah law in the future? Perhaps it may, though if Americans have been paying attention to Egypt and the experience Egyptians, particularly Coptic Christians, have endured with the Muslim Brotherhood, perhaps not. The Muslim Brotherhood is a sophisticated political organization that is patient and will use any political means when possible but will also use violence if necessary. Coptic Christians are a minority of 10% in Egypt and they have suffered. More than 40 Coptic Christian churches were burned to the ground in Egypt at the hands of the Brotherhood in the few weeks after President Morsi was toppled in a coup, as was a Franciscan School (Hendawi, 2013). What the Muslim Brotherhood desires wherever it exists is what it has largely achieved in Egypt under Morsi: total control of all institutions, a theocracy, and Shariah law.

Briefly, Shariah means *straight path,* which sounds fine until one realizes that it means that under Shariah law everything is meant to be controlled by an Islamic state with Islam and Shariah the answer for everything, including worship only to Allah, with Islam controlling family relations, inheritance, commerce, property law, civil (tort) law, criminal law, administration, taxation, constitution, international relations, war, and even ethics.

Shariah law is brutal, repressive, totalitarian, intolerant. It condones repressive treatment of women, homosexuals, Jews and other infidels, apostates, and petty criminals, among others, and it is not compatible with democracy. This legal code is quite thorough, as Lawrence Wright (2011) notes, covering how to respond to someone sneezing, whether

▲ Many countries, especially Great Britain have turned to Closed Circuit Television systems. This photo demonstrates the capability of one system in a hotel.

© Can Stock Photo Inc. / CarolinaSmith

it is permissible to wear gold jewelry, and even has prescribed punishments for adultery and drinking alcohol, and it is "untainted by Western Influence" and modernity.

How does the Muslim Brotherhood propose to take over or change the United States of America? It will do this by establishing numerous institutions that do not seem dangerous, such as educational, social, economic, and scientific institutions and the establishment of mosques, schools, clinics, shelters, and clubs; most universities have Muslim student associations. In the report *Shariah: The Threat to America: An Exercise in Competitive Analysis (Report of Team B II)*, a number of authors from the Center for Security Policy (2010) make a convincing case that just as with Egypt, the Muslim Brotherhood, the Council for American-Islamic Relations (founded by members of Hamas, a terrorist group), and others plan to change America fundamentally with the end goal of making it an Islamic state. The authors agree that it is possible but unlikely due to the fact that Muslims make up less than 1% of the population in the United States. The United States is less tolerant than Europe, which may be losing or lost in the next few decades, but the battle continues in the United States with perhaps a dozen states considering laws prohibiting Shariah law, but as one author wryly notes, a takeover is about as likely as a zombie apocalypse.

The Muslim Brotherhood and the Wahhabists are making some small progress in the United States and, unfortunately, they have a road map. One of the authors has a book on the shelf titled *The Amish and the State* (Kraybill, 1993) that painstakingly documents how the Amish changed things in the United States to suit themselves, through legislation and court battles at the state and federal level since 1925. The Amish have gained concessions from almost every state and the federal government to practice their religion and lifestyle when it comes to education and schooling, health care, and even not driving vehicles but buggies. Recently, the State of Oklahoma passed a constitutional amendment prohibiting the state's courts from weighing or using Shariah law; it was struck down in a federal court (Gershman, 2013). Everything is going according to the plan. Even the Saudis are helping, funding nearly 80% of all new mosques, including large ones (Stakelbeck, 2010), built in this country, and their form of Islam is Wahhabi Islam, a radical version.

Europe, as noted earlier, does have some difficulties with unassimilated immigrants, many practicing Islam. These large numbers of Muslims in Europe have resulted in some areas seeing the proclamation of a "Shariah Controlled Zone" using flyers that state that "Islamic Rule Enforced" and also proclaiming there will be no gambling, no music or concerts, no porn or prostitution, no drugs or smoking, and no alcohol (Camber, 2011). Mark Steyn, (2006) in his book *American Alone: The End of the World as We Know It*, does predict that demography is destiny and in Europe, Muslim immigration and astonishing birth rates will, according to Steyn, within a generation, set Europe on a path to becoming Islamic. Steyn is convincing with such nuggets as the Muslim population of Rotterdam is 40%, the most popular name for a baby boy in Belgium is Mohammed, and it is the fifth most popular name in Great Britain. Steyn is also not a fan of multiculturalism, diversity, or tolerance, and believes that it is a certain way to destroy a nation. To see the pervasive wave that may be coming to Europe, realize that Belgium, for example, will see Muslims in the majority by 2030, and watch the video called *Welcome Belgistan* (n.d.; see also Web Resources, below).

The paragraphs above consider whether the United States would fundamentally change at some point to a more Islamic orientation. The authors consider that unlikely but expect some concessions, though minor. Europe, on the other hand, does have some serious issues with demography, declining birth rates, and a lack of assimilation. Mark Steyn may well be correct about Europe.

Technology Will Keep Us Safe and Free?

The next area of concern going forward is technology, including the massive National Security Agency (NSA) and its surveillance programs that seem to be much more pervasive and intrusive than believed. The NSA has a budget that is perhaps $10 billion annually, and it has 30,000 employees although with contractors, such as the now infamous Edward Snowden, probably more. The NSA director, General Keith Alexander, also heads U.S. Cyber Command, five directorates, several administrative directorates, and three operational centers. NSA does everything from generating codes to signal intelligence on a massive scale with multiple programs for various purposes including intercepting foreign communications, cracking codes, helping track down terrorists, and defending U.S. interests against cyber attacks. While it may keep us safe, it very definitely infringes on our privacy. According to the *Wall Street Journal*, the NSA has a significant capability to spy on Americans:

> The National Security Agency—which possesses only limited legal authority to spy on U.S. citizens—has built a surveillance network that covers more Americans' Internet communications than officials have publicly disclosed, current and former officials say. The system has the capacity to reach roughly 75% of all U.S. Internet traffic in the hunt for foreign intelligence, including a wide array of communications by foreigners and Americans. In some cases, it retains the written content of emails sent between citizens within the U.S. and also filters domestic phone calls made with Internet technology, these people say. (Gorman & Valentino-Devries, 2013)

The NSA has oversight from a federal court called the Foreign Intelligence Surveillance Court (FISA) but that has not deterred it from efforts to snoop. A recently declassified report noted that the FISA Court was very unhappy that for a three-year period of time the NSA was collecting domestic communications without proper privacy safeguards.

The NSA does all of this at massive facilities in the United States in Maryland, Hawaii, Colorado, Texas, Georgia, and Utah. To give an idea of the size of the NSA facilities and capabilities, people in Utah have been following the construction of the NSA Utah Data Center, a $1.5 billion project at Camp William near Bluffdale, Utah. The Utah Data Center will be 1 million square feet in size and it is estimated that it can store one thousand trillion gigabytes of data (one thousand megabytes is a gigabyte, for perspective). That facility is big but while not as large, another $792 million center is being built in Maryland; that means the NSA facilities are seven times bigger than the Pentagon (Sternstein, 2013). Of course generating and storing that much data (the term for that is actually "Big Data") is worthless unless one can do something with it, so President Obama announced in March of 2012 a $200 million project on data mining the Big Data (Semerad, 2013). The Department of Defense spends $250 million a year on this and the Defense Advanced Research Projects Agency spends another $25 million on this as well.

▲ The National Security Agency (NSA) is a huge bureaucracy though many know little about it. One of the authors trained some NSA agents periodically. They would always say they worked for the federal government; or it was concluded, No Such Agency.

Last, when it comes to technology, and while still in their infancy, the use of drones and especially armed drones constitutes a frightening expansion of government power, especially with a current executive in President Obama who can rationalize the sanctioning of a drone strike on an American citizen. This is clearly an area of concern, and an editorial in the *Washington Post* asked when drone strikes would end, especially on American citizens:

> After U.S. citizen Anwar al-Awlaki was killed in a September drone attack, the ACLU's deputy legal director rightly said, "It is a mistake to invest the President—any President—with the unreviewable power to kill any American whom he deems to present a threat to the country." (Miller, 2011)

The American Civil Liberties Union and the Center for Constitutional Right are two groups thankfully taking the matter to court in a case called *Al-Aulaqi v. Panetta* (the groups charged that the U.S. government's killings of U.S. citizens Anwar Al-Aulaqi, Samir Khan, and 16-year-old Abdulrahman Al-Aulaqi in Yemen in 2011 were illegal). The case was dismissed by a federal court in 2014, with U.S. District Judge Rosemary Collyer deciding that courts must defer to Congress and the administration in this area of jurisprudence (Petterson, 2014). In the Stanford/New York University report *Living Under Drones*, one finds this: "From June 2004 through mid-September 2012, available data indicate that drone strikes killed 2,562–3,325 people in Pakistan, of whom 474–881 were civilians, including 176 children. *The Bureau of Investigative Journalism* reports that these strikes also injured an additional 1,228–1,362 individuals (International Human Rights and Conflict Resolution Clinic, n.d.).

The report alleges that the administration fails to acknowledge that many of the drone strikes result in "collateral damage" and the drone strikes continue, especially in Yemen. Some have suggested that the use of lethal drones on such a wide scale reflects President's Obama's preference for drones killing terrorists versus capturing them and having to send them to Guantanamo, the U.S. military prison in Cuba; as of the date of the article, May 2, 2013, approximately 4,700 people had been killed in approximately 300 drone strikes in four countries (Roberts, 2013). The Guantanamo military prison in all likelihood will never close as there are prisoners there that the United States regards as so dangerous they will never release them (Crowley, 2013), and indeed it may even be impossible to try them (though if they will never be released, why try them?). Drones will continue to be an issue as they move into the hands of law enforcement and even homeland security.

There are other technologies as well that may be useful in fighting terrorists going forward. Geographic Information Systems have been used to analyze crime patterns and more specifically to actually predict offender behaviors and locations of future crimes. This is based on the notion that a few offenders are very busy, termed frequent offenders, committing the bulk of crime, often returning to the same or nearby locations, usually within just a mile of their residence (Croisdale, 2012). Researchers are now employing the same analysis for terrorists and terrorist attacks, but it's just beginning. An article in *Popular Science* in 2011 is revealing. It recounts a project by geography students at the University of California at Los Angeles predicting the location of Osama bin Laden using this approach and commercial satellite imagery. The students did not actually get it right, but for a limited effort using simple, non-military or CIA, software and equipment, they predicted his location within 278 miles (Kvinta, 2011).

Another innovation that will ensure safety is the marriage of Closed Circuit Television (CCTV) with software and complex algorithms. CCTV amasses a very large amount of information. Going through it requires a huge number of man hours, often simply to find out what occurred in the past. The Thales Group (n.d.) and others, such as SAAB with its SAFE system, in the United Kingdom now have video and data fusion capabilities that are incredible. Again, CCTV was once used after the fact to determine what happened. Now, CCTV can be programmed to spot real-time intrusions, patterns, out-of-place packages or shipments, and walking styles. It integrates facial recognition software, reads license plates, and identifies colors. The technology is in place in Mexico City, Mecca, and a number of sports stadiums.

The next era in safety and security, particularly in air travel, will be biometrics. Biometrics refers to a "process by which a person's unique physical and other traits can be detected and recorded by an electronic device or system for the means of confirming identity" ("Biometrics," n.d.). Examples of this would be DNA or the scan of a human iris. This area has exploded. As an example, between 2001 and 2011 there were 633 patents issued for facial recognition. The FBI is boldly developing what is called Next Generation Identification that is incredible, and it will include facial recognition technology that can be used to examine any public photo or video, looking for specific individuals and then pairing them with biographic data from another database.

Some predict that biometrics will replace airport boarding passes. The more advanced biometrics come from Israel, which also uses behavioral profiling. Israel recently deployed what it calls a Unipass system at its main airport where travelers provide a fingerprint, have their photograph recorded for a facial recognition database, and then receive a smart card with their information on it. Intel Corporation recently purchased another company called IDesia Biometrics from Israel (Savitz, 2012) that has developed a biometric system that captures an electronic signal generated by a person's heartbeat to create an electronic, unique individual biometric signature. This is really uncharted territory, especially when it comes to privacy protections. The government and security officials are so eager for some of this new technology, much of it characterized as Remote Biometric Identification or RBI, that they have not addressed privacy concerns or constitutional protections. RBI simple means a system that does the following, "RBI allows the government to ascertain the identity of (1) multiple people; (2) at a distance; (3) in a public space; (4) absent notice and consent; and (5) in a continuous and on-going manner" (Donohue, 2012).

The author of the above quote, Laura Donohue, also notes that Congress has been silent about legislation protecting privacy, especially when it comes to RBI. This is a fluid area and one where innovations may cause a loss of privacy and liberty. One hopes that Congress wakes up.

The Fifth Wave of Terrorism?

While examining the technology that one hopes will keep us safe from terrorism, the question remains, when will terrorism end? Experts mentioned earlier suggest that the current wave of religion-based terrorism will last 10 to 20 more years. After that, will there be a fifth wave? One scholar believes there will be and that it has already begun (Kaplan, 2011). It eerily fits with the warning earlier from the DNI on the situation with food and water capacity and availability shrinking in places such as Africa and Asia. Some background is needed.

In 2010, in the journal *Foreign Policy*, Jeffrey Gettleman wrote an article called "Africa's Forever Wars" in which he decried the seemingly endless conflicts in Africa, where he sees, as the *New York Times*' East Africa bureau chief, "un-wars" where there is not soldier-versus-soldier conflict but soldier-versus-civilian conflict. This terror is practiced by predators committing stunning atrocities, such as hundreds of thousands of brutal sexual assaults in the eastern Congo so sadistically that victims are incontinent for life; and he says terror has become an end, not a means. He also notes that at the time of the writing of the article in almost half of the 53 African countries there was an active conflict or a recently ended one. Gettleman sees no solution, but he may have been on to something.

Addressing where the United States is when it comes to the battle with terrorists (which he prefers to the term war on terror), Jeffrey Kaplan suggests that the country is in for more of the current wave of religious-based terrorism but suggests that perhaps there is already a Fifth Wave:

> My own contribution to this expanding body about terrorism concerns the emergence of a new form of tribalism in places like Africa, where increasingly vicious or outright genocidal conflicts are taking place in Sudan, Uganda, and the Democratic Republic of the Congo, and for which Rapoport's Four Waves fails to account. My Fifth Wave theory finds the ultimate goal of each of these movements to be nothing less than the creation of a new and perfected people, making all who do not belong to the respective groups and the "reconstituted Golden Age" subject to the intent of extermination. The signature weapon of the Fifth Wave is rape. (Kaplan, 2008)

Arguing that Africa is really not that bad and is actually improving, another scholar claims that large-scale wars have declined by about half from earlier decades, but admits there are still political struggles surrounding such issues as electoral clashes and violence associated with access to "livelihood resources such as land and water."

Earlier in this section, the authors cited DNI Clapper and his and the intelligence community's concerns about emerging conflicts in areas where water and adequate food may be issues that encourage terrorism. The authors believe that he is correct to be concerned and concur that in Asia and Africa, with decreasing water, terrorism and conflict are inevitable and may indeed constitute a Fifth Wave.

 ## Summary

This section reviewed the status of our battle with terror groups such as al Qaeda, concluding that they have not gone away but may not be an imminent threat. They will constitute a threat for 10 to 20 more years. It examined current threats as represented to a Senate Select Committee that makes clear the growing concern about cyber attacks, as well as addressing al Qaeda and related groups, coming to similar conclusions from other experts. The DNI does express concern as well about WMD threats from Iran and North Korea.

The Middle East's current volatile situations was reviewed, noting that the outcome is not clear and may not be for a decade, although it seems that Egypt will not immediately become an Islamic state. The ideology and designs of the Muslim Brotherhood were reviewed again and its impact and that of Islam on Europe and the United States was examined with the depressing fact emerging that Europe will likely undergo a major shift to a more Islamic orientation.

The NSA, an intrusive yet needed agency, and its technologies were presented, including the new area of biometrics that shows promise but may also impact privacy. This is uncharted territory and safeguards for privacy protection have yet to catch up, especially when it comes to Remote Biometric Identification or RBI. It is worth explaining again what this means:

> RBI allows the government to ascertain the identity of (1) multiple people; (2) at a distance; (3) in a public space; (4) absent notice and consent; and (5) in a continuous and on-going manner. (Donohue, 2012, p. 415)

Soon, if it is not already doing it, the government can identify anyone using remote biometrics and may be able to do so without the person being aware of it. This is an area where constant vigilance is warranted on the NSA, on authorities, and yes, even contractors. The NSA and other agencies and their emerging technologies may save us while impairing our privacy and freedom, including the use of drones. Drones as weapons are controversial as mentioned above, and Congress currently seems willing to allow this to proceed. No one knows the future of drones but they are problematic and soon will be in the hands of many, including law enforcement and federal authorities, for surveillance and possibly as weapons. Drones have already been used by the FBI for surveillance, and the FBI maintains it does not need a warrant to use them (Dinan, 2013).

Finally, the authors presented the theory of a "Fifth Wave of Terrorism" and, based on the theory by Kaplan and examining the situation in the DNI briefing, believe it is possible there is now a new wave of terror.

 ## Key Points

- ◆ Predicting the future is difficult, but it is clear that for many the coming decades will be challenging with regard to water and food in many areas, a sure source of conflict.
- ◆ Al Qaeda remains a threat, if not an immediate one, with numerous groups affiliated with it.

- The Middle East is unstable and likely to remain that way for years to come. The Arab Spring has turned into an Arab Winter. The winter will last for a decade.
- Cyber threats receive more attention, especially the infrastructures controlled by computers.
- Technology may keep us safe, but it is intrusive and compromises freedom and privacy. Any widespread use of killer drones should give one pause, especially in light of the decision to kill American citizens using drones.
- The United States, Europe, and Islam will need to confront the issue of Shariah law, although it may be inevitable in the future.
- One author suggests there may be a "fifth wave of terrorism in the future" based on tribalism and genocidal actions in certain countries, many in Africa.

© Can Stock Photo Inc. / Sangoiri

▲ An aircraft that could by brought down by hacking into the on-board computers, something not inconceivable.

KEY TERMS

Cybersecurity

Department of Homeland Security (DHS)

Director of National Intelligence (DNI)

Fifth Wave of Terrorism Theory

National Counterterrorism Center

National Security Agency (NSA)

National Security Special Event

Remote Biometric Identification (RBI)

Transportation Security Agency (TSA)

DISCUSSION QUESTIONS

1. Given all of the technology employed to keep the United States safe and secure, discuss how safe you currently feel, on a scale of one (very unsafe) to ten (completely safe).

2. Should the NSA, with approval from a court, be able to keep every telephone conversation you have had in the last year, as well as every text message sent and received?

3. Do you believe the evidence is beginning to demonstrate there is a Fifth Wave of terrorism? Yes or no? Support your answer with evidence from reputable sites, not Wikipedia.

4. Should the government be required to get a search warrant before conducting Remote Biometric Surveillance on someone? For a day? For a month? For a year?

5. Based on material presented, do you believe that cyber threats have been adequately addressed by the federal government?

WEB RESOURCES

DNI Clapper, Current Threat Assessment: http://www.intelligence.senate.gov/130312/clapper.pdf

FBI Director's take on Cyber Crime: http://www.fbi.gov/news/speeches/combating-threats-in-the-cyber-world-outsmarting-terrorists-hackers-and-spies

For what is possible with CCTV software, see this: https://www.thalesgroup.com/en/worldwide/big-data/smarter-data-safer-public

Meet Dr. Schneck, the new cyber person at DHS: http://fcw.com/articles/2013/08/19/schneck-dhs.aspx

This think tank believes Shariah is coming to America; the report is here: http://www.centerforsecuritypolicy.org/upload/wysiwyg/article%20pdfs/Shariahh%20-%20The%20Threat%20to%20America%20%28Team%20B%20Report%29%20Web%20Version%2009302010.pdf

To see what the Islamic presence is in some European cities, watch this on Belgistan: http://www.cbn.com/tv/1509282970001

READING 24

The following journal article is presented in its entirety. It is a look at the relationship between various factors and the collapse of regimes such as Egypt and Tunisia, but not China. Read the article and answer the questions at the end of the article.

From the Arab Spring to the Chinese Winter

The Institutional Sources of Authoritarian Vulnerability and Resilience in Egypt, Tunisia, and China

Steve Hess

 ## Introduction

On 17 December 2010, Mohamed Bouazizi, an unknown street vendor in the small Tunisian city of Sidi Bouzid, triggered a succession of events that have fundamentally altered the political trajectory of not only his home country of Tunisia, but also created powerful reverberations across the Middle East and North Africa (MENA) region. Frustrated with repeated mistreatment at the hands of local officials, Bouazizi set himself aflame. Tapping into all manner of grievances held by Tunisia's citizens, this incident sparked protests across the country so intense that they quickly brought about the destabilization of the regime and soon drove the long-standing autocrat, Zine al-Abidine Ben Ali, from power. As Ben Ali fled into exile, the Tunisian uprising triggered protests in countries across the MENA region (including Egypt, Libya, Syria, Yemen, and Bahrain) now known as the 'Arab Spring.' Within months, Hosni Mubarak's regime had also collapsed. At the time of writing, protests in Syria have raged on for more than a year despite increasingly violent regime crackdowns.

As suggested by Blake Hounshell (2011), the sudden, unpredicted collapse of seemingly durable autocracies in Tunisia and Egypt has raised important questions about the conventional wisdom on authoritarian resilience.

Consequently, the next question is 'Who is next?' For a number of commentators and academics, one common answer has been another presumably durable autocracy: China. In one noteworthy investigation of this question, Jay Ulfelder (2011) applied a combination of 17 statistical indicators for popular unrest from 163 countries and found that as of 2011, cases such as Egypt, Tunisia, Syria, and Libya appeared in the top 26 countries that were expected to experience nonviolent rebellion. Perhaps most surprisingly, China surpassed high-risk countries such as Iran and Egypt, ranking *first* among all 163 countries in its likelihood of experiencing regime-destabilizing unrest (Ulfelder, 2011). Ulfelder explained, 'China reportedly experiences tens of thousands of scattered protests, riots, and strikes each year, but many observers of that country's politics dismiss those events as background noise in an otherwise well-managed political system.' Citing recent riots and protests in Guangdong and Inner Mongolia, he suggested the People's Republic of China (PRC) might be 'riper for nonviolent rebellion than many China watchers believe' (Ulfelder, 2011).

The following article considers two prevailing clusters of explanations for the recent breakdown of autocracies in Tunisia and Egypt and considers their relevance for assessing the likelihood of authoritarian breakdown in China.

SOURCE: *International Political Science Review*, *34*(3), 254–272. © The Author(s) 2013

The first section centers on the social drivers of unrest in the cases of Tunisia and Egypt: youth unemployment, socioeconomic inequality, official corruption, and an increasingly 'tech-savvy' population, while the second explores capacity-centered explanations for the strength and resilience of authoritarian regimes. This analysis finds that in spite of China's much more impressive economic performance, many factors often identified as the drivers of unrest in the MENA region have, in fact, also been present in China. Even in a context of growth, which might imply a greater satisfaction with the regime, social unrest has been on the rise. Additionally, prior to 2011, Tunisia and especially Egypt were rated as highly durable according to the primary elements of authoritarian capacity.

If divergence in repressive capacity played a critical role in determining the different political outcomes in China and the MENA cases, greater work remains to be done in the direction of cleanly identifying the sources of vulnerability within seemingly high-capacity regimes. Moving in this direction, the third section of this article explores a critical intervening variable that has received little attention in comparisons drawn between China and the Arab Spring autocracies: the dramatic difference in state centralization among these cases. Observing that uncoordinated 'parochial' protests aimed at local, community-specific grievances (rather than the regime itself) have emerged as the prevailing mode of contention in the PRC, the decentralized state structure of the Chinese state appears to deny popular claimants a unifying target to mobilize around and with it the political opportunities needed to develop forms of popular contention coordinated and sustained on a national level.

 Social Drivers of Protest

Many analyses of the Arab Spring have emphasized the bottom-up factors driving Middle Eastern discontent. Looking at the participants involved in mass street demonstrations and their specific grievances, these interpretations have suggested unrest in the MENA region is most closely linked to socioeconomic inequities, the perception of official corruption, a large youth bulge, and the diffusion of modern communications technologies. In an observation that has given Beijing pause, many of these factors are also present in contemporary Chinese society and have fueled a growing problem of social instability across the country. The political scientist Suisheng Zhao (2011) has written that 'The Jasmine Revolution that began in North Africa early 2011 frightened the Chinese government because China faces social and political tensions caused by rising inequality, injustice, and corruption.'

Economic Performance

Any comparison of the social forces driving unrest in the Middle East and China must begin with a discussion of the diverging economic trajectories of these two regions. Over the past three decades, China has undergone dynamic economic development, its GDP growth maintaining an average annual rate of 10.0 percent from 1978 to 2010 (World Bank, 2012). During the same period, MENA economies have grown at a comparatively sluggish rate of 3.59 percent, with Egypt (5.10 percent) and Tunisia (4.52 percent) only modestly outperforming the region as a whole (World Bank, 2012). Clearly, this variation in overall economic growth is no trivial consideration. As noted by Huntington (1991), Diamond and Linz (1989), and Bermeo (1990), in autocracies lacking the kind of legitimacy provided by democratic procedures, the maintenance of political power is heavily dependent on economic performance criteria. The large–N quantitative research of Przeworski and Limongi has provided some general support for this hypothesis, demonstrating that autocracies with high per capita incomes have been remarkably highly resilient to collapse (1997: 159–60).

Breaking in some respects with these findings, Geddes (1999) and, more recently, Ulfelder (2005: 311–34) have found that poor long-term economic performance in itself has little to do with destabilizing autocratic regimes. While low growth 'is never good news,' only sudden and severe economic crises in the short term seem to be capable of destabilizing otherwise-resilient personalist and single-party autocracies. Since the 1970s, single-party autocracies that have broken down have experienced average declines in per capita income of 4 percent in the year before their respective political transitions (Geddes, 1999: 134–6). This pattern has clearly been evident in cases such as the Philippines in 1986 and Indonesia in 1998, where regime breakdown was preceded by steep and sudden economic declines.

Considering that the Arab Spring uprisings have emerged in the wake of the global financial crisis of 2008, a sudden economic downswing seems to be a likely source of the internal discontent that emerged in many MENA states.

However, according to World Bank data, the cases in question experienced tepid growth, not outright collapse, in the lead up to the turbulent year of 2011. Tunisia's GDP per capita increased 3.5 percent in 2008, 2.0 percent in 2009, and 2.6 percent in 2010, while Egypt's grew by 5.3 percent in 2008, 2.9 percent in 2009, and 3.3 percent in 2010. These rates were quite unimpressive in comparison with China's rapid per capita growth rates of 9.0 percent in 2008, 8.6 percent in 2009, and 9.8 percent in 2010 (World Bank, 2012). On the other hand, they were not akin to the kind of deep economic crises that have typically destabilized autocratic regimes.

China's sustained growth amid global economic turmoil may well have bolstered the regime's performance legitimacy and dampened short-term pressures for regime change. But in the greater East Asian region in particular, economic achievement has not had a straightforward causal relationship with the resilience of autocratic regimes. In Singapore and Malaysia, economic success has walked hand in hand with durable single-party rule. Severe economic crises helped terminate the regimes of Suharto in Indonesia and Marcos in the Philippines. Taiwan and South Korea also enjoyed strong economic growth in the 1980s, but nevertheless experienced growing domestic challenges that culminated in political transitions.

Importantly, China's growth has come with serious social consequences. Unemployment, inequality, and official corruption (factors that helped fuel popular discontent in the MENA region) have also appeared in China at high levels and fueled increasingly frequent outbreaks of social unrest. Social and demographic trends, not the least being its now (as of 2010) shrinking labor force, are working against China's ability to sustain near double-digit GDP growth rates in the coming decades (Chang, 2012; Goldstone, 2011a). Consequently, it has become increasingly important to look beyond China's current rates of economic growth and consider if other factors may also be at play in sustaining regime resilience.

Angry, Unemployed Youths

Writing in *Foreign Policy* in early 2011, Ellen Knickmeyer described the 'chronically unemployed twenty-somethings' across the Middle East and North Africa as the 'Arab world's youth army', a group she identified as the leading social force behind the uprisings and political upheavals spreading across the region. In Tunisia and Egypt, high fertility rates had produced a substantial youth bulge, such that by 2005 some 56.1 percent and 62.7 percent of citizens, respectively, were under the age of 30. Meanwhile, in China, family-planning policies since the early 1980s have kept the youth cohort to a smaller size. In 2005, only 45.3 percent of the population was under 30 (Leahy, 2007: 87–90). If the youth were the critical foot soldiers of the 2011 social unrest in the Middle East, any emergent Chinese protest movement would have a comparably smaller social base to draw recruits from.

In Egypt and Tunisia, this large, educated youth cohort became increasingly frustrated with its poor job prospects, which likely played a major role in fueling recent unrest. Many of these educated youth joined the growing ranks of the *hittistes* (Arab slang for 'those who lean against walls') (Knickmeyer, 2011). In Tunisia and Egypt, official unemployment in 2005 had reached rates of 14.2 percent and 11.2 percent, respectively, leaving many youths frustrated with the political and economic status quo and the time needed to plot and organize anti-regime collective actions.

Official data has suggested a rosier picture in China, where despite a reported growth in unemployment, the official unemployment rate has remained comparatively low at 4.1 percent (United Nations Development Program, 2010). By many accounts, these figures understate the problem, as they reflect only the number of citizens who have formally registered for unemployment benefits (*China Labor Bulletin*, 2007; Giles et al., 2005: 168). Data from household surveys has suggested that the actual unemployment rate has been much higher, reaching figures as high as 14.0 percent or 20.0 percent since the late 1990s (Giles et al., 2005: 163). According to the estimates of Giles et al., unemployment has been highest among young workers between the ages of 16 and 30 and permanent urban residents, with young urban residents having an estimated unemployment rate of 24.3 percent in 2000 (2005: 163).

The impact of youth unemployment might also be enhanced by the disproportionately poor job prospects of college graduates. The number of graduates unable to find work has expanded rapidly from 750,000 in 2003 to 1.2 million in 2005 and almost 2 million in 2009, that is, about 32 percent of all graduates (Zhao and Huang, 2010: 2). According to some recent research, these official numbers are also likely to understate the problem (Zhao and Huang, 2010: 2). Based on the observed increase in the involvement of white-collar workers and university students in many recent protest actions (Chen, 2009: 87–106),[1] growing frustrations among this critical and highly trained cohort of unemployed youths could potentially be a driving force behind social unrest heading forward.

Inequality

As noted by a number of commentators, widening socioeconomic inequality has also helped drive the Arab Spring protests in Tunisia and Egypt (Knickmeyer, 2011). Seeing their poverty and material hardship contrasted with that of those better connected to the regime, working-class citizens joined educated youths as they poured into the streets in the regime-toppling protests of 2011. In protests preceding the Arab Spring in Tunisia and Egypt, popular grievances centered on the regime's inability to curtail rising food prices or provide other basic services (Ottaway and Hamzawy, 2011: 2–6). In Egypt, more than 1000 protest actions took place from 1998 to 2004. After the implementation of economic liberalization policies, which cut social services and government spending, protests increased by 200 percent, amounting to 250 in 2004 alone. In April 2008, as many as a half-million Egyptians participated in more than 400 actions, including a general strike centering in al-Mahalla al-Kubra and involving tens of thousands of state workers, youth activists, and professionals. Meanwhile, in Tunisia, materially aggrieved citizens organized collective actions against a mining company in 2008, which soon came to involve protestors demonstrating against rising inflation and unemployment in other parts of the country (Ottaway and Hamzawy, 2011: 2–6).

Data on socioeconomic inequality suggests the gap between the rich and poor has widened even further in China. In 2001, Egypt's Gini index for the distribution of family inequality was reported at 34.4 (90th most unequal of 136 countries), whereas Tunisia's was estimated at 40.0 (61st most unequal) in 2005. China's figure of 41.5 (52nd most unequal) in 2007 (CIA World Factbook, 2011) indicates that socioeconomic inequality has widened dramatically during the deepening of economic reforms from the mid-1990s to early 2000s. At this time, state-provided social services were dramatically cut and large state-owned

enterprises were restructured or privatized, leading many recently laid-off factory workers to join the growing ranks of the unemployed and the poor (Wang, 2006: 252–8). At a time in which a record 1.11 million Chinese have become millionaires (Kroll, 2011), Huang (2008: 246–50) estimates that the country's poorest citizens have actually experienced absolute declines in living standards. Meanwhile, a 2004 household survey revealed that 71.7 percent of Chinese respondents considered income in the country to be either 'somewhat large' or 'too large' and held the view that the 'rich get richer, [while the] poor get poorer' (Whyte, 2010: 306–7). In short, in the midst of growing national prosperity, working-class Chinese have seen living standards drop in relative and, more recently, absolute terms.

In a similar vein to the economic protests reported in Egypt and Tunisia over the past decade, materially driven 'subsistence crises' have fueled growing social unrest among China's working class (Feng, 2000: 41–63; Tong and Lei, 2010: 490). A 2004 survey indicated that nearly 73 percent of public officials in China considered 'income distribution' to be their greatest concern (People's Daily, 2004). Frustrations over extreme inequality have also percolated upward from the lowest rungs of society into the middle class. As illustrated by the survey data of Brockmann et al. (2009), even the 'winners' of market reforms have experienced declining levels of subjective well-being. These 'frustrated achievers' have become increasingly dissatisfied as they have witnessed an even narrower slice of Chinese society achieve dramatically higher levels of relative affluence (Brockmann et al., 2009: 387–405). Clearly, following the economically motivated protests in Egypt identified in the late 1990s and mid-2000s, issues of socioeconomic injustice have also become a troubling source of social instability in present-day China.

Corruption

By many accounts, popular discontent in the MENA region has been further inflamed by the pervasive presence of official corruption. Stuart Levey (2011) has pointed out that official corruption has been a 'key grievance' driving protests throughout the Arab world—a reality highlighted by the recent trials of Ben Ali and Mubarak for corrupt practices ranging from money laundering to drug trafficking. If corruption in itself was a driver of political instability in the Arab Spring, then Beijing has reason for concern. According to the 2010 Corruption Perceptions Index (CPI), a measure of the overall extent of corruption as perceived by foreign and domestic country experts and business leaders, China received a score of 3.5, placing it 78th in the world out of 178 participating countries. Tunisia and Egypt were scored at 4.3 and 3.1, respectively, placing them at the ranks of 59th and 98th overall (Transparency International, 2011). The Pew Forum's Global Attitudes Project and *The China Survey* have also both reported high levels of official corruption. According to the former, 78 percent of Chinese respondents considered corrupt officials to be a 'very' or 'moderately big' problem (Pew Global Attitudes Survey in China, 2008); the latter found that 67.5 percent of respondents viewed official corruption to be a 'serious' problem (see Harmel and Yeh, 2011: 7). In short, much like their counterparts in Tunisia and Egypt, average Chinese consider corruption to be a widespread and very serious problem.

Scholars have been in general agreement that corruption is present at high levels in China, but not on the broader ramifications of this problem. Minxin Pei (2006) has argued that pervasive 'decentralized predation' by state agents will gradually result in economic stagnation and malaise in China. Francis Fukuyama (2010: 35) has argued that China's rule of law is 'good enough' to contain corruption and sustain economic growth. Kellee Tsai (2007: 6–11) has even found that corruption and other adaptive informal practices have been essential to China's successful economic transition.

While the broader impact of corruption on China's economic vibrancy remains in dispute, it is clear that the perception of official corruption has appeared as a driver of social unrest with growing frequency. As revealed in Chen's analysis of mass incidents reported in Chinese news, protests directly motivated by land seizures, unpaid wages, or factory layoffs are often closely linked to complaints over cadre corruption (2009: 90–95). These 'anger-venting' mass incidents have revealed the groundswell of frustration against corrupt and abusive local officials that has marked many of China's communities (Fewsmith, 2008). Often involving the destruction of the property and offices of local governments, these mass protests (in a noteworthy parallel to the case of Mohamed Bouazizi in Tunisia) have been sparked by specific incidents of officials abusing citizens.

Two recent major mass incidents have been triggered by actions such as local security forces' mistreatment of a female street vendor in Guangdong and the mysterious death of an anticorruption activist in Hubei Province (McLaughlin, 2011). While these anger-venting mass incidents have typically been restricted to specific communities, involving no visible coordination or linkages across localities, they have a common theme: they all represent popular backlashes against perceived official malfeasance. As noted by Ben Heineman (2011), 'this corruption—both in the sense of officials/cadres taking money illicitly or in the arbitrary use of "law" for personal ends—only increases, in turn, the pressure for protests.' Much like the now-toppled regimes in Tunisia and Egypt, the Chinese government has not effectively reined in the persistent problem of official corruption, an outcome that adds significantly to the country's likelihood of experiencing destabilizing outbursts of social unrest.

This analysis has revealed that while China has a significantly smaller youth cohort, it has comparable levels of inequality and official corruption. It may also have a similar problem of unemployment, which is most pronounced among the college-educated. Importantly, these drivers of discontent have helped fuel the rapid growth of social unrest in China. As reported by the Ministry of Public Security, 'mass incidents' in the country increased from 10,000 in 1994 to 87,000 in 2005 (Kahn, 2006), with some accounts suggesting the number had accelerated to 180,000 by 2010 (Orlik, 2011). Clearly, as reported by Zhao (2011), Ulfelder (2011), and others, social discontent is very much on the rise in China. Yet in an important deviation from the MENA cases, the prevailing mode of contention in China has been the parochial protest, aimed at specific local functionaries of the Chinese Communist Party (CCP), but not the regime itself—an outcome discussed at some length below.

Authoritarian Capacity

In many explanations of the Arab Spring, commentators have followed the conventional wisdom embraced by researchers of authoritarianism in academia, which centers on the capacity of autocratic regimes to maintain elite cohesion while also stamping out popular challenges. In this view, authoritarian breakdown is less associated with the tactics, grievances, or organization of popular protests and more with the top-down deficiencies of regimes themselves. In one such interpretation of the Arab Spring, Jack Goldstone (2011b) has argued:

> Although such regimes often appear unshakable, they are actually highly vulnerable, because the very strategies they use to stay in power make them brittle, not resilient. It is no coincidence that although popular protests have shaken much of the Middle East, the only revolutions to succeed so far—those in Tunisia and Egypt—have been against modern sultans.

In other words, the primary driving factor behind the Arab Spring lay in the preexisting structural weaknesses of the regimes themselves. The regimes of Egypt and Tunisia were slowly being corroded by their personalism and lack of effective institutional mechanisms for maintaining longterm internal cohesion among elites and control over society. However, as admitted by Goldstone, this 'degree of . . . weakness is often visible only in retrospect,' appearing after a regime has fallen in the face of popular challenges (2011b: 8–16). This is demonstrated by the many ways in which the MENA regimes, especially Egypt, received extraordinarily high marks in terms of their degree of 'capacity.' As noted by Tarek Masoud, up to the turbulent year of 2011, Egypt in particular was known in academic circles as an 'exemplar of something we called "durable authoritarianism"—a new breed of modern dictatorship that had figured out how to tame the political, economic, and social forces that routinely did in autocracy's lesser variants' (2011: 22–34). The durability of this brand of dictatorship was based in an explanatory variable often used in studies of authoritarianism: the overarching quality of 'authoritarian capacity.' This variable involves three primary elements: coercive capacity, political capacity, and discretionary control over the economy (Way, 2008: 55–69). On all three criteria, the autocracies of Egypt, Tunisia, and China have exhibited high levels of strength, indicating the need to add greater precision to this approach.

Internal Security Forces

The first of these elements involves an effective internal security force, which enhances a regime's 'coercive capacity'—its ability to 'prevent or crack down on opposition protest' (Levitsky and Way, 2010: 57). In most respects, China is thought to have a powerful and effective coercive apparatus. The military and internal security forces have demonstrated their willingness to apply violence when requested—both at Tiananmen in 1989 and in thousands of lesser-known cases (Thompson, 2001: 63–83). Moreover, in the post-Tiananmen years, the country's internal security apparatus, the People's Armed Police (PAP), has expanded to more than 1 million personnel (supplementing local police forces), has an annual budget of nearly US$2 billion, and has received advanced training and equipment for crowd dispersal (Sun and Wu, 2009: 107–28).

In much the same manner, both Egypt and Tunisia also received high marks in their degree of coercive capacity. Citing the exceptional 'robustness' of the coercive apparatuses of these Middle Eastern regimes, Bellin (2004: 139–57) and Brownlee (2010: 468–89) considered this capacity to be an important reason for the remarkable resilience of dictatorships in the region. Heading into 2011, Egypt in particular was flush with roughly US$1.3 billion in annual military aid from the USA, its internal security forces included in excess of 1.4 million personnel, and these forces had repeatedly demonstrated their willingness and capability to crack down on all manner of regime opponents, ranging from moderate politicians to radical militants (Cook, 2009: 3).

Beyond the impressive ability to harass and intimidate opponents as well as crack down on demonstrators found in all three regimes, a lower-intensity form of repression (Levitsky and Way, 2010: 58), media control, takes a central position in discussions of the Arab Spring uprisings. Clearly, new social media vehicles, such as Twitter and Facebook, played a critical role in enabling opposition activists in these two MENA cases to express publicly criticisms of their respective governments and to organize massive anti-regime street demonstrations.

If social media and Internet usage challenge existing modes of authoritarian control, then China's current regime has serious cause for concern. Chinese society has been affected by the rapid diffusion of these technologies in a similar fashion to the MENA region at large. In Egypt and Tunisia, Internet penetration rates exploded from 0.7 percent and 1.0 percent, respectively, in 2000 to 21.1 percent and 33.4 percent, respectively, in 2009. Similarly, in China, Internet penetration has accelerated from 1.7 percent of the population using the web in 2000 to 28.7 percent in 2009 (Internet World Stats, 2012).

Of course, as noted by Lynch (2011) and Morozov (2011), Internet technology and social media present a double-edged sword to opposition activists in authoritarian contexts. These technologies not only create new opportunities for anti-regime activists, but can also help authoritarian regimes to monitor, silence, and even distract these oppositionists as well as their supporters (Lynch, 2011: 305–6; Morozov, 2011). Observers noting Beijing's 'Great Firewall' which blocks politically sensitive websites, its army of 50,000 government web censors, and its innovative technique of 'crowdsourcing' Internet monitoring to private '50-centers' who are paid small amounts to report on their fellow web-users have often held up China as an exemplar of media control and censorship (Diamond, 2010: 74). Commentators such as James Fallows (2011) have argued that the ingenuity of China's Internet censorship, which he brands 'flexible repression,' has in many ways helped it avert an Arab-Spring-style uprising.

While certainly reasonable, upon closer inspection weighing the media-control capacity of various autocracies is difficult—particularly during an era of 'authoritarian learning' in which autocracies are quick to adopt the best practices for quelling unrest from their nondemocratic counterparts (Silitski, 2006). After all, the regimes in Egypt and Tunisia were both noted for their extensive filtering of web content, blocking of opposition websites, revocation of press licenses for media outlets that published politically sensitive news, and use of the Internet and social media sites to conduct surveillance of political and social activists (Deibart et al., 2010: 537–44, 581–8). In the end, while these efforts were clearly insufficient in preventing enterprising activists from using these technologies to mobilize anti-regime collective actions in Tunisia and Egypt, Chinese 'netizens' have also proven adept at using text-message and web technology to circumvent state censorship to expose official malfeasance and challenge public officials.

Such online activism has even spilled over into the coordination of mass protest actions, such as a 2007 demonstration against the construction of a chemical plant in Xiamen that involved as many as 20,000 participants (Lim, 2007). While the Chinese regime has maintained an impressive coercive capacity, contentious collective action has nevertheless been frequent and intense. But unlike in Egypt and Tunisia, this contention has been sporadic, localized, and targeted at corrupt subnational officials, not the national regime perse. In addition, while calls to organize a 'Jasmine Revolution' modeled on the Arab Spring successfully evaded censors and were distributed widely across China's cyberspace, few citizens responded with enthusiasm or appeared at designated public gatherings (Ramzy, 2012). In contrast to Egypt and Tunisia, these recent episodes in China suggest that while communications technologies have certainly helped mobilize popular contention, citizens have overwhelmingly used these tactics not to challenge the regime itself, but rather to confront local cadres based on local grievances.

Economic Control

A second element of authoritarian capacity, a regime's discretionary control over the economy, is critical to funding a robust and highly trained security apparatus as well as supplying rents that can be distributed to supporters in exchange for lasting loyalty (Way, 2008: 55–69). It reduces the risk that private economic interests might provide much-needed financial support to potential regime opponents. This state control over economic resources limits the demands the state must make in extracting revenue from the public and enables the regime to reward loyalists, buy off potential challengers, and starve opponents (Way, 2008: 64–5).

Of course, when considering the distribution of economic resources within the MENA region, it must first be noted that among leading oil-exporting countries, such as Saudi Arabia, Iran, the United Arab Emirates, Kuwait, Qatar, and Bahrain, no autocratic regime collapsed during the turbulent year of 2011. Libya, which was toppled in large part through a NATO military intervention, stands as the lone exception (Ross, 2011; US Energy Information Administration, 2011). Clearly, an economy that is highly dependent upon oil exports offers major benefits to an autocratic ruler, as is often noted by proponents of the 'resource curse' hypothesis, and those countries most strongly affected by Arab Spring protests were among the poorest in the MENA region in terms of oil resources.

In both Egypt and Tunisia, scholars have commented that while these authoritarian regimes lacked the oil reserves of many of their neighbors, maintaining control over their economies remained a priority. While they implemented some liberalizing reforms, these regimes by and large had steadfastly sustained their discretionary grip over the economy (Bellin, 2004: 139), which largely contributed to the creation of a 'rent-seeking urban bourgeoisie and landed elite with no interest in democracy or political participation' (King, 2007: 434). In much the same way, over the past three decades the CCP has also relaxed its controls over the state socialist economy. Much like their counterparts in the MENA region, China specialists have generally not seen market reforms as eroding the regime's capacity for maintaining political and social control. Rather, leading researchers such as Tsai (2007), Solinger (2008), and Chen and Dickson (2010) have found that the emerging social forces that were expected to challenge the regime's grip on power, namely an increasingly politically assertive bourgeoisie, have achieved their material economic gains in concert with the status quo political regime. They have consequently demonstrated a remarkably high level of support for it.

Political Institutions

Another element that enhances the durability of authoritarian regimes is the presence of a powerful, highly institutionalized political party (Brownlee, 2007: 42–3; Geddes, 1999: 135; Magaloni, 2008: 715–42). Many commentaries written immediately after the Arab Spring argued that the 'sultanistic' nature of authoritarian rule in Egypt and Tunisia, compared with the better institutionalized single-party rule of the CCP, made these regimes vulnerable to internal divisions and collapse (Fukuyama, 2011; Goldstone, 2011b: 8–16). According to the large-N empirical work of Geddes (1999: 131–2) and Brownlee (2007), single-party regimes have far exceeded their personalist counterparts in longevity. Presumably, if Ben Ali and Mubarak ran their regimes as personal fiefdoms, relying primarily on kinsmen and cronies for their support, this made them much more exposed to popular challenges and internal defections than China's better institutionalized single-party regime.

However, variation in institutional capacity among these three cases is not so transparent. Only a cursory examination of research on Egypt and Tunisia reveals that until 2011 students of authoritarianism believed these countries were bolstered by effective institutional mechanisms, ranging from hegemonic parties to nominally democratic elections. Brownlee (2007), King (2007: 433–59), Lust (2009), and Blaydes (2011: 2–5) identified political institutions in Egypt and Tunisia (that is, hegemonic party apparatuses, legislatures, and elections) as critical regime supports that helped manage internal divisions and co-opt potential challengers. The Chinese regime has also received high marks for its institutionalization, with scholars noting the CCP's highly specialized institutions and smooth, merit-based methods for cadre advancement (Nathan, 2003: 6–17; Shambaugh, 2009). Meanwhile, others have inspected internal party documents (Gilley, 2003: 19–22; Zong, 2002) and studied patterns in cadre promotion (Shih et al., 2012: 166–87) to suggest that it is personal and faction-based connections that drive advancement in the party, not objective performance criteria. Supporting this view, the spectacular and intrigue-laced collapse of Bo Xilai in recent months has revealed serious fractures and divisions underlying China's otherwise well-managed political system. These findings suggest that while China clearly appears to have an impressive repressive capacity in many respects, the opacity of the political system may mask vulnerabilities that do exist. Egypt and Tunisia are instructive in that the fragility of these seemingly durable autocracies only became apparent after they were challenged by coordinated, national-level protest movements. Consequently, studies of authoritarian resilience should ask not only whether a regime is strong or weak in terms of capacity, but whether elements of the regime tend to facilitate or inhibit national-level contention from emerging in the first place.

 The Missing Variables: Centralization and Modes of Contention

This analysis does not suggest that the PRC is on the verge of collapse. Rather, when looking at the dynamics of state-society interactions within China, several sources of its resilience have almost uniformly been overlooked in comparative studies of authoritarianism. Speaking to a deeper problem in the current literature on authoritarianism, scholars have tended to focus on elite-level factors such as competition between majority and minority factions within regimes (see Geddes, 1999) and to ignore protest actors as independent agents of popular mobilization in their own right. As noted by Eric McGlinchey (2009: 124–5), 'The existing literature, perhaps understandably given its focus on institutional weakness, [has] overemphasized state variables while underemphasizing the causal role of social opposition movements.' Sharing a similar sentiment, Bunce and Wolchik have been 'extremely skeptical that structural factors alone' can explain popular revolutions, and emphasize the need for an integrated approach that appreciates the causal role of popular oppositions as agents of authoritarian breakdown: 'Put simply, structure, agency, and process are all important' (2009: 70).

First, by leaving social opposition movements outside their analyses and focusing on authoritarian states as entities that can be characterized as having high or low 'strength' or 'capacity', students of authoritarianism have missed an important consideration that has taken central importance in the parallel subfield of contentious politics: the way in which the state and other political environments can act as 'structure[s] of political opportunities' that create certain 'constraints or open avenues' for different kinds of individual and group political actions, greatly influencing the manner of their political behavior (Eisinger, 1973: 11–12). Presumably, particular structures of opportunities might alternatively facilitate or inhibit the development of sustained, large-scale, high-participant forms of popular contention seen on the streets of Tunis and Cairo's Tahrir Square.

Second, the treatment of popular protests as residual phenomena deriving from authoritarian weakness has meant that researchers have often failed to disaggregate different modes of popular contention that tend to appear in these various political opportunity structures. Of particular relevance to this comparison between the cases of Tunisia and Egypt, on the one hand, and China, on the other, is the distinction between 'national' and 'parochial' forms of popular contention. As best articulated in the writings of Sidney Tarrow and Charles Tilly, parochial forms of contention are framed around material and issue-specific grievances, lack broad and coordinated coalitions of social actors who are based in diverse societal and economic sectors and geographic localities, and target particular and usually local officials or layers of the state. National forms of contention or movements are framed in general and inclusive terms that incorporate outside groups, coordinated across many previously unconnected sites and social actors, and united against a single unifying target, such as a national government or leader (Tilly and Tarrow, 2007: 31–4).

Of importance to studying comparative cases of authoritarian resilience or breakdown, parochial forms of contention are fragmented and uncoordinated on a national scale. Since they can be dealt with on a case-by-case basis, they generally present a manageable challenge to authoritarian regimes—including those that tend to rate as low-capacity regimes. These latter, nationally coordinated movements can quickly overwhelm even the highest capacity regimes, such as the Shah's Iran, the Philippines under Marcos, regimes across Central and Eastern Europe, and even the seemingly durable Mubarak and Ben Ali dictatorships in Egypt and Tunisia, leading to elite fragmentation and collapse.

An investigation that treats the state as a structure of political opportunities for particular forms of popular contention would begin by identifying common features of these regimes and their respective state structures that might enable popular claimants rapidly to mobilize diverse segments of society in concerted action against their respective governments. One important feature that immediately comes to attention is the extraordinarily high degree of state centralization in both Egypt and Tunisia, which contrasts dramatically with China's decentralized state structure. In this context, centralization refers to the distribution of 'responsibility for planning, management and resource raising and allocation' between 'the central government and its agencies [and] the lower levels of government' (Work, 2002: 5) and, of particular importance in autocratic regimes, the distribution of power over a state's coercive resources and discretion over decisions related to the suppression of popular opponents.

Yet, while a half-century ago Arthur Maass (1959: 9) pointed out that concerns related to the 'distribution and division of governmental power' have been fundamental issues in political science dating back to the time of Aristotle, few recent scholars have considered how major variation on this most basic element of an authoritarian state's structure might influence its vulnerability to popular challenges from below. Moreover, as noted by Falleti (2010: 6–7), we are seeing a growing number of countries around the world experimenting with dramatic decentralizing reforms. This shift toward greater decentralization has extended to nondemocracies, which, according to World Bank (2011) data, distributed an average of 14.9 percent of all government expenditure at the subnational level in the 1970s and 1980s, a figure that surged in the 1990s to 32.8 percent by 1999. At a point in history in which more and more autocracies have decentralized, it has become important to explore how this most basic change to state structures may impact the way in which popular actors interact with and contest the authoritarian state.

While research on the subject of centralization and decentralization has been extensive on the global level, driven by the issue's emphasis on international financial institutions such as the World Bank and International Monetary Fund, it has been relatively limited in the MENA region. Available research, mostly on topics such as development and governance, immediately reveals an extremely high level of centralization in many MENA countries, including Egypt and Tunisia (Amin and Ebel, 2006; Tosun and Yilmaz, 2008). This is reflective of the policy choices made by MENA regimes, which unlike many countries in Latin America, Africa, East Asia, and Eastern Europe, did not undertake the decentralizing reforms promoted by the World Bank and other international financial institutions in the 1980s and 1990s (World Bank, 2008).

Looking at available data on subnational shares of public expenditure and revenue, intergovernmental transfers, and the distribution of authority over personnel management and internal security forces, researchers in the mid- to late 2000s described the cases of Tunisia and Egypt as extraordinarily centralized in terms of functional and coercive state power (Boex, 2011; Tosun and Yilmaz, 2008: 11–12; United Nations Development Program, Program on Governance in the Arab Region, 2009). In fact, writing in 2006, Amin and Ebel noted that 'one cannot [even] track how Egypt compares with other countries because *all spending is carried out by central entities;* thus expenditure flow data gets reported by ministry sector rather than by function' (2006: 3, emphasis added). In their overall analysis of the Egyptian state, these

authors specifically described Egypt as having 'one of the most centralized public sector systems in the world' (Amin and Ebel, 2006: 9).

Compared with the MENA cases of Tunisia and Egypt, China has been an ideal case of what can be termed 'decentralized authoritarianism' (Landry, 2008). In Tunisia and Egypt, subnational governments respectively controlled 12.1 percent and 15.6 percent of all government expenditure (Tosun and Yilmaz, 2008: 27). In China, an average of 54.84 percent of fiscal expenditure was spent at the subnational level from 1995 to 1998 (World Bank, 2011). The degree of decentralization in post-Maoist China has been extremely remarkable. As noted by Landry (2008: 6), from 1972 to 2000 nondemocracies' average subnational share of state expenditure was 17.76 percent. The country, in other words, has been three times as decentralized as the average authoritarian regime, and nearly four times as decentralized as its counterparts in the MENA region.

This variation in measures of state centralization has likely played an important role in facilitating the outbreak of nationwide forms of popular contention in settings such as Egypt and Tunisia, while promoting only parochial forms of contention in decentralized states such as China. According to Alexis de Tocqueville (1955: 76), the extraordinary centralization of successive governments proved critical in bringing about the revolutionary upheavals that shook France during the late 18th and early 19th centuries. More recently, scholars of contentious politics have argued that the growing centralization associated with the birth of the modern national state was intimately linked to the appearance of nationwide movements of popular contention. According to Sidney Tarrow (1994: 72):

> As the activities of national states expanded and penetrated society, they also caused the targets of collective action to shift from private and local actors to national centers of decision-making. The national state not only centralized the targets of collective action; it involuntarily provided a fulcrum for . . . standard forms of collective action.

Instead of taking localized, particular, and bifurcated actions that were restricted to subnational targets, involved direct action, and were carried out by specific social groups, contentious repertoires evolved into genuine national protest movements (Tarrow, 1994: 6; Tilly, 1993: 272). Following the logic of this presumed linkage between state centralization and the appearance of national forms of protest, acts of popular contention in more decentralized authoritarian settings are expected to emerge in a recurrently fragmented and localized form, failing to undergo the process of 'upward scale shift' to national protest movements (Tilly and Tarrow, 2007: 95). According to terminology applied by Tarrow (1994), this limited mode of protest does not diffuse a general, 'modular repertoire' of contention throughout a national society. Additionally, it lacks broad, coordinated coalitions of social actors who are based in diverse societal and economic sectors and geographic localities (Tarrow, 1994: 72).

When comparing World Bank (2011) Government Finance Statistics (GFS) data on fiscal decentralization (available for the years 1972 to 2000) with historical cases of authoritarian breakdown during the 'third wave of democracy', one can observe a general global pattern linking high levels of state centralization with sudden instances of regime collapse (Huntington, 1991; Ulfelder, 2005: 327–30). Among the 45 historical autocracies with available data, the proportion of government expenditure at the subnational level has averaged 18 percent. For the 36 regimes that have collapsed since the 1970s, the share of subnational spending during their years under autocratic rule averaged 17 percent. There is a great deal of variation within this group, including a subgroup of highly centralized autocracies such as Marcos's Philippines (11.8 percent subnational spending) and Suharto's Indonesia (11.9 percent subnational spending). Preceding the Arab Spring by decades, these two centralized regimes (as well as a cascade of communist regimes in Central and Eastern Europe, such as Poland, Hungary, Czechoslovakia, and Romania) broke down in the face of national bursts of popular protest.

Meanwhile, the smaller group of 10 surviving autocracies had an average subnational expenditure of 26 percent. Among these survivors, there is a large degree of variation in the measure of fiscal decentralization. These cluster into three general groupings: a highly centralized pair of resource-rich states in the Middle East, that is Iran (3.0 percent) and Bahrain (3.2 percent); three very diverse autocracies in the medium-range in terms of decentralization, that is Malaysia (19.1 percent), Zimbabwe (18.1 percent), and Azerbaijan (24.1 percent); and five more-decentralized states, that is Tajikistan (30.9 percent), Kazakhstan (31.4 percent), Belarus (32.6 percent), Russia (38.1 percent), and China (54.8 percent). In short, in a global-historical sense, all else being equal, decentralized regimes (such as China, Kazakhstan, Russia, Tajikistan, and Belarus) have outpaced their more centralized neighbors. China has avoided nationalized waves of protest that have affected neighboring countries such as Indonesia and the Philippines. Meanwhile, Kazakhstan, Russia, Tajikistan, and Belarus have resisted the wave of 'color revolutions' that toppled regimes across the former Soviet Union in the 2000s.

A closer look at China, the world's most decentralized autocracy, provides evidence that autocracies with very low levels of centralization deny popular claimants the opportunity to use an intrusive, centralized state as a common, unifying target for mobilizing national-level contention. Meanwhile, this structure creates opportunities for localized, parochial forms of contention. In a pattern noted by a number of leading specialists in Chinese contentious politics, the decentralized state structure of the PRC has encouraged popular claimants to take parochial collective actions aimed at corrupt and abusive local officials (Cai, 2008: 411–32; Lee, 2007; O'Brien and Li, 2006). The research of Yongshun Cai has suggested that by granting local officials greater authority within their jurisdiction, the center avoids blame for local authorities' official misdeeds and their use of repression (2008: 415). This reality grants the center a degree of plausible deniability when acts of State violence occur at the local level—helping the national state preserve its legitimacy even when coercion is used against protestors.

As noted by O'Brien and Li (2006: 27), gaps between the center and local authorities in the Chinese state provide a 'structural opening' that local claimants can capitalize upon in launching collective actions aimed at addressing their grievances. By framing their protests against local cadres as defending the laws and regulations promulgated by the center, these 'rightful resisters' can seek allies within the state or in the wider public, such as media outlets, and avoid the accusation that they are antistate or unpatriotic (O'Brien and Li, 2006: 23). By framing their contention in such a way, claimants develop repertoires of contention and bases of support that are conducive to local and particularized, but not national anti-system modes of action. As noted by Tilly and Tarrow (2007), effective social movement campaigns require a 'social movement base'. This refers to 'movement organizations, networks, participants, and the accumulated cultural artifacts, memories, and traditions that contribute to social movement campaigns' (Tilly and Tarrow, 2007: 114). In a decentralized state structure, in which localized acts of contention often prove effective and more broadly coordinated efforts are extraordinarily risky to the participants involved, claimants tend to construct bases oriented around particular and specific, not national and inclusive concerns, and parochial, not national collective action becomes the prevailing mode of contention.

In widely publicized outbreaks of unrest in China, elements of parochial forms of contention have prevailed. High-profile outbreaks of protest that have diffused across regions, such as the November 2008 strike of taxi drivers that originated in Chongqing and the rapid succession of strikes in auto-parts factories across China's industrial Southeast from May to July 2010, have involved scattered acts of contention that targeted subnational officials and governments, involved no observable cross-regional coordination, and were resolved through interactions between subnational authorities and strike organizers within their various jurisdictions (*China Labor Bulletin*, 2010; Hess, 2009: 61–77; Richburg, 2010; *Straits Times*, 2010). In major cases of not-in-my backyard-style environmental contention, such as the demonstrations in Xiamen during 2006 and Dalian during 2011 against paraxylene plants (which involved as many as 20,000 participants in the former case), protestors have targeted municipal officials and been successful in compelling them to relocate polluting industries outside of their local communities (Bradsher, 2011; Lim, 2007). These cases support the findings of Cai (2008) in his study of 78 incidents of popular contention in China from 1995 to 2006. In more than 80 percent of these incidents, the state response of either repressing protestors or extending concessions was determined entirely at the local level. In another ten cases, provincial authorities intervened, leaving only five cases in which the central government became directly involved (Cai, 2008: 420).

Conclusions

These findings suggest that the variation in state centralization between Tunisia and Egypt, on the one hand, and China, on the other, helps explain why popular protests prompted authoritarian breakdown in the former, but have not affected the resilience of the regime in the latter. In all three cases discussed in this article, authoritarian regimes have demonstrated high levels of capacity: they have had powerful and effective coercive apparatuses, highly institutionalized hegemonic parties, and comparable levels of discretionary control over the economy. These factors led scholars to the conclusion that all three regimes were extraordinarily durable—an assessment that has since quickly fallen apart with the fall of Ben Ali and Mubarak. Moreover, all three regimes have struggled with the problem of social instability, which was driven in no small part by high levels of corruption, inequality, and youth unemployment. While China has had a smaller cohort of 20-somethings than the cases of Egypt and Tunisia and higher rates of overall economic growth, this has not prevented the outbreak of frequent and intense outbursts of social unrest, which have expanded rapidly in number over the past several decades (Kahn, 2006).

While fragmented outbursts of social unrest transformed into protests coordinated and sustained on the national level in Egypt and Tunisia, protests in China have remained scattered and oriented around local and limited issues and targeted against subnational officials. The divergence on this outcome, that is, on the prevailing *mode* of protest (national contention in the Arab Spring cases and parochial contention in China), helps explain why the PRC has continued to endure and not faced nationwide protest movements despite the frequency of protest actions overall within its borders. These findings suggest that the literature on authoritarian resilience can be enhanced by asking not only if regimes are 'high' or 'low' capacity, but also how particular states' structures can impede or facilitate the appearance of national-level protest movements as autocratic regimes interact with popular claimants.

Note

1. This notion of middle-class discontent is also supported by a number of recent high-profile protest actions. These include ones against polluting chemical plants at Xiamen in 2007 and Dalian in 2011 that involved tens of thousands of protesters and the well-known protest against a proposed magnetic train line in Shanghai during 2008, which were overwhelmingly middle-class affairs.

References

Amin K and Ebel R (2006) *Egyptian intergovernmental relations and fiscal decentralization diagnostics and an agenda for reform*. World Bank Policy Note, Egypt Public Expenditure Review. Available at: http://lgi.osi.hu/cimg/0/l/3/2/l/Egypt_Decentralizatiion._World_Bank._2006_Ebel_Amin.pdf (accessed 21 July 2011).

Bellin E (2004) The robustness of authoritarianism in the Middle East: Exceptionalism in comparative perspective. *Comparative Politics* 36(2): 139–57.

Bermeo N (1990) Rethinking regime change. *Comparative Politics* 22(3): 359–77.

Blaydes L (2011) *Elections and Distributive Politics in Mubarak's Egypt*. New York: Cambridge University Press.

Boex J (2011) *Democratization in Egypt: The potential role of decentralization*. Urban Institute Center on International Development and Governance Policy Brief, February. Available at: http://www.urban .org/ uploadedpdf/412301-Democratization-in-Egypt.pdf (accessed 21 July 2011).

Bradsher K (2011) China moves swiftly to close chemical plant after protests. *New York Times*, 14 August. Available at: http://www .nytimes.com/2011/08/15/world/asia/15dalian.html (accessed 10 October 2011).

Brockmann H, Delhey J, Welzel C, et al. (2009) The China puzzle: Falling happiness in a rising economy. *Journal of Happiness Studies* 10(4): 387–405.

Brownlee J (2007) *Authoritarianism in an Age of Democracy*. Cambridge: Cambridge University Press.

Brownlee J (2010) Unrequited moderation: Credible commitments and state repression in Egypt. *Studies in Comparative International Development* 45(4): 468–89.

Bunce V and Wolchik S (2009) Getting real about real causes. *Journal of Democracy* 20(1): 69–73.

Cai Y (2008) Power structure and regime resilience: Contentious politics in China. *British Journal of Political Science* 38(3): 411–32.

Chang G (2012) China's zero-growth economy. *Forbes*, 11 March. Available at: http://www.forbes.com/ sites/gordonchang/2012/03/11/ chinas-zero-growth-economy/ (accessed 15 April 2012).

Chen CJ (2009) Growing social unrest and emergent protest groups in China. In: Hsiao H-H and Lin C (eds) *Rise of China: Beijing's Strategies and Implications for the Asia-Pacific*. New York: Routledge, pp. 87–106.

Chen J and Dickson B (2010) *Allies of the State: China's Private Entrepreneurs and Political Change.*

Cambridge, MA: Harvard University Press. *China Labor Bulletin* (2007) Unemployment in China. 14 December. Available at: http:// www.clb.org.hk/en/node/100060#3b (accessed 16 August 2011).

China Labor Bulletin (2010) Guangdong ponders another increase in the minimum wage. 30 November. Available at: http://www.clb .org.hk/en/node/100940 (accessed 17 February 2011).

CIA World Factbook (2011) Country reports. Available at: https:// www.cia.gov/library/publications/the-world-factbook/ (accessed 29 August 2011).

Cook S (2009) *Political instability in Egypt*. Council on Foreign Relations Contingency Planning Memorandum 4, August. New York and Washington, DC: Council on Foreign Relations Press.

Deibart R, et al. (2010) *Access Controlled: The Shaping Power, Rights and Rule in Cyberspace*. Cambridge, MA: MIT Press.

De Tocqueville A (1955 [1856]) *The Old Regime and the French Revolution*. Gilbert (trans.). Garden City, NY: Doubleday.

Diamond L (2010) Liberation technology. *Journal of Democracy* 21(3): 69-83.

Diamond L and Linz J (1989) Introduction: Politics, society, and democracy in Latin America. In: Diamond L and Linz J (eds) *Democracy in Developing Countries: Latin America*. Boulder, CO: Lynne Rienner.

Eisinger P (1973) The conditions of protest behavior in American cities. *American Political Science Review* 67(1): 11–28.

Falleti T (2010) *Decentralization and Subnational Politics in Latin America*. Cambridge: Cambridge University Press.

Fallows J (2011) Arab Spring, Chinese winter. *The Atlantic*, September. Available at: http://www.theatlantic.com/magazine/ archive/2011/09/arab-spring-chinese-winter/8601/ (accessed 10 January 2012).

Feng C (2000) Subsistence crises, managerial corruption and labor protests in China. *China Journal* 44: 41–63.

Fewsmith J (2008) An 'anger-venting' mass incident catches the attention of China's leadership. *China Leadership Monitor* 26. Available at: http://www.hoover.org/publications/china-leadership-monitor/ article/5673 (accessed 1 September 2011).

Fukuyama F (2010) Transitions to the rule of law. *Journal of Democracy* 21(1): 33–44.

Fukuyama F (2011) Is China next? *Wall Street Journal*, 12 March. Available at: http://online.wsj.com/article/ SB1000142405274870 3560404576188981829658442.html (accessed 14 July 2011).

Geddes B (1999) What do we know about democratization after twenty years? *Annual Review of Political Science 2 :* 115–44.

Giles J, et al. (2005) What is China's true unemployment rate? *China Economic Review* 16(2): 149–70.

Gilley B (2003) The limits of authoritarian resilience. *Journal of Democracy* 14(1): 18–26.

Goldstone J (2011a) Rise of the TIMBIs. *Foreign Policy, 2* December. Available at: http://www.foreignpolicy.com/articles/2011/12/ 02/rise_the_timbis? (accessed 15 April 2012).

Goldstone J (201lb) Understanding the revolutions of 2011. *Foreign Affairs* 90(3): 8–16.

Harmel R and Yeh Y-Y (2011) Corruption and government satisfaction in authoritarian regimes: The case of China. *Annual Meeting of the American Political Science Association*, Seattle, USA, 1–4 September 2011.

Heineman B (2011) In China, corruption and unrest threaten autocratic rule. *The Atlantic*, 29 June. Available at: http://www.theatlantic .com/international/archive/2011/06/in-china-corruption-and-unrest-threaten-autocratic-rule/241128/ (accessed 1 September 2011).

Hess S (2009) Deliberative institutions as mechanisms for managing social unrest: Jlie case of the 2008 Chongqing taxi strike. *China: An International Journal* 7(2): 61–77.

Hounshell B (2011) Dark crystal: Why didn't anyone predict the Arab revolutions? *Foreign Policy*, 15 July. Available at: http://www .foreignpolicy.com/articles/2011/06/20/dark_crystal (accessed 15 July 2011).

Huang Y (2008) *Capitalism with Chinese Characteristics: Entrepreneurship and the State*. Cambridge: Cambridge University Press.

Huntington S (1991) *The Third Wave.* Norman: University of Oklahoma Press.

Internet World Stats (2012) Internet usage and telecommunications reports. Available at: http://www.internetworldstats.com/ (accessed 4 January 2012).

Kahn J (2006) Pace and scope of protest in China accelerated in '05 *New York Times,* 20 January. Available at: http://www.nytimes .com/2006/01/20/international/asia/20china.html (accessed 25 October 2010).

King S (2007) Sustaining authoritarianism in the Middle East and North Africa. *Political Science Quarterly* 122(3): 433–59.

Knickmeyer E (2011) The Arab world's youth army. *Foreign Policy,* 27 January. Available at: http://www.foreignpolicy.com/articles/ 2011/01/27/the_arab_world_s_youth_army (accessed 19 August 2011).

Kroll L (2011) Record number of millionaires. *Forbes,* 31 May. Available at: http://www.forbes.com/sites/luisakroll/2011/05/31/record-number-of-millionaires/ (accessed 16 August 2011).

Landry P (2008) *Decentralized Authoritarianism in China.* New York: Cambridge University Press.

Leahy E (2007) *The shape of things to come.* Population Action International Report, 11 April. Available at: http://www.popula tionaction.org/Publications/Report/The_Shape_of_Things_to_ Come/SOTC.pdf (accessed 19 August 2011).

Lee CK (2007) *Against the Law: Labor Protests in China's Rustbelt and Sunbelt.* Berkeley: University of California Press.

Levey S (2011) Fighting corruption after the Arab Spring. *Foreign Affairs,* 16 June. Available at: http://www.foreignaffairs.com/ articles/67895/stuart-levey/fighting-corruption-after-the-arab-spring (accessed 30 August 2011).

Levitsky S and Way L (2010) *Competitive Authoritarianism: Hybrid Regimes after the Cold War.* New York: Cambridge University Press.

Lim B (2007) Thousands protest against S. China chemical plant. *Reuters,* 1 June. Available at: http://www.reuters.com/article/ 2007/06/01/idUSPEKl12258 (accessed 22 February 2011).

Lust E (2009) Competitive clientelism in the Middle East. *Journal of Democracy* 20(3): 122–35.

Lynch M (2011) After Egypt: The limits and promise of online challenges to the authoritarian Arab state. *Perspectives on Politics* 9(2): 301–10.

Maass A (ed.) (1959) *Area and Power: A Theory of Local Government.* Glencoe, IL: Free Press.

McGlinchey E (2009) Central Asian protest movements: Social forces or state resources? In: Wooden A and Stefes C (eds) *The Politics of Transition in Central Asia and the Caucasus.* London: Routledge.

McLaughlin KE (2011) Protests grip China: But is it just the usual springtime unrest? *Global Post,* 17 June. Available at: http://www .globalpost.com/dispatch/news/regions/asia-pacific/china/ l10617/china-protests (accessed 1 September 2011).

Magaloni B (2008) Credible power-sharing and the longevity of authoritarian rule. *Comparative Political Studies* 41(4–5): 715–42.

Masoud T (2011) The upheavals in Egypt and Tunisia: The road to (and from) Liberation Square. *Journal of Democracy* 22(3): 20–34.

Morozov E (2011) *The Net Delusion.* New York: Perseus Books.

Nathan A (2003) Authoritarian resilience. *Journal of Democracy* 14(1): 6–17.

O'Brien K and Li L (2006) *Rightful Resistance in Rural China.* Cambridge: Cambridge University Press.

Orlik T (2011) Unrest grows as economy booms. *Wall Street Journal,* 26 September. Available at: http://online.wsj.com/article/SB1000 1424053111903703604576587070600504108.html (accessed 6 October 2011).

Ottaway M and Hamzawy A (2011) *Protest movements and political change in the Arab world.* Carnegie Endowment for International Peace Policy Outlook, 28 January. Available at: http://carnegieen dowment.org/files/OttawayHamzawy_Outlook_Jan11_Protest Movements.pdf (accessed 25 August 2011).

Pei M (2006) *China's Trapped Transition.* Cambridge, MA: Harvard University Press.

People's Daily (2004) Income distribution reform top concern: Survey among Chinese officials. 30 November. Available at: http://english .peopledaily.com.cn/200411/30/eng20041130_165611.html (accessed 29 August 2011).

Pew Global Attitudes Survey in China (2008) *The Chinese celebrate their roaring economy as they struggle with its costs.* 22 July. Available at: http://pewglobal.org/files/pdf/261.pdf (accessed 1 September 2011).

Przeworski A and Limongi F (1997) Modernization: Theories and facts. *World Politics* 49(2): 155–83.

Ramzy A (2012) Simmering discontent: The biggest challenge to social harmony. *Time,* 7 June. Available at: http://world.time.com/2012/06/ 07/chinas-simmering-discontent-the-biggest-challenge-to-social-harmony/ (accessed 12 June 2012).

Richburg K (2010) In China, unrest spreads as more workers rally. *Washington Post,* 11 June. Available at: http://www.washington post.com/wp-dyn/content/article/2010/06/11/AR2010061102012 .html (accessed 17 February 2011).

Ross M (2011) Will oil drown the Arab Spring? *Foreign Affairs,* September–October. Available at: http://www.foreignaffairs.com/ articles/68200/michael-l-ross/will-oil-drown-the-arab-spring (accessed 11 January 2012).

Shambaugh D (2009) *China's Communist Party: Atrophy and Adaptation.* Berkeley: University of California Press.

Shih V, et al. (2012) Getting ahead in the Communist Party. *American Political Science Review* 106(1): 166–87.

Silitski V (2006) *Contagion deterred: Preemptive authoritarianism in the former Soviet Union (the case of Belarus).* CDDRL Working Paper No. 66. Available at: http://iis-db.stanford.edu/pubs/21152/ Silitski_No_66.pdf (accessed 31 December 2011).

Solinger D (2008) Business groups: For or against the regime? In: Gilley B and Diamond L (eds) *Political Change in China: Comparisons with Taiwan.* Boulder, CO: Lynne Rienner.

Straits Times (2010) Dozens hurt in factory clash. 10 June. Available at: http://www.straitstimes.com/BreakingNews/Asia/Story/STI Story_538209.html (accessed 17 February 2011).

Sun I and Wuy (2009) The role of the People's Armed Police in Chinese policing. *Asian Criminology* 4(2): 107–28.

Tarros (1994) *Power in movement: social movements, collective action, and politics.* Cambridge, UK: Cambridge University Press.

Thompson M (2001) To shoot or not to shoot: Post-totalitarianism in China and Eastern Europe. *Comparative Politics* 34(1): 63–83.

Tilly C (1993) Contentious repertoires in Great Britain, 1758–1834. *Social Science History* 17(2): 253–80.

Tilly C and Tarrow S (2007) *Contentious Politics.* Boulder, CO: Paradigm.

Tong Y and Lei S (2010) Large-scale mass incidents and government responses in China. *International Journal of Chinese Studies* 1(2): 487–508.

Tosun MS and Yilmaz S (2008) *Centralization, decentralization, and conflict in the Middle East and North Africa.* World Bank Policy Research Working Paper 4774 (November). Washington, DC: World Bank.

Transparency International (2011) Corruption Perceptions Index: 2010 results. Available at: http://www.transparency.org/policy_research/surveys_indices/cpi/2010/results (accessed 19 August 2011).

Tsai K (2007) *Capitalism Without Democracy.* Ithaca, NY: Cornell University Press.

Ulfelder J (2005) Contentious collective action and the breakdown of authoritarian regimes. *International Political Science Review* 26(3): 311–34.

Ulfelder J (2011) Crystal clear: Yes, rows of numbers can help predict revolutions. You just have to know where to look. *Foreign Policy,* 22 June. Available at: http://www.foreignpolicy.com/articles/2011/06/22/crystal clear (accessed 15 April 2012).

United Nations Development Program (2010) International Human Development Indicators. 4 November. Available at: http://hdr.undp.org/en/statistics/ (accessed 19 August 2011).

United Nations Development Program, Program on Governance in the Arab Region (2009) Country theme, local government. Available at: http://www.pogar.org/ (accessed 21 July 2011).

US Energy Information Administration (2011) Top world oil producers: 2010. Available at: http://www.eia. gov/countries/index.cfm (accessed 11 January 2012).

Wang S (2006) Openness and inequality: The case of China. In: Dittmer L and Liu G (eds) *China's Deep Reform: Domestic Politics in Transition.* Lanham, MD: Rowman and Littlefield, pp. 252–8.

Way L (2008) The real causes of the color revolutions. *Journal of Democracy* 19(3): 55–69.

Whyte M (2010) Fair versus unfair: How do Chinese citizens view current inequalities? In: Oi JC, Rozelle S and Zhou X (eds) *Growing Pains: Tensions and Opportunity in China's Transformation.* Stanford, CA: Shorenstein Center.

Work R (2002) Overview of decentralization worldwide: A stepping stone to improved governance and human development. In: *2nd International Conference on Decentralization,* Manila, Philippines, 25–27 July.

World Bank (2008) Decentralization in client countries. Available at: http://www.dpwg-lgd.org/cms/upload/pdf/WB-Dez.pdf (accessed 30 April 2012).

World Bank (2011) Fiscal decentralization indicators. Available at: http://www.worldbank.org/publicsector/decentralization/fiscalindicators.htm (accessed 4 October 2011).

World Bank (2012) World Development Indicators: 2011. Available at: http://data.worldbank.org/ (accessed 12 April 2012).

Zhao L and Huang Y (2010) *Unemployment problem of China's youth.* East Asian Institute Background Brief No. 523, 28 April. Singapore: East Asian Institute.

Zhao S (2011) The China model and the authoritarian state. *East Asia Forum,* 31 August. Available at: http://www.eastasiaforum.org/2011/08/31/the-china-model-and-the-authoritarian-state/ (accessed 1 September 2011).

Zong H (2002) *Disidai (The Fourth Generation).* New York: Mirror Books.

Steve Hess is Assistant Professor of Political Science at the University of Bridgeport's International College. His research, which focuses on contentious politics and authoritarian governance in Eastern and Central Asia as well as Chinese foreign policy, has previously appeared in *Asian Survey, Central Asian Survey,* and *Problems of Post-Communism.*

REVIEW QUESTIONS

1. What is the primary factor that allows the Chinese government to retain power?

2. Why did scholars conclude, wrongly, that Tunisia's and Egypt's regimes were durable, only to see them collapse?

3. Is there a notion of middle-class discontent in China?

4. In this section, one of the predictions on the future is that China will shrink and stop growing. If that occurs, should the Chinese government worry, based on this article? Why or why not?

GLOSSARY

Abu Sayyaf Group: A violent, Islamic, al Qaeda–linked terrorist group operating in the Philippines. It split from the Moro National Liberation Front and cooperates at times with Jemaah Islamiyah from Indonesia. It uses kidnapping for ransom, bombings, assassinations, and extortion for financial gain and to further its jihadist aims for an Islamic state in the region.

Action Direct: Left-wing terrorist group operating in France in the 1970s and 1980s.

Acquisition: The ability of terrorist groups to acquire or gain control of the materials needed for a weapon of mass destruction.

Afghanistan: Country reference. See map in Section 8.

Africa: Geographic reference. See map in Section 6.

Algeria: A country in North Africa. See map in Section 6

Allied Powers: The alliance of Britain, France, the United States, Russia, Japan, and Italy that defeated Germany and its ally, the Ottoman Empire, in World War I.

Al Qaeda: "The base," a terrorist organization founded by Abdullah Yusuf Azzam in Afghanistan along with Osama bin Laden to funnel manpower and funds to the mujahedeen battling the Soviet Union.

Al Qaeda affiliates: Groups around the world that follow the ideology of al Qaeda and may engage in some coordination with the original group, but are not under its direct control. Some initial affiliates, such as the Islamic State of Iraq and Syria have since been repudiated by al Qaeda for being too brutal.

Al Qaeda in the Arabian Peninsula (AQAP): Al Qaeda affiliate in Yemen.

Al Qaeda in the Lands of the Islamic Maghreb: Al Qaeda affiliate in Algeria.

Al-Shabbab: Islamic fundamentalist group operating in Somalia; conducts many of its attacks in neighboring Kenya.

Anarchism: A political ideology in the 19th century seeking to overthrow the existing system of governance and replace it with one governed by natural law rather than man-made institutions.

Animal Liberation Front (ALF): A leaderless, cell-like organization allied with People for the Ethical Treatment of Animals (PETA), engaged in terror acts to stop the exploitation of animals, specifically animal testing of products.

Anti-terrorism, Crime and Security Act of 2000: An anti-terrorism law in Great Britain that became the first permanent statute to deal with terrorism. It included domestic terrorism and greatly broadened the definition of what constitutes terrorism. It also gave authorities greatly expanded power to search and detain suspects.

Armed Islamic Group: Group formed in 1992 in Algeria. Its goal was to overthrow the Algerian government and replace it with an Islamic state.

Assassins: Muslim terrorist group active in Persia and the Middle East from approximately 1090 to 1272. Its goal—to purify the practice of Islam—dictated the nature of its attacks and led it to become one of the earliest religious/suicide groups, similar to Hamas and al Qaeda of today.

Aum Shinrikyo: A cult-like organization in Japan that conducted a savage attack on the Tokyo subway system using sarin, a dangerous nerve gas. Eventually all involved were captured and prosecuted.

Ayman al-Zawahari: Current leader of al Qaeda. Succeeded to the position after bin Laden's death.

Bacteria: Living organisms that are present in nature and capable of living outside the human body. Some of the more well-known bacteria that pose a bioterrorism concern include typhoid, brucellosis, plague, tularemia, and anthrax.

Basque Fatherland and Liberty (ETA): Long-standing terrorist group operating out of the Basque region of Spain and with the goal of creating an independent Basque state.

Basques: Ethnic group living in the mountains on the border of France and Spain with their own distinct language and culture.

Belfast Agreement: A 1998 agreement between Britain and Northern Ireland that established shared governance over Northern Ireland and revamped Northern Ireland's criminal justice system, which had been plagued with abuses and overzealousness throughout the "Troubles."

Beslan: Town in Russia where Chechen terrorists took a school hostage in 2011. Almost 300 people were killed, most of whom were children.

Bhutto, Benazir: One-time civilian prime minister of Pakistan.

bin Laden, Osama: Founder and leader of al Qaeda.

Bioengineering: The process of modifying or creating bacteria and other organisms in a laboratory. The vast majority of bioengineering efforts focus on designing new cures for serious diseases and making full use of new technologies to develop effective vaccines and treatments for a variety of health concerns. This technology in the hands of terrorists or rogue states could allow such groups to more easily conduct a biological terrorist attack using bioengineering to design strains of diseases that are resistant to current treatments or that combine various pathogens to create more lethal and harder-to-treat agents.

Biological agents: Dangerous bacteria used to create weapons of mass destruction.

Black Panthers: Founded by Huey Newton and Bobby Seale in Oakland, California, in 1966, originally called the Black Panther Party for Self-Defense. They were a Marxist organization hoping to establish Marxism using mobilizing communities to rebel against the established government. With aggressive policing by the federal government, their efforts were quickly checked.

Black Tigers: The suicide bombers belonging to the Liberation Tigers of Tamil Elam (LTTE).

Boko Haram: Particularly violent Islamic fundamentalist group operating in Northern Nigeria. Its goal is to overthrow the Nigerian government and create an Islamic state over the areas in its control. This group is most famous for its 2014 abduction of 200 school girls.

Centers for Disease Control and Prevention (CDC): The main U.S. public health agency in the event of a bioterrorist attack or release of an infectious disease.

Central Powers: The alliance of the Ottoman Empire, Germany, and Austria-Hungary during World War I.

Chechnya: Region of Russia in the Caucasus Mountains that has been agitating for independence since the days of Imperial Russia.

Chemical agents: Chemical substances used to make chemical weapons.

Chemical Facility Anti-Terrorism Standards in the Code of Federal Regulations (CFATS): Regulations that provide a process whereby facilities submit information and security plans to the Department of Homeland Security, which reviews and approves the plans, and then inspects their implementation.

Code of Federal Regulations (CFR): Collection of all federal regulatory provisions that implement federal legislation and policy.

Communist Combatant Cells: Left-wing terrorist group that briefly operated out of Belgium in the 1980s.

Comprehensive conventions: International conventions that cover a wide array of issues or behaviors by states and even non-state actors.

Conventional terrorism: Attacks using conventional explosives or other devices or tactics that do not depend on chemical, biological, radiological, or nuclear agents.

Conventions: International agreements among a number of sovereign states.

Critical infrastructure protection: The protection of key assets and industries—such as power generation, transportation, and computer systems—vital to the day-to-day operation of a state and the daily lives of its citizens.

Cybersecurity: Measures taken to protect computers or computer systems from attack, compromise, or hacking. This is a relatively new term, originating in 1994, needed because of the burgeoning Internet.

Databases: Large collections of data stored for rapid search and retrieval on a computer, usually a server or servers; much of the information is sensitive, such as social security numbers, medical records, and so on, necessitating cybersecurity, a term and a job unheard of a couple of decades ago.

Delivery: The ability of a terrorist group to use a weapon of mass destruction in a way that causes the most casualties and spreads over the largest area.

Dozier, James: Deputy Chief of Staff of NATO's Southern European headquarters in Italy. General Dozier was kidnapped by the Red Brigades and held for 42 days until he was rescued by a specially trained counterterrorism squad.

Department of Homeland Security (DHS): Established in 2002, combining 22 different federal agencies with a role in providing security for the United States, it has grown to be the third largest federal agency with 240,000 employees.

Director of National Intelligence (DNI): This position was recommended following the 9/11 attacks on the United States that demonstrated that the 17 various intelligence agencies failed to cooperate well together. The Director of National Intelligences lacks the authority that the Director of Homeland Security enjoys, which some consider problematic.

Earth Liberation Front (ELF): An autonomous cell-like organization dedicated to protecting the environment and the Earth. With religious-like fervor the movement, which began in Great Britain, has spread internationally, with actions designed to preclude activities such as logging.

Egypt: Country reference. See map in Section 6.

Egyptian Islamic Jihad (EIJ): A group that merged with al Qaeda in 2001 through marriage and ideology. EIJ's leader, Ayman Al-Zawahari, became bin Laden's number two in the organization and succeeded bin Laden upon bin Laden's death.

Enemy combatants: Members or associates of the Taliban or al Qaeda—whether U.S. or non-U.S. citizens—caught in Afghanistan and Pakistan and often held at the U.S. Naval facility in Guantanamo Bay, Cuba.

Executive orders: Orders issued by the president that can have the force of law. They are designed to promote or implement policies that do not require Congressional legislation or approval.

Fatah: The major political party of the Palestinians, founded in 1965 as the Palestinian National Liberation Movement. Fatah is a reverse acronym for "conquest by means of jihad"; Fatah's flag shows a grenade and crossed rifles superimposed on a map of Israel. Many offshoots of Fatah have been involved in terrorism. Despite this, the United States has been supportive, though there is a current dilemma and the current Fatah charter does not mention Israel. After much conflict with Hamas, the two parties reconciled in 2015. Hamas is a designated foreign terrorist group, according to the United States.

Fatwa: A finding by an expert in Islamic religious law concerning a dispute. Osama bin Laden issued two aimed at mobilizing Muslims around the world to target the West and the United States.

Federal Emergency Management Agency (FEMA): A federal agency once known for being effective and nimble, absorbed by the Department of Homeland Security just prior to the disastrous response to Hurricane Katrina.

Federally Administered Tribal Area: Area in North Pakistan, bordering Afghanistan, mostly controlled by the Taliban. Only limited Pakistani government resources or authority are present in the area.

Fifth Wave of Terrorism Theory: A theory by Jeffrey Kaplan that posits that terrorism will continue, particularly in some areas of Africa, and will have as a goal the destruction of groups of people that are unlike their aggressors.

Fissile material: A type of radioactive isotope. Radioactive isotopes are drawn from elements having the same number of protons but different numbers of neutrons. Fissile material is a necessary component of nuclear weapons.

Four Waves of Terrorism: The organizing principle set out by David Rapaport to categorize terrorism into four eras from the 19th century to the present.

Freedom Fighter: A term fraught with difficulty and subjectivity, depending on the individual, the cause, supporters, politics, and ideological beliefs. Terrorists, believing they are correct, would regard themselves as freedom fighters.

Gama'a Islamiyya (IG): Indigenous to Egypt, IG was formed in the 1970s as a loosely organized group whose goal it was to create an Islamic state, replacing the sovereign state of Egypt. Its most famous attack took place against a major tourist attraction when IG bombed an Egyptian archeological site in 1997.

Golden Temple: The holiest site in the Sikh religion, it is located in the city of Amritsar. In 1984, radical Sikhs calling for independence for the Punjab region took refuge in the temple. The harsh response by Indian troops led to the assassination of Prime Minister Indira Gandhi, which was followed by four days of anti-Sikh rioting.

Great Terror: A time during the French Revolution when members of the nobility and non-noble political opponents of Robespierre and the Committee for Public Safety were sent to the guillotine.

Grenzschutzgruppe 9 (GSG 9): Specially trained counterterrorism force in Germany formed after the attack on the Munich Olympics in 1972. Still in operation today and is one of the most respected of such groups worldwide.

Hamas: A Sunni political party particularly effective in the Gaza Strip, a designated Foreign Terrorist Organization, pioneer in the use of suicide bombings during Intifadas (uprisings in Palestinian Territories and Israel). Currently it works jointly with the Fatah party in the Palestinian Territories.

Haqqani Network: One of the most active Taliban affiliates; its leaders have operated from Pakistan since early in the Afghan war.

Hezbollah: A Shia political party based in Lebanon; it is designated a Foreign Terrorist Organization. It is an effective social services agency, a government within a government, and often perpetrates acts of terror and war in the service of Iran.

Homeland Security Act of 2002: A 187-page bill that created the Department of Homeland Security. It covers the organization, scope, and functions of the department. What it failed to do was to preclude excessive oversight by Congress with over 100 committees and sub-committees having jurisdiction. This creates havoc for the agency, but little has been done to resolve this.

Hussein, Saddam: Ruler of Iraq from 1979 until overthrown by the U.S. invasion in 2003.

Ideology: A set of opinions or beliefs that guides an individual or group, usually strongly held.

Improvised Explosive Device(s) (IED): An effective but often crude explosive device, containing a trigger, an initiator or fuse, explosive material, and a power source in a container. One of the authors has seen examples in Israel that ranged from a bicycle, to a flower pot, to a fiberglass rock.

Indefinite detention: A term that describes the results of a battle in the U.S. Congress. Citizens and non-citizens may be indefinitely detained if deemed a threat, such as being affiliated with al Qaeda or a similar group. Other countries, such as Great Britain, have also used this, though it is controversial, and may apply to not only terror suspects but also to immigrants. As of 2015, Great Britain remains the only country with this policy but it is increasingly being criticized and will review this.

India: Country reference. See map in Section 8.

Inter-Services Intelligence Directorate (ISI): Pakistan's intelligence service, which has worked closely with the United States since 2001. Many officials believe the ISI has continued to support Islamic fundamentalist groups such as the Taliban on both sides of the Pakistan border for its own purposes.

Iran: Country reference. See map in Section 6.

Iraq: Country reference. See map in Section 6.

Islamic State of Iraq and Syria (ISIS): Islamic fundamentalist group that began life as al Qaeda's affiliate in Iraq but later moved to Syria, was repudiated by al Qaeda for its brutality, and began to seize major cities and territories in Iraq and Syria in 2014.

Islamization: The evolution of the Chechen independence movement from a nationalist movement to an Islamic fundamentalist one, partly based on the direct involvement of al Qaeda.

Jaish-e-Mohammed (JEM): Group founded by an Islamic extremist by the name of Masood Azhar following his release from prison in India in early 2000. The aim of JEM is to unite Kashmir and Pakistan and expel all foreign troops from Afghanistan.

Jannah, Mohammed Ali: Founder of the state of Pakistan as part of the breakup of Britain's Indian Empire.

Japanese Red Army: A far-left terrorist group in Japan, founded by Fusako Shigenobu who then went to Lebanon in 1971 with some members and founded the Palestinian branch. It was never large, but it had close ties to the Popular Front for the Liberation of Palestine. Notably brutal, it killed several of its own members in a training camp in the mountains before police raided the place. The founder helped plan an attack on the Lod airport in Tel Aviv in 1972. She is serving prison time, as are most of the other members.

Jihadists: Muslims who advocate or participate in jihad; *jihad* means to struggle and this is where the difficulty begins. Moderate Muslims claim jihad is simply a personal inner struggle to be a good person. Radical Muslims claim jihad means they must secure the entire world as a Muslim enclave.

Karzai, Hamid: First president of Afghanistan after the overthrow of the Taliban by U.S. and Northern Alliance forces.

Kashmir: Region claimed by both India and Pakistan. See map in Section in 8.

Khmer Rouge: A radical communist movement that came to power in Cambodia and quickly degenerated into a genocidal one with 1.5 million to 2 million persons dead; most of those murdered were from professional, technical, or religious groups or wore glasses.

Ku Klux Klan (KKK): Founded in the United States in 1866, the Klan soon became prominent in most Southern states, as a check against federal power that members believed was punishing the South. The Klan's outrages were such that the Congress passed legislation that limited its power in the late 1870s. It did have two major revivals, following WWI and after the civil rights movement in the late 1960s. It still exists but as a minor group.

Kurdish Workers Party (PKK): Name once used by a group founded in 1978 by Abdullah Olacan. Its goal was a Marxist-Leninist state for the Kurdish people situated in the region shared by Turkey, Iran, Syria, Armenia, and Iraq. Today it is known as the Kongra Gel.

Kurds: Ethnic group whose geographic origins span the mountainous region that forms the border between and encompasses Turkey, Iran, Iraq, and Syria. Turkey is home to almost half of the world's 30 million Kurds.

Lashkar-e-Tayibba (LeT): Islamic fundamentalist group originally operating in Kashmir. It is believed responsible for the 2006 attack against Western targets in Mumbai, India, and for planning further attacks in the United States and the West.

Liberation Tigers of Tamil Elam (LTTE): Known as the Tamil Tigers. One of the longest operating separatist groups, it tried without success to create an independent state of Tamil in Northern Sri Lanka. It is known for its suicide attacks and its use of women fighters.

Libya: Country reference. See map in Section 6.

Marxism-Leninism: A political ideology that combines the broad socialist theories of Marx with the application of those theories by Lenin in his efforts to transform Russia into a communist state. In the 20th century, this philosophy, in part, spawned several left-wing terrorist groups operating in Western Europe.

Middle East: Regional reference. See map in Section 6.

Moro, Aldo: Former prime minister of Italy who was captured and killed by the Red Brigades.

Mine Resistant Ambush Protected (MRAP): Vehicles requested by the Department of Defense that are able to withstand IEDs. Several vendors produced versions of this vehicle, costing anywhere from $500,000 to a million dollars. Approximately 12,000 were manufactured. They did work; now a number of U.S. police departments have requested them.

Mujahedeen: Insurgent guerrilla groups founded in response to the Soviet invasion of Afghanistan in 1979 to harass and drive out the Soviets. The *mujahedeen* believed they were answering a call to jihad against the godless enemies of Islam. These groups were often funded and armed by a Cold War–driven United States, which saw a communist-run Afghanistan as the worst possible outcome.

Mumbai: City in India that was the site of simultaneous attacks against several Western targets throughout the city by members of the terrorist group Lashkar-e-Tayibba (LeT) in 2006. Former name was Bombay.

Muslim Brotherhood: A group founded in Egypt in 1928 by Hassan al Banna to return purity to the practice of Islam and to create a single Muslim nation, originally through education and reform. It has been at various times both a terrorist group and a "school of thought" that has influenced the development of Islamic groups all over the world with varying levels of radicalization.

Muslim National Liberation Front: An Islamic terrorist group in the Philippines. It merged with the Moro Islamic Liberation Front in 2007. In 2014 it signaled that it would accept a peace accord with the Philippine government that gives them some autonomy.

Nabka: Palestinian term meaning "catastrophe"; refers to the day the Palestinians fled prior to the 1948 war between Israel and a number of Arab countries.

National Counterterrorism Center (NCTC): Established by Presidential Executive Order 13354 in August 2004, later codified by the Intelligence Reform and Terrorism Prevention Act of 2004. It is charged with encouraging interagency cooperation and tracks terrorists and terrorist groups.

National Liberation Army, Colombia (ELN): A Marxist terrorist group in Colombia more centered in urban areas than the larger group, FARC. It opposes its government, despises the United States, and uses drug production and terror to further its goal of a Marxist state. It has been in decline recently; the current Colombian government hopes to reach a solution with it soon. It is designated as a terrorist organization by the U.S. Department of State.

National Liberation Front (FLN): Group formed in Algeria in the 1950s to target French officials, police, and military posts. After few results, it turned to terrorizing and killing French civilians with the aim of independence from French control. It's best known for its brief control of the main city of Algiers and the battle between the French forces and FLN fighters that ensued.

National Security Agency: Established in 1954 by President Harry Truman to continue advances into signal intelligence and code-breaking that greatly aided the United States and Allies in World War II.

National Security Special Event: In the Presidential Protection Act of 2000, language was included that allowed the Secret Service to become the lead agency. Now the Director of Homeland Security may designate an event a National Security Special Event with the Secret Service in charge of security and the FBI in charge of intelligence.

Naxalites: Popular term for what has been known since 2004 as the Communist Party of India. It follows the Maoist version of that ideology and targets police stations, factories, Indian government officials, and multinational corporations.

New terrorism: The terrorism faced today, often with a religious motivation, and the accompanying fear that the Internet and access to weapons of mass destruction make it more deadly. This term is controversial and many suggest it is not helpful.

Northern Alliance: Collection of militias and tribal forces opposed to Taliban control of Afghanistan. Its leader was killed just days before the U.S. invaded Afghanistan in 2001 with the goal of overthrowing the Taliban and capturing Osama bin Laden.

November 17: Left-wing group operating in Greece beginning in the 1970s. November 17 sought three goals: (1) an end to the U.S. military presence in Greece; (2) the removal of the Turkish presence in Cyprus; and (3) an end to Greece's ties to the North Atlantic Treaty Organization (NATO) and the European Union (EU).

Objective legal environment: Actual legislation, regulations, and judicial decisions dealing with a particular policy area or issue.

Office of the Director of National Intelligence (ODNI): The centers, offices, and projects or the various intelligence agencies "managed" by the Director of National Intelligence. They work independently and collaboratively.

Omar, Mullah: Leader of the Afghan Taliban. He harbored Osama bin Laden and al Qaeda until the U.S. invasion in 2001.

Oslo Accords: A 1993 accord between Israeli and Palestinian leaders that generally saw Israel agree to return some land, Gaza and Jericho, for Palestinian recognition of Israel's right to exist. It began a lengthy process of negotiation but ended in 2002 with attacks forcing then-Prime Minister Ariel Sharon to declare that further progress was impossible.

Ottoman Empire: Muslim empire that dominated the Middle East, Eastern Europe, and parts of the Mediterranean until defeated by British, French, Italian, and American forces during World War I.

Pakistan: Country reference. See map in Section 7.

Palestinian Authority: Governing body of the Palestinian territories, currently led by Fatah and recently joined by Hamas. Its most recent effort is to get the United Nations to recognize the 1967 borders that existed prior to that war and have Israel leave in two years.

Pan American Flight 103: American airliner downed by Libyan agents while traveling from London to New York. A bomb made of a plastic explosive was placed in checked luggage. All 259 passengers and crew on board the aircraft were killed along with 11 people on the ground in Lockerbie, Scotland, when the plane exploded above the town.

Perceived legal environment: The understanding by practitioners in a particular policy area of how the objective legal environment should be interpreted. Recent research found a difference between the objective legal environment and the perceived legal environment in the public health arena.

Persia: Once one of the dominant empires in the world; currently known as Iran.

Plantation of Ulster: Established in the late 16th/early 17th century, the Plantation of Ulster allowed English settlers to take over the most prosperous agrarian section of Ireland, the North, and push out the Irish peasants. This move set the tone for a division of territory that ran along religious as well as geographic lines.

Plutonium: A radioactive element that does not occur in nature but is manufactured. Plutonium isotope 239 (PL-239) is the type of plutonium most often used in nuclear weapons. The facilities, money, expertise, and resources needed to produce plutonium are probably far outside the capability range of non-state actors.

Posse Comitatus Act of 1878: Prohibits the use of the U.S. military in domestic law enforcement activities.

Prabhakaran, Vellupillai: Leader of the Tamil Tigers (see Liberation Tigers of Tamil Elam).

Precursor chemicals: Chemicals that when used separately are harmless, or even beneficial, but in combination with others can be used to make chemical weapons.

Presidential Directive: A type of executive order.

Provos: Name often used to refer to the Provisional Irish Republican Army and its offshoots. Their goal was to create a Catholic state in Northern Ireland and to end British control of the region. They also engaged in a religious struggle with the Protestant residents of the region.

Qadhafi, Muammar: Former dictator of Libya, deposed by Libyan militias, with the help of U.S. and Western airstrikes in 2011. He fled to his home town and was killed by a mob while trying to escape.

Qutb, Sayyid: Founder of a particularly militant wing of the Muslim Brotherhood. He espoused violence to achieve the goals of the Brotherhood. He was jailed by the Egyptian government for attempting to overthrow the government in 1954 and was hanged in 1966.

Radiological agents: Agents used to create radiological weapons, which are weapons based on conventional explosives but designed to spread radioactive material.

RAND Database of Worldwide Terrorism Incidents: A database created by the RAND Corporation that ranges from 1968 to 2009; it is interactive, fully searchable, and comprehensive.

Real ID: An effort, in response to the 2005 REAL ID Act, that attempts to set standard federal guidelines for the issuance of driver's licenses in an effort to ensure their validity and security. It was spurred in part by the fears of terrorist sleeper cells hiding in U.S. suburbia, but also by broader concerns over U.S. immigration policy.

Real Irish Republican Army (RIRA): Group that claims to be the successor to the old Provisional Irish Republican Army (see Provos) and has engaged in some relatively minor attacks in Northern Ireland. In 2011, the RIRA was linked to at least seven attacks on Northern Irish office buildings, government facilities, banks, and the Police Service of Northern Ireland (PSNI) targets using improvised explosive devices (IEDs).

Red Army Faction: Left-wing group operating out of Germany beginning in the early 1970s. It was often called the Baader Meinhof Gang after two of its founders. The group was known for bombings, assassinations, and kidnappings and had its hand in a number of criminal enterprises used to fund its activities. A Marxist-Leninist group, it targeted both U.S. and German targets as part of its opposition to the ills of capitalism.

Red Brigades: Left-wing group operating in Italy from the late 1960s through the 1990s.

Regional conventions: International conventions where the signatories are from the same region of the world.

Religion: The belief in and worship of a superhuman controlling power, especially a personal God or gods. The heart of many conflicts is religion.

Remote Biometric Identification (RBI): Biometrics are measurements of the body, such as a fingerprint or a DNA sample. Remote Biometric Identification means that someone, such as the FBI, can do this using facial recognition and tracking movements through Closed Circuit Television, as an example. This is an area that Congress has largely ignored.

Revolutionary Armed Forces of Colombia (FARC): Large Marxist terror group in Colombia, operating primarily in rural areas, often controlling large areas of the country. They oppose intervention in the country by the United States, desire a Marxist government, and fund their operations with drugs, kidnappings, ransom, and extortion. The government of Colombia is currently negotiating with them.

Risk assessment: Uses quantitative, technologically advanced analysis of actual risks from disasters or crisis. The unpredictability of a terrorist attack makes risk assessment difficult.

Risk perception: The subjective, intuitive feelings individuals hold as to their own personal safety, whether accurate or not.

Royal Ulster Constabulary (RUC): Created by the British and consisting mostly of Northern Irish Protestants. The RUC was meant to serve as the police force for Northern Ireland, but it was soon universally hated by the Catholic residents of the region for its harsh and abusive tactics. It was a favorite target of the Provos.

Sarin: A manmade toxin originally developed as a pesticide in Germany in 1938.

Saudi Arabia: Country reference. See map in Section 6.

Saud, Ibn: Leader of the Saud family who united the tribes of the Arabian Peninsula and became the first king of modern Saudi Arabia.

Sectoral conventions: International conventions designed to deal with a specific policy area or problem, such as conventions against terrorist acts against aviation.

Sikhs: The Sikhs are a 500-year-old religious group that embodies aspects of both Islam and Hinduism. They reside mostly in the Punjab region of India. Radicalized Sikhs have used terrorist attacks and violence to try to gain independence for the region.

Sri Lanka: Country reference. See map in Section 8.

State sponsor of terrorism: Country that has repeatedly provided support for international terrorism is designated by the State Department as a state sponsor. Currently these include Cuba, Sudan, Iran, and Syria.

Sudan: Country reference. See map in Section 6.

Suicide bombing: A bomb attack by a terrorist knowing he or she will be killed in the attack. Also called homicide bombing.

Sovereign citizen movement: According to the FBI, these are ant-government extremists who do not believe they have to answer to the federal government. The FBI raises the concern that law enforcement officials may be harmed when stopping an individual or individuals who are members of the movement.

Taliban: Islamic fundamentalist movement consisting of a number of militias that gained control of Afghanistan after the civil war that followed the expulsion of the Soviet forces. While no longer in control of Afghanistan, the Taliban still operates in both Afghanistan and Pakistan and is considered a significant threat to both governments.

Tehrik-e-Taliban Pakistan (TTP): Composed of various militant tribes, the TTP was formed in 2007 by Baitullah Mehsud to oppose the Pakistan military in the Federally Administered Tribal Areas. The TTP is also suspected of involvement in the death of Prime Minister Benazir Bhutto and may have directed a plot to bomb Times Square in New York City in May 2010.

Terrorism: There are hundreds of definitions, but most have the following elements: It is an act of violence, against innocents, with a political goal, carried out to capture media attention, with the purpose of instilling fear.

Thuggees: A group operating in India from around the 7th century to the 19th century that targeted and strangled travelers as a sacrifice to the Hindu Goddess of Death.

Toxins: In contrast to bacteria and viruses, toxins are not considered living organisms. Instead, they are non-living chemical substances produced as by-products of living organisms, such as the venom produced by rattlesnakes. Botulism and ricin (sometimes used in assassinations) are also part of the toxin family.

Transnational terrorism: Terrorism by groups whose ideology, motivations, and operational base are not tied to one country, but may span several countries or even regions.

Transportation Security Agency (TSA): Created following 9/11 as part of the Department of Homeland Security. It is in charge of protecting the entire transportation system of the United States. Most believe it is simply involved with airline security but it also covers rail systems and ports.

Treaties: Agreements between or among countries. Treaties among large numbers of countries are often called Conventions.

The "Troubles": Term used to refer to the violence between Catholics and the British-backed Protestants in Northern Ireland that began in 1969 and ran through to the early 21st century.

Túpac Amaru Revolutionary Movement, Peru (MRTA): A small, traditional Marxist-Leninist group with a goal to overthrow the leadership of Peru. It was not very effective but it did manage to take over the Japanese ambassador's residence in Lima, Peru, in 1996 and hold 72 persons hostage for four months until they were rescued by Peruvian forces in April of 1997; the rescue resulted in the deaths of all 14 MRTA members and one hostage.

TWA Flight 847: American airliner captured by Hezbollah terrorists in 1985 while traveling from Athens to Rome. American hostages were held in Beirut for over two weeks before finally being released. A Navy diver on the plane—Robert Stethem—was killed.

Typologies of terrorism: A means to classify and hopefully find meaning in determining the rationale and motives of various types of terrorist individuals and groups.

Ul-Haq, Zia: Pakistani general who took control of Pakistan in 1977. He immediately instituted Islamic law and revised the educational curriculum to promote an Islamic national identity. This all-encompassing move from the secular to the religious might have sparked a rebellion among Pakistan's secular elite but for the Soviet invasion of Afghanistan.

United Self Defense Force, Colombia (AUC): A para-military organization in Colombia originally formed to counter Marxist terrorist groups. It rapidly became equally opportunistic, engaging in the drug trade while fighting the left-wing insurgents. Following the death of its founder in 2004, the group disbanded in 2006.

Universal conventions: Conventions that involve a large number of countries spanning multiple regions or even continents.

Uranium: A naturally occurring radioactive element; the ore contains 7% uranium. It is enriched, a process that enhances or increases the amount of uranium relative to other materials in the ore. Isotope U-235 is the most commonly used for weapons.

USNORTHCOM: United States Northern Command. Formed after 9/11 to bolster and better coordinate U.S. homeland defense capabilities.

USS *Cole*: U.S. naval vessel bombed by al Qaeda while moored to a fuel dock in the harbor of Yemen's capital in 2000.

Vehicle Borne Improvised Explosive Device (VBIED): Similar to an IED but using a vehicle of some sort that can carry a much more explosive material. This would have an effective explosive device, containing a trigger, an initiator or fuse, explosive material, and a power source, and the container would be a vehicle. As an example, a truck bomb was the source of the explosive at the 1983 Marine Barracks bombing in Lebanon that killed 241 personnel.

Viruses: Viruses, though considered living organisms, cannot live outside a host (plant, animal, or human) and are dependent on their ability to replicate inside a host for their long-term survival.

Wahhabism: Rigid and conservative interpretation of Islam that controls religious life in Saudi Arabia. The Saud family is often blamed for the spread of radical Islamic terrorism because of its use of its financial resources to spread this version of Islam worldwide.

Weaponization: The ability of a terrorist group to modify a bacterial, chemical, or nuclear agent into its most lethal form before delivery.

Weather Underground: A small but violent student radical group that managed to bomb the U.S. Capitol and planned to bomb military headquarters.

Yemen: Country reference. See map in Section 6.

Zealots: A Jewish terrorist group operating around 70 A.D. in an effort to gain independence from Rome. In addition to Roman officials, the group also targeted Jewish leaders it believed to be oppressive or corrupt.

REFERENCES

"Artimeo" have founded the Movadef. (2013, June 5). Retrieved July 4, 2013, from elcomercio.pe: http://elcomercio.pe/actualidad/1585464/noticia-terroristaartemio-habria-fundado-movadef

18 USC § 1385—Use of Army and Air Force as posse comitatus. (n.d.). Retrieved August 13, 2013, from www.law.cornell.edu: http://www.law.cornell.edu/uscode/text/18/1385

1991: Bomb kills India's former leader Rajiv Gandhi. (n.d.). Retrieved from *BBC: On This Day:* http://news.bbc.co.uk/onthisday/hi/dates/stories/may/21/newsid_2504000/2504739.stm

40 killed; Shining Path guerrillas shut down much of Lima. (1992, July 26). *New York Times.*

63 lynchings in 1921; Tuskegee Institute gives 4,096 as total of mob victims since 1885. (1922, January 1). *New York Times.*

9/11 Commission Report. (2004). *9/11 Commission Report.* Retrieved June 26, 2013, from Ch. 8: http://govinfo.library.unt.edu/911/report/911Report_Ch8.htm

9/11 Commission, The. (2004, July 22). *9/11 Commission report,* The Retrieved June 20, 2013, from *govinfo.library:* http://govinfo.library.unt.edu/911/report/911Report_Exec.pdf

Abdullah Sungkar. (n.d.). Retrieved July 12, 2012, from GlobalJihad.net: http://www.globaljihad.net/view_page.asp?id=533

Abel, T. (1951). The sociology of concentration camps. *Social Forces, 30*(2), 150–155.

Abrams, N. (2008). *Anti-terrorism and criminal enforcement* (3rd ed.). St. Paul, MN: Thomson-West.

ACLU. (n.d.). *Al-Aulaqi v. Panetta: Lawsuit challenging targeted killings.* Retrieved August 23, 2013, from aclu.org: https://www.aclu.org/national-security/al-aulaqi-v-panetta

Agape International. (n.d.). *Aims-Story.* Retrieved July 19, 2012, from agapewebsite: http://agapewebsite.org/aims-story/

Alberto Fujimori. (n.d.). In *Infoplease online encyclopedia.* Retrieved from http://www.infoplease.com/encyclopedia/people/fujimori-alberto.html

Aldo Moro snatched at gunpoint. (1978, March 16). Retrieved from BBC: *On This Day.* http://news.bbc.co.uk/onthisday/hi/dates/stories/march/16/newsid_4232000/4232691.stm

Almasmari, H., Fitch A., & Nissenbaum, D. (2015, March 23). Chaos in Yemen stymies U.S. counterterror operations: Group that controls capital and northern region tries to extend its reach. *Wall Street Journal (Online).* Retrieved from http://search.proquest.com.proxy.lib.csus.edu/newsstand

al-Mujahed, A., & Morris, L. (2015, March 23). Rebel's defiance could tip Yemen into civil war. *Washington Post,* p. A10.

Al-Qaeda attacks: Killing in the name of Islam. (2011, May 6). *The Economist.* Retrieved June 20, 2013, from economist.com: http://www.economist.com/blogs/dailychart/2011/05/al-qaeda_attacks

Al-Qaida timeline: Plots and attacks. (n.d.). Retrieved May 2, 2013, from NBC News: http://www.nbcnews.com/id/4677978/ns/world_news-hunt_for_al_qaida/t/al-qaida-timeline-plots-attacks/#.UYLsQrWG3cw

Altiok, T. (2009, January). In defense of goods: Research validates simulation's role in port security. *Industrial Engineer, 41*(1), 334–337.

Ambinder, M. (2013, August 14). *What the NSA's massive org chart (probably) looks like.* Retrieved August 21, 2013, from defenseone.com: http://www.defenseone.com/ideas/2013/08/what-nsas-massive-org-chart-probably-looks/68642/

Ambrose, P. (2005, November 17). A path of destruction. Retrieved July 1, 2013, from *themorningnew.org:* http://www.themorningnews.org/article/a-path-to-destruction

American Civil Liberties Union. (n.d.). Retrieved July 25, 2013, from Patriot Act: http://www.aclu.org/timelines/post-911-surveillance

American Experience, The. (n.d.). *The rise of the Ku Klux Klan.* Retrieved January 15, 2013, from http://www.pbs.org/wgbh/americanexperience/features/general-article/grant-kkk/+

Andaya, B. W. (n.d.). *Introduction to Southeast Asia: History, Geography and Livliehood.* Retrieved 26 2012, June, from asiasociety.org: http://asiasociety.org/countries/traditions/introduction-southeast-asia

Anderson, R. H. (1989, June 5). Ayatollah Ruhollah Khomeini, 89, relentless founder of Iran's Islamic Republic. *New York Times,* p. B5.

Animal Enterprise Terrorism Act (AETA), The. (n.d.). Retrieved February 14, 2013, from Center for Constitutional Rights: http://

ccrjustice.org/learn-more/faqs/factsheet%3A-animal-enterprise-terrorism-act-%28aeta%29

Anti-terrorism, Crime and Security Act 2001. (n.d.). Retrieved August 12, 2011, from The National Archives: http://www.legislation.gov.uk/ukpga/2001/24/contents

Anzalone, C. (2013). The Nairobi attack and Al-Shabab's media strategy. *CTC Sentinel, 6*(10), 1–6.

Applebaum, A. (2011, January 2). *Homeland Security hasn't made us safer.* Retrieved August 6, 2013, from http://www.foreignpolicy.com/articles/2011/01/02/unconventional_wisdom?page=0,2

Arab League agrees to create joint military force. (2015, March 29). *BBC News.* Retrieved from http://www.bbc.com/news/world-middle-east-32106939

Archick, K. (2013). *U.S.-EU cooperation against terrorism* (CRS Report No. RS 22030). Washington, DC: Congressional Research Service.

Arnaz, F. (2012, June 27). *Web Hacking Is Latest Trend in Indonesia's Terrorism Funding: Police.* Retrieved July 13, 2012, from Jakarta Globe: http://www.thejakartaglobe.com/lawandorder/web-hacking-is-latest-trend-in-indonesias-terrorism-funding-police/526735

Associated Press. (2002, October 12). U.S. Congress, "Authorization for the use of military force against Iraq." *New York Times,* p. A1.

Associated Press. (2013, August 9). 12 in Yemen die in strikes by U.S. drones. Retrieved August 23, 2013, from nytimes.com: http://www.nytimes.com/2013/08/09/world/middleeast/12-in-yemen-die-in-strikes-by-us-drones.html

At home in Peru's nastiest cell-block. (2007, September 27). Retrieved July 2, 2013, from *peruthisweek:* http://archive.peruthisweek.com/news/2589#c331

Ba'asyir get 15 years in prison. (2011, June 16). *Jakarta Post* (National ed.). Retrieved July 12, 2012, from http://www.thejakartapost.com/news/2011/06/16/ba%E2%80%99asyir-gets-15-years-prison.html

Background note: North Korea. (2012, April 4). Retrieved August 1, 2012, from U.S. State Department: http://www.state.gov/r/pa/ei/bgn/2792.htm

Badawi, N. (2013, March 19). US lawmakers seek to "redirect" Egypt aid. Retrieved May 28, 2013, from *dailynewsegypt.com:* www.dailynewsegypt.com/201303/19us-lawmakers-seek-to-redirect/egypt/aid/

Baer, S. (2003). *Peru's MRTA: Tupac Amaru Revolutionary Movement.* New York, NY: Rosen Publishing Group.

Baker, A. (2011, May 12). Why we're stuck with Pakistan. *Time.* Retrieved from http://www.time.com/time/magazine/article/0,9171,2071131.00.html

Baker, A. (2013, May 27). The YouTube war. *Time,* pp. 38–40.

Baker, A. (2014, January 20). A nightmare returns. *Time,* pp. 30–35.

Balko, R. (2013, July 19). Rise of the warrior cop: Is it time to reconsider the militarization of American policing? *Wall Street Journal,* p. C1.

BBC News. (2013). Syria chemical attack: What we know. Retrieved from http://www.bbc.co.uk/news/world-middle-east-23927399

Behind the Houthi insurgency. (2015, January 21). *New York Times,* p. 5.

Beittel, J. S. (2012, November 28). *Colombia: Background, U.S. relations, and interests.* Retrieved June 28, 2013, from fas.org: http://www.fas.org/sgp/crs/row/RL32250.pdf

Belgistan? Sharia showdown looms in Brussels. (n.d.). Retrieved August 23, 2013, from *cbn.com:* http://www.cbn.com/tv/ 1509282970001

Benko, R. (2013, March 11). 1.6 billion rounds of ammo for Homeland Security? It's time for a national conversation. Retrieved August 21, 2013, from *Forbes.com:* http://www.forbes.com/sites/ralphbenko/2013/03/11/1-6-billion-rounds-of-ammo-for-homeland-security-its-time-for-a-national-conversation/

Bennett, B. (2012, February 23). "Sovereign citizen" movement now on FBI's radar. Retrieved February 21, 2013, from *Los Angeles Times:* http://articles.latimes.com/2012/feb/23/nation/la-na-terror-cop-killers-20120224

Bentley, E. (2012). Homeland security law and policy. In K. G. Logan & J. D. Ramsay (Eds.), *Introduction to homeland security* (pp. 19–47). Boulder, CO: Westview Press.

Berko, A., & Erez, E. (2005, December). "Ordinary people" and "death work": Palestinian suicide bombers as victimizers and victims. *Violence and Victims, 20*(6), 603–623.

Berman, D. K. (2013, July 23). Daddy, could you tell me what a truck driver was? *Wall Street Journal,* p. B1.

Biometrics. (n.d.). Retrieved August 22, 2013, from *reference.com:* http://dictionary.reference.com/browse/biometrics

Birkland, T. A. (1997). *After disaster: Agenda setting, public policy, and focusing events.* Washington, DC: Georgetown University Press.

Birkland, T. A. (2001). *An introduction to the policy process: Theories, concepts, and models of public policy making.* Armonk, NY: M. E. Sharpe.

Blakeley, R. (2007). Bringing the state back into terrorism studies. *European Political Science, 6,* 228.

Bland, B. (2012, June 21). *Indonesia sentences last Bali bomber.* Retrieved July 13, 2012, from Financial Times: http://www.ft.com/intl/cms/s/0/f904dc72-bb7d-11e1-9436-00144feabdc0.html#axzz20XOqbnlF

Bloom, M. M. (2003). Ethnic conflict, state terror, and suicide bombing in Sri Lanka. *Civil Wars, 6,* 54–84.

Bloomberg Business Week News. (2012, July 17). *Myanmar's Suu Kyi to travel to US to receive award.* Retrieved July 18, 2012, from businessweek.com: http://www.businessweek.com/ap/2012-07-17/myanmars-suu-kyi-to-travel-to-us-to-receive-award

Bodansky, Y. (2007). *Chechen jihad: Al Qaeda's training ground and the next wave of terror.* New York, NY: HarperCollins.

Bombing of Air India Flight 182, The. (2006, September 25). Retrieved September 23, 2012, from *CBC News:* http://www.cbc.ca/news/background/airindia/bombing.html

Bombing of the King David hotel, The. (n.d.). Retrieved March 23, 2012, from *etzel.org:* http://www.etzel.org.il/english/ac10.htm

Bradley, M. (2012, June 29). Morsi promises to free "Blind Sheik" from U.S. prison. Retrieved May 28, 2013, from online.wsj.com:

http://online.wsj.com/article/SB1000142405270230356150457749705302635 6034.html

Bradley, M., & Abdellatif, R. (2013, July 4). *Egyptian military ousts President Morsi.* Retrieved July 5, 2013, from online.wsj.com: http://online.wsj.com/article/SB10001424127887323899704578583319518313964.html

Bradley, M., & El-Ghobashy, T. (2013, August 22). Mubarak to be placed under house arrest after release. *Wall Street Journal*, p. A10.

Bridges, T. (2005, November 7). Fujimori arrested in Chile. Retrieved July 2, 2013, from *Los Angeles Times:* http://www .latinamericanstudies.org/peru/fujimori-arrested.htm

Brief introduction to Falun Dafa. (n.d.). Retrieved July 5, 2012, from Falun Dafa website: http://www.falundafa.org/eng/home.html

Buchanan, A. S. (2000). *Peace with justice: A history of the Israeli-Palestinian Declaration of Principles on Interim Self-Government Arrangement.* Basingstoke, UK: Macmillan.

Bueno de Mesquita, B. (2009). The predictioneer's game: Using the logic of brazen self-interest to see and shape the future. New York, NY: Random House.

Burgonio, T. J. (2012, June 30). Filing charges not enough, rights group tells Philippine gov't. Retrieved June 30, 2012, from *Philippine Daily Inquirer:* http://newsinfo.inquirer.net/221035/filing-charges-not-enough-rights-group-tells-philippine-gov%E2%80%99t

Burns, R. G. (2013). *Policing: A modular approach.* Upper Saddle River, NJ: Pearson Education.

Burt, J.-M. (2006, October). "Quien Habla Es Terrorista": The political use of fear in Fujimori's Peru. *Latin America Research Review, 41*(3), 32–61.

Bush, G. W. (2001, September 20). *Transcript of President Bush's address to a joint session of Congress on Thursday night, September 20, 2001.* Retrieved July 24, 2013, from cnn.com: http://archives .cnn.com/2001/US/09/20/gen.bush.transcript/

Byman, D. (2008). Iran, terrorism, and weapons of mass destruction. *Studies in Conflict & Terrorism, 31*, 169–181.

Calabresi, M., Crowley, M., & Newton-Small, J. (2013, March 11). The path to war. *Time*, pp. 20–25.

Camber, R. (2011, July 28). "No porn or prostitution": Islamic extremists set up Sharia law controlled zones in British cities. Retrieved August 22, 2013, from *dailymail.co.uk:* http://www .dailymail.co.uk/news/article-2019547/Anjem-Choudary-Islamic-extremists-set-Sharia-law-zones-UK-cities.html

Campion-Smith. (2012, September 7). *Canada closes Iran Embassy, expels remaining Iranian diplomats.* Retrieved October 14, 2012, from thestar.com: http://www.thestar.com/news/canada/politics/article/1252838—canada-closes-iran-embassy-expels-remaining-iranian-diplomats

Canada rejects anti-terror laws. (2007, February 28). Retrieved October 6, 2012, from *BBC News:* http://news.bbc.co.uk/2/hi/americas/6403241.stm

Canadian Parliament. (2001). *House Government Bill 36 C.* Retrieved October 6, 2012, from Parliament of Canada: http://www.parl .gc.ca/LegisInfo/BillDetails.aspx?Bill=C36&Mode=1&Parl=37&Ses=1&View=10

Canadian Security Intelligence Service. (n.d.). *Terrorism.* Retrieved September 13, 2012, from Canadian Security Intelligence Service: http://www.csis-scrs.gc.ca/prrts/trrrsm/index-eng.asp

Capella, M. B., & Sahliyeh, E. (2007). Suicide terrorism: Is religion a critical factor? *Security Journal, 20*, 267–283.

Capron, R. (2012, June 20). Associate Dean, Business Affairs. (T. A. Capron, Interviewer)

Carafano, J. J., & Weitz, R. (2005, December 7). The truth about FEMA: Analysis and proposals. Retrieved July 9, 2013, from *heritage.org:* http://www.heritage.org/research/reports/2005/12/the-truth-about-fema-analysis-and-proposals

Carlos Castaño murder: Nine men sentenced in Colombia. (2011, March 17). BBC. Retrieved June 28, 2013, from BBC.com: http://bbc.co.uk/news/world-latin-america-11400959

Carson, J. V., LaFree, G., & Dugan, L. (2012, March 14). Terrorist and non-criminal acts by radical environmental and animal rights groups in the United States 1970 to 2007. *Terrorism and Political Violence, 24*(2), 295–319.

Carter, J. (2006). *Palestine peace: Not apartheid.* New York, NY: Simon & Schuster.

Catherwood, C. (2006). *A brief history of the Middle East.* Philadelphia, PA: Running Press.

Center for Security Policy. (2010, September 10). *Shariah: The threat to America: An exercise in competitive analysis (Report of Team B II) [Paperback].* Retrieved August 20, 2013, from centerforsecurity policy.org: http://www.centerforsecuritypolicy.org/upload/wysiwyg/article%20pdfs/Shariah%20-%20The%20Threat%20to%20America%20%28Team%20B%20Report%29%20Web%20Version%2009302010.pdf

Centers for Disease Control and Prevention. (n.d.). *Facts about sarin.* Retrieved August 7, 2012, from: http://www.bt.cdc.gov/agent/sarin/basics/facts.asp

Central Intelligence Agency. (2012). *CIA World Factbook.* Retrieved from https://www.cia.gov/library/publications/the-world-factbook/index.html

Chailand, G., & Blin, A. (2007). Zealots and assassins. In G. Chailand & A. Blin (Eds.), *The history of terrorism* (pp. 55–78). Berkeley: University of California Press.

Chalamers, D. M. (1981). *Hooded Americanism: The history of the Ku Klux Klan.* Durham, NC: Duke University Press.

Challands, S. (2006, January 27). Hamas: A brief history of the rise to power. Retrieved June 3, 2012, from *CTV News:* http://www.ctv.ca/CTVNews/Specials/20060127/hamas_chronology_060127/

Chalmers, D. M. (1981). *Hooded Americanism* (3rd ed.). Durham, NC: Duke University Press.

Chao, L. (2012, June 2). Google tips off users in China. *Wall Street Journal*, p. B1.

Charlton, L. (1984, November 1). *Assassination in India: A leader of will and force; Indira Gandhi, born to politics, left her own imprint on India*. Retrieved October 5, 2012, from NY Times: http://www.nytimes.com/learning/general/onthisday/bday/1119.html

Charters, D. A. (2008, October). The (Un) peaceable kingdom? Terrorism and Canada before 9/11. *IRPP Policy Matters, 9*(4), 27.

Chavez, L. (1985, February 5). *Power is cut during papal visit; Peruvian rebel group suspected.* Retrieved July 3, 2013, from nytimes.com: http://www.nytimes.com/1985/02/05/world/power-is-cut-during-papal-visit-peruvian-rebel-group-suspected.html?n=Top%2fReference%2fTimes%20Topics%2fOrganizations%2fR%2fRoman%20Catholic%20Church

Chaze, W. I., Dudney, R. S., Wallace, J. N., Shapiro, J. P., & Mullen, D. (1985, July 1). Reagan's hostage crisis. *U.S. News and World Report.*

Chertoff, M. (2011). 9/11: Before and after. Retrieved July 5, 2013, from *Journal of the Navy Post Graduate School Center for Defense and Homeland Security*: http://www.hsaj.org/?fullarticle=7.2.13

China leader urges resistance against Western forces. (n.d.). Retrieved July 5, 2012, from http://news.yahoo.com/http://news.yahoo.com/china-leader-urges-resistance-against-western-forces-223528649.html

China's population: The most surprising demographic crisis. (2011, May 5). Retrieved from *economist.com*: http://www.economist.com/node/18651512/print

China's urban population higher than rural areas. (2011, January 18). Retrieved July 7, 2012, from sfgate.com: http://www.sfgate.com/world/article/China-s-urban-population-higher-than-rural-areas-2595957.php#photo-2085598

Choi, S. W. (2010). Fighting terrorism through the rule of law? *Journal of Conflict Resolution, 54*, 940–966.

CIA World Factbook. (n.d.). Retrieved from https://www.cia.gov/library/publications/the-world-factbook/

Clapper, J. R. (2013, March 12). *Statement for the record: Worldwide threat assessment.* Retrieved August 16, 2013, from intelligence.senate.gov: http://www.intelligence.senate.gov/130312/clapper.pdf

Clark, C. (2013, April 22). *The Tehran factor: New wrinkle added to Canadian counter-terrorism efforts.* Retrieved April 23, 2013, from theglobeandmail.com: http://www.theglobeandmail.com/news/politics/the-tehran-factor-new-wrinkle-added-to-canadian-counter-terrorism-efforts/article11492175/

Clark, C. S. (2012a, October 12). *Homeland Security's fusion center lambasted in Senate report.* Retrieved August 8, 2013, from *govexec.com:* http://www.govexec.com/defense/2012/10/homeland-securitys-fusion-centers-lambasted-senate-report/58535/

Clark, C. S. (2012b, September 1). Lifting the lid. Retrieved August 14, 2013, from *govexec.com:* http://www.govexec.com/magazine/features/2012/09/lifting-lid57807/

Clayton, M. (2013, May 22). Terror watch lists: Are they working as they should? *Christian Science Monitor*, pp. 1–6.

Clinton, H. (2012, April 4). The future of Burma is neither clear nor certain, Secretary Clinton says. Retrieved July 19, 2012, from *diplonews.com*: http://www.diplonews.com/feeds/free/5_April_2012_14.php

Clymer, A. (2010, June 29). Robert C. Byrd, a pillar of the Senate, dies at 92. Retrieved January 21, 2013, from *New York Times:* http://www.nytimes.com/2010/06/29/us/politics/29byrd.html?pagewanted=all&_r=0

Cobb, N. (1982, October 18). Where have you gone, Bobby Seale? *The Boston Globe*, p. 1.

Cobb, R. W., & Elder, C. D. (1983). *Participation in American politics: The dynamics of agenda building* (2nd ed.). Baltimore, MD: Johns Hopkins University Press.

Coburn, T. (2012, December). Safety at any price: Assessing the impact of Homeland Security spending in U.S. cities. Retrieved August 9, 2013, from *coburn.senate.gov:* http://www.coburn.senate.gov/public/index.cfm?a=Files.Serve&File_id=b86fdaeb-86ff-4d19-a112-415ec85aa9b6

Colello, T. (1987). *Syria: A country study.* Washington, DC: Government Printing Office.

Collins, J. (2012, November 232). *Sovereign citizens inspire rise in real estate filings.* Retrieved February 21, 2013, from OCRegister.com: http://www.ocregister.com/articles/sovereign-378636-county-mortgage.html

Colombia's armed groups. (2013, February 8). Retrieved June 28, 2013, from *BBC.com:* http://www.bbc.co.uk/news/world-latin-america-12779945

Combs, C. C. (2006). *Terrorism in the twenty-first century* (4th ed.). Upper Saddle River, NJ: Pearson.

Commission of Inquiry Into the Investigation of the Bombing of Air India Flight 182. (2010, June 17). *Final report.* Retrieved October 5, 2012, from Library and Archives of Canada: http://epe.lac-bac.gc.ca/100/206/301/pco-bcp/commissions/air_india/2010-07-23/www.majorcomm.ca/en/reports/finalreport/default.htm

Community Security Trust, United Kingdom, The. (2010). *Terrorist incidents against Jewish communities and Israeli citizens abroad 1968–2010.* Retrieved January 20, 2012, from http://www.thecst.org.uk/docs/CST%20Terrorist%20Incidents%201968%20-%202010.pdf

Congress of the United States. (2001, October 26). *Patriot Act.* Retrieved July 5, 2013, from fincen.gov: http://www.fincen.gov/statutes_regs/patriot/

Congressional Budget Office. (2012, January). Comparing the compensation of federal and private-sector employees. Retrieved August 5, 2013, from *cbo.gov:* http://www.cbo.gov/sites/default/files/cbofiles/attachments/01-30-FedPay.pdf

Corrin, A. (2012, July 9). Cyber warfare: New battlefield, new rules. Retrieved July 9, 2012, from *Federal Computer Week:* http://fcw.com/Articles/2012/07/15/FEAT-Inside-DOD-cyber-warfare-rules-of-engagement.aspx?Page=1

Costa, K. J. (2001, April). *Hart-Rudman calls for homeland defense.* Retrieved March 14, 2013, from Air Force Magazine: http://www.airforce-magazine.com/MagazineArchive/Pages/2001/April%202001/0401hartrud.aspx

Country guides: Colombia. (n.d.). Retrieved June 24, 2013, from *washingtonpost.com:* http://www.washingtonpost.com/wp-srv/world/countries/colombia.html

Court hands Shining Path leader life sentence. (2013, June 18). *Al Jazeera*. Retrieved July 3, 2013, from aljazeera.com: http://www.aljazeera.com/news/americas/2013/06/2013681296617928.html

Coyne, T. (2012, December 6). North Van "eco-terrorism" suspect to plead guilty to some, not all, charges. *The North Shore Outlook*. Vancouver, British Columbia, Canada.

Crandall, R. (2011, August–September). Closing argument: Requiem for the FARC. *Survival, 53*(4), 233–240.

Crenshaw, M. (2009). Intimations of mortality or production lines? *Political Psychology, 30*(3), 360.

Crepeau, F. (2011, February 3). Anti-terrorism measures and refugee law challenges in Canada. *Refugee Survey Quarterly, 29*(4), 3.

Croisdale, T. (2012, October 12). Frequent offenders and crime mapping using GIS. (T. A. Capron, Interviewer)

Crowley, M. (2013, June 10). Why Gitmo will never close. *Time*, pp. 44–48.

Crowley, M. (2014a, June 30). Iraq's eternal war. *Time*, pp. 28–34.

Crowley, M. (2014b, September 29). Coalition of the wary. *Time*, pp. 24–27.

Cunningham, W. G., Jr. (2003). *Terrorism: Concepts, causes, and conflict resolution*. Fort Belvoir, VA: Defense Threat Reduction Agency.

Danzig, R., Sageman, M., Leighton, T., Hough, L., Yuki, H., Kotani, R., & Hosford, Z. M. (2011). *Aum Shinrikyo insights into how terrorists develop biological and chemical weapons*. Washington, DC: Center for New American Security. Retrieved August 7, 2012, from http://www.cnas.org/aumshinrikyo

Dark winter. (n.d.). Retrieved February, 10, 2012, from *UPMC Center for Biosecurity:* http://www.upmc-biosecurity.org/website/events/2001_darkwinter/

De Castro, R. C. (2010). Abstract of counter-insurgency in the Philippines and the global war on terror. Examining the dynamics of the twenty-first. *European Journal of East Asian Studies, 9*(1), 135–160.

Defense Intelligence Agency. (n.d.). About. Retrieved July 9, 2013, from *dia.mil:* http://www.dia.mil/about/

Defense.gov. (2013, June 20). *Casualty*. Retrieved June 20, 2013, from defense.gov/news/casualty: http://www.defense.gov/news/casualty.pdf

Demarest, J. M. (2013, May 8). *Responding to the cyber threat*. Retrieved August 19, 2013, from fbi.gov: http://www.fbi.gov/news/testimony/responding-to-the-cyber-threat

Department of Homeland Security. (2008, April 11). Retrieved April 4, 2011, from Homeland Security: http://www.dhs.gov/xabout/history/editorial_0133.shtm

Department of Homeland Security. (n.d.). About DHS. Retrieved July 26, 2013, from *dhs.gov:* http://www.dhs.gov/about-dhs

Department of Homeland Security. (n.d.). Retrieved April 5, 2011, from Homeland Security: http://www.dhs.gov/xabout/responsibilities.shtm

Department of State, Bureau of Counterterrorism. (2012, January 27). *Foreign terrorist organizations*. Retrieved June 27, 2012, from http://www.state.gov/j/ct/rls/other/des/123085.htm

Desloges, A. (2011, February 28). *International terrorism in Canada: Facilitating factors*. Retrieved October 14, 2012, from *Centre for Foreign Policy Studies:* http://centreforforeignpolicystudies.dal.ca/pdf/gradsymp11/Desloges.pdf

DHS. (n.d.). Preventing terrorism results. Retrieved August 8, 2013, from *dhs.gov:* https://www.dhs.gov/topic/preventing-terrorism-results

Dinan, S. (2013, July 29). *FBI says it doesn't need warrant to use drones*. Retrieved August 23, 2013, from washingtontimes.com: http://www.washingtontimes.com/news/2013/jul/29/fbi-says-it-doesnt-need-warrant-use-drones/

Donohue, L. K. (2012). Technological leap, statutory gap, and constitutional abyss: Remote Biometric Identification comes of age. *Minnesota Law Review, 97*, 408–559. Retrieved from http://scholarship.law.georgetown.edu/cgi/viewcontent.cgi?article=2043&context=facpub

Drum, K. (2013, October 20). The Benghazi controversy explained. *Mother Jones*. Retrieved from http://www.motherjones.com/print/201931

Dycus, S., Banks, W. C., & Raven-Hansen, P. (2007). *Counterterrorism law*. New York, NY: Aspen Publishers.

Early, B. R. (2006, Winter). Larger than a party, yet smaller than a state: Locating Hezbollah's place within Lebanon's state and society. *World Affairs, 168*(3), 125.

Editorial Board. (2013, August 19). *Thank Obama's ego for the carnage and crisis in Egypt*. Retrieved August 20, 2013, from investors.com: http://news.investors.com/ibd-editorials-081913-668027-egypt-crisis-result-of-obama-arrogance-and-incompetence.htm

Editorial Board. (2013, July 23). DHS hinders reform push. Retrieved August 12, 2013, from *azcentral.com:* http://www/azcentral.com/opinions/articles/20130723dhs-hinders-reform-push.html

Editorial: Memos show Gorelick involvement in "wall." (2004, April 29). Retrieved June 18, 2013, from washingtontimes.com: http://www.washingtontimes.com/news/2004/apr/29/20040429-122228-6538r/print/

Elsea, J. K. (2014). *The Military Commissions Act of 2009: Overview and legal issues* (CRS Report No. R41163). Washington, DC: Congressional Research Service.

Embassy of Indonesia, Oslo. (2012, June 28). Indonesia's cultural diversity unparalleled. Retrieved June 28, 2012, from *indonesai-oslo.no/:* http://indonesia-oslo.no/indonesias-cultural-diversity-unparalleled/

Emerson, S., & Duffy, B. (1990). *The fall of Pan Am 103: Inside the Lockerbie investigation*. New York, NY: Putnam.

Entous, A., & Gorman, S. (2012, November 15). Petraeus's battle in final days—CIA chief weakened by criticism of his role In Libya crisis, then affair. *Wall Street Journal*, p. A1.

Escobar, G. (1996, December 19). Peruvian guerrillas hold hundreds hostage: Ambassadors among those detained; Rebels demand comrades' freedom. *New York Times*, p. A1.

Exec. Order No. 13231. (2001). *Critical infrastructure protection in the information age*. Retrieved August 19, 2014, from http://fas.org/irp/offdocs/eo/eo-13231.htm

Execution of Aum founder likely postponed. (2012, June 12). Retrieved August 8, 2012, from *The Daily Yomiuri:* http://www.yomiuri.co.jp/dy/national/T120604003901.htm

Explaining Boko Haram, Nigeria's Islamist insurgency. (2014, November 11). *New York Times.* Retrieved December 10, 2014, from http://www.nytimes.com/2014/11/11/world/africa/boko-haram-in-nigeria.html

Fair, C. C., Crane, K., Chivvis, S., Puri, S., & Spirtas, M. (2009). *Pakistan: Can the United States secure an insecure state?* Santa Monica, CA: RAND Project Air Force Report.

Falkenrath, R. A. (2001). Problems of preparedness: U.S. readiness for a domestic terrorist attack. *International Security, 25,* 147–186.

Falkenrath, R. A. (2001). U.S. readiness for a domestic terrorist attack. *International Security, 25,* 147–186.

Falkenrath, R. A., Newman, R. D., & Thayer, B. A. (2000). *America's Achilles heel: Nuclear, biological, and chemical terrorism and covert attack.* Cambridge, MA: MIT Press.

Fatah and Hamas sign landmark reconciliation deal. (2011, April 5). Retrieved June 5, 2012, from *France 24 International News:* http://www.france24.com/en/20110504-fatah-hamas-sign-landmark-reconciliation-deal-palestine-cairo

FBI. (2010, March 8). *InfraGard: A partnership that works.* Retrieved August 23, 2013, from fbi.gov: http://www.fbi.gov/news/stories/2010/march/infragard_030810

FBI's Counterterrorism Analysis Section. (2011, September). *FBI Law Enforcement Bulletin.* Retrieved February 18, 2013, from Sovereign Citizens: A Growing Domestic Threat to Law Enforcement: http://www.fbi.gov/stats-services/publications/law-enforcement-bulletin/september-2011/sovereign-citizens

Fear of missing out; security in Colombia. (2013). *The Economist,* 35–36.

Federal Bureau of Investigation. (2010, April 10). *Domestic terrorism: The sovereign citizens movement.* Retrieved February 18, 2013, from Federal Bureau of Investigation: http://www.fbi.gov/news/stories/2010/april/sovereigncitizens_041

Federal Bureau of Investigation. (2011, August 3). *FBI Federal Bureau of Investigation Albuquerque Division.* Retrieved August 3, 2011, from http://www.fbi.gov/albuquerque/priorities

Federal Bureau of Investigation. (n.d.). *Intelligence collection disciplines (INTs).* Retrieved August 14, 2013, from fbi.gov: http://www.fbi.gov/about-us/intelligence/disciplines

Federal Bureau of Investigation. (n.d.). *Terror hits home: The Oklahoma City bombing.* Retrieved February 13, 2013, from Famous Cases: http://www.fbi.gov/about-us/history/famous-cases/oklahoma-city-bombing

Federal Transit Administration. (2013, July 2). TSA commends 16 mass transit and rail agencies for highest security levels. Retrieved August 9, 2013, from *fta.dot.gov:* http://bussafety.fta.dot.gov/2013/07/tsa-commends-16-mass-transit-and-rail-agencies-for-highest-security-levels/

Federman, J. (2009, June 15). *Netanyahu peace speech: Israeli prime minister appeals to Arab leaders for peace.* Retrieved June 5, 2012, from *Huffington Post:* http://www.huffingtonpost.com/2009/06/14/netanyahu-peace-speech-is_n_215337.html

Fennel, T., & Chauvin, L. (1997). Rescue in Lima: A bold assault sends a message to terrorists. *Macleans,* p. 38.

Ferran, L. (2010, February 10). *Extremists in "patriot" movement calling Joe Stack a hero.* Retrieved February 21, 2013, from abcnews.go.com: http://abcnews.go.com/TheLaw/patriot-movement-calling-joe-stack-hero/story?id=9889443

Fetini, A. (2009, June 8). A brief history of Hizballah. Retrieved June 3, 2012, from *time.com:* http://www.time.com/time/world/article/0,8599,1903301,00.html

Fitzpatrick, B., & Norby, M. (2013, January/February). Colombia, the riven land. *New Internationalist,* pp. 46–48.

Flanigan, S. T. (2008). Nonprofit services provision by insurgent organizations: The cases of Hizballah and the Tamil Tigers. *Studies in Conflict and Terrorism, 31,* 499–519.

Fletcher, D. (2009, February 17). A brief history of the Khmer Rouge. Retrieved July 19, 2012, from *time.com:* http://www.time.com/time/world/article/0,8599,1879785,00.html

Fletcher, H. (2012, June 19). *Aum Shinrikyo.* Retrieved August 8, 2012, from Council on Foreign Relations cfr.org: http://www.cfr.org/japan/aum-shinrikyo/p9238

Foreign Policy. (2012, July). The 2012 Failed State Index: Interactive map and rankings. Retrieved 16, from http://www.foreignpolicy.com/failed_states_index_2012_interactive

Foroohar, R. (2011, June 27). Turkey's man of the people. *Time,* pp. 36–38.

France 24. (2010, December 29). Retrieved March 19, 2012, from International News 24/7: http://www.france24.com/en/20101229-israel-approves-2011-2012-budget

Frankel, G. (2004, November 3). *Dutch filmmaker shot, stabbed in Amsterdam.* Retrieved Marcj 5, 2013, from *Chicago Tribune:* http://articles.chicagotribune.com/2004-11-03/news/0411030299_1_gogh-dutch-filmmaker-muslims-and-non-muslims

French, C. (2010, January 18). *"Toronto 18" mastermind gets life for bomb plot.* Retrieved September 20, 2012, from Reuters: http://www.reuters.com/article/2010/01/18/us-security-sentence-idUSTRE60H3PN20100118

Fried, A., & Chinnareddy, A. (2012, July 2). *3 key trends in China.* Retrieved July 7, 2012, from http://www.hsp.com/blog/2012/7/3-key-trends-china

Friedman, B. H. (2011). Managing fear: The politics of homeland security. *Political Science Quarterly,* 77–106.

Friedman, T. L. (1989). *From Beirut to Jerusalem.* New York, NY: Farrar Straus & Giroux.

Fromkin, D. (2009). *A peace to end all peace: The fall of the Ottoman Empire and the creation of the modern Middle East.* New York, NY: Henry Holt.

Full text of Netanyahu speech to AIPAC. (2012, March 5). Retrieved June 6, 2012, from *algemeiner.com:* http://www.algemeiner.com/2012/03/05/full-text-of-netanyahu-speech-to-aipac-2012/

Fund for Peace. (2012). *Failed states.* Retrieved August 20, 2012, from Failed States: http://www.foreignpolicy.com/failed_states_index_2012_interactive

Galafaro, C. (2013, January 18). *Suspect in St. John deputies' shooting was "paranoid," witness testifies.* Retrieved February 21, 2013, from nola Times Picayune: http://www.nola.com/crime/index.ssf/2013/01/witness_testifies_that_suspect.html

Galicki, Z. (2005). International law and terrorism. *American Behavioral Scientist, 48,* 743–757.

Ganor, B. (2010). *Is one man's terrorist another man's freedom fighter?* Retrieved August 19, 2011, from International Institute for Counter-terrorism: http://www.ict.org.il/ResearchPublications/tabid/64/Articlsid/432/Default.aspx

Gans, J. J. (2012, October 10). *"This is 50-50": Behind Obama's decision to kill bin Laden.* Retrieved June 21, 2013, from atlantic.com: http://www.theatlantic.com/international/archive/2012/10/this-is-50-50-behind-obamas-decision-to-kill-bin-laden/263449/

GAO Report. (2013, June 27). *Progress and challenges in DHS implementation and.* Retrieved August 12, 2013, from www.gao.gov: http://www.gao.gov/assets/660/655540.pdf

Gardels, N. (2012, September 18). *Ayaan Hirsi Ali on anger in the Muslim world over video: Don't apologize.* Retrieved March 11, 2013, from Huffington Post: http://www.huffingtonpost.com/nathan-gardels/ayaan-hirsi-ali-on-anger_b_1893278.html

Garfinklle, A. M. (1991, October). On the origin, meaning, use and abuse of a phrase. *Middle Eastern Studies, 27*(4), 539–550.

Garson, G. D. (2006). Securing the virtual state: Recent developments in privacy and security. *Social Science Computer Review, 24,* 489–496.

Gatehouse, J. (2011, May 16). The world's most hated terrorist. *Maclean's, 124*(18), pp. 152–157.

Gedalyahu, T. B. (2011, May 8). Prime minister visits grave of Entebbe hero Yoni Netanyaho. *Israeli National News.* Retrieved from http://www.israelnationalnews.com/News/News.aspx/143972#.T5HdUdl62So

Geisler, N. L., & Saleeb, A. (2003). *Answering Islam: The Crescent in light of the Cross.* Grand Rapids, MI: Baker Books.

Gershman, J. (2013, August 16). *Oklahoma ban on Sharia law unconstitutional, US judge rules.* Retrieved August 20, 2013, from wsj.com: http://blogs.wsj.com/law/2013/08/16/oklahoma-ban-on-sharia-law-unconstitutional-us-judge-rules/

Gerson, A., & Adler, J. (2001). *The price of terror.* New York, NY: HarperCollins.

Gettleman, J. (2010, March/April). *Africa's forever wars: Why the continent's conflicts never end.* Retrieved August 22, 2013, from http://www.foreignpolicy.com/articles/201002/22/africas_forever_wars

Gettleman, J. (2012, July 1). Gunmen attack 2 churches in Kenya, killing at least 15 people. *New York Times.* Retrieved from http://www.nytimes.com/2012/07/02/world/africa

Gideon, B. (2004). *The boundaries of modern Palestine, 1840–1947.* New York, NY: Routledge Curzon Taylor & Francis Group.

Gil-Alana, L. A., & Barros, C. P. (2010). A note on the effectiveness of anti-terrorist policies: Evidence from ETA. *Conflict Management and Peace Science, 27*(1), 28–46.

Gilbert, M. (1998). *Israel: A history.* New York, NY: William Morrow.

Global Terrorism Database. (2012). Retrieved from http://www.start.umd.edu/gtd/

Goldenberg, S. (2008, March 3). US plotted to overthrow Hamas after election victory. Retrieved May 30, 2012, from *The Guardian World News*: http://www.guardian.co.uk/world/2008/mar/04/usa.israelandthepalestinians

Goldhill, O. (2013, August 16). *Breast implant explosives could be used in terrorist attack.* Retrieved August 23, 2013, from telegraph.co.uk: http://www.telegraph.co.uk/news/uknews/terrorism-in-the-uk/10247069/Breast-implant-explosives-could-be-used-in-terrorist-attack.html

Goldman, A., & Gearan, A. (2014, January 15). Senate report: Attacks on U.S. compounds in Benghazi could have been prevented. *Washington Post.* Retrieved from http://www.washingtonpost.com/world/national-security

Gorman, S., Barrett, D., & Valentino-Devries, J. (2013, August 22). Secret spy court raps NSA. *Wall Street Journal,* p. A1.

Gorman, S., & Valentino-Devries, J. (2013, August 21). NSA reaches deep into U.S. to spy on Net. *Wall Street Journal,* p. A1.

Government Accountability Office. (2013, March 7). *A better defined and implemented national strategy is needed to address persistent challenges.* Retrieved August 12, 2013, from http://www.gao.gov: http://www.gao.gov/products/GAO-13-462T

Government plans to eradicate house churches. (2012, May 10). Retrieved July 9, 2012, from

Gradstein, L. (2009, August 5). Abbas's Fatah party holds 1st convention in two decades. *Washington Post.* Retrieved June 5, 2012, from: http://www.washingtonpost.com/wp-dyn/content/article/2009/08/04/AR2009080403114.html

Green, L. (2011, January 20). Christianity in China. Retrieved July 7, 2012, from *FoxNews.com*: http://www.foxnews.com/world/2011/01/20/christianity-china/

Greenwald, G. (2013, May 17). Washington gets explicit: Its "war on terror" is permanent. Retrieved August 17, 2013, from *theguardian.com*: http://www.theguardian.com/commentisfree/2013/may/17/endless-war-on-terror-obama

Greenwald, G., & Ball, J. (2013, June 20). *The top secret rules that allow NSA to use US data without a warrant.* Retrieved June 21, 2013, from *The Guardian:* http://www.guardian.co.uk/world/2013/jun/20/fisa-court-nsa-without-warrant

Gregory, K. (2009, August 22). Shining Path, Tupac Amaru (Peru, leftists). Retrieved July 1, 2013, from http://www.cfr.org/peru/shining-path-tupac-amaru-peru-leftists/p9276#p1

Guardian.co.uk.world.datablog. (2010, February 1). Retrieved August 21, 2012, from http://www.guardian.co.uk/world/datablog/2010/feb/01/united-nations-population-world-data

Guled, A. (2014, September 7). Somali extremist group names new leader after deadly U.S. airstrike. *Washington Post,* p. A09.

Gursky, E., Inglesby, T. V., & O'Toole, T. (2008). Anthrax 2001: Observations on the medical and public health response. In R. D. Howard & J. F. Forest (Eds.), *Weapons of mass destruction and terrorism* (pp. 390–411). New York, NY: McGraw-Hill.

Gurule, J., & Corn, G. S. (2011). *Principles of counterterrorism law*. St. Paul, MN: West-Thomson Reuters.

Hacker, F. J. (1976). *Crusaders, criminals, crazies: Terror and terrorism in our time*. New York, NY: W. W. Norton.

Hamas Covenant 1988. (2008, June 3). *Yale Law School Lillian Goldman Law Library*. Retrieved from: http://avalon.law.yale.edu/20th_century/hamas.asp

Hamzawy, A., Ottoway, M., & Brown, N. J. (2007, February). *Policy outlook: What Islamists need to be clear about: The case of the Egyptian Muslim Brotherhood*. Washington, DC: Carnegie Endowment for International Peace.

Hanson, S. (2008, January 11). Colombia's right-wing paramilitaries and splinter groups. Retrieved June 25, 2013, from *Council on Foreign Relations*: http://www.cfr.org/colombia/colombias-right-wing-paramilitaries-splinter-groups/p15239

Hanson, S. (2009, August 19). FARC, ELN: Colombia's left-wing guerrillas. Retrieved June 25, 2013, from *Council on Foreign Relations*: http://www.cfr.org/colombia/farc-eln-colombias-left-wing-guerrillas/p9272

Hanson, V. D. (2013, August 13). *The mother of all scandals*. Retrieved August 14, 2013, from victorhanson.com: http://victorhanson.com/wordpress/?p=6331#more-6331

Harrington, E. (2013, March 5). TSA sealed $50-million sequester-eve deal to buy new uniforms. Retrieved August 12, 2013, from *cnsnews.com*: http://cnsnews.com/news/article/tsa-sealed-50-million-sequester-eve-deal-buy-new-uniforms

Harrison, R. (2012, June 17). Cyber urgency needed: Complacency leaves U.S. vulnerable. Retrieved July 9, 2012, from *defensenews.com*: http://www.defensenews.com/print/article/20120617/DEFFEAT05/306170006/Cyber-Urgency-Needed

Hearn, K. (2012, April 27). Peru guerrillas set aside rebellion for drug money. *Washington Times*.

Hendawi, H. (2013, August 17). *Egypt: Islamists hit Coptic Christian churches, torch Franciscan school*. Retrieved August 19, 2013, from huffingtonpost.com: http://www.huffingtonpost.com/2013/08/17/christians-in-egypt_n_3773991.html

Herzog, C. (2004). The *Arab-Israeli wars: War and peace in the Middle East*. New York, NY: Vintage.

Hess, S. (2013, February 13). From the Arab Spring to the Chinese Winter: The institutional sources of authoritarian vulnerability in Egypt, Tunisia and China. *International Political Science Review, 34*(3), 254–272.

Heupel, M. (2007). Adapting to transnational terrorism: The UN Security Council's evolving approach to terrorism. *Security Dialogue, 28*, 477–499.

Hevesi, D. (1989, August 23). Huey Newton symbolized the rising black anger of a generation [Obituary]. *New York Times*, p. B7.

Hezbollah. (2012, June 4). *New York Times*. Retrieved June 4, 2012, from http://topics.nytimes.com/top/reference/timestopics/organizations/h/hezbollah/index.html

Hinderocker, S. (2010, November 1). *Investigate this*. Retrieved March 28, 2011, from Power Line Blog: http://www.powerlineblog.com/archives/2010/11/027580.php

History crash course #65: The state of Israel. (n.d.). Retrieved March 15, 2012, from *The Jewish website: aish.com*: http://www.aish.com/jl/h/48961671.html

Hjelmgaard, K. (2013, 21 August). Syria opposition claims hundreds dead in "gas" attacks. Retrieved August 21, 2013, from *usatoday.com*: http://www.usatoday.com/story/news/world/2013/08/21/syria-poisonous-gas-attack/2680089/

Hodgson, M. G. S. (1955). *The order of Assassins*. The Hague, Netherlands: Moulton Press.

Hoffman, B. (2006). *Inside terrorism*. New York, NY: Columbia University Press.

Holt, M. (2012, September 18). After 20 years, the "Blind Sheikh" is back in the news; Here's what you need to know about him. Retrieved May 23, 2013, from *theblaze.com*: www.theblaze.com/stories/2012/09/18/after-20-years-in-prison-the-blind-sheikh-is-back-in-the-news-heres-what-you-need-to-know-about-html/

Homeland Security Act of 2002. PL 107-296, 116. Stat. 2135.

Homeland Security Presidential Decision Directive 4. 2002. *National strategy to combat weapons of mass destruction*. Retrieved August 18, 2014, from http://fas.org/irp/offdocs/nspd/nspd-17.html

Homeland Security Presidential Decision Directive 9. (2004). *Defense of United States agriculture and food*. Retrieved August 19, 2014, from as.org/irp/offdocs/nspd/hspd-9.html

Hosseinian, Z. (2013, June 15). Iran's new president hails "victory of moderation." *Reuters News Service*. Retrieved from http://www.reuters.com/assets/print?aid=USBRE95C1E120130615

Houben, V. J. (2003, July). Southeast Asia and Islam. Annals of the American Academy of Political and Social Science, 588, 149–170.

How North Korean children are taught to hate the "American b*******" at kindergarten. (2012, June 23). Retrieved July 2, 2012, from http://www.dailymail.co.uk/news/article-2163817/How-North-Korean-children-taught-hate-American-b——kindergarten.html

Human Rights Watch. (2012, January). *Cambodia 2012*. Retrieved July 20, 2012, from Human Rights Watch *hrw.org*: http://www.hrw.org/sites/default/files/related_material/cambodia_2012.pdf

Hurtubise, S. (2013, July 24). *Experts: Obamacare will lead to massive spying on US health records*. Retrieved August 19, 2013, from dailycaller.com: http://dailycaller.com/2013/07/24/experts-obamacare-will-lead-to-massive-spying-on-u-s-health-records/

Hussain, Y. (2013, June 21). Colombia crude puts heat on oil sands. Retrieved June 24, 2013, from *business.financialpost.com*: http://business.financialpost.com/2013/06/21/colombia-crude-puts-heat-on-oil-sands/? Isa=db&-d444

Idoiaga, G. E. (2006). *The Basque conflict: New ideas and prospects for peace* (United States Institute for Peace Special Report No. 161). Washington, DC: United States Institute for Peace.

Indira Gandhi's death remembered. (2009). Retrieved from http://news.bbc.co.uk/2/hi/south_asia/8306420.stm

Insight Crime. (n.d.). ELN. Retrieved June 27, 2013, from *insightcrime.org*: http://www.insightcrime.org/groups-colombia/eln

Institute for Economics and Peace. (2012). *Fact sheet: 2012 Global Peace Index*. Retrieved July 18, 2012, from http://www

.visionofhumanity.org/wp-content/uploads/2012/06/2012GPI-Fact-Sheet2.pdf

Intelligence.gov. (2011, March 14). Retrieved March 14, 2011, from Intelligence.gov: http://www.intelligence.gov/about-the-intelligence-community/

International Crisis Group. (2002, December 11). *Indonesia backgrounder: How the Jemaah Islamiyah terrorist network operates* (Asia Report No. 43). Retrieved July 12, 2012, from Asia Report No. 43: http://www.seasite.niu.edu/indonesian/islam/ICG-Indonesia%20Backgrounder%20JI.pdf

International Human Rights and Conflict Resolution Clinic of Stanford Law School (Stanford Clinic) and the Global Justice Clinic at New York University School of Law (NYU Clinic). (n.d.). Living under drones. Retrieved August 23, 2013, from livingunderdrones.org: http://www.livingunderdrones.org/report/

International Human Rights and Conflict Resolution Clinic Stanford Law, Global Justice Clinic NYU School of Law. (2012, September). *Living under drones: Death, injury and trauma to civilians from US drone practices in Pakistan.* Retrieved July 5, 2013, from livingunderdrones.org: http://livingunderdrones.org/wp-content/uploads/2012/10/Stanford-NYU-living-under-drones.pdf

International Institute for Counter-Terrorism. (2011, March 25). Retrieved March 25, 2011, from http://www.ict.org.il/

Iran-backed group claims credit for murdering Jewish family as "natural response." (2011, March 15). Retrieved from WorldTribune.com: http://www.worldtribune.com/worldtribune/WTARC/2011/me_palestinians0285_03_15.html

Israel today. (2009, April 28). PBS. Retrieved from http://www.israeltoday.co.il/NewsItem/tabid/178/nid/18682/Default.aspx?archive=article_title

Israel used chocs to poison Palestinian. (2006, May 8). Retrieved April 20, 2012, from *Sydney Morning Herald:* http://www.smh.com.au/news/World/Israel-used-chocs-to-poison-Palestinian/2006/05/08/1146940441701.html

Israel war of independence. (n.d.). Retrieved March 16, 2012, from http://www.zionism-israel.com/his/Israel_war_independence_1948_timeline.htm

Israeli Ministry of Foreign Affairs. (2010, November 28). *Facts about Israel: The people.* Retrieved March 17, 2012, from http://www.mfa.gov.il/MFA/Facts+About+Israel/People/SOCIETY.htm

Israeli Ministry of Foreign Affairs. (n.d.). *Israel in brief.* Retrieved March 13, 2012, from http://www.mfa.gov.il/MFA/Facts+About+Israel/Israel+in+Brief/ISRAEL+IN+BRIEF.htm?DisplayMode=print

Jacobs, A. (2009, April 27). China still presses crusade against Falun Gong. *The New York Times.* Retrieved July 5, 2012, from http://www.nytimes.com/2009/04/28/world/asia/28china.html

Jacobsen, P. D., Wasserman, J., Botoseneanu, A., Silverstein, A., & Wu, H. W. (2012). The role of law in public health preparedness: Opportunities and challenges. *Journal of Health Politics, Policy, and Law, 37,* 297–328.

Jenkins, B. M. (1980). The study of terrorism: Definitional problems. *RAND Corporation.* Retrieved August 3, 2011, from RAND Corporation: http://www.rand.org/pubs/papers/P6563.html

Jenkins, B. M. (2006). The new age of terrorism. *RAND Corporation.* Retrieved August 11, 2011, from RAND Corporation: http://www.rand.org/pubs/reprints/RP1215

Jenkins, B. M. (2012). *Al Qaeda in its third decade: Irreversible decline or imminent victory?* Santa Monica, CA: RAND Corporation.

Johnson, D. (1981). *American law enforcement: A history.* St. Louis, MO: Forum Press.

Johnson, J. (2014). *Remarks by Secretary of Homeland Security Jeh Johnson at the American Bar Association annual convention—As prepared for delivery* (August 9). Retrieved August 22, 2014, from http://www.dhs.gov/news-releases/press-releases

Johnson, R. (2013, April). The role of communication in the prevention of terrorist attacks. Retrieved August 14, 2013, from *CIP Report:* http://tuscany.gmu.edu/centers/cip/cip.gmu.edu/wp-content/uploads/2013/06/April2013_PartershipsInformationSharing.pdf

Johnston, D. (1995, April 20). At least 31 are dead, scores are missing after car bomb attack in Oklahoma City wrecks 9-story federal office building. Retrieved February 22, 2013, from *New York Times:* http://www.nytimes.com/learning/general/onthisday/big/0419.html

Jones, B. (2005, April 26). Wrongdoing for a cause? Animal Liberation Front is focus of law enforcement in probe of harassment, theft from lab executive. *Newsday* [Nassau and Suffolk Ed.], p. A08.

Jones, R. (2012, April 3). Hamas leader admits "Palestinian" identity is invented. Retrieved April 3, 2012, from *Israel Today Magazine:* http://www.israeltoday.co.il/News/tabid/178/nid/23179/language/en-US/Default.aspx

Jones, S. G. (2013, July 18). *Re-examining the al Qa'ida threat to the United States: Global al Qaeda: Affiliates, objectives, and future challenges.* Retrieved August 16, 2013, from house.gov: http://docs.house.gov/meetings/FA/FA18/20130718/101155/HHRG-113-FA18-Wstate-JonesS-20130718.pdf

Joshua, M. (2012, June 18). Sinai attacks show risks in Israel. Retrieved June 19, 2012, from *online.wsj.com:* http://online.wsj.com/article/SB10001424052702303379204577474852043709024.html?mod=googlenews_wsj

Joshua, P. (2008, November 12). In Peru, a rebellion reborn. *Washington Post,* p. A12.

Kadeer, R. (2012, July 3). China's second Tibet. *Wall Street Journal,* p. A13.

Kageyama, Y. (2012, July 31). Japan's pro-bomb voices rise as nuke power debated. Retrieved August 3, 2012, from *Huffington Post:* http://www.huffingtonpost.com/huff-wires/20120731/as-japan-the-nuclear-option/

Kaplan, J. (2008). Terrorism's fifth wave: A theory, a conundrum and a dilemma. *Perspectives on Terrorism.* Retrieved March 16, 2015, from http://www.terrorismanalysts.com/pt/index.php/pot/article/view/26

Kaplan, J. (2011, Fall). The new/old terrorism. *Phi Kappa Phi Forum,* pp. 1–6.

Kaplan, R. (2010). *Monsoon: The Indian Ocean and the future of American power*. New York, NY: Random House.

Karimi, N. (2012, November 21). *Iran long-range missile technology sent to Gaza, commander says*. Retrieved February 26, 2013, from Honolulu Star Advertiser: http://www.staradvertiser.com/news/breaking/180358621.html?id=180358621

Kasmeri, S. A. (2011, July). *The North Atlantic Treaty Organization and the European Union's common security and defense policy: Intersecting trajectories*. Retrieved July 13, 2012, from Strategic Studies Institute: http://www.strategicstudiesinstitute.army.mil/pdffiles/PUB1078.pdf

Kaufman, J. (1989, August 27). Huey Newton died violently—as did the Black Panthers. *Boston Globe*, A 25.

Kennedy, P. J. (2009). *From Mujahideen to mainstream: The evolution of Hezbollah* (Unpublished thesis). Georgetown University, Washington, D.C.

Kenny, C. (2013, July 15). The case for abolishing the DHS. Retrieved August 6, 2013, from *businessweek.com*: http://www.businessweek.com/articles/2013-07-15/the-case-for-abolishing-the-dhs

Kenny, J. (2012, June 25). Ten years of TSA. *The New American*, pp. 21–23.

Kerner Commission Report. (1969). *Report of the National Advisory Commission on Civil Disorders*. New York, NY: Bantam Books. Retrieved 23 2013, January, from Eisenhower Foundation: http://www.eisenhowerfoundation.org/docs/kerner.pdf

Kershner, I., & EL-Khodar, T. (2007, June 17). Abbas moves to reassert Fatah authority. Retrieved June 2, 2012, from *New York Times*: http://www.nytimes.com/2007/06/17/world/middleeast/17cnd-mideast.html?_r=1

Kershner, I., & EL-Khodary, T. (2009, January 3). Israeli troops launch attack on Gaza. Retrieved June 2, 2012, from *New York Times*: http://www.nytimes.com/2009/01/04/world/middleeast/04mideast.html?pagewanted=all

Kershner, I., & Graham, B. (2008, June 19). Gaza cease-fire takes hold. Retrieved June 2, 2012, from *New York Times*: http://www.nytimes.com/2008/06/19/world/africa/19iht-20mideast.13828230.html

Kessler, G. (2003, January 12). U.S. decision on Iraq has puzzling past: Opponents of war wonder when, how policy was set. *Washington Post*, p. A1.

Kifner, J. (1998, May 2). Eldridge Cleaver, Black Panther who became G.O.P. conservative, is dead at 62 [Obituary]. *New York Times*, p. 8.

Kimmerling, B. (2006). The continuation of Israeli-Palestinian conflict by "academic" means: Reflections on the problematiques of publishing books. *Contemporary Sociology: A Journal of Reviews, 35*(5), 447–449.

Kine, P. (2012, May 25). *China's thuggish para-police*. Retrieved July 7, 2012, from *thediplomat.com*: http://thediplomat.com/2012/05/25/chinas-thuggish-para-police/

King, R. (2013, July 31). NSA chief Keith Alexander speaks about PRISM at Black Hat. Retrieved August 7, 2013, from *blogs.wsj.com*: http://blogs.wsj.com/cio/2013/07/31/general-keith-alexander-speaks-about-prism-at-black-hat/

Kingdon, J. W. (1995). *Agendas, alternatives, and public policies* (2nd. ed.). New York, NY: Longman.

Kirchick, J., & Ahmari, S. (2012, May 30). *We are all Persian grammarians now*. Retrieved June 6, 2012, from *the americaninterst.com*: http://www.the-american-interest.com/article.cfm?piece=1261

Kirkpatrick, D. D. (2013, January 14). Morsi's slurs against Jews stir concern. Retrieved March 4, 2013, from *New York Times*: http://www.nytimes.com/2013/01/15/world/middleeast/egypts-leader-morsi-made-anti-jewish-slurs.html?hp

Kirkpatrick, D. D., & Cowell, A. N. (2013, August 19). *Egyptian court is said to order that Mubarak be released*. Retrieved August 19, 2013, from *New York Times*: http://www.nytimes.com/2013/08/20/world/middleeast/egypt.html?pagewanted=all&_r=0

Kissinger, H. (2014). *World order*. New York, NY: Penguin Press.

Knickerbocker, B. (2011, May 2). Obama: Al Qaeda leader Osama bin Laden killed by US forces. *Christian Science Monitor*, n.p.

Knickmeyer, E. (2011, January 27). *The Arab world's youth army*. Retrieved August 19, 2013, from foreignpolicy.com: http://www.foreignpolicy.com/articles/2011/01/27/the_arab_world_s_youth_army

Krämer, G. (2002). *A history of Palestine: From the Ottoman conquest to the founding of Israel* (G. A. Harman, Trans.). Munchen, Bavaria, Germany: Verlag C. H. Beck oHG.

Kraul, C., & Leon, A. (2013, June 7). Peru's leader rejects Fujimori's plea for pardon. Retrieved July 1, 2013, from *Los Angeles Times*: http://www.latimes.com/news/world/worldnow/la-fg-wn-peru-fujimori-pardon-20130607,0,2397275.story

Kraul, C., & Rotella, S. (2008, October 22). Drug probe finds Hezbollah link. Retrieved June 29, 2013, from *Los Angeles Times*: http://articles.latimes.com/2008/oct/22/world/fg-cocainering22

Kraybill, D. B. (Ed.). (1993). *The Amish and the state*. Baltimore, MD: Johns Hopkins University Press.

Kreiger, H. L. (2012, March 23). Iran may activate US Hezbollah cells after strike. Retrieved June 4, 2012, from *Jerusalem Post*: http://www.jpost.com/International/Article.aspx?id=263098

Kronenwetter, M. (2004). *Terrorism: A guide to events and documents*. Westport, CT: Greenwood Press.

Kronstadt, K. A. (2008). Terrorist attacks in Mumbai, India, and implications for U.S. interests (Congressional Research Service report no. R40087). Washington, DC: Government Printing Office.

Kruglanski, A. W., Chen, X., Dechensne, M., Fishman, S., & Orehek, E. (2009a). Fully committed: Suicide bombers' motivation and the quest for personal significance. *Political Psychology, 30*(3), 331–357.

Kruglanski, A. W., Chen, X., Dechensne, M., Fishman, S., & Orehek, E. (2009b). Yes, no, and maybe in the world of terrorism. *Political Psychology, 30*(3), 401–417.

Kuntzel, M. (2008). Suicide terrorism and Islam. *American Foreign Policy Interests, 30*(4), 227–232.

Kvinta, P. (2011, October). Hunting bin Laden. *Popular Science*, pp. 25–27. Retrieved from *Popular Science*. Retrieved from http://

www.popsci.com/technology/article/2011-09/how-college-class-tracked-down-osama-bin-laden

Kyodo News. (2010, July 17). Terrorist Shigenobu loses appeal. Retrieved August 1, 2012, from *Japan Times:* http://www.japantimes.co.jp/text/nn20100717b1.html

Lady GaGa "devastated" as Indonesia concert cancelled. (2012, May 28). Retrieved July 14, 2012, from BBC News Asia: http://www.bbc.co.uk/news/world-asia-18224783

Lahoud, N. (2010). *The jihadis' path to self-destruction.* New York, NY: Columbia University Press.

Lake, E., & Seper, J. (2011, June 16). Under new leader, al Qaeda issues hit list. Retrieved June 13, 2013, from *washingtontimes.com:* http://www.washingtontimes.com/news/2011/jun/16/bin-laden-deputy-al-zawahri-named-al-qaeda-leader/?page=all

Landler, M. A. (2011, May 20). Obama seeks end to the stalemate on Mideast talks. *New York Times,* p. A1.

Lane, B. (2012, June 28). Terror expert calls for ban on Indonesian militants. Retrieved July 13, 2012, from *The Australian.com:* http://www.theaustralian.com.au/higher-education/terror-expert-calls-for-ban-on-indonesian-militants/story-e6frgcjx-1226409497354

Laquer, W. (1999). *The new terrorism.* New York, NY: Oxford University Press.

Laquer, W. (2003). *No end to war: Terrorism in the 21st century.* New York, NY: Continuum Press.

Laquer, W. A. (Ed.). (2008). *The Israel-Arab reader: A documentary history of the Middle East conflict* (7th Updated and Rev. ed.). New York, NY: Penguin.

Leheny, D. (2010). Terrorism risks and counterterrorism cost in Post-9/11 Japan. *Japan Forum,* pp. 219–237.

Lerhe, E. (2009, March). "Connecting the dots" and the Canadian counter-terrorism efforts: Steady progress or technical, bureaucratic, legal, and political failure. Retrieved October 16, 2012, from *Canadian Defence & Foreign Affairs Institute* (www.cdfai.org): http://www.cdfai.org/PDF/Connecting%20the%20Dots%20and%20the%20Canadian%20Counter-terrorism%20Effort.pdf

Levant, E. (2012, October 2). Out of Guantanamo and into a Canadian prison. Retrieved October 14, 2012, from *online.wsj.com:* http://online.wsj.com/article/SB10000872396390444592404578028600592805828.html

Levitt, G. M. (1988). *Democracies against terror: The Western response to state-supported terrorism.* Washington, DC: Center for Strategic and International Studies.

Levitt, M. (2011). A history of violence: Is there anyone who still doubts that Iran is a terrorist state? *Foreign Policy.* Retrieved from http://foreignpolicy.com/2011/10/12/a-history-of-violence/

Lewis, B. (1995). *The Middle East: A brief history of the last 2,000 years.* New York, NY: Scribner.

Lewis, B. (2009, October 23). *The Iranian difference.* Washington, DC: Association for the Study of the Middle East and Africa.

Lewis, J. E. (2004, May 18). Testimony: Animal rights extremism and ecoterrorism. Retrieved February 7, 2013, *FBI.gov:* http://www.fbi.gov/news/testimony/animal-rights-extremism-and-ecoterrorism

Lipin, M. (2012, July 5). China's Xinjiang province a "police state" 3 years after riots. Retrieved from *Voice of America News:* http://www.voanews.com/content/uighur-china-xingiang-rights/1364092.html

Little, D. (2004, November). Mission impossible: The CIA and the cult of covert action in the Middle East. *Diplomatic History, 28*(5), 663–701.

Loiko, S. L., & McDonnell, P. J. (2013, March 1). Russia charges that U.S. aid helps Syrian "extremists." Retrieved March 4, 2013, from http://www.latimes.com/news/world/worldnow/la-fg-wn-russia-criticizes-aid-syrian-rebels-20130301,0,7206321.story

Long, B. (2003, March 20). Bush: "No outcome except victory." Retrieved July 25, 2013, from *cnn.com:* http://www.cnn.com/2003/WORLD/meast/03/19/sprj.irq.war.bush/

Lum, T. (2011). *Human rights in China and U.S. policy.* Washington, DC: Congressional Research Service. Retrieved July 5, 2012

Lum, T. (2012, April 5). *Republic of the Philippines and U.S. interests.* Retrieved July 17, 2012, from fas.org: http://www.fas.org/sgp/crs/row/RL33233.pdf

Lyman, P. N., & Morrison, J. S. (2004). The terrorist threat in Africa. *Foreign Affairs, 83*(1), 75–86.

Madeira, J. L. (2012, June 12). McVeigh's execution, eleven years later. Retrieved February 22, 2013, from *Huffington Post:* http://www.huffingtonpost.com/jody-lynee-madeira/mcveighs-execution-11-yea_b_1587786.html

Mandel, H. (2012, June 10). Amnesty International criticizes cutting Tibet off from world. Retrieved July 5, 2012, from *examiner.com:* http://www.examiner.com/article/amnesty-international-criticizes-tibet-off-from-world

Manning, T. (2013, June 25). Written testimony. Retrieved August 12, 2013, from *dhs.gov:* https://www.dhs.gov/news/2013/06/25/written-testimony-fema-senate-homeland-security-and-governmental-affairs

Manuel, J. (2013, May 1). The long road to recovery: Environmental health impacts of Hurricane Sandy. Retrieved August 12, 2013, from *Environmental Health Perspective:* http://ehp.niehs.nih.gov/121-a152/

Maoist Shining Path rebels killed four people and exploded bombs across Peru to mark the ninth anniversary. (1989, May 18). *Los Angeles Times,* p. 2.

Martin, G. (2011). *Terrorism and homeland security.* Thousand Oaks, CA: Sage

Maulia, E. (2014, December 30). Indonesia comes of age in 2014 elections. Retrieved March 13, 2015, from *JakartaGlobe.com:* http://thejakartaglobe.beritasatu.com/news/%EF%BB%BFindonesia-comes-age-2014-elections/

May, C. (2012, May 10). The real Palestinian refugee problem. *National Review Online.* Retrieved May 10, 2012, from http://www.nationalreview.com/articles/299512/real-palestinian-refugee-problem-clifford-d-may#

McCarthy, A. C. (2013, May 25). Obama's cynical war speech: Islamic supremacism is not based on a lie. Retrieved June 21, 2013, from *National Review Online:* http://www.nationalreview.com/article/349313/obama-s-cynical-war-speech

McCullough, M. (2005, June 21). Bio tech fights back on animal rights. *Philadelphia Inquirer*.

McCurry, J. (2008, December 12). Founder of Japan's Red Army in final appeal for freedom. Retrieved August 1, 2012, from *Guardian.Co.uk*: http://www.guardian.co.uk/world/2008/dec/13/japan-fusako-shigenobu-red-army

McDougall, A. (2009). State power and its implications for civil war Colombia. *Studies in Conflict and Terrorism, 32*, 322–345.

McNicoll, G. (2005). Demographic future of East Asian regional integration. In T. Pempel (Ed.), *Remapping East Asia: The construction of a region* (pp. 54–75). Ithaca, NY: Cornell University Press.

Meir Amit Intelligence and Information Center. (2011, March 25). *Home*. Retrieved March 25, 2011, from Meir Amit Intelligence and Information Center: http://www.terrorism-info.org.il/site/home/default.asp

Menon, R. (2012, April 16). How to handle North Korea: Don't just do something, stand there. *Huffington Post*. Retrieved July 2, 2012, from http://www.huffingtonpost.com/rajan-menon/north-korea_b_1425168.html

Metz, H. C. (Ed.). (1992). *Saudi Arabia: A country study*. Washington, DC: Government Printing Office.

Metz, H. C. (Ed.). (1994). *Algeria: A country study*. Washington, DC: Government Printing Office.

Meyer, J. (2007, July 22). Congressman eyes the Chiquita case; The Justice Department took four years to file a criminal case. *Los Angeles Times*, p. A19.

Meyer, J., & Kraul, C. (2007, June 29). U.S. firms linked to Colombia militias. *Los Angeles Times*, p. A19.

MI5 or MI6? (n.d.). Retrieved April 2, 2013, from *mi5.gov*: https://www.mi5.gov.uk/careers/working-at-mi5/working-with-mi6-and-gchq/mi5-or-mi6.aspx

Milewski, T. (2007, June 28). Symbols and suits: Sikh extremism enters politics. Retrieved October 5, 2012, from *CBC News*: http://www.cbc.ca/news/background/sikh-politics-canada/index.html

Milewski, T. (2010, June 23). Excusing the inexcusable: After a quarter-century, Canada says sorry for botching the Air India investigation. Retrieved October 5, 2012, from *CBC News*: http://www.cbc.ca/news/canada/story/2010/06/23/f-air-india-25th.html

Miller, D. (Ed.). (1991). *The Blackwell encyclopedia of political thought*. Oxford, UK: Blackwell.

Miller, G. (2013, June 30). A trail of inaccuracy about NSA programs. Retrieved August 14, 2013, from *washingtonpost.com*: http://articles.washingtonpost.com/2013-06-30/world/40292346_1_programs-clapper-jr-remark

Miller, H. I. (2012, August 5). *The Nuking of Japan Was a Military and Moral Imperative*. Retrieved April 20, 2015, from forbes.com: http://www.forbes.com/sites/henrymiller/2014/08/05/the-nuking-of-japan-was-a-military-and-moral-imperative

Miller, H. I., & Conko, G. (2012, September 5). Rachel Carson's deadly fantasies. Retrieved May 22, 2013, from *forbes.com*: www.forbes.com/sites/miller/2012/09/05rachel-carsons-deadly-fantasies/print/

Miller, P. D. (2011, November 17). *When will the U.S. drone war end?* Retrieved August 23, 2013, from washingtonpost.com: http://www.washingtonpost.com/opinions/when-will-the-us-drone-war-end/2011/11/15/gIQAZ677VN_story.html

Miniter, R. (2003). *Losing bin Laden*. Washington, DC.: Regnery Publishing.

Mintz, A., & Brule, D. (2009). Methodological issues in studying suicide terrorism. *Political Psychology, 30*(3), 365–371.

Modern history. (n.d.). Retrieved March 17, 2012, from *mideastweb.org*: http://www.mideastweb.org/abumazen.htm

Modern Israel and the Israeli-Palestinian conflict: A brief history. (n.d.). *MidEast Web*. Retrieved March 17, 2012, from http://mideastweb.org/briefhistory-oslo.htm

Moens, A. (1991). President Carter's advisers and the fall of the Shah. *Political Science Quarterly, 106*(2), 211–237.

Mohamed Morsi. (n.d.). Retrieved March 4, 2013, from *New York Times* Topics: http://topics.nytimes.com/top/reference/timestopics/people/m/mohamed_morsi/index.html?inline=nyt-per

Molinski, D. (2013, January 15). Colombian rebel group steps up violence. *Wall Street Journal*, p. A13.

Montgomery, D. (2002, April 20). An unclenched fist; Black Panthers recall the bad times and salute the good that grew from them. *The Washington Post*, p. C1.

Mozingo, J. (2006, September 6). A thin line on animal rights; Dr. Jerry Vlasak stays carefully in the world of medicine while serving as a spokesman for extremists who threaten laboratory researchers. *Los Angeles Times*, p. B1.

Mugabe lables NATO a "terrorist group" over Libya. (2011, August 8). BBC News Africa. Retrieved August 8, 2011, from BBC News Africa: http://www.bbc.co.uk/news/world-africa-14452522

Munoz, S. S. (2013, June 29). Colombian land deals probed. *Wall Street Journal*, p. A11.

Murphy, K. (2001, June 9). The Nation; Environmental group forms an incendiary core, FBI says; sabotage: With attacks rising, concern grows that a new hierarchy is turning the Earth Liberation Front into a terrorist organization. *Los Angeles Times*, p. A12.

Myanmar's Suu Kyi to travel to US to receive award. (2012, July 17). *Bloomberg Business Week*. Retrieved July 18, 2012, from: http://www.businessweek.com/ap/2012-07-17/myanmars-suu-kyi-to-travel-to-us-to-receive-award

Myers, S. L. (2007, November 27). Framework set by Palestinians and Israelis for peace talks. Retrieved June 2, 2012, from *New York Times*: http://www.nytimes.com/2007/11/27/washington/27cnd-prexy.html?pagewanted=all

Nadler, J. L., Markey, E. J., & Thompson, B. G. (2012, June 27). Cargo, the terrorists' Trojan horse. *New York Times*, p. A27.

Nagle, D. (2002, April 19). USS Cole rejoins the fleet. Retrieved June 17, 2013, from *navy.mil*: http://www.navy.mil/submit/display.asp?story_id=1415

Nakashima, E. (2010, November 3). Control of intelligence budget will shift. *The Washington Post*, p. A2.

Nakashima, E., & Wilson, S. (2013, June 11). ACLU sues over NSA surveillance program. Retrieved August 14, 2013, from *washingtonpost.com:* http://articles.washingtonpost.com/2013-06-11/politics/39893547_1_surveillance-program-clapper-jr-aclu

Napolitano, J. (2013, August 19). *Appointment of new Deputy Under Secretary for Cybersecurity.* Retrieved August 19, 2013, from dhs.gov: http://www.dhs.gov/blog/2013/08/19/appointment-new-deputy-under-secretary-cybersecurity

Nash, N. C. (1992, September 22). Lima Journal; Shining Path women: So many and so ferocious. *New York Times.* Retrieved from http://www.nytimes.com/1992/09/22/world/lima-journal-shining-path-women-so-many-and-so-ferocious.html

National Consortium for the Study of Terrorism and Responses to Terrorism. (2009, July 17). *Press Release.* Retrieved July 12, 2012, from Background on Terrorism and Public Opinion in Indonesia: http://www.start.umd.edu/start/media/Background_on_Terrorism_Public_Opinion_Indonesia.pdf

National Counterterrorism Center. (2009, April 30). *2008 report on terrorism.* Retrieved February 18, 2011, from www.nctc.gov: http://docs.google.com/viewer?a=v&q=cache:E24Gz7hVV_0J:www.fbi.gov/stats-services/publications/terror_08.pdf+definition+of+terrorism&hl=en&gl=us&pid=bl&srcid=ADGEESjbWbufK_Lq8rkQ0aFqh7nGt3x0eNd3Le_lT5eYAtgLViRS5hHOt8CjdW3mLvcm-M8Icbfa9wMvdvjg25A077FIzK6xd

National Counterterrrrorism Center. (2011, December 27). *Foreign terrorist organizations.* Retrieved July 12, 2012, from *Counterterrorism 2014 Calendar:* http://www.nctc.gov/site/other/fto.html

National Counterterrorism Center. (2011). *National Counterterrorism Center 2010 report on terrorism.* Washington, DC. Retrieved January 20, 2012, from http://www.nctc.gov/

National Counterterrorism Center. (2012). *Counterterrorism 2012 calendar.* Retrieved from http://www.nctc.gov

National Institute of Justice. (2014). *Terrorism.* Retrieved July 28, 2014, from http://www.nij.gov/topics/crime/terrorism/Pages/welcome.aspx#note1

National Security Agency. (n.d.). National Security Agency: Sixty years of defending our nation. Retrieved August 13, 2013, from *nsa.gov:* http://www.nsa.gov/about/cryptologic_heritage/60th/book/NSA_60th_Anniversary.pdf

NBC News. (2006, April 16). Lessons learned, and not learned, 11 years later: Survivors fear impact of 1995 Oklahoma City bombing being forgotten. Retrieved February 15, 2013, from *nbcnews.com:* http://www.nbcnews.com/id/12343917/#.UR7O06U4vcw

Nemeth, C. P. (2010). *Homeland security: An introduction to principles and practices.* Boca Raton, FL: CRC Press.

Net Bible. (n.d.). Retrieved March 20, 2012, from bible.org: http://net.bible.org/#!bible/1+Chronicles+11:2

Neuman, W. (2012, May 27). Peru forced to confront deep scars of civil war. *New York Times,* p. A6.

Nieto, W. A. (2011, September). Give war a chance—Revisited the price to pay: The military and terrorism in Peru. *Defence Studies, 11*(3), 517–540.

Nixon, R. (2013, August 6). T.S.A. expands duties beyond airport. *New York Times,* p. A11.

Nordland, R. (2014, September 22). After rancor, Afghans agree to share power. *New York Times,* p. A1.

No sex please, we're Japanese: Country heads for extinction as survey reveals young people shunning marriage. (2011, November 28). Retrieved August 3, 2012, from *Daily Mail Reporter:* http://www.dailymail.co.uk/news/article-2067327/Japan-heads-extinction-survey-reveals-young-people-shunning-marriage.html

Nossiter, A., & Kirkpatrick, D. D. (2014, May 8). Abduction of girls an act not even al Qaeda can condone. *New York Times,* p. 5.

Nuclear aftershocks. (2012, January 17). Retrieved from *Frontline:* http://www.pbs.org/wgbh/pages/frontline/nuclear-aftershocks

Nuclear Threat Initiative. (2012). *Country profiles: Pakistan.* Retrieved from www.nti.org/country-profiles/pakistan

Obama, B. (2013, February 12). Presidential policy directive—Critical infrastructure security and resilience PPD 21. Retrieved August 12, 2013, from *whitehouse.gov:* http://www.whitehouse.gov/the-press-office/2013/02/12/presidential-policy-directive-critical-infrastructure-security-and-resil

Obama, B. (2013, May 23). Remarks of President Barack Obama. Retrieved June 21, 2013, from *whitehouse.gov:* http://www.whitehouse.gov/the-press-office/2013/05/23/remarks-president-barack-obama

Office for Victims of Crime. (2000, October). *Responding to terrorism victims: Oklahoma City and beyond.* Retrieved July 22, 2013, from http://www.ojp.usdoj.gov/ovc/publications/infores/respterrorism/welcome.html

Oklahoma Department of Civil Emergency Management After Action Report Alfred P. Murrah Federal Building Bombing, The. (n.d.). Retrieved July 23, 2013, from Office of Emergency Management: http://www.ok.gov/OEM/documents/Bombing%20After%20Action%20Report.pdf

Oklahoma Emergency Management. (n.d.). *The Oklahoma Department of Civil Emergency: After action report Alfred P. Murrah Federal Building 19 April 1995 in Oklahoma City, Oklahoma.* Retrieved February 22, 2013, from Office of Emergency Management: http://www.ok.gov/OEM/documents/Bombing%20After%20Action%20Report.pdf

Operation Cast Lead. (n.d.). *Global Security.org.* Retrieved June 2, 2012, from http://www.globalsecurity.org/military/world/war/operation-cast-lead.htm

Oren, A. (2012, June 10). *Israel's settlement policy could trigger a third intifada, experts warn Netanyahu.* Retrieved June 18, 2012, from http://www.haaretz.com/news/diplomacy-defense/israel-s-settlement-policy-could-trigger-a-third-intifada-experts-warn-netanyahu.premium-1.435377

Oren, M. (2011, June 6). Remembering six says in 1967. *ForeignPolicy.com.* Retrieved March 14, 2012, from http://www.foreignpolicy.com/articles/2011/06/06/remembering_six_days_in_1967

Origins of Sikhism. (2009). *BBC Religions.* Retrieved from http://www.bbc.co.uk/religion/religions/sikhism/history/history

O'Rourke, L. A. (2008, August 2). Behind the woman behind the bomb. *New York Times*. Retrieved February 11, 2012, from http://www.nytimes.com/2008/08/02/opinion/02orourke.html?pagewanted=all

O'Rourke, L. A. (2009). What's special about female suicide terrorism? *Security Studies, (18)*, 681–718.

OSAC. (n.d.). *Indonesia 2012 crime and safety report: Jakarta*. Retrieved July 13, 2012, from osac.gov: https://www.osac.gov/Pages/ContentReportDetails.aspx?cid=12568

Otis, J. (2013, June 11). Political fallout for Colombia's peace talks with FARC rebels. Retrieved June 25, 2013, *from world.time.com:* http://world.time.com/2013/06/11/political-fallout-for-colombias-peace-talks-with-farc-rebels/

Ott, J. (2011, September 12). Assessing Risk aviation security since 9/11. *Aviation Week & Space Technology, 173*(32), 48.

Owens, J. E. (2009, June-September). Congressional acquiescence to presidentialism in the US "war on terror." *The Journal of Legislative Studies, 15*(2,3), 147–190.

Painter, W. L. (2013). *Issues in homeland security policy for the 113th Congress* (CRS Report No. R42985). Washington, DC: Congressional Research Service.

Palmer, M. (2003). Breaking the real axis of evil: How to oust the world's last dictators by 2025. Lanham, MD: Rowman & Littlefield.

Pangi, R. (2008). Consequence management in the 1995 sarin attacks on the Japanese subway system. In R. D. Howard & J. F. Forest (Eds.), *Weapons of mass destruction and terrorism* (pp. 429–458). New York, NY: McGraw-Hill.

Parlade, A. G., Jr. (2006, June 16). *An analysis of the communist insurgency in the Philippines*. Retrieved July 17, 2012, from DTIC: http://www.dtic.mil/cgi-bin/GetTRDoc?AD=ADA463770

Parry, A. (1976). *Terrorism: From Robespierre to Arafat*. New York, NY: Vanguard Press.

Parsons, D. (1986, September 24). Police learn '60s riot experience has relevance in '80s. Retrieved January 23, 2013, from *LA Times*: http://articles.latimes.com/1986-09-24/news/vw-8927_1_police-departments

Partnership for Public Service, The. (2011, August). Securing the future: Management lessons of 9/11. Retrieved August 6, 2013, from *ourpublicservice.org*: http://www.ourpublicservice.org/OPS/publications/viewcontentdetails.php?id=164

Paul, R. (2012, June 14). *S. 3303*. A bill to require.... Retrieved August 9, 2013, from *govtrack.us*: http://www.govtrack.us/congress/bills/112/s3303

Payton, L. (2012, June 20). Sweeping immigration changes to give new power to minister. Retrieved October 14, 2012, from *cbcnews*: http://www.cbc.ca/news/politics/story/2012/06/20/pol-new-law-deport-foreigners-crimes.html

Peace Pledge Union. (n.d.). *Cambodia 1975*. Retrieved July 2012, 2012, from ppu.org: http://www.ppu.org.uk/genocide/g_cambodia1.html

Penketh, A. (2005, October 27). Iran's leader says Jewish state "should be wiped from map." Retrieved March 4, 2013, from *The Independent*: http://www.independent.co.uk/news/world/middle-east/irans-leader-says-jewish-state-should-be-wiped-from-map-8037358.html

Peru greets New Years Eve in darkness as rebels sabotage power sources. (1988, January 2). *Los Angeles Times*, p. 20.

Peru snares elusive Shining Path leader. (2003, July 8). *The Australian*, p. 14.

Petrou, M. (2011, January 17). Iran: His brand of extremism is under fire from political and religious opponents. *Maclean's, 124(1)*, 40–42.

Philippine government, rebels try to break impasse. (2012, June 17). norwaynews.com/en. (2012, June 17). *Philippine government, rebels try to break impasse*. Retrieved July 17, 2012, from norwaynews.com.en: http://www.norwaynews.com/en/~view.php?72X8b54IQ84824x285Qpe844OR3883SS76Azh353Pd28

Phl gov't, MILF conclude 29th formal peace talks. (2012, July 18). *Phl gov't, MILF conclude 29th formal peace talks*. Retrieved July 18, 2012, from philstar.com: http://www.philstar.com/Article.aspx?articleId=828987&publicationSubCategoryId=200

Platt, T. (1982). *The iron fist and the velvet glove: An analysis of the U.S. police* (3rd ed.). San Francisco, CA: Crime and Social Justice Associates.

Ploughshares. (2011, April 30). Philippines-CPP/NPA (1969—first combat deaths). Retrieved July 17, 2012, from ploughshares.ca: http://www.ploughshares.ca/content/philippines-cppnpa-1969-first-combat-deaths

Plummer, C. (2012). Failed states and connections to terrorist activity. *International Criminal Justice Review, 22*, 416–449.

Population Reference Bureau. (n.d.). *Southeast Asia highlights*. Retrieved June 27, 2012, from prb.org: http://www.prb.org/DataFinder/Geography/Data.aspx?loc=391

Preston, T. (2007). *From lambs to lions: Future security relationships in a world of biological and nuclear weapons*. Lanham, MD: Rowman & Littlefield.

Price, H., Jr. (1977). The strategy and tactics of revolutionary terrorism. *Comparative Studies in Society and History, 19*, 52–65.

Proposal for Department of Homeland Security. (2002, June). Retrieved April 4, 2011, from Proposal for Department of Homeland Security: http://www.dhs.gov/xlibrary/assets/book.pdf

Rabasa, A., & Chalk, P. (2012). Non-traditional threats and maritime domain awareness in the tri-border area of Southeast Asia: The coast watch system of the Philippines. Retrieved July 10, 2012, from *rand.org*: http://www.rand.org/pubs/occasional_papers/OP372.html

RAND Corporation. (2011, March 25). About the RAND Database of Worldwide Terrorism Incidents. Retrieved March 25, 2011, from The Rand Terrorism Database: http://www.rand.org/nsrd/projects/terrorism-incidents/about.html

Rapoport, D. C. (1984). Fear and trembling: Terrorism in three religious traditions. *American Political Science Review, 78*(3), 658–677.

Rapoport, D. C. (2002). The four waves of rebel terror and September 11. *Anthropoetics, 8*(Spring/Summer). Retrieved from http://www.anthropoetics.ucla.edu/ap0801/terror.htm

Rapoport, D. C. (2004). The four waves of terrorism. In A. K. Cronin & J. M. Ludes (Eds.), *Attacking terrorism: Elements of a grand strategy* (pp. 46–73). Washington, DC: Georgetown University Press.

Rapoport, D. C. (2006, June 5). *The four waves of modern terrorism.* Retrieved August 19, 2013, from Burkle Center for International Relations: http://www.international.ucla.edu/burkle/article.asp?parentid=47153

Rastogi, N. (2008, March 28). *Why does China care about Tibet?* Retrieved July 5, 2012, from slate.com: http://www.slate.com/articles/news_and_politics/explainer/2008/03/why_does_china_care_about_tibet.html

Read, J. M. (2009). Charting a course through radical Islam: Origins, rise, transformation. *Defence Studies*, 269–305.

Red Brigades suspects to be tried. (2004, October 20). *BBC.* Retrieved from http://news.bbc.co.uk/go/pr/fr/-/2/hi/europe/3758326.stm

Reidel, B. (2008). Pakistan and terror: The eye of the storm. *Annals of the American Academy of Political and Social Science, 618*, 31.

Religion and the Comunist Party. (2011, February 11). Retrieved July 7, 2012, from *economist.com:* http://www.economist.com/node/21547287

Results in brief: Marine Corps implementation of the urgent universal needs process for mine resistant ambush protected vehicles. (2008). DoD Inspector General Report No. D-2009-030 (Project No. D-2009-030). Retrieved March 2, 2012, from http://www.dodig.osd.mil/Audit/reports/fy09/09-030%20RIB.pdf

Rettig, M. (2013, February 12). Incomplete intelligence reform: Why the U.S. intelligence community needs an empowered ODNI. Retrieved August 14, 2013, from *diplomaticourier.com:* http://www.diplomaticourier.com/news/topics/security/1250-incomplete-intelligence-reform-why-the-u-s-intelligence-community-needs-an-empowered-odni

Rhodes, J. (1999, January 1). Fanning the flames of racial discord: The national press and the Black Panther Party. *Harvard International Journal of Press/Politics, 4*(95), 96–118.

Richardson, J. (2011, January 3). *US cable leaks' collateral damage in Zimbabwe.* Retrieved February 12, 2011, from guardian.co.uk: http://www.guardian.co.uk/commentisfree/cifamerica/2011/jan/03/zimbabwe-morgan-tsvangirai. Copyright Guardian News & Media Ltd 2011. Reprinted with permission.

Riechel, P. L. (1988). Southern slave patrols as a transitional police type. *American Journal of Police, 7*(1), 51–77.

Roberts, D. (2013, May 2). *US drone strikes being used as alternative to Guantánamo, lawyer says.* Retrieved August 23, 2013, from theguardian.com: http://www.theguardian.com/world/2013/may/02/us-drone-strikes-guantanamo

Rochlin, J. (2011). Plan Colombia and the revolution in military. *Review of International Studies, 37*, 715–740.

Romano, L. (1998, May 28). Fortier gets 12 years in bombing case. Retrieved February 22, 2013, from *Washington Post:* http://www.washingtonpost.com/wp-srv/national/longterm/oklahoma/stories/fortier052898.htm

Romero, S. (2009, December 11). World Briefing/The Americas; Ecuador: Report says U.S. aided attack on rebels. Retrieved June 28, 2013, from *nytimes.com:* http://query.nytimes.com/gst/fullpage.html?res=9C06E5D6143EF932A25751C1A96F9C8B63

Ronen, G. (2011, August 26). Arut Sheva Israel National News 7. Retrieved August 26, 2011, from http://www.israelnationalnews.com/: http://www.israelnationalnews.com/News/News.aspx/147116#.TlQUIV3bAxE

Ross, J. I. (2006). *Political terrorism: An interdisciplinary approach.* New York, NY: Peter Lang.

Rudd, M. (2011, January 11). An ex-Weather Underground radical on the Tucson shootings and political violence. Retrieved January 24, 2013, from *Washington Post:* http://www.washingtonpost.com/wp-dyn/content/article/2011/01/14/AR2011011405029.html

Rudoren, J., & Fares, A. (2012, May 20). Palestinians sign deal to set up elections. Retrieved June 5, 2012, from *New York Times:* http://www.nytimes.com/2012/05/21/world/middleeast/hamas-and-fatah-agree-in-cairo-to-begin-work-on-elections.html

Ruiz, B. (2001). *The Colombian civil war.* Jefferson, NC: McFarland.

Rummel, R. J. (1994). Statistics of democide: Genocide and mass murder since 1900 (Vol. 2: Macht und Gesellschaft series). Munster, Germany: LIT.

Ryan, J. R., & Glarum, J. F. (2008). *Biosecurity and bioterrorism: Containing and preventing biological threats.* Amsterdam, the Netherlands: Butterworth-Heinemann.

Sachar, H. M. (2010). *A history of Israel from the rise of Zionism to our time* (3rd ed.). New York, NY: Alfred A. Knopf.

Sadler, S. (2013, June 18). Threat, risk and vulnerability: The future of the TWIC program. Retrieved from *dhs.gov:* http://www.dhs.gov/news/2013/06/18/written-testimony-tsa-house-homeland-security-subcommittee-border-and-maritime

Safire, W. (2002, January 20). The way we live now: 01-20-02: On language; Homeland. *New York Times Magazine, 12.*

Saiget, R. (2012, June 4). *China rounds up activists on Tiananmen anniversary.* Retrieved June 26, 2012, from tibetsun.com: http://www.tibetsun.com/archive/2012/06/04/china-rounds-up-activists-on-tiananmen-anniversary/

Salaheddin, S., & Karam, Z. (2013, August 19). *Syria refugee crisis: More than 20,000 have fled war-torn country since last Thursday.* Retrieved August 19, 2013, from huggingtonpost.com: http://www.huffingtonpost.com/2013/08/19/syria-refugee-crisis_n_3778358.html

Sanders, E. (2011, January 26). Leaked documents show Palestinians ready to deal at 2008 peace talks. *Los Angeles Times.* Retrieved June 5, 2012, from http://articles.latimes.com/2011/jan/26/world/la-fg-palestinian-papers-20110126

Sato, S. (2012, July 12). *Khmer tribunal "a battle against time": Japan judge.* Retrieved July 19, 2012, from AFP: http://news.yahoo.com/khmer-tribunal-battle-against-time-japan-judge-171934953.html

Savage, C., Wyatt, E., & Baker, P. (2013, June 6). U.S. confirms gathering of web data overseas. *New York Times,* p. A1.

Savitz, E. (2012, July 2). *Intel reportedly buys Israeli startup IDesia Biometrics*. Retrieved August 22, 2013, from Forbes.com: http://www.forbes.com/sites/ericsavitz/2012/07/02/intel-reportedly-buys-israeli-startup-idesia-biometrics/

Scham, P. (n.d.). Traditional narratives of Israeli and Palestinian history: Distillation. *jewishvirtuallibrary.org*. Retrieved 5 2012, April, from http://www.jewishvirtuallibrary.org/jsource/History/narratives.html

Schanzer, J. (2012, May 31). US taxpayers aid to UNRWA tops $10 billion. *The Hill*. Retrieved June 6, 2012, from http://thehill.com/blogs/congress-blog/foreign-policy/230249-us-taxpayers-aid-to-unrwa-tops-10-billion

Schemo, D. J. (1997, April 23). Peru held hostage: Four months of torment ends abruptly. *New York Times*. Retrieve from http://www.nytimes.com/1997/04/23/world/peru-held-hostage-four-months-of-torment-ends-abruptly.html

Schemo, D. J. (1997, January 4). A born revolutionary's path to a living "tomb" in Peru. *New York Times*. Retrieved from http://www.nytimes.com/1997/01/04/world/a-born-revolutionary-s-path-to-a-living-tomb-in-peru.html

Scheuer, M. (2004). *Imperial hubris: Why the West is losing the war on terror*. Washington, DC: Potomac Books.

Scheuer, M. (2007, April 17). *Prepared statement for U. S. House of Representatives: "Extraordinary rendition in U.S. counterterrorism policy: The impact on transatlantic relations."* Retrieved from https://fas.org/irp/congress/2007_hr/rendition.pdf

Schimmel, K. S. (2012). Protecting the NFL/militarizing the homeland: Citizen soldiers and urban resilience in post-9/11 America. *International Review for the Sociology of Sport, 47*(3), 338–357.

Schmid, A. P., Jongman, A. J., & Stohl, M. (2005). *Political terrorism: A new guide to actors, authors, concepts, databases, theories and literature* (updated, expanded ed.). New Brunswick, NJ: Transaction.

Schmitt, E. (2015, March 23). Out of Yemen, U.S. is hobbled in terror fight. *New York Times*. Retrieved from http://search.proquest.com/docview/1665092888?accountid=10358

Schnelle, S. (2012, January). Abdullah Azzam, ideologue of jihad: Freedom fighter or terrorist? *Journal of Church and State, 54*(4), 625–647.

Schwartz, M. J. (2012, November 14). Obama secret order authorizes cybersecurity strikebacks. Retrieved August 12, 2013, from *informationweek.com:* http://www.informationweek.com/government/security/obama-secret-order-authorizes-cybersecur/240134945

Schwirtz, M. (2011, July 25). After killings, unease in Norway, where few police carry guns. *New York Times*. Retrieved July 28, 2011, from New York Times: http://www.nytimes.com/2011/07/26/world/europe/26police.html

Secretary of Defense Robert Gates, D. N. (2010, February 1). DoD news briefing on the release of the financial year 2010 defense budget. Mine Resistant Ambush Protected Vehicle Program. Available at: strategicstudiesinstitute.army.mil/files/SSI-pub-1064.epub

Seditionist and sovereignty movements in the USA. (2014). Retrieved from *angelfire.com:* http://www.angelfire.com/nv/micronations/usa.html

Semerad, T. (2013, June 29). *NSA in Utah: Mining a mountain of data*. Retrieved August 21, 2013, from sltrib.com: http://www.sltrib.com/sltrib/news/56515678-78/data-nsa-http-www.html.csp

Shahar, Y. (n.d.). The history of Iranian sponsored terrorism. Retrieved March 15, 2013, from *higginsctc.org:* http://www.higginsctc.org/terrorism/Iraniansponsoredterrorism.htm

Shay, C. (2009, October 1). *A brief history of Abu Sayaf*. Retrieved July 16, 2012, from time.com: http://www.time.com/time/world/article/0,8599,1927124,00.html

Shea, D. A. (2014). *Implementation of chemical facility anti-terrorism standards (CFATS): Issues for Congress* (CRS Report No. R43346). Washington, DC: Congressional Research Service.

Shining Path. (n.d.). In *Infoplease online encyclopedia*. Retrieved from http://www.infoplease.com/encyclopedia/history/shining-path.html

Shirley, W. D. (2012, June). *When activism is terrorism: Special interest politics and state repression of the animal rights movement* (Doctoral dissertation, University of Oregon). UMI Dissertation Express, Number: 3523394. Available at http://dissexpress.umi.com/dxweb/results.html?QryTxt=&By=&Title=when+activism+is+terrorism&pubnum=3523394

Silke, A. (2008). Holy warriors: Exploring the psychological processes of jihadi radicalization. *European Journal of Criminology, 5*, 99–123.

Simpson, P. (2012, January 17). *Chinas urban population exceed rural for first time ever*. Retrieved July 7, 2012, from telegraph.co.uk: http://www.telegraph.co.uk/news/worldnews/asia/china/9020486/Chinas-urban-population-exceeds-rural-for-first-time-ever.html

Sinclair, A. (2003). *An anatomy of terror*. London: Pan Books.

Slovic, P. (1987, April 17). Perception of risk. *Science, 236*, pp. 280–285.

Slovic, P. (2004). What's fear got to do with it? Its affect we need to worry about. *Missouri Law Review, 69*, 971–990.

Slovic, P., Finucane, M. L., Peters, E., & MacGregor, D. G. (2004). Risk as analysis and risk as feelings: Some thoughts about affect, reason, risk, and rationality. *Risk Analysis, 24*, 311–322.

Smith, A. (2011). Improvised explosive devices in Iraq, 2003–2009: A case of operational surprise and institutional response. *Strategic Studies Institute*. Carlisle, PA 17013-5046: Director, Strategic Studies Institute, U.S. Army War College, 632 Wright Ave, Carlisle, PA 17013-5046. Retrieved March 2, 2012, from http://www.strategicstudiesinstitute.army.mil/pubs/display.cfm?pubid=1064

Smith, J. F. (1989, September 11). Drug war stepped up in Colombia and Peru, U.S. pilots, agents join in assaults on Andean cocaine labs. *Los Angeles Times* [Home ed.], p. A2.

Smith, M. (2013, February 18). '93 WTC plotter Ramzi Yousef wants contact ban lifted. Retrieved June 232, 2013, from *cnn.com:* http://www.cnn.com/2013/02/17/us/terrorist-prison

Snyder, R. A. (1988). *Negotiating with terrorists: TWA 847*. Washington, DC: Institute for the Study of Diplomacy.

Solomon, E. (2013, December 2). Syria's death toll hits nearly 126,000. *Reuters News Service*. Retrieved from http://www.reuters.com/assets/print?aid=USBRE9B1OES20131202

Southern Poverty Law Center. (n.d.). *Sovereign citizens movement*. Retrieved February 21, 2013, from Southern Poverty Law Center: http://www.splcenter.org/get-informed/intelligence-files/ideology/sovereign-citizens-movement

South Korea profile. (2012, July 3). *BBC News Asia*. Retrieved July 5, 2012, from bbc.com: http://www.bbc.co.uk/news/world-asia-pacific-15292674

Spindlove, J. R., & Simonsen, C. E. (2010). *Terrorism today: The past, the players, the future* (4th ed.). Upper Saddle River, NJ: Prentice Hall.

Spindlove, J. R., & Simonsen, C. E. (2013). *Terrorism today: The past, the players, and the future (5th ed.)*. Upper Saddle River, NJ: Pearson.

Staff of the KlanWatch Project of the Southern Poverty Law Center. (2011). Ku Klux Klan: A history of racism and violence (R. Baudouin, Ed.). Retrieved January 15, 2013, from *Southern Poverty Law Center*: http://www.splcenter.org/get-informed/publications/ku-klux-klan-a-history-of-racism

Stakelbeck, E. (2010, August 22). *Mega mosque plans target America's heartland*. Retrieved August 20, 2013, from http://www.cbn.com/cbnnews/us/2010/August/Mega-Mosque-Plans-Target-Americas-Heartland/

START. (2011, January 24). *Suicide attack at Moscow airport* (Background report). College Park, MD: National Consortium for the Study of Terrorism and Responses to Terrorism.

Steffan, M. (2013, March 11). Drunken quarrel between friends sparks Pakistan biggest religious riot since 2009. Retrieved March 11, 2013, from *blog.christianitytoday.com*: http://blog.christianitytoday.com/ctliveblog/archives/2013/03/drunken-quarrel-between-friends-sparks-pakistan-biggest-religious-riot-since-2009.html

Stenersen, A. (2009). Are the Afghan Taliban involved in international terrorism? *CTC Sentinel, 2*(9), 1–4.

Sternstein, A. (2013, February 12). *Obama's cyber executive order lays foundation for mandatory regulations*. Retrieved August 12, 2013, from http://www.nextgov.com/cybersecurity/2013/02/obamas-cyber-executive-order-lays-foundation-mandatory-regulations/61267/?oref=ng-channelriver

Sternstein, A. (2013, July 25). *NSA's new spy facilities are seven times bigger than the Pentagon*. Retrieved August 21, 2013, from nextgov.com: http://www.nextgov.com/cloud-computing/2013/07/nsas-big-dig/67412/?oref=d-river

Steyn, M. (2006). *America alone: The end of the world as we know it*. Washington, DC: Regnery Publishing.

Stier, M. (2012, March 22). Building one DHS: Why is employee morale so low? Retrieved August 7, 2013, from *homeland.house.gov/hearing*: http://homeland.house.gov/hearing/subcommittee-hearing-building-one-dhs-why-employee-morale-low

Stimson, H. L. (1947, February). The decision to use the atomic bomb. *Harpers Magazine, 194*(1161), pp. 97–107.

Stone, H. (2013, January 23). Peru puts Shining Path leader on trial. Retrieved July 3, 2013, from *Council on Hemispheric Affairs*: http://www.coha.org/the-pan-american-post-peru-puts-shining-path-leader-on-trial/

Story of Maher Arar rendtion to torture, The. (n.d.). Retrieved October 16, 2012, from Center for Constitutional Rights: http://ccrjustice.org/files/rendition%20to%20torture%20report.pdf

Straus, S. (2012). Wars do end! Changing patterns of political violence in sub-Saharan Africa. *African Affairs, 111*(443), 179–201.

Study of Terrorism and Responses to Terrorism. (2013, February 12). *Animal Liberation Front*. Retrieved from Global Terrorism Database: http://www.start.umd.edu/tops/terrorist_organization_profile.asp?id=14

Sunstein, C. R. (2003). Terrorism and probability neglect. *The Journal of Risk and Uncertainty, 26*, 121–136.

Suro, R. (1987, February 6). *Italians see links to Syria in 1985 airport attack*. Retrieved March 5, 2014, from http://www.nytimes.com/1987/02/06/world/italians-see-links-to-syria-in-1985-airport-attack.html

Sweetman, B., Mackenzie, C., & Eshel, D. (2012). Better security through technology. *Aviation Week, 174*(41), 4–5.

Sylves, R. (2008). *Disaster policy and politics: Emergency management and homeland security*. Washington, DC: CQ Press.

Syria. (2011, August 31). Retrieved August 2011, 2011, from *New York Times* News International Countries and Territories Syria: http://topics.nytimes.com/top/news/international/countriesandterritories/syria/index.html

Taddeo, V. (2010). U.S. response to terrorism: A strategic analysis of the Afghanistan campaign. *Journal of Strategic Security, 3*(2), 27–38.

Teves, O. (2012, July 10). *Suspected militants in Southern Philippines kill 6*. Retrieved July 16, 2012, from sfgate.com: http://www.sfgate.com/news/article/Suspected-militants-in-southern-Philippines-kill-6-3697724.php

Thakur, R. (2011). Delinking destiny from geography: The changing balance of India-Pakistan relations. *India Quarterly: A Journal of International Affairs, 67*, 197–212.

Thales Group. (n.d.). *Thales' intelligent video analysis unlocks the potential of CCTV*. Retrieved August 21, 2013, from thalesgroup.com: https://www.thalesgroup.com/en/worldwide/big-data/smarter-data-safer-public

Theocracy. (n.d.). Retrieved February 26, 2013, from *merriam-webster.com*: http://www.merriam-webster.com/dictionary/theocracy

Thiessen, M. A. (2011, December 8). Iran responsible for 1998 U.S. embassy bombings. Retrieved June 8, 2013, from washingtonpost.com: http://articles.washingtonpost.com/2011-12-08/opinions/35285776_1_al-qaeda-sophisticated-bombs-nuclear-weapons

Thompson, B. G. (2010). A legislative prescription for confronting 21st-century risks to the homeland. *Harvard Journal on Legislation*, pp. 277–326.

Thompson, G. (2008, April 22). *Court ban on Jemaah Islamiyah*. Retrieved July 12, 2012, from abc.net: http://www.abc.net.au/am/content/2008/s2223796.htm

Tibet profile timeline. (2012, May 28). *BBC News*. Retrieved July 5, 2012, from BBC News: http://www.bbc.co.uk/news/world-asia-pacific-17046222

Timeline: History of a revolution. (2009, August 31). Retrieved April 16, 2012, from *alJazeera.com*: http://www.aljazeera.com/programmes/plohistoryofrevolution/2009/07/200974133438561995.html

Timeline: History of a revolution. (2009, July). *AlJazeera*. Retrieved May 24, 2012, from http://www.aljazeera.com/programmes/plohistoryofrevolution/2009/07/200972094351911191.html

Tobin, J. (2012, February 3). Iran threatens Israel with destruction, but the *Times* doesn't hear it. *commentarymagazine.com*. Retrieved June 6, 2012, from: http://www.commentarymagazine.com/2012/02/03/iran-khamenei-threat-israel-new-york-times/

Top Khmer Rouge leaders guilty of crimes agains humanity. (2014, August 7). Retrieved March 13, 2015, from *bbc.com*: http://www.bbc.com/news/world-asia-28670568

Totten, M. (2013, August 15). *The truth about Egypt*. Retrieved August 19, 2013, from worldaffairsjournal.org: http://www.worldaffairsjournal.org/blog/michael-j-totten/truth-about-egypt

Tran, M. (2010, May 26). US journalist granted parole in Peru after 15 years in jail. Retrieved July 4, 2013, from *guardian.co.uk*: http://www.guardian.co.uk/world/2010/may/26/peru-frees-american-lori-berenson

Travis, L. E., & Langworthy, R. H. (2008). *Policing in America: A balance of forces* (4th ed.). Upper Saddle River, NJ: Pearson Education.

Trent, C. (2012, December 5). AUC. Retrieved June 27, 2013, from *colombiareports.com*: http://colombiareports.com/auc/

Trimble, J. (1985, July 1). For Israelis, new strains and decisions. *U.S. News and World Report*.

Truth and Reconciliation Commission. (2003, August 25). *Final report of the Truth and Reconciliation Commission: Vol. VI. Crimes and human rights violations*. Retrieved July 2, 2013, from cverdad.org.pe: http://www.cverdad.org.pe/ingles/ifinal/conclusiones.php

Tucker, J., & Kadlec, R. (2001, Spring). Infectious disease and international security. *Strategic Review*, 12–20.

Turgut, P. (2013, March 21). New day for the Kurds: Will Olacon's declaration bring peace with Turkey? *Time*. Retrieved from http://content.time.com/time/magazine

UNIFIL. (2012, June 6). Retrieved June 6, 2012, from United Nations Interim Forces in Lebanon: http://unifil.unmissions.org/Default.aspx?tabid=1501

U.S. Census Bureau. (2015). Los Angeles County Quickfacts. Retrieved March 3, 2015, from http://quickfacts.census.gov/qfd/states/06/06037.html

U.S. Commission on National Security/21st Century. (2001, February 15). *Road map for national security: The phase III report of the U.S. Commission on National Security/21st Century*. Retrieved July 9, 2013, from au.af.mil: http://www.au.af.mil/au/awc/awcgate/nssg/phaseIIIfr.pdf

U.S. Department of Defense. (2014). *Department of Defense dictionary of military and associated terms* (Joint Publication 1-02).

Retrieved July 28, 2014, from http://www.dtic.mil/doctrine/dod_dictionary

U.S. Department of State. (1987). *Patterns of global terrorism 1986*. Retrieved from http://www.state.gov

U.S. Department of State. (1996). *Patterns of global terrorism 1995*. Retrieved from http://www.state.gov

U.S. Department of State. (1997). *Patterns of global terrorism 1996*. Retrieved from http://www.state.gov

U.S. Department of State. (1998). *Patterns of global terrorism 1997*. Retrieved from http://www.state.gov

U.S. Department of State. (2011). *Country reports on terrorism 2010*. Retrieved from http://www.state.gov

U.S. Department of State. (2012). *Country reports on terrorism 2011*. Retrieved from http://www.state.gov

U.S. Department of State. (2013). *Country reports on terrorism 2012*. Retrieved from http://www.state.gov

U.S. Department of State. (n.d.). Foreign terrorist organizations. Retrieved from http://www.state.gov/j/ct/rls/other/des/123085.htm

U.S. Food and Drug Administration. (2014). *FDA Food Safety Modernization Act*. Retrieved August 19, 2014, from http://www.fda.gov/Food/GuidanceRegulation/FSMA

U.S. General Accounting Office. (1992). *National security: The use of presidential directives to make and implement U.S. policy* (GAO Report No. GAO/NSIAD-92-72). Washington, DC: General Accounting Office.

U.S. general rescued from Red Brigade. (1982, January 28). *BBC: On This Day*. Retrieved from http://news.bbc.co.uk/onthisday/hi/dates/stories/january/28

U.S. Government Accountability Office. (2003). *Public health response to anthrax incidents of 2001* (GAO-04-152). Washington, DC: Government Printing Office.

U.S. House of Representatives. (2003). Committee on Government Reform. *Project BioShield: Contracting for the health and security of the American public*. 108th Cong. (April 4).

U.S. House of Representatives. (2005a). Committee on Homeland Security. Subcommittee on Emergency Preparedness, Science, and Technology. *Project BioShield: Linking bioterrorism threats and countermeasures*. 109th Cong. (July 12).

U.S. House of Representatives. (2005b). Committee on Government Reform. *One year later: Evaluating the effectiveness of Project BioShield*. 109th Cong. (July 14).

U.S. State Department. (2012, January 20). *Background note: Indonesia*. Retrieved July 11, 2012, from state.gov: http://www.state.gov/r/pa/ei/bgn/2748.htm

United Nations. (2001). *On threats to international peace and security caused by terrorist acts*, (S/Res/1373). Retrieved August 25, 2014, from http://www.un.org/en/sc/documents/search.shtml

United Nations. (2010). *Measures to eliminate international terrorism*. New York, NY: United Nations.

United Nations. (2011, April 5). Resolution 51/210: Measures to eliminate international terrorism. Retrieved from *United Nations*,

General Assembly: http://www.un.org/documents/ga/res/51/a51r210.htm

United Nations General Assembly Resolution 181. (1947, November 29). Retrieved March 15, 2012, from United Nations: http://daccess-dds-ny.un.org/doc/RESOLUTION/GEN/NR0/038/88/IMG/NR003888.pdf?OpenElement

United States House of Representatives. (2001). Committee on Government Reform. Subcommittee on National Security, Veteran's Affairs, and International Relations. *Biological warfare defense: Vaccine research and development program.* 107th Cong. (October 23).

United States, Department of State. (2011, 22 September). Terrorist designation of HAMAS operative Muhammad Hisham Muhammad Isma'il Abu Ghazala [Press release]. Retrieved March 19, 2012, from http://www.state.gov/r/pa/prs/ps/2011/09/173352.htm

Vick, K. (2013, April 8). Turkey's triumphs. *Time.* Retrieved from http://content.time.com/time/magazine

Vick, K. (2013, July 22). Street rule. *Time,* pp. 29–35

Vick, K. (2014a, February 3). The largest camp. *Time,* pp. 26–27.

Vick, K. (2014b, November 3). Why Kobani matters. *Time,* pp. 30–33.

Vidino, L. (2011). *Radicalization, linkage, and diversity: Current trends in terrorism in Europe.* Santa Monica, CA: RAND National Defense Research Institute.

Vidino, L. (2011a, March 6). Five myths about the Muslim Brotherhood. *Washington Post.* Retrieved from http://www.rand.org/commentary/2011/03/06/wp.html

Vidino, L. (2011b). *Radicalization, linkage, and diversity: Current trends in terrorism in Europe.* Santa Monica, CA: RAND National Defense Research Institute.

Vigo, M. (2013, February 4). *Peru government says it is committed to evaluating Fujimori pardon.* Retrieved July 5, 2013, from http://peruthisweek.peru.com/news-peru-government-says-it-is-committed-to-evaluating-fujimori-pardon-13539

Villeneuve, N. (2009, December 13). Thoughts on critical infrastructure protection. Retrieved August 16, 2013, from *nartv. org:* http://www.nartv.org/2009/12/13/thoughts-on-critical-infrastructure-protection/

Voice of the Martyrs, The. http://www.persecution.net/cn-2012-05-10.htm

von Luebk, C. (2011, April 2). *Inequality: How much longer can elites hide their privileges from view?* Retrieved July 17, 2012, from Inside Indonesia: http://www.insideindonesia.org/edition-104-apr-june-2011/inequality-10041444

Wallis, R. (2001). *Lockerbie: The story and lessons.* Westport, CT: Praeger.

Wang, A. (2012, July 2). *Chinese corruption comes in staggering sums.* Retrieved July 7, 2012, from theepochtimes: http://www.theepochtimes.com/n2/china-news/chinese-corruption-comes-in-staggering-sums-259995.html

Wartzman, R. (2008, April 10). Peter Drucker's winning team. Retrieved April 10, 2013, from *businessweek.com:* http://www.businessweek.com/stories/2008-04-10/peter-druckers-winning-teambusinessweek-business-news-stock-market-and-financial-advice

Weaver, M. A. (1993, April 12). The trail of the sheikh. *The New Yorker,* p. 71.

Weinburger, S. (2010). Intent to deceive. *Nature, 465,* 412–415.

Wells, M. (2012, September 20). Peru: As Shining Path's political arm grows, government clamps down. *Christian Science Monitor.*

Wenkel, D. (2007). Palestinians, Jebusites, and evangelicals. *The Middle East Quarterly,* 49–56.

White, J. (2012). *Terrorism and homeland security (7th ed.).* Belmont, CA: Wadsworth-Cengage Learning.

White, J. R. (2009). *Terrorism and homeland security* (6th ed.). Belmont, CA: Wadsworth.

White, J. R. (2012). *Terrorism and homeland security* (7th ed.). Belmont, CA: Wadsworth-Cengage Learning.

Whitlock, C. (2005, December 21). Hijacker sought by U.S. released. *The Washington Post.* Retrieved from http://www.washingtonpost.com/wp-dyn/content/article/2005/12/20/AR2005122001615.html

Who are Hamas? (2009, January 4). Retrieved May 30, 2012, from *BBC News:* http://news.bbc.co.uk/2/hi/1654510.stm

Wilkinson, T. (2011, June 4). Many ghosts haunt Peru's election. *Los Angeles Times,* A3.

Williams, D., & Edwards, A. (2013, June 6). Austria withdraws peacekeepers from Golan Heights as Bin Laden's former Al Qaeda number two urges Sunni Muslims to fight in Syria civil war. Retrieved June 12, 2013, from dailymail.co.uk/news: http://www.dailymail.co.uk/news/article-2336802/Overthrow-Assad-set-Islamic-rule-Bin-Ladens-Al-Qaeda-number-urges-Sunni-Muslims-fight-Syria-civil-war.html

Williams, H., & Murphy, P. V. (1990, January). *The Evolving Strategy of Police: A minority view.* Retrieved January 15, 2013, from National Criminal Justice Reference Service, Perspective of Policing: https://www.ncjrs.gov/pdffiles1/nij/121019.pdf

Wills, D. C. (2003). *The first war on terrorism: Counterterrorism policy during the Reagan administration.* Lanham, MD: Rowman & Littlefield.

Winthrop, J. (1997, July). The Oklahoma City bombing: Immediate response authority and other military assistance to civil authority (MACA). Retrieved July 22, 2013, from *Homeland Security Digital Library:* http://www.hsdl.org/?view&did=463752

Wolf, J. (2013, July 7). *Scanning cargo containers is more important than scanning emails.* Retrieved August 10, 2013, from atlanticreview.org: http://atlanticreview.org/archives/1573-Scanning-Cargo-Containers-is-More-Important-than-Scanning-Emails.html

Woo, A. (2012, June 19). *Ajaib Singh Bagri and Ripudaman Singh Malik.* Retrieved October 5, 2012, from *The Globe and Mail:* http://www.theglobeandmail.com/news/british-columbia/bc-court-upholds-perjury-conviction-for-air-india-bomber/article4427686/

Woodhouse, L. (2012, December 17). *Earth Liberation Front Activist Released After 7 Years in "Little Guantanamo."* Retrieved February

14, 2013, from Huffington Post: http://www.huffingtonpost.com/leighton-woodhouse/earth-liberation-front-ac_b_2295064.html

World Future Society. (n.d.). Forecasts from *The Futurist* magazine. Retrieved August 16, 2013, from *World Future Society:* http://www.wfs.org/Forecasts_From_The_Futurist_Magazine

Wright, L. (2006). *The looming tower: Al-Qaeda and the road to 9/11.* New York, NY: Alfred A. Knopf.

Wright, L. (2011). *The looming tower: Al-Qaeda and the road to 9/11* (Rev. ed.). New York, NY: Vintage Books.

Wright, R. (2014, January 27). A new beginning in Tehran. *Time,* pp. 37–41.

Yeoman, B., & Hogan, B. (2002). Airline insecurity. In A. J. Cigler (Ed.), *Perspectives on terrorism* (p. 112). Boston, MA: Houghton Mifflin.

Yerger, D. B. (2011). The Economic Costs of 9/11 on the U.S. *Phi Kappa Phi Forum, 91*(3), 12–13.

Zahran, M. (2012). Jordan Is Palestinian. *Middle East Quarterly,* 3–12.

Zeine, Z. N. (1973). *The emergence of Arab nationalism.* Delmar, NY: Caravan Books.

Zenn, J. (2012, July 12). *Runaway radicals in Indonesia.* Retrieved July 14, 2012, from atimes.com: http://www.atimes.com/atimes/Southeast_Asia/NG12Ae02.html

Zmaneh, R. (2012, November 12). *Canadian Court Freezes on Iranian Properties.* Retrieved November 2, 2012, from Eurasia Review: http://www.eurasiareview.com/02112012-canadian-court-freezes-on-iranian-properties/

Appendix A

Religions in Their Own Words

The Thirteen Foundations of Jewish Belief

Rabbi Boruch Clinton

The source for the following summary of each of these thirteen principles is Maimonides' introduction to the tenth chapter of the Mishnaic tractate, Sanhedrin.

1. The existence of the world or any part of it is dependent upon the existence of the single, unique Creator. But the existence of this Creator, the Master of the universe, is not dependent on anything.

2. This principle is known as "**Yichud**"—**G-d's Oneness**. There is only one G-d. He is unique, and is without any divisions. There is nothing in the universe with which we can compare His Oneness. This aspect of G-d's existence is clear from the verse "Hear, oh Israel, G-d is our Lord, G-d is one." (Deut. 6 4).

3. G-d has no body or any physical aspect, nor is His Power that of a physical body. This principle builds on the logic of the previous one. If G-d were to have a body, it would limit Him to the confines of that body, and therefore He would not be infinite and incomparable in the same way. There are many places in the Bible where G-d is described as "stretching out His hand" (or doing some similar physical action). These are only figures of speech (anthropomorphisms). They are sublime actions couched in words that humans can understand.

4. G-d has always existed and always will. He is eternal. Again, if this weren't true, and G-d would be limited (by time) and he would no longer be "infinite."

5. **Idolatry**: There is no individual or power besides G-d whom it is fitting to worship or serve. Even to worship (or attribute independent power to) intermediaries (like angels, other human beings, stars or planets) is forbidden. Such worship is in the category of idolatry. G-d created the universe and every single one of its parts, it is to Him that we owe all of our gratitude and subservience.

6. The sixth principle of the Rambam is "**Nevu'a**"—**prophecy**. G-d grants prophecy to people who have perfected their personal character and who follow all the commandments of the Torah. Prophecy does not come to unlearned and unprepared people.

7. The prophecy of our teacher Moses (through whom the Torah was transmitted at Mt Sinai) was greater than all other prophecy in four ways:

   1. It was not "heard" through any intermediary (i.e. an angel, a cloudy vision) but was directly from G-d.

   2. It was always given while Moshe was wide awake, in complete control of his faculties.

   3. Moshe was not overcome with shaking and dread as were other prophets, but was calm and alert.

   4. Moshe had the incredible ability to summon prophecy at will. Other prophets had to prepare and wait until G-d chose to appear.

8. **The giving of the Torah**: The words of the Written Torah (the "Five Books of Moses") are the true and completely accurate words of G-d. The words in the Torah were dictated by G-d to Moshe. In essence, the verse "Shema Yisroel . . ."—(Hear O Israel . . .) is just as meaningful to us as the lists of names and places written in the Torah. They all come from G-d and there is great, limitless wisdom to be found in every word. G-d also taught Moshe how to carry out the commandments found in the Written Torah; these G-d-given explanations form part of what we call the Oral Torah.

9. Since the entire Torah comes from G-d, one may not add to it or subtract from it (i.e. add or subtract commandments. For instance to say that there is no commandment of Tefillin . . .).

10. G-d is aware of all our actions and does not ignore them.

11. There is **reward and punishment** for our actions in this world.

12. The Messiah (lit "anointed one"), a descendant of King David, will come, and could come at any time. He will be wiser than King Soloman and possess a level of prophecy close to that of Moses.

13. There will eventually be a revival of the dead.

Rabbi Boruch Clinton teaches at the Ottawa Torah Institute yeshiva high school and Machon Sarah high school for girls (both in Ottawa, Canada). You may reach him with comments and questions at bclinton@torah.org.

SOURCE: Copyright © 2000 by Rabbi Boruch Clinton and Project Genesis, Inc. Retrieved at http://www.torah.org

What We Believe

Arizona Community Church

The Bible

The Bible is inspired by God and is the only infallible, authoritative Word of God. *2 Timothy 3:15-17; 1 Peter 1:23-25; Hebrews 4:12; John 20:31.*

God

There is One God eternally existent in Three Persons: Father, Son and Holy Spirit. *Matthew 3:16-17; 28:19; Luke 1:35; 1 John 5:7; 2 Corinthians 13:14*

Jesus Christ

Jesus Christ is God, *John 1:1-2, 14*; came in the flesh through the virgin birth, *Isaiah 7:14; Matthew 1:23*; lived a sinless life, *Hebrews 4:15*; performed miracles, *John 2:1, Luke 1:1-4*; died for the sins of the world, *Romans 3:23-2; 1 Corinthians 15:3; Colossians 1:14; 1 John 1:7; Hebrews 10:19*; bodily rose from the dead, *Matthew 28:6; Romans 1:4, 8:1; 1 Corinthians 15*; ascended to Heaven and intercedes for us with the Father, *Luke 24:50-51; 1 John 2:1-2; Hebrews 7:25*; will return in power and glory to judge all creation and to establish an eternal kingdom. *Acts 1:11; Titus 2:13; Revelation 1:7, 11:15; Philippians 2:9-11*

Our Need for Salvation through Jesus Christ

We believe that, in Adam's sin of disobedience to the revealed will and word of God, he incurred the penalty of spiritual and physical death, became subject to the wrath of God, became inherently corrupt, and utterly incapable of choosing or doing that which is acceptable to God, apart from Divine grace *Genesis 2:17; Romans 5:12-19; Ephesians 2:1-3, 4:18-19, 6:10-19; Mark 7:20-23; Revelation 12:9; 1 Peter 5:3*

Becoming a Christian-How, and the Benefits

The sole condition for receiving everlasting life is faith alone in the Lord Jesus Christ, who died a substitutionary death on the cross for humanity's sin and rose bodily from the dead. All who believe are assured that their sins are forgiven, *Colossians 1:14; 2:13-15; the believer becomes a child of God, John 1:12*; a new life with Christ begins, 2 Corinthians 5:17; John 10:10; and eternity in Heaven is assured. *John 11:25-25, 17:3; 2 Corinthians 5:1-10; 1 Thessalonians 4:13-18*; Salvation is a gift from God. *Ephesians 2:8-10*

God. The Holy Spirit

The Holy Spirit indwells the Christian and is actively ministering in the world today enabling and empowering the living of a life pleasing to God; He gives gifts to every believer for building up the Body of Christ (the Church). *Romans 8:9; 1 Corinthians 6:19, 12:1-14; Galatians 5:22-23; Ephesians 5:18; Romans 12:1-8; 1 Peter 4: 1-10.*

The Eternal Future

All people will be resurrected, the Christian to everlasting life in Heaven and eternal rewards; the non-Christian to everlasting death in Hell and eternal punishment. *John 5:24-29; 1 Corinthians 15:20-23; Revelation 20:11-15*

The Body of Christ-The Church

There is spiritual unity of believers in Jesus Christ; many churches but one Lord, *Ephesians 4:4*. The Church is made up of all who put their faith in Jesus Christ, and is to give praise to God, build up believers, and proclaim the Gospel. *Matthew 16:16-18, 28:19-20; Acts 2:42-47; Acts 1:8; 1 Peter 2:9-10*

Marriage

The only legitimate marriage sanctioned by God is the joining of one man and one woman in a single, exclusive union, as taught in Scripture.   We believe marriage is a picture of the relationship of Christ and the Church and is to be a lifelong, covenant relationship between a man and a woman based on love, respect, mutual submission and personal sacrifice.   We believe God intended sexual intimacy to be enjoyed only within the context of the male and female marriage relationship; and that all other sexual relations outside of marriage are inconsistent with the teachings of the Bible.   Gen. 1:27-28; Mark 10:2-12; Ephesians 5:21-33; Hebrews 13:4; Jude 1:7

SOURCE: Retrieved from http://www.azcc.org/mission-beliefs/

Welcome to Submission.org

Editors of Submission.org

Submission or Islam in the Arabic language is a meaning or a description rather than a name or a title. It describes the state of mind of anyone who recognizes God's absolute authority, and reaches a conviction that God alone possesses all power; no other entity possesses any power or control independent of Him. The logical consequence of such a realization is to devote one's life and one's worship absolutely to God alone.

So, Submission (or Islam in Arabic language) is a spiritual state of mind and not a title of a religion that belongs to a specific group of people. ANYONE who submits and worships one God without idolizing other entities is a Submitter by definition (Muslim in Arabic language).

This state of mind basically conforms with God's one and only message He delivered to man-kind through all of His messengers since Noah; worship God alone and avoid idolatry. Based on that, one can safely conclude that the message of Islam or Submission has been in existence way before the time of prophet Muhammad and way before Quran. All God's messengers, since Noah, devoted their lives and worship to God alone and were Submitters to Him alone.

[Final Testament/Quran 21:25] We did not send any messenger before you except with the inspiration: "There is no god except Me; you shall worship Me alone."

[Final Testament/Quran 16:36] We have sent a messenger to every community, saying, "You shall worship God, and avoid idolatry. . . . . "

[Final Testament/Quran 10:71] Recite for them the history of Noah. He said to his people, "O my people, if you find my position and my reminding you of God's revelations too much for you, then I put my trust in God. You should get together with your leaders, agree on a final decision among yourselves, then let me know it without delay.

[Final Testament/Quran 10:72] "If you turn away, then I have not asked you for any wage. My wage comes from God. I have been commanded to be a submitter."

[Final Testament/Quran 2:133] Had you witnessed Jacob on his death bed; he said to his children, "What will you worship after I die?" They said, "We will worship your god; the god of your fathers Abraham, Ismail, and Isaac; the one god. To Him we are submitters."

[Final Testament/Quran 3:52] When Jesus sensed their disbelief, he said, "Who are my supporters towards God?" The disciples said, "We are God's supporters; we believe in God, and bear witness that we are submitters."

With this understanding in mind, and how Submission to God alone has been the only message God delivered through all His messengers, we can easily conclude that it is indeed the only religion that has been delivered, authorized, and approved by God. This is stated clearly in God's Final Testament:

[Final Testament/Quran 3:19] The only religion approved by God is "Submission." Ironically, those who have received the scripture are the ones who dispute this fact, despite the knowledge they have received, due to jealousy. For such rejectors of God's revelations, God is most strict in reckoning.

[Final Testament/Quran 3:85] Anyone who accepts other than Submission as his religion, it will not be accepted from him, and in the Hereafter, he will be with the losers.

[Final Testament/Quran 21:92] Your congregation is but one congregation, and I alone am your Lord; you shall worship Me alone.

[Final Testament/Quran 21:93] However, they divided themselves into disputing religions. All of them will come back to us (for judgment).

[Final Testament/Quran 21:94] As for those who work righteousness, while believing, their work will not go to waste; we are recording it.

By the will of God, this site is made available to shed the light on, define and promote God's message to mankind since Noah; Submission to Him alone. It is our mission to present the whole truth we know about the religion of Submission (Islam in Arabic language).

Knowledge is the cure for one's ignorance.

"And you shall know the truth, and the truth shall make you free."

[Bible, John 8:32]

[Final Testament/Quran 17:36] "You shall not accept any information, unless you verify it for yourself. I have given you the hearing, the eyesight, and the brain, and you are responsible for using them."

[Final Testament/Quran 20:114] . . . say, "My Lord, increase my knowledge."

SOURCE: Retrieved from http://submission.org/index.html

INDEX

Page references with boxes, charts, figures, and tables are labeled as (box), (chart), (fig.), and (table), respectively.

ABOUT THE AUTHORS

Timothy A. Capron graduated from the local university and cleverly avoided the draft by graduating from Marine Corps Officer Candidate School at Quantico, Virginia. Following graduation, he then attended the Marine Corps Basic School for Officers at Quantico, and the Army Artillery Basic Officer Course at Ft. Sill, Oklahoma. He was then assigned to Okinawa and Viet Nam and Hawaii. Leaving active duty, he attended graduate school, completing studies in 1979 with a PhD in criminal justice from Sam Houston State's College of Criminal Justice. He then taught in a college program for a year and returned to Active Duty with the Marine Corps in 1980, transferring to the Army later that year. He enjoyed numerous assignments as a military police officer and a nuclear weapons officer, attending scores of schools and courses, including the Army Management Staff College. He retired from the military in 1995 as a Lieutenant Colonel, worked for ITT Systems as a manager working for the Defense Nuclear Information Center, until the latter part of 1998, relocating to California. He is a professor at Sac State, having taught there since 2000. He is lifetime member of the Academy of Criminal Justice Sciences and a member of the Association for the Study of the Mideast and Africa. His latest textbook is: Capron, Timothy A. (with Charles H. McCaghy, J.D. Jamieson and Sandra Carey), *Deviant Behavior: Crime, Conflict and Interest Groups, 8th edition,* (2008). He is also a program advisor for the California State Mitigation Assessment Response Team (SMART).

Stephanie B. Mizrahi received her law degree from McGeorge School of Law in 1994 and served as a prosecuting attorney in Washington and Oregon prior to receiving her master's in criminal justice from Washington State University in 2001, and her PhD in criminal justice from Washington State University in 2014. Stephanie's topic for her dissertation was terrorism policy in relation to chemical and biological terrorism.From 1986-1991, Dr. Mizrahi served in the Central Intelligence Agency. She spent the majority of her time as an analyst in the Office of European Analysis with a year's rotation in the Counterterrorism Center (CTC). She is also an officer in the Western Association of Criminal Justices and has served on several committees for the academy of criminal justice sciences. She teaches in the division of criminal justice at California State University, Sacramento. Her teaching and research interests center around law, terrorism, homeland security and emergency management policy, and curriculum development in homeland security and emergency management programs. She has also served as the conference coordinator and on several committees with the CSU council on emergency management and homeland security (CEMHS).